ART INSTITUTE OF LAS VEGAS
2350 CORPORATE CIRCLE DRIVE
HENDERSON, NV 89074−7737

Video Compression and Communications

Video Compression and Communications

From Basics to H.261, H.263, H.264, MPEG4 for DVB and HSDPA-Style Adaptive Turbo-Transceivers

Second Edition

L. Hanzo, P. J. Cherriman and J. Streit

All of
University of Southampton, UK

IEEE PRESS

IEEE Communications Society, Sponsor

John Wiley & Sons, Ltd

Copyright © 2007   John Wiley & Sons Ltd, The Atrium, Southern Gate, Chichester,
West Sussex PO19 8SQ, England

Telephone   (+44) 1243 779777

Email (for orders and customer service enquiries): cs-books@wiley.co.uk
Visit our Home Page on www.wiley.com

All Rights Reserved. No part of this publication may be reproduced, stored in a retrieval system or transmitted in any form or by any means, electronic, mechanical, photocopying, recording, scanning or otherwise, except under the terms of the Copyright, Designs and Patents Act 1988 or under the terms of a licence issued by the Copyright Licensing Agency Ltd, 90 Tottenham Court Road, London W1T 4LP, UK, without the permission in writing of the Publisher. Requests to the Publisher should be addressed to the Permissions Department, John Wiley & Sons Ltd, The Atrium, Southern Gate, Chichester, West Sussex PO19 8SQ, England, or emailed to permreq@wiley.co.uk, or faxed to (+44) 1243 770620.

Designations used by companies to distinguish their products are often claimed as trademarks. All brand names and product names used in this book are trade names, service marks, trademarks or registered trademarks of their respective owners. The Publisher is not associated with any product or vendor mentioned in this book. All trademarks referred to in the text of this publication are the property of their respective owners.

This publication is designed to provide accurate and authoritative information in regard to the subject matter covered. It is sold on the understanding that the Publisher is not engaged in rendering professional services. If professional advice or other expert assistance is required, the services of a competent professional should be sought.

Other Wiley Editorial Offices

John Wiley & Sons Inc., 111 River Street, Hoboken, NJ 07030, USA

Jossey-Bass, 989 Market Street, San Francisco, CA 94103-1741, USA

Wiley-VCH Verlag GmbH, Boschstr. 12, D-69469 Weinheim, Germany

John Wiley & Sons Australia Ltd, 42 McDougall Street, Milton, Queensland 4064, Australia

John Wiley & Sons (Asia) Pte Ltd, 2 Clementi Loop #02-01, Jin Xing Distripark, Singapore 129809

John Wiley & Sons Canada Ltd, 22 Worcester Road, Etobicoke, Ontario, Canada M9W 1L1

Wiley also publishes its books in a variety of electronic formats. Some content that appears in print may not be available in electronic books.

IEEE Communications Society, Sponsor
COMMS-S Liaison to IEEE Press, Mostafa Hashem Sherif

Library of Congress Cataloging-in-Publication Data

Hanzo, Lajos, 1952-
    Video Compression and Communications : from basics to H.261, H.263,
H.264, MPEG4 for DVB and HSDPA-style adaptive turbo-transceivers / L. Hanzo,
P. J. Cherriman and J. Streit – 2nd ed.
        p. cm.
    Includes bibliographical references and index.
    ISBN 978-0-470-51849-6 (cloth)
1. Video compression. 2. Digital video. 3. Mobile communication systems.
I. Cherriman, Peter J., 1972- II. Streit, Jürgen, 1968- III.
Title.
    TK6680.5.H365 2007
    006.6'–dc22

2007024178

British Library Cataloguing in Publication Data

A catalogue record for this book is available from the British Library

ISBN 978-0-470- 51849-6 (HB)

Typeset by the authors using LaTeX software.
Printed and bound in Great Britain by Antony Rowe Ltd, Chippenham, England.
This book is printed on acid-free paper responsibly manufactured from sustainable forestry in which at least two trees are planted for each one used for paper production.

Contents

About the Authors   xvii

Other Wiley and IEEE Press Books on Related Topics   xix

Preface   xxi

Acknowledgments   xxiii

1 Introduction   1
- 1.1 A Brief Introduction to Compression Theory   1
- 1.2 Introduction to Video Formats   2
- 1.3 Evolution of Video Compression Standards   5
  - 1.3.1 The International Telecommunications Union's H.120 Standard   8
  - 1.3.2 Joint Photographic Experts Group   8
  - 1.3.3 The ITU H.261 Standard   11
  - 1.3.4 The Motion Pictures Expert Group   11
  - 1.3.5 The MPEG-2 Standard   12
  - 1.3.6 The ITU H.263 Standard   12
  - 1.3.7 The ITU H.263+/H.263++ Standards   13
  - 1.3.8 The MPEG-4 Standard   13
  - 1.3.9 The H.26L/H.264 Standard   14
- 1.4 Video Communications   15
- 1.5 Organization of the Monograph   17

I Video Codecs for HSDPA-style Adaptive Videophones   19

2 Fractal Image Codecs   21
- 2.1 Fractal Principles   21
- 2.2 One-dimensional Fractal Coding   23
  - 2.2.1 Fractal Codec Design   27
  - 2.2.2 Fractal Codec Performance   28

2.3	Error Sensitivity and Complexity				32
2.4	Summary and Conclusions				33

3 Low Bitrate DCT Codecs and HSDPA-style Videophone Transceivers — 35

- 3.1 Video Codec Outline 35
- 3.2 The Principle of Motion Compensation 37
  - 3.2.1 Distance Measures 40
  - 3.2.2 Motion Search Algorithms 42
    - 3.2.2.1 Full or Exhaustive Motion Search 42
    - 3.2.2.2 Gradient-based Motion Estimation 43
    - 3.2.2.3 Hierarchical or Tree Search 44
    - 3.2.2.4 Subsampling Search 45
    - 3.2.2.5 Post-processing of Motion Vectors 46
    - 3.2.2.6 Gain-cost-controlled Motion Compensation 46
  - 3.2.3 Other Motion Estimation Techniques 48
    - 3.2.3.1 Pel-recursive Displacement Estimation 49
    - 3.2.3.2 Grid Interpolation Techniques 49
    - 3.2.3.3 MC Using Higher Order Transformations 49
    - 3.2.3.4 MC in the Transform Domain 50
  - 3.2.4 Conclusion 50
- 3.3 Transform Coding 51
  - 3.3.1 One-dimensional Transform Coding 51
  - 3.3.2 Two-dimensional Transform Coding 52
  - 3.3.3 Quantizer Training for Single-class DCT 55
  - 3.3.4 Quantizer Training for Multiclass DCT 56
- 3.4 The Codec Outline 58
- 3.5 Initial Intra-frame Coding 60
- 3.6 Gain-controlled Motion Compensation 60
- 3.7 The MCER Active/Passive Concept 61
- 3.8 Partial Forced Update of the Reconstructed Frame Buffers .. 62
- 3.9 The Gain/Cost-controlled Inter-frame Codec 64
  - 3.9.1 Complexity Considerations and Reduction Techniques . 65
- 3.10 The Bit-allocation Strategy 66
- 3.11 Results 67
- 3.12 DCT Codec Performance under Erroneous Conditions 70
  - 3.12.1 Bit Sensitivity 70
  - 3.12.2 Bit Sensitivity of Codec I and II 71
- 3.13 DCT-based Low-rate Video Transceivers 72
  - 3.13.1 Choice of Modem 72
  - 3.13.2 Source-matched Transceiver 73
    - 3.13.2.1 System 1 73
      - 3.13.2.1.1 System Concept 73
      - 3.13.2.1.2 Sensitivity-matched Modulation 74
      - 3.13.2.1.3 Source Sensitivity 74
      - 3.13.2.1.4 Forward Error Correction 75
      - 3.13.2.1.5 Transmission Format 75

		3.13.2.2	System 2	78
			3.13.2.2.1 Automatic Repeat Request	78
		3.13.2.3	Systems 3–5	79
	3.14	System Performance		80
		3.14.1	Performance of System 1	80
		3.14.2	Performance of System 2	83
			3.14.2.1 FER Performance	83
			3.14.2.2 Slot Occupancy Performance	85
			3.14.2.3 PSNR Performance	86
		3.14.3	Performance of Systems 3–5	87
	3.15	Summary and Conclusions		89

4 Very Low Bitrate VQ Codecs and HSDPA-style Videophone Transceivers — 93

4.1	Introduction		93
4.2	The Codebook Design		93
4.3	The Vector Quantizer Design		95
	4.3.1	Mean and Shape Gain Vector Quantization	99
	4.3.2	Adaptive Vector Quantization	100
	4.3.3	Classified Vector Quantization	102
	4.3.4	Algorithmic Complexity	103
4.4	Performance under Erroneous Conditions		105
	4.4.1	Bit-allocation Strategy	105
	4.4.2	Bit Sensitivity	106
4.5	VQ-based Low-rate Video Transceivers		107
	4.5.1	Choice of Modulation	107
	4.5.2	Forward Error Correction	109
	4.5.3	Architecture of System 1	109
	4.5.4	Architecture of System 2	111
	4.5.5	Architecture of Systems 3–6	112
4.6	System Performance		113
	4.6.1	Simulation Environment	113
	4.6.2	Performance of Systems 1 and 3	114
	4.6.3	Performance of Systems 4 and 5	115
	4.6.4	Performance of Systems 2 and 6	117
4.7	Joint Iterative Decoding of Trellis-based Vector-quantized Video and TCM		118
	4.7.1	Introduction	118
	4.7.2	System Overview	120
	4.7.3	Compression	120
	4.7.4	Vector Quantization Decomposition	121
	4.7.5	Serial Concatenation and Iterative Decoding	121
	4.7.6	Transmission Frame Structure	122
	4.7.7	Frame Difference Decomposition	123
	4.7.8	VQ Codebook	124
	4.7.9	VQ-induced Code Constraints	126
	4.7.10	VQ Trellis Structure	127
	4.7.11	VQ Encoding	129

		4.7.12	VQ Decoding	130

        4.7.12 VQ Decoding 130
        4.7.13 Results 132
  4.8   Summary and Conclusions 136

5 Low Bitrate Quad-tree-based Codecs and HSDPA-style Videophone Transceivers   139

  5.1   Introduction 139
  5.2   Quad-tree Decomposition 139
  5.3   Quad-tree Intensity Match 142
        5.3.1   Zero-order Intensity Match 142
        5.3.2   First-order Intensity Match 144
        5.3.3   Decomposition Algorithmic Issues 145
  5.4   Model-based Parametric Enhancement 148
        5.4.1   Eye and Mouth Detection 149
        5.4.2   Parametric Codebook Training 151
        5.4.3   Parametric Encoding 152
  5.5   The Enhanced QT Codec 153
  5.6   Performance and Considerations under Erroneous Conditions 154
        5.6.1   Bit Allocation 155
        5.6.2   Bit Sensitivity 157
  5.7   QT-codec-based Video Transceivers 158
        5.7.1   Channel Coding and Modulation 158
        5.7.2   QT-based Transceiver Architectures 159
  5.8   QT-based Video-transceiver Performance 162
  5.9   Summary of QT-based Video Transceivers 165
  5.10 Summary of Low-rate Video Codecs and Transceivers 166

II High-resolution Video Coding   171

6 Low-complexity Techniques   173

  6.1   Differential Pulse Code Modulation 173
        6.1.1   Basic Differential Pulse Code Modulation 173
        6.1.2   Intra/Inter-frame Differential Pulse Code Modulation 175
        6.1.3   Adaptive Differential Pulse Code Modulation 177
  6.2   Block Truncation Coding 177
        6.2.1   The Block Truncation Algorithm 177
        6.2.2   Block Truncation Codec Implementations 180
        6.2.3   Intra-frame Block Truncation Coding 180
        6.2.4   Inter-frame Block Truncation Coding 182
  6.3   Subband Coding 183
        6.3.1   Perfect Reconstruction Quadrature Mirror Filtering 185
                6.3.1.1   Analysis Filtering 185
                6.3.1.2   Synthesis Filtering 188
                6.3.1.3   Practical QMF Design Constraints 189
        6.3.2   Practical Quadrature Mirror Filters 191

		6.3.3	Run-length-based Intra-frame Subband Coding 195
		6.3.4	Max-Lloyd-based Subband Coding 198
	6.4	Summary and Conclusions 202

7 High-resolution DCT Coding   205

   7.1 Introduction 205
   7.2 Intra-frame Quantizer Training 205
   7.3 Motion Compensation for High-quality Images 209
   7.4 Inter-frame DCT Coding 215
         7.4.1 Properties of the DCT Transformed MCER 215
         7.4.2 Joint Motion Compensation and Residual Encoding 222
   7.5 The Proposed Codec 224
         7.5.1 Motion Compensation 225
         7.5.2 The Inter/Intra-DCT Codec 226
         7.5.3 Frame Alignment 227
         7.5.4 Bit-allocation 229
         7.5.5 The Codec Performance 230
         7.5.6 Error Sensitivity and Complexity 233
   7.6 Summary and Conclusions 235

III   H.261, H.263, H.264, MPEG2 and MPEG4 for HSDPA-style Wireless Video Telephony and DVB   237

8 H.261 for HSDPA-style Wireless Video Telephony   239

   8.1 Introduction 239
   8.2 The H.261 Video Coding Standard 239
         8.2.1 Overview 239
         8.2.2 Source Encoder 240
         8.2.3 Coding Control 242
         8.2.4 Video Multiplex Coder 243
                 8.2.4.1 Picture Layer 244
                 8.2.4.2 Group of Blocks Layer 245
                 8.2.4.3 Macroblock Layer 247
                 8.2.4.4 Block Layer 247
         8.2.5 Simulated Coding Statistics 250
                 8.2.5.1 Fixed-quantizer Coding 251
                 8.2.5.2 Variable Quantizer Coding 252
   8.3 Effect of Transmission Errors on the H.261 Codec 253
         8.3.1 Error Mechanisms 253
         8.3.2 Error Control Mechanisms 255
                 8.3.2.1 Background 255
                 8.3.2.2 Intra-frame Coding 256
                 8.3.2.3 Automatic Repeat Request 257
                 8.3.2.4 Reconfigurable Modulations Schemes 257
                 8.3.2.5 Combined Source/Channel Coding 257

	8.3.3	Error Recovery 258
	8.3.4	Effects of Errors 259
		8.3.4.1 Qualitative Effect of Errors on H.261 Parameters 259
		8.3.4.2 Quantitative Effect of Errors on a H.261 Data Stream . . . 262
		8.3.4.2.1 Errors in an Intra-coded Frame 263
		8.3.4.2.2 Errors in an Inter-coded Frame 265
		8.3.4.2.3 Errors in Quantizer Indices 267
		8.3.4.2.4 Errors in an Inter-coded Frame with Motion Vectors 268
		8.3.4.2.5 Errors in an Inter-coded Frame at Low Rate . . . 271

- 8.4 A Reconfigurable Wireless Videophone System 272
  - 8.4.1 Introduction 272
  - 8.4.2 Objectives 273
  - 8.4.3 Bitrate Reduction of the H.261 Codec 273
  - 8.4.4 Investigation of Macroblock Size 274
  - 8.4.5 Error Correction Coding 275
  - 8.4.6 Packetization Algorithm 278
    - 8.4.6.1 Encoding History List 278
    - 8.4.6.2 Macroblock Compounding 279
    - 8.4.6.3 End of Frame Effect 281
    - 8.4.6.4 Packet Transmission Feedback 282
    - 8.4.6.5 Packet Truncation and Compounding Algorithms 282
- 8.5 H.261-based Wireless Videophone System Performance 283
  - 8.5.1 System Architecture 283
  - 8.5.2 System Performance 286
- 8.6 Summary and Conclusions 293

9 Comparative Study of the H.261 and H.263 Codecs   295
- 9.1 Introduction 295
- 9.2 The H.263 Coding Algorithms 297
  - 9.2.1 Source Encoder 297
    - 9.2.1.1 Prediction 297
    - 9.2.1.2 Motion Compensation and Transform Coding 297
    - 9.2.1.3 Quantization 298
  - 9.2.2 Video Multiplex Coder 298
    - 9.2.2.1 Picture Layer 300
    - 9.2.2.2 Group of Blocks Layer 300
    - 9.2.2.3 H.261 Macroblock Layer 301
    - 9.2.2.4 H.263 Macroblock Layer 302
    - 9.2.2.5 Block Layer 305
  - 9.2.3 Motion Compensation 306
    - 9.2.3.1 H.263 Motion Vector Predictor 307
    - 9.2.3.2 H.263 Subpixel Interpolation 308
  - 9.2.4 H.263 Negotiable Options 309
    - 9.2.4.1 Unrestricted Motion Vector Mode 309
    - 9.2.4.2 Syntax-based Arithmetic Coding Mode 310

| | | 9.2.4.2.1 | Arithmetic coding . . . . 311 |
| | 9.2.4.3 | Advanced Prediction Mode 312

		9.2.4.3.1	Four Motion Vectors per Macroblock 313
		9.2.4.3.2	Overlapped Motion Compensation for Luminance 313
	9.2.4.4	P-B Frames Mode 315

9.3 Performance Results 318
  9.3.1 Introduction 318
  9.3.2 H.261 Performance 319
  9.3.3 H.261/H.263 Performance Comparison 322
  9.3.4 H.263 Codec Performance 325
    9.3.4.1 Gray-Scale versus Color Comparison 325
    9.3.4.2 Comparison of QCIF Resolution Color Video 328
    9.3.4.3 Coding Performance at Various Resolutions 328
9.4 Summary and Conclusions 335

10 H.263 for HSDPA-style Wireless Video Telephony 339

10.1 Introduction 339
10.2 H.263 in a Mobile Environment 339
  10.2.1 Problems of Using H.263 in a Mobile Environment 339
  10.2.2 Possible Solutions for Using H.263 in a Mobile Environment 340
    10.2.2.1 Coding Video Sequences Using Exclusively Intra-coded Frames 341
    10.2.2.2 Automatic Repeat Requests 341
    10.2.2.3 Multimode Modulation Schemes 341
    10.2.2.4 Combined Source/Channel Coding 342
10.3 Design of an Error-resilient Reconfigurable Videophone System 343
  10.3.1 Introduction 343
  10.3.2 Controlling the Bitrate 343
  10.3.3 Employing FEC Codes in the Videophone System 345
  10.3.4 Transmission Packet Structure 346
  10.3.5 Coding Parameter History List 347
  10.3.6 The Packetization Algorithm 349
    10.3.6.1 Operational Scenarios of the Packetizing Algorithm . . . 349
10.4 H.263-based Video System Performance 352
  10.4.1 System Environment 352
  10.4.2 Performance Results 354
    10.4.2.1 Error-free Transmission Results 354
    10.4.2.2 Effect of Packet Dropping on Image Quality 354
    10.4.2.3 Image Quality versus Channel Quality without ARQ 356
    10.4.2.4 Image Quality versus Channel Quality with ARQ 357
  10.4.3 Comparison of H.263 and H.261-based Systems 359
    10.4.3.1 Performance with Antenna Diversity 361
    10.4.3.2 Performance over DECT Channels 362
10.5 Transmission Feedback 367
  10.5.1 ARQ Issues 371

        10.5.2 Implementation of Transmission Feedback 371
                10.5.2.1 Majority Logic Coding 372
  10.6 Summary and Conclusions 376

11 MPEG-4 Video Compression   379
  11.1 Introduction 379
  11.2 Overview of MPEG-4 380
        11.2.1 MPEG-4 Profiles 380
        11.2.2 MPEG-4 Features 381
        11.2.3 MPEG-4 Object-based Orientation 384
  11.3 MPEG-4: Content-based Interactivity 387
        11.3.1 VOP-based Encoding 389
        11.3.2 Motion and Texture Encoding 390
        11.3.3 Shape Coding 393
                11.3.3.1 VOP Shape Encoding 394
                11.3.3.2 Gray-scale Shape Coding 396
  11.4 Scalability of Video Objects 396
  11.5 Video Quality Measures 398
        11.5.1 Subjective Video Quality Evaluation 398
        11.5.2 Objective Video Quality 399
  11.6 Effect of Coding Parameters 400
  11.7 Summary and Conclusion 404

12 Comparative Study of the MPEG-4 and H.264 Codecs   407
  12.1 Introduction 407
  12.2 The ITU-T H.264 Project 407
  12.3 H.264 Video Coding Techniques 408
        12.3.1 H.264 Encoder 409
        12.3.2 H.264 Decoder 410
  12.4 H.264 Specific Coding Algorithm 410
        12.4.1 Intra-frame Prediction 410
        12.4.2 Inter-frame Prediction 412
                12.4.2.1 Block Sizes 412
                12.4.2.2 Motion Estimation Accuracy 413
                12.4.2.3 Multiple Reference Frame Selection for Motion
                          Compensation 414
                12.4.2.4 De-blocking Filter 414
        12.4.3 Integer Transform 415
                12.4.3.1 Development of the 4×4-pixel Integer DCT 416
                12.4.3.2 Quantization 419
                12.4.3.3 The Combined Transform, Quantization, Rescaling, and
                          Inverse Transform Process 420
                12.4.3.4 Integer Transform Example 421
        12.4.4 Entropy Coding 423
                12.4.4.1 Universal Variable Length Coding 424

		12.4.4.2 Context-based Adaptive Binary Arithmetic Coding 424
		12.4.4.3 H.264 Conclusion 425
	12.5	Comparative Study of the MPEG-4 and H.264 Codecs 425
		12.5.1 Introduction 425
		12.5.2 Intra-frame Coding and Prediction 425
		12.5.3 Inter-frame Prediction and Motion Compensation 426
		12.5.4 Transform Coding and Quantization 427
		12.5.5 Entropy Coding 427
		12.5.6 De-blocking Filter 427
	12.6	Performance Results 428
		12.6.1 Introduction 428
		12.6.2 MPEG-4 Performance 428
		12.6.3 H.264 Performance 430
		12.6.4 Comparative Study 433
		12.6.5 Summary and Conclusions 435

13 MPEG-4 Bitstream and Bit-sensitivity Study   437

    13.1 Motivation 437
    13.2 Structure of Coded Visual Data 437
        13.2.1 Video Data 438
        13.2.2 Still Texture Data 439
        13.2.3 Mesh Data 439
        13.2.4 Face Animation Parameter Data 439
    13.3 Visual Bitstream Syntax 440
        13.3.1 Start Codes 440
    13.4 Introduction to Error-resilient Video Encoding 441
    13.5 Error-resilient Video Coding in MPEG-4 441
    13.6 Error-resilience Tools in MPEG-4 443
        13.6.1 Resynchronization 443
        13.6.2 Data Partitioning 445
        13.6.3 Reversible Variable-length Codes 447
        13.6.4 Header Extension Code 447
    13.7 MPEG-4 Bit-sensitivity Study 448
        13.7.1 Objectives 448
        13.7.2 Introduction 448
        13.7.3 Simulated Coding Statistics 449
        13.7.4 Effects of Errors 452
    13.8 Chapter Conclusions 457

14 HSDPA-like and Turbo-style Adaptive Single- and Multi-carrier Video Systems   459

    14.1 Turbo-equalized H.263-based Videophony for GSM/GPRS 459
        14.1.1 Motivation and Background 459
        14.1.2 System Parameters 460
        14.1.3 Turbo Equalization 462
        14.1.4 Turbo-equalization Performance 465

    14.1.4.1 Video Performance 467
    14.1.4.2 Bit Error Statistics 469
  14.1.5 Summary and Conclusions 472
14.2 HSDPA-style Burst-by-burst Adaptive CDMA Videophony: Turbo-coded Burst-by-burst Adaptive Joint Detection CDMA and H.263-based Videophony 472
  14.2.1 Motivation and Video Transceiver Overview 472
  14.2.2 Multimode Video System Performance 477
  14.2.3 Burst-by-burst Adaptive Videophone System 480
  14.2.4 Summary and Conclusions 484
14.3 Subband-adaptive Turbo-coded OFDM-based Interactive Videotelephony . . 485
  14.3.1 Motivation and Background 485
  14.3.2 AOFDM Modem Mode Adaptation and Signaling 486
  14.3.3 AOFDM Subband BER Estimation 487
  14.3.4 Video Compression and Transmission Aspects 487
  14.3.5 Comparison of Subband-adaptive OFDM and Fixed Mode OFDM Transceivers 488
  14.3.6 Subband-adaptive OFDM Transceivers Having Different Target Bitrates 492
  14.3.7 Time-variant Target Bitrate OFDM Transceivers 498
  14.3.8 Summary and Conclusions 504
14.4 Burst-by-burst Adaptive Decision Feedback Equalized TCM, TTCM, and BICM for H.263-assisted Wireless Videotelephony 506
  14.4.1 Introduction 506
  14.4.2 System Overview 507
    14.4.2.1 System Parameters and Channel Model 509
  14.4.3 Employing Fixed Modulation Modes 512
  14.4.4 Employing Adaptive Modulation 514
    14.4.4.1 Performance of TTCM AQAM 515
    14.4.4.2 Performance of AQAM Using TTCM, TCC, TCM, and BICM 518
    14.4.4.3 The Effect of Various AQAM Thresholds 519
  14.4.5 TTCM AQAM in a CDMA system 520
    14.4.5.1 Performance of TTCM AQAM in a CDMA system 522
  14.4.6 Conclusions 525
14.5 Turbo-detected MPEG-4 Video Using Multi-level Coding, TCM and STTC . 526
  14.5.1 Motivation and Background 526
  14.5.2 The Turbo Transceiver 527
    14.5.2.1 Turbo Decoding 529
    14.5.2.2 Turbo Benchmark Scheme 531
  14.5.3 MIMO Channel Capacity 531
  14.5.4 Convergence Analysis 534
  14.5.5 Simulation Results 539
  14.5.6 Conclusions 543
14.6 Near-capacity Irregular Variable Length Codes 543
  14.6.1 Introduction 543

		14.6.2	Overview of the Proposed Schemes 544

        14.6.2   Overview of the Proposed Schemes 544
                 14.6.2.1   Joint Source and Channel Coding 545
                 14.6.2.2   Iterative Decoding 547
        14.6.3   Parameter Design for the Proposed Schemes 549
                 14.6.3.1   Scheme Hypothesis and Parameters 549
                 14.6.3.2   EXIT Chart Analysis and Optimization 550
        14.6.4   Results 552
                 14.6.4.1   Asymptotic Performance Following Iterative Decoding
                            Convergence 553
                 14.6.4.2   Performance During Iterative Decoding 554
                 14.6.4.3   Complexity Analysis 555
        14.6.5   Conclusions 557
14.7   Digital Terrestrial Video Broadcasting for Mobile Receivers 558
        14.7.1   Background and Motivation 558
        14.7.2   MPEG-2 Bit Error Sensitivity 559
        14.7.3   DVB Terrestrial Scheme 570
        14.7.4   Terrestrial Broadcast Channel Model 572
        14.7.5   Data Partitioning Scheme 573
        14.7.6   Performance of the Data Partitioning Scheme 579
        14.7.7   Nonhierarchical OFDM DVBP Performance 589
        14.7.8   Hierarchical OFDM DVB Performance 594
        14.7.9   Summary and Conclusions 600
14.8   Satellite-based Video Broadcasting 601
        14.8.1   Background and Motivation 601
        14.8.2   DVB Satellite Scheme 602
        14.8.3   Satellite Channel Model 604
        14.8.4   The Blind Equalizers 605
        14.8.5   Performance of the DVB Satellite Scheme 607
                 14.8.5.1   Transmission over the Symbol-spaced Two-path
                            Channel 608
                 14.8.5.2   Transmission over the Two-symbol Delay Two-path
                            Channel 614
                 14.8.5.3   Performance Summary of the DVB-S System 614
        14.8.6   Summary and Conclusions on the Turbo-coded DVB System 621
14.9   Summary and Conclusions 622
14.10 Wireless Video System Design Principles 623

Glossary   **625**

Bibliography   **635**

Index   **659**

Author Index   **667**

About the Authors

Lajos Hanzo (http://www-mobile.ecs.soton.ac.uk) FREng, FIEEE, FIET, DSc received his degree in electronics in 1976 and his doctorate in 1983. During his 30-year career in telecommunications he has held various research and academic posts in Hungary, Germany and the UK. Since 1986 he has been with the School of Electronics and Computer Science, University of Southampton, UK, where he holds the chair in telecommunications. He has co-authored 15 books on mobile radio communications totalling in excess of 10 000 pages, published about 700 research papers, acted as TPC Chair of IEEE conferences, presented keynote lectures and been awarded a number of distinctions. Currently he is directing an academic research team, working on a range of research projects in the field of wireless multimedia communications sponsored by industry, the Engineering and Physical Sciences Research Council (EPSRC) UK, the European IST Programme and the Mobile Virtual Centre of Excellence (VCE), UK. He is an enthusiastic supporter of industrial and academic liaison and he offers a range of industrial courses. He is also an IEEE Distinguished Lecturer of both the Communications Society and the Vehicular Technology Society (VTS). Since 2005 he has been a Governer of the VTS. For further information on research in progress and associated publications please refer to http://www-mobile.ecs.soton.ac.uk

Peter J. Cherriman graduated in 1994 with an M.Eng. degree in Information Engineering from the University of Southampton, UK. Since 1994 he has been with the Department of Electronics and Computer Science at the University of Southampton, UK, working towards a Ph.D. in mobile video networking which was completed in 1999. Currently he is working on projects for the Mobile Virtual Centre of Excellence, UK. His current areas of research include robust video coding, microcellular radio systems, power control, dynamic channel allocation and multiple access protocols. He has published about two dozen conference and journal papers, and holds several patents.

Jürgen Streit received his Dipl.-Ing. Degree in electronic engineering from the Aachen University of Technology in 1993 and his Ph.D. in image coding from the Department of Electronics and Computer Science, University of Southampton, UK, in 1995. From 1992 to 1996 Dr Streit had been with the Department of Electronics and Computer Science working in the Communications Research Group. His work led to numerous publications. Since then he has joined a management consultancy firm working as an information technology consultant.

Other Wiley and IEEE Press Books on Related Topics[1]

- R. Steele, L. Hanzo (Ed): *Mobile Radio Communications: Second and Third Generation Cellular and WATM Systems*, John Wiley and IEEE Press, 2nd edition, 1999, ISBN 07 273-1406-8, 1064 pages

- L. Hanzo, F.C.A. Somerville, J.P. Woodard: *Voice Compression and Communications: Principles and Applications for Fixed and Wireless Channels*, John Wiley and IEEE Press, 2001, 642 pages

- L. Hanzo, P. Cherriman, J. Streit: *Wireless Video Communications: Second to Third Generation and Beyond*, John Wiley and IEEE Press, 2001, 1093 pages

- L. Hanzo, T.H. Liew, B.L. Yeap: *Turbo Coding, Turbo Equalisation and Space-Time Coding*, John Wiley and IEEE Press, 2002, 751 pages

- J.S. Blogh, L. Hanzo: *Third-Generation Systems and Intelligent Wireless Networking: Smart Antennas and Adaptive Modulation*, John Wiley and IEEE Press, 2002, 408 pages

- L. Hanzo, C.H. Wong, M.S. Yee: *Adaptive Wireless Transceivers: Turbo-Coded, Turbo-Equalised and Space-Time Coded TDMA, CDMA and OFDM Systems*, John Wiley and IEEE Press, 2002, 737 pages

- L. Hanzo, L-L. Yang, E-L. Kuan, K. Yen: *Single- and Multi-Carrier CDMA: Multi-User Detection, Space-Time Spreading, Synchronisation, Networking and Standards*, John Wiley and IEEE Press, June 2003, 1060 pages

- L. Hanzo, M. Münster, T. Keller, B-J. Choi: *OFDM and MC-CDMA for Broadband Multi-User Communications, WLANs and Broadcasting*, John Wiley and IEEE Press, 2003, 978 pages

---

[1]For detailed contents and sample chapters please refer to http://www-mobile.ecs.soton.ac.uk

- L. Hanzo, S-X. Ng, T. Keller, W.T. Webb: *Quadrature Amplitude Modulation: From Basics to Adaptive Trellis-Coded, Turbo-Equalised and Space-Time Coded OFDM, CDMA and MC-CDMA Systems*, John Wiley and IEEE Press, 2004, 1105 pages

- L. Hanzo, T. Keller: *An OFDM and MC-CDMA Primer*, John Wiley and IEEE Press, 2006, 430 pages

- L. Hanzo, F.C.A. Somerville, J.P. Woodard: *Voice and Audio Compression for Wireless Communications*, 2nd edition, John Wiley and IEEE Press, 2007, 880 pages

- L. Hanzo, P.J. Cherriman, J. Streit: *Video Compression and Communications: H.261, H.263, H.264, MPEG4 and HSDPA-Style Adaptive Turbo-Transceivers*, John Wiley and IEEE Press, 2007, 680 pages

- L. Hanzo, J. Blogh, S. Ni: *HSDPA-Style FDD Versus TDD Networking: Smart Antennas and Adaptive Modulation*, John Wiley and IEEE Press, 2007, 650 pages

Preface

Against the backdrop of the fully-fledged third-generation wireless multimedia services, this book is dedicated to a range of topical wireless video communications aspects. The transmission of multimedia information over wireline based links can now be considered a mature area, even Digital Video Broadcasting (DVB) over both terrestrial and satellite links has become a mature commercial service. Recently, DVB services to handheld devices have been standardized in the DVB-H standard.

The book offers a historical perspective of the past 30 years of technical and scientific advances in both digital video compression and transmission over hostile wireless channels. More specifically, both the entire family of video compression techniques as well as the resultant ITU and MPEG video standards are detailed. Their bitstream is protected with the aid of sophisticated near-capacity joint source and channel coding techniques. Finally, the resultant bits are transmitted using advanced near-instantaneously adaptive High Speed Downlink Packet Access (HSDPA) style iterative detection aided turbo transceivers as well as their OFDM-based counterparts, which are being considered for the Third-Generation Partnership Project's Long-Term Evolution i(3GPP LTE) initiative.

Our hope is that the book offers you - the reader - a range of interesting topics, sampling - and hopefully without "gross aliasing errors", the current state of the art in the associated enabling technologies. In simple terms, finding a specific solution to a distributive or interactive video communications problem has to be based on a compromise in terms of the inherently contradictory constraints of video-quality, bitrate, delay, robustness against channel errors, and the associated implementational complexity. Analyzing these trade-offs and proposing a range of attractive solutions to various video communications problems are the basic aims of this book.

Again, it is our hope that the book underlines the range of contradictory system design trade-offs in an unbiased fashion and that you will be able to glean information from it in order to solve your own particular wireless video communications problem. Most of all however we hope that you will find it an enjoyable and relatively effortless reading, providing you with intellectual stimulation.

Lajos Hanzo, Peter J. Cherriman, and Jürgen Streit
School of Electronics and Computer Science
University of Southampton

Acknowledgments

We are indebted to our many colleagues who have enhanced our understanding of the subject, in particular to Prof. Emeritus Raymond Steele. These colleagues and valued friends, too numerous to be mentioned, have influenced our views concerning various aspects of wireless multimedia communications. We thank them for the enlightenment gained from our collaborations on various projects, papers and books. We are grateful to Jan Brecht, Jon Blogh, Marco Breiling, Marco del Buono, Sheng Chen, Clare Sommerville, Stanley Chia, Byoung Jo Choi, Joseph Cheung, Peter Fortune, Sheyam Lal Dhomeja, Lim Dongmin, Dirk Didascalou, Stephan Ernst, Eddie Green, David Greenwood, Hee Thong How, Thomas Keller, Ee-Lin Kuan, W. H. Lam, C. C. Lee, Soon-Xin Ng, M. A. Nofal, Xiao Lin, Chee Siong Lee, Tong-Hooi Liew, Matthias Muenster, Vincent Roger-Marchart, Redwan Salami, David Stewart, Jeff Torrance, Spyros Vlahoyiannatos, William Webb, John Williams, Jason Woodard, Choong Hin Wong, Henry Wong, James Wong, Lie-Liang Yang, Bee-Leong Yeap, Mong-Suan Yee, Kai Yen, Andy Yuen, and many others with whom we enjoyed an association.

We also acknowledge our valuable associations with the Virtual Centre of Excellence in Mobile Communications, in particular with its chief executive, Dr Walter Tuttlebee, as well as other members of its Executive Committee, namely Dr Keith Baughan, Prof. Hamid Aghvami, Prof. John Dunlop, Prof. Barry Evans, Prof. Mark Beach, , Prof. Peter Grant, Prof. Steve McLaughlin, Prof. Joseph McGeehan and many other valued colleagues. Our sincere thanks are also due to the EPSRC, UK for supporting our research. We would also like to thank Dr Joao Da Silva, Dr Jorge Pereira, Bartholome Arroyo, Bernard Barani, Demosthenes Ikonomou, and other valued colleagues from the Commission of the European Communities, Brussels, Belgium, as well as Andy Aftelak, Andy Wilton, Luis Lopes, and Paul Crichton from Motorola ECID, Swindon, UK, for sponsoring some of our recent research. Further thanks are due to Tim Wilkinson at HP in Bristol for funding some of our research efforts.

We feel particularly indebted to Chee Siong Lee for his invaluable help with proofreading as well as co-authoring some of the chapters. Jin-Yi Chung's valuable contributions in Chapters 11–13 are also much appreciated. The authors would also like to thank Rita Hanzo as well as Denise Harvey for their skillful assistance in typesetting the manuscript in Latex. Similarly, our sincere thanks are due to Jenniffer Beal, Mark Hammond, Sarah Hinton and a number of other staff members of John Wiley & Sons for their kind assistance throughout the preparation of the camera-ready manuscript. Finally, our sincere gratitude is due to the numerous authors listed in the Author Index, as well as to those, whose work was not cited due to space limitations. We are grateful for their contributions to the state of the art, without their contributions this book would not have materialized.

Lajos Hanzo, Peter J. Cherriman, and Jürgen Streit
School of Electronics and Computer Science
University of Southampton

Chapter 1

Introduction

1.1 A Brief Introduction to Compression Theory

The ultimate aim of data compression is the removal of redundancy from the source signal. This, therefore, reduces the number of binary bits required to represent the information contained within the source. Achieving the best possible compression ratio requires not only an understanding of the nature of the source signal in its binary representation, but also how we as humans interpret the information that the data represents.

We live in a world of rapidly improving computing and communications capabilities, and owing to an unprecedented increase in computer awareness, the demand for computer systems and their applications has also drastically increased. As the transmission or storage of every single bit incurs a cost, the advancement of cost-efficient source-signal compression techniques is of high significance. When considering the transmission of a source signal that may contain a substantial amount of redundancy, achieving a high compression ratio is of paramount importance.

In a simple system, the same number of bits might be used for representing the symbols "e" and "q". Statistically speaking, however, it can be shown that the character "e" appears in English text more frequently than the character "q". Hence, on representing the more-frequent symbols with fewer bits than the less-frequent symbols we stand to reduce the total number of bits necessary for encoding the entire information transmitted or stored.

Indeed a number of source-signal encoding standards have been formulated based on the removal of predictability or redundancy from the source. The most widely used principle dates back to the 1940s and is referred to as Shannon–Fano coding [2, 3], while the well-known Huffman encoding scheme was contrived in 1952 [4]. These approaches, however, have been further enhanced many times since then and have been invoked in various applications. Further research will undoubtedly endeavor to continue improving upon those techniques, asymptotically approaching the information theoretic limits.

Digital video compression techniques [5–9] have played an important role in the world of wireless telecommunication and multimedia systems, where bandwidth is a valuable commodity. Hence, the employment of video compression techniques is of prime importance

Table 1.1: Image Formats, Their Dimensions, and Typical Applications

Resolution	Dimensions	Pixel/s at 30 frames/s	Applications
Sub-QCIF	128×96	0.37 M	Handheld mobile video and
QCIF	176×144	0.76 M	videoconferencing via public phone networks
CIF	352×288	3.04 M	Videotape recorder quality
CCIR 601	720×480	10.40 M	TV
4CIF	704×576	12.17 M	
HDTV 1440	1440×960	47.00 M	Consumer HDTV
16CIF	1408×1152	48.66 M	
HDTV	1920×1080	62.70 M	Studio HDTV

in order to reduce the amount of information that has to be transmitted to adequately represent a picture sequence without impairing its subjective quality, as judged by human viewers. Modern compression techniques involve complex algorithms which have to be standardized in order to obtain global compatibility and interworking.

1.2 Introduction to Video Formats

Many of the results in this book are based on experiments using various resolution representations of the "Miss America" sequence, as well as the "Football" and "Susie" sequences. The so-called "Mall" sequence is used at High Definition Television (HDTV) resolution. Their spatial resolutions are listed in Table 1.1 along with a range of other video formats.

Each sequence has been chosen to test the codecs' performance in particular scenarios. The "Miss America" sequence is of low motion and provides an estimate of the maximum achievable compression ratio of a codec. The "Football" sequence contains pictures of high motion activity and high contrast. All sequences were recorded using interlacing equipment. *Interlacing* is a technique that is often used in image processing in order to reduce the required bandwidth of video signals, such as, for example, in conventional analog television signals, while maintaining a high frame-refresh rate, hence avoiding flickering and video jerkiness. This is achieved by scanning the video scene at half the required viewing-rate — which potentially halves the required video bandwidth and the associated bitrate — and then displaying the video sequence at twice the input scanning rate, such that in even-indexed video frames only the even-indexed lines are updated before they are presented to the viewer. In contrast, in odd-indexed video frames only the odd-indexed lines are updated before they are displayed, relying on the human eye and brain to reconstruct the video scene from these halved scanning rate even and odd video fields. Therefore, every other line of the interlaced frames remains un-updated.

For example, for frame 1 of the interlaced "Football" sequence in Figure 1.1 we observe that a considerable amount of motion took place between the two recoding instants of each

1.2. INTRODUCTION TO VIDEO FORMATS

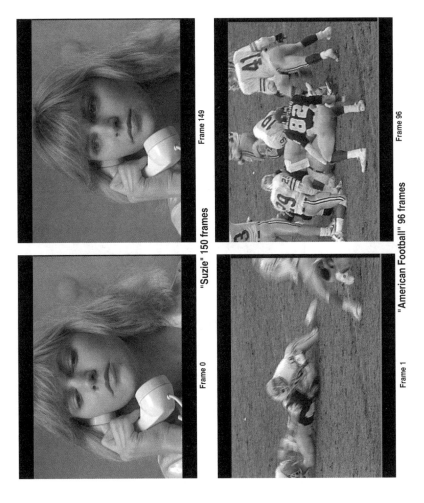

Figure 1.1: 4CIF video sequences.

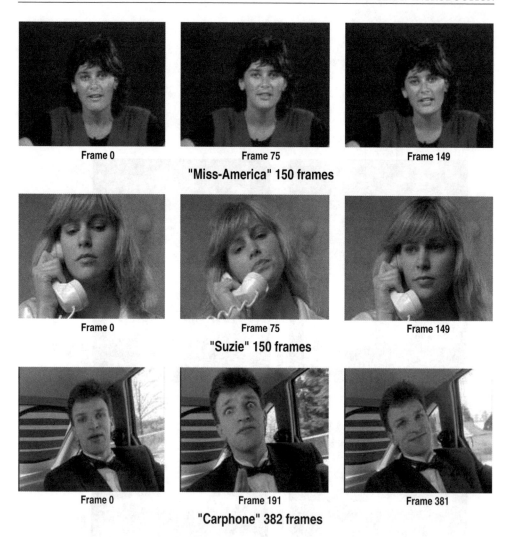

Figure 1.2: QCIF video sequences.

frame, which correspond to the even and odd video fields. Furthermore, the "Susie" sequence was used in our experiments in order to verify the color reconstruction performance of the proposed codecs, while the "Mall" sequence was employed in order to simulate HDTV sequences with camera panning. As an example, a range of frames for each QCIF video sequence used is shown in Figure 1.2. QCIF resolution images are composed of 176×144 pixels and are suitable for wireless handheld videotelephony. The 4CIF resolution images are suitable for digital television, which are 16 times larger than QCIF images. A range of frames for the 4CIF video sequences is shown in Figure 1.1. Finally, in Figure 1.3 we show a range of frames from the 1280×640-pixel "Mall" sequence. However, because the 16CIF resolution is constituted by 1408×1152 pixels, a black border was added to the sequences before they were coded.

We processed all sequences in the YUV color space [10] where the incoming picture information consists of the luminance (Y) plus two color difference signals referred to as chrominance U (Cr_u) and chrominance V (Cr_v). The conversion of the standard Red–Blue–Green (RGB) representation to the YUV format is defined in Equation 1.1:

$$\begin{bmatrix} Y \\ U \\ V \end{bmatrix} = \begin{pmatrix} 0.299 & 0.587 & 0.114 \\ -0.146 & -0.288 & 0.434 \\ 0.617 & -0.517 & -0.100 \end{pmatrix} \begin{pmatrix} R \\ G \\ B \end{pmatrix}. \qquad (1.1)$$

It is common practice to reduce the resolution of the two color difference signals by a factor of two in each spatial direction, which inflicts virtually no perceptual impairment and reduces the associated source data rate by 50%. More explicitly, this implies that instead of having to store and process the luminance signal and the two color difference signals at the same resolution, which would potentially increase the associated bitrate for color sequences by a factor of three, the total amount of color data to be processed is only 50% more than that of the associated gray-scale images. This implies that there is only one Cr_u and one Cr_v pixel for every four luminance pixels allocated.

The coding of images larger than the QCIF size multiplies the demand in terms of computational complexity, bitrate, and required buffer size. This might cause problems, considering that a color HDTV frame requires a storage of 6 MB per frame. At a frame rate of 30 frames/s, the uncompressed data rate exceeds 1.4 Gbit/s. Hence, for real-time applications the extremely high bandwidth requirement is associated with an excessive computational complexity. Constrained by this complexity limitation, we now examine two inherently low-complexity techniques and evaluate their performance.

1.3 Evolution of Video Compression Standards

Digital video signals may be compressed by numerous different proprietary or standardized algorithms. The most important families of compression algorithms are published by recognized standardization bodies, such as the International Organization for Standardization (ISO), the International Telecommunication Union (ITU), or the Motion Picture Expert Group (MPEG). In contrast, proprietary compression algorithms developed and owned by a smaller interest group are of lesser significance owing to their lack of global compatibility and interworking capability. The evolution of video compression standards over the past half-a-century is shown in Figure 1.4.

As seen in the figure, the history of video compression commences in the 1950s. An analog videophone system had been designed, constructed, and trialled in the 1960s, but it required a high bandwidth and it was deemed that using the postcard-size black-and-white pictures produced did not substantially augment the impression of telepresence in comparison to conventional voice communication. In the 1970s, it was realized that visual identification of the communicating parties may be expected to substantially improve the value of multi-party discussions and hence the introduction of videoconference services was considered. The users' interest increased in parallel to improvements in picture quality.

Figure 1.3: 16CIF "Mall" video sequence.

1.3. EVOLUTION OF VIDEO COMPRESSION STANDARDS

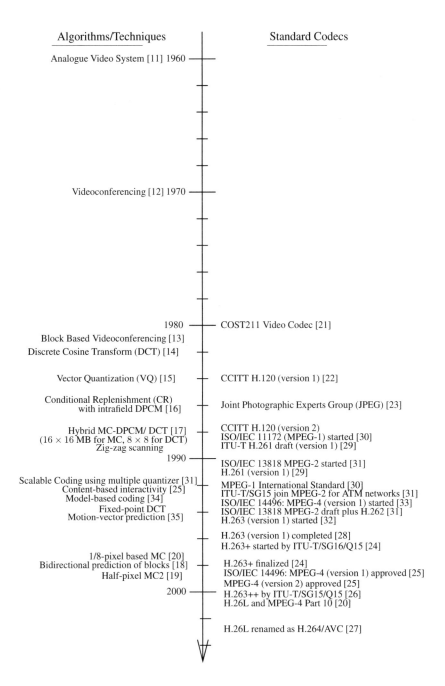

Figure 1.4: A brief history of video compression.

Video coding standardization activities started in the early 1980s. These activities were initiated by the International Telegraph and Telephone Consultative Committee (CCITT) [34], which is currently known as the International Telecommunications Union — Telecommunication Standardisation Sector (ITU-T) [22]. These standardization bodies were later followed by the formation of the Consultative Committee for International Radio (CCIR; currently ITU-R) [36], the ISO, and the International Electrotechnical Commission (IEC). These bodies coordinated the formation of various standards, some of which are listed in Table 1.2 and are discussed further in the following sections.

1.3.1 The International Telecommunications Union's H.120 Standard

Using state-of-the-art technology in the 1980s, a video en**co**der/**dec**oder (codec) was designed by the Pan-European Cooperation in Science and Technology (COST) project 211, which was based on Differential Pulse Code Modulation (DPCM) [59, 60] and was ratified by the CCITT as the H.120 standard [61]. This codec's target bitrate was 2 Mbit/s for the sake of compatibility with the European Pulse Code Modulated (PCM) bitrate hierarchy in Europe and 1.544 Mbit/s for North America [61], which was suitable for convenient mapping to their respective first levels of digital transmission hierarchy. Although the H.120 standard had a good spatial resolution, because DPCM operates on a pixel-by-pixel basis, it had a poor temporal resolution. It was soon realized that in order to improve the image quality without exceeding the above-mentioned 2 Mbit/s target bitrate, less than one bit should be used for encoding each pixel. This was only possible if a group of pixels, for example a "block" of 8×8 pixels, were encoded together such that the number of bits per pixel used may become a non-integer. This led to the design of so-called block-based codecs. More explicitly, at 2 Mbit/s and at a frame rate of 30 frames/s the maximum number of bits per frame was approximately 66.67 kbits. Using black and white pictures at 176×144-pixel resolution, the maximum number of bits per pixel was 2 bits.

1.3.2 Joint Photographic Experts Group

During the late 1980s, 15 different block-based videoconferencing proposals were submitted to the ITU-T standard body (formerly the CCITT), and 14 of these were based on using the Discrete Cosine Transform (DCT) [14] for still-image compression, while the other used Vector Quantization (VQ) [15]. The subjective quality of video sequences presented to the panel of judges showed hardly any perceivable difference between the two types of coding techniques. In parallel to the ITU-T's investigations conducted during the period of 1984–1988 [23], the Joint Photographic Experts Group (JPEG) was also coordinating the compression of static images. Again, they opted for the DCT as the favored compression technique, mainly due to their interest in progressive image transmission. JPEG's decision undoubtedly influenced the ITU-T in favoring the employment of DCT over VQ. By this time there was worldwide activity in implementing the DCT in chips and on Digital Signal Processors (DSPs).

1.3. EVOLUTION OF VIDEO COMPRESSION STANDARDS

Table 1.2: Evolution of Video Communications

Date	Standard
1956	AT&T designs and construct the first Picturephone test system [37]
1964	AT&T introduces Picturephone at the World's Fair, New York [37]
1970	AT&T offers Picturephone for $160 per month [37]
1971	Ericsson demonstrates the first Trans-Atlantic videophone (LME) call
1973 Dec.	ARPAnet packet voice experiments
1976 March	Network Voice Protocol (NVP), by Danny Cohen, USC/ISI [38]
1981 July	Packet Video Protocol (PVP), by Randy Cole, USC/ISI [39]
1982	CCITT (predecessor of the ITU-T) standard H.120 (2 Mbit/s) video coding, by European COST 211 project [22]
1982	Compression Labs begin selling $250,000 video conference (VC) system, $1,000 per hour lines
1986	PictureTel's $80,000 VC system, $100 per hour lines
1987	Mitsubishi sells $1,500 still-picture phone
1989	Mitsubishi drops still-picture phone
1990	TWBnet packet audio/video experiments, portable video players (pvp) (video) from Information Science Institute (ISI)/Bolt, Beranek and Newman, Inc. (BBN) [40]
1990	CCITT standard H.261 ($p \times 64$) video coding [29]
1990 Dec.	CCITT standard H.320 for ISDN conferencing [41]
1991	PictureTel unveils $20,000 black-and-white VC system, $30 per hour lines
1991	IBM and PictureTel demonstrate videophone on PC
1991 Feb.	DARTnet voice experiments, Voice Terminal (VT) program from USC/ISI [42]
1991 June	DARTnet research's packet video test between ISI and BBN. [42]
1991 Aug.	University of California, Berkeley (UCB)/Lawrence Berkeley National Laboratories (LBNL)'s audio tool vat releases for DARTnet use [42]
1991 Sept.	First audio/video conference (H.261 hardware codec) at DARTnet [42]
1991 Dec	dvc (receive-only) program, by Paul Milazzo from BBN, Internet Engineering Task Force (IETF) meeting, Santa Fe [43]
1992	AT&T's $1,500 videophone for home market [37]
1992 March	World's first Multicaster BackBONE (MBone) audio cast (vat), 23rd IETF, San Diego
1992 July	MBone audio/video casts (vat/dvc), 24th IETF, Boston
1992 July	Institute National de Recherche en Informatique et Automatique (INRIA) Videoconferencing System (ivs), by Thierry Turletti from INRIA [44]
1992 Sept.	CU-SeeMe v0.19 for Macintosh (without audio), by Tim Dorcey from Cornell University [45]
1992 Nov.	Network Video (nv) v1.0, by Ron Frederick from Xerox's Palo Alto Research Center (Xerox PARC), 25th IETF, Washington DC
1992 Dec.	Real-time Transport Protocol (RTP) v1, by Henning Schulzrinne [46]
1993 April	CU-SeeMe v0.40 for Macintosh (with multipoint conferencing) [47]
1993 May	Network Video (NV) v3.2 (with color video)
1993 Oct.	VIC Initial Alpha, by Steven McCanne and Van Jacobson from UCB/LBNL
1993 Nov.	VocalChat v1.0, an audio conferencing software for Novell IPX networks

Table 1.2: Continued

Date	Standard
1994 Feb.	CU-SeeMe v0.70b1 for Macintosh (with audio), audio code by Charley Kline's Maven [47]
1994 April	CU-SeeMe v0.33b1 for Windows (without audio), by Steve Edgar from Cornell [47]
1995 Feb.	VocalTec Internet Phone v1.0 for Windows (without video) [48]
1995 Aug.	CU-SeeMe v0.66b1 for Windows (with audio) [47]
1996 Jan.	RTP v2, by IETF avt-wg
1996 March	ITU-T standard H.263 ($p \times 8$) video coding for low bitrate communication [28]
1996 March	VocalTec Telephony Gateway [49]
1996 May	ITU-T standard H.324 for Plain Old Telephone System (POTS) conferencing [50]
1996 July	ITU-T standard T.120 for data conferencing [51]
1996 Aug.	Microsoft NetMeeting v1.0 (without video)
1996 Oct.	ITU-T standard H.323 v1, by ITU-T SG 16 [52]
1996 Nov.	VocalTec Surf&Call, the first Web to phone plugin
1996 Dec.	Microsoft NetMeeting v2.0b2 (with video)
1996 Dec.	VocalTec Internet Phone v4.0 for Windows (with video) [48]
1997 July	Virtual Room Videoconferencing System (VRVS), Caltech-CERN project [53]
1997 Sept.	Resource ReSerVation Protocol (RSVP) v1 [54]
1998 Jan.	ITU-T standard H.323 v2 [55]
1998 Jan.	ITU-T standard H.263 v2 (H.263+) video coding [24]
1998 April	CU-SeeMe v1.0 for Windows and Macintosh (using color video), from Cornell University, USA [47]
1998 May	Cornell's CU-SeeMe development team has completed their work [47]
1998 Oct.	ISO/IEC standard MPEG-4 v1, by ISO/IEC JTC1/SC29/WG11 (MPEG) [25]
1999 Feb.	Session Initiation Protocol (SIP) makes proposed standard, by IETF musicwork group [56]
1999 April	Microsoft NetMeeting v3.0b
1999 Aug.	ITU-T H.26L Test Model Long-term (TML) project, by ITU-T SG16/Q.6 (VCEG) [20]
1999 Sept.	ITU-T standard H.323 v3 [57]
1999 Oct.	Network Address Translation (NAT) compatible version of iVisit, v2.3b5 for Windows and Macintosh
1999 Oct.	Media Gateway Control Protocol (MGCP) v1, IETF
1999 Dec.	Microsoft NetMeeting v3.01 service pack 1 (4.4.3388)
1999 Dec.	ISO/IEC standard MPEG-4 v2
2000 May	Columbia SIP user agent sipc v1.30
2000 Oct.	Samsung releases the first MPEG-4 streaming 3G (CDMA2000-1x) video cell phone
2000 Nov.	ITU-T standard H.323 v4 [58]
2000 Nov.	MEGACO/H.248 Protocol v1, by IETF megaco-wg and ITU-T SG 16
2000 Dec.	Microsoft NetMeeting v3.01 service pack 2 (4.4.3396))
2000 Dec.	ISO/IEC Motion JPEG 2000 (JPEG 2000, Part 3) project, by ISO/IEC JTC1/SC29/WG1 (JPEG) [23]

1.3. EVOLUTION OF VIDEO COMPRESSION STANDARDS

Table 1.2: Continued

Date	Standard
2001 June	Windows XP Messenger supports the SIP
2001 Sept.	World's first Trans-Atlantic gallbladder surgery using a videophone (by surgeon Lindbergh)
2001 Oct.	NTT DoCoMo sells $570 3G (WCDMA) mobile videophone
2001 Oct.	TV reporters use $7,950 portable satellite videophone to broadcast live from Afghanistan
2001 Oct.	Microsoft NetMeeting v3.01 (4.4.3400) on XP
2001 Dec.	Joint Video Team (JVT) video coding (H.26L and MPEG-4 Part 10) project, by ITU-T SG16/Q.6 (VCEG) and ISO/IEC JTC1/SC29/WG 11 (MPEG) [20]
2002 June	World's first 3G video cell phone roaming
2002 Dec.	JVT completes the technical work leading to ITU-T H.264 [27]
2003	Wireless videotelephony commercialized

1.3.3 The ITU H.261 Standard

During the late 1980s it became clear that the recommended ITU-T videoconferencing codec would use a combination of motion-compensated inter-frame coding and the DCT. The codec exhibited a substantially improved video quality in comparison with the DPCM-based H.120 standard. In fact, the image quality was found to be sufficiently high for videoconferencing applications at 384 kbits/s and good quality was attained using 352×288-pixel Common Intermediate Format (CIF) or 176×144-pixel Quarter CIF (QCIF) images at bitrates of around 1 Mbit/s. The H.261 codec [29] was capable of using 31 different quantizers and various other adjustable coding options, hence its bitrate spanned a wide range. Naturally, the bitrate depended on the motion activity and the video format, hence it was not perfectly controllable. Nonetheless, the H.261 scheme was termed as a $p \times 64$ bits/s codec, $p = 1, \ldots, 30$ to comply with the bitrates provided by the ITU's PCM hierarchy. The standard was ratified in late 1989.

1.3.4 The Motion Pictures Expert Group

In the early 1990s, the Motion Picture Experts Group (MPEG) was created as Sub-Committee 2 of ISO (ISO/SC2). The MPEG started investigating the conception of coding techniques specifically designed for the storage of video, in media such as CD-ROMs. The aim was to develop a video codec capable of compressing highly motion-active video scenes such as those seen in movies for storage on hard disks, while maintaining a performance comparable to that of Video Home System (VHS) video-recorder quality. In fact, the basic MPEG-1 standard [30], which was reminiscent of the H.261 ITU codec [29], was capable of accomplishing this task at a bitrate of 1.5 Mbit/s. When transmitting broadcast-type distributive, rather than interactive of video, the encoding and decoding delays do not constitute a major constraint, one can trade delay for compression efficiency. Hence, in contrast to the H.261 interactive codec, which had a single-frame video delay, the MPEG-

1 codec introduced the bidirectionally predicted frames in its motion-compensation scheme.

At the time of writing, MPEG decoders/players are becoming commonplace for the storage of multimedia information on computers. MPEG-1 decoder plug-in hardware boards (e.g. MPEG magic cards) have been around for a while, and software-based MPEG-1 decoders are available with the release of operating systems or multimedia extensions for Personal Computer (PC) and Macintosh platforms.

MPEG-1 was originally optimized for typical applications using non-interlaced video sequences scanned at 25 frames/s in European format and at 29.9 frames/s in North American format. The bitrate of 1.2 to 1.5 Mbits/s typically results in an image quality comparable to home Video Cassette Recorders (VCRs) [30] using CIF images, which can be further improved at higher bitrates. Early versions of the MPEG-1 codec used for encoding interlaced video, such as those employed in broadcast applications, were referred to as MPEG-1+.

1.3.5 The MPEG-2 Standard

A new generation of MPEG coding schemes referred to as MPEG-2 [8, 31] was also adopted by broadcasters who were initially reluctant to use any compression of video sequences. The MPEG-2 scheme encodes CIF-resolution codes for interlaced video at bitrates of 4–9 Mbits/s, and is now well on its way to making a significant impact in a range of applications, such as digital terrestrial broadcasting, digital satellite TV [5], digital cable TV, digital versatile disc (DVD) and many others. Television broadcasters started using MPEG-2 encoded digital video sequences during the late 1990s [31].

A slightly improved version of MPEG-2, termed as MPEG-3, was to be used for the encoding of HDTV, but since MPEG-2 itself was capable of achieving this, the MPEG-3 standards were folded into MPEG-2. It is foreseen that by the year 2014, the existing transmission of NTSC format TV programmes will cease in North America and instead HDTV employing MPEG-2 compression will be used in terrestrial broadcasting.

1.3.6 The ITU H.263 Standard

The H.263 video codec was designed by the ITU-T standardization body for low-bitrate encoding of video sequences in videoconferencing [28]. It was first designed to be utilized in H.323-based systems [55], but it has also been adopted for Internet-based videoconferencing.

The encoding algorithms of the H.263 codec are similar to those used by its predecessor, namely the H.261 codec, although both its coding efficiency and error resilience have been improved at the cost of a higher implementational complexity [5]. Some of the main differences between the H.261 and H.263 coding algorithms are listed below. In the H.263 codec, half-pixel resolution is used for motion compensation, whereas H.261 used full-pixel precision in conjunction with a smoothing filter invoked for removing the high-frequency spatial changes in the video frame, which improved the achievable motion-compensation efficiency. Some parts of the hierarchical structure of the data stream are now optional in the H.263 scheme, hence the codec may be configured for attaining a lower data rate or better error resilience. There are four negotiable options included in the standard for the sake of potentially improving the attainable performance provided that both the encoder and decoder are capable of activating them [5]. These allow the employment of unrestricted motion vectors, syntax-based arithmetic coding, advanced prediction modes as well as both forward- and backward-frame prediction. The latter two options are similar to the MPEG codec's Predicted (P) and Bidirectional (B) modes.

1.3.7 The ITU H.263+/H.263++ Standards

The H.263+ scheme constitutes version 2 of the H.263 standard [24]. This version was developed by the ITU-T/SG16/Q15 Advanced Video Experts Group, which previously operated under ITU-T/SG15. The technical work was completed in 1997 and was approved in 1998. The H.263+ standard incorporated 12 new optional features in the H.263 codec. These new features support the employment of customized picture sizes and clock frequencies, improve the compression efficiency, and allow for quality, bitrate, and complexity scalability. Furthermore, it has the ability to enhance the attainable error resilience, when communicating over wireless and packet-based networks, while supporting backwards compatibility with the H.263 codec. The H.263++ scheme is version 3 of the H.263 standard, which was developed by ITU-T/SG16/Q15 [26]. Its technical content was completed and approved in late 2000.

1.3.8 The MPEG-4 Standard

The MPEG-4 standard is constituted by a family of audio and video coding standards that are capable of covering an extremely wide bitrate range, spanning from 4800 bit/s to approximately 4 Mbit/s [25]. The primary applications of the MPEG-4 standard are found in Internet-based multimedia streaming and CD distribution, conversational videophone as well as broadcast television.

The MPEG-4 standard family absorbs many of the MPEG-1 and MPEG-2 features, adding new features such as Virtual Reality Markup Language (VRML) support for 3D rendering, object-oriented composite file handling including audio, video, and VRML objects, the support of digital rights management and various other interactive applications.

Most of the optional features included in the MPEG-4 codec may be expected to be exploited in innovative future applications yet to be developed. At the time of writing, there are very few complete implementations of the MPEG-4 standard. Anticipating this, the developers of the standard added the concept of "Profiles" allowing various capabilities to be grouped together.

As mentioned above, the MPEG-4 codec family consists of several standards, which are termed "Layers" and are listed below [25].

- Layer 1: Describes the synchronization and multiplexing of video and audio.

- Layer 2: Compression algorithms for video signals.

- Layer 3: Compression algorithms for perceptual coding of audio signals.

- Layer 4: Describes the procedures derived for compliance testing.

- Layer 5: Describes systems for software simulation of the MPEG-4 framework.

- Layer 6: Describes the Delivery Multimedia Integration Framework (DMIF).

1.3.9 The H.26L/H.264 Standard

Following the finalization of the original H.263 standard designed for videotelephony, which was completed in 1995, the ITU-T Video Coding Experts Group (VCEG) commenced work on two further developments, specifically, a "short-term" effort including adding extra features to the H.263 codec resulting in Version 2 of the standard and a "long-term" effort aiming at developing a new standard specifically designed for low-bitrate visual communications. The long-term effort led to the draft "H.26L" standard, offering a significantly better video compression efficiency than the previous ITU-T standards. In 2001, the ISO MPEG recognized the potential benefits of H.26L and the Joint Video Team (JVT) was formed, including experts from both the MPEG and the VCEG. The JVT's main task was to develop the draft H.26L model1 into a full international standard. In fact, the outcome of these efforts turned out to be two identical standards, namely the ISO MPEG-4 Part 10 scheme of MPEG-4 and the ITU-T H.264 codec. The official terminology for the new standard is Advanced Video Coding (AVC), although, it is widely known by its old working title of H.26L and by its ITU document number H.264 [62].

In common with earlier standards, such as the MPEG-1, MPEG-2, and MPEG-4 schemes, the H.264 draft standard does not explicitly define an unambiguous coding standard. Rather, it defines the syntax of an encoded video bitstream and the decoding algorithm for this bitstream. The basic functional elements, such as motion prediction, transformation of the motion-compensated error residual, and the quantization of the resultant DCT coefficients as well as their entropy encoding are not unlike those of the previous standards, such as MPEG-1, MPEG-2, MPEG-4, H.261, H.263, etc. The important advances found in the H.264 codec occur in the specific implementation of each functional element. The H.264 codec is described in more detail in Section 12.2.

This book reports on advances attained during the most recent years of the half-a-century history of video communications, focussing on the design aspects of wireless videotelephony, dedicating particular attention to the contradictory design aspects portrayed in Figure 1.5.

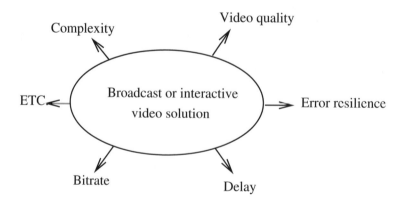

Figure 1.5: Contradictory system design requirements of various video communications systems.

1.4 Video Communications

Video communication over rate-limited and error-prone channels, such as packet networks and wireless links, requires both a high error resilience and high compression. In the past, considerable efforts have been invested in the design and development of the most efficient video compression schemes and standards. For the sake of achieving high compression, most modern codecs employ motion-compensated prediction between video frames, in order to reduce the temporal redundancy, followed by a spatial transform invoked for reducing the spatial redundancy. The resultant parameters are entropy-coded, in order to produce the compressed bitstream. These algorithms provide high compression, however, the compressed signal becomes highly vulnerable to error-induced data losses, which is particularly detrimental when communicating over best-effort networks. In particular, video transmission is different from audio transmission because the dependency across successive video frames is much stronger owing to the employment of inter-frame motion-compensated coding. In this book, a network-adaptive source-coding scheme is proposed for dynamically managing the dependency across packets, so that an attractive trade-off between compression efficiency and error resilience may be achieved. Open standards such as the ITU H.263 [63], H.264 [27] and the ISO/IEC MEPG-4 video codecs [25] were invoked in our proposed schemes.

To address the challenges involved in the design of wireless video transmissions and video streaming, in recent years the research efforts of the community have been directed particularly towards communications efficiency, error resilience, low latency, and scalability [5, 64, 65]. In video communications, postprocessing is also applied at the receiver side for the sake of error concealment and loss recovery. The achievable subjective quality was also improved by an adaptive deblocking filter in the context of the H.264/MPEG-4 video codec [66]. A range of techniques used to recover the damaged video frame areas based on the characteristics of image and video signals have been reviewed in [67]. More specifically, spatial-domain interpolation was used in [68] to recover an impaired macroblock; transform-domain schemes were used to recover the damage inflicted by partially received DCT coefficients, as presented in [69–72]. Temporal-domain schemes interpolate the missing information by exploiting the inherent temporal correlation in adjacent frames. Application examples include, for example, interpolated motion compensation [73, 74] and state recovery [75]. More specifically, the conventional video compression standards employ an architecture, which we refer to as a single-state architecture, because, for example, they have a prediction loop assisted with a single state constituted by the previous decoded frame, which may lead to severe degradation of all subsequent frames, until the corresponding state is reinitialized in the case of loss or corruption. In the state recovery system proposed by Apostolopoulos [75], the problem of having an incorrect state or that of encountering error propagation at the decoder is mitigated by encoding the video into multiple independently decodable streams, each having its own prediction process and state, such that if one stream is lost, the other streams can still be used for producing usable video sequence. Other schemes, such as the temporal smoothness method [76], the coding mode recovery scheme of [72, 76], and the Displaced Frame Difference (DFD) as well as the Motion Vector (MV) recovery management of [77–80] have also resulted in substantial performance. These schemes can also be combined with layered coding, as suggested in [81, 82].

The development of flexible, near-instantaneously adaptive schemes capable of maintaining a perceptually attractive video quality regardless of the channel quality encountered is one of the contributions of this book. Recently, significant research interests have also been devoted to Burst-by-Burst Adaptive Quadrature Amplitude Modulation (BbB-AQAM) transceivers [83, 84], where the transceiver reconfigures itself on a near-instantaneous basis, depending on the prevalent perceived wireless channel quality. Modulation schemes of different robustness and different data throughput have also been investigated [85, 86]. The BbB-AQAM principles have also been applied to Joint Detection Code Division Multiple Access (JD-CDMA) [83, 87] and OFDM [88]. A range of other adaptive video transmission schemes have been proposed for the sake of reducing the transmission delay and the effective of packet loss by Girod and co-workers [89, 90].

Video communication typically requires higher data transmission rates than other sources, such as, audio or text. A variety of video communications schemes have been proposed for increasing the robustness and efficiency of communication [91–95]. Many of the recent proposals employ Rate Distortion (R-D) optimization techniques for improving the achievable compression efficiency [96–98], as well as for increasing the error resilience, when communicating over lossy networks [99, 100]. The goal of these optimization algorithms is to jointly minimize the total video distortion imposed by both compression and channel effects, subject to a given total bitrate constraint. A specific example of recent work in this area is related to intra/inter-mode switching [101, 102], where intra-frame coded macroblocks are transmitted according to the prevalent network conditions for mitigating the effects of error propagation across consecutive video frames. More specifically, an algorithm has been proposed in [102–104] for optimal intra/inter-mode switching, which relies on estimating the overall distortion imposed by quantization, error propagation, and error concealment.

A sophisticated channel coding module invoked in a robust video communication system may incorporate both Forward Error Correction (FEC) and Automatic Re-transmission on Request (ARQ), provided that the ARQ-delay does not affect "lip-synchronization". Missing or corrupted packets may be recovered at the receiver, as long as a sufficiently high fraction of packets is received without errors [5, 105, 106]. In particular, Reed–Solomon (RS) codes are suitable for this application as a benefit of their convenient features [107, 108]. FEC is also widely used for providing Unequal Error Protection (UEP), where the more vulnerable bits are protected by stronger FEC codes. Recent work has addressed the problem of how much redundancy should be added and distributed across different by prioritized data partitions [108–112]. In addition to FEC codes, data randomization and interleaving have also been employed for providing enhanced protection [75, 113, 114]. ARQ techniques incorporate channel feedback and employ the retransmission of erroneous data [5, 115–118]. More explicitly, ARQ systems use packet acknowledgments and timeouts for controlling which particular packets should be retransmitted. Unlike FEC schemes, ARQ intrinsically adapts to the varying channel conditions and hence in many applications tends to be more efficient. However, in the context of real-time communication and low-latency streaming the latency introduced by ARQ is a major concern. Layered or scalable coding, combined with transmission prioritization, is another effective approach devised for providing error resilience [73, 109, 119–121]. In a layered scheme, the source signal is encoded such that it generates more than one different significance group or layer, with the base layer containing the most essential information required for media reconstruction at an acceptable quality, while the enhancement layer(s) contains information that may be invoked for reconstruction at an enhanced quality. At high packet loss rates, the more-important, more strongly protected layers can still be recovered, while the less-important layers might not. Commonly used layered techniques may be categorized into temporal scalability [122], spatial scalability [123, 124], Signal-to-Noise Ratio (SNR) scalability [25], data partitioning [27], or any combinations of these. Layered scalable coding has been widely employed for video streaming over best-effort networks, including the Internet and wireless networks [121, 125–128]. Different layers can be transmitted under the control of a built-in prioritization mechanism without network support, such as the UEP scheme mentioned above, or using network architectures capable of providing various different Quality of Service (QoS) [129–132]. A scheme designed for optimal intra/inter-mode selection has recently been proposed for scalable coding, in order to limit the inter-frame error propagation inflicted by packet losses [133]. Another scheme devised for adaptive bitrate allocation in the context of scalable coding was presented in [134]. Layered scalable coding has become part of various established video coding standards, such as the members of the MPEG [25, 30, 31] and H.263+ codec family [24].

Dogan *et al.* [135] reported promising results on adopting the MPEG-4 codec for wireless applications by exploiting the rate control features of video transcoders and combined them with error-resilient General Packet Radio Service (GPRS) type mobile access networks.

The employment of bidirectionally predicted pictures during the encoding process is capable of substantially improving the compression efficiency, because they are encoded using both past and future pictures as references. The efficiency of both forward and backward prediction was studied as early as 1985 by Musmann *et al.* [136]. Recent developments have been applied to the H.264/MPEG codecs, amongst others by Flierl and Girod [137] and Shanableh and Ghanbari [138]. In order to achieve even higher compression in video coding, Al-Mualla, Canagarajah and Bull [139] proposed a fast Block Matching Motion Estimation (BMME) algorithm referred to as the Simplex Minimization Search (SMS), which was incorporated into various video coding standards such as the H.261, H.263, MPEG-1, and MPEG-2 for the sake of both single or multiple reference aided motion estimation.

A plethora of video coding techniques designed for the H.264 standard have been proposed also in the excellent special issues edited by Luthra, Sullivan and Wiegand [140]. At the time of writing there are numerous ongoing research initiatives with the objective of improving the attainable video transmission in wireless environments. Wenger [141] discussed the transmission of H.264 encoded video over IP networks while Stockhammer *et al.* [142] studied the transmission of H.264 bitstreams over wireless environments. Specifically, the design of the H.264 codec specifies a video coding layer and a network adaptation layer, which facilitate the transmission of the bitstream in a network-friendly fashion. As for the wireless networking area, a recent publication of Arumugan *et al.* [143] investigated the coexistence of 802.11g WLANs and high data rate Bluetooth-enabled consumer electronic devices in both indoor home and office environments.

Further important contributions in the area of joint source and channel coding entail the development of a scheme that offers the end-to-end joint optimization of source coding and channel coding/modulation over wireless links, for example those by Thobaben and Kliewer [144] or Murad and Fuja [145].

1.5 Organization of the Monograph

- **Part I** of the book is dedicated to video compression basics. These introductory topics are revised in the context of a host of fixed but arbitrarily programmable rate video codecs based on fractal coding, on the DCT, on VQ codecs, and quad-tree-based codecs. These video codecs and their associated Quadrature Amplitude Modulated (QAM) video systems are described in Chapters 2–5.

- **Part II** of the book is focussed on high-resolution video coding, encompassing Chapters 6 and 7.

- **Part III** of the book entails Chapters 8–14, which characterize the H.261 and H.263 video codecs, constituting important representatives of the family of hybrid DCT codecs. Hence, the associated findings of these chapters can be readily applied in the context of other hybrid DCT codecs, such as the MPEG family, including the MPEG-1, MPEG-2 and MPEG-4 codecs. Chapters 8–14 also portray the interactions of these hybrid DCT video codecs with HSDPA-style near-instantaneously reconfigurable multimode QAM transceivers.

Chapter 11 provides an overview of the MPEG-4 video codec. This chapter will assist the reader in following our further elaborations in the forthcoming chapters. Chapter 12 has been divided into two parts, the first half provides an overview of the H.264 video codec, while the second part is constituted by a comparative study of the MPEG-4 and H.264 video codecs.

Video compression often invokes Variable Length Coding (VLC). All of the standard video codecs have adopted these techniques, because they are capable of dramatically reducing the bitrate. However, VLC techniques render the bitstream vulnerable to transmission errors. In Chapter 13 we show the effects of transmission errors on the MPEG-4 codec, quantifying the sensitivity of the various bits.

The book concludes in Chapter 14 by offering a range of system design studies related to wideband burst-by-burst (BbB) adaptive TDMA/TDD, OFDM, and CDMA interactive as well as distributive mobile video systems and their performance characterization over highly dispersive transmission media. More specifically, both H.263/H.264 as well as MPEG-4 compression-based interactive videophone schemes are proposed and investigated, using BbB adaptive High Speed Downlink Packet Access (HSDPA) style iterative detection-aided turbo transceivers. Amongst a range of other system design examples, Coded Modulation-Aided JD-CDMA video transceivers are investigated, which are capable of near-instantaneously dropping as well as increasing their source coding rate and video quality under transceiver control as a function of the near-instantaneous channel quality.

Several MPEG-4 compression based videophone schemes are also studied. First of all, we investigate an Iterative Parallel Interference Cancellation (PIC) aided CDMA MPEG-4 videophone scheme. Following this study, we investigated a novel turbo-detected unequal error protection MPEG-4 videophone scheme using a serially concatenated convolutional outer code, trellis-coded modulation-based inner code, and space–time coding. We also proposed a simple packetization scheme, where we partitioned the MPEG-4 bitstream into two bit-sensitivity classes and assigned an Unequal Protection Scheme (UEP).

Part I

Video Codecs for HSDPA-style Adaptive Videophones

Chapter 2

Fractal Image Codecs

2.1 Fractal Principles

Fractal image codecs have attracted considerable interest in recent years through the valuable contributions, theses, of Barnsley [146], Beaumont [147], Jacquin [148], and Monroe [149, 150]. Fractals are geometrical objects with endless self-similar details, resulting in efficient parametric image representations [146]. An often-quoted example for self-similarity is a fern, the leaves of which can be viewed as small ferns, and the same is valid for the leaf of the leaf, and so on, as seen in Figure 2.1.

This self-similarity expressed in technical terms is intra-frame redundancy, which can be removed by expressing regions of the image to be encoded as transformed versions of other image segments using so-called *contractive affine* transforms. This transform can be expressed in two dimensions (2D) [146], as follows:

$$\begin{bmatrix} X \\ Y \end{bmatrix} = \begin{bmatrix} r\cos\phi & -s\sin\theta \\ r\sin\phi & s\cos\theta \end{bmatrix} \begin{bmatrix} x \\ y \end{bmatrix} + \begin{bmatrix} X_0 \\ Y_0 \end{bmatrix} \qquad (2.1)$$

$$= \begin{bmatrix} A & B \\ C & D \end{bmatrix} \begin{bmatrix} x \\ y \end{bmatrix} + \begin{bmatrix} X_0 \\ Y_0 \end{bmatrix} \qquad (2.2)$$

or, alternatively, as:

$$\begin{aligned} X &= (r\cos\phi)x - (s\sin\theta)y + X_0 \\ &= Ax + By + X_0 \\ Y &= (r\sin\phi)x + (s\cos\theta)y + Y_0 \\ &= Cx + Dy + Y_0. \end{aligned}$$

These transforms allow us to represent each pixel (X, Y) of a 2D two-tone object to be encoded as a transformed version of the pixels x, y of another object by scaling the x and y coordinates using the factors $(r, s) < 1$, rotating them by angles (ϕ, θ), as well as linearly translating the original object in both directions by (X_0, Y_0). This contractivity of the affine transformations, which is imposed by setting $(r, s) < 1$, is important, for these

Figure 2.1: Affine transformation example.

transformations are invoked repetitively a high number of times in order to be able to resolve arbitrarily fine details in images. The effect of this contractive affine transformation is shown in Figure 2.1 using the example of a fern. Note, however, that from an image-coding point of view, it is preferable to find a transformation that translates each pixel of an object into the corresponding pixels of another object in a single object-transformation step, rather than carrying out the transformation on a pixel-by-pixel basis, since this would incur a high computational complexity and require storing all the transformation parameters for the sake of facilitating the inverse mapping. These remarks will be augmented later in this chapter.

For nonbinary gray-scale images, a third dimension must be introduced in order to represent the luminance information. The simple principle of fractal coding requires us first to partition the original image into small regions, which are referred to as range blocks (RB) that perfectly tile the original image. This is particularly necessary for large image segments, for would be difficult to find other similar segments, which could be used for their encoding. The larger the image segment, the more difficult to find matching objects. Then a pool of domain blocks (DB) is defined in Figure 2.2, which can be thought of as a two-dimensional (2D) codebook. Each image segment constituted by the legitimate domain blocks is tentatively contractive affine transformed to the specific RB about to be encoded. The fidelity of the match expressed typically in terms of the Mean Squared Error (MSE) is remembered, and the coordinates of the best matching DB for each RB are stored and transmitted to the decoder for image reconstruction. This process is illustrated in Figure 2.2, and the specific techniques to be used will be detailed later.

The affine transform associated with the best MSE match is described by means of its Iterated Function System (IFS) code [146] given by $(r, s, \phi, \theta, X_0, Y_0)$, which can be used at the decoder to reconstruct the original image. The technique of image reconstruction will be highlighted after exemplifying the concept of IFS codes with reference to our previous "fern example", whose IFS code is given in Table 2.1. A more detailed discussion on one-dimensional fractal coding is given in Section 2.2.

Returning to the problem of image reconstruction from the IFS codes, the Random Collage Theorem [146] must be invoked, which states that from any arbitrary initial state the image can be reconstructed by the iterative application of the contractive affine transform

2.2. ONE-DIMENSIONAL FRACTAL CODING

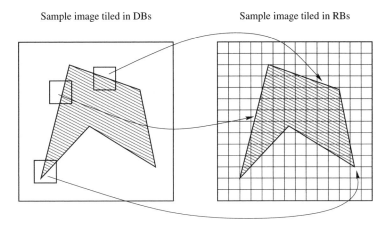

Figure 2.2: Example of mapping three RBs on DBs.

Table 2.1: The IFS Code of a Fern

A	B	C	D	X_0	Y_0
0	0	0	0.16	0.0	0
0.2	−0.26	0.23	0.22	0.0	1.6
−0.15	0.28	0.26	0.24	0.0	0.44
0.85	0.04	−0.04	0.85	0.0	1.6

described by the IFS code. As previously mentioned, contractivity is required in order to consistently reduce the size of the transformed objects, and thus resolve more and more fine detail to ensure algorithmic convergence. Algorithmic details of the above techniques will be augmented in the next section, first in the context of a one-dimensional fractal coding example, before delving into details of 2D fractal video coding.

2.2 One-dimensional Fractal Coding

One-dimensional signals can be acoustic signals or line-by-line scanned images. In order to allow reduced-complexity processing, the signal waveform is divided into short one-dimensional segments, constituting the range blocks. The Collage Theorem requires that the range blocks (RBs) perfectly tile the waveform segment to be encoded. Because of the required contractivity for each RB, a longer domain block (DB) has to be found, which has a similar waveform shape. Finally, each DB is mapped to the RB using a specific affine transform. A set of legitimate contractive affine transforms is as follows:

- Scale the amplitude of the block.
- Add an offset to the amplitude.

Table 2.2: IFS Codes for the Example of Figure 2.4

RB	DB	Scaling	Offset	Rotation
a	A	0.5	−0.25	0
b	A	0.5	0.25	0
c	B	0.5	0.25	0
d	A	0.25	0.125	180

- Reflect the waveform of the block about the x-axis or y-axis.
- Rotation by 180 degrees.

The larger the set of legitimate transforms of the block, the better the match becomes. However, an increasing number of possible transformations results in a higher number of IFS codes and ultimately in a higher bitrate, as well as higher computational complexity.

The encoding technique is explained with reference to Figure 2.3, following the Beaumont approach [147]. At the top of the figure, a segment of an image scanline luminance is divided into eight RBs. In order to encode the first RB, a similar pattern in the signal has to be found. In general, it is convenient to use a DB size, that is an integer multiple of the RB size, and here we opted for a double-length DB. Observe in the figure that RB1 could be represented by DB4 reasonably well, if DB4 were appropriately rotated and scaled using contractive affine transformations as follows. After contracting the length of DB4 to the RB size, as portrayed in Figures 2.3(a) and (b), the waveform segment has to be amplitude-scaled and vertically shifted, as displayed in Figure 2.3(c), before it is finally rotated in Figure 2.3(d) in order to produce a waveform closely matching RB1. The amplitude-scaling, offset, and reflection parameters represent the IFS code of RB1. By calculating these values for all RBs, the IFS code of the scanline is fully described.

Let us now consider the more specific example of Figure 2.4, for which we can readily compute the associated IFS codes. The description of RB "a" by the help of the appropriately transformed RB "A" is illustrated step-by-step in Figure 2.5, resulting in the IFS codes given in the first line of Table 2.2. Specifically, DB "A" is first subjected to "contraction" by a factor of 2, as seen at the consecutive stages of Figure 2.5, and then its magnitude is scaled by 0.5. Lastly, an offset of −0.25 is applied, yielding exactly RB "a". Similar fractal transformation steps result in the IFS codes of Table 2.2 for the remaining RBs, namely, RB b, c, and d of Figure 2.4. The decoder has to rely on the IFS codes of each RB in order to reproduce them. However, the decoder is oblivious of the DB pool of the encoder, which was used to generate the IFS codes. Simple logic would suggest that a repeated application of the fractal transforms conveyed by the IFS codes and applied to the DBs is the only way of reproducing the RBs. *The surprising fact is that repeated application of the IFS-coded contractive affine transforms to an arbitrary starting pattern will converge on the encoded RB, which was stated before as the Random Collage Algorithm.*

The application of the Random Collage Algorithm is demonstrated in Figure 2.6. As seen in the figure, the decoding operation commences with a simple "flat" signal, since the original DBs are not known to the decoder. According to Table 2.2, the first RB can

2.2. ONE-DIMENSIONAL FRACTAL CODING

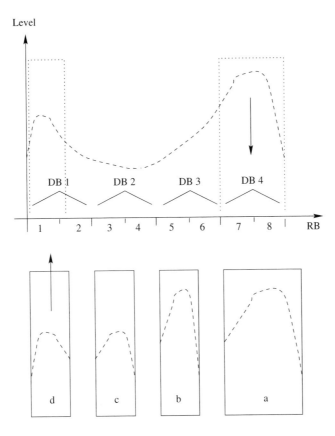

Figure 2.3: One-dimensional fractal coding.

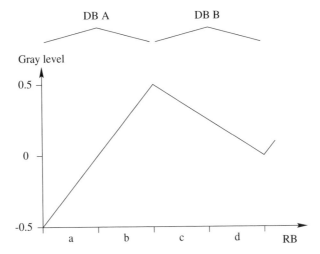

Figure 2.4: DBs and RBs for IFS code computation example for which the IFS codes are given in Table 2.2.

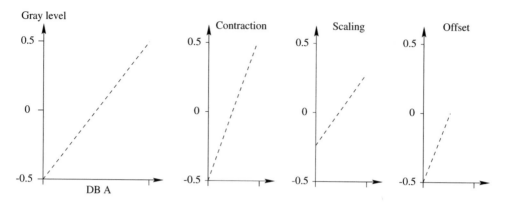

Figure 2.5: Contractive affine transform steps for the IFS code computation example related to RB "a" of Figure 2.4 for which the IFS codes are given in Table 2.2.

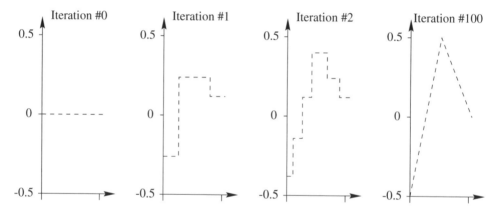

Figure 2.6: Application of the one-dimensional random collage algorithm to the example of Figure 2.4 for which the IFS codes are given in Table 2.2.

be obtained by scaling the decoder's DB A by 0.5 and shifting it by -0.25. During the first iteration, the scaling does not have any influence, the initialization with zero makes all DBs "flat", thus, at this point, every RB contains a constant DC level, corresponding to the offset information. In the second step, the DBs are already different. DB A now contains a step function. It is scaled and shifted again, and after this second iteration the original waveform is already recognizable. With every iteration the number of steps approximating the ramp-signal increases, whereas the step size shrinks. In this example, the encoder was able to find a perfect match for every RB, and hence after an infinite number of iterations the decoded signal yields the source signal.

Having considered the basic fractal coding principles, let us now examine a range of fractal codec designs contrived for QCIF "head-and-shoulders" videophone sequences.

2.2.1 Fractal Codec Design

Previously proposed fractal codec designs were targeted at high-resolution images, having large intra-frame domain-block pools [148, 149], which will be introduced later in this chapter. Following the approaches proposed by pioneering researchers of the field, such as Barnsley [146], Jacquin [148], Monroe [149, 150], Ramamurthi and Gersho [15], and Beaumont [147], in this study we explored the range of design trade-offs available using a variety of different head-and-shoulders fractal videophone codecs (Codecs A-E) and compared their complexity, compression ratio, and image quality using QCIF images.

In fractal image coding, the QCIF image to be encoded is typically divided into 4-by-4 or 8-by-8 picture elements (pel) based on nonoverlapping range blocks (RB), which perfectly tile the original image [146]. Other block sizes can also be used, but perfect tiling of the frames must be ensured. Every RB is then represented by the contractive affine transformation [149] of a larger, typically quadruple-sized domain block (DB) taken from the same frame of the original image. In general, the larger the pool of domain blocks, the better the image quality, but the higher the computational complexity and bitrate, requiring a compromise. For gray-scale coding of 2D images, a third dimension, representing the brightness of the picture, must be added before affine transformation takes place. Jacquin suggested that for the sake of reduced complexity it is attractive to restrict the legitimate affine transforms to the following manipulations [148]:

1. Linear translation of the block.
2. Rotation of the block by 0, 90, 180, and 270 degrees.
3. Reflection about the diagonals, vertical, and horizontal axis.
4. Luminance shift of the block.
5. Contrast scaling of the block.

The Collage Theorem [146] and practical contractivity requirements facilitate the following (DB, RB) size combinations: $(16 \times 16, 8 \times 8), (16 \times 16, 4 \times 4), (8 \times 8, 4 \times 4)$, and our codecs attempt to match every RB with every DB of the same frame, allowing rotations by $0°, 90°, 180°$, or $270°$. Adopting the MSE expression of

$$\text{MSE} = \sqrt{\frac{1}{p^2} \sum_{i=1}^{p} \sum_{k=1}^{p} (X_{ik} - (aY_{ik} - b))^2} \quad (2.3)$$

as a block-matching distortion measure, where p is the RB size, the optimum contrast scaling factor a and luminance shift b for the contracted DBs Y_{ik} and RBs X_{ik} can be derived by minimizing the MSE defined above leading to:

$$0 = \sum_{i=1}^{p} \sum_{k=1}^{p} (aY_{ik}^2 - bY_{ik} - X_{ik}Y_{ik}) \quad (2.4)$$

$$0 = \sum_{i=1}^{p} \sum_{k=1}^{p} (X_{ik} - aY_{ik} + b), \quad (2.5)$$

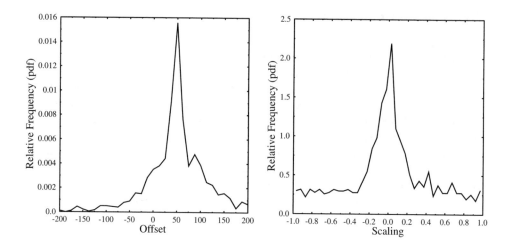

Figure 2.7: PDF of contrast scaling and luminance shift.

where the first derivative of Equation 2.3 is set to zero with respect to a and b, respectively. This yields a set of two equations with two variables, which finally leads to:

$$b = \frac{\sum_{i=1}^{p}\sum_{k=1}^{p} X_{ik}Y_{ik} \sum_{i=1}^{p}\sum_{k=1}^{p} Y_{ik} - \sum_{i=1}^{p}\sum_{k=1}^{p}(Y_{ik}^2 \sum_{i=1}^{p}\sum_{k=1}^{p} X_{ik})}{p^2 \sum_{i=1}^{p}\sum_{k=1}^{p} Y_{ik}^2 - (\sum_{i=1}^{p}\sum_{k=1}^{p} Y_{ik})^2} \quad (2.6)$$

$$a = \frac{\sum_{i=1}^{p}\sum_{k=1}^{p} X_{ik}}{\sum_{i=1}^{p}\sum_{k=1}^{p} Y_{ik}} + \frac{p^2}{\sum_{i=1}^{p}\sum_{k=1}^{p} Y_{ik}} \cdot b. \quad (2.7)$$

The optimum contrast-scale factor a and luminance-shift b have to be found for all the RBs, and their values have to be quantized for transmission to the decoder.

2.2.2 Fractal Codec Performance

In order to identify the range of design trade-offs five different codecs, following the philosophy in [15, 146–150] Codecs A–E, were simulated and compared in Table 2.3. The DB indices and the four different rotations of $0°, 90°, 180°$, and $270°$ were gray-coded in order to ensure that a single bit error results in the neighboring codewords, while the luminance shift b and contrast scaling a were Max-Lloyd quantized using 4 bits each. The PDFs of these quantities are portrayed in Figure 2.7.

Comparison of the three basic schemes, Codecs A–C featured in Table 2.3, suggested that a RB size of 4×4 used in Codecs B and C was desirable in terms of image quality, having a peak signal-to-noise ratio (PSNR) of 5–6 dB higher than Codec A which had a RB size of 8×8. The PSNR is the most frequently used objective video-quality measure defined as

$$PSNR = 10 \log_{10} \frac{\sum_{i=1}^{p}\sum_{j=1}^{q} \max^2}{\sum_{i=1}^{p}\sum_{j=1}^{q}(f_n(i,j) - \tilde{f}_n(i,j))^2} \quad (2.8)$$

which we will often use throughout this book. However, Codec A had an approximately four times higher compression ratio or lower coding rate expressed in bits/pels (bpp). Codec C of

2.2. ONE-DIMENSIONAL FRACTAL CODING

Table 2.3: Comparison of Five Fractal Codecs

Codec	DB Size	RB Size	Classi-fication	Split	PSNR (dB)	Rate (bpp)
A	16	8	None	No	31	0.28
B	16	4	None	No	35	1.1
C	8	4	None	No	37	1.22
D	16	8/4	Twin	Yes	36	0.84
E	16	8/4	Quad	Yes	29	1.0

Table 2.4: Classified Block Types and Their Relative Frequencies in Codec E

Block Type	Description / Edge Angle	Frequency (%)
Shade	No significant gradient	18.03
Midrange	Moderate gradient, no edge	46.34
Edge	Steep gradient, edge detected	26.39 (total)
	0 deg	2.66
	5 deg	1.85
	90 deg	5.69
	135 deg	3.18
	180 deg	3.92
	225 deg	2.96
	270 deg	3.92
	315 deg	2.22
Mixed	Edge angle ambiguous	9.24

Table 2.3 had four times more DBs than Codec B, which yielded quadrupled block-matching complexity, but the resulting PSNR improvement was limited to about 1–2 dB and the bitrate was about 20% higher due to the increased number of DB addresses. These findings were also confirmed by informal subjective assessments.

In order to find a compromise between the four times higher compression ratio of the 8×8 RBs used in Codec A and the favorable image quality of the 4×4 RBs of Codec B and C, we followed Jacquin's [148] suggestions of splitting inhomogeneous RBs in two, three, or four subblocks. Initially, the codec attempted to encode an 8×8 RB and calculated the MSE associated with the particular mapping. If the MSE was above a certain threshold, the codec split up the block into four nonoverlapping subblocks. The MSE of these subblocks was checked against the error threshold individually, and if necessary one or two 4×4 subblocks were coded in addition. However, for three or four poorly matching subblocks, the codec stored only the transforms for the four small quarter-sized subblocks. This splitting technique was used in Codecs D and E of Table 2.3.

In addition to Jacquin's splitting technique [148], the subjectively important edge representation of Codecs D and E was improved by a block classification algorithm originally suggested by Ramamurthi and Gersho [15], which was then also advocated by Jacquin [148]. Accordingly, the image blocks were divided into four classes:

1. Shade blocks taken from smooth areas of an image with no significant gradient.

2. Midrange blocks having a moderate gradient but no significant edge.

3. Edge blocks having a steep gradient and containing only one edge.

4. Mixed blocks with a steep gradient that contains more than one edge, and hence the edge orientation is ambiguous.

Codec D used a basic twin-class algorithm, differentiating only between shade and nonshade blocks, whereas Codec E used the above quad-class categorization. The relative frequencies of all registered subclasses were evaluated following the above approach [15,148] for Codec E using our QCIF sequences, which are shown in Table 2.4.

In Codecs D and E after the classification of all DBs and RBs normal coding ensued, but with the advantage that the codec predetermined the angle the DB had to be rotated by and it attempted to match only blocks of the same class. If, for example, a RB was classified as an edge block with a certain orientation, the codec exploited this by limiting the required search to the appropriate DB pool. Furthermore, shade blocks were not fully encoded; only their mean was transmitted to the decoder, yielding a significant reduction in complexity and bitrate. Interestingly, we found the quad-class Codec E to perform worse than the twin-class codec D. This was due to classification errors and to the smaller size of the DB pools in case of Codec E, which often were too limited to provide a good DB match.

A comparison of the five QCIF videophone codecs is presented in Table 2.3. Codecs A–C represent basic fractal codecs with no RB classification or splitting. When comparing Codecs A and B using RB sizes of 8×8 and 4×4, respectively, the compression ratio of Codec A is four times higher, but its PSNR is 5 dB lower at similar complexity. The PSNR versus frame index performance of codecs A and B is portrayed in Figure 2.8. The typical subjective image quality achieved by Codec B is portrayed in Figure 2.9 as a function of the Random Collage Theorem's iteration index. It was found that five approximations are sufficient to achieve pleasant communications image quality at a PSNR in excess of 33 dB. The 1 dB PSNR advantage of Codec C does not justify its quadruple complexity. Codecs D and E employ twin- or quad-class block classification combined with RB splitting, if the MSE associated with a particular mapping is above a certain threshold. Interestingly, the less complex Codec D has a higher compression ratio and higher image quality. The lower performance of codec E is attributed to block classification errors and to the limited size of the DB pool provided by our QCIF images.

The RB bit-allocation scheme of Codecs A and B is identical, but the lower-compression Codec B has four times as many RBs. The specific bit assignment of a RB in Codecs A

2.2. ONE-DIMENSIONAL FRACTAL CODING

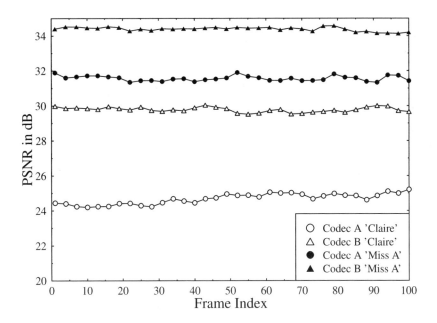

Figure 2.8: PSNR versus frame index performance of Codecs A and B for the "Claire" and "Miss America" sequences.

Table 2.5: Bit-allocation per RB for Codecs A and B

Bit Index	Parameter
1–2	Rotation
3–6	DB X coordinate
7–10	DB Y coordinate
11–14	Scaling
15–18	Offset

and B is shown in Table 2.5. Bits 1–2 represent four possible rotations, bits 3–6 and 7–10 are the X and Y domain-block coordinates, respectively, while bits 11–14 are the Max-Lloyd quantized scaling gains and bits 15–18 represent offset values used in the Random Collage Algorithm. However, Codec A uses $22 \times 18 = 3968 \times 8$ pels RBs associated with a rate of $R = 18/64 \approx 0.28$ bits/pel, while Codec B has $44 \times 38 = 15844 \times 4$ RBs, which is associated with a quadrupled bitrate of $R = 18/16 \approx 1.1$ bits/pel. The number of bits per frame becomes $396 \cdot 18 = 7128$ bits/frame for Codec A and 28512 for Codec B, corresponding to bitrates of 71.28 kbit/s and 285.12 kbit/s, respectively, at a scanning rate of 10 frames/s. The associated PSNR values are about 31 and 35 dB, respectively. The compression ratio of all of the fractal codecs was more modest than that of the Discrete

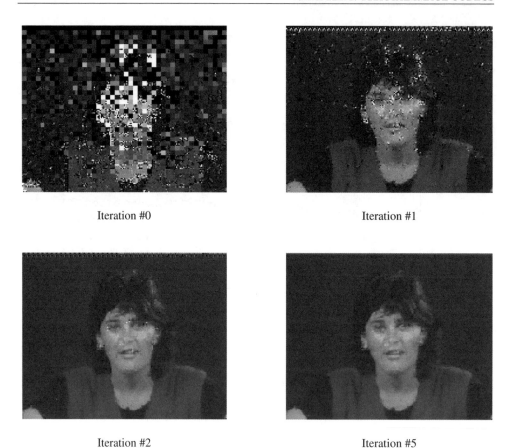

Iteration #0   Iteration #1

Iteration #2   Iteration #5

Figure 2.9: Demonstration of fractal image reconstruction from an arbitrary initial picture by the Random Collage Algorithm as a function of iteration index and the typical fractal video quality.

Cosine Transform (DCT), Vector Quantization (VQ), and Quad-tree (QT) codecs, which will be discussed in Chapters 3, 4, and 5, respectively.

2.3 Error Sensitivity and Complexity

Both Codecs A and B were subjected to bit-sensitivity analysis by consistently corrupting each bit of the 18-bit frame and evaluating the Peak Signal to Noise Ratio (PSNR) degradation inflicted, which is defined by Equation 2.8. These results are shown for both codecs in Figure 2.10. Observe from the figure that the significance of the specific coding bits can be explicitly inferred from the PSNR degradations observed. Generally, Codec A is more vulnerable to channel errors, since the bits for each block carry the information for $8 \times 8 = 64$ pixels rather than $4 \times 4 = 16$ pixels in the case of Codec B. Very vulnerable to channel errors

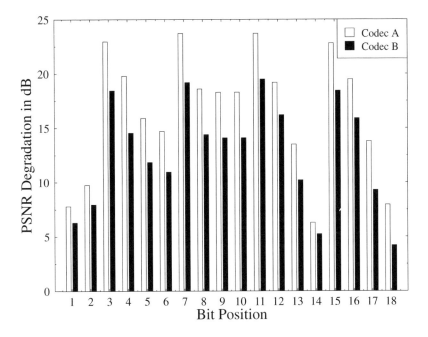

Figure 2.10: Bit sensitivities versus Bit Index for Codecs A and B using the bit-allocation of Table 2.3.

are the MSBs of the DB coordinates, the scaling and offset values. These are about three times more sensitive than the corresponding LSBs. A more detailed analysis of the error sensitivities and a matching wireless transceiver design can be found in [151]. The PSNR versus BER degradation due to random bit errors, which is characteristic of the codec's behavior over wireline-based AWGN channel is depicted in Figure 2.11. On the basis of subjective video-quality degradation a 2 dB PSNR degradation, is deemed acceptable, which requires the transceiver to operate at a BER of about 10^{-3}.

Fractal codecs suffer from a high computational demand. This is due to the evaluation of Equations 2.7 and 2.6 for every block. The number of Flops per DB-RB match is proportional to $7 \cdot p^2$, leading to 99 DB $\times$ $7 \cdot 8^2$ $\times$ 396 RB $>$ 17.5 Mflop for Codecs A and B. The computational load has to be multiplied with the frame rate, which leads to a huge computational demand compared to other compression techniques.

2.4 Summary and Conclusions

A range of QCIF fractal video codecs was studied in terms of image quality, compression ratio, and complexity. Two fixed-rate codecs, Codecs A and B, were also subjected to bit-sensitivity analysis. We found that for a pleasant image quality, the rather small block size of 4×4 pixels is necessary. This leads to a relatively high bitrate of 71 kbit/s. Furthermore, the computational demand is very high. We note furthermore that fractal coding is not amenable

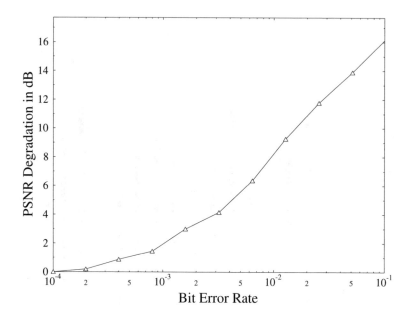

Figure 2.11: PSNR degradation versus bit error rate for Codec A.

to motion-compensated inter-frame coding, since the motion compensation removes temporal redundancy in subsequent frames, which results in a strongly decreased intra-frame correlation. It was found that the remaining intra-frame correlation was insufficient to perform fractal coding. The following chapters will demonstrate that motion compensation in conjunction with a range of other compression techniques assists us in achieving high compression ratios.

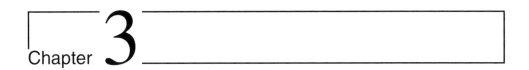

Low Bitrate DCT Codecs and HSDPA-style Videophone Transceivers

3.1 Video Codec Outline

In video sequence coding, such as videotelephony, a combination of temporal and spatial coding techniques is used in order to remove the "predictable or redundant" image contents and encode only the unpredictable information. How this can be achieved is the subject of the forthcoming chapters. These techniques guarantee a reduced bitrate and hence lead to bitrate economy. Let us begin our discussion with the simple encoder/decoder (codec) structure of Figure 3.1.

Assuming that the 176×144-pixel Quarter Common Intermediate Format (QCIF) ITU standard videophone sequence to be encoded contains a head-and-shoulder video clip, the consecutive image frames f_n and f_{n-1} typically do not exhibit dramatic scene changes. Hence, the consecutive frames are similar, a property that we refer to as being correlated. This implies that the current frame can be approximated or predicted by the previous frame, which we express as $f_n \approx \hat{f}_n = f_{n-1}$, where $\hat{f}_n$ denotes the nth predicted frame. When the previous frame f_{n-1} is subtracted from the current one, namely f_n, due to the speaker's movement, this prediction typically results in a "line-drawing-like" difference frame, as shown in Figure 3.2. Most areas of this difference frame are "flat", having values close to zero, and the variance or second moment of it is significantly lower than that of the original frame. We have removed some of the predictable components of the video frame.

This reduced variance difference signal, namely, $e_n = f_n - f_{n-1}$, is often referred to as Motion Compensated Error Residual (MCER) since the associated frame-differencing effectively attempts to compensate for the motion of the objects between consecutive video frames, yielding a reduced-variance MCER. Thus, e_n requires a reduced coding rate, that is, a lower number of bits in order to represent it with a certain distortion than f_n. The MCER

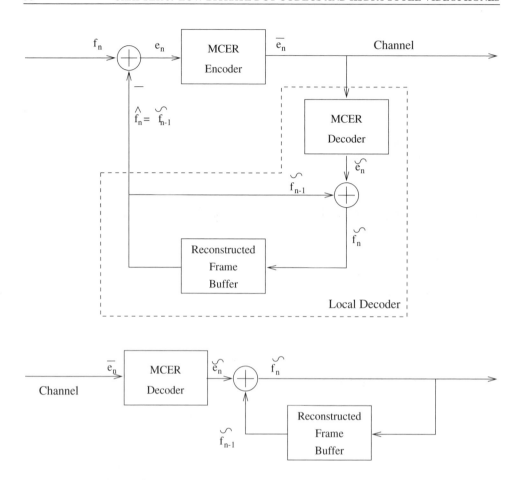

Figure 3.1: Basic video codec schematic using frame-differencing.

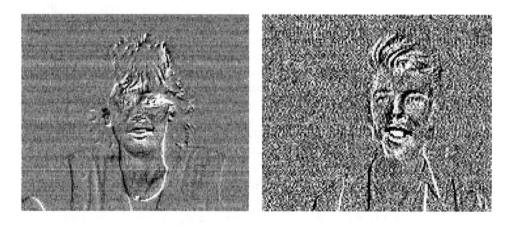

Figure 3.2: Videophone frame difference signals for two sequences.

3.2. THE PRINCIPLE OF MOTION COMPENSATION

$e_n = f_n - f_{n-1}$ can then be encoded as $\bar{e}_n$, with the required distortion using a variety of techniques, which constitute the subject of the following chapters. The quantized or encoded MCER signal $\bar{e}_n$ is then transmitted over the communications channel. In order to reproduce the original image $f_n = e_n + f_{n-1}$ exactly, knowledge of e_n would be necessary, but only its quantized version, $\bar{e}_n$ is available at the decoder, which is contaminated by the quantization distortion introduced by the MCER encoder, inflicting the reconstruction error $\Delta e_n = e_n - \bar{e}_n$. Since the previous undistorted image frame f_{n-1} is not available at the decoder, image reconstruction has to take place using their available approximate values, namely, $\bar{e}_n$ and $\tilde{f}_{n-1}$, giving the reconstructed image frame as follows:

$$\tilde{f}_n = \tilde{e}_n + \tilde{f}_{n-1}, \tag{3.1}$$

where in Figure 3.1 we found that the locally decoded MCER residual $\tilde{e}_n$ is an equal-valued, noiseless equivalent representation of $\bar{e}_n$. This equivalence will be further elaborated during our forthcoming discussions. The above operations are portrayed in Figure 3.1, where the current video frame f_n is predicted by $\hat{f}_n = \tilde{f}_{n-1}$, an estimate based on the previous reconstructed frame $\tilde{f}_{n-1}$. Observe that (ˆ) indicates the predicted value and (˜) the reconstructed value, which is contaminated by some coding distortion. Note furthermore that the encoder contains a so-called local decoder, which is identical to the remote decoder. This measure ensures that the decoder uses the same reconstructed frame $\tilde{f}_{n-1}$ in order to reconstruct the image, as the one used by the encoder to generate the MCER e_n. Note, however, that in case of transmission errors in $\tilde{e}_n$, the local and remote reconstructed frames become different, which we often refer to as being misaligned. This phenomenon leads to transmission error propagation through the reconstructed frame buffer, unless countermeasures are employed.

To be more explicit, instead of using the previous original frame, the so-called locally decoded frame is used in the motion compensation, where the phrase "locally decoded" implies decoding it at the encoder (i.e., where it was encoded). This local decoding yields an exact replica of the video frame at the distant decoder's output. This local decoding operation is necessary, because the previous original frame is not available at the distant decoder, and so without the local decoding operation the distant decoder would have to use the reconstructed version of the previous frame in its attempt to reconstruct the current frame. The absence of the original video frame would lead to a mismatch between the operation of the encoder and decoder, a phenomenon that will become clearer during our further elaborations.

3.2 The Principle of Motion Compensation

The simple codec of Figure 3.1 used a low-complexity temporal redundancy removal or motion compensation technique, often referred to as frame-differencing. The disadvantage of frame-differencing is that it is incapable of tracking more complex motion trajectories, where different objects, for example, the two arms of a speaker, move in different directions. This section briefly summarizes a range of more efficient motion compensation (MC) techniques, discusses their computational complexity and efficiency, and introduces the notation to be used.

Image sequences typically exhibit spatial redundancy within frames and temporal redundancy in consecutive frames. Motion compensation attempts to remove the latter.

If there is no drastic change of scenery, consecutive frames often depict the same video objects, but some of them may be moving in different directions. The motivation behind block-based motion compensation is to analyze the motion trajectory and to derive a set of descriptors for the changes. These descriptors characterize the location, shape, and movement of a moving object. In the ideal case, when no new information is introduced in the video scene, consecutive frames can be fully described by their appropriately motion-translated predecessors. Apart from tracking motion in the two spatial dimensions, some techniques, such as model-based coding [152, 153], also interpolate the third, temporal dimension of a given sequence so that the movement is represented by a three-dimensional vector and all three image dimensions are reproduced. Although promising results can be achieved using these techniques [154, 155], their real-time implementation has been hindered due to the associated high complexity.

In an often used practical MC implementation, the image frame is divided in a number of perfectly tiling 2×2 to 16×16-pixel blocks, which are then slid over a certain search area of Figure 3.4 surrounding the corresponding location in the previous reconstructed frame $\tilde{f}_{n-1}$, in order to identify the specific position from which each block has originated. The corresponding motion vectors (MV) describing the motion translation, also known as displacement vectors, are two dimensional and are typically restricted to integer multiples of the pixel separation, although we note that at a later stage subpixel resolution will also be considered in the context of the H.263 standard video codec. The motion vector identified this way is then usually applied to the entire group of pixels within a block, so that the predicted current block constituted by the appropriately motion translated, previously reconstructed block is subtracted from the current block about to be encoded, as portrayed in Figure 3.4. Again, the result of this operation is referred to as the Motion Compensated Error Residual (MCER). These operations are reflected in the modified codec schematic of Figure 3.3.

An MC algorithm operating with subpixel accuracy has been proposed [156]. The above motion prediction algorithm is based on two consecutive frames, that impose a minimum encoding delay of one frame duration. We refer to the previous reconstructed frame $\tilde{f}_{n-1}$, rather than the uncoded original input frame f_n during the derivation of the motion vectors, because both the encoder and the decoder must use an identical reference frame, to which they apply the motion vectors (MV) during the image reconstruction. However, the original input frame f_n is not available at the remote decoder. These issues are elaborated on in more depth later in this chapter.

Formulating the above algorithm slightly more rigorously, we find that the entire $p \times q$-sized frame f_n

$$f_n = \{(x, y) | x \epsilon [1 \ldots p], y \epsilon [1 \ldots q]\} \tag{3.2}$$

is divided into $n = (p/b)(q/b)$ smaller, $b \times b$ pixel rectangles defined as:

$$b_n = \{(x, y) | x, y \epsilon [1 \ldots b]\}. \tag{3.3}$$

Again, this process is depicted in Figure 3.4. Then for each of the n blocks, the most similar block within the stipulated search area of the previous reconstructed or locally decoded frame, $\tilde{f}_{n-1}$ has to be found. Recall from our previous arguments that the previous locally decoded, rather than the previous original frame, has to be used in this operation, because the previous original frame is unavailable at the remote decoder. Then the vector emerging from a given pixel — for example, at the center of the block under consideration and pointing

3.2. THE PRINCIPLE OF MOTION COMPENSATION

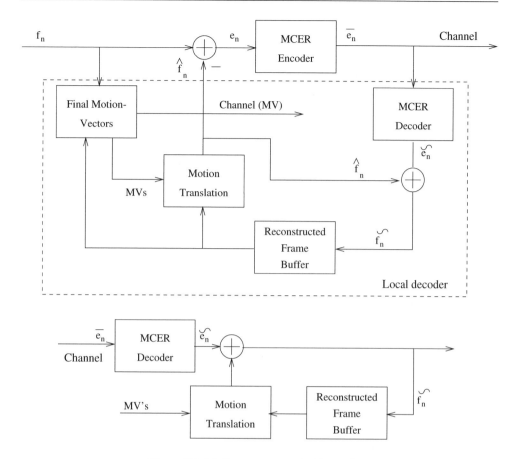

Figure 3.3: Motion-compensated video codec.

to the corresponding pixel of the most similar block within the search area of the previous reconstructed block — is defined as the motion vector $\vec{M} = M(m_x, m_y)$. The similarity criterion or distance measure used in this block-matching process has not yet been defined. A natural choice is to use the mean squared error (MSE) between the block for which the MVs are sought and the momentarily tested, perfectly overlapping block of pixels within the previous reconstructed frame $\tilde{f}_{n-1}$. This topic will be revisited in more depth in the next section.

The motion-compensated prediction error residual (MCER) is also often termed a displaced frame difference (DFD). Depending on the input sequence, the MCER typically contains a zero mean signal of nonstationary nature. Figure 3.5 depicts the two-dimensional autocorrelation function (ACF) of a typical MCER and its probability density function (PDF). The ACF of the MCER suggests that there is residual spatial correlation or "predictability" over a number of adjacent pixels in both vertical and horizontal directions. Since the ACF values are above 0.5 for displacements of $X_0, Y_0 < 3$, this residual spatial correlation can be exploited using a variety of coding schemes. The PDF of the MCER signal in Figure 3.5 also suggests that most of the values of the MCER are within a range of $[-10, 10]$, assuming that

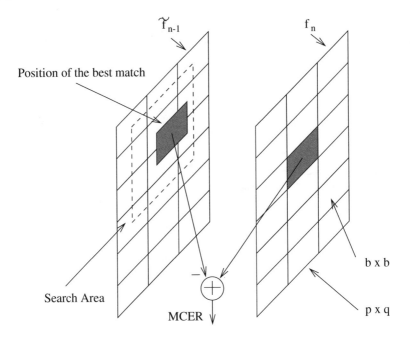

Figure 3.4: The process of motion compensation.

the source frame luminance values are sampled with an 8-bit resolution within the range of [0..255]. However, when using the lossy coding scheme of Figure 3.3 (i.e., schemes that do not perfectly reconstruct the original image), the quantization distortion associated with the MCER encoder will accumulate in the local reconstructed frame buffer, and it will alter the statistical characteristics of the MCER, reducing the correlation and resulting in a more widely spread PDF.

3.2.1 Distance Measures

At the time of writing, there is insufficient understanding of the human visual perception. Many distance criteria have been developed in order to assist in pattern matching or image processing research, but most of them fail under various circumstances. Here we introduce the three most commonly used distance measures of image processing.

The widely accepted Mean Square Error (MSE) criterion is defined as:

$$M_{mse} = MIN \sqrt{\sum_{i=1}^{b} \sum_{j=1}^{b} (f_n(x+i, y+j) - f_{n-1}(x+i-m_x, y+j-m_y))^2}, \quad (3.4)$$

where b denotes the block size, x and y represent the top left corner of the block under consideration, and m_x, m_y constitute the coordinates of the motion vector $\vec{M}$. Equation 3.4 evaluates the luminance difference of the given $b \times b$ blocks f_n and f_{n-1} on a pixel-by-pixel basis, sums the squared difference over the entire block, and after taking the square root of

3.2. THE PRINCIPLE OF MOTION COMPENSATION

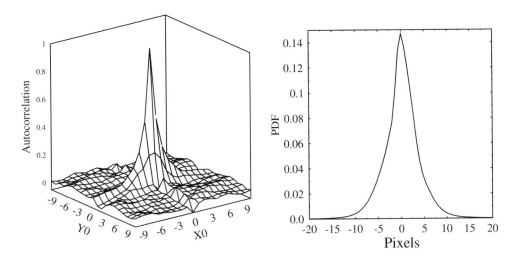

Figure 3.5: Two-dimensional autocorrelation function and the PDF of the MCER.

the sum, the MC scheme finds the position, where this expression has a minimum over the search area of Figure 3.4.

Two simplifications of this criterion are used in MC. The mean absolute difference (MAD) criterion is defined as the sum of the absolute differences rather than its second moment:

$$M_{mad} = MIN \sum_{i=1}^{n} \sum_{k=1}^{n} \mid f_n(x+i, y+j) - f_{n-1}(x+i-m_x, y+j-m_y) \mid . \quad (3.5)$$

The pel difference classification [157] (PDC) criterion of Equation 3.6 compares the matching error for every pel location of the block with respect to a preset threshold and classifies the pel as a "matching" or "mismatching" pel, yielding:

$$M_{pdc} = MIN \sum_{i=1}^{n} \sum_{k=1}^{n} T \mid f_n(x+i, y+j) - f_{n-1}(x+i-m_x, y+j-m_y) \mid, \quad (3.6)$$

where T denotes the threshold function defined as:

$$T(y) = \begin{cases} 1 & \text{if } y > 0 \\ 0 & \text{otherwise} . \end{cases} \quad (3.7)$$

This algorithm was contrived for real-time implementations, as its thresholding operation is readily implementable in hardware. However, the threshold must be set carefully and eventually adjusted for different types of image sequences. Other variants of the mentioned algorithms are also feasible. For example, for each of the above-mentioned matching criteria, a "fractional" measure may be defined that would consider only a fraction of the pixels within the two blocks under consideration. Accordingly, a 50% fractional criterion would consider only every other pixel, while halving the matching complexity.

As in Figure 3.6, the Peak Signal to Noise Ratio (PSNR) will serve as the objective quality measure, which is defined as:

$$PSNR = 10 \log_{10} \frac{\sum_{i=1}^{p} \sum_{j=1}^{q} \max^2}{\sum_{i=1}^{p} \sum_{j=1}^{q} (f_n(i,j) - \tilde{f}_n(i,j))^2}, \quad (3.8)$$

as opposed to the conventional SNR of:

$$SNR = 10 \log_{10} \frac{\sum_{i=1}^{p} \sum_{j=1}^{q} f(i,j)^2}{\sum_{i=1}^{p} \sum_{j=1}^{q} (f_n(i,j) - \tilde{f}_n(i,j))^2}. \quad (3.9)$$

The PSNR is a derivative of the well-known Signal to Noise Ratio (SNR), which compares the signal energy to the error energy as defined in Equation 3.9. The PSNR compares the maximum possible signal energy of $\max^2 = 225^2$ to the noise energy, which was shown to result in a higher correlation with the subjective quality perception of images than the conventional SNR. Figure 3.6 compares the performance for the distance criteria defined above for a codec using motion compensation only, that is, allocating zero bits for the encoding of the MCER, which is equivalent to setting MCER $= 0$ during the frame reconstruction. Observe in Figure 3.6 that even when no MC is used — corresponding to simple frame-differencing — the PSNR is around 25 dB for the limited-motion "Miss America" video sequence. In addition, we derived another measure from the MSE criterion, where only 50% of the pels in each block are considered. We expect the MSE criterion to lead to the best possible results, since the PSNR, is like the MSE criterion, is a mean squared distance function. Figure 3.6 also underlines the relatively poor performance of the PDC algorithm.

3.2.2 Motion Search Algorithms

Having introduced a number of pattern matching criteria, in this section we briefly highlight the so-called optimum full-search motion compensation algorithm and its derivatives, while comparing their complexities. The gain-cost controlled motion compensation algorithm to be introduced in Section 3.2.2.6 will be used in all of our compression algorithms presented throughout Part II of the book.

3.2.2.1 Full or Exhaustive Motion Search

The full search MC algorithm determines the associated matching criterion for every possible motion vector within the given search scope and selects that particular motion vector, which results in the lowest matching error. This was portrayed in Figure 3.4. Thus, a search within a typical motion search window of 16×16 pixels requires 256 block comparisons. Using the MSE criterion as quality measure and a block size of 8×8 pels, the full search (FS) requires 13 million floating point operations (Mflop) per QCIF frame. This constitutes a computational load, which we cannot afford in typical real-time applications, such as, for example, in hand-held mobile videotelephony. Note, however, that the matching complexity is a quadratic function of the search window size. For image sequences with moderate motion activity or for sequences scanned at a high frame rate, the search window can be reduced to the size of ± 2 adjacent pixel positions, without significant performance penalty, leading to

3.2. THE PRINCIPLE OF MOTION COMPENSATION

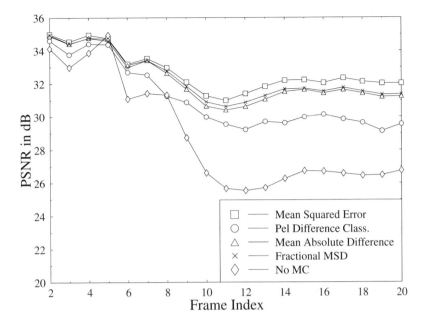

Figure 3.6: PSNR versus frame index performance comparison for various distance measures while encoding the MCER with zero bits, when using the "Miss America" sequence.

a computational demand below one MFlop for the same frame size. The computational complexity associated with a single motion compensated block b_n corresponds to two additions and one multiplication for the b^2 elements of each block, which have to be executed for each of the s^2 positions within the search window $s \times s$, when using the MSE criterion of Equation 3.4, yielding:

$$Com_{fs} = s^2(3b^2 - 1). \qquad (3.10)$$

3.2.2.2 Gradient-based Motion Estimation

The disadvantage of the exhaustive search algorithm is its complexity. The number of search steps can be reduced efficiently, if the image sequence satisfies the gradient constraint equation [158].

Let $I(x, y, t)$ be the luminance of the pixel (x, y) at the time instant t. The motion of the pixel in each spatial direction is defined as $m_x(x, y)$ and $m_y(x, y)$, respectively. At the time instant $t + \delta t$, the pixel will have moved to the location $(x + \delta x, y + \delta y)$. Using the motion vector defined as:

$$\vec{M} = M(m_x, m_y) = \frac{dx}{dt}\vec{e_x} + \frac{dy}{dt}\vec{e_y} = m_x \cdot \vec{e_x} + m_y \cdot \vec{e_y} \qquad (3.11)$$

where $\vec{e_x}$ and $\vec{e_y}$ represent the corresponding unit vectors along the x- and y-axis, we approximate the new location of the reference pixel as $(x + \delta t \cdot m_x, y + \delta t \cdot m_y)$, assuming that motion translation during the short time interval δt is linear. With these assumptions we obtain:

$$I(x,y,t) = I(x + \delta t \cdot m_x, y + \delta t \cdot m_y, t + \delta t), \qquad (3.12)$$

which physically implies that the luminance of the pixel at a new position in the current frame is the same as the luminance of the reference pixel in the previous frame.

The Taylor expansion of the right-hand side of Equation 3.12 leads to:

$$\begin{aligned} I(x,y,t) = I(x,y,t) &+ \frac{\delta I(x,y,t)}{\delta x}\frac{dx}{dt} + \frac{\delta I(x,y,t)}{\delta y}\frac{dy}{dt} \\ &+ \frac{\delta I(x,y,t)}{\delta t} + e(x,y,t), \end{aligned} \qquad (3.13)$$

where $e(x,y,t)$ contains the higher order components of the Taylor expansion. For small δt values, that is, when $\delta t \to 0$ we have $e(x,y,t) \to 0$, and upon neglecting $e(x,y,t)$, we arrive at the gradient constraint equation:

$$\frac{\delta I(x,y,t)}{\delta x}m_x + \frac{\delta I(x,y,t)}{\delta y}m_y + \frac{\delta I(x,y,t)}{\delta t} = 0. \qquad (3.14)$$

A ramification of this equation is that an iterative pattern matching algorithm, which minimizes the MSE between segments in consecutive frames, as it is carried out in MC, will converge if Equation 3.14 is satisfied. All the iterative MC algorithms to be introduced in this section, such as the hierarchical search or the subsampling search, assume that Equation 3.14 holds for the given image sequence.

3.2.2.3 Hierarchical or Tree Search

The so-called tree search algorithm proposed by Jain and Jain [158, 159] drastically reduces the complexity of MC, while maintaining a high-quality motion prediction. Assuming that the DFD function exhibits a dominant minimum associated with a certain MV, the search algorithm visits in the first iteration only a fraction of all possible positions in the search window.

The algorithm initially estimates the block motion vector as $(0,0)$. A set of eight correction vectors arranged on a grid, as seen in Figure 3.7, are in turn added to the estimated motion vector and a new value of the MCER is computed each time. The correction vector, which yields the lowest value of the MCER so far computed, is added to the estimated motion vector, in order to obtain a more refined estimate. The size of the grid of correction vectors is then halved. The process is repeated until the final grid size is 3·3, at which point the algorithm terminates with an estimate of the best motion vector. When comparing this suboptimum algorithm with the optimal full search, the motion-compensated error energy is typically only 10% higher, while the search complexity is significantly reduced, a definitely worthwhile trade-off.

For example, the full search in a search window of 16×16 pels requires 256 comparisons, whereas the hierarchical search already converges after 27 comparisons. As seen in Figure 3.8, the complexity reduction is achieved without a significant loss in terms of PSNR performance. The set of curves displayed in Figure 3.8 demonstrates the performance of

3.2. THE PRINCIPLE OF MOTION COMPENSATION

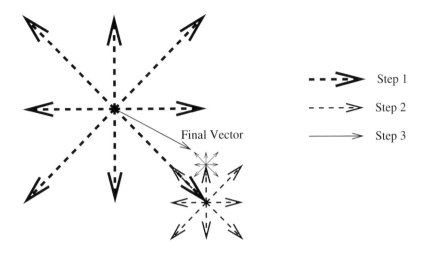

Figure 3.7: Suboptimum MV tree search.

various search algorithms applied to the "Miss America" head-and-shoulder video sequence. A total of 50 active motion vectors was assumed out of the 396 8×8 QCIF blocks, while for the remaining 346 blocks frame-differencing was used. The associated MCER was not transmitted to the decoder at all, which was equivalent to the assumption of MCER $= 0$. This crude assumption represented a worst-case codec performance, which was found adequate for assessing the performance of the various MC schemes.

3.2.2.4 Subsampling Search

The subsampling search algorithm [160] operates similarly to the tree search. However, instead of reducing the search window, the whole search area is subsampled by a given factor. The search in the lower resolution area is less complex since the number of possible locations and the block size are reduced according to the subsampling factor. Once the best location minimizing the MCER has been found in the subsampled domain, the resolution will be increased step-by-step, invoking the full-resolution search in the last stage of operations. The advantage is that the search window can be limited around the best position found in the previous step. The procedure continues until the original resolution is reached.

As an example, for a block size of 16×16 pels, both the incoming and previous reconstructed frame would be subsampled by a factor of 4 in each spatial direction. The subsampling factor of 4 means that a distance of 1 pel equals a distance of 4 pels in the original frames. At the lowest resolution level, full search is applied using a search window of 3×3 pixels. In the next step the resolution is doubled in each spatial direction, and again, a full search using a search window of 3×3 pels is applied around the previously found optimum position, minimizing the MCER. The same procedure applied once again will result in the suboptimum vector in the original search window.

The complexity is even lower than that of the hierarchical search (HS), since the same number of block comparisons is performed at reduced block sizes. However, prior to any computations, the area covered by the search window must be subsampled to all required

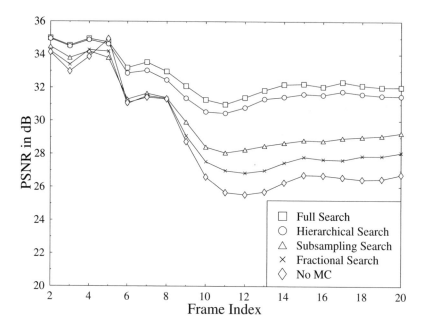

Figure 3.8: PSNR versus frame index performance comparison for various MC search algorithms using the "Miss America" sequence, where "No MC" refers to simple frame-differencing.

spatial resolutions. For some codecs, as, for example, quad tree codecs [160], which constitute the subject of Chapter 5, the image has to be subsampled anyway and hence no additional complexity is incurred. Although the search procedure is similar to the hierarchical search, it does not reach its performance.

3.2.2.5 Post-processing of Motion Vectors

When using very small block sizes of 4×4 or even smaller, the motion vector field can be interpreted as a subsampled field of the "true motion" of each pixel [156]. Thus, the "true motion field" can be derived by generating an individual MV for each of the pixels on the basis of the nearest four MVs using interpolation. This is true if the correlation of adjacent motion vectors is sufficiently high and the gradient constraint of Equation 3.14 is satisfied. If Equation 3.14 is not satisfied, the MV interpolation may increase the MCER energy. Simulations revealed that for the commonly used block size of 8×8 pixels, only around 10% of the motion vectors derived by motion field interpolation resulted in an MCER gain, (i.e., a further reduction of the MCER energy). However, the reduction of the MCER energy due to this method remained generally below 3%. This marginal improvement is insufficient to justify the additional complexity of the codec.

3.2.2.6 Gain-cost-controlled Motion Compensation

Motion compensation reduces the MCER energy at the cost of additional complexity and channel capacity demand for the motion vectors. The contribution of each vector toward

3.2. THE PRINCIPLE OF MOTION COMPENSATION

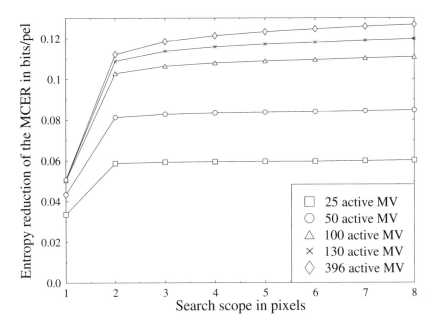

Figure 3.9: Entropy reduction versus search scope in case of limiting the number of active MVs per 8×8 block for QCIF head-and-shoulder video sequences.

the MCER entropy reduction strongly depends on the movement associated with each vector. Comparing the MCER energy reduction with the requirement for the storage of the additional MVs leads to the concept of gain-cost controlled MC.

For most of the blocks the MCER entropy reduction, where the entropy was defined in Chapter 1 of [161], does not justify the additional transmission overhead associated with the MV. Figure 3.9 reveals this entropy reduction for various numbers of active MVs based on QCIF frames containing a total of 396 8×8 blocks. In order to evaluate the potential MC gain due to every block's motion vector, we quantified the MCER energy reduction in Equation 3.15, where b_{0n} and $b_{0(n-1)}$ denote the block under consideration in the current and previous frames, respectively, and the MC gain is measured with respect to the MCER of $[b_{0n}(i,j) - b_{0(n-1)}(i,j)]$ generated by simple frame-differencing:

$$MC_{gain} = \sum_{i=1}^{b}\sum_{j=1}^{b}(b_{0n}(i,j) - b_{0(n-1)}(i,j))^2$$
$$- \sum_{i=1}^{b}\sum_{j=1}^{b}(b_{0n}(i,j) - b_{0(n-1)}(i+m_x, j+m_y))^2. \quad (3.15)$$

In Figure 3.9 we selected only the most efficient vectors for each experiment. All low-gain motion-passive MVs were set to [0, 0], and simple frame-differencing was used at the motion-passive locations. Surprisingly, our results suggested that the most important 25% of the MVs (i.e., using about 100 active MVs) resulted in 80–90% of the possible entropy reduction.

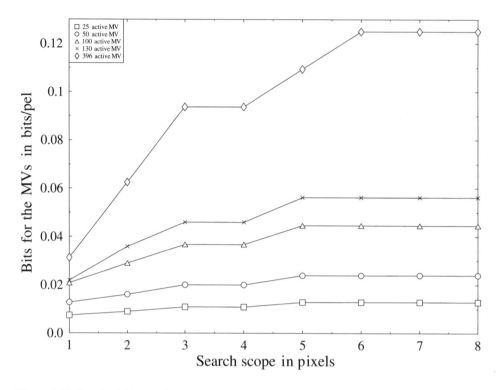

Figure 3.10: Required bitrate of active MVs versus MC search-scope for different number of active 8×8 blocks in QCIF head-and-shoulder videophone sequences.

The disadvantage of the motion passive/active concept is the additional requirement of indices or tables necessary to identify the active vectors. In a conceptually simple approach, we can assign a 1-bit motion activity flag to each of the 396 8×8 blocks, which constitutes a transmission overhead of 396 bits for each QCIF frame. An efficient alternative technique will be introduced in Section 3.7 in order to further compress this 396-bit binary motion activity table and hence to reduce the number of bits required. Figure 3.10 reveals the required coding rate of the MVs, when applying the above-mentioned efficient activity table compression method. From Figures 3.9 and 3.10 we concluded that the MC search window size can be limited to ± 2 adjacent pixels for limited-dynamic head-and-shoulder videophone sequences, while the adequate number of active motion vectors is below 100 in the case of QCIF-sized images. When using a higher proportion of active MVs, the number of bits assigned to their encoding will reduce the number of bits available for the encoding of the MCER and inevitably reduce the image quality in the case of a fixed bitrate budget.

3.2.3 Other Motion Estimation Techniques

Various modifications of the above-mentioned algorithms have been published in the literature [162–164]. Many of them are iterative simplifications of the full search algorithm and are based on the gradient constraint criterion of Equation 3.14. Other techniques improve

3.2. THE PRINCIPLE OF MOTION COMPENSATION

the accuracy of the motion prediction by interpolating the given images to a higher resolution and defining the MVs at a sub-pel accuracy [165]. Another classical approach to motion compensation is represented by the family of the pel-recursive techniques [166, 167], which will be described in Section 3.2.3.1. The grid interpolation techniques introduced quite recently are rooted in model-based techniques, and they will be highlighted in Section 3.2.3.2.

3.2.3.1 Pel-recursive Displacement Estimation

This technique predicts the motion vectors for the current frame from previous frames. Thus, no additional information must be transmitted for the motion vectors because the prediction is based on frames, which are available at both the transmitter and the receiver at the same time. Therefore, it is affordable to carry out the prediction for every single pixel rather than grouping them into blocks. The algorithms observe the motion for each pel over the last frames and derive from these vectors the speed and orientation of the movement. These values are then used to estimate the motion vectors for the current frame. The problem associated with these algorithms is that they fail at the edges of movement, which causes a motion-compensated error residual with strongly increased nonstationary characteristics. This phenomenon limits the performance of this technique and that of the subsequent image-coding steps. Another principal problem of these so-called backward estimation techniques is that they require perfect alignment between the operations of the transmitter and receiver. A tiny misalignment between the transmitter and receiver will result in a different motion estimation at both ends and induce further, rapidly accumulating misalignment. Hence, this motion compensation technique is not suitable for mobile communications over high error rate channels.

3.2.3.2 Grid Interpolation Techniques

The so-called grid interpolation technique is based on block deformation rather than block movement. This technique divides the video frames into perfectly tiling blocks using a regular grid pattern, as seen at the left-hand side of Figure 3.11. This so-called wire frame, the image objects, and their underlying texture are appropriately deformed in order to reconstruct the next frame, which is portrayed at the right-hand side of Figure 3.11. The deformation of each patch of texture is obtained by an affine transform, which was the subject of Chapter 2 on fractal coding. This transform is defined by the motion of each grid crossing. Hence, only the motion vectors of the grid crossings have to be transmitted. This technique has been employed with some success in [168, 169]. Its drawback is associated with the required interpolation of points. This is because the affine transform will not map every pixel on a valid position of the destination frame. This increases the complexity dramatically, especially during the iterative motion estimation step of the encoding process.

3.2.3.3 MC Using Higher Order Transformations

Instead of deforming the grid, we may apply MVs, which are a function of the location within the blocks. The nth order geometric transformation of:

$$mv_x(u,v) = \alpha_0 + \alpha_1 u + \cdots \alpha_n u^n + \beta_0 + \beta_1 v + \cdots \beta_n v^n$$
$$mv_y(u,v) = \gamma_0 + \gamma_1 u + \cdots \gamma_n u^n + \delta_0 + \delta_1 v + \cdots \delta_n v^n \qquad (3.16)$$

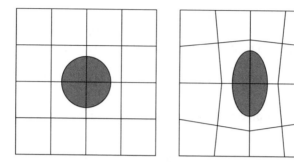

Figure 3.11: Grid interpolation example.

allows simultaneous translation, rotation, and change of scale [170], and it has $4(n+1)$ different parameters given by $\alpha_0 \ldots \alpha_n, \beta_0 \ldots \beta_n, \gamma_0 \ldots \gamma_n$ and $\delta_0 \ldots \delta_n$.

This is a very general approach to MC, since the motion-compensated objects may differ in their appearance, yet guarantee a low MCER for relatively large block sizes of 16×16 and larger [171]. Equation 3.16 can be simplified, allowing shifting, scaling, and translation only [172], reducing it to a total of three variables. In practical terms, however, the additional complexity and bitrate requirement of the algorithm outweigh its advantages.

3.2.3.4 MC in the Transform Domain

Block-based MC often results in *blocking artifacts* [173], which are subjectively objectionable. This problem can be mitigated by using blocks having fuzzy block borders, which can be created by interleaving the pixels of adjacent blocks according to a number of different mapping rules that are characterized by jigsaw puzzle-like block borders [174]. This kind of block shape reduces the likelihood of artifacts at the cost of a reduced MC performance. An elegant way of combatting this problem is to apply MC in a transform domain, for example, using complex wavelets [175] or the lapped orthogonal transform (LOT) [176, 177]. In the LOT, the images are transformed and then divided into overlapping blocks, hence mitigating the time-domain artifacts.

3.2.4 Conclusion

In this chapter, we have reviewed a range of block-matching motion estimation techniques and analyzed their performance in the case of various parameters. We found that upon using a rather limited proportion of active motion vectors we were able to reduce the DFD energy significantly. Another result was that small search window sizes of 4×4 pixels were adequate for maintaining a low complexity and near-optimum performance in case of slow-motion QCIF head-and-shoulder videophone sequences, such as the "Miss America" sequence. Such a small search window allowed us to use the highest complexity exhaustive search method, while keeping the technique's computational demand within the requirements of real-time applications. This is naturally valid only for low-activity head-and-shoulder images. Finally, we concluded that backward estimation techniques, such as pel-recursive motion compensation, are not suitable for communications over high error rate mobile channels.

3.3 Transform Coding

3.3.1 One-dimensional Transform Coding

As is well-known from Fourier theory, signals are often synthesized by so-called orthogonal basis functions, a term, which will be augmented during our further discourse. Specifically, when using Fourier transforms, an analog time-domain signal, which can be the luminance variation along a scanline of a video frame or a voice signal, can be decomposed into its constituent frequencies.

For signals such as the above-mentioned video signal representing the luminance variation along a scanline of a video frame, orthogonal series expansions can provide a set of coefficients, which equivalently describe the signal concerned. We will make it plausible that these equivalent coefficients may become more amenable to quantization than the original signal.

For example, for a one-dimensional time-domain sample sequence $\{x(n), 0 \leq n \leq N-1\}$ a unitary transform is given in vectorial form by $\underline{X} = \underline{\underline{A}}\underline{x}$, which can also be expressed in a less compact scalar form as [10]:

$$X(k) = \sum_{n=0}^{N-1} a(k,n) \cdot x(n), \quad 0 \leq k \leq N-1 \tag{3.17}$$

where the transform is referred to as unitary, if $\underline{\underline{A}}^{-1} = \underline{\underline{A}}^{*T}$ holds. The associated inverse operation requires us to invert the matrix $\underline{\underline{A}}$ and because of the unitary property, we have $\underline{x} = \underline{\underline{A}}^{-1}\underline{X} = \underline{\underline{A}}^{*T}\underline{X}$, yielding [10]:

$$x(n) = \sum_{k=0}^{N-1} X(k) a^*(k,n), \quad 0 \leq k \leq N-1 \tag{3.18}$$

which gives a *series expansion* of the time-domain sample sequence $x(n)$ in the form of the *transform coefficients* $X(k)$. The columns of $\underline{\underline{A}}^{*T}$, that is, the vectors $\underline{a}_k^* \triangleq \{a^*(k,n), 0 \leq n \leq N-1\}$, are the basis vectors of $\underline{\underline{A}}$, or the *basis vectors of the decomposition*. According to the above principles, the time-domain signal $x(n)$ can be equivalently described in the form of the *decomposition* in Equation 3.18, where the *basis functions* $a^*(k,n)$ are weighted by the transform coefficients $X(k)$ and then superimposed on each other, which corresponds to their summation. The transform-domain weighting coefficients $X(k)$ can be determined from Equation 3.17.

The transform-domain coefficients $X(k); k = 0 \cdots N-1$ often give a more "compact" representation of the time-domain samples $x(n)$, implying that if the original time-domain samples $x(n)$ are correlated, then in the transform domain most of the signal's energy is concentrated in a few transform-domain coefficients. To elaborate a little further — according to the Wiener-Khintchine theorem — the AutoCorrelation Function (ACF) and the Power Spectral Density (PSD) are Fourier transform pairs. Because of the Fourier-transformed relationship of the ACF and PSD, it is readily seen that a slowly decaying autocorrelation function, which indicates a predictable signal $x(n)$ in the time domain is associated with a PSD exhibiting a rapidly decaying low-pass nature. Therefore, in case of correlated time-domain $x(n)$ sequences, the transform-domain coefficients $X(k)$ tend to be statistically small for high frequencies, that is, for high-coefficient indices and exhibit large magnitudes for low-frequency transform-domain coefficients, that is, for low transform-domain indices. This concept will be exposed in a little more depth below, but for a deeper exposure to these issues the reader should consult Jain's excellent book [10].

3.3.2 Two-dimensional Transform Coding

The one-dimensional signal decomposition can also be extended to two-dimensional (2D) signals, such as 2D image signals of a video frame, as follows [10]:

$$X(k,l) = \sum_{m=0}^{N-1}\sum_{n=0}^{N-1} x(m,n) \cdot a_{k,l}(m,n) \quad 0 \leq k,l \leq N-1 \tag{3.19}$$

$$x(m,n) = \sum_{k=0}^{N-1}\sum_{l=0}^{N-1} X(k,l) \cdot a^*_{k,l}(m,n) \quad 0 \leq m_1 n \leq N-1 \tag{3.20}$$

where $\{a^*_{k,l}(m,n)\}$ is a set of discrete two-dimensional basis functions, $X(k,l)$ are the 2D transform-domain coefficients, and $\underline{X} = \{X(k,l)\}$ constitutes the transformed image.

As in the context of the one-dimensional transform, the two-dimensional (2D) time-domain signal $x(m,n)$ can be equivalently described in the form of the decomposition in Equation 3.20, where the 2D basis functions $a^*_{k,l}(m,n)$ are weighted by the coefficients $X(k,l)$ and then superimposed on each other, which again corresponds to their summation at each pixel position in the video frame. The transform-domain weighting coefficients $X(k,l)$ can be determined from Equation 3.19.

Once a spatially correlated image block $x(m,n)$ of, for example, $N \times N = 8 \times 8$ pixels is orthogonally transformed using the discrete cosine transform (DCT) matrix $\underline{A}$ defined as [10, 178]:

$$A_{mn} = \frac{2c(m)c(n)}{N} \sum_{i=0}^{N}\sum_{j=0}^{N} b(i,j) \cos\frac{(2i+1)m\pi}{2N} \cos\frac{(2j+1)n\pi}{2N}$$

$$c(m) = \begin{cases} \frac{1}{\sqrt{2}} & \text{if } (m=0) \\ 1 & \text{otherwise} \end{cases} \tag{3.21}$$

the transform-domain image described by the DCT coefficients [10, 178] can be quantized for transmission to the decoder. *The rationale behind invoking the DCT is that the frequency-domain coefficients $X(k,l)$ can typically be quantized using a lower number of bits than the original image pixel values $x(m,n)$*, which will be further augmented during our forthcoming discourse.

For illustration's sake, the associated two-dimensional 8×8 DCT [10, 178] *basis-images are portrayed in Figure 3.12*, where, for example, the top left-hand corner represents the zero horizontal and vertical spatial frequency, since there is no intensity or luminance change in any direction across this basis image. Following similar arguments, the bottom right corner corresponds to the highest vertical and horizontal frequency, which can be represented using 8×8 basis images, since the luminance changes between black and white between adjacent pixels in both the vertical and horizontal directions. Similarly, the basis image in the top

3.3. TRANSFORM CODING

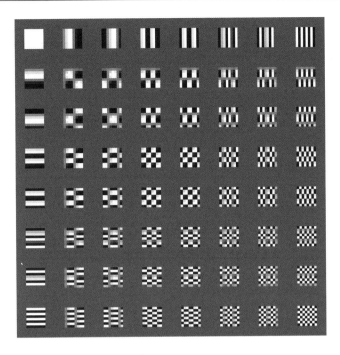

Figure 3.12: 8×8 DCT basis images. ©A. Sharaf [179].

right-hand corner corresponds to the highest horizontal frequency but zero vertical frequency component, and by contrast, the bottom left image represents the highest vertical frequency but zero horizontal frequency. *In simple terms, the decomposition of Equations 3.19 and 3.20 can be viewed as finding the required weighting coefficients $X(k, n)$ in order to superimpose the weighted versions of all 64 different "patterns" in Figure 3.12 for the sake of reconstituting the original 8×8 video block.* In other words, each original 8×8 video block is represented as the sum of the 64 appropriately weighted 8×8 basis image.

It is plausible that for blocks over which the video luminance or gray shade does not change dramatically (i.e., at a low spatial frequency), most of the video frame's conveyed energy is associated with these low spatial frequencies. Hence, the associated low-frequency transform-domain coefficients $X(k, n)$ are high, and the high-frequency coefficients $X(k, n)$ are of low magnitude. In contrast, if there is a lot of fine detail in a video frame, such as in a finely striped pattern or in a checkerboard pattern, most of the video frame's conveyed energy is associated with high spatial frequencies. Most practical images contain more low spatial frequency energy than high-frequency energy. This is also true for those motion-compensated video blocks, where the motion compensation was efficient and hence resulted in a flat block associated with a low spatial frequency. For these blocks therefore most of the high-frequency DCT coefficients can be set to zero at the cost of neglecting only a small fraction of the video block's energy, residing in the high spatial frequency DCT coefficients. In simple terms this corresponds to gentle low-pass filtering, which in perceptual terms results in a slight blurring of the high spatial-frequency image fine details.

In other words, upon exploiting that the human eye is rather insensitive to high spatial frequencies, in particular when these appear in moving pictures, the spatial frequency-domain block is amenable to data compression. Again, this can be achieved by more accurately quantizing and transmitting the high-energy, low-frequency coefficients, while typically coarsely representing or masking out the low-energy, high-frequency coefficients. We note, however that in motion-compensated codecs there may be blocks along the edges of moving objects where the MCER does not retain the above-mentioned spatial correlation and hence the DCT does not result in significant energy compaction. Again, for a deeper exposure to the DCT, and for an example the reader is referred to [10, 178].

In critical applications, such as medical image compression, these high-frequency coefficients carry the perceptually important edge-related video information and so must be retained. On the same note, our discussion of high-quality DCT-based coding will show that if the original image contains a significant amount of high-frequency energy and hence there is less adjacent pixel correlation in the image, then this manifests itself in terms of a less concentrated set of DCT coefficients. For such pictures, the DCT typically achieves less energy compaction.

A range of orthogonal transforms have been suggested in the literature for various applications [180]. The DCT is widely used for image compression, since its compression ratio approaches that of the optimum Karhunen-Loeve (KL) transform [10], while maintaining a significantly lower implementational complexity. This is due to the fact that the KL transform's basis vectors used in the transformation are data dependent, and so they have to be computed for each data segment before transformation can take place. The simple reason for the energy compaction property of orthogonal transforms, in particular that of the DCT, is that for correlated time-domain sequences, which have a flat-centered autocorrelation function, the spatial frequency-domain representative tends to be of a low-pass nature.

The efficiency of the DCT transform has constantly attracted the interest of researchers and has led to a wide variety of DCT and DCT-based codecs. Recently, many attractive results have been published using adaptive methods [181–183] and hybrid coding [184–186]. These approaches are typically associated with a fluctuating, time-variant compression ratio, requiring buffering methods in order to allow a constant bitrate. The following approach presents a fixed-rate DCT scheme, which can dispense with adaptive feedback buffering while still guaranteeing a fixed frame rate at a constant user-selectable bitrate.

As argued earlier, most of the energy of the 8×8 block is concentrated in the low horizontal and low vertical frequency regions corresponding to the top left-hand corner region of its two-dimensional spectrum. We capitalize on this characteristic by assigning higher resolution quantizers in the frequency regions, where we locate the highest energy, whereas lower energy frequency regions are quantized with coarser quantizers or are completely neglected by setting them to zero. The issue of choosing the appropriate frequency regions for quantization and the appropriate quantizer resolution are the most crucial design steps for a DCT codec.

Next, we describe in Sections 3.3.3 and 3.3.4, how we derived a set of quantizers for the DCT transformed blocks of the MCER, which are specially tailored for very low bitrates. Then Sections 3.4–3.11 reveal the codec structure and its performance under ideal channel conditions. Finally, in Section 3.12, we examine the sensitivity to channel errors and propose two configurations for fixed and mobile video communicators.

3.3. TRANSFORM CODING

Table 3.1: Bit-allocation Tables for the Fixed-rate DCT Coefficient Quantizers

3	2	1	0	0	0	0	0
2	1	0	0	0	0	0	0
1	0	0	0	0	0	0	0
0	0	0	0	0	0	0	0
0	0	0	0	0	0	0	0
0	0	0	0	0	0	0	0
0	0	0	0	0	0	0	0
0	0	0	0	0	0	0	0

3.3.3 Quantizer Training for Single-class DCT

Initially, we sought to determine which of the DCT coefficients had to be retained and quantized and which could be ignored or set to zero, while not inflicting severe image-quality degradation. We commenced by setting some of the DCT coefficients to zero, whereas the remaining coefficients were left unquantized. On the basis of the energy compaction property of the DCT [10], the unquantized coefficients were typically the ones closest to the top left corner of the matrix. This location represents the direct current (DC) and low-frequency region. We found that in order to maintain the minimum required videophone quality, about six DCT coefficients had to be retained.

In a second step, we assigned the appropriate number of reconstruction levels and the reconstruction levels themselves, which were required in order to maintain the target bitrate, which in our case was around 10 kbit/s. A particular design constraint was that the number of reconstruction levels had to be an integer power of two. That is, we required $n_{rec} = 2^k \mid k = \{1, 2, 3, \ldots\}$, so that each coefficient was quantized to k bits. Initially, we allocated 8 bits for the DC coefficient, 4 bits for each of the positions $[0, 1]$ and $[1, 0]$ on the $[8, 8]$-dimensional frequency plane, and 2 bits for each of the positions $[2, 1], [1, 1]$, and $[1, 2]$. Appropriately trained Max-Lloyd quantizers [187] were used to achieve minimum distortion quantization. We obtained the appropriate training data for the Max-Lloyd quantizers by recording the DCT coefficients of the MCER for the 30 highest-energy blocks of each QCIF frame for a 100 frame long motion-compensated sequence. This selection of training blocks considers the fact that we generally incorporate an active/passive classification scheme into the final codec. Therefore the training is also based only on blocks, which are likely to be selected by an active/passive classification scheme for quantization. The samples for each coefficient were collected, and the corresponding Max-Lloyd quantizers were derived from the training data. Perceptually pleasant videophone quality was achieved, when quantizing every 8×8 pixel block to a total of 22 bits. In order to achieve the maximum possible compression ratio attainable using this technique, while maintaining an adequate image quality, we iteratively reduced the number of reconstruction levels until we reached the required image quality versus bitrate trade-off deemed appropriate for QCIF videophony. At this stage, the entire block was quantized to a total of 10 bits, which leads to the bit-allocation depicted in Table 3.1.

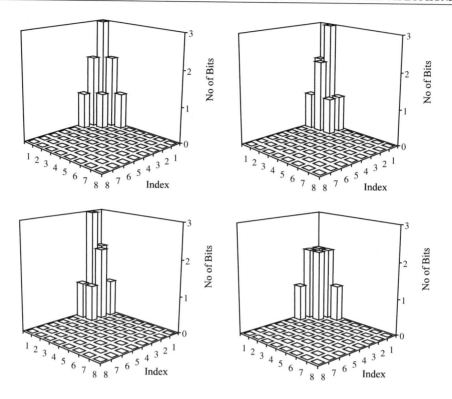

Figure 3.13: Bit-allocation for the Quad-class DCT quantizers.

3.3.4 Quantizer Training for Multiclass DCT

In the previous section, we demonstrated how we derived a set of quantizers, which achieved the maximum compression while providing an adequate quality for hand-held video communications. Noticeable degradation was caused at the edges of motion trajectories or at areas of high contrast. This was because for uncorrelated MCER sequences the DCT actually results in energy expansion rather than energy compaction [160]. Furthermore, if there is more correlation in one particular spatial direction than the other, the center of the spectral-domain energy peak moves toward the direction of higher correlation in the time domain. This happens if the block contains, for example, a striped texture. As a consequence, we added two sets of quantizers which shall cater for blocks with higher horizontal or vertical correlation. A fourth quantizer was added, which caters for the case of decreased energy compaction. Thus, its bit allocation is more widely spread than that of the other quantizers. The bit-allocation for all four quantizers is depicted in Figure 3.13 across the 8×8 frequency-domain positions.

After fixing the quantizer resolutions, we had to find an appropriate training sequence for each set of quantizers. The quantizer training required a training sequence exhibiting similar statistical properties to the actual discrete cosine transformed MCER, which has to be quantized by the quantizers to be designed. This required an initial tentative quantizer

3.3. TRANSFORM CODING

for the design of the actual quantizer. This is because the design of the MCER encoder in Figure 3.1 influences the future distribution of the MCER e_n through the reconstruction frame buffer. These issues constitute the subject of our forthcoming deliberations. Hence, again, a classification algorithm was needed in order to assign each of the training blocks to one of those categories. To simulate the effect of the actual quantizer about to be designed on the MCER, which is then used to generate the quantizer training set, we decided to use a simple initial tentative quantization function, which we will refer to as a pseudo-quantizer, producing a randomly quantized quantity $\tilde{q}$ from the unquantized variable q by contaminating q using an additive, quantizer-resolution dependent fraction of the unquantized variable itself. Specifically, we attempted to simulate each quantizer by applying Equation 3.22,

$$\tilde{q} = q \times \left(1.0 + 1.5 \frac{(-1)^{RAND}}{n}\right), \qquad (3.22)$$

where q is the quantity to be quantized, $\tilde{q}$ represents its pseudo-quantized value, n equals the number of reconstruction levels, and *RAND* is a random integer. Clearly, this pseudo-quantizer simulates the quantization error by randomly adding or subtracting a fraction of itself from the unquantized quantity q. The absolute factor of 1.5 was found empirically and should be in the range of $[1.3\ldots1.7]$. The magnitude of the added distortion is inversely proportional to the number of quantization levels n.

Each quantizer training block was DCT transformed, pseudo-quantized, and then inverse-transformed back to the time domain, while tentatively using all four sets of quantizer seen in Figure 3.13. The unquantized coefficients corresponding to the specific quantizer of Figure 3.13 leading to the best reconstruction in the MSE sense were collected separately for each class. Finally, the Max-Lloyd quantizers for each energy distribution class were derived from each of these classified training sets. We found that the initial set of quantizers performed well, since a retraining of those quantizers upon using the newly generated quantizers instead of the pseudo-quantizer resulted only in marginal further performance improvement.

Figure 3.14 reveals the codec's PSNR performance for a single-, dual-, and quad-class DCT codec at a constant bitrate of 1000 bits per frame, or 10 kbit/s at a video frame scanning rate of 10 frames/s. The dual-class codec was using the quantizer sets derived for high and normal energy compaction, which are depicted in the top left and bottom right corner of Figure 3.13. Although the dual-class scheme requires one and the quad-class scheme two more bits per DCT transformed block for the encoding of the energy distribution classifier and therefore the number of active blocks to be encoded — given the 1000 bits per frame bitrate constraint — must be reduced, if the overall bitrate is kept constant, the perceived image quality increased significantly and the objective PSNR also improved by up to 0.7 dB due to the classification technique. The relative frequency of selecting any one of the four quantizers varies from frame to frame and depends on the specific encoded sequences. The average relative frequency of the four quantizers was evaluated using the "Miss America" sequence for both the twin- and the quad-class DCT schemes, which is shown in Table 3.2. Observe that the highest relative frequency bottom entry of the quad-class DCT codec in Table 3.2 is associated with an extended horizontal correlation, which is also confirmed by the autocorrelation function of the typical MCER, as seen in Figure 3.5.

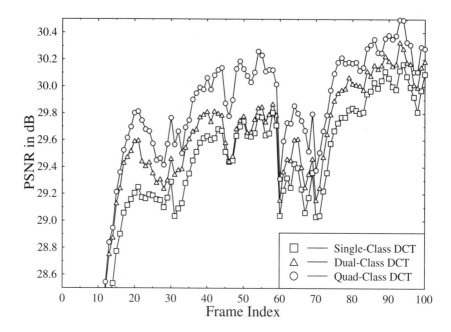

Figure 3.14: Performance comparison of multiclass DCT schemes.

Table 3.2: Relative Frequency of the Various Quantizers in the Multiclass DCT Codecs in the "Miss America" Sequence

Quantizer Set Type	Av. Rel. Freq. Quad-class DCT	Av. Rel. Freq. Dual-class DCT
Very High Correlation	14.7%	39%
High Correlation	29.3%	61%
Increased Vertical Correlation	24.0%	—
Increased Horizontal Correlation	32.0%	—

3.4 The Codec Outline

In order to achieve a time-invariant compression ratio associated with a constant encoded video rate, the codec is designed to switch between two modes of operation as depicted in Figure 3.15. This is necessary because the DCT-based inter-frame codec assumes that the previous reconstructed frame is known to both the encoder and decoder. At the commencement of communications, the reconstructed frame buffers of the encoder and decoder are empty. Thus, the MCER energy for the first frame would be very high, and it would be the first frame to be encoded. The quantizers we derived using the philosophy of Section 3.3 were not trained for this particular case, and so the DC-coefficient quantizer would be driven to saturation. Therefore, we transmit a single independent intra-coded

3.4. THE CODEC OUTLINE

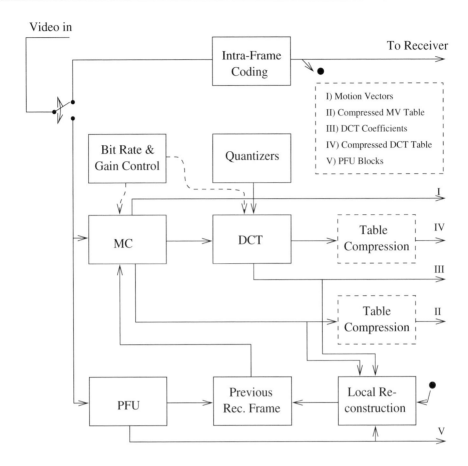

Figure 3.15: Schematic of the multiclass DCT codec.

frame during the initial phase of communication in order to replenish the reconstructed frame buffers. Once switched to inter-frame mode, any further mode switches are optional and are required only if a drastic change of video scene occurs.

In the inter-frame mode of Figure 3.15 the incoming video sequence is first motion compensated, and then the resulting MCER is encoded. The active/passive decisions carried out during both the MC and DCT coding stages are cost/gain controlled within the allocated constant, but programmable bitrate budget as it was highlighted in Section 3.2.2.6. An optional table compression algorithm of Figure 3.15, which will be described at a later stage, is also incorporated in order to reduce the amount of redundancy associated with the transmission of the 396-bit motion- and DCT activity tables. Finally, a partial forced update (PFU) algorithm is included in Figure 3.15 to ensure that a realignment of the encoder's and decoder's reconstructed buffers becomes possible, while operating under high error-rate channel conditions. The following sections elaborate further on the individual blocks of Figure 3.15.

Table 3.3: Block Sizes Used in the Startup Phase of the Intra-frame DCT Codec. The Block Size Along the Fringes of the QCIF Frames was Enlarged Slightly in Order to Perfectly Tile the Frame and to Generate the Exact Required Number of Bits per Frame

Bits per Frame	Block Size
500	14 × 14
800	12 × 12
1000	10 × 10
1200	10 × 10
1500	9 × 9

3.5 Initial Intra-frame Coding

The objective of the intra-frame mode of Figure 3.15 is to provide an initial frame for both the encoder's and decoder's reconstructed frame buffer in order to ensure that the initial MCER frame does not saturate the DC-coefficient quantizer. Constrained by the fixed bitrate budget of about 1000 bits/frame, the 176 × 144-pixel QCIF frame can only be represented very coarsely, at a resolution of about $1000/(176 \times 144) \simeq 0.04$ bits/pixel. This can be achieved by dividing the frame into perfectly tiling blocks and coarsely quantizing the average of each block. It was found appropriate to quantize each block average to 16 levels, which are equidistantly distributed between the absolute pixel values of 52 and 216. Since the entire intra-frame coded image must be encoded within the limits of the per-frame bitrate budget, the block size had to be adjusted accordingly. The resulting block sizes for various target bitrates are depicted in Table 3.3. When the intra-frame coded block size did not tile the QCIF format perfectly, the codec increased the block size for those particular blocks, which were situated along the fringes of the frame.

3.6 Gain-controlled Motion Compensation

The motion compensation process of Figure 3.15 is divided into two steps. In the first step the codec determines a motion vector for each of the 8 × 8 sized blocks. The search window is set to 4 × 4 pixels, which proved to be efficient in removing redundancies between consecutive frames at a cost of 4 bits per MV in the context of our QCIF head-and-shoulder videophone sequences. Since we concluded in Section 3.2.4 that MC achieves a significant MCER gain only for a fraction of the blocks, the codec stores the potential MCER energy gain for every block. The gain of a particular block B at location (x_0, y_0) in the video frame f is defined as:

$$G_{mc} = \sum_{i=1}^{b}\sum_{j=1}^{b}(f_n(x_0+i, y_0+j) - f_{n-1}(x_0+i, y_0+j))^2$$
$$- \sum_{i=1}^{b}\sum_{j=1}^{b}(f_n(x_0+i, y_0+j) - f_{n-1}(x_0+i+m_x, y_0+j+m_y))^2. \quad (3.23)$$

The bitrate control algorithm selects the specific number of most efficient MVs according to the available bitrate budget. The remaining MVs are set to zero, before subtracting the current motion translated reconstructed block from the incoming one, which corresponds to simple frame-differencing, as far as the "motion-passive" blocks are concerned. At this stage we have to further clarify our active/passive block terminology in order to avoid future confusion. Our later discussion will refer to those blocks where MVs were allocated as motion-active, while to those blocks, where MV = 0 and hence frame-differencing is used, as motion-passive. On a similar note, blocks where the MCER is encoded are termed MCER-active and those where the MCER is low, and hence it is set to zero, are referred to as MCER-passive.

3.7 The MCER Active/Passive Concept

Section 3.3 introduced the philosophy of energy compaction and demonstrated how this property can be exploited in order to reduce the number of encoded bits by using the DCT. In most cases, it is sufficient to transmit a total of 10 to 12 bits for an 8×8 pixel block in order to maintain adequate videophone quality, which corresponds to a compression ratio of (25 384 pixels $\times$8 bits/pixel)/(396 blocks $\times$12 bits) ≈ 42, when neglecting the motion vectors. Further compression can be achieved when the codec intelligently allocates bits to those image regions, where it is most beneficial. From the analysis of the MCER, we know that large sections of the MCER frame are flat and there is no need to allocate bits to these MCER-passive areas. Therefore, a protocol is required in order to identify the location of MCER-active blocks, which we encode and that of the MCER-passive ones, which we neglect.

The most conceptually obvious solution would be to transmit the index of every active block. In case of QCIF images and a block size of 8×8, there are 396 blocks, which requires 9 bits to store one MCER-active block index. Assuming a total of perhaps 100 active blocks, 900 bits/frame would be necessary only to identify their locations. A more promising approach is to establish an active/passive table, which for each of the 396 blocks contains a 1-bit flag marking the corresponding block either as MCER-active or MCER-passive. This becomes advantageous compared to the previous method, if there are more than 44 active blocks. Bearing in mind that the active probability for very low bitrate applications is usually below 15%, we note that most of the entries of the active-passive table will be marked as passive. This implies predictability or redundancy and opens further compression possibilities, such as run-length or entropy-encoding.

We decided to scan the active-passive flags on a line-by-line basis and then packetized a certain number of them into a single symbol, which were then Huffman codes, as highlighted in [161]. It was expected that most symbols would contain a low number of 1 and a high number of 0 flags. This *a priori* knowledge could then be exploited to assign short Huffman code words to the high-probability symbols and long ones to the low-probability symbols. We found that the optimum number of bits per symbol was around five. This bit-to-symbol conversion exploits most of the latent redundancy inherent in the bit-by-bit representation, while preventing us from optimizing the codewords of the symbol-based Huffman codec to a certain activity rate. Figure 3.16 reveals the advantage of this active/passive table compression technique in case of QCIF frames and 8×8 sized blocks. However, this technique does not exploit the fact that increased motion activity often covers a group of several blocks rather

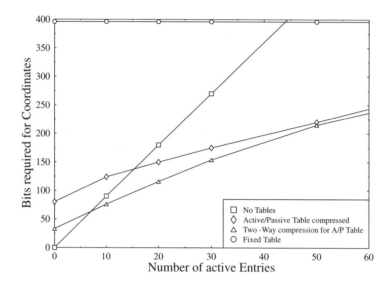

Figure 3.16: Comparison for various active/passive classifier encoding strategies in QCIF frames for 8×8 blocks.

than a single block implying that the uncompressed 396 activity flags are likely to be clustered in active and passive islands.

Therefore, we developed another approach in order to further compress the active/passive table. The motion activity flags were grouped into blocks of two by two, covering four original 8×8 pixel blocks. For those groups that did contain active motion flags, we assigned a 4-bit activity symbol, reflecting its activity-flag contents. These symbols were then Huffman coded. This concept of only transmitting information regarding those groups of vectors, which were active, required a second active/passive table. This second table reflected which of the grouped motion activity flags were set to active. It contained $396/4 = 99$ entries, which we again packetized to three entries per symbol, a value that was found to guarantee best coding efficiency, and these entries were then Huffman encoded. This method was found superior to the previous technique, especially, when the active rate was below 10% or 40 blocks per frame, as demonstrated by Figure 3.16. In this range of block activity ratio, on the average we needed about 5 bits per active entry in order to convey the activity information to the decoder. This corresponds to a bitrate reduction of nearly 50%, when compared to the 9-bit "indexing" and "uncompressed table" techniques of Figure 3.16. Further potential bitrate economies can be achieved upon invoking the error-resilient positional coding (ERPC) principle [188].

3.8 Partial Forced Update of the Reconstructed Frame Buffers

Let us now consider the partial forced update block of Figure 3.15. Communications over nonideal channels is prone to errors, which is particularly true in hostile mobile environments

3.8. PARTIAL FORCED UPDATE

characterized by Rayleigh-fading channels. As a consequence of inter-frame coding, the video quality of the reconstructed frames is impaired not only in the affected frames, but also in consecutive frames. The errors result in a misalignment between the encoder's local reconstructed buffer and the equivalent reconstructed frame buffer at the receiver side. This misalignment will persist if no measures are implemented to mitigate the effects of errors. A potentially suitable remedy is to apply a so-called leakage technique, which de-weights the reconstructed frame decoded from the received information using a leakage factor [187]. Hence, this procedure inflicts some image degradation under error-free channel conditions in order to allow for the effects of channel errors to decay.

In a specific implementation of the leakage technique, the video frame or segments of the frame are periodically luminance scaled. A typical leakage factor of, for example, 0.9 would mean that the brightness of every pixel in the reconstructed frame buffer would be multiplied by 0.9, resulting in a slowly fading luminance of consecutive frames. This then would result in an increased MCER and a slightly more moderate coding performance, while improving the codec's robustness. The gradually fading segments would therefore remove the effects of channel errors over a period of time. This simple technique renders the PDF of the DC coefficient to become less highly peaked in the center, which inevitably leads to a wider dynamic range and hence typically requires more reconstruction levels for the corresponding quantizer, unless a higher coding distortion is acceptable. The extent to which the quantizers have to be adapted to the modified input data depends on the actual value of the leakage factor, which in turn depends on prevailing BER. Higher BERs require a faster luminance scaling or fading, that is, lower valued leakage factors.

Another often used procedure is to invoke a so-called forced updating technique, which forcibly realigns the reconstructed frame buffers at regular instants using typically a compromise value, which again allows the decoder to gradually taper the effects of errors. In order to incorporate the highest possible degree of flexibility, we opted for a combination of the above-mentioned leakage and replenishment techniques. Before inter-frame encoding of the MCER ensues, we select a predetermined, bitrate dependent number of 8×8 pixel blocks in the incoming frame buffer, for which we determine the coarsely quantized average, as we detailed in Section 3.5. Then every pixel of the selected blocks in the reconstructed buffers is first scaled down by a certain factor $(1 - l)$. Then the 4-bit quantized average is superimposed, which is scaled by the leakage factor l. This partial forced updating (PFU) scheme guarantees that the average MCER energy does not change and the DC-quantizers of the subsequent DCT codec are not overloaded. The leakage factor values used in our approach are typically between 0.3 and 0.5 in order to ensure that the PFU blocks do not become blurred in the next frame. Since this Partial Forced Update (PFU) constitutes a completely separate encoding step, the subsequent encoding of the same block in the following MC and DCT encoding steps may prevent blurring completely.

The PSNR degradation inflicted by this PFU technique under error-free conditions is portrayed in Figure 3.17, which appears particularly low as long as the proportion of PFU blocks is low. When a PFU rate of about 10%, corresponding to 40 blocks is applied, the PSNR degrades by about 1 dB at a concomitantly improved codec robustness. Observe furthermore in the figure that in case of a modest proportion of replenished blocks, perhaps up to 5%, the codec is able to reduce the MCER energy and hence a slightly improved PSNR performance is achieved.

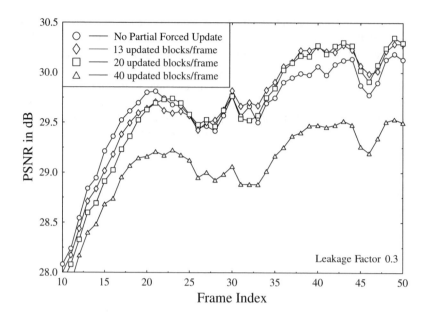

Figure 3.17: PSNR degradation as a consequence of invoking partial forced updates.

We made use of the same quantizer, which was invoked for the intra-frame coding, but the size of the updated blocks is fixed to 8×8 pixels. The number of updated blocks per frame and the leakage factor determine the inflicted intentional video impairment due to PFU and the codec's ability to recover from erroneous conditions, that is, the codec's robustness. In our experiments we found that the number of updated blocks per frame had to be below 25, as also evidenced by Figure 3.17.

3.9 The Gain/Cost-controlled Inter-frame Codec

In the previous sections, we have described the components of the coding scheme in Figure 3.15. This section reports on how we combined the introduced components in order to create a coding scheme for QCIF images at bitrates around 10 kbit/s.

The motion detector attempts to minimize the MCER between the incoming and the local reconstructed frame buffers. As a result of the conclusions of Section 3.2.4, the search window is limited to a size of 4×4 pixels, which limits the search scope to -2 in the negative and $+1$ in the positive horizontal/vertical spatial directions. The motion detector stores the best motion vector for each of the 396 blocks, as well as the corresponding gain, as defined in Section 3.6. Now the bitrate control unit in Figure 3.15 marks inefficient motion vectors as passive and determines the corresponding active/passive tables as described in Section 3.7, computing also the resulting overall bitrate requirement for the MVs. The codec then estimates how many encoded MVs can be accommodated by the available bitrate budget and relegates some of the active motion vectors to the motion-passive class. Then the bitrate requirement is updated and compared to the bitrate limit. This deactivation of motion vectors continues until we reach the predetermined bitrate for the MC.

Finally, motion compensation takes place for the motion-active blocks, while for the passive blocks simple frame-differencing is applied. The MCER is passed on to the DCT codec, which has a similar active/passive classification regime to that of the motion compensation scheme. Explicitly, each block is transformed to the frequency domain, quantized and then transformed back to the time domain in order to assess the potential benefit of the DCT-based coding in terms of overall PSNR or MSE contribution. The best set of quantizers is found by a full search — that is, by invoking all four available quantizers and evaluating their MSE performance. Now, the bitrate control unit in Figure 3.15 determines the number of available bits for the DCT codec remaining from the total budget for the frame after reserving the required capacity for the partial forced update, the motion compensation, and the frame alignment word (FAW). The FAW is a 22-bit unique word, which allows the decoder to resynchronize at the commencement of the next video-frame using correlation techniques after losing synchronization. Again, the active/passive DCT tables are determined, and blocks attaining low MSE gains are not encoded in order to meet the overall bitrate requirement. In order to emphasize the subjectively more important central eye and lip region of the screen, the codec incorporates the option of scaling the DCT gains and allows for the codec to gradually improve its image representation in its central section. This is particularly important when operating at 5–8 kbit/s or during the first transmitted frames, when the codec builds up fine details, to reach its steady-state video quality, commencing from the coarsely quantized intra-frame coded initial state.

3.9.1 Complexity Considerations and Reduction Techniques

The number of multiplications required for the direct discrete cosine transformation of the $b \times b$ matrix B is proportional to b^4 [10]. The dimensionality of the problem can be reduced to a more realistic complexity order, which is proportional to b^3 if we invoke the separable transform: $B_{dct} = TBT^t$, where T is the $b \times b$ unitary transformation matrix we defined by Equation 3.21 and T^t is the transpose of T. A single matrix multiplication of a $b \times b$ matrix requires $b^2(2b-1)$ floating point operations (Flops). In case of our 8×8 blocks ($b = 8$), 1920 Flops are necessary to evaluate one transformation. If we apply gain-controlled quad-class DCT, it is necessary to transform each block back to the time domain after choosing one of the four quantizers. If we apply the MSE criterion to select the best of the possible four quantizers, we require a total of $4b^2(2b-1) + 4(3b^2-1) \simeq 8b^3 + 12b^2$ Flops per block, which is the equivalent of 1.9 Mflops per QCIF frame. This can be reduced by about 50% upon exploiting some regularities in the matrix operations, by reducing the complexity to the order of $b^2 \log_2 b^2 = 384$ Flops per transformation, at the cost of increased data handling steps [189], or by using the so-called fast DCT [178]. This may be important for a direct implementation in silicon.

A much more efficient way of reducing the computational complexity becomes possible using a block classification algorithm prior to any encoding steps. Similar to our experience in Section 3.2.2.6 as regards MC, we found that a substantial proportion of the MCER contains flat blocks, which are unlikely to be selected by a gain-cost controlled codec. Hence, prior to the actual encoding, the encoder determines the energy contents of all blocks. Only the blocks exhibiting high energy content, that is, blocks carrying important information, are considered

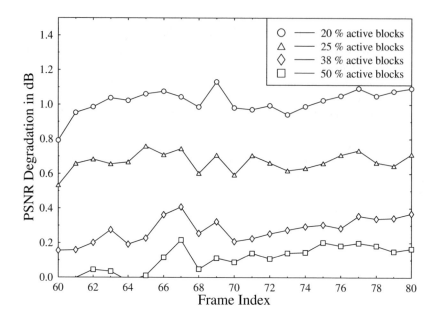

Figure 3.18: PSNR degradation caused by the preselection of active blocks in an attempt to reduce the codec's complexity.

for the MC and DCT encoding operations. In case of the MC, the encoder first determines the frame difference, and then the motion prediction focuses on the locations where significant movement took place. After carrying out the MC step, a similar preclassification phase is invoked for the DCT. The degradation caused by the above suboptimum preselection phase is depicted in Figure 3.18 for a range of active block proportions between 20 and 50%. These results reveal that the complexity can be reduced by about 50% without noticeable PSNR impairment, when compared to the original scheme. When the proportion of active blocks is as low as 20%, the preclassification is more likely to be deceived than in the case of retaining a higher proportion of blocks. When assuming, for example, a 38% block activity rate, the codec exhibits a complexity of 2 Mflop per frame or 20 Mflops, which is comparable to that of the standard so-called Advanced Multi-Rate AMR speech codec [190] of the 3G systems [83], hence constituting a realistic real-time implementational complexity at the time of writing.

3.10 The Bit-allocation Strategy

Let us now consider the bit-allocation scheme of our codec portrayed in Figure 3.15. The bitstream transmitted to the decoder consists of bits for:

- The block averages for the partial forced update.
- The table for the active MVs.

- The active MVs.
- The table for the active DCT blocks.
- The active DCT coefficients.

In our experiments we found that the best subjective and objective quality was achieved when the number of active blocks for the MC and DCT was roughly the same, although not necessarily the same blocks were processed by the two independent active/passive classification schemes. Since the 4-bit motion vectors require a lower channel capacity than the 12-bit quantized DCT coefficients, we decided to earmark between $\frac{1}{2}$ and $\frac{2}{3}$ of the available bitrate budget to the DCT activity table and DCT coefficients, while the remaining bits were used for the MC and PFU. The PFU is usually configured to refresh 22 out of the 396 blocks in each frame. Therefore, $4 \times 22 = 88$ bits were reserved for the PFU. The actual number of encoded DCT blocks and MVs depends on the selected bitrate. It typically varied between 30 and 50 for bitrates between 8 and 12 kbit/s at a scanning rate of 10 frames/s.

The output of the codec contains two classes of bits — namely, the entropy-encoded MC and DCT activity tables on one hand, and the motion vectors, quantized DCT coefficients, and the partial forced updated blocks on the other. The first class of information, due to the encoding procedure, is extremely vulnerable against any corruption. A corrupted bit is likely to create a code associated with a different length, and as a result the entire frame may have to be dropped. Because this high vulnerability to channel errors is unacceptable in wireless applications, we also contrived another, more robust codec, which sacrifices coding efficiency and abandons the run-length coding concept for the sake of improved error resilience. A typical bit-allocation example for the variable-length compressed Codec I and for the more error-resilient Codec II is given in Table 3.4.

Table 3.4: Bit-allocation of the Variable-length Compressed Codec I at 8 kbit/s and for the More Error Resilient Codec II at 11.3 kbit/s

Codec	FAW	PFU	MV Ind.	MV	DCT Ind.	DCT	Total
Codec I	22	22×4	—	< 350	—	< 350	800
Codec II	22	22×4	30×9	30×4	30×9	30×12	1130

Later in this chapter, we will refer to the previously introduced codec discussed above as Codec I. In contrast, Codec II does not take advantage of the compression capabilities of run-length coding. In Codec II, we decided to transmit the index of each active DCT block and MV requiring 9 bits to identify one of the 396 indices using the enumerative method. The increased robustness of the codec is associated with an approximately 35% increased bitrate. As Figure 3.19 reveals, Codec I at 8 kbit/s achieves a similar quality to that of Codec II at 11.3 kbit/s.

3.11 Results

The performance of the variable-length compressed Codec I was tested at various bitrates in the range of 5 to 15 kbit/s. All PSNR versus frame index results presented in Figures 3.21

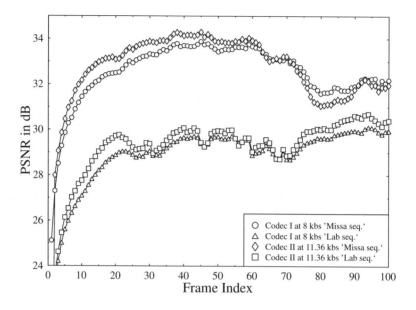

Figure 3.19: PSNR versus frame index performance for Codec I at 8 kbit/s and Codec II at 11.3 kbit/s.

Figure 3.20: Frame 87 of the "Lab" sequence, original (left) and 8kbit/s DCT encoded (right). Both Sequences encoded at various bitrates can be viewed under the WWW address http://www-mobile.ecs.soton.ac.uk

and 3.22 were obtained at a constant frame rate of 10 frames/s and without partial forced update. The results for the more error-resilient Codec II are similar at a 35% higher bitrate. A typical frame of the "Lab" sequence, which was one of our locally recorded high activity sequences, is depicted in Figure 3.20.

A frame rate of 10 frames/s is sufficiently high for head-and-shoulder images, as we do not expect very high motion activity. The codec builds up its steady-state video quality

3.11. RESULTS

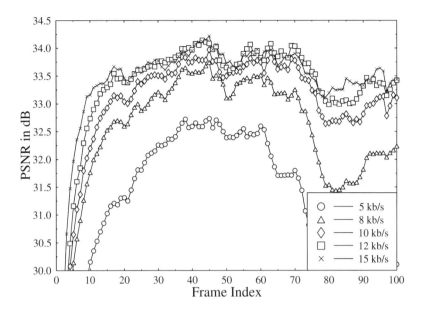

Figure 3.21: PSNR versus frame index performance of the "table-compressed" DCT Codec I at various bitrates for the "Miss America" sequence.

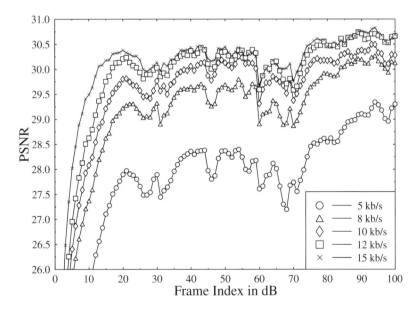

Figure 3.22: PSNR versus frame index performance DCT Codec I at various bitrates for the "Lab" sequence.

following the intra-frame initialization phase after about 15 frames or 1.5 seconds, which does not constitute an annoyance. The drop of the PSNR performance between frames 80 and 90 is due to the fact that "Miss America" moves her whole body, and, especially when operating at 5 kbit/s, the codec cannot instantaneously cope with the drastically increased motion activity. However, as soon as the motion activity surge decays, the PSNR curve recovers from its previous loss of quality. The fact that the image quality does not significantly increase, when using the higher bitrates of 15 kbit/s or 12 kbit/s compared to the 10 kbit/s scenario, demonstrates how well the DCT scheme performs at low bitrates around 10 kbit/s. The codec achieves a similar PSNR to the ITU H261 scheme presented in [191] and [192], while operating at twice the frame rate and maintaining a similar bitrate, which implies a factor 2 higher compression ratio. The improved performance is attributable mainly to the intelligent cost-gain quantization invoked throughout the coding operation while maintaining a moderate complexity of about 20 Mflops.

3.12 DCT Codec Performance under Erroneous Conditions

The codecs presented here are suitable for a wide range of applications, including videotelephony over both fixed and wireless networks, surveillance and remote sensing etc. For example, mobile videophony at a transmission rate comparable to that of the standard so-called Advanced Multi-Rate AMR speech codec [190] of the 3G systems [83] becomes realistic.

Over error-free transmission media, when for example Automatic Repeat Request (ARQ) techniques can be invoked, the proposed codecs could be employed without PFU. Such a channel would provide the best performance and compression, since no bits are dedicated to PFU. However, for most practical channels the PFU must be invoked. In practical scenarios, it is essential to analyze the sensitivities of the various encoded bits to channel errors in order to be able to protect more vulnerable bits with stronger error correction codes and to design appropriate source-sensitivity matched error protection schemes.

3.12.1 Bit Sensitivity

Shannon's ideal source encoder is constituted by an ideal entropy codec, as argued in Chapter 1 of [161], where corrupting a single bit results in an undecodable codeword. The output bits of Shannon's ideal source codec are hence equally sensitive to bit errors. The proposed codecs still retain some residual redundancy and exhibit unequal bit sensitivities. In particular, the run-length (RL) encoded tables of Codec I are very sensitive to bit errors and need more protection than others. A common solution to this problem is to divide the bitstream in two or more sensitivity classes and assign a different forward error correction (FEC) code to each of the classes. In case of Codec I, the sensitivity of the RL-coded bits is difficult to quantify, for their corruption affects all other bits in the same transmission burst. Various procedures are available in order to quantify the bit sensitivity of the remaining non-RL encoded bits of Codec I and all bits of Codec II. In [193] a single bit of a video-coded frame was consistently corrupted, and the inflicted image peak signal-to-noise ratio (PSNR) degradation was observed. Repeating this method for all bits of a frame provided the

3.12. DCT CODEC PERFORMANCE UNDER ERRONEOUS CONDITIONS

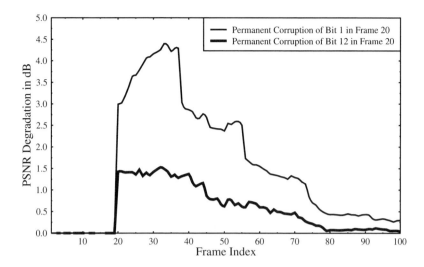

Figure 3.23: PSNR degradation profile for Bits 1 and 12 of the DCT Codecs I and II in Frame 2, where Bit 1 is the MSB of the PFU segment, while Bit 12 is a MV bit, as seen in Figure 3.24.

required sensitivity figures, and on this basis bits having different sensitivities were assigned matching FEC codes. This technique does not take adequate account of the phenomenon of error propagation across image frame boundaries. Here we used the method suggested in [194], where each bit of the same type is corrupted in the current frame and the PSNR degradation for the consecutive frames due to the error event in the current frame is observed. As an example, Figure 3.23 depicts the PSNR degradation profile in case of corrupting Bit No. 1 of a transmission frame, namely the Most Significant Bit (MSB) of the PFU segment in frame 20 of the "Miss America" sequence, and Bit No. 12, one of the MV bits. The observable PSNR degradation over its whole duration is accumulated, averaged over all occurrences of that particular type of bit, and then weighted by the corresponding relative frequency of each bit under consideration in order to obtain the PSNR degradation sensitivity measures portrayed in Figure 3.24.

3.12.2 Bit Sensitivity of Codec I and II

For Additive White Gaussian Noise (AWGN) channels we propose forward error correction (FEC). Both convolutional and block codes are feasible. Once the FEC scheme fails to remove the transmission errors, we have to differentiate between two possible error events. If the run-length encoded tables are corrupted, it is likely that a codeword of a different length is generated and the decoding process becomes impossible. This error is detectable because the decoded frame length and the preset frame length differ. Hence, a single bit error can force the decoder to drop an entire frame. If, however, one of the PFU, DCT, or MV bits is corrupted, the decoder is unable to determine that a corruption took place, but a maximum of two 8×8 blocks is affected by a single bit error. The difference in vulnerability for the run-length and non-run-length encoded bits is highlighted in Figure 3.25. If the whole bitstream

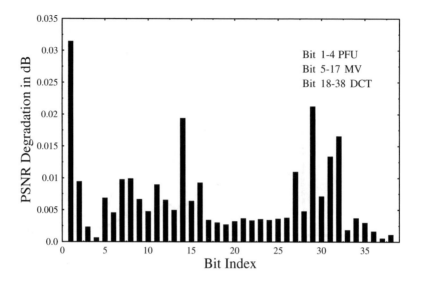

Figure 3.24: Integrated bit sensitivities for the error-resilient DCT Codec II.

of Codec I is subjected to bit corruption, a BER of $2 \cdot 10^{-4}$ is sufficient to inflict unacceptable video degradation. If, however, the bit corruption only affects the non-run-length encoded bits, while the RL-coded bits remain intact, the codec can tolerate BERs up to $2 \cdot 10^{-3}$. In reference [195], we proposed an appropriate transmission scheme, which takes advantage of this characteristic. These schemes will be highlighted in Section 3.13. Another way of imposing the error resilience of Codec I would be the application of error resilient positional coding [188].

As evidenced by Figure 3.25, the absence of run-length encoded bits increases the error resilience of Codec II by an order of magnitude. Therefore, Codec II suits, for example, mobile applications over Rayleigh-fading channels. Further issues of unequal protection FEC and ARQ schemes were discussed in [195] and will be highlighted in Section 3.13. Having designed and analyzed our proposed DCT-based video codec, let us now consider some of the associated systems aspects in the context of a wireless videophone system.

3.13 DCT-based Low-rate Video Transceivers

3.13.1 Choice of Modem

The factors affecting the choice of modulation for a particular system were discussed in [196], and so we will not consider modulation aspects. Suffice to say here that the differentially coded noncoherent Star-QAM modems described in [196] exhibit typically lower complexity than their pilot symbol-assisted (PSA) coherent counterparts, but as we have shown, they inflict a characteristic 3 dB differential coding SNR penalty over AWGN channels, which also persists over Rayleigh channels. A further SNR penalty is imposed by the reduced minimum distance property of the rotationally symmetric differentially encoded StarQAM

3.13. DCT-BASED LOW-RATE VIDEO TRANSCEIVERS

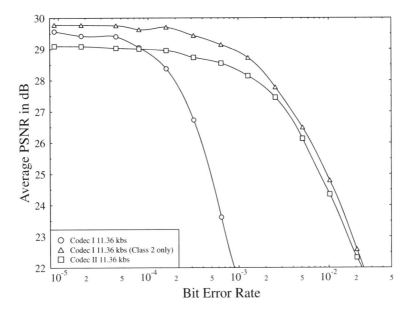

Figure 3.25: PSNR degradation versus BER for Codec I and II using random errors, demonstrating that the more robust class 2 bits of Codec I can tolerate a similar BER to Codec II.

constellation. Hence, Star-QAM requires higher SNR and SIR values than the slightly more complex coherent PSA schemes. Therefore in our proposed video transceivers second-order switched-diversity aided coherent Pilot Symbol Assisted Modulation (PSAM) using the maximum-minimum distance square QAM constellation was employed.

3.13.2 Source-matched Transceiver

3.13.2.1 System 1

3.13.2.1.1 System Concept The system's schematic is portrayed in Figure 3.26, where the source-encoded video bits generated by the video encoder are split in two bit-sensitivity classes, and sensitivity matched channel coding/modulation is invoked. This schematic will be detailed throughout this section. A variety of video codecs were employed in these investigations, which were loosely based on the candidate codecs presented in Table 3.4, but their bitrates were slightly adjusted in order to better accommodate the bit-packing constraints of the binary BCH FEC codecs employed. The run length-compressed Codec I of Table 3.4 had a bitrate of 8 kbit/s, and Codec II was programmed to operate at 11.36 kbit/s, while a derivative of Codec I, Codec I-bis generated a rate of 8.52 kbit/s, as seen in Table 3.5, along with a range of other system features. The proposed system was designed for mobile videotelephony, and it had two different modes of operation, namely 4-level and 16-level quadrature amplitude modulation (QAM), which were the subject of [196]. Our intention was to contrive a system, in which the more benign propagation environment of indoor cells would benefit from the prevailing higher signal-to-noise ratio (SNR) by using bandwidth-efficient 16QAM, thereby requiring only half the number of transmission packets compared to 4QAM. When the portable station (PS) is handed over to an outdoor microcell or roams in

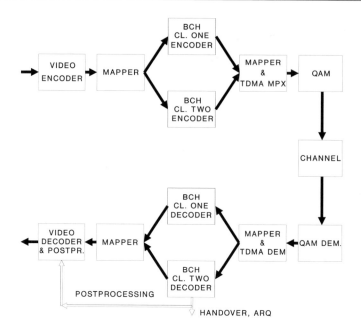

Figure 3.26: DCT-based videophone system schematic.

a lower SNR region toward the edge of a cell, the base station (BS) instructs the PS to lower its number of modulation levels to 4, in order to maintain an adequate robustness under lower SNR conditions. Let us now focus our attention on specific details of System 1, which was designed to accommodate the bitstream of the error-resilient Codec II.

3.13.2.1.2 Sensitivity-matched Modulation As shown in Figure 5.2 of [196], 16-level pilot symbol-assisted quadrature amplitude modulation (16-PSAQAM) provides two independent 2-bit subchannels having different bit error rates (BER). Specifically, the BER of the *higher integrity C1 subchannel* is a factor two to three times lower than that of *the lower quality C2 subchannel*. Both subchannels support the transmission of 2 bits per symbol. This implies that the 16-PSAQAM scheme inherently caters for sensitivity-matched protection, which can be fine-tuned using appropriate FEC codes to match the source codec's sensitivity requirements. This property is not retained by the 4QAM scheme, since it is a single-subchannel modem, but the required different protection for the source-coded bits can be ensured using appropriately matched channel codecs.

3.13.2.1.3 Source Sensitivity In order to find the appropriate FEC code for our video codec, its output stream was split in two equal sensitivity classes, namely, Class One and Class Two, according to our findings in Figure 3.24. Note that *the notation Class One and Two introduced here for the more and less sensitive video bits is different from the higher and lower integrity C1 and C2 modulation channels*. Then the PSNR degradation inflicted by both the Class One and Two video bits as well as the average PSNR degradation was evaluated for a range of BER values in Figure 3.27 using randomly distributed bit errors. These results

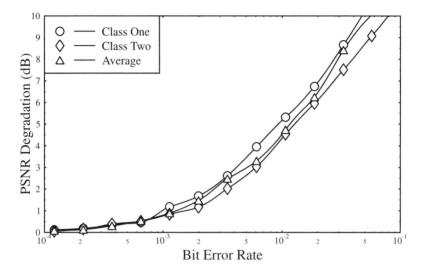

Figure 3.27: PSNR versus BER degradation of Codec II for Class One and Two bits.

showed that a lower BER was required by the Class One bits than by the Class Two bits. More explicitly, about a factor two lower BER was required by the Class One bits than by the Class Two bits in order for the video system to limit the PSNR degradations to 1–2 dB. These integrity requirements conveniently coincided with the integrity ratio of the C1 and C2 subchannels of our 16-PSAQAM modem, as seen in Figure 5.2 of [196]. We therefore can apply the same FEC protection to both the Class One and Two video source bits and direct the Class One bits to the C1 16-PSAQAM subchannel, and the Class Two bits to the C2 subchannel.

3.13.2.1.4 Forward Error Correction Both convolutional and block codes can be successfully used over mobile radio links, but in our proposed scheme we have favored the binary Bose-Chaudhuri-Hocquenghem (BCH) codes introduced in Chapter 3 and 4 of [86]. Suffice to say that BCH codes combine a good burst-error correction capability with reliable error detection, a facility useful for invoking image post-enhancement in monitoring the channel's quality and in controlling handovers between traffic cells. The preferred $R = 71/127 \approx 0.56$-rate BCH(127,71,9) code can correct nine errors in a block of 127 bits, an error correction capability of about 7.1%. The number of channel-coded bits per image frame becomes $1136 \times 127/71 = 2032$, while the bitrate becomes 20.32 kbit/s at an image frame rate of 10 frames/s.

3.13.2.1.5 Transmission Format The transmission packets are constructed using one Class One BCH(127,71,9) codeword, and one Class Two BCH(127,71,9) codeword. A stronger BCH(127,50,13) codeword is allocated to the packet header, which conveys user-specific identification (ID) and dedicated control information, yielding a total of 381 bits per packet. In the case of 16QAM, these codewords are represented by 96 symbols, and after adding 11 pilot symbols using a pilot spacing of $P = 10$ as well as four ramp symbols to ensure smooth power amplifier ramping, the resultant 111-symbol packets are transmitted over the radio channel. Eight such packets represent an entire QCIF image frame. Thus the signaling rate becomes 111 symb/12.5 ms $\approx$ 9 kBd. When using for example a time division multiple access (TDMA) channel bandwidth of 200 kHz, for the sake of illustration, as in the Pan-European second-generation mobile radio system known as GSM [197] and employing a Nyquist filtering (see [196]) modulation excess bandwidth of 38.8%, the signaling rate becomes 144 kBd. This allows us to accommodate $144/9 = 16$ video users, or eight speech and eight video slots for videotelephony. For example, the number of voice/videophone users accommodated in a bandwidth of 200 kHz then coincides with the number of full-rate speech users supported by the GSM system [197]. In contrast, the 3G systems [83] have an approximately 5 MHz bandwidth and a chip-rate of 3.84 MChip/s. Thus, depending on the spreading factor employed, they are capable of potentially accommodating a significantly higher number of users, although naturally, the video coding rate only depends on the video source codec. The combined voice/video user bandwidth depends also on the speech coding rate and on the channel codec used.

When the prevailing channel SNR does not allow 16QAM communications, 4QAM must be invoked by the HSDPA-style [198] reconfigurable transceiver. In this case, the 381-bit packets are represented by 191 two-bit symbols, and after adding 30 pilot symbols and four ramp symbols, the packet-length becomes 225 symb/12.5 ms, yielding a signaling rate of 18 kBd. In this case, the number of "video only" users supported by System 1 becomes 8 in the 200 KHz bandwidth used in our example. Thus the number of combined voice/video users would be reduced to four. The system also facilitates mixed-mode operation, where 4QAM users must reserve two slots in each 12.5 ms TDMA frame near the fringes of the traffic cell, while in the central section of the cell, 16QAM users will require only one slot per frame in order to maximize the number of users supported. Assuming an equal proportion of 4QAM and 16QAM users, the average number of users per carrier becomes 12. The equivalent user bandwidth of the 4QAM PSs is 200 kHz/8 = 25 kHz, while that of the 16QAM users is 200 kHz/16 = 12.5 kHz.

For very high quality indoor mobile channels or for conventional telephone lines, 64-QAM can be invoked, which further reduces the required bandwidth at the cost of a higher channel SNR demand. However, the packet format of this mode of operation is different from that of the 16QAM and 4QAM modes and so requires a different slot length. The 381-bit payload of the packet is represented by 64 six-bit symbols, and four ramp symbols are added along with 14 pilot symbols, which corresponds to a pilot spacing of $P = 5$. The resulting 82-symbol/12.5 ms packets are transmitted at a signaling rate of about 6.6 kBd, which allows us to host 22 videophone users. The equivalent user bandwidth in our 200 KHz-bandwidth example becomes 200 kHz/22 $\approx$ 9.1 kHz.

These features of the 16QAM/4QAM System 1, along with the characteristics of a range of other systems introduced in the next section, are summarized in Table 3.5.

Clearly, the required video signaling rate and bandwidth are comparable to those of most state-of-the-art mobile radio speech links, which renders our scheme attractive for mobile videotelephony in the framework of existing second- and forthcoming third-generation mobile radio systems, where an additional physical channel can be provided for the video stream. This rate can also be readily accommodated by conventional telephone subscriber

3.13. DCT-BASED LOW-RATE VIDEO TRANSCEIVERS

Table 3.5: Summary of DCT-based Videophone System Features. * ARQs in Systems 2, 3, and 4 are Activated by Class One Bit Errors, but not by Class Two [195]. ©IEEE 1995, Hanzo, Streit

Feature	System 1	System 2	System 3	System 4	System 5
Video Codec	Codec II	Codec I-bis	Codec II	Codec I	Codec I
Video rate (kbps)	11.36	8.52	11.36	8	8
Frame Rate (fr/s)	10	10	10	10	10
C1 FEC	BCH(127,71,9)	BCH(127,50,13)	BCH(127,71,9)	BCH(127,50,13)	(BCH(127,50,13)
C2 FEC	BCH(127,71,9)	BCH(127,92,5)	BCH(127,71,9)	BCH(127,50,13)	(BCH(127,50,13)
Header FEC	BCH(127,50,13)	BCH(127,50,13)	BCH(127,50,13)	BCH(127,50,13)	(BCH(127,50,13)
FEC-coded Rate (kbps)	20.32	15.24	20.32	20.32	20.32
Modem	4/16-PSAQAM	4/16-PSAQAM	4/16-PSAQAM	4/16-PSAQAM	4/16-PSAQAM
Re-transmitted	None	Cl. One	Cl. One & Two	Cl. One & Two	None
User Signal. Rate (kBd)	18 or 9	6.66	18 or 9	18 or 9	18 or 9
System Signal. Rate (kBd)	144	144	144	144	144
System Bandwidth (kHz)	200	200	200	200	200
No. of Video Users	8 or 16	(21-2)=19	6 or 14	6 or 14	8 or 16
Eff. User Bandwidth (kHz)	25 or 12.5	10.5	33.3 or 14.3	33.3 or 14.3	25 or 12.5
Min. AWGN SNR (dB)	7 or 15	15	6 or 13	5 or 11	8 or 12
Min. Rayleigh SNR (dB)	15 or 20	25	7 or 20	7 or 14	15 or 16

loops or cordless telephone systems, such as the European DECT and CT2 or the Japanese PHS and the American PACS systems of Table 1.1 in [322].

3.13.2.2 System 2

In this section we contrive various transceivers, which are summarized in Table 3.5, in order to expose the underlying system design trade-offs. We again emphasize that the effects of transmission errors are particularly objectionable, if the run-length coded activity table bits of Table 3.4 are corrupted. Therefore, in System 2 of Table 3.5, which was designed to incorporate Codec I, the more sensitive run-length (RL) coded activity table bits of Table 3.4 are protected by the powerful binary Bose-Chaudhuri-Hocquenghem BCH(127,50,13) codec, while the less vulnerable remaining bits are protected by the weaker BCH(127,92,5) code. Note that the overall coding rate of $R = (50 + 92)/(127 + 127) \approx 0.56$ is identical to that of System 1, but the RL-coded Class One video bits are more strongly protected. At a fixed coding rate this inevitably assumes a weaker code for the protection of the less vulnerable Class Two video bits. The 852 bits/100ms video frame is encoded using six pairs of such BCH code words, yielding a total of $6 \cdot 254 = 1524$ bits, which is equivalent to a bitrate of 15.24 kbit/s.

As in System 1, the more vulnerable run-length and BCH(127,50,13) coded Class One video bits are then transmitted over the higher integrity C1 16QAM subchannel. The less sensitive BCH(127,92,5) coded Class Two DCT coefficient bits are conveyed using the lower-integrity C2 16QAM subchannel. This arrangement is favored in order to further emphasize the integrity differences of the BCH codecs used, which is necessitated by the integrity requirements of the video bits.

The transmission burst is constructed by adding an additional BCH(127,50,13) codeword for the packet header conveying the user identifier (ID) as well as control information. The resulting 381 bits are again converted to 96 16QAM symbols, and 11 pilots as well as 4 ramp-symbols are added. In System 2 six such $96 + 11 + 4 = 111$-symbol packets represent a video frame; hence, the single-user signaling rate becomes 666 symb/100 ms, which corresponds to 6.66 kBd. This allows us to accommodate now Integer[144 kBd/6.66] = 21 such video users, if no timeslots are reserved for packet re-transmissions. This number will have to be reduced in order to accommodate Automatic Repeat Requests (ARQs).

3.13.2.2.1 Automatic Repeat Request ARQ techniques have been successfully used in data communications [199–202] in order to render the bit- and frame-error rate arbitrarily low. However, because of their inherent delay and the additional requirement for a feedback channel required by message acknowledgments, they have not been employed in interactive speech or video communications. In state-of-the-art wireless systems, however, such as the third-generation systems of [83] there exists a full duplex control link between the BS and PS, which can be used for acknowledgments. The short TDMA frame length ensures a low packet delay and acknowledgment latency. Thus ARQ can be invoked.

In System 2, when the more powerful BCH codec conveying the more sensitive run-length coded Class One bits over the C1 16QAM subchannel is overloaded by channel errors, we re-transmit these bits only, using robust 4QAM. For the first transmission attempt (TX1), we use contention-free Time Division Multiple Access (TDMA). If an ARQ request occurs, the re-transmitted packets will have to vie for a number of earmarked timeslots similarly to

3.13. DCT-BASED LOW-RATE VIDEO TRANSCEIVERS

Packet Reservation Multiple Access (PRMA), which was introduced in [196]. The intelligent base station (BS) detects these events of packet corruption and instructs the portable stations (PS) to re-transmit their packets during the slots dedicated to ARQ packets. Reserving slots for ARQ packets reduces the number of video users supported, depending on the prevailing channel conditions, as we will show in Section 3.14.

Although the probability of erroneous packets can be reduced by allowing repeated re-transmissions, there is a clear trade-off between the number of maximum transmission attempts and the BCH-coded frame error rate (FER). In order to limit the number of slots required for ARQ attempts, which potentially reduce the number of video users supported, in System 2 we invoke ARQ only, if the more sensitive run-length coded Class One bits transmitted via the C1 16-PSAQAM channel and protected by the BCH(127,50,13) codec are corrupted. Furthermore, we re-transmit only the Class One bits, but in order to ensure a high success rate, we use 4-PSAQAM, which is more robust than 16-PSAQAM. Since only half of the information bits are re-transmitted, they can be accommodated within the same slot interval and same bandwidth, as the full packet. If there are only C2 bit errors in the packet, the corrupted received packet is not re-transmitted, which implies that typically there will be residual Class Two errors. In order to limit the number of slots dedicated to re-transmissions, we limited the number of transmission attempts to three, which implies that a minimum of two slots per frame must be reserved for ARQ. In order to maintain a low system complexity, we dispense with any contention mechanism and allocate two timeslots to that particular user, whose packet was first corrupted within the TDMA frame. Further users cannot therefore invoke ARQ since there are no more unallocated slots. A further advantage is that in possession of three copies of the transmitted packet, majority decisions can be invoked as regards all video bits, if all three packets become corrupted. The basic features of System 2 designed to accommodate the run-length compressed Codec I-bis are also summarized in Table 3.5.

3.13.2.3 Systems 3–5

In order to explore the whole range of available system-design trade-offs, we have contrived three further systems, — Systems 3–5 of Table 3.5.

System 3 of Table 3.5 uses the same video and FEC codecs as well as modems as System 1, but it allows a maximum of three transmission attempts in the case of C1 BCH(127,71,9) decoding errors. However, in contrast to System 2, which invokes the more robust 4QAM for ARQs, both the Class One and Two video bits are re-transmitted using the original modem mode. If there are only C2 errors, no ARQ is invoked. Employing ARQ in System 1 constitutes a further trade-off in terms of reducing the number of subscribers supported by two, while potentially improving the communications quality at a certain BER or allowing an expansion of the range of operating channel SNRs toward lower values as seen in the last two rows of Table 3.5 for Systems 1 and 3.

System 4 employs the run-length coded source compression scheme referred to as Codec I, having a bitrate of 8 kbit/s or 800 bits per 100 ms video frame. This system follows the philosophy of System 3, but the Class One and Two video bits were protected by the more powerful BCH(127,50,13) code instead of the BCH(127,71,9) scheme as portrayed in row 5 of Table 3.5. The slightly reduced video rate of 8 kbit/s was imposed in order to accommodate the BCH(127,50,13) code in both 16QAM subchannels, while maintaining the

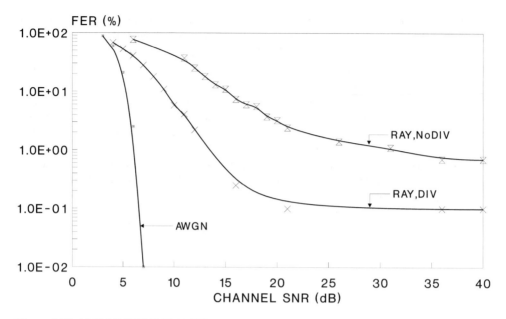

Figure 3.28: 4QAM BCH(127,71,9) FER versus channel SNR performance of the 18 kBd mode of operation of the DCT codec based System 1 of Table 3.5 over AWGN and Rayleigh (RAY) channels with diversity (DIV) and without diversity (NoDIV) [195]. ©IEEE 1995, Hanzo, Streit.

same 20.32 kbit/s overall FEC-coded video rate, as Systems 1 and 3. System 4 will allow us to assess whether it is a worthwhile complexity investment to introduce run-length coding in the slightly higher-rate, but more error-resilient, Codec II in order to reduce the source bitrate and whether the increased error sensitivity of the lower-rate Codec I can be compensated for by accommodating the more complex and more powerful BCH(127,50,13) codec.

In order to maximize the number of video subscribers supported, the performance of System 4 can also be studied without ARQ techniques. We will refer to this scheme as **System 5**. Again, these system features are summarized in Table 3.5. Having designed the video transceivers, we present their performance results in the next section.

3.14 System Performance

3.14.1 Performance of System 1

In our experiments the signaling rate was 144 kBd, while the propagation frequency and the vehicular speed were 1.8 GHz and 30 mph, respectively. For pedestrian speeds, the fading envelope fluctuates less dramatically, and so our experimental conditions constitute an urban worst-case pedestrian scenario.

Here we characterize the performance of the transceiver in terms of the BCH(127,71,9) coded Frame Error Rate (FER) versus channel Signal-to-Noise Ratio (SNR), as portrayed in Figures 3.28 and 3.29 in the case of the 4QAM and 16QAM modes of operation of System 1.

3.14. SYSTEM PERFORMANCE

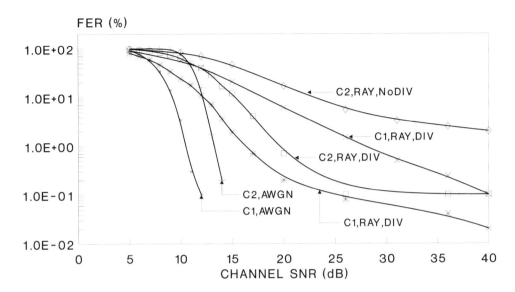

Figure 3.29: 16QAM C1 and C2 BCH(127,71,9) FER versus channel SNR performance of the 9 kBd mode of operation of the DCT codec based System 1 over AWGN and Rayleigh (RAY) channels with diversity (DIV) and without diversity (NoDIV) [195]. ©IEEE 1995, Hanzo, Streit.

In these figures we displayed the FER over both AWGN and Rayleigh channels, in case of the Rayleigh channel both with and without diversity. Note that for near-unimpaired video quality, the FER must be below 1% but preferably below 0.1%. This requirement is satisfied over AWGN channels for SNRs in excess of about 7 dB for 4QAM. In the case of 16QAM and AWGN channels, the C1 and C2 FERs are reduced to about 0.1% for SNRs above 13 and 15 dB, respectively. Observe in the figures that over Rayleigh-fading (RAY) channels with diversity (DIV) the corresponding FER values are increased to about 15 dB for 4QAM and 20 dB for 16QAM, while without diversity (NoDIV) further increased SNR values are necessitated.

The overall video PSNR versus channel SNR (ChSNR) performance of System 1 is shown in Figures 3.30 and 3.31 for the 4QAM and 16QAM modes of operation, respectively. The PSNR versus ChSNR characteristics of the 6.6 kBd 64QAM arrangement are also given for the sake of completeness in Figure 3.32. Observe in the above PSNR versus channel SNR figures that the AWGN performance was also evaluated without forward error correction (FEC) coding in order to indicate the expected performance in a conventional AWGN environment, such as telephone or satellite channels without FEC coding.

Because of its limited bandwidth efficiency gain, high SNR requirement, and incompatible slot structure, we recommend the 64QAM system only for those applications, where the bandwidth is at absolute premium, and in our further discourse we favor the 16QAM/4QAM modes of System 1. The corresponding figures suggest that best performance was achieved over AWGN channels with FEC, requiring a channel SNR (ChSNR) of about 15 and 7 dB in the case of the 16QAM and 4QAM modes of operation, respectively, in order to achieve an unimpaired image quality associated with a PSNR value of about 34 dB. Without FEC

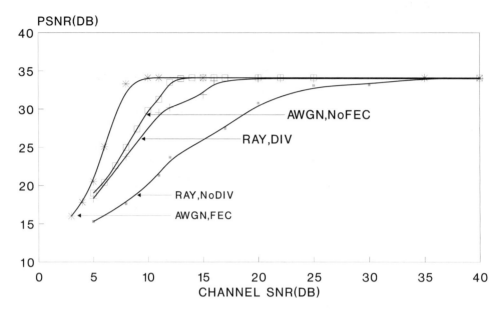

Figure 3.30: 4QAM PSNR versus channel SNR performance of the DCT codec based System 1 in its 18 kBd mode of operation over various channels [195]. ©IEEE 1995, Hanzo, Streit.

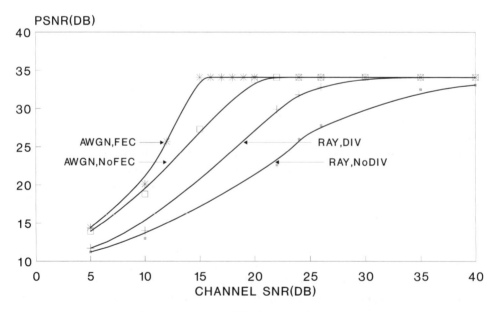

Figure 3.31: 16QAM PSNR versus channel SNR performance of the DCT codec-based System 1 of Table 3.5 in its 9 kBd mode of operation over various channels [195]. ©IEEE 1995, Hanzo, Streit.

3.14. SYSTEM PERFORMANCE

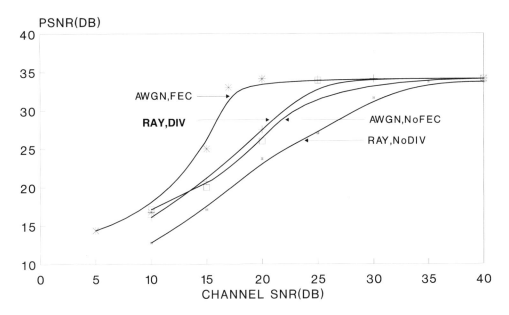

Figure 3.32: 64QAM PSNR versus channel SNR performance of the DCT codec-based System 1 of Table 3.5 in its 6.6 kBd mode of operation over various channels [195]. ©IEEE 1995, Hanzo, Streit.

coding over AWGN channels, these SNR values had to be increased to about 20 and 12 dB, respectively. Over Rayleigh channels with second-order diversity, the system required ChSNR values of about 15 and 25 dB in the 4QAM and 16QAM modes in order to reach an image PSNR within 1 dB of its unimpaired value of 34 dB. This 1 dB PSNR degradation threshold will be used in all scenarios for characterizing the near-unimpaired image quality. Finally, without diversity over Rayleigh channels, SNRs of about 25 and 33 dB were needed for near-unimpaired PSNR performance in the 4QAM and 16QAM modes.

3.14.2 Performance of System 2

3.14.2.1 FER Performance

In order to evaluate the overall video performance of System 2, 100 frames of the "Miss America" (MA) sequence were encoded and transmitted over both the best-case additive white Gaussian noise (AWGN) channel and the worst-case narrowband Rayleigh-fading channel. The BCH(127,50,13) and BCH(127,92,5) decoded frame error rate (FER) was evaluated for both the C1 and C2 bits after the first transmission attempt (TX1) over AWGN and Rayleigh channels with and without second-order diversity, as seen in Figures 3.33–3.35. These figures also portray the C1 FER after the second (TX2) and third (TX3) transmission attempts, which were carried out using 4QAM, in order to maximize the success rate of the C1 bits, representing the vulnerable run-length coded activity table encoding bits of Table 3.4.

Over AWGN channels, a C1 FER of less than 1% can be maintained for channel SNRs in excess of about 5 dB, if three transmission attempts are allowed, although at such low SNRs

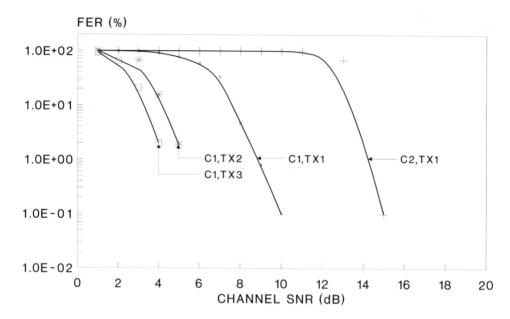

Figure 3.33: BCH(127,50,13) and BCH(127,92,5) FER versus channel SNR performance of the DCT codec-based System 2 of Table 3.5 over AWGN channels [195]. ©IEEE 1995, Hanzo, Streit.

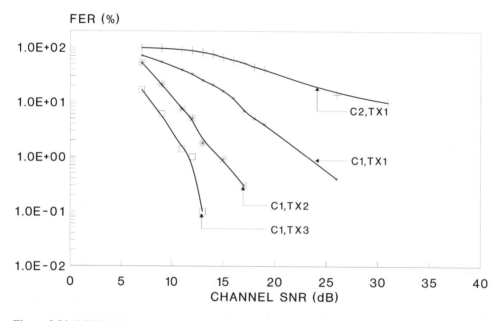

Figure 3.34: BCH(127,50,13) and BCH(127,92,5) FER versus channel SNR performance of the DCT codec-based System 2 of Table 3.5 over Rayleigh channels without diversity [195]. ©IEEE 1995, Hanzo, Streit.

3.14. SYSTEM PERFORMANCE

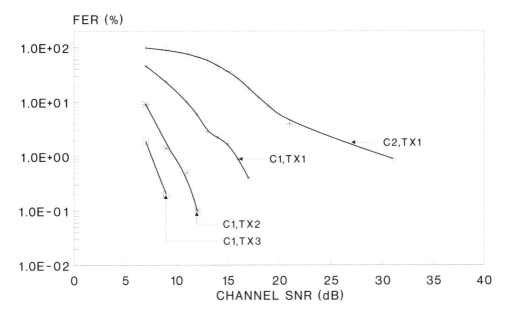

Figure 3.35: BCH(127,50,13) and BCH(127,92,5) FER versus channel SNR performance of the DCT codec-based System 2 of Table 3.5 over Rayleigh channels with diversity [195]. ©IEEE 1995, Hanzo, Streit.

the C2 bit errors are inflicting an unacceptably high video degradation. The corresponding C2 FER over AWGN channels becomes sufficiently low for channel SNRs above about 14–15 dB in order to guarantee unimpaired video communications, which is significantly higher than that required by the C1 subchannel. This ensures that under normal operating conditions the C1 bits are never erroneous. Over the Rayleigh channel, but without diversity, a channel SNR of about 12 dB was required with a maximum of three transmissions in order to reduce the C1 FER below 1% or FER $= 10^{-2}$, as shown in Figure 3.34. However, the C2 FER curved flattened out for high SNR values, which resulted in a severe "leakage" of erroneous C2 bits, and this yielded a somewhat impaired video performance. When diversity reception was used, the minimum required SNR value necessary to maintain a similar C1 FER was reduced to around 10 dB, while the C2 FER became adequately low for SNRs in excess of about 20–25 dB, as demonstrated by Figure 3.35.

3.14.2.2 Slot Occupancy Performance

As mentioned earlier, the ARQ attempts require a number of reserved timeslots for which the re-transmitting MSs have to contend. When the channel SNR is too low, there is a high number of re-transmitted packets contending for too low a number of slots. The slot occupancy increase versus channel SNR performance, which was defined as the percentage of excess transmitted packets, is portrayed for a range of scenarios in Figure 3.36. For SNR values in excess of about 10, 15, and 25 dB, when using 16QAM over AWGN as well as Rayleigh channels with and without diversity, respectively, the slot occupancy was increased

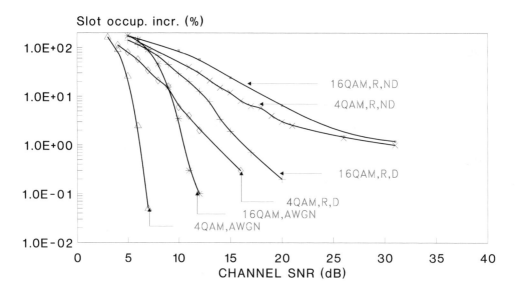

Figure 3.36: Slot occupancy increase versus channel SNR performance of the proposed DCT-based video transceivers of Table 3.5 over AWGN and Rayleigh channels with (D) and without (ND) diversity [195]. ©IEEE 1995, Hanzo, Streit.

due to re-transmissions only marginally. Therefore, reserving two timeslots per frame for a maximum of two re-transmission attempts ensures a very low probability of packet collision during ARQ operations, while reducing the number of subscribers supported by two. In a simplistic approach, this would imply that for a channel SNR value, where the FER is below 1% and assuming 20 users, the reserved ARQ slots will be occupied only in about every fifth frame. However, we cannot earmark fewer than two slots for two additional transmission attempts. The 4QAM slot occupancy is even lower at a given channel SNR than that of the 16QAM schemes, as suggested by Figure 3.36.

3.14.2.3 PSNR Performance

The PSNR versus ChSNR performance of System 2 is characterized by Figure 3.37. Observe that over AWGN channels ChSNR values in excess of 15 dB are required for unimpaired video performance. Over Rayleigh channels with diversity, about 20 dB ChSNR is necessitated for an unimpaired PSNR performance, while without diversity the PSNR performance seriously suffers from the leaking Class Two bit errors. Overall, the 6.6 kBd System 2 has a lower robustness than the 9 kBd System 1, since its behavior is predetermined by the initially transmitted and somewhat impaired Class Two video bits, which were protected by the weaker BCH(127,92,5) code. Recall that System 1 used the BCH(127,71,9) code in both the C1 and C2 subchannels. In fact, the performance of the 6.6 kBd System 2 is more similar to that of the 6.6 kBd 64QAM system characterized in Figure 3.32, which does not use ARQ. We round that re-transmission attempts invoked in order to improve the integrity of the initially received Class One bits only and to ensure an adequate integrity for these vulnerable bits, but without enhancing the quality of the initial 16QAM C2 subchannel

3.14. SYSTEM PERFORMANCE

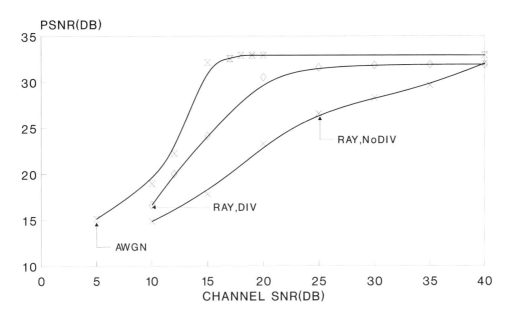

Figure 3.37: PSNR versus channel SNR performance of the 6.6 kBd DCT-based video System 2 over various channels [195]. ©IEEE 1995, Hanzo, Streit.

failed. In other words, System 2 cannot outperform System 1. Furthermore, System 2 is inherently more complex than System 1 and only marginally more bandwidth efficient. Therefore, in contriving the remaining systems we set out to improve the noted deficiencies of System 2.

3.14.3 Performance of Systems 3–5

Our experience with System 2 suggested that it was necessary to re-transmit both the Class One and Two bits if the overloading of the C1 FEC codec indicated poor channel conditions. This plausible hypothesis was verified using System 3, which is effectively the ARQ-assisted System 1. This allowed us to assess the potential benefit of ARQs in terms of the minimum required channel SNR, while its advantages in terms of FER reduction were portrayed in Figures 3.33–3.35. The corresponding PSNR curves of System 3 are plotted in Figures 3.38 and 3.39 for its 18 kBd 4QAM and 9 kBd 16QAM modes, respectively. Re-transmission was invoked only if the C1 FEC decoder was overloaded by a preponderance of channel errors, but in these cases both the C1 and C2 subchannels were re-transmitted. Comparison with Figures 3.30 and 3.31 revealed substantial ChSNR reduction over Rayleigh channels, in particular without diversity. This was because, in case of a BCH frame error, by the time of the second or third transmission attempt, the channel typically emerged from a deep fade. Over AWGN channels, the channel conditions during any further ARQ attempts were similar to those during the previous ones. Hence, ARQ offered more limited ChSNR reduction. The minimum required ChSNR values for the 4QAM mode over AWGN and Rayleigh channels with and without diversity are 7, 8, and 13 dB, while value for the 16QAM mode are 13, 18, and 27 dB, respectively, as also shown in Table 3.5.

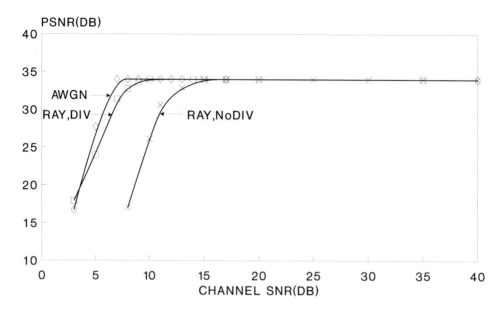

Figure 3.38: PSNR versus channel SNR performance of the 18 kBd 4QAM mode of the DCT codec-based System 3 over various channels using three transmission attempts [195]. ©IEEE 1995, Hanzo, Streit.

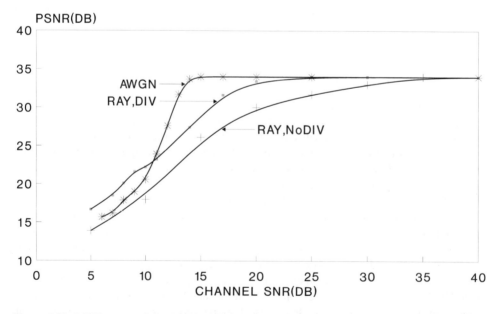

Figure 3.39: PSNR versus channel SNR performance of the 9 kBd 16QAM mode of the DCT codec-based System 3 using three transmission attempts over various channels [195]. ©IEEE 1995, Hanzo, Streit.

3.15. SUMMARY AND CONCLUSIONS

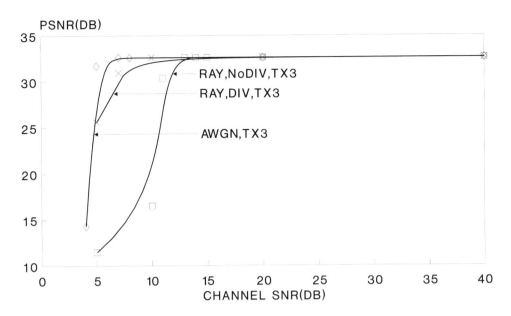

Figure 3.40: PSNR versus channel SNR performance of the 18 kBd 4QAM mode of the DCT codec-based System 4 using three transmission attempts over various channels [195]. ©IEEE 1995, Hanzo, Streit.

Employment of ARQ was more crucial in System 4 than in System 3, because the corrupted run-length coded activity tables of the video codecs of Table 3.4 would inflict severe quality degradations for the whole video frame. The corresponding PSNR curves are portrayed for the 18 and 9 kBd 4QAM and 16QAM operating modes in Figures 3.40 and 3.41 over various channels, which can be contrasted with the results shown for System 5 without ARQ in Figures 3.42 and 3.43. Again, over Rayleigh channels the ChSNR requirement reductions due to ARQ are substantial, in particular without diversity, where the received signal typically emerges from a fade by the time ARQ takes place. Over AWGN channels the benefits of ARQ are less dramatic but still significant. This is because during re-transmission each packet faces similar propagation conditions, as during its first transmission. The required ChSNR thresholds for near-perfect image reconstruction in the 4QAM mode of System 4 are about 6, 8, and 12 dB over AWGN and Rayleigh channels with and without diversity, which are increased to 11, 14, and 27 dB in the 16QAM mode. In contrast, System 5 necessitates ChSNRs of 8, 13, and 25 dB, as well as 12, 16 and 27 dB under the previously stated conditions over 4QAM and 16QAM, respectively. The required ChSNR values are summarized in Table 3.5.

3.15 Summary and Conclusions

In this chapter, initially DCT-based QCIF video codecs were designed, which were then incorporated into various videophone transceivers. A range of bandwidth-efficient, fixed-rate mobile videophone transceivers have been presented, which retain the features summarized

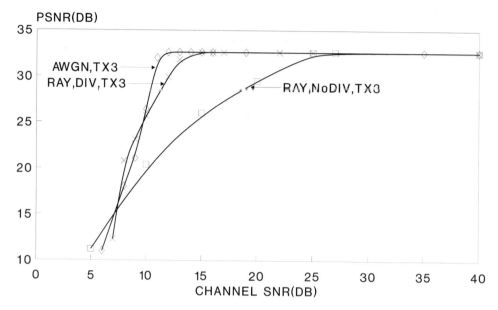

Figure 3.41: PSNR versus channel SNR performance of the 9 kBd 16QAM mode of the DCT codec-based System 4 using three transmission attempts over various channels [195]. ©IEEE 1995, Hanzo, Streit.

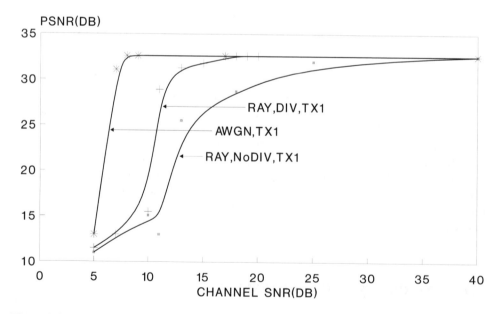

Figure 3.42: PSNR versus channel SNR performance of the 18 kBd 4QAM mode of the DCT codec-based System 5 over various channels [195]. ©IEEE 1995, Hanzo, Streit.

3.15. SUMMARY AND CONCLUSIONS

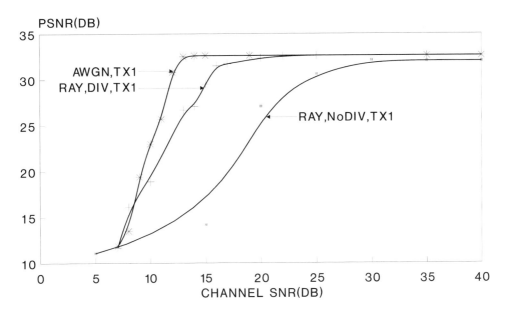

Figure 3.43: PSNR versus channel SNR performance of the 9 kBd 16QAM mode of the DCT codec-based System 5 over various channels [195]. ©IEEE 1995, Hanzo, Streit.

in Table 3.5. The video source rate can be fixed to any arbitrary value in order to be able to accommodate the videophone signal by a single speech channel of the 3G systems [203], for example at rates between 6.7 and 13 kbit/s. This rate is comparable to that of the standard so-called Advanced Multi-Rate AMR speech codec [190] of the 3G systems, requiring an additional speech channel for the transmission of the video information.

In System 1 the 11.36 kbit/s Codec II of Table 3.4 was used. After BCH(127,71,8) coding, the channel rate becomes 20.32 kbit/s. When using an adaptive transceiver, which can invoke 16QAM and 4QAM depending on the channel conditions experienced, the signaling rate becomes 9 and 18 kBd, respectively. Hence, for example, 16 or 8 videophone users can be accommodated in the GSM system's bandwidth of 200 kHz, which implies video user bandwidths of 12.5 and 25 kHz, respectively. For transmission over line-of-sight AWGN channels, SNR values of about 15 and 7 dB are required when using 16QAM and 4QAM, respectively, in order to maintain unimpaired video PSNR values of about 34 dB. An increased channel SNR of about 20 and 15 dB is needed in the diversity-assisted Rayleigh scenario.

In System 2 we have opted for the 8.52 kbit/s Codec I-bis videophone scheme, maintaining a PSNR of about 33 dB for the "Miss America" sequence. The source-coded bitstream was sensitivity-matched binary BCH(127,50,13) and BCH(127,92,5) coded and transmitted using pilot assisted 16QAM. Because of the lower source-coded rate of 8.52 kbit/s of Codec I-bis, the single-user signaling rate of System 2 was reduced to 6.66 kBd, allowing us to accommodate $21 - 2 = 19$ videotelephone users in the 200 kHz GSM bandwidth used for exemplifying the number of users accommodated. If the signal-to-interference ratio (SIR) and signal-to-noise ratio (SNR) values are in excess of about 15 dB and 25 dB over the AWGN as well as diversity-assisted nondispersive Rayleigh-fading channels, respectively, pleasant videophone quality is maintained. The implementation complexity of System 1 is lower than that of System 2, while System 2 can accommodate more users, although it is less robust, as also demonstrated by Table 3.5. This is because only the more vulnerable C1 bits are re-transmitted.

On the basis of our experience with System 2, the fully ARQ-assisted System 3 was contrived, which provided a better image quality and a higher robustness, but was slightly less bandwidth-efficient than Systems 1 and 2 due to reserving two timeslots for ARQ. Furthermore, the question arose as to whether it was better to use the more vulnerable run-length coded but reduced-rate Codec I of Table 3.4 with stronger and more complex FEC protection, as in Systems 4 and 5, or to invoke the slightly higher rate Codec II with its weaker and less complex FEC. In terms of robustness, System 5 proved somewhat more attractive than System 4, although the performances of the non-ARQ-based System 1 and System 5 are rather similar.

Overall, using schemes similar to the proposed ones, mobile videotelephony may be implemented over existing mobile speech links, such as the standard AMR speech coding rate [190] of the 3G systems, requiring an additional speech channel for the transmission of the video information. Naturally, at a given total number of physical channels the number of combined voice/video users becomes half of the "voice-only" users.

Our future work in this field will be targeted at improving the complexity/quality balance of the proposed schemes using a variety of other video codecs, such as parametrically assisted quad-tree and vector quantized codecs. A further important research area to be addressed is devising reliable HSDPA-style [198] transceiver reconfiguration algorithms. In the next chapter, we examine a range of vector-quantized video codecs and videophone systems.

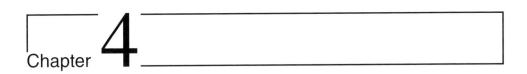

Chapter 4

Very Low Bitrate VQ Codecs and HSDPA-style Videophone Transceivers

4.1 Introduction

Vector quantization (VQ) is a generalization of scalar quantization [204]. Extensive studies of vector quantizers have been performed by many researchers using VQ for analysis purposes [205, 206] or for image compression. In the latter field of applications various advances, such as adaptive [207–209], multistage [210–213], or hybrid coding [214–216], have been suggested. In contrast to these contributions, we will show that VQ may be rendered attractive even when using large vectors (Section 4.2) and small codebooks (< 512) in the codec's inter-frame mode. Furthermore, we will compare the computationally demanding full search VQ with the mean-shape gain VQ to be highlighted in Section 4.3.1, with adaptive VQ in Section 4.3.2, and with classified VQ in Section 4.3.3. Finally, we propose complexity reduction techniques and focus on bit-sensitivity issues. The chapter concludes with the system-design study of a VQ-codec based videophone system similar to the DCT-based system of the previous chapter. In this chapter we adopt a practically motivated approach to vector quantization; for a deeper exposure to VQ principles, the book by Gersho and Gray [204] is recommended.

4.2 The Codebook Design

The codebook design is a crucial issue for every VQ design. Unfortunately, there is no practical algorithm that leads to the optimal codebook C for a given training sequence S. Practical algorithms are usually based on a two-stage method. First, an initial codebook is derived, which is then improved by a second, so-called generalized Max-Lloyd algorithm [204].

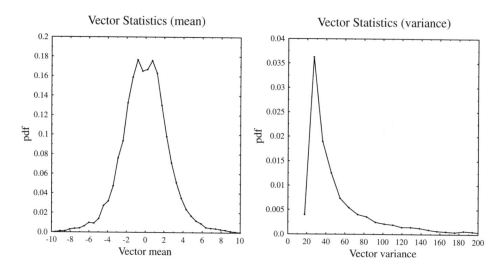

Figure 4.1: Statistical analysis of the training sequence.

The choice of the VQ training sequence is also critical, since it should contain video data, which are characteristic of a wide range of inputs. In our experiments we calculated the Displaced Frame Difference (DFD) signal for various input sequences, while only allowing for 35 active motion vectors out of the 396 8×8 blocks of a Quarter Common Intermediate Format (QCIF) frame to be generated. This implies that the VQ will be used in very low bitrate image coding, where an active/passive block classification is necessary. The DFD energy was analyzed, and the 35 blocks containing the highest energy were copied into the training sequence. The size of the training sequence was chosen to be sufficiently extensive in order to guarantee a sufficient statistical independence between the codebook and the training sequence. The statistical properties of the training sequence are portrayed in terms of the Probability Density Function (PDF) of its mean and variance in Figure 4.1. The mean and variance are highly peaked between $[-2\ldots 2]$ and $[20\ldots 40]$, respectively. Note in the figure that the narrow, but highly peaked, PDF of the DFD mean underlines the efficiency of MC, since in the vast majority of cases its mean is close to zero and its variance implies a low DFD energy.

Initially the codebook was designed using the so-called "pruning" method [204], where a small codebook was derived from a large codebook or from a training sequence. The algorithm was initialized with a single-vector codebook. Then the vectors of the training sequence were compared to those in the codebook using the component-wise squared and summed distance as a similarity criterion. If this distance term exceeds a given threshold, the vector is classified as new and accordingly copied into the codebook. Hence, the threshold controls the size of the codebook, which contains all "new" vectors. In our case this method did not result in an adequate codebook, since the performance of a large codebook of 512 8×8-pixel entries, which had an unacceptably high complexity, was found to be poor.

In a second attempt, we invoked the pairwise nearest neighbor (PNN) algorithm [204]. This approach shrinks a given codebook step-by-step until a desired size is achieved. Initially, each training vector is assigned to a separate cluster, which results in as many clusters as the

4.3. THE VECTOR QUANTIZER DESIGN

Algorithm 1 *The pairwise nearest neighbor algorithm summarized here reduces the initial training set to the required codebook size step by step, until a desired size is achieved.*

1. Assign each training vector to a so-called cluster of vectors.
2. Evaluate the potential distortion penalty associated with tentatively merging each possible pairs of clusters.
3. Carry out the actual merging of that specific pair of clusters, which inflicts the lowest excess distortion due to merging.
4. Repeat Steps 2 and 3, until the required codebook size is arrived at.

number of training vectors. Then for two candidate clusters the distortion penalty incurred by merging these two clusters is determined. This is carried out for all possible pairs of clusters, and finally the pair with the minimum distortion penalty is merged with a single cluster. This process continues until the codebook is shrunk to the desired size. The algorithm's complexity increases drastically with the codebook size. This technique became impractical for our large training set. We therefore simplified this algorithm by limiting the number of tentative cluster combinations as follows. Instead of attempting to merge each possible pair of clusters, where the number of combinations exhibits a quadratic expansion with increasing codebook size, in our approach a single cluster is preselected and tentatively combined with all the others in order to find the merging pair associated with the lowest distortion. The PNN procedure is summarized in Algorithm 1. Following this technique, the codebook size was initially reduced, and then the full PNN algorithm was invoked in order to create a range of codebooks having sizes between 4 and 1024. In the second step of the codebook generation, we used the generalized Max-Lloyd algorithm [204] instead of the PNN algorithm in order to enhance our initial codebooks, which resulted in a slight reduction of the average codebook distortion. This underlines the fact that our initial codebooks used by the PNN algorithm were adequate and the suboptimum two-stage approach caused only a negligible performance loss. As an example, our sixteen-entry codebook constituted by 8×8 vectors is depicted in Figure 4.2, where the vector components were shifted by 127 and multiplied by 30 in order to visually emphasize their differences. The corresponding 128-entry codebook is portrayed in Figure 4.3.

4.3 The Vector Quantizer Design

After creating codebooks having a variety of sizes, we had to find the best design for the vector quantizer itself. The basic concept of the image codec is the same as that described in Section 3.4, which is depicted in Figure 4.4.

Since the first frame used in inter-frame coding is unknown to the decoder, we initialize the buffers of the local and remote decoder with the coarsely quantized intracoded frame we defined in Section 3.5. Once switched to the inter-frame mode, the optional PFU of Section 3.8 may periodically update the local and remote buffers, before the gain-controlled motion compensation of Section 3.6 takes place. The resulting MCER is then passed to the VQ codec. The VQ determines the closest codebook entry for each of the 396 8×8 blocks

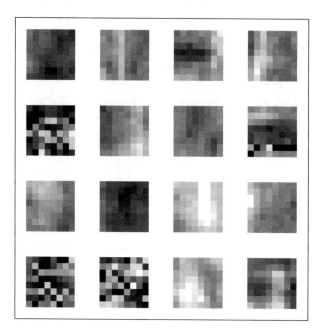

Figure 4.2: Enhanced sample codebook with sixteen 8×8 vectors.

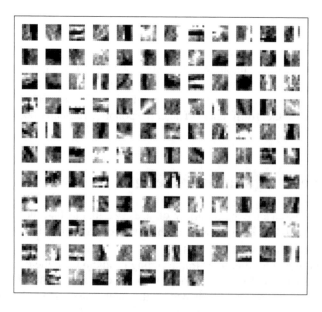

Figure 4.3: Enhanced sample codebook with 127 8×8 vectors.

4.3. THE VECTOR QUANTIZER DESIGN

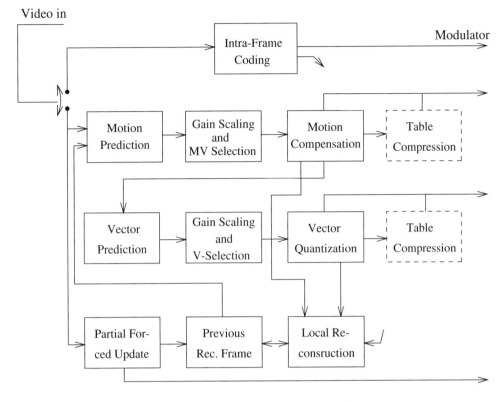

Figure 4.4: Basic schematic of the VQ codec.

in the MCER frame and calculates their potential encoding gain, which is defined in Equation 4.1:

$$VQ_{gain} = \sum_{i=1}^{8}\sum_{j=1}^{8}(b_n(i,j))^2 - \sum_{i=1}^{8}\sum_{j=1}^{8}(b_n(i,j) - C_m(i,j))^2, \qquad (4.1)$$

where b_n denotes the block to be encoded and C_m denotes the best codebook match for the block b_n. A second gain- and bitrate-controlled algorithm decides which of the VQ blocks are active. The VQ indices are stored and transmitted using the activity table concept of Section 3.7. Since we experienced similar VQ block activities to those of the DCT-based codec in Section 3.7, we were able to use the same Huffman codes for the active/passive tables. The VQ frame was also locally decoded and fed back to the motion compensation scheme for future predictions.

As mentioned earlier, a coarsely quantized intra-frame is transmitted at the beginning of the encoding procedure and at drastic changes of scene as highlighted in Section 3.5. Recall that the DCT codec of Section 3.4 used a 12 bit per encoded 8×8 block. Furthermore, we found that in the context of the DCT-based codec of Section 3.4 the reconstructed 8×8 block means could range up to $+/-40$. In contrast, the block means represented in the VQ 256-codebook, for example, covers only the interval of $[-4.55\ldots 6.34]$. This explains why

Figure 4.5: Subjective performance of the VQ at 8 kbit/s, 256-codebook, 10 frames/s, frame 87 of the "Lab" sequence, PSNR 30.33 dB; left: original; right: frame coded using 800 bits. A range of video sequences encoded at various bitrates can be viewed under the WWW address http://www-mobile.ecs.soton.ac.uk

we sometimes experience a better PSNR performance when the PFU is employed, for it helps to restore the original block means.

Figure 4.5 demonstrates the impressive performance at 8 kbit/s and 10 frames/s, but it also reveals granular noise around the mouth and some blockiness in the smooth regions of the background. The codec's failure in the mouth region can be explained by the fact that at this particular location new information is introduced and the MC fails. The following VQ fails to reconstruct the lip movement and results in a delayed motion of the mouth. Surprisingly good is the reproduction of the sharp edges along the waistcoat and the collar of the speaker. Subjectively, the VQ coded frame appears to be slightly sharper than the frame of the 8 kbit/s DCT codec in Figure 3.20, and the background seems to be more adequately reproduced, although some disturbing artifacts are visible. This is partially due to the fact that a quantized DCT causes a bandwidth limitation, whereas the VQ codec maintains a better spatial resolution.

The VQ codec outperformed the DCT scheme in terms of subjective and objective quality at the cost of a higher complexity, an issue to be elaborated on explicitly during our further discourse. For each of the 396 blocks in a QCIF frame, the VQ searches through the entire codebook and compares every possible match by summing the squared differences for each pixel within the codebook entry block and the block to be quantized. Since the block size is fixed to 8×8, it is readily seen that 191 Flops are necessary to perform one block match. In the case of the 128-sized codebook, a total of 396 blocks $\times 128$ codebook entries $\times$ 191 Flops = 13 Mflops have to be executed. The 256-sized codebook requires 26 Mflops. The worst-case DCT complexity of 3.8 Mflops seems to be more modest compared to the higher complexity of the VQ.

In order to explore the range of design trade-offs, we tested several VQ schemes, in which we incorporated adaptive codebooks, classified VQ [15], and mean-shape gain VQ. Before commenting on these endeavors, let us first report our findings as regards the practical range of codebook sizes. As mentioned, we opted for a fixed, but programmable, bitrate scheme and

4.3. THE VECTOR QUANTIZER DESIGN

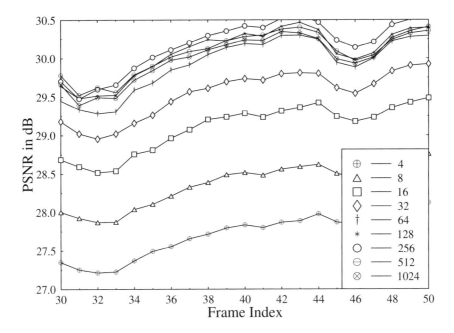

Figure 4.6: PSNR performance for various VQ codebook sizes in the range of 4–1024 entries at a constant bitrate of 1000 bits/frame or 10 kbit/s.

varied the codebook size between 4 and 1024. As Figure 4.6 reveals, the best peak signal-to-noise ratio (PSNR) performance was achieved using codebook sizes in the range of 128 and 512. For these investigations, we used a locally recorded high-activity head-and-shoulders videophone sequence, which we refer to as the "Lab" sequence, since the well-known low-activity "Miss America" (MA) sequence was inadequate for evaluating the VQ performance. Observe that the 256-entry codebook results in the best PSNR performance. This corresponds to a VQ data rate of 0.125 bit per pixel or 8 bit per 8×8 block. This is because increasing the codebook (CB) size beyond 256 requires more bits for the CB address, which inevitably reduces the number of active blocks under the constraint of a fixed bitrate budget. However, a codebook size of 128 is preferable, it halves the codec's complexity without significantly reducing the video quality. The corresponding visually enhanced codebook is depicted in Figure 4.3.

4.3.1 Mean and Shape Gain Vector Quantization

In an attempt to increase the VQ codec's video quality without increasing the codebook size, we also experimented with a so-called mean- and variance-normalized VQ scheme. The approach is based on normalizing the incoming blocks to zero mean and unity variance. As seen in Figure 4.7, the Mean and Shape Gain Vector Quantizer (MSVQ) codec first removes the quantized mean of a given block, and then it normalizes the block variance to near-unity variance by its quantized variance. These two operations require $3n^2 - 2$ Flops per block. Given a certain CB size for the mean and variance, multiplying the CB entry with

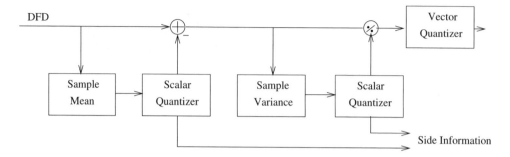

Figure 4.7: Block diagram for a Mean Shape Gain Vector Quantizer.

the quantized gain and adding the quantized mean values can be viewed as increasing the variety of the CB entries, which is equivalent to virtually expanding the codebook size by the product of the number of reconstruction levels of the mean and variance quantizers. Hence, a 2 bit quantizer for both the mean and variance would virtually expand the codebook to sixteen times its original size. The appropriate reconstruction levels were obtained by employing a Max-Lloyd quantizer [187].

We assessed the performance of the MSVQ using codebook sizes of 64 and 256 and quantizers, which ranged from 1 to 32 reconstructions levels, while maintaining a constant bitrate. Using a 1-level MSVQ scheme corresponds to the baseline VQ with no scaling at all. As demonstrated by Table 4.1, the MSVQ codec's performance was slightly degraded in comparison to the baseline codec, except for the 64-entry codebook with a 2-level mean and gain quantizer, where a minor improvement was achieved. This implies that the additional reconstruction precision of the MSVQ increased the bitrate requirement for each block, allowing fewer blocks to be encoded and finally resulting in a worse overall performance at a constant bitrate. As a comparison, it should be noted that the 64-entry/2-level mean- and gain-scaled scheme has an almost identical performance to that of the 256-entry, 1-level arrangement, while retaining a lower complexity. Following our previous arguments, the 64-entry/2-level mean- and gain-scaled arrangement virtually expands the size of the 64-entry CB by a factor of 4 to 256 — justifying a similar performance to that of the 256-entry VQ scheme, which dispensed with invoking the MSVQ principle. From these endeavors we concluded that our codebooks obtained by the modified PNN algorithm are close to the optimum codebooks; therefore, the virtual codebook expansion cannot improve the performance significantly. The advantage of the 64-entry MSVQ scheme was that we maintained the performance of the 256-entry VQ arrangement with a complexity reduction of around 70%. Having reviewed the MSVQ principles in this section, let us now consider adaptive vector quantization.

4.3.2 Adaptive Vector Quantization

In adaptive vector quantization (AVQ) the codebook is typically updated with vectors, which occur frequently but are not represented in the codebook. Unfortunately, the increased flexibility of the method is often associated with an increased vulnerability against channel errors. Our adaptive approach is depicted in Figure 4.8. The codec is based on two codebooks

4.3. THE VECTOR QUANTIZER DESIGN

Table 4.1: Average PSNR versus Number of Quantization Levels for the MSVQ at a Constant Bitrate of 8kbit/s Using a 64- and 256-Entry Codebook (CB) and an Identical Number of Quantization Levels for the Mean- and Gain-scaling Scheme

No. of Quant. Levels	1	2	4	8	16	32
Av. PSNR for 64-entry CB	28.22	28.35	27.09	26.61	26.64	26.62
Av. PSNR for 256-entry CB	28.60	28.16	27.68	27.46	27.56	27.49

implemented as first-in-first-out (FIFO) pipelines, one of which is referred to as the active codebook and the other one as passive. The VQ itself has access only to the active codebook and can choose any of the vectors contained in that codebook. Vectors selected by the VQ codec are then written back to the beginning of the pipeline. After each encoding step, a certain number of vectors are taken from the end of the pipeline and fed back to the beginning of the passive codebook. The same number of vectors are taken from the end of the pipeline in the passive codebook and entered at the beginning of the active codebook. This method forces a continuous fluctuation between the two codebooks but allows frequently used vectors to remain active. The principle is vulnerable to channel errors, but a forced realignment of the codebooks at both the encoder and at the remote decoder at regular intervals is possible, increasing the codec's robustness.

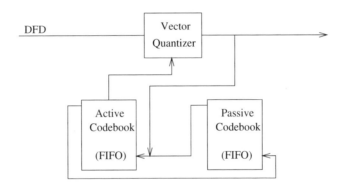

Figure 4.8: Block diagram of the adaptive VQ.

Our performance evaluation experiments were again based on the "Lab" sequence and the 256-sized codebook, which was split into an active codebook of 64 vectors and a remaining passive codebook of 192 vectors. The variable in this scenario was the number of changed vectors after each encoding step, which was fixed to 4, 8, 16, and 32 vectors per step. The corresponding PSNR versus frame index results are depicted in Figure 4.9 along with the performance of the standard 256-entry VQ. As in all other experiments, the comparison was based on a constant bitrate, specifically at 1000 bits/frame. Using this concept, we were able to increase the performance of the 64-entry VQ to a level similar to that of the standard 256-entry VQ, when applying a high vector exchange rate of 32 vectors per step. The lower vector exchange rates still improved the performance, when compared to a standard 64-entry VQ,

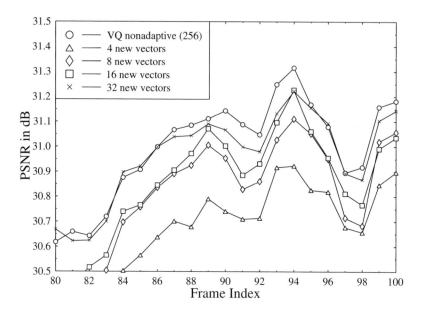

Figure 4.9: PSNR performance of VQ with and without adaptive codebooks at a constant bitrate of 10 kbit/s.

but the PSNR gain was more modest, amounting to about 0.2 dB. Hence, adaptive VQ with a high vector replenishment rate of 32 vectors is also an attractive candidate to reduce the complexity of the VQ at a concomitant increased vulnerability to channel errors. Following the above brief notes on adaptive vector quantization, let us now consider classified vector quantization as another means of reducing the codec's complexity.

4.3.3 Classified Vector Quantization

Classified vector quantization (CVQ) [15] has been introduced as a means of reducing the VQ complexity and preserving the edge integrity of intra-frame coded images. As the block diagram of Figure 4.10 outlines, a CVQ codec is a combination of a classifier and an ordinary VQ using a series of codebooks. The incoming block is classified into one of n classes, and thus only the corresponding reduced-size codebook c_n is searched, thereby reducing the search complexity. In [15] the image blocks to be vector quantized were classified as so-called shade, midrange, mixed, and edge blocks. The edge blocks were further sorted according to their gradient across the block. Specifically, blocks that exhibited no significant luminance gradient across the 8×8-pixel area were classed as shade-blocks. In contrast, those that had a strong gradient were deemed to be edge-blocks. Midrange blocks had moderate gradients but no definite edge, while mixed blocks were typically those that did not fit the above categories. The same principle was invoked in our fractal codec in Chapter 2, and the relative frequencies of the associated blocks were summarized in Table 2.4.

In the present case the VQ's input was the DFD signal, for which the above classification algorithm is not applicable. In our approach a smaller codebook, derived from the same

4.3. THE VECTOR QUANTIZER DESIGN

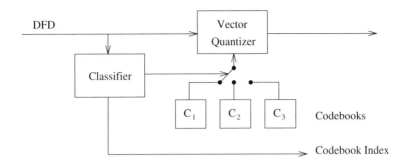

Figure 4.10: Block diagram of a classified VQ.

training sequence as the unclassified codebook C, can be viewed as a set of "centroids" for the codebook C. Hence, codebook C can be split into n codebooks c_n by assigning each vector in C to one of the centroids. The encoding procedure then consists of a two-stage VQ process and may be seen as a tree-structured VQ (TSVQ) [15]. The classifier is a VQ in its own right, using a codebook filled with the n centroids. Once the closest centroid has been found, the subcodebook containing the associated vectors is selected for the second VQ step. We carried out various experiments, based on the optimum 256-entry codebook, with classifiers containing $n = 16$, 64, and 128 centroids, as revealed in Figure 4.11. The PSNR performance loss due to using this suboptimum two-stage approach is less than 0.3 dB. This is surprising since the number of block comparisons was reduced from 256 for the standard VQ to about 64 for the CVQ. The complexity reduction cannot be exactly quantified, because the codebooks c_n do not necessarily contain N/n vectors.

4.3.4 Algorithmic Complexity

Although VQ has been shown [204] to match or outperform any other schemes that map an incoming vector onto a finite number of states, often other techniques such as transform coding (MPEG/H261) are preferred. This is because of the complexity of VQ, which increases drastically with increasing block and codebook sizes. The previous sections have shown that VQ is feasible for even such a large block size as 8×8 pixels using relatively small codebooks. This section will show that the amalgamation of various simplifications allows us to reduce the complexity of an entire codec down to 10 Mflops at a frame rate of 10 frames/s. We commence with a focus on the nearest neighbor rule, which has to be invoked for every possible MCER-block/codebook vector combination.

If we define a block as in Equation 3.3, then the MSE is defined as

$$d_{MSE} = \sum_{i=1}^{b} \sum_{j=1}^{b} (x(i,j) - y(i,j))^2, \qquad (4.2)$$

where $\vec{x}$ is the input block and $\vec{y}$ is the codebook vector. The search for the codebook vector, which results in the lowest MSE is known as the nearest neighbor rule. Equation 4.2 can be

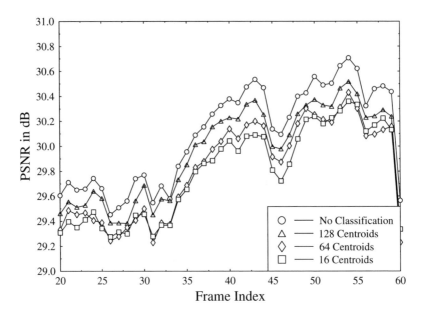

Figure 4.11: PSNR versus frame index performance of classified VQ codecs when using the "Lab" sequence at 10 kbit/s.

expanded as:

$$d_{MSE} = \sum_{i=1}^{b}\sum_{j=1}^{b} x(i,j)^2 + \sum_{i=1}^{b}\sum_{j=1}^{b} y(i,j)^2 - 2\sum_{i=1}^{b}\sum_{j=1}^{b} x(i,j)y(i,j). \quad (4.3)$$

The first term of Equation 4.3 depends only on the input vector; therefore it is constant for every comparison, not affecting the nearest neighbor decision [217]. The second term depends only on the codebook entries, and so it can be determined prior to the encoding process. The last term of Equation 4.3 depends on both the incoming vector and the codebook entries. The evaluation of Equation 4.3 implies a complexity of $2b^2+1$ Flops, whereas that of Equation 4.2 is associated with $3b^2-1$ Flops, corresponding to a complexity reduction of about 30%.

When using a CVQ, the average number of codebook comparisons for each MCER can be reduced from 256 to 32. In addition the block preclassification technique presented in Section 3.9.1 may be employed, which limits the search to those blocks that contain high MCER energy. The VQ appeared to be less sensitive to such active block limitations (as evidenced by comparing Figures 4.12 and 3.18) than the previously studied DCT codec. When the preclassification is applied to both the MC and VQ at an activity rate of 25%, the total preclassification complexity increases the codec complexity by 0.81 MFlop per video frame. The overall VQ process complexity becomes about 1 Mflop/video frame or 10 Mflops at 10 frames/s. The overall PSNR penalty compared to the optimum full search scheme is typically less than 1 dB.

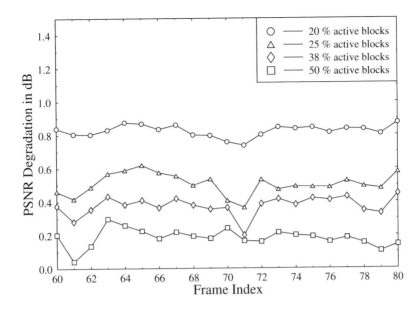

Figure 4.12: PSNR degradation performance of active/passive classified VQ codecs at 10 kbit/s upon using different fractions of active blocks.

4.4 Performance under Erroneous Conditions

As observed in Section 3.12, an increased data compression ratio typically results in an increased vulnerability to channel errors. Since the codec's vulnerability increases drastically upon the introduction of run-length coding, we again design two codecs as it will become explicit during our forthcoming elaborations.

4.4.1 Bit-allocation Strategy

Both codecs are completely reconfigurable and allow bitrate adjustments after every transmitted frame. This is a very useful characteristic, since mobile channels typically exhibit time-variant channel capacities. We now focus our attention on two fixed configurations at 1131 bits per frame, although other bitrates and configurations are possible. The schemes are based on the low-complexity CVQ with a preselection of blocks.

The transmission burst of Codec I was composed of the bits of the FAW, PFU, the MC, and VQ bits. The 22-bit FAW is inserted to support the video decoder's operation in order to regain synchronous operation after loss of frame synchronization, as we saw in Chapter 3 in the context of the DCT-based codec. The PFU was set to periodically refresh 22 blocks per video frame. Therefore, every 18 frames or 1.8 seconds, the update refreshes the same blocks of the 396-block QCIF frame. This periodicity is signaled to the decoder by transmitting the inverted FAW. The gain-cost-controlled MC algorithm was configured so that the resulting bitrate contribution of the MVs in the QCIF frame did not exceed 500 bits. The remaining bits of the 1131-bit long transmission burst was devoted to the VQ. These parameters are

Table 4.2: The 11.31 kbit/s VQ Codec Bit-allocation Tables Both With and Without VLC

Codec	FAW	PFU	MV Ind.	MV	VQ Ind.	VQ	Padding	Total
I	22	22×4	—	< 500 VLC	—	VLC	< 10	1131
II	22	22×4	38×9	38×4	31×9	31×8	0	1131

reflected in Table 4.2. Because it is impossible to generate exactly 1131 bits per QCIF frame, we allow up to 10 bits for padding in order to achieve a constant bitrate of 11.31 kbit/s.

Recall that in the previous chapter Codecs I and II had an identical video quality, but their bitrates differed by about 30% and hence we had an 8 kbit/s and an 11.36 kbit/s DCT codec. Here our philosophy is slightly different, for we exploit the VLC principle in order to compress an approximately 30% higher original bitrate to 11.31 kbit/s and to support a higher video quality by the more error-sensitive Codec I. Logically, the identical-rate 11.31 kps Codec II then exhibits a lower video quality but a higher error resilience. Thus, in Codec I the indices of the active MV and VQ blocks have to be transmitted. The index of MV and VQ blocks ranges from 1 to 396, requiring 9 bits each, so that a total of $9 + 4 = 13$ bits for each MV and $9 + 8 = 17$ bits for each VQ block are required, as seen in Table 4.2. The use of indices instead of the run-length encoded activity tables is associated with redundancy, which manifested itself in a decreased compression ratio of about 30% to 40% or a performance loss of 1 to 2 dB, depending on the actual bitrate.

4.4.2 Bit Sensitivity

In order to prepare the codec for source-sensitivity matched protection, the codec bits were subjected to bit-sensitivity analysis. Here we have to separate those parts of the transmission burst that are VLC and those that are not. If a VLC bit is corrupted, the decoding process is likely to lose synchronization and the decoding of the frame becomes impossible. The decoder detects these bits and reacts by dropping the corrupted transmission burst and keeping the current contents of the reconstructed frame buffers. Therefore, the VLC bits of Codec I are classified as very sensitive.

In the case of non-VLC bits, we distinguish between various sensitivities. Using the same approach as in Section 3.12.1, we determined the integrated and averaged PSNR degradation for each non-VLC bit type, which we portrayed in Figure 4.13.

The VQ-index sensitivity depends on the codebook contents, as well as on the order of the codebook entries. We decreased the sensitivity of the VQ indices by reordering the codebook entries using the simulated annealing algorithm [218]. This technique was invoked several times in order to prevent the algorithm from being trapped in a suboptimum local minimum. The improved robustness is revealed by comparing the two sets of PSNR degradation bars associated with the corresponding bit positions 27–34 in Figure 4.13.

The behavior of both codecs is portrayed in Figure 4.14. Under erroneous conditions here, we inflicted random bit errors and measured the PSNR performance at various BERs. As we expect, Codec I outperforms codec II at low BERs, for it has a higher error-free performance. However, since Codec I comprises vulnerable VLC bits, it breaks down for BERs exceeding

4.5. VQ-BASED LOW-RATE VIDEO TRANSCEIVERS

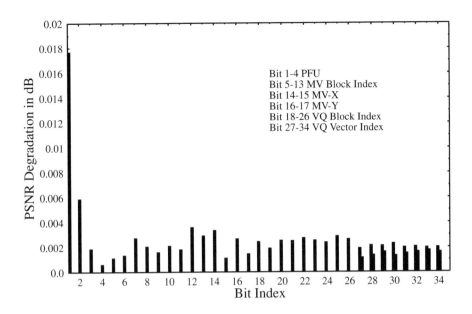

Figure 4.13: Integrated bit sensitivities for the non-VLC Codec II.

$2 \cdot 10^{-4}$. In order to emphasize that this high vulnerability is due to the VLC bits, Figure 4.14 also depicts the performance of Codec I when corruption only affects the non-VLC parts of the transmission burst. In case of Codec II, the breakdown BER is at around $2 \cdot 10^{-3}$.

As a result of Figure 4.14, we propose Codec I for channels at high-channel SNRs and for channels with AWGN characteristics. The more robust Codec II suffers about one dB PSNR degradation under error-free conditions, while exhibiting an order of magnitude higher error resilience than Codec I.

The full investigation of these VQ codecs over mobile channels was discussed in [219] and the results of these will be outlined in the next section.

4.5 VQ-based Low-rate Video Transceivers

4.5.1 Choice of Modulation

In recent years the increasing teletraffic demands of mobile systems have led to compact frequency reuse patterns, such as micro- and picocellular structures, which exhibit more benign propagation properties than conventional macrocells. Specifically, because of the low-power constraint and smaller coverage area, the transmitted power is typically channelled into street canyons. This mitigates both the dispersion due to far-field, long-delay multipath reflections and co-channel interferences. This is particularly true for indoor channels, where interferences are further mitigated by the room partitions. Under these favorable propagation conditions, multilevel modulation schemes have been invoked in order to further increase the teletraffic delivered, as in the American IS-54 [220] and the Japanese Digital Cellular (JDC)

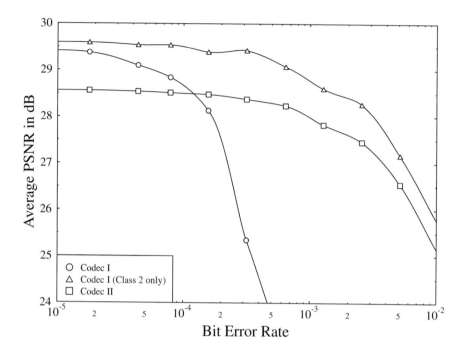

Figure 4.14: PSNR versus BER for the VLC-based Codec I and for the non-VLC Codec II.

systems [221] of Table 9.1 in [161]. Hence, similarly to the DCT-based videophone systems of Section 3.13, we incorporated the VQ-based codecs designed earlier in this chapter in coherent pilot symbol assisted (PSA) 4- and 16QAM modems, which were the subjects of [196].

Again, similarly to the DCT-based videophones of Section 3.13, we created a range of HSDPA-style [198] reconfigurable source-sensitivity matched videophone transceivers, which have two modes of operation — namely, a more robust, but less bandwidth efficient four-level Quadrature Amplitude Modulation (4QAM) mode of operation for outdoors applications. For the sake of illustration, we consider accommodating for example eight users in a bandwidth of 200 kHz, as in the GSM system [197] and a less robust but more bandwidth efficient 16QAM mode supporting 16 video users in indoor cells or near the BS. In contrast, the 3G systems [83] have an approximately 5MHz bandwidth and a chip-rate of 3.84 MChip/s. Thus, depending on the spreading factor employed, they are capable of potentially accommodating a significantly higher number of users, although naturally, the video coding rate only depends on the video source codec.

In other words, indoor cells exploit the prevailing higher signal-to-noise ratio (SNR) and signal-to-interference ratio (SIR), thereby allowing these cells to invoke 16QAM and thereby requiring only half as many TDMA slots as 4QAM. When the portable station (PS) is roaming in a lower-SNR outdoor cell, the intelligent base station (BS) reconfigures the system in order to operate at 4QAM. The philosophy of these systems is quite similar to that of Section 3.13. Let us now consider the choice of error correction codecs.

4.5.2 Forward Error Correction

For convolutional codes to operate at their highest achievable coding gain, it is necessary to invoke soft-decision information. Furthermore, in order to be able to reliably detect decoding errors, they are typically combined with an external, concatenated error detecting block code, as in the 3G systems, for example [83]. As mentioned in Chapters 3 and 4 of [86] high minimum-distance block codes possess an inherently reliable error detection capability, which is a useful means of monitoring the channel's quality in order to activate Automatic Repeat Requests (ARQ), handovers, or error concealment. In the proposed systems we have opted for binary Bose-Chaudhuri-Hocquenghem (BCH) codes, which were characterized in Chapters 3 and 4 of [86]. Let us now consider the specific video transceivers investigated.

In contrast to the 11.31 kbit/s codecs of Table 4.2, we followed the philosophy of our candidate DCT-based videophone schemes of Section 3.13 and designed a more error-resilient, 11.36 kbit/s codec, namely, Codec II, and a lower-rate variable-length coded scheme, the 8 kbit/s Codec I. Although in Table 4.2 both the Type I and II codecs had a bitrate of 11.31 kbit/s, because of their programmable nature in these videophone systems we configured Codec I to operate at 8 kbit/s in order to derive comparable results with our similar DCT-based schemes. The system's schematic was the same as that for the DCT-based systems given in Figure 3.26.

In Systems 1, 3, and 6, which are characterized in Table 4.3, we used the $R = 71/127 \approx 0.56$-rate BCH(127,71,9) code in both 16QAM subchannels, which, in conjunction with the 11.36 kbit/s video Codec II, resulted in $1136 \times 127/71 = 2032$ bits/frame and a bitrate of 20.32 kbit/s at an image frame rate of 10 frames/s. These system features are summarized in Table 4.3.

In contrast, in Systems 2, 4, and 5 we employed the 8 kbit/s video Codec I, which achieved this lower rate essentially due to invoking RL coding, in order to compress the motion- and VQ-activity tables. Since erroneous RL-coded bits corrupt the entire video frame, their protection is crucial. Hence, in Systems 2, 4, and 5 we decided to use the stronger BCH(127,50,13) code, which after FEC coding yielded the same 20.32 kbit/s bitrate as the remaining systems. This will allow us to assess in Section 4.6, whether it is a worthwhile complexity investment in terms of increased system robustness to use vulnerable RL coding in order to reduce the bitrate and then accommodate a stronger FEC code at the previous 20.32 kbit/s overall bitrate.

Source Sensitivity: Before FEC coding, the output of Codec II was split into two equal sensitivity classes, Class One and Two. The bitstream of Codec II was split into two sensitivity classes according to our findings in Figure 4.13. Note, again, that the notation Class One and Two introduced here for the more and less sensitive video bits is different from the higher and lower integrity C1 and C2 modulation channels. In the case of Codec I, all RL encoded bits were assigned to C1 and then the remaining bits were assigned according to their sensitivities to fill the capacity of the C1 subchannel, while relegating the more robust bits to C2. Let us now consider the architecture of System 1.

4.5.3 Architecture of System 1

Transmission Format: Similarly to the equivalent DCT-codec based systems of Table 3.5, the transmission packets of System 1 are constituted by a Class One BCH(127,71,9)

Table 4.3: Summary of Vector-quantized Video System Features * ARQs in Systems 2, 3, and 4 Are Activated by Class One Bit Errors But Not by Class Two [219] ©IEEE 1997, Streit, Hanzo

Feature	System 1	System 2	System 3	System 4	System 5	System 6
Video Codec	Codec II	Codec I	Codec II	Codec I	Codec I	Codec II
Video rate (kbps)	11.36	8	11.36	8	8	11.36
Frame Rate (fr/s)	10	10	10	10	10	10
C1 FEC	BCH(127,71,9)	BCH(127,50,13)	BCH(127,71,9)	BCH(127,50,13)	BCH(127,50,13)	BCH(127,71,9)
C2 FEC	BCH(127,71,9)	BCH(127,92,5)	BCH(127,71,9)	BCH(127,50,13)	BCH(127,50,13)	BCH(127,71,9)
Header FEC	BCH(127,50,13)	BCH(127,50,13)	BCH(127,50,13)	BCH(127,50,13)	BCH(127,50,13)	BCH(127,50,13)
FEC-coded Rate (kbps)	20.32	20.32	20.32	20.32	20.32	20.32
Modem	4/16-PSAQAM	4/16-PSAQAM	4/16-PSAQAM	4/16-PSAQAM	4/16-PSAQAM	4/16-PSAQAM
Retransmitted	None	Cl. One	Cl. One & Two	Cl. One & Two	None	Cl. One
User Signal. Rate (kBd)	18 or 9	9	18 or 9	18 or 9	18 or 9	9
System Signal. Rate (kBd)	144	144	144	144	144	144
System Bandwidth (kHz)	200	200	200	200	200	200
No. of Users	8-16	(16-2)=14	6-14	6-14	8-16	(16-2)=14
Eff. User Bandwidth (kHz)	25 or 12.5	14.3	33.3 or 14.3	33.3 or 14.3	33.3 or 14.3	14.3
Min. AWGN SNR (dB) 4/16QAM	5/11	11	4.5/10.5	6/11	8/12	12
Min. Rayleigh SNR (dB) 4/16QAM	10/22	15	9/18	9/17	13/19	17

codeword transmitted over the C1 16QAM subchannel, a Class Two BCH(127,71,9) codeword conveyed by the C2 subchannel, and a stronger BCH(127,50,13) codeword for the packet header conveying the MS ID and user-specific control information. The generated 381-bit packets are represented by 96 16QAM symbols. Then 11 pilot symbols are inserted according to a pilot spacing of $P = 10$, and 4 ramp symbols are concatenated, over which smooth power amplifier ramping is carried out in order to mitigate spectral spillage into adjacent frequency bands. Eight 111-symbol packets represent an entire video frame, and so the signaling rate becomes 111 symb/12.5 ms $\approx$ 9 kBd. Again, for the sake of illustration, we may assume the simple scenario of a time division multiple access (TDMA) channel bandwidth of 200 kHz, such as in the Pan-European second-generation mobile radio system known as GSM and a Nyquist modulation excess bandwidth of 38.8%. The resultant total signaling rate becomes 144 kBd. This allows us to accommodate $144/9 = 16$ video users, which coincides with the number of so-called half-rate speech users supported by the GSM system [197] of Table 9.1 in [161]. In contrast, the 3G system [83] have an approximately 5 MHz bandwidth and a chip-rate of 3.84 MChip/s. Thus, depending on the spreading factor employed, they are capable of potentially accommodating a significantly higher number of users, although naturally, the video coding rate only depends on the video source codec.

If the channel conditions degrade and 16QAM communications cannot be supported, the BS instructs the system to switch to 4QAM, which requires twice as many timeslots. The 381-bit packets are now conveyed by 191 4QAM symbols, and after adding 30 pilot symbols and four ramp symbols the packet length becomes 225 symb/12.5 ms, resulting in a signaling rate of 18 kBd. Hence, the number of videophone users supported by System 1 is reduced to 8. The system also facilitates mixed-mode operation, where 4QAM users must reserve two slots in each 12.5 ms TDMA frame toward the fringes of the cell, while in the central section of the cell 16QAM users will only require one slot per frame in order to maximize the number of users supported. Assuming an equal proportion of 4QAM and 16QAM users, the average number of users per carrier becomes 12. The equivalent user bandwidth of the 4QAM PSs is 200 kHz/8 = 25 kHz, while that of the 16QAM users, is 200 kHz/16 = 12.5 kHz. The characteristics of the whole range of our candidate systems are highlighted in Table 4.3.

4.5.4 Architecture of System 2

As already noted in Section 4.5.2, the philosophy behind System 2 was to contrive a scheme that allowed an assessment of the value of Codec I in a system's context. Specifically, we sought to determine whether it pays dividends in robustness terms to invest complexity and hence reduce the 11.36 kbit/s rate of Codec II to 8 kbit/s and then accommodate a more powerful but more complex BCH(127,50,13) code within the same 20.32 kbit/s bitrate budget. The sensitive RL and BCH(127,50,13) coded Class One bits are then conveyed over the lower BER C1 16QAM subchannel. Furthermore, in order to ensure the error-free reception of the RL-coded Class One bits, Automatic Repeat Request (ARQ) is invoked, if the C1 BCH(127,50,13) code indicates their corruption. Similarly to System 1, System 2 can also accommodate 16 timeslots for the 9 kBd 4QAM users, but if we want to improve the integrity of the RL-coded video bits by ARQ, some slots will have to be reserved for the ARQ packets. This proportionately reduces the number of users supported, as the channel conditions degrade and requires more ARQ attempts. The underlying trade-offs will be analyzed in Section 4.6.

Automatic Repeat Request: ARQ techniques have been successfully used in data communications [199–202] in order to render the bit and frame error rate arbitrarily low. However, because of their inherent delay and the additional requirement for a feedback channel for message acknowledgment, they have not been employed in interactive speech or video communications. In the operational wireless systems, however, such as the 3G systems of [83] there exists a full duplex control link between the BS and PS, which can be used for acknowledgments, and the short TDMA frame length ensures a low packet delay. Hence ARQ can be invoked.

System 2 was contrived so that for the first transmission attempt (TX1) we use contention-free Time Division Multiple Access (TDMA) and 16QAM to deliver the video packet. If the Class One bits transmitted over the C1 16QAM subchannel are corrupted and an ARQ request occurs, only the Class One bits are re-transmitted using the "halved-capacity" but more robust 4QAM mode of operation. These packets have to contend for a number of earmarked timeslots, similarly to thePacket Reservation Multiple Access (PRMA) aided networks [196], as also highlighted in Chapter 6 of [161]. If, however, there are only C2 bit errors in the packet, it is not re-transmitted. Since the Class One bits are delivered by the lower BER C1 16QAM subchannel, often they may be unimpaired when there are Class Two errors. To strike a compromise between the minimum required channel SNR and the maximum number of users supported, we limited the number of transmission attempts to three. Hence, two slots per frame must be reserved for ARQ. For the sake of low system complexity, we dispense with any contention mechanism and allocate two timeslots to that particular user, whose packet was first corrupted within the TDMA frame. Further users cannot therefore invoke ARQ, because there are no more unallocated slots. A further advantage of this scheme is that since there are three copies of the transmitted packet, majority decisions can be invoked on a bit-by-bit basis, if all three packets became corrupted. The basic features of System 2 designed to accommodate Codec I are also summarized in Table 4.3.

4.5.5 Architecture of Systems 3–6

With the aim of exploring the whole range of design options, we have created four other systems, namely, Systems 3–6.

System 3 is similar to System 1, for it is constructed of the 11.36 kbit/s Codec II, BCH(127,71,9) FEC codecs, and reconfigurable 16QAM/4QAM modems, but it relies on ARQ assistance. In contrast to System 2, where only the Class One bits were re-transmitted using 4QAM, here both the Class One and Two bits are re-transmitted using the same modulation scheme, as during the first attempt. System 3 also limits the maximum number of transmission attempts to three in case of C1 BCH(127,71,9) decoding errors and therefore reduces the number of users by two but improves the transmission integrity. In the 16QAM and 4QAM modes, 14 and 6 users can be supported, respectively.

System 4 is similar to System 2, for it employs the 8 kbit/s Codec I in conjunction with the robust BCH(127,50,13) FEC codecs and the 16QAM/4QAM modems, but instead of Class One only re-transmission, it invokes full packet re-transmissions, if the Class One bits are corrupted. The number of video users is again reduced to 14 and 6 in the 16QAM and 4QAM modes, respectively. **System 5** is a derivative of System 4, which dispenses with ARQ and hence can serve 16/8 users in the 16QAM/4QAM modes of operation, respectively. Finally, **System 6** is related to System 1, for it is constructed of the same components but allows

ARQs in the same fashion as System 2, where in the case of Class One errors only these bits are re-transmitted using the previously introduced 16QAM/4QAM/4QAM TX1/TX2/TX3 regime. Therefore, System 6 can support 14 users.

4.6 System Performance

4.6.1 Simulation Environment

Before commenting on the system's performance, let us consider the simulation environment and the system parameters. The 11.36 and 8 kbit/s VQ video codecs, the previously mentioned BCH codecs, and pilot-assisted diversity, and ARQ-aided 4QAM and 16QAM modems were simulated, including the AWGN and Rayleigh channel. The receiver carried out the inverse functions of the transmitter. Simple switched diversity was implemented. The criterion for deciding on which diversity channel to decode was the minimum fading-induced phase shift of the pilot symbols with respect to their transmitted phase rather than the maximum received signal power, since this has resulted in slightly better BER performance. As already mentioned in Section 3.13.2.2, the ARQ scheme, which was invoked in various situations in the different systems of Table 4.3, was controlled by the error detection capability of the BCH codes used. This was possible, since the error detection capability was extremely reliable in case of the powerful codes used and hence allowed us to dispense with Cyclic Redundancy Coding (CRC).

In interactive communications ARQ techniques have not been popular because of their delay and the required feedback channel for message acknowledgment. In intelligent low-delay picocells, however, it is realistic to assume a near-instantaneous full duplex control link between the BS and PS, for example, for ARQ acknowledgments, which in our case were always received error freely. In order to compromise between traffic capacity and robustness, only two slots per frame were reserved for ARQ, limiting the number of transmission attempts to three, where we allocated these two timeslots to that particular user, whose packet was first corrupted within the TDMA frame. Since in the ARQ-assisted arrangements re-transmission took place within the last two slots of the same TDMA frame, no extra delay was inflicted. As regards the video packet of the last user in the frame, whose information was transmitted during the third slot from the end of the frame, it is not realistic, however, to expect a near-instantaneous acknowledgment even in low-delay picocells, implying the lack of ARQ assistance for this user. The specific criterion for invoking ARQ depended on which system of Table 4.3 was used.

Our results refer to the previously mentioned signaling rate of 144 kBd, while the propagation frequency and the vehicular speed were 1.8 GHz and 30 mph, respectively. These results are also comparable to those of the similar DCT-codec-based performance figures of Table 3.5. In the case of lower walking speeds, the signal's envelope fades more slowly. Therefore the pilots provide a better fading estimate, and the PSAM modem has a lower BER. On the other hand, in slow-fading conditions the fixed-length BCH codecs have to face a somewhat more pessimistic scenario, since in some codewords — despite interleaving — there may be a high number of errors, and the error correction capability may be overloaded. In contrast, in some codewords there may be only a low number of errors, which does not fully exploit the BCH codec's correction power. Throughout these experiments noise-limited, rather than interference-limited, operation was assumed, which is a realistic assumption in

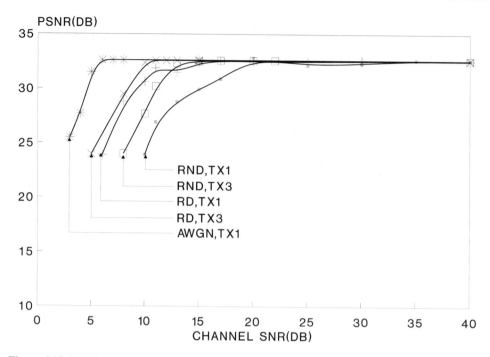

Figure 4.15: PSNR versus channel SNR performance of the VQ codec-based video Systems 1 and 3 of Table 4.3 in their 4QAM 18 kBd mode of operation over various channels [219]. ©IEEE 1997, Streit, Hanzo.

partitioned indoor picocells or in the six-cell microcellular cluster used, for example, in the Manhatten model of Chapter 17 in [222]. If, however, the decoded BCH frame error rate due to either interference or noise, which is monitored by the BS, becomes excessively high for 16QAM communications, the BS may instruct the system to switch to 4QAM, thereby reducing the number of users supported. The consideration of these reconfiguration algorithms is beyond the scope of this chapter. Having summarized the simulation conditions, let us now concentrate on the performance aspects of the candidate systems.

4.6.2 Performance of Systems 1 and 3

Recall from Table 4.3 that a common feature of Systems 1, 3, and 6 was that these all used the 11.36 kbit/s video codec associated with the BCH(127,71,9) FEC codec, although Systems 3 and 6 used ARQ in addition in order to enhance their robustness at the price of reserving two more timeslots and hence reducing the number of users supported. The PSNR versus channel SNR (ChSNR) performance of System 1 is portrayed in Figures 4.15 and 4.16 for its 4QAM/16QAM 17.2/9 kBd modes of operation, respectively. Observe in these figures that the system's performance was evaluated for the best-case additive white Gaussian noise (AWGN) channel and the worst-case Rayleigh (R) channel, both with diversity (D) and with no diversity (ND), using one transmission attempt (TX1). However, in order to be able to

4.6. SYSTEM PERFORMANCE

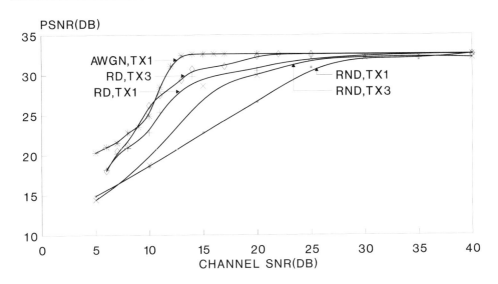

Figure 4.16: PSNR versus channel SNR performance of the VQ codec-based video Systems 1 and 3 of Table 4.3 in their 9 kBd mode of operation over various channels [219]. ©IEEE 1997, Streit, Hanzo.

assess the benefits of ARQ in terms of required ChSNR reduction, Figures 4.15 and 4.16 also display the corresponding results for System 3, using three transmission attempts (TX3).

Each system was deemed to deliver nearly unimpaired video if its PSNR was reduced by less than 1 dB owing to channel impairments. These corner-SNR values are tabulated for ease of reference in Table 4.3 for each system studied. Considering System 1 first, in its 4QAM/18kBd mode of operation over AWGN channels a channel SNR of about 5 dB was required, which had to be increased to around 10 dB and 18 dB for the RD, TX1, and RND, TX1 scenarios. When opting for the doubled bandwidth efficiency but less robust 16QAM/9kBd mode of System 1, the corresponding ChSNR values were 12, 23 and 27 dB, respectively. These corner-SNR values are also summarized in Table 4.3. This trade-off becomes explicit in Figure 4.17.

Comparing the above values in Figure 4.15 to the corresponding ChSNRs of System 3 reveals an approximately 6 dB reduced corner-SNR, when using 4QAM with no diversity, but this gain significantly eroded in the inherently more robust diversity-assisted scenario. In the 16QAM/9kBd mode, Figure 4.16 shows approximately 3 dB ARQ gain with no diversity, which is increased to about 4–5 dB, when using the diversity-aided arrangement.

4.6.3 Performance of Systems 4 and 5

Systems 4 and 5 combine the lower rate, but more error-sensitive, RL-coded 8 kbit/s Codec I with the more robust BCH(127,50,13) codec. These systems are more complex to implement, and hence we can assess whether the system benefited in robustness terms. The corresponding PSNR versus ChSNR curves are plotted in Figures 4.18 and 4.19 for the 4QAM/18kBd and the 16QAM/9kBd modes of operation, respectively. Considering the 4QAM/17.2 kBd mode of System 4 first, where a maximum of three transmission attempts are used, if any of the

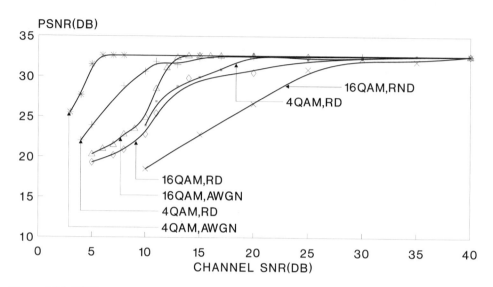

Figure 4.17: PSNR versus channel SNR performance of the VQ codec-based video System 1 of Table 4.3 in its 4QAM/16QAM 18/9 kBd modes of operation over various channels [219]. ©IEEE 1997, Streit, Hanzo.

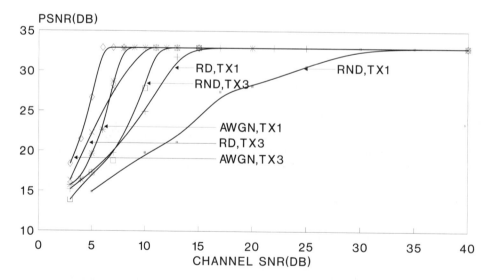

Figure 4.18: PSNR versus channel SNR performance of the 4QAM/18kBd modes of the VQ codec-based video Systems 4 and 5 of Table 4.3 over various channels [219]. ©IEEE 1997, Streit, Hanzo.

4.6. SYSTEM PERFORMANCE

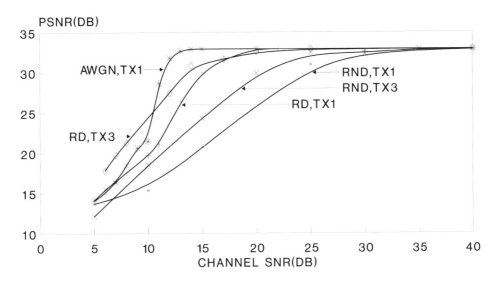

Figure 4.19: PSNR versus channel SNR performance of the 16QAM/9kBd mode of the VQ codec-based video Systems 4 and 5 of Table 4.3 over various channels [219]. ©IEEE 1997, Streit, Hanzo.

C1 or C2 bits are corrupted, an AWGN performance curve similar to that of System 3 is observed. Over Rayleigh channels with diversity, about 9 dB ChSNR is necessitated for near-unimpaired performance, increasing to around 11 dB with no diversity. In the 16QAM mode, the AWGN performance is similar to that of System 3. The ChSNR over Rayleigh channels with diversity must be in excess of about 17 dB, increasing to around 24 dB without diversity. As expected, System 5 is typically less robust than the ARQ-assisted System 4. It is interesting to observe that the RL-coded Systems 4 and 5, despite more robust FEC coding, tend to be less robust than the corresponding schemes without RL-coding, such as System 1 and 3, even though the latter systems also exhibit a reduced complexity. Another powerful solution for the RL-coded codecs is simply to drop corrupted packets and not update the corresponding video frame segment. This, however, requires a feedback acknowledgment flag in order to inform the encoder's local reconstruction frame buffer as to this packet dropping event. This issue will be discussed in great detail in Section 10.3.6 in the context of the ITU's H.263 codec.

4.6.4 Performance of Systems 2 and 6

The situation is slightly different when only the Class One bits are re-transmitted in case of erroneous decoding. These systems were contrived to constitute a compromise between robustness and high user capacity. For the RL-coded System 2, it is vital to invoke ARQ, whereas for System 6 it is less critical. However, System 6 uses the weaker BCH(127,71,9) code. The PSNR results of Figure 4.20 reveal that due to the stronger FEC code, System 2 is more robust than System 6. However, for all studied scenarios the degradation of System 2 was more rapid with decreasing ChSNR values than that of System 6. The minimum required ChSNR values are given in Table 4.3.

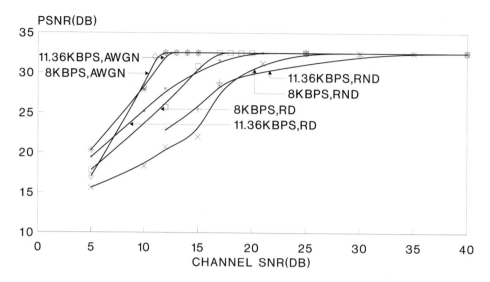

Figure 4.20: PSNR versus channel SNR performance of the VQ codec-based video Systems 2 and 6 of Table 4.3 [219]. ©IEEE 1997, Streit, Hanzo.

As expected, in all systems, regardless of the mode of operation invoked, best performance was always achieved over AWGN channels, followed by the diversity-assisted Rayleigh (RD) curves, while the least robust curves represented the system's performance with no diversity (ND). The ARQ attempts did improve the performance of all systems, but the lowest ARQ gains were offered over AWGN channels, since due to the random error distribution in every new attempt, the video packet was faced with similar channel conditions, as during its first transmission. The highest ARQ gains were typically maintained over Rayleigh channels with no diversity (ND), since during re-transmission, the signal envelope had a fair chance of emerging from a deep fade, increasing the decoding success probability.

4.7 Joint Iterative Decoding of Trellis-based Vector-quantized Video and TCM[1]

R. Maunder, J. Kliewer, S.X. Ng, J. Wang, L-L. Yang and L. Hanzo

4.7.1 Introduction

Shannon's source and channel separation theorem [223] states that under certain idealized conditions, source and channel coding can be performed in isolation without any loss in performance. This motivated the designs of the Vector Quantization (VQ) based [224] video transmission system in [225] and the MPEG4-based [226] system of [227]. However,

---

[1]R. G. Maunder, J. Kliewer, S. X. Ng, J. Wang, L-L. Yang and L. Hanzo: Joint Iterative Decoding of Trellis-based Vector-quantized Video and TCM, to appear in IEEE Transactions on Wireless Communications

4.7. JOINT ITERATIVE DECODING OF TRELLIS-BASED VQ VIDEO AND TCM 119

Shannon's findings only apply under a number of idealiztic assumptions. These assumptions have limited validity for practical video transmission systems communicating over realistic mobile-radio channels [224]. This motivates the application of joint source and channel coding techniques in wireless video transmission, as exemplified by [228].

Let us now summarize the novelty of this section in contrast to the state-of-the-art. In [228], joint source and channel coding techniques were applied at the bit-level to mitigate channel-induced distortion within an MPEG-4 [226] coded video sequence. More specifically, knowledge of the residual correlation of bits within the MPEG-4 bitstream was exploited to assist their recovery upon reception. In our approach, joint source and channel coding techniques are applied at the Video Block (VB)-level within a specially designed novel proprietary video codec, which offers the advantage of directly assisting the recovery of the transmitted video information, rather than of its intermediate bit-based representation.

The design of the proposed video transmission system extends the serial concatenation and iterative decoding proposals of [229], in a manner similar to that of [230]. More explicitly, a novel trellis-decoded VQ-based video codec is serially concatenated with the separate In-phase, Quadrature-phase (IQ) interleaved TCM codec of [231]. In the receiver, VQ- and TCM-decoding are performed iteratively.

The video codec of the proposed VQ-TCM system is designed to maintain a simple VQ-based transmission frame structure. The rules governing the formation of the legitimate bit-sequences of a transmission frame are referred to as the *VQ-induced code constraints* in this section. The video codec was specifically designed to impose VQ-induced code constraints that may be completely described by a novel VQ trellis structure, reminiscent of the symbol-level trellis structure of [232]. This is in contrast to the video transmission systems of [225] and [227], in which the code constraints imposed by the video codec did not lend themselves to a trellis-based description.

As we will outline in Section 4.7.10, the employment of the proposed VQ trellis structure represents the consideration of all legitimate transmission frame permutations. More explicitly, we perform optimal Minimum Mean-Squared-Error (MMSE) VQ encoding of the video sequence in a novel manner that is reminiscent of Trellis-Coded Quantization (TCQ) [233]. In addition, the employment of the proposed VQ trellis structure during VQ decoding guarantees the recovery of legitimate — although not necessarily error-free — video information. Hence, unlike in the video transmission systems of [225] and [227], the proposed video decoder is never forced to discard information. During VQ decoding, a novel modification of the Bahl-Cocke-Jelinek-Raviv (BCJR) [234] algorithm provided in [235] is performed on the basis of the proposed VQ trellis structure. This allows the exploitation of the VQ-induced code constraints to assist in joint iterative VQ- and TCM-decoding. In addition, this allows the unconventional soft *A Posteriori* Probability (APP) based [236] MMSE reconstruction of the transmitted video frames.

This section is organized as follows. In Section 4.7.2, the proposed VQ-TCM system is introduced. Section 4.7.6 describes the VQ-TCM system's transmission frame structure and the VQ-induced code constraints. In addition, the complete description of the VQ-induced code constraints by the proposed VQ trellis structure is discussed in Section 4.7.6. The employment of the VQ trellis structure in VQ-encoding and -decoding is described in Sections 4.7.11 and 4.7.12, respectively. In Section 4.7.13, both low- and high-latency implementations of the proposed VQ-TCM system are introduced and their video reconstruction quality is assessed. Finally, the performance of the proposed joint video and channel coding approach is compared to that of the video transmission systems of [225] and [227].

4.7.2 System Overview

In this section we introduce the proposed VQ-TCM video transmission system, which employs a joint video and channel coding philosophy, as introduced in Section 4.7.1. The video codec of this system achieves compression by employing low-complexity frame differencing and VQ of the resultant Frame Difference (FD) [224]. These are chosen in favor of the more advanced but higher complexity compression techniques, such as sub-pixel-accuracy MPEG4-style Motion Compensation (MC) or the Discrete Cosine-Transform (DCT) [224]. This ensures that a simple transmission frame structure is maintained, as we will detail in Section 4.7.6. Rather than performing an excessive-complexity single VQ-encoding or -decoding operation for the entire video frame as usual, a decomposition into a number of smaller VQ operations is performed. This approach is associated with a significant computational complexity reduction, as will be shown in Section 4.7.10. The proposed video codec is protected by the serially concatenated TCM codec of [231] and joint iterative decoding is employed in the receiver. These issues are further detailed in the following sub-sections with reference to the proposed VQ-TCM system's transmitter and receiver schematics seen in Figures 4.21 and 4.22, respectively.

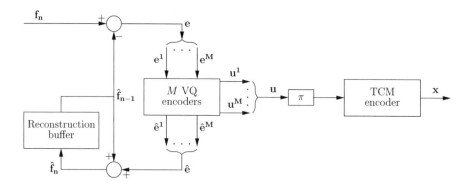

Figure 4.21: The proposed VQ-TCM system's transmitter.

4.7.3 Compression

In the VQ-TCM system's transmitter of Figure 4.21, the previous reconstructed video frame $\hat{f}_{n-1}$ is employed as a prediction for the current video frame f_n. The FD $e = f_n - \hat{f}_{n-1}$ has a lower variance than the current video frame f_n. Hence, the transmission of the FD e to convey the current video frame f_n provides compression.

As will be discussed in Section 4.7.6, VQ is employed to represent the FD e in a compact bit-sequence form, namely the transmission frame u. Hence, the VQ offers compression. This compression is termed "lossy" because the VQ approximates the FD e by the reconstructed FD $\hat{e}$, as shown in Figure 4.21. The reconstructed FD $\hat{e}$ is employed to obtain the current

4.7. JOINT ITERATIVE DECODING OF TRELLIS-BASED VQ VIDEO AND TCM

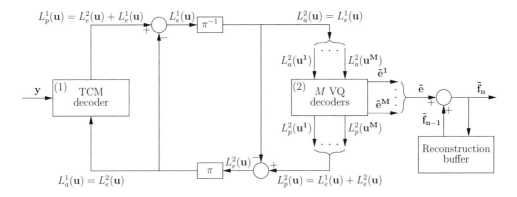

Figure 4.22: The proposed VQ-TCM system's receiver.

reconstructed video frame $\hat{\mathbf{f}}_n = \hat{\mathbf{e}} + \hat{\mathbf{f}}_{n-1}$, which is stored for the sake of providing the prediction of the next video frame.

4.7.4 Vector Quantization Decomposition

In the proposed VQ-TCM system's transmitter and receiver, VQ-encoding and -decoding are performed, respectively. As will be detailed in Sections 4.7.7 and 4.7.10, we decompose each FD **e** into M number of smaller sub-frames, which may be represented by VQs operating in parallel. This decomposition implies the corresponding decomposition of both the reconstructed FD **ê** and the transmission frame **u**, as indicated by the braces { and } in Figure 4.21. In the transmitter, M number of separate VQ operations are employed. Each VQ encoder approximates the FD sub-frame $\mathbf{e}^m$ by the reconstructed FD sub-frame $\hat{\mathbf{e}}^m$, where $m \in [1\ldots M]$ denotes the sub-frame index. In addition, the transmission sub-frame $\mathbf{u}^m$ is generated, as will be detailed in Section 4.7.11. The bit-based transmission frame **u** is constituted by the concatenation of the set of these M number of transmission sub-frames.

4.7.5 Serial Concatenation and Iterative Decoding

In the proposed VQ-TCM system, the video codec is serially concatenated with the TCM codec of [231]. In the transmitter, the current video frame $\mathbf{f}_n$ is conveyed by the reconstructed FD **ê**, where the latter is generated using VQ encoding and is represented by the transmission frame **u**, as discussed in Section 4.7.3. The transmission frame **u** is interleaved in the block π of Figure 4.21 and TCM encoded to generate the channel's input symbols **x**. These are transmitted over the channel and are received as the channel's output symbols **y**, as shown in Figure 4.22. Finally, TCM- and VQ-decoding are employed to recover the reconstructed FD estimate $\tilde{\mathbf{e}}$. This allows the recovery of the current video frame estimate $\tilde{\mathbf{f}}_n = \tilde{\mathbf{e}} + \tilde{\mathbf{f}}_{n-1}$, as shown in Figure 4.22.

In the receiver, Soft-In Soft-Out (SISO) TCM- and VQ-decoding are performed iteratively, as shown in Figure 4.22. Soft information, represented in the form of Logarithmic Likelihood-Ratios (LLRs) [236], is exchanged between the TCM and VQ iterative decoding stages for the sake of assisting each other's operation. With each successive decoding iteration, the reliability of this soft information increases, until iterative decoding convergence is achieved. In Figure 4.22, $L(\cdot)$ denotes the LLRs of the specified bits, where the superscript (1) denotes TCM decoding and (2) represents VQ decoding. In addition, a subscript denotes the role of the LLRs, with a, p and e indicating *a priori*, *a posteriori* and extrinsic information, respectively.

During each decoding iteration, *a priori* LLRs $L_a(\mathbf{u})$ are provided for each iterative decoding stage, as shown in Figure 4.22. Naturally, in the case of TCM decoding, the channel's output symbols $\mathbf{y}$ are also exploited. Note that the *a priori* LLRs $L_a(\mathbf{u})$ are obtained from the most recent operation of the other decoding stage, as will be highlighted below. In the case of the first decoding iteration no previous VQ decoding has been performed. In this case, the *a priori* LLRs $L_a^1(\mathbf{u})$ provided for TCM decoding are all zero, corresponding to a probability of 0.5 for both "0" and "1". Each iterative decoding stage applies the BCJR [234] algorithm, as described in [231] and Section 4.7.12 for TCM- and VQ-decoding, respectively. The result is the generation of the *a posteriori* LLRs $L_p(\mathbf{u})$, as shown in Figure 4.22.

During iterative decoding, it is necessary to prevent the re-use of already-exploited information, since this would limit the attainable iteration gain [237]. This is achieved following each decoding stage by the subtraction of $L_a(\mathbf{u})$ from $L_p(\mathbf{u})$, as shown in Figure 4.22. Following VQ decoding, we arrive at the extrinsic LLRs $L_e^2(\mathbf{u})$. In the case of TCM decoding, the LLRs $L_e^1(\mathbf{u})$ additionally contain information extracted from the channel's output symbols $\mathbf{y}$. It is these sets of LLRs that provide the *a priori* LLRs for the next iteration of the other decoding stage. De-interleaving, indicated by the block π^{-1} in Figure 4.22, is applied to $L_e^1(\mathbf{u})$ in order to generate $L_a^2(\mathbf{u})$. Similarly, interleaving is applied to $L_e^2(\mathbf{u})$ for the sake of providing $L_a^1(\mathbf{u})$, as shown in Figure 4.22. These interleaving and de-interleaving operations are necessary for the sake of mitigating the correlation of consecutive LLRs, before forwarding them to the next iterative decoding stage [236]. As always, the interleaver's ability to provide this desirable statistical independence is related to its length.

As stated in Section 4.7.4, M number of separate VQ decoding processes are employed in the proposed VQ-TCM system's receiver. Similarly to the decomposition of the bit-based transmission frame $\mathbf{u}$, the *a priori* LLRs $L_a^2(\mathbf{u})$ are decomposed into M number of sub-frames, as shown in Figure 4.22. This decomposition is accompanied by the corresponding decomposition of the *a posteriori* LLRs $L_p^2(\mathbf{u})$ and the reconstructed FD estimate $\tilde{\mathbf{e}}$. Each VQ decoder is provided with the *a priori* LLR sub-frame $L_a^2(\mathbf{u}^\mathbf{m})$ and generates the *a posteriori* LLR sub-frame $L_p^2(\mathbf{u}^\mathbf{m})$, where $m \in [1\ldots M]$. In addition, each of the M number of VQ decoding process recovers the reconstructed FD sub-frame estimate $\tilde{\mathbf{e}}^\mathbf{m}$, as will be detailed in Section 4.7.12.

4.7.6 Transmission Frame Structure

As stated in Section 4.7.2, each video frame $\mathbf{f_n}$ is conveyed between the proposed VQ-TCM system's transmitter and receiver by means of a single bit-based transmission frame $\mathbf{u}$. This comprises the concatenation of M number of VQ-based transmission sub-frames. Again, the formation of legitimate bit-sequences within these transmission sub-frames is governed by simple VQ-induced code constraints, which are imposed by the decomposition of FDs into sub-frames and by the specific nature of the VQ codebook to be outlined in Section 4.7.8. These VQ-induced code constraints may be described by a novel VQ trellis structure. These issues are discussed in the following sub-sections and are described with the aid of

4.7. JOINT ITERATIVE DECODING OF TRELLIS-BASED VQ VIDEO AND TCM

an example that continues throughout Figures 4.23, 4.24, 4.25 and 4.26. It should be noted that the generalization of this example is straightforward.

4.7.7 Frame Difference Decomposition

As stated in Section 4.7.4, the FD **e**, the reconstructed FD **ê** and the reconstructed FD estimate **ẽ** are decomposed into M number of sub-frames, as detailed in the example of Figure 4.23.

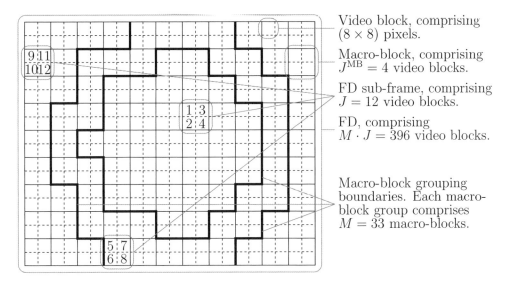

Figure 4.23: Example of selecting $J = 12$ (8×8)-pixel VBs from a (176×144)-pixel FD to provide one of the $M = 33$ FD sub-frames.

For the sake of implementational simplicity, the FD decomposition is designed to yield FD sub-frames with statistical properties that are similar to each other. This allows the allocation of an equal number of bits, namely I, to each transmission sub-frame $\mathbf{u}^m = \{u_i^m\}_{i=1}^I$, where $u_i^m \in \{0, 1\}$ and $i \in [1 \ldots I]$ is the bit index. In addition, the same codebook may be employed for the M number of VQ-encoding and -decoding processes.

The proposed video codec operates on a block-based philosophy. In this section a *video block* (VB) has the dimensions of (8×8) pixels and is defined as the smallest unit of video information that is considered in isolation. Each of the M number of FD sub-frames $\mathbf{e}^m$ comprises a unique combination of J number of (8×8)-pixel VBs of the FD **e**. Hence, the FD **e** comprises $M \cdot J$ number of VBs in total. In the example of Figure 4.23, each of the $M = 33$ FD sub-frames in the (176×144)-pixel FD comprises $J = 12$ of the $M \cdot J = 396$ (8×8)-pixel VBs, which are shown with dashed boundaries.

We proceed by decomposing the FD **e** into a set of perfectly tiling *macro-blocks* (MBs). Each MB has the dimensions of $(J_x^{\mathrm{MB}} \times J_y^{\mathrm{MB}}) = J^{\mathrm{MB}}$ number of VBs, where $J = 12$ is an integer multiple of J^{MB}. In the example of Figure 4.23, MBs are shown with solid boundaries and have the dimensions of $(J_x^{\mathrm{MB}} \times J_y^{\mathrm{MB}}) = (2 \times 2)$ VBs, giving $J^{\mathrm{MB}} = 4$. The MBs are then assigned to $J/J^{\mathrm{MB}} = 3$ different groups on the basis of the distance between their centre and the FD centre, with each group comprising $M = 33$ MBs. This results in

$J/J^{\text{MB}} = 3$ quasi-concentric MB groups, as indicated by the thick boundaries in Figure 4.23. A pseudo-random selection of one MB from each of the $J/J^{\text{MB}} = 3$ groups is performed for each FD sub-frame $\mathbf{e}^{\mathbf{m}}$. It is the $J = 12$ (8×8)-pixel VBs identified by this pseudo-random MB selection that constitute the FD sub-frame $\mathbf{e}^{\mathbf{m}}$. We note that a pre-determined fixed seed is employed for the sake of allowing identical pseudo-random MB selections to be made independently by both the video encoder and decoder.

Each of the $J = 12$ (8×8)-pixel VBs e_j^m, that constitute the FD sub-frame $\mathbf{e}^{\mathbf{m}} = \{e_j^m\}_{j=1}^J$, is allocated a VB index $j \in [1 \ldots J]$. These $J = 12$ VB indices are allocated *amongst* the $J/J^{\text{MB}} = 3$ pseudo-randomly selected MBs in a quasi-radial ordering. In this way, the $J^{\text{MB}} = 4$ lowest-valued VB indices are allocated to the MB nearest to the FD centre, as seen in Figure 4.23. Similarly, the MB nearest to the FD perimeter is assigned the $J^{\text{MB}} = 4$ highest-valued VB indices. The allocation of VB indices *within* MBs should be made with specific consideration of the VQ codebook employed, as will be detailed in Section 4.7.9. Again, Figure 4.23 exemplifies the indices allocated to the $J = 12$ VBs of a FD sub-frame.

As mentioned, the FD sub-frames have statistical properties that are similar to each other, however it should be noted that each of the $J/J^{\text{MB}} = 3$ constituent MBs is likely to exhibit different statistical properties. For example, the MB that is allocated the lowest-valued VB indices in each FD sub-frame can be expected to exhibit a high level of video activity, since this MB is located near the centre of the FD. In contrast, a low level of video activity can be expected for the MB that is assigned the highest-valued VB indices.

The reconstructed FD sub-frame $\hat{\mathbf{e}}^{\mathbf{m}} = \{\hat{e}_j^m\}_{j=1}^J$ employs the same VB selection and indexing as the corresponding FD sub-frame $\mathbf{e}^{\mathbf{m}} = \{e_j^m\}_{j=1}^J$. The same applies to the reconstructed FD sub-frame estimate $\tilde{\mathbf{e}}^{\mathbf{m}} = \{\tilde{e}_j^m\}_{j=1}^J$.

In contrast to the FD decomposition described above, an alternative approach could comprise a selection of $J = 12$ *adjacent* VBs for each FD sub-frame. Although this alternative would permit the exploitation of correlation between adjacent VBs, each FD sub-frame would have different statistical properties. As a result, the corresponding transmission sub-frames would require different individual bit-allocations on a demand-basis. In addition, each transmission sub-frame's different bit-allocation would require a separate VQ-encoder and -decoder design. Finally, the transmission of side information would be required for the sake of signalling this allocation between the transmitter and the receiver. Hence, this alternative approach was discarded for the sake of maintaining simplicity.

4.7.8 VQ Codebook

As stated in Section 4.7.2, each FD sub-frame $\mathbf{e}^{\mathbf{m}}$ is approximated by the reconstructed FD sub-frame $\hat{\mathbf{e}}^{\mathbf{m}}$ and represented by the bit-based transmission sub-frame $\mathbf{u}^{\mathbf{m}}$ on the basis of the same VQ codebook. This comprises K number of VQ tiles addressed by their Reversible Variable-Length Coding (RVLC) based [238] codebook index. These are known to both the video encoder and decoder. An example of a $K = 5$-entry VQ codebook is provided in Figure 4.24. In the VQ tiles of this figure, dark pixels indicate negative FD values and light pixels represent positive FD values.

As exemplified in Figure 4.24, the K number of VQ tiles in the VQ codebook may have different dimensions, which must be multiples of the (8×8)-pixel VB dimensions. This allows the adequate representation of both small areas of high video activity and large areas of low video activity. In this way, coding efficiency is maintained. The VQ codebook entry

4.7. JOINT ITERATIVE DECODING OF TRELLIS-BASED VQ VIDEO AND TCM

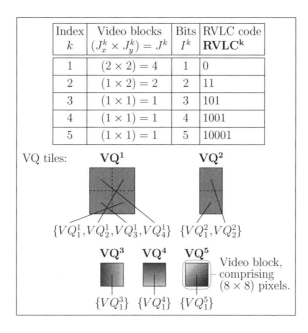

Figure 4.24: Example of a $K = 5$-entry VQ codebook.

with index $k \in [1 \ldots K]$ is associated with the VQ tile $\mathbf{VQ^k}$. This has the dimensions of $(J_x^k \times J_y^k) = J^k$ number of (8×8)-pixel VBs, as exemplified in Figure 4.24. However, a VQ tile's dimensions must not exceed the MB dimensions of $(J_x^{\mathrm{MB}} \times J_y^{\mathrm{MB}}) = (2 \times 2)$ VBs, as defined in Section 4.7.7. Each VB $VQ_{j^k}^k$, from the set of J^k VBs that constitute the VQ tile $\mathbf{VQ^k} = \{VQ_{j^k}^k\}_{j^k=1}^{J^k}$, is allocated an index $j^k \in [1 \ldots J^k]$. Note that these indices may be arbitrarily allocated. To emphasize this point, a random allocation of VB indices is employed in Figure 4.24, as exemplified by the VQ tiles $\mathbf{VQ^1}$ and $\mathbf{VQ^2}$.

In addition, the VQ codebook entry index k is represented by the RVLC code $\mathbf{RVLC^k}$, as exemplified in Figure 4.24. Each RVLC code $\mathbf{RVLC^k} = \{RVLC_{i^k}^k\}_{i^k=1}^{I^k}$ comprises I^k number of bits having values of $RVLC_{i^k}^k \in \{0,1\}$, where $i^k \in [1 \ldots I^k]$ is the bit index. The employment of RVLC codes instead of their more efficient, higher coding-rate alternatives, such as Huffman codes [239], will be justified in Section 4.7.9.

The VQ codebook should be designed by considering the statistical properties of the FD sub-frames. This can be achieved by employing the Linde-Buzo-Gray algorithm [240] to design the VQ tile set and a Huffman coding based algorithm [238] to design the RVLC code set.

Recall from Section 4.7.7 that all FD sub-frames exhibit similar statistical properties, but different statistical properties are exhibited by each of the $J/J^{\mathrm{MB}} = 3$ MBs that constitute the FD sub-frames. In an alternative approach, a separate VQ codebook could be designed and employed to model the statistical properties of each of these MBs. This alternative approach would allow the achievement of a higher coding efficiency. However, it was discarded for the sake of simplicity.

4.7.9 VQ-induced Code Constraints

Let us now elaborate on how the employment of VQ imposes code constraints that govern the formation of legitimate bit-sequences within the transmission sub-frames. In Figure 4.25 an example of a VQ-based $J = 12$-block reconstructed FD sub-frame $\hat{\mathbf{e}}^m$ is provided. Here, we employed the FD decomposition example of Figure 4.23 and the $K = 5$-entry VQ codebook example of Figure 4.24. The corresponding $I = 17$-bit transmission sub-frame $\mathbf{u}^m$ is subject to the VQ-induced code constraints outlined below.

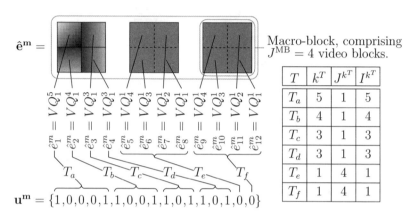

Figure 4.25: Example of a reconstructed FD sub-frame, comprising $J = 12$ VBs selected as exemplified in the FD decomposition of Figure 4.23, and the corresponding $I = 17$-bit transmission sub-frame. These comprise the entries k^T, where $T \in [T_a \ldots T_f]$, from the VQ codebook example of Figure 4.24.

As stated in Section 4.7.7, each reconstructed FD sub-frame $\hat{\mathbf{e}}^m = \{\hat{e}_j^m\}_{j=1}^J$ comprises $J = 12$ (8×8)-pixel VBs. These $J = 12$ VBs constitute $J/J^{MB} = 3$ MBs, as shown in Figure 4.25. Each of these MBs comprises an appropriate tessellation of VQ tiles from the K-entry VQ codebook. The tiles $\mathbf{VQ}^k = \{VQ_{j^k}^k\}_{j^k=1}^{J^k}$ may cover regions of $(J_x^k \times J_y^k) = J^k$ number of (8×8)-pixel VBs. Note that, in Figure 4.25 each of the entries k^T, where $T \in [T_a \ldots T_f]$, invoked from the VQ codebook example of Figure 4.24 provides J^{k^T} number of (8×8)-pixel VBs for the reconstructed FD sub-frame $\hat{\mathbf{e}}^m$.

A specific constraint is imposed that restricts the positioning of the tile $\mathbf{VQ}^k$ in legitimate tessellations. Specifically, the J^k number of VBs in $\hat{\mathbf{e}}^m = \{\hat{e}_j^m\}_{j=1}^J$ that are represented by the tile $\mathbf{VQ}^k$ must have consecutive VB indices j. Note that this is true in all cases in Figure 4.25. This VQ tile positioning constraint is imposed to allow the formation of the novel trellis structure to be described in Section 4.7.10. In order that the video-degradation imposed by this constraint is minimized, the VQ tiles and the allocation of VB indices within the MBs should be jointly designed, as stated in Section 4.7.7. In view of this, the presence of the vertical VQ tile $\mathbf{VQ}^2$ in the VQ codebook example of Figure 4.24 motivates the vertical allocation of consecutive VB indices within each MB, as seen in Figure 4.23.

The employment of the VQ tile $\mathbf{VQ}^k$ is expressed using J^k number of mappings of the form $\hat{e}_j^m = VQ_{j^k}^k$, as exemplified in Figure 4.25. With reference to the examples of

4.7. JOINT ITERATIVE DECODING OF TRELLIS-BASED VQ VIDEO AND TCM

Figures 4.23 and 4.24, we note that the resultant values of $j \in [1 \ldots J]$ and $j^k \in [1 \ldots J^k]$ in such mappings are dependent on the specific positioning of the VQ tile.

Each of the VQ tiles $\mathbf{VQ^k}$ that comprise the reconstructed FD sub-frame $\hat{\mathbf{e}}^m$ is associated with an RVLC code $\mathbf{RVLC^k}$, where k is the VQ codebook entry index. These RVLC codes are concatenated in the order of the VB indices j that are associated with the employment of the corresponding VQ tiles in the reconstructed FD sub-frame $\hat{\mathbf{e}}^m = \{\hat{e}_j^m\}_{j=1}^J$. This is exemplified in Figure 4.25, in which each of the entries k^T, where $T \in [T_a \ldots T_f]$, invoked from the VQ codebook example of Figure 4.24 is represented by I^{k^T} number of bits. In the proposed video codec this concatenation is constrained to having a total length of $I = 17$ bits, since it constitutes the $I = 17$-bit transmission sub-frame $\mathbf{u}^m$.

It follows from the above discussions that each $I = 17$-bit transmission sub-frame must represent a legitimate tessellation of VQ tiles, comprising $J = 12$ VBs in total. These VQ-induced code constraints represent redundancy within the bit-based transmission sub-frames. Note that the degree of this redundancy is dependent on the distance properties of the RVLC code set employed. As will be outlined in Section 4.7.13, the employment of RVLC codes rather than Huffman codes is justified, because their additional redundancy is desirable.

4.7.10 VQ Trellis Structure

As described in Section 4.7.9, the formation of legitimate bit-sequences within each of the bit-based transmission sub-frames is governed by the VQ-induced code constraints. During VQ-encoding and -decoding we consider only legitimate $I = 17$-bit transmission sub-frame permutations. Since satisfying this constraint is non-trivial, the employment of a trellis structure is proposed to describe the complete set of VQ-induced code constraints. These VQ-induced code constraints depend on the FD decomposition and the VQ codebook employed, as described in Section 4.7.9. Hence, the design of the proposed VQ trellis structure also depends on these aspects of the proposed video codec's operation. The VQ trellis structure example of Figure 4.26 describes the complete set of VQ-induced code constraints that correspond to the FD decomposition example of Figure 4.23 and to the $K = 5$-entry VQ codebook example of Figure 4.24. In this example, each reconstructed FD sub-frame $\hat{\mathbf{e}}^m$ comprises $J = 12$ VBs and each transmission sub-frame $\mathbf{u}^m$ comprises $I = 17$ bits.

The proposed VQ trellis structure comprises a set of transitions between trellis states, as exemplified in Figure 4.26. A novel block-based modification of the Variable Length Coding (VLC) symbol-level trellis structure described in [232] is employed. Whilst the bit index axis of [232] is retained in the proposed VQ trellis structure, the VLC symbol index axis of [232] is replaced by a VB index axis. In contrast to [232], transitions are permitted to skip a number of consecutive indices along this axis.

During a VQ-encoding or -decoding operation, there are a number of instances when it is possible to employ each VQ codebook entry, as will be elaborated on below. The proposed VQ trellis structure represents each of these legitimate possibilities with a transition, as exemplified in Figure 4.26. In this way, the proposed VQ trellis structure describes the complete set of VQ-induced code constraints. Each transition T represents a possible employment of the VQ codebook entry having the index $k^T \in [1 \ldots K]$. This is associated with a unique combination of (a) the positioning of the VQ tile $\mathbf{VQ^{k^T}}$ within the reconstructed FD sub-frame $\hat{\mathbf{e}}^m$ and (b) the corresponding positioning of the associated

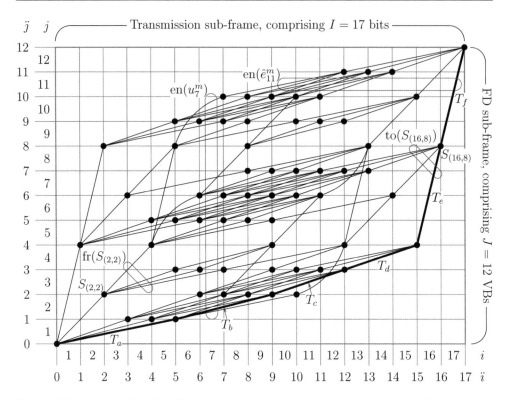

Figure 4.26: Example of a VQ trellis structure for the FD decomposition example of Figure 4.23 and for the $K = 5$-entry VQ codebook example of Figure 4.24, where we have $J = 12$ and $I = 17$. Trellis states occur between the consideration of specific bits and VBs, hence bit state indices $\ddot{i}$ and VB state indices $\ddot{j}$ occur between the bit indices i and the VB indices j, respectively.

RVLC code $\mathbf{RVLC}^{k^T}$ within the transmission sub-frame $\mathbf{u}^m$. Note that for reasons to be discussed below, (a) is subject to the legitimate VQ tile positioning constraint that was stated in Section 4.7.9.

Each trellis state $S_{(\ddot{i},\ddot{j})}$ in the proposed VQ trellis structure represents the progress made at a particular point during the VQ-encoding or -decoding operation. This point occurs immediately after the consideration of the first $\ddot{j}$ number of VBs in the reconstructed FD sub-frame $\hat{\mathbf{e}}^m$ and the first $\ddot{i}$ number of bits in the transmission sub-frame $\mathbf{u}^m$. Here, $\ddot{j} \in [0 \ldots J]$ denotes a VB state index. These occur between the VB indices $j \in [1 \ldots J]$ that were introduced in Section 4.7.7. Likewise, bit state indices $\ddot{i} \in [0 \ldots I]$ occur between the bit indices $i \in [1 \ldots I]$, as shown in Figure 4.26.

Each transition T represents the employment of the VQ codebook entry with index $k^T \in [1 \ldots K]$ immediately after reaching a particular point during the VQ-encoding or -decoding operation. This point is identified by the state indices $\ddot{i}^T \in [0 \ldots I]$ and $\ddot{j}^T \in [0 \ldots J]$. Hence, the transition T emerges from the trellis state $S_{(\ddot{i}^T,\ddot{j}^T)}$. The VQ tile $\mathbf{VQ}^{k^T}$ provides J^{k^T} number of VBs for the reconstructed FD sub-frame $\hat{\mathbf{e}}^m$, as stated in Section 4.7.9. In addition, the RVLC code $\mathbf{RVLC}^{k^T}$ provides I^{k^T} number of bits for the transmission sub-frame $\mathbf{u}^m$.

4.7. JOINT ITERATIVE DECODING OF TRELLIS-BASED VQ VIDEO AND TCM

Hence, the transition T merges into the trellis state $S_{(iT+J^{k^T},jT+I^{k^T})}$. Note that the VQ tile positioning constraint described in Section 4.7.9 is necessary to ensure that each transition is continuous with respect to the VB index axis. Also note that a particular employment of a VQ codebook entry is only possible if the associated transition T contributes to a legitimate transition path between the trellis states $S_{(0,0)}$ and $S_{(I,J)}$. This condition is satisfied if at least one transition path exists between the trellis states $S_{(0,0)}$ and $S_{(iT,jT)}$ and between the trellis states $S_{(iT+I^{k^T},jT+J^{k^T})}$ and $S_{(I,J)}$.

Note that the reconstructed FD sub-frame and transmission sub-frame examples of Figure 4.25 correspond to the bold trellis path in Figure 4.26. Here, each transition $T \in [T_a \ldots T_f]$ corresponds to the similarly labelled employment of a VQ codebook entry in Figure 4.25. With reference to Figure 4.25, observe that each transition $T \in [T_a \ldots T_f]$ encompasses I^{k^T} number of bit indices and J^{k^T} number of VQ indices. Whilst the trellis path considered comprises six transitions, it should be noted that the trellis structure of Figure 4.26 additionally contains trellis paths comprising seven transitions. In general, each transition path in the proposed VQ trellis structure comprises a varying number of transitions.

Note furthermore that a single VQ trellis structure considering $J \cdot M = 396$ VBs and $I \cdot M = 561$ bits would contain more transitions than the combination of $M = 33$ VQ trellis structures, each considering $J = 12$ VBs and $I = 17$ bits. This justifies the decomposition of the VQ-encoding and -decoding operations of an entire FD into $M = 33$ less complex trellis-based VQ operations, as described in Section 4.7.4. This decomposition is associated with a reduced grade of VQ encoding freedom and hence a reduced video reconstruction quality. However, this slight video degradation is insignificant compared to the resultant computational complexity reduction.

For the benefit of the following sections, we now introduce some trellis-transition set notation, which is exemplified in Figure 4.26. The set of all transitions that encompasses (en) the VB $j \in [1 \ldots J]$ of the reconstructed FD sub-frame $\hat{\mathbf{e}}^m$ is termed $\text{en}(\hat{e}_j^m)$. Furthermore, $\text{en}(\hat{e}_j^m = VQ_{j^k}^k)$ is the specific sub-set of $\text{en}(\hat{e}_j^m)$ that maps the VB $j^k \in [1 \ldots J^k]$ of the VQ tile $\mathbf{VQ}^k$ onto $\hat{e}_j^m$. Note that some of these sub-sets may be empty. This is a consequence of the VQ tile positioning constraint described in Section 4.7.9. The set of all transitions that encompasses the bit $i \in [1 \ldots I]$ of the transmission sub-frame $\mathbf{u}^m$ is termed $\text{en}(u_i^m)$. In addition, $\text{en}(u_i^m = b)$ is the specific sub-set of $\text{en}(u_i^m)$ that maps the bit value $b \in \{0,1\}$ onto u_i^m. Note that for a particular transition T, we have $RVLC_{i-iT}^{k^T} = b$ if $T \in \text{en}(u_i^m = b)$. Finally, the set of transitions emerging from (fr) the trellis state $S_{(\bar{i},\bar{j})}$ is denoted as $\text{fr}(S_{(\bar{i},\bar{j})})$, whilst the set merging into (to) that trellis state is represented as $\text{to}(S_{(\bar{i},\bar{j})})$.

4.7.11 VQ Encoding

In the proposed VQ-TCM system's transmitter of Figure 4.21, VQ encoding is performed separately for each of the M number of FD sub-frames. Each VQ encoder operates on the basis of the proposed VQ trellis structure described in Section 4.7.10 and exemplified in Figure 4.26. Since it describes the complete set of VQ-induced code constraints, the employment of the proposed VQ trellis structure represents the consideration of every legitimate FD sub-frame encoding. This allows us to find the MMSE approximation of the FD sub-frame $\mathbf{e}^m$. The result is the optimal reconstructed FD sub-frame $\hat{\mathbf{e}}^m$ and the corresponding bit-based transmission sub-frame $\mathbf{u}^m$.

We quantize the video sequence in a novel manner, which is reminiscent of TCQ [233], but considers the tessellation of potentially differently sized VQ tiles. The philosophy of Viterbi decoding [237] is employed, with a survivor path being selected at each trellis state $S_{(\tilde{i},\tilde{j})}$. This selection yields the $\tilde{i}$-bit encoding of the first $\tilde{j}$ number of VBs in the FD sub-frame $\mathbf{e^m}$ that introduces the lowest possible cumulative video distortion $D(S_{(\tilde{i},\tilde{j})})$.

As stated in Section 4.7.10, each transition T in the proposed VQ trellis structure represents the employment of the VQ codebook entry with index k^T during VQ encoding. This corresponds to employing the VQ tile $\mathbf{VQ^{k^T}}$ to represent a total of J^{k^T} number of (8×8)-pixel VBs of the FD sub-frame $\mathbf{e^m}$. The distortion $d(T)$ associated with the transition T is the sum of the squared difference between $\mathbf{VQ^{k^T}}$ and the corresponding J^{k^T} number of VBs of $\mathbf{e^m}$.

The survivor path at the trellis state $S_{(\tilde{i},\tilde{j})}$ is deemed to be that associated with the specific merging transition $T \in \mathrm{to}(S_{(\tilde{i},\tilde{j})})$ having the minimum cumulative video distortion. This is calculated as $D(T) = d(T) + D(S_{(i^T, j^T)})$, where $D(S_{(0,0)}) = 0$. Having determined the survivor path at the trellis state $S_{(I,J)}$, the MMSE VQ encoding of the FD sub-frame $\mathbf{e^m}$ has been found. The reconstructed FD sub-frame $\mathbf{\hat{e}^m}$ is formed as the tessellation of the VQ tiles associated with the survivor path transitions. In addition, the I-bit transmission sub-frame $\mathbf{u^m}$ is formed as the concatenation of the associated RVLC codes, as described in Section 4.7.9.

4.7.12 VQ Decoding

In the proposed VQ-TCM system's receiver of Figure 4.22, VQ decoding is performed separately for each of the M number of I-bit transmission sub-frames. Each of the M number of VQ decoders operates on the basis of the proposed VQ trellis structure. The recovery of legitimate – although not necessarily error-free – video information is therefore guaranteed, since the proposed VQ trellis structure describes the complete set of VQ-induced code constraints. As stated in Section 4.7.9, these VQ-induced code constraints impose redundancy within each transmission sub-frame. This redundancy may be exploited to assist during VQ decoding by invoking the BCJR [234] algorithm on the basis of the proposed VQ trellis structure. In addition, residual redundancy is exhibited by the transmission sub-frame $\mathbf{u^m}$ owing to the simplicity of the proposed video codec. This may be exploited by the BCJR algorithm with consideration of the statistical properties of FD sub-frames, as will be detailed below.

Each VQ decoder is provided with the *a priori* LLR sub-frame $L_a^2(\mathbf{u^m})$ and generates the *a posteriori* LLR sub-frame $L_p^2(\mathbf{u^m})$, as stated in Section 4.7.5. In addition, each VQ decoder recovers the optimal MMSE-based reconstructed FD sub-frame estimate $\mathbf{\tilde{e}^m}$. This is achieved using soft APP-based reconstruction, as follows.

A novel block-based modification of the BCJR algorithm of [235] is invoked in the proposed VQ trellis structure. This obtains an APP for each transition, giving the probability that this specific transition was in the survivor path during the corresponding VQ encoding operation, as described in Section 4.7.11. These APPs are calculated as

$$P(T|\mathbf{y}) = \frac{1}{C_1} \cdot \alpha(S_{(i^T, j^T)}) \cdot \gamma(T) \cdot \beta(S_{(i^T + I^{k^T}, j^T + J^{k^T})}). \tag{4.4}$$

The specific value of the normalization factor $C_1 = p(\mathbf{y})$ in (4.4) may be ignored in this application, since we are only concerned with the ratios of *a posteriori* transition

4.7. JOINT ITERATIVE DECODING OF TRELLIS-BASED VQ VIDEO AND TCM

probabilities, as we will show below. The terms $\alpha(S_{(i^T, j^T)})$, $\gamma(T)$ and $\beta(S_{(i^T+I^{k^T}, j^T+J^{k^T})})$ consider the probability of the requisite trellis activity before, during and after the occurrence of the transition T, respectively.

Specifically, the term $\gamma(T)$ in (4.4) is calculated as

$$\gamma(T) = \frac{P(k^T)}{C_2(S_{(i^T, j^T)})} \cdot \prod_{i^k=1}^{I^{k^T}} P(u_{i^T+i^k}^m = RVLC_{i^k}^{k^T} | y), \quad (4.5)$$

where $P(u_i^m = b | y)$ is the *a priori* probability that the transmission sub-frame bit u_i^m has a value of $b \in \{0, 1\}$. This is obtained from the *a priori* LLR $L_a^2(u_i^m) = \ln(\frac{P(u_i^m=0|y)}{P(u_i^m=1|y)})$ [236]. Furthermore, $P(k)$ in (4.4) is the probability of occurrence for the VQ codebook entry with index k. This is obtained based on knowledge of the statistical properties of the FD sub-frames. As stated in Section 4.7.7, different statistical properties are associated with each of the J/J^{MB} number of MBs constituting a FD sub-frame. Hence, different values of $P(k)$ are employed, depending on the specific MB that T is constituent of. Finally, the normalization factor $C_2(S_{(i,j)}) = \sum_{T \in \text{fr}(S_{(i,j)})} P(k^T)$ was proposed in [235] for the sake of ensuring that (4.5) represents a true probability.

The BCJR algorithm's forward recursion emerging from the trellis state $S_{(0,0)}$ is employed to obtain the values of

$$\alpha(S_{(i,j)}) = \sum_{T \in \text{to}(S_{(i,j)})} \gamma(T) \cdot \alpha(S_{(i^T, j^T)}), \quad (4.6)$$

where $\alpha(S_{(0,0)}) = 1$. Similarly, a backward recursion from the trellis state $S_{(I,J)}$ is employed to obtain the values of

$$\beta(S_{(i,j)}) = \sum_{T \in \text{fr}(S_{(i,j)})} \gamma(T) \cdot \beta(S_{(i^T+I^{k^T}, j^T+J^{k^T})}), \quad (4.7)$$

where $\beta(S_{(I,J)}) = 1$.

Having determined the *a posteriori* transition probabilities, the method of [235] is employed for obtaining bit-based soft outputs. This is facilitated by the employment of the normalization factor $C_2(S_{(i^T, j^T)})$ during the calculation of $\gamma(T)$, as described above. For each of the I number of bits in the transmission sub-frame $\mathbf{u^m} = \{u_i^m\}_{i=1}^I$, we consider a cross-section of the trellis structure, which is perpendicular to the bit index axis at the particular index i. The associated *a posteriori* LLR is calculated as

$$L_p^2(u_i^m) = \ln \left(\frac{\sum_{T \in \text{en}(u_i^m=0)} P(T|\mathbf{y})}{\sum_{T \in \text{en}(u_i^m=1)} P(T|\mathbf{y})} \right). \quad (4.8)$$

The recovery of the reconstructed FD sub-frame estimate $\tilde{\mathbf{e}}^\mathbf{m}$ is performed on an individual block-by-block basis. A soft APP-based output is obtained for each of the J number of (8×8)-pixel VBs in $\tilde{\mathbf{e}}^\mathbf{m}$. Again, a novel modification of the method of [235] is employed for obtaining these block-based soft outputs. For each of the J number of VBs in the reconstructed FD sub-frame $\hat{\mathbf{e}}^\mathbf{m} = \{\hat{e}_j^m\}_{j=1}^J$, we consider a cross-section of the trellis structure. In analogy to the generation of the previously mentioned bit-based soft outputs, this cross-section is now perpendicular to the VB index axis at the particular index j. The APP of the VB $\hat{e}_j^m$ being provided by a particular one of the J^k number of VBs in the VQ tile $\mathbf{VQ^k} = \{VQ_{j^k}^k\}_{j^k=1}^{J^k}$ is calculated as

$$P(\hat{e}_j^m = VQ_{j^k}^k|\mathbf{y}) = \frac{\sum_{T \in \text{en}(\hat{e}_j^m = VQ_{j^k}^k)} P(T|\mathbf{y})}{\sum_{T \in \text{en}(\hat{e}_j^m)} P(T|\mathbf{y})}. \quad (4.9)$$

Note that some of the sets $\text{en}(\hat{e}_j^m = VQ_{j^k}^k)$ may be empty, as described in Section 4.7.10. In this case, the corresponding APP is zero-valued.

Each of the J number of (8×8)-pixel VBs that constitute the reconstructed FD sub-frame estimate $\tilde{\mathbf{e}}^m = \{\tilde{e}_j^m\}_{j=1}^J$ is obtained using optimal MMSE estimation according to

$$\tilde{e}_j^m = \sum_{k=1}^K \sum_{j^k=1}^{J^k} P(\hat{e}_j^m = VQ_{j^k}^k|\mathbf{y}) \cdot VQ_{j^k}^k. \quad (4.10)$$

The Log-MAP algorithm [241] is employed to reduce the computational complexity of VQ decoding. Specifically, the above mentioned calculations are performed in the logarithmic domain by employing the Jacobian approximation with an eight-entry table-lookup correction factor [237]. This reduces the number of multiplications required by the BCJR algorithm. In addition, the T-BCJR algorithm [242] is employed to prune insignificant transitions from the proposed VQ trellis structure, where the transition paths passing through the trellis state $S_{(\tilde{i},\tilde{j})}$ are pruned during the forward recursion if

$$\frac{\alpha(S_{(\tilde{i},\tilde{j})})}{\sum_{i'=0}^I \alpha(S_{(i',\tilde{j})})} < 0.001. \quad (4.11)$$

In the next section we consider the performance of the proposed VQ-TCM system, where the above-mentioned complexity-reduction methods were observed to impose no significant performance degradation.

4.7.13 Results

In this section we assess the performance of the proposed VQ-TCM system. We transmitted 250 video frames of the "Lab" video-sequence [224]. This 10 frames/s gray-scale head-and-shoulders (176×144)-pixel Quarter Common Intermediate Format (QCIF) video sequence exhibits a moderate level of video activity.

The proposed video codec was designed to achieve an attractive trade-off between the conflicting design requirements associated with bitrate, video reconstruction quality and computational complexity. The 396-block FDs were decomposed into $M = 33$ FD sub-frames, each comprising $J = 12$ (8×8)-pixel VBs. This was performed exactly as exemplified in Figure 4.23. However, in contrast to our simplified example of Figure 4.26 using $I = 17$ bits per sub-frame, each transmission sub-frame comprised $I = 45$ bits in the system implemented. A $K = 512$-entry VQ codebook was employed. This comprised the five VQ tiles shown in Figure 4.24 and an additional 507 single-VB VQ tiles. The corresponding RVLC codes were designed to have a minimum free distance of two [238]. The coding rate

4.7. JOINT ITERATIVE DECODING OF TRELLIS-BASED VQ VIDEO AND TCM

of the proposed video codec is defined here as the ratio of its entropy to its average RVLC code length. This may be estimated as the area beneath the inverted VQ decoding EXIT characteristic [236] provided in Figure 4.27, which will be discussed in more detail below. This gives a coding rate of $R_{VQ} \approx 0.666$.

Two VQ-TCM schemes associated with different latencies were employed. The first scheme imposed a low latency equal to the duration of a single video frame, namely 0.1 s at 10 frames/s. This is suitable for realtime interactive videotelephony applications. In this scheme the length of each transmission frame **u** and that of the interleaver π equals $M \cdot I = 1485$ bits. The second VQ-TCM scheme had a high latency of 50 video frames, i.e. 5 s at 10 frames/s. This is suitable for non-realtime video streaming and wireless-Internet download applications. Here, 50 transmission frames **u** are concatenated before interleaving, giving an interleaver length of $50 \cdot M \cdot I = 74250$ bits. Note that both schemes have a video encoded bitrate of 14.85 kbit/s.

The same TCM codec was employed in both VQ-TCM schemes. This terminated $R_{TCM} = 3/4$-rate TCM codec employed a coding memory of 6, $M_{TCM} = 16$-level QAM modulation and IQ-interleaving [231] for transmission over an uncorrelated narrowband Rayleigh fading channel. The bandwidth efficiency of the proposed VQ-TCM system is given by $\eta = R_{VQ} \times R_{TCM} \times \log_2(M_{TCM}) = 2.00$ bit/s/Hz without a Nyquist excess bandwidth. More explicitly, we assume ideal Nyquist filtering and ignore the code termination symbols added by the TCM encoder. Note that at $\eta = 2$ bit/s/Hz the uncorrelated Rayleigh-fading channel's capacity limit for 16QAM is $E_b/N_0 = 3.96$ dB [243], where $E_b/N_0 = SNR/\eta$ is the Signal to Noise Ratio (SNR) per bit.

Again, in Figure 4.27 we provide the EXIT characteristics [244] for TCM decoding at various E_b/N_0 values. These can be seen to achieve unity extrinsic mutual information I_E^1 for unity *a priori* mutual information I_A^1. In addition, Figure 4.27 provides the inverted VQ decoding EXIT characteristic. Similarly to TCM decoding, VQ decoding achieves unity extrinsic mutual information I_E^2 for unity *a priori* mutual information I_A^2. This was found to be the benefit of the employment of an RVLC code set having a minimum free distance of two.

In addition, Figure 4.27 provides EXIT trajectories [244] for both the low- and high-latency VQ-TCM schemes at $E_b/N_0 = 6$ dB. The low-latency trajectory can be seen to deviate from the EXIT characteristics and fails to converge to the desired unity mutual information. In contrast, the high-latency trajectory closely matches these EXIT characteristics and converges to unity mutual information. The improved performance of the high-latency scheme is a benefit of its longer interleaver [236], justified by the reasons noted in Section 4.7.5. Note that iterative decoding convergence to unity mutual information is associated with the achievement of an infinitesimally low video degradation. The proposed VQ-TCM system supports this for $E_b/N_0 \geq 5.25$ dB, as shown in Figure 4.27. With the advent of a sufficiently long interleaver and of a sufficiently high number of decoding iterations, infinitesimally low video degradation could be achieved at $E_b/N_0 = 5.25$ dB. This is just 1.29 dB from the proposed VQ-TCM system's channel capacity limit of 3.96 dB.

In Figure 4.28, the video reconstruction quality of both the low- and high-latency VQ-TCM schemes is assessed after a number of decoding iterations and for a range of channel E_b/N_0 values. We employ the Peak-Signal to Noise Ratio (PSNR) [224] as the objective video reconstruction quality metric. In this application, a PSNR of 29.5 dB is associated with an aesthetically pleasing video reconstruction quality. The proposed VQ-TCM system exhibited substantial iteration gains, as shown in Figure 4.28. After eight decoding iterations

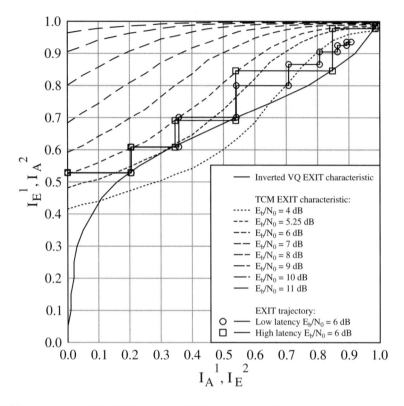

Figure 4.27: The proposed VQ-TCM system's EXIT chart for the "Lab" video sequence and a 3/4-rate TCM scheme.

an E_b/N_0 iteration gain of 4.34 dB was achieved by the low-latency scheme at a PSNR of 29.5 dB. In the case of the high-latency scheme, the corresponding iteration gain was 5.61 dB. The high-latency scheme was seen to outperform the low-latency scheme, regardless of the number of decoding iterations. This is in agreement with the findings of the EXIT chart analysis of Figure 4.27, as discussed above. For the low-latency scheme, a PSNR higher than 29.5 dB was achieved after eight decoding iterations for $E_b/N_0 > 7.00$ dB, as shown in Figure 4.28. This is 3.04 dB from the proposed VQ-TCM system's channel capacity limit of 3.96 dB. In the case of the high-latency scheme, this is achieved for $E_b/N_0 > 5.75$ dB, which is just 1.79 dB from the channel's capacity limit.

We now compare the performance of the proposed VQ-TCM system with two benchmarkers. Similarly to the proposed VQ-TCM system, the IQ-TCM-VLC system of [225] is also VQ-based and is hence referred to as the *VQ-based benchmarker* here. In contrast, the IQ-TCM-RVLC system of [227] is MPEG4-based and is referred to here as the *MPEG4-based benchmarker*. Similarly to the proposed VQ-TCM system, the benchmarkers employ both RVLC and TCM. However, they adopt Shannon's source and channel coding separation philosophy [223], with video decoding being performed independently of iterative RVLC- and TCM-decoding. Since the benchmarkers have the same latency, coding rate and bitrate as the low-latency VQ-TCM scheme, their direct comparison is justified.

4.7. JOINT ITERATIVE DECODING OF TRELLIS-BASED VQ VIDEO AND TCM

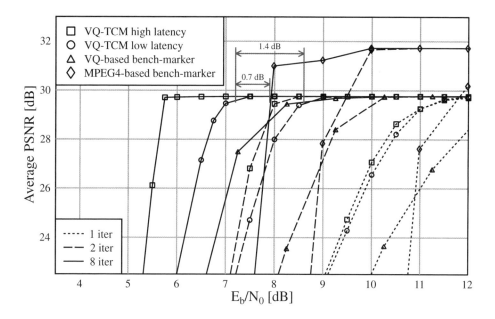

Figure 4.28: PSNR performance of the proposed VQ-TCM system as well as of the VQ- and MPEG4-based benchmarkers for the "Lab" video sequence, when communicating over an uncorrelated narrowband Rayleigh fading channel.

In Figure 4.28, the low-latency VQ-TCM scheme can be seen to outperform the VQ-based benchmarker, regardless of the number of decoding iterations. Although the MPEG4-based benchmarker achieves a slightly higher error-free video reconstruction quality than the proposed VQ-TCM system, it is outperformed by the scheme advocated at low values of E_b/N_0. At a PSNR of 29.5 dB, the proposed VQ-TCM system offers a consistent improvement over both the VQ- and MPEG4-based benchmarkers, which amounts to about 1.4 dB and 0.7 dB, respectively. This is a benefit of the iterative exploitation of the video codec's code constraints, as outlined in Section 4.7.12. In addition, this approach guarantees the recovery of legitimate video information. Hence, the proposed VQ-TCM system is never forced to discard any video information. In contrast, both the benchmarkers of [225] and [227] drop video information, when the iterative decoding process does not recover legitimate video information, reducing the attainable video reconstruction quality.

The improved performance offered by the proposed VQ-TCM system is achieved at the cost of an increased computational complexity. A simple metric of computational complexity considers the number of BCJR trellis transitions encountered during iterative decoding. Note that this ignores the complexity contribution imposed by trellis pruning and video reconstruction. Although an equal number of BCJR transitions are employed during TCM decoding in the proposed VQ-TCM system and in the benchmarkers, the complexity of the respective VQ- and RVLC-decoding operations are different. More explicitly, in the proposed VQ-TCM system, VQ decoding employs a channel-condition-dependent number of BCJR transitions, which is typically equal to the number of trellis transitions encountered during TCM decoding. In contrast, RVLC decoding has approximately a quarter of the number of trellis transitions, in both benchmarkers. Hence, the proposed VQ-TCM system has a computational complexity that may be deemed approximately 1.6 times that of the benchmarkers.

4.8 Summary and Conclusions

Following the comparative study of a variety of VQ-based constant-rate video codecs, an 8 and an 11.36 kbit/s scheme were earmarked for further study in the context of various systems. It was demonstrated that the proposed VQ codecs slightly outperformed the identical-rate DCT codecs. The various studied reconfigurable VQ-codec-based mobile videophone arrangements are characterized by Table 4.3. The video codecs proposed are programmable to any arbitrary transmission rate in order to host the videophone signal by conventional mobile radio speech channels, such as the standard AMR speech coding rate [190] of the 3G systems, requiring an additional speech channel for the transmission of the video information.

In Systems 1, 3, and 6 of Table 4.3, the 11.36 kbit/s Codec II was invoked. The BCH(127,71,9) coded channel rate was 20.32 kbit/s. Upon reconfiguring the 16QAM and 4QAM modems depending on the BCH decoded frame error rate due to the prevailing noise and interference conditions experienced, signaling rates of 9 and 17.2 kBd, respectively, can be maintained. For example, depending on the noise and interference conditions, between 8 and 16 video users can be accommodated in the GSM bandwidth of 200 kHz, which implies minimum and maximum user bandwidths of 12.5 and 25 kHz, respectively. When configuring the video codec to operate at the same bitrate as the AMR speech codec, the wireless system becomes capable of supporting half the combined voice-video users in comparison to "voice-only" users.

In Systems 2, 4, and 5 of Table 4.3, the RL-coded 8 kbit/s Codec I performed similarly to Codec II in terms of PSNR, but it was more error sensitive. The lower rate allowed us, and the lower robustness required us, to use the stronger BCH(127,50,13) FEC codecs. Further system features are summarized in Table 4.3.

As expected, the 4QAM/17.2kBd mode of operation of all systems is more robust than the 16QAM/9kBd mode, but it can only support half the number of users. Explicitly, without ARQ, eight or sixteen users can be supported in the 4QAM/17.2kBd or 16QAM/9kBd modes. When ARQ is used, these numbers are reduced by two in order to reserve slots for re-transmitted packets. A compromise scheme is constituted by Systems 2 and 6, which re-transmit only the corrupted Class One bits in case of decoding errors and halve the number of bits per modulation symbol, facilitating higher integrity communications for fourteen users. Both Systems 2 and 6 slightly outperform in terms of robustness the 16QAM modes of Systems 1 and 5, but not their 4QAM mode, since the initially erroneously received 16QAM Class Two bits are not re-transmitted and hence remain impaired. Systems 2, 4, and 5, which are based on the more sensitive Codec I and the more robust BCH(127,50,13) scheme, tend to have a slightly more robust performance with decreasing channel SNRs than their corresponding counterparts using the Codec II and BCH(127,71,9) combination, namely Systems 6, 3 and 1. However, once the error correction capability of the BCH(127,50,13) codec is overloaded, the corresponding PSNR curves decay more steeply, which is explained by the violently precipitated RL-coded bit errors. This becomes quite explicit in Figure 4.20. Clearly, if best SNR performance is at a premium, the Codec I and BCH(127,50,13)

4.8. SUMMARY AND CONCLUSIONS

combination is preferable, while in terms of low complexity, the Codec I and BCH(127,71,9) arrangement is more attractive.

Careful inspection of Table 4.3 provides the system designer with a vast plethora of design options, while documenting their expected performance. Many of the listed system configuration modes can be invoked on a time-variant basis and so do not have to be preselected. In fact, in a true "software radio" [245], these modes will become adaptively selectable in order to comply with the momentary system optimization criteria.

The most interesting aspects are highlighted by contrasting the minimum required channel SNR values with the systems' user capacity, in particular in the purely 4QAM- and entirely 16QAM-based scenarios. System 2 is an interesting scheme, using 4QAM only during the Class One ARQ attempts, thereby ensuring a high user capacity, although the 11 and 15 dB minimum channel SNR requirements are valid only for one of the fourteen users, whose operation is 4QAM/ARQ assisted. The remaining thirteen users experience a slightly extended error-free operating SNR range in comparison to the 16QAM System 1 mode due to the employment of the more robust BCH(127,50,13) Class One codec. The comparison of System 1 to the otherwise identical, but ARQ-aided, System 3 suggests that over AWGN channels only very limited SNR gains can be attained, but over Rayleigh channels up to 4 dB SNR reduction can be maintained in the 16QAM mode. In general, the less robust the modem scheme, the higher the diversity and AGC gains, in particular over Rayleigh channels. Hence we concluded that both the diversity and ARQ gains are substantial and that the slightly reduced user capacity is a price worth paying for the increased system robustness. This is slightly surprising perhaps, but it is often beneficial to opt for the lower-rate, higher-sensitivity Codec I, if the additional system complexity is acceptable, since due to the stronger accommodated BCH(127,50,13) codec, an extended operating range is maintained. All in all, in terms of performance, System 4 constitutes the best arrangement, with a capability of supporting between 6 and 14 users at minimum channel SNRs between 9 and 17 dB. Here we refrain from describing all potentially feasible systems; this is left for the interested reader to explore.

In summary, the proposed systems are directly suitable for combined voice/videotelephony by halving the number of "voice-only" channels of the operation wireless systems, such as the 3G systems of [83]. Our future work is aimed at improving the complexity/quality characteristics of these systems using parametrically enhanced coding. In Chapter 5, we will concentrate on quad-tree-based coding techniques and on matching multimode videophone transceivers.

In the last section of this chapter we have considered the concept of joint video and channel coding. The proposed video codec imposed VQ-induced code constraints, which were described by a novel VQ trellis structure. This VQ trellis structure was the basis of the proposed MMSE VQ-encoding and -decoding processes. In the latter case, a novel BCJR algorithm was employed to facilitate the iterative exchange of soft information with the serially concatenated TCM decoder. The proposed VQ-TCM system was shown to support error-free video reconstruction at 1.29 dB from the Rayleigh-channel's capacity limit of 3.96 dB.

The joint video and channel coding approach of the proposed VQ-TCM system was shown to consistently outperform two powerful benchmarkers, both employing the Shannonian source and channel separation philosophy [223]. However, this performance improvement was found to accrue at the cost of a 1.6 times increase in computational complexity. Our further research will consider the application of the techniques introduced here to standard video codecs.

Chapter 5

Low Bitrate Quad-tree-based Codecs and HSDPA-style Videophone Transceivers

5.1 Introduction

Quad-tree decomposition is a technique known from the field of image analysis [10]. A practical quad-tree structured intra-frame codec was presented by Strohbach [160], after which Zhang [246] proposed a codec suitable for ISDN applications. Other efforts concentrated on intra-frame coding [247–250], achieving bitrates as low as 0.25 bpp. These codecs share the characteristic that the images are initially segmented into equal-sized blocks, which are then subjected to quad-tree structured coding, generating variable-sized segments that can be represented, for example, by a constant luminance level at the cost of a low loss of sharpness or resolution. In this chapter, we design a low-rate quad-tree decomposed videophone codec, which in its cost-gain quantized approach is similar to the VQ- and DCT-based codecs of the previous two chapters. Furthermore, the codec incorporates a model-based parametric coding enhancement option for the subjectively critical eye-mouth region of head-and-shoulder videophone sequences. Lastly, the proposed codec will be incorporated in a multimode video transceiver, similar to those of the VQ- and DCT-based transceivers of the previous two chapters.

5.2 Quad-tree Decomposition

Quad-trees (QT) represent a subclass of region growing techniques [10] in which the image (in our case the MCER), is described by the help of variable-size sectors characterized by similar features, for example, by similar gray levels. The stylized MCER frame portrayed in

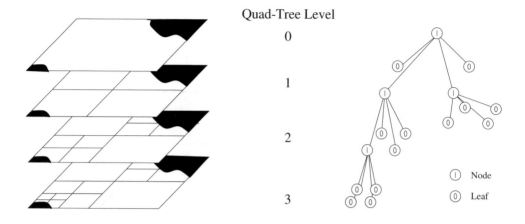

Figure 5.1: Regular QT decomposition example, the corresponding quad tree, and its quad-tree code.

Figure 5.1 is described in terms of two sets of parameters: the structure or spatial distribution of similar regions and their gray levels. This figure is described in more depth later in this chapter. Note that the information describing the QT structure is potentially much more sensitive to bit errors than the gray level coding bits.

Before QT decomposition takes place, the 176×144 pixel QCIF MCER frame can be optionally divided into smaller, for example, 16×16 pixel rectangular blocks, perfectly tiling the frame. This initial decomposition may prove useful if the original image frame is not rectangular, although it may hamper coding efficiency by limiting the size of the largest homogeneous blocks to a maximum of 16×16 pixels. Creating the QT regions is a recursive operation that can follow either a top-down or a bottom-up approach. For example, according to the latter approach and considering each individual pixel, two or more neighbors are merged together, providing a certain similarity criterion is satisfied. This criterion may, for instance be, a similar gray level. This merging procedure is repeated until no more regions satisfy the merging criterion, and hence no more merging is possible.

Similarly, the QT regions can be obtained in a top-down approach, as portrayed in Figure 5.1, in which the MCER is initially divided into four sections, since the MCER frame is too inhomogeneous to satisfy the similarity criterion. Then two of the quadrants become homogeneous, whereas two of them need a number of further splitting or decomposition operations in order to satisfy the similarity criterion. This splitting operation can be continued, until the pixel level is reached and no further splits are possible. The variable-length QT-code at the bottom of the figure will be derived after elaborating on the similarity criterion.

Denoting the gray levels of the quadrants of a square by $m1 \ldots m4$, we compute their mean according to $m = (m_1 + m_2 + m_3 + m_4)/4$. If the absolute difference of all four pixels and the mean gray level is less than the system parameter σ, then these pixels satisfy

5.2. QUAD-TREE DECOMPOSITION

the merging criterion. A simple merging criterion can be formulated as follows [160]:

$$(|m - m_1| < \sigma) \cap (|m - m_2| < \sigma) \cap (|m - m_3| < \sigma)$$
$$\cap (|m - m_4| < \sigma) = \text{True}, \tag{5.1}$$

where $\cap$ represents the logical *AND* operation.

If the system parameter σ is reduced, the matching criterion is expected to become more stringent and hence less merging will take place, which is likely to increase the required encoding rate at a concomitant improvement of the MCER's representation quality. In contrast, an increased σ value is expected to allow more merging to take place and therefore reduce the bitrate. Note, however, that the number of bits generated, and accordingly the bitrate, is inherently variable. In order to circumvent this problem and to introduce a gain-cost-controlled approach, we will later derive Algorithm 2.

If the merging criterion is satisfied, the mean gray level m becomes the gray level of the merged quadrant in the next generation, and so on. At this stage it is important to note that the quality control threshold σ does not need to be known to the QT decoder. Therefore, the image representation quality can be rendered position-dependent within the video frame being processed, which allows perceptual weighting to be applied to the subjectively important image sections, such as the eyes and lips, without increasing the complexity of the decoder or the transmission rate.

Pursuing the top-down QT decomposition approach of Figure 5.1, we now derive the variable-length QT-code given at the bottom of the Figure. The QCIF MCER frame constitutes a node in the QT which is associated with a binary 1 in this example. After QT splitting, this *node* gives rise to four further nodes, which are classified on the basis of the similarity criterion. Specifically, if all the pixels at this level of the QT differ from the mean m by less than the threshold σ, then they are considered to be a *leaf node* in the QT which is denoted by a binary 0. Hence they do not have to be subjected to further similarity tests; they can simply be represented by the mean value m. In our example, a four-level QT decomposition was used (0–3), but the number of levels and/or the similarity threshold σ can be arbitrarily adjusted in order to achieve the required image quality and/or bitrate target.

If, however, the pixels constituting the current node to be classified differ by more than the threshold σ, the pixels forming the node cannot be adequately represented by their mean m, and thus they must be further split until the threshold condition is met. This repetitive splitting process can be continued until there are no more nodes to split, since all the *leaf nodes* satisfy the threshold criterion. Consequently, the QT structure describes the contours of similar gray levels in the frame difference signal. The derivation of the QT-code now becomes explicit from Figure 5.1, where each *parent-node* is flagged with a binary one classifier, while the *leaf nodes* are denoted by a binary zero classifier and the flags are read from top to bottom and left to right. Observe in the example of Figure 5.1 that the location and size of 13 different blocks can be encoded using a total of 17 bits.

For the sake of completeness, we note that an alternative decomposition technique is constituted by the Binary Tree (BT) decomposition. The segmentation for a typical frame of the "Lab" sequence is exemplified in Figure 5.2, where the QT structure is portrayed with and without the overlaid MCER frame and the original video frame. In the eye and lip regions a more stringent similarity match was required than in the background, which led to a finer tree decomposition. In our experience, typically similar performance is achieved using BT-based arrangements and QT codecs. Hence only a brief exposure to BT is offered here.

Figure 5.2: Quad-tree segmentation example with and without overlaid MCER and the original video frame. This decomposition example can be viewed in motion under http://www-mobile.ecs.soton.ac.uk/

Logically, in BT decomposition every BT splitting step leads only to two subblocks. Therefore, typically twice as many splitting steps are necessary, to reach the same small block sizes, as in the case of QT codecs. Furthermore, the decomposition process is not as regular as for QT codecs. A parent-block may be divided both horizontally and vertically. However, in our experiments we did not allow several consecutive splits to take place in the same dimension, since this would require an extra 1-bit signaling bit. Instead, we invoked consecutive vertical/horizontal or horizontal/vertical decompositions, which allowed us to avoid the transmission of additional tree information in order to describe whether horizontal or vertical splitting was used.

In the case of high image-quality requirements, the simple block average representation of the QT decomposed blocks can be substituted by more sophisticated techniques, such as vector quantization (VQ), discrete cosine transformed (DCT) [185], or subband (SB) decomposed representations. Naturally, increasing the number of hierarchical levels in the QT decomposition leads to blocks of different sizes and applying VQ or DCT to blocks of different sizes increases the codec's complexity. Hence, the employment of these schemes for more than two to three hierarchical levels becomes impractical. Having reviewed the principles of QT decomposition, let us now consider the design of our candidate codecs.

5.3 Quad-tree Intensity Match

With low implementational complexity in mind, we contrived two candidate codecs. In the first one we used the above mentioned zero-order mean- or average-based decomposition, whereas in the second one we attempted to model the luminance intensity profile over the block by a first-order linear approximation corresponding to a luminance plane sloping in both x and y directions.

5.3.1 Zero-order Intensity Match

In order to achieve a fixed bitrate, we limited the number of bits per QCIF frame to 1000, which corresponds to a transmission rate of 10 kbit/s at a frame rate of 10 frames/s. The associated compression ratio is 176×144 pixels $\times 8$ bits/pixel/1000 bits ≈ 203, corresponding

5.3. QUAD-TREE INTENSITY MATCH

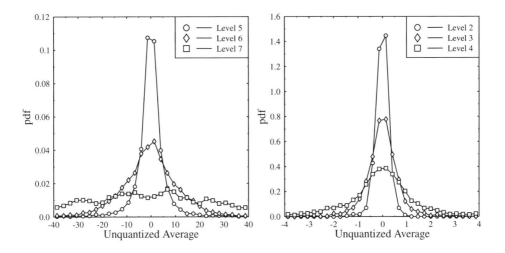

Figure 5.3: Probability density function of the block averages at Levels 2 to 7.

to a relative coding rate of about 0.04 bit/pixel. The statistical evaluation of the Probability Density Functions (PDF) of the average values m of the variable-sized blocks portrayed in Figure 5.3 revealed that for various block sizes quantizers having different mean values and variances were required. As the mean-value PDFs revealed, the mean of the blocks toward the top of the QT, namely, at QT Levels 2–4, which cover a large picture area, is more likely to be close to zero than the mean of smaller blocks. This fact exhibited itself in a more highly peaked and hence less spread PDF for the smaller picture areas, such as, for example, those at QT-levels 2–4 in Figure 5.3. Observe that for clarity of visualization the horizontal PDF scales of Levels 2–4 and 5–7 are about an order of magnitude different. The mean of the smallest blocks at QT Levels 5–7 tends to fluctuate over a wide range, yielding a near-uniform PDF for Level 7. Table 5.1 summarizes the intervals over which the block means fluctuate, as the block size is varied. Therefore, it was necessary to design separate quantizers for each QT hierarchy level.

With this objective in mind, a two-stage quantizer training was devised. First the unquantized mean values for each QT-decomposed block size were recorded using a training sequence, based on MCER frames, in order to derive an initial set of quantizers. During the second, true training stage, this initial set of quantizers was used tentatively in the QT codec's operation, in order to record future unquantized block averages generated by the codec operated at the chosen limited, constant bitrate, namely, at 1000 bits/frame. This two-stage approach was necessary for obtaining realistic training data for the Max-Lloyd training [187] of the final quantizers.

In order to achieve the best codec performance under the constraint of generating 1000 bits per frame, we derived codebooks for a range of different numbers of reconstruction levels. Figure 5.4 characterizes the codec's performance for various quantizers, ranging from 2- to 64-level schemes. Observe in the figure that the two-level and four-level quantizers were found to have the best performance. This was because the 10 kbit/s bitrate limit severely restricted the number of hierarchical levels in the QT decomposition process, when more

Table 5.1: Maximum Quantizer Ranges at Various Hierarchy Levels of the QT Codec

Level	0	1	2	3	4	5	6
Block Size	176 × 144	88 × 72	44 × 36	22 × 18	11 × 9	5/6 × 4/5	2/3 × 2/3
Range +/−	0.22/0.36	0.57/0.79	2.9/2.5	8.5/7.7	20.6/20.7	57.6/50.4	97.3/86.6

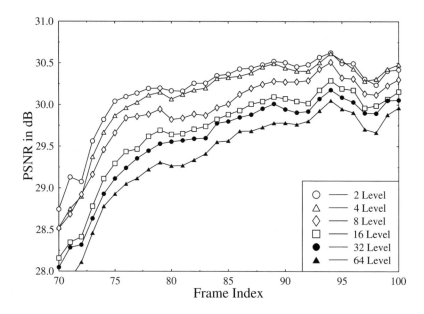

Figure 5.4: Performance for various QT codec quantizers at a constant bitrate of 1000 bits/frame.

bits were allocated for quantizing the block averages. Hence, the codec failed to adequately decompose the inhomogeneous regions due to allocating an excessive number of bits to the quantization of their mean values.

5.3.2 First-order Intensity Match

Following our endeavors using the zero-order model of constant-luminance MCER representation within the QT decomposed subblocks, we embarked on studying the performance of the first-order luminance-intensity approximation corresponding to a sloping luminance plane across the MCER QT-block, which was defined by:

$$b(x,y) = a_0 + a_{x1}x + a_{y1}y. \tag{5.2}$$

Specifically, the block's luminance is approximated by a plane sloping in both x and y directions where the coefficients a_{x1} and a_{y1} are characteristic of the luminance slope.

Table 5.2: PSNR Performance Comparison for Zero- and First-order QT Models at Various Constant Bitrates

Approximation	8 kbit/s	10 kbit/s	12 kbit/s
Zero-order PSNR (dB)	27.10	28.14	28.46
First-order PSNR (dB)	25.92	27.65	27.82

The squared error of this linear approximation is given by:

$$e = \sum_{x=1}^{b}\sum_{y=1}^{b}(b(x,y) - (a_0 + a_{x1}x + a_{y1}y))^2 \qquad (5.3)$$

where the constants a_0, a_{x1}, and a_{y1} are determined by setting the partial derivative of Equation 5.3 with respect to all three variables to zero and solving the resulting three-dimensional problem for all decomposition steps. In order to assess the performance of this scheme, Max-Lloyd quantizers [187] were designed for each constant, and the PSNR performance of a range of quantizers using different numbers of quantization levels was tested. We found that the increased number of quantization bits required for the three different coefficient quantizers was too high to facilitate a sufficiently fine QT-based MCER decomposition given the fixed bitrate budget imposed. Thus, the PSNR performance of the first-order scheme became inferior to that of the zero-order model under the constraint of a fixed bitrate. Specifically, Table 5.2 reveals that the PSNR performance of the first-order model is at least 0.5 dB worse than that of the simple zero-order model. Consequently, for our further experiments the latter scheme was preferred.

5.3.3 Decomposition Algorithmic Issues

The crucial element of any QT or BT codec is the segmentation algorithm controlled by the similarity threshold σ. Most codecs presented in the recent literature [160, 246, 249] invoke the previously outlined thresholding process in order to decide whether a tree should be further decomposed. This principle results in an approximately time-invariant video quality and a fluctuating time-variant bitrate. This bitrate fluctuation can be smoothed by applying adaptive buffer feedback techniques and controlling the similarity threshold σ [160]. Since our proposed videophone system was designed for constant-rate mobile radio channels, we prefer a constant selectable bitrate scheme. A further requirement in very low bitrate video communications is following the principle of gain-cost control, which is difficult to achieve in the framework of QT schemes, since the gain of a potential decomposition becomes known only after the decomposition took place.

In our preferred approach, the codec develops the QT or BT structure down to a given maximum number of decomposition levels and then determines the gain of each decomposition step, as summarized in Algorithm 2. Exposition of the algorithm is further aided by referring to Figure 5.1. The decomposition gain referred to in Step 2 of the algorithm is defined as the difference between the mean squared video reconstruction error of the parent block and the total MSE contribution associated with the sum of its four child blocks.

Algorithm 2 *This algorithm adaptively adjusts the required QT resolution, the number of QT description bits, and the total number of encoding bits required in order to arrive at the target bitrate.*

1. Develop the full tree from minimum to maximum number of QT levels (e.g., 2--7).
2. Determine the decomposition gains associated with all decomposition steps for the full QT.
3. Determine the average decomposition gain over the full set of leaves.
4. If the potentially required number of coding bits is more than twice the target number of bits for the frame, then delete all leaves having less than average gains and repeat Step 3.
5. Otherwise delete leaves on an individual basis, starting with the lowest gain leaf, until the required number of bits is attained.

The question that arose was whether the potential gain due to the current additional decomposition step should be weighted by the actual area of the blocks under consideration. Toward the bottom of the QT the subblocks represent small image sections, while toward the top the QT blocks correspond to larger areas. Our simulations showed that for most of the cases such a weighting of the gains was disadvantageous. This somewhat surprising result was an indirect consequence of the constant bitrate requirement, since the adaptive codec often found it more advantageous to resolve some of the fine image details at an early stage — despite the restricted size of the QT-segment, if the bitrate budget allowed it — and then save bits during future frames over the area in question.

Returning to Algorithm 2, we find that the purpose of Steps 3 and 4 is to introduce a bitrate-adaptive, computationally efficient way of pruning the QT to the required resolution. This allows us to incorporate an element of cost-gain quantized coding, while arriving at the required target bitrate without many times tentatively decomposing the image in various ways in an attempt to find the optimum fixed bit-allocation scheme. The algorithm typically encountered four to five such fast QT pruning recursions before branching out to Step 5, which facilitated a slower converging fine-tuning phase during the bit-allocation optimization.

Algorithm 2 allowed us to eliminate the specific child blocks or leaves from the tree that resulted in the lowest decomposition gains. During this QT pruning process, the elimination of leaves converted some of the nodes to leaves, which were then considered for potential elimination during future coding steps. Therefore, the list of leaves associated with the lowest decomposition gains had to be updated before each QT pruning step. During the fast pruning phase — constituted by Steps 3 and 4 — in a computationally efficient but suboptimum approach, we deleted leaves on the basis of comparing them to the average gain of the entire set of leaves, rather than deleting the leaves associated with the lowest decomposition gain one-by-one. All leaves with a gain lower than the average gain were deleted in a single step if the potentially required number of coding bits was more than twice the target number. The slower one-by-one pruning constituted by Step 5 was then invoked before concluding the bit-allocation in order to fine-tune the number of bits allocated. This suboptimum deletion

5.3. QUAD-TREE INTENSITY MATCH

process was repeated until the tree was pruned to the required size and the targeted number of coding bits was allocated.

In an attempt to explore any eventual residual redundancy or predictability within the QT, which could lead to further potential bitrate reduction, we examined the tree structure. The philosophy of the following procedure was to fix the minimum and maximum tree depth. The advantage of such a restriction is twofold, since both the number of bits required for the QT and the complexity can be reduced. Under this QT structure limitation, bits can be saved at both the top and bottom of the tree. If a certain minimum depth of the tree is inherently assumed, the tree information of the top levels becomes *a priori* knowledge. The same happens as regards the bottom level if the tree depth is limited. In that case all decomposed subblocks at the last level are leaf nodes or, synonymously, child nodes, implying that no further splitting is allowed to take place. There is no need to transmit this *a priori* information. Hence, the decomposition complexity is limited since the tree needs to be developed only for the tree branches, which do not constitute *a priori* knowledge.

In Table 5.3 we list the average number of nodes associated with each tree level. Observe from the table that $15.3/16 = 95.6\%$ of the time the QT depth exceeds three levels. It is therefore, advantageous to restrict the minimum tree depth to three without sacrificing coding efficiency. Note from Table 5.3 that only about $52.9/16384 = 0.32\%$ of the potentially possible Level 7 leaves are ever decomposed. Thus, if we restrict the maximum tree depth to six for our QCIF codec, the codec's complexity is reduced, which is associated with a concomitant average PSNR reduction from 33.78 dB to 32.79 dB for the first 100 "Miss America" frames.[1]

Table 5.3: Average Number of Quad-tree Nodes and Leaves per Hierarchical Level within the Tree at 1200 Bits per Frame

QT Level	0	1	2	3	4	5	6	7
Av. No. of nodes/leaves	1	4	15.3	28.4	47.6	66.9	72.6	52.9
Max. No. of nodes/leaves	1	4	16	64	256	1024	4096	16384

As seen from Table 5.3, limiting the QT depth to six levels increases the number of Level 6 leaves by the number of relegated Level 7 leaves, which now have to be allocated additional reconstruction level coding bits. On the other hand, because of this restriction the $(72.6 + 52.9) = 125.5$ Level 6 leaves do not have to be specifically flagged as leaves in the variable-length QT code, which results in bitrate savings. When using a four-level quantizer for the QT luminance and a bitrate budget of 1000 bits/frame, on the basis of simple logic one would expect a similar performance with and without Level 7 decomposition, since we could save 125.5 QT description bits, due to our *a priori* knowledge that these are all leaf nodes. These bits then could be invested to quantize $INT[125.5/2] = 62$ newly created leaf nodes using 2 bits per luminance value. Nonetheless, the 1 dB PSNR reduction encountered due to removing Level 7 from the QT suggested that the adaptive codec reacted differently,

---

[1]The "Miss America" sequence encoded at 11.36 kbit/s can be viewed under the following WWW address: http://www-mobile.ecs.soton.ac.uk

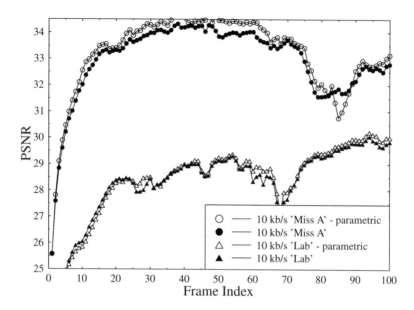

Figure 5.5: PSNR performance of the QT codec with and without parametrical enhancement at 10 kbit/s.

although a complexity reduction was achieved. Hence, if video quality was at premium, the seventh QT level was necessary to develop important fine details in the frame.

The achievable PSNR performance of the proposed QT codec at various bitrates is characterized by Figure 5.5 in the case of the "Lab" and "Miss America" sequences. Furthermore, PSNR versus frame-index performance of the model-based parametric coding enhancement (to be introduced in Section 5.4) is also portrayed in the figure, although its objective PSNR effects appear to be mitigated by the relatively small area of the eyes and lips, where it is applied. (A clearer indication of its efficiency will be demonstrated in Figure 5.12, where its PSNR improvement will be related to its actual area.) Having described the proposed QT codec, let us now discuss the proposed parametric enhancement scheme.

5.4 Model-based Parametric Enhancement

Although in videophony the high-quality representation of the subjectively most important eye and lip regions is paramount, owing to the paucity of bits imposed by the system's bandlimitation this cannot always be automatically ensured. This is because the QT cannot be fully developed down to the pixel level, that is, to Level 7, even if the eye and lip regions were easily and unambiguously detectable and the total bitrate budget of 1000 bits would be invested in these regions. When using 1000 bits/frame, only too small a fraction of the 25 384-pixel frame could be adequately modeled. Hence, in these highly active spots, we propose to employ model-based parametric coding enhancement. The penalty associated with this technique is that the decoder needs to store the eye and lip codebooks. Alternatively, these codebooks must be transmitted to the decoder during the call-setup phase.

5.4. MODEL-BASED PARAMETRIC ENHANCEMENT

Algorithm 3 *This algorithm summarizes the parametric coding enhancement steps.*

1. Generate the binary image using smoothing, frame-differencing, and thresholding (see Figure 5.6).
2. Identify the axis of symmetry yielding the maximum number of symmetric pixels (see Equation 5.4).
3. Identify the position of eyes, nostrils, lips, and nose using the scaled template (see Figure 5.7).
4. Find the best matching eye and lip codebook entries and send their position, luminance shift, and codebook index.

Note, however, that the proposed technique is general and that it can be invoked in conjunction with any other coding technique, such as the DCT- and VQ-based codecs of the previous two chapters.

In the next two sections we describe our approach to the eye and mouth detection and the parametric codebook or database training, while Section 5.4.3 highlights the encoding process.

5.4.1 Eye and Mouth Detection

Eye and mouth detection has been the aim of various research interests such as model-based coding, vision-assisted speech recognition [251], and lip-reading [252]. A reliable detection of the eye and mouth location is the most crucial step for all of the above techniques, and a simple procedure is proposed in the following detection steps, which are also summarized in Algorithm 3.

Step 1: It is critical for the reliable operation of the proposed eye-mouth detection algorithm that the face is the largest symmetrical object in the image frame and that its axis of symmetry is vertical. Initially, we generate a black and white two-tone image from the incoming video frame in order to detect this symmetry and to simplify the detection process. This two-tone image is free from gradual brightness changes but retains all edges and object borders. This was achieved by smoothing the picture using a simple two-dimensional averaging Finite Impulse Response (FIR) filter and then subjecting the smoothed image to frame-differencing and thresholding. Our experiments were based on 3×3- and 5×5-order simple two-dimensional averaging filters, where all 9 or 25 filter coefficients were set to $\frac{1}{9}$ and $\frac{1}{25}$, respectively, in order to preserve energy. The filtered image was then subtracted from the incoming frame and finally thresholded, which led to a binary image $f_{bin}(x, y)$, similar to the one shown at the left-hand side (LHS) of Figure 5.6. A threshold of 8.0 has been found suitable for this operation. This concludes the first step of the parametric coding algorithm summarized in Algorithm 3.

Step 2: The next step is to find the axis of symmetry for the face. Assuming tentatively that the axis of symmetry is x_0, symmetry to this axis is tested by counting the number of symmetric pixels in the two-tone image. The specific x_0 value yielding the highest number of symmetric pixels is then deemed to be the axis of symmetry. Once this axis is known, the pixel-symmetric two-tone image $f_{sym}(x, y)$ at the right-hand side (RHS) of Figure 5.6 is

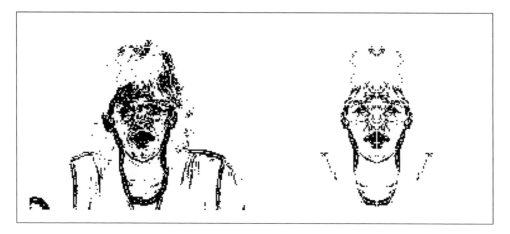

Figure 5.6: Binary (LHS) and binary-symmetric (RHS) frame of the "Miss America" sequence generated by Steps 1 and 2 of Algorithm 3.

generated from the binary image $f_{bin}(x, y)$ as follows:

$$f_{sym}(x, y) = f_{bin}(2x_0 - x, y) f_{bin}(x, y). \tag{5.4}$$

Step 3: The pixel-symmetric image is then used to localize the eyes and the mouth. Initially, we attempted to locate the eye and the mouth as the two most symmetrical objects, assuming a given axis of symmetry x_0 in the symmetric frame $f_{sym}(x, y)$ at the RHS of Figure 5.6. This technique resulted in a detection probability of around 80% for the "Miss America" and "Lab" sequences. An object was deemed to be correctly detected if its true location in the original frame was detected with a precision of +/- 4 pixels. In most of the cases, the location of the eyes was correctly identified, but often the algorithm erroneously detected the chin as the mouth.

In a refined approach, we contrived a more sophisticated template, which consisted of separate areas for the eyes and the nose and a combined area for the nostrils and the mouth, as seen at the LHS of Figure 5.7. We expect many "contrast pixels" in the binary symmetric image $f_{sym}(x, y)$ at the location of the eyes, mouth, and nostrils, while the nose is not conspicuously represented in Figure 5.6. The template was scalable in the range of 0.8 to 1.1 in order to cater for a range of face sizes and distances measured from the camera. This template was then vertically slid along the previously determined axis of symmetry, and at each position the number of symmetric pairs of contrast pixels appearing at both sides of the axis of symmetry within the template overlaid on the binary symmetric image $f_{sym}(x, y)$ of Figure 5.6 was determined. An exception was the nose rectangle, where the original luminance change was expected to be gradual, leading to no contrast pixels at all. This premise was amalgamated with our identification procedure by reducing the total number of symmetric contrast pixels detected within the confines of the template by the number of such contrast pixels found within the nose rectangle. Clearly, a high number of symmetric contrast pixels in the currently assumed nose region weakened the confidence that the current template position was associated with the true position. Finally, the template location resulting in the highest number of matching symmetric pixels was deemed to be the true position of the template.

5.4. MODEL-BASED PARAMETRIC ENHANCEMENT

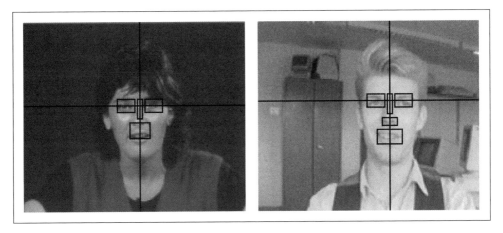

Figure 5.7: Image frames with overlaid initial (left) and improved (right) templates used in Algorithm 3.

Although through these measures the eye and lip detection probability was increased by another 10%, in some cases the nostrils were mistaken for the mouth. We further improved the template by adding a separate rectangular area for the nostrils, as depicted at the RHS of Figure 5.7. This method reached a detection probability of 97%.

In our previous endeavors, the eye and lip detection technique outlined was tested using image sequences, where the speaker keeps a constant distance from the camera and so the size of the face remains unchanged. However, the algorithm can adapt to the more realistic situation of encountering a time-variant face size by scaling the template. The magnification of the template was allowed to vary in the range of 0.8 to 1.1 due to limiting the associated detection complexity. The algorithm was tested using various head-and-shoulder sequences, and we found that it performed well, as long as the speaker was sufficiently close to the camera. Even the fact that a speaker wore glasses did not degrade the detection performance, although difficulties appeared in the case of individuals with a beard. In these situations, the parametric QT codec enhancement had to be disabled on grounds of reduced MSE performance, which was signaled to the decoder using a one-bit flag. Again, the above parametric coding steps are summarized in Algorithm 3; Step 4 will be discussed in more detail later in this chapter.

5.4.2 Parametric Codebook Training

A critical issue as regards the parametric codec's subjective performance is the training of the eye and lip codebooks. Large codebooks have better performance and higher complexity than small ones. The previously described eye and lip identification algorithm was also invoked for training the codebooks, which can then be manually edited in order to remove redundant entries and thus reduce the codebook search complexity. Initially, we generated a single codebook for the eyes and derived the second eye codebook by mirroring the captured codebook entries. As this could lead to pattern-matching problems, when the head was inclined, we proceeded with separate codebooks for each eye. A sample codebook of 16 eye

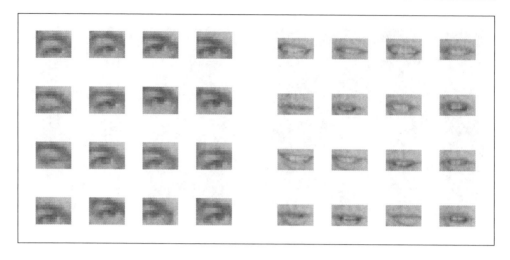

Figure 5.8: Eye and lip codebooks for the parametric coding of the "Lab" sequence.

and lip entries for the "Lab" sequence is portrayed in Figure 5.8, where the contrast of the entries was enhanced for viewing convenience. Let us now consider the parametric encoding algorithm itself.

5.4.3 Parametric Encoding

Step 4: The parametric encoder detects the eyes and mouth using the template-matching approach described in Algorithm 2. Once the template position is detected, the codec attempts to fit every codebook entry into the appropriate locations by sliding it over a window of $+/-3$ pixels in each spatial direction. The MSE for each matching attempt corresponding to the full search of the codebook and to all the legitimate positions is compared, and the best-matching entry is chosen. A luminance shifting operation allows the codec to appropriately adjust the brightness of the codebook entries as well. This is necessary, because the brightness during codebook generation and encoding may differ. An example of the subjective effects of parametric coding and enhancement (PC) is portrayed in Figure 5.9, where we attempted to match the eye and mouth entries of the "Miss America" sequence into the "Lab" sequence in order to assess the codec's performance outside the training sequence. This scenario would represent the worst-case situation, when a prestored parametric codebook is used, which does not contain entries from the current user. A better alternative is to train the codebook during the call setup phase. Observe that there are no annoying artifacts, although the character of the face appears somewhat different. Hence, if the system protocols allow, it is preferable to opt for codebook training before the commencement of the communication phase, which would incur a slight call setup delay.

The optional parametric coding (PC) steps were included here for completeness also in the QT-based video codec's schematic in Figure 5.10, if PC affected the video quality advantageously in MSE terms. If the parametric enhancement was deemed to be successful, then for each modeled object the exact position, luminance shift, and codebook index were transmitted. This required, including the 1-bit PC enable flag, a total of 70 bits, as detailed

5.5. THE ENHANCED QT CODEC

Figure 5.9: Parametric coding example: The uncoded "Lab" sequence with the superimposed entries from the "Miss America" eye/lip codebook (LHS) and the original frame (RHS). Both sequences encoded at various bitrates can be viewed under the WWW address http://www-mobile.ecs.soton.ac.uk

Table 5.4: Bit-allocation for the Parametric QT Codec Enhancement

Type	Location of the Object	Lumin. Shift	Codeb. Entry	PC Flag	Total
No. of bits	$3 \times 15 = 45$	$3 \times 4 = 12$	$3 \times 4 = 12$	1	69

in Table 5.4. Assuming arbitrary, independent eye and lip locations, for each object a 15-bit position identifier was required, yielding a total of 45 bits. When using sixteen independent luminance shifts for the three objects, a total of 12 bits was necessary, similarly to the object codebook indices. This 70-bit segment contained some residual redundancy, since the object positions and luminance shifts are correlated both to each other and to their counterparts in consecutive frames, which would allow us to reduce the above 45-bit location identifier to around 32 bits and the total number of bits to about 57, but this further compression potential was not exploited here. Let us now consider the features of our parametrically enhanced QT codec.

5.5 The Enhanced QT Codec

The parametrically enhanced QT codec of Figure 5.10 is based on the same MC compensated approach, as described in the respective sections of Chapter 3. Here we focus on the parametric coding and QT-specific issues. After initializing the reconstructed frame buffers with the intra-coded frame, the PFU of Figure 5.10 step may be invoked, after which the optimal parametric eye and mouth detection takes place. The PC is optional for two reasons. First, PC requires *a priori* knowledge or training of the PC codebooks before the call setup;

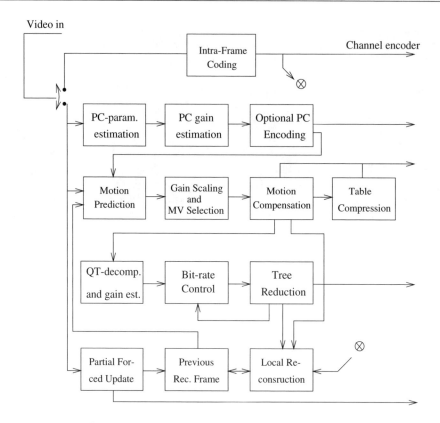

Figure 5.10: Parametrically enhanced QT codec schematic.

and second, the eye and mouth detection might be unsuccessful. In those cases, the encoder sends a single-bit flag in order to signal to the decoder that the PC is disabled. The success of PC is determined by the overall MSE reduction. An increase of the PSNR in the eye–mouth region enabled the PC, whereas a PSNR decrease was interpreted as a false identification of the eye–mouth region and hence the PC-flag sent to the decoder indicated that no PC was in operation. The motion compensation (see Section 3.6) was then invoked, and the MCER was passed to the previously described QT codec.

5.6 Performance and Considerations under Erroneous Conditions

The objective Peak Signal to Noise Ratio (PSNR) at various bitrates of the codec is portrayed in Figure 5.11 both with and without parametric coding enhancement. Observe that, due to the relatively small, 470-pixel area of the eye and lip regions and because of the optional employment of the parametric enhancement, the overall objective PSNR video quality improvement appears more limited than its subjective improvements exemplified

5.6. PERFORMANCE AND CONSIDERATIONS UNDER ERRONEOUS CONDITIONS

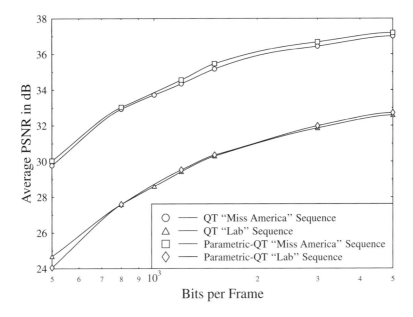

Figure 5.11: PSNR versus the number of bits per frame performance of the proposed QT codec with and without parametric enhancement.

in Figure 5.12.[2] The objective PSNR improvement due to the parametric enhancement is de-weighted by the factor of the template-to-frame area ratio, namely, by $470/(176 \times 144) = 0.0185$. The PSNR improvement of the template area becomes more explicit in Figure 5.13, which is the equivalent of Figure 5.5 related to this smaller 470-pixel image frame section where the PC enhancement is active. Observe in Figure 5.5 that for the "Miss America" sequence PSNR values in excess of 34 dB are possible.

Compared to the codecs based on the VQ and the DCT of the previous two chapters, we found that the QT-based codec required a higher bitrate. The previously described codecs supported bitrates as low as 8 kbit/s. Figure 5.11 reveals that the minimum required bitrate for the QT codec lies around 10 kbit/s.

5.6.1 Bit Allocation

We presented two versions of the DCT and VQ codecs introduced in the previous two chapters. Codec I employed run-length encoded activity tables for the active blocks, while Codec II used simpler, but error-resilient, active block indexing. Explicitly, Codec II exhibited a higher error resilience at the cost of a 30 to 40% higher bitrate. Since the tree code for the QT representation is a variable-length code, its vulnerability is as high as that of the run-length encoded bitstreams of the Type I codecs in Chapters 3 and 4. Here we propose a single QT-based codec, where the MVs are encoded using the active/passive concept from Section 3.7.

---

[2]Both sequences encoded at various bitrates can be viewed under the WWW address http://www-mobile.ecs.soton.ac.uk

Figure 5.12: Comparison of the QT codec without (left) and with (right) PC at 10kbit/s. Both sequences encoded at various bitrates can be viewed under the WWW address http://www-mobile.ecs.soton.ac.uk

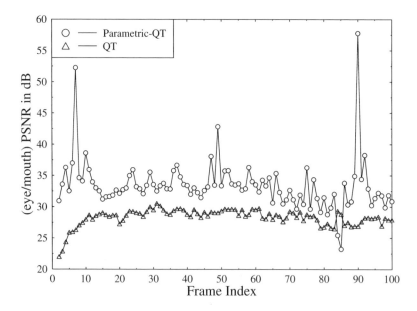

Figure 5.13: PSNR versus frame index performance for the parametrically enhanced Codec 1 in the 470-pixel PC template region using the "Lab" sequence.

5.6. PERFORMANCE AND CONSIDERATIONS UNDER ERRONEOUS CONDITIONS

Table 5.5: Bit-allocation Table for the 11.36 kbit/s QT-codec Using Optional PC Enhancement

Parameter	FAW	MC	PC	PFU	QT	Total
No. of bits	16	< 500	1 or 70	80	> 500	≤ 1136

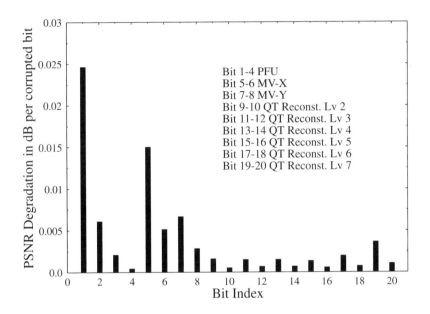

Figure 5.14: Integrated bit sensitivities of the QT codec.

The typical bit-allocation of the codec is summarized in Table 5.5. A total of less than 500 bits is usually allocated to the motion compensation activity table and the motion vectors, 1 or 70 bits are earmarked for parametric coding, 80 bits for partial forced updates, 16 bits for the FAW, while a minimum of 500 bits should be assigned to variable-length QT coding using the previously highlighted QT pruning, as well as to the four-level luminance mean quantization. Let us now briefly consider the bit error sensitivity of the QT codec.

5.6.2 Bit Sensitivity

We employed the technique described in Section 3.12.1 to resolve the integrated bit sensitives for each bit, which are depicted in Figure 5.14. Note that the sensitivity of the four PFU bits and that of the $2 + 2 = 4$ MV bits follow the natural trend from the Most Significant Bits (MSB) toward the Least Significant Bits (LSB). As Figure 5.14 demonstrates, the bit sensitivity of the QT reconstruction level bits increases toward smaller block sizes or deeper QT levels. This is because the average luminance of smaller blocks has a more spread distribution, as we have seen in Figure 5.3, and hence their quantization levels are more sparsely allocated, which results in an increased vulnerability. The variable-length quad-tree code itself can be interpreted as run-length-encoded information, since a corrupted bit affects the entire tree-decoding process.

In [253] we reported our results on a video transceiver designed for the previously described QT codecs, which is discussed next.

5.7 QT-codec-based Video Transceivers

5.7.1 Channel Coding and Modulation

Trellis-coded modulation (TCM) or block-coded modulation (BCM) [222, 254] uses modulation constellation expansion, such as from 2 bits/symbol 4-QAM to 4 bits/symbol 16-QAM, in order to accommodate the parity bits of a half-rate FEC codec without expanding the bandwidth of the transmitted signal. This constellation expansion is invoked in order to "absorb" the channel-coding bits while maintaining the same signaling rate. TCM schemes usually achieve higher coding gain over AWGN channels than isolated coding and modulation, but this is not easily achieved over Rayleigh-fading channels, where coding gains are usually achieved only at quite high channel SNR values. This is a consequence of the hostile channel's tendency to overwhelm the error correction capability of the incorporated FEC scheme, which then often precipitates more errors than the number of errors without FEC. This is particularly true for the inherently reduced-resilience multilevel QAM schemes of [196], which are capable of accommodating the increased number of bits due to TCM. Furthermore, the bits generated by our DCT, VQ, and QT video codecs have unequal bit sensitivities, and therefore unequal protection channel coding [193] must be invoked. In the spirit of [193, 222, 255], here we employ source-sensitivity matched unequal protection joint-source/channel-coding and modulation schemes. The QT-coded video bits are mapped in two different protection classes, namely, Class One and Two. Both streams are binary Bose-Chaudhuri-Hocquenghem (BCH) coded, as detailed in [86] and transmitted using 4- or 16QAM, as described in [196]. We employed simple Time Division Multiple Access (TDMA).

Similarly to our videophone systems designed for the DCT- and VQ-based video codecs in the previous two chapters, in order for the transmitted signal to fit into a GSM-like bandwidth of 200 kHz, while using a Nyquist excess bandwidth of 38.8% corresponding to a Nyquist rolloff factor of 0.388, the signaling rate was limited to 144 kBd. Again, the philosophy of the HSDPA-style [198] reconfigurable source-sensitivity matched transceivers employed complied with that of the DCT- and VQ-based transceivers of the previous two chapters. Coherent detection was achieved using pilot symbol-assisted modulation (PSAM), where channel sounding symbols with known phase and magnitude were inserted in the transmitted TDMA burst in order to inform the receiver about the channel's estimated momentary phase rotation and attenuation. The inverse complex-domain pilot-assisted de-rotation and de-attenuation can then be applied to the data symbols in order to remove the effects of the fading channel.

After extracting the pilot symbols from the received TDMA burst, a complex channel attenuation and phase rotation estimate must be derived for each received symbol, using the simple linear or polynomial interpolation techniques highlighted in [196]. As in the context of the previous two chapters on DCT- and VQ-based HSDPA-style [198] reconfigurable source-sensitivity matched video transceivers, we found that the maximum-minimum distance rectangular 16QAM constellation exhibits two independent 2-bit subchannels, having different BERs. The lower integrity C2 subchannel showed a factor two to three times higher BER than the higher quality C1 subchannel. Although these integrity differences can be fine-tuned using different BCH codes in the subchannels, for our system it was found appropriate to maintain this integrity ratio for the Class One and Class Two QT-coded bits.

5.7.2 QT-based Transceiver Architectures

The QT-based video codecs employed in our videophone transceiver were based on the coding concepts previously outlined earlier in this chapter, and their bit-allocation scheme was similar to that of Table 5.5. In order to comply with the bit-packing constraints of the BCH codecs employed in the videophone transceivers, QT Codec 1 was configured to operate at 11.36 kbit/s, while its slightly modified version, QT Codec 1-bis, generated a bitrate of 11 kbit/s. Again we note that in contrast to the previous Codec I and Codec II schemes of the DCT- and VQ-based systems of the previous two chapters, which represented the more bandwidth-efficient and the more error-resilient codecs, respectively, the QT codec is inherently more vulnerable. This is indicated by our different notation, where we refrained from using the convention Codec I, opting instead for Codec 1 and Codec 1-bis. The video transceiver's schematic is similar to that shown in Figure 3.26.

For the QT-based video codecs Codec 1 and Codec 1-bis, four different transceivers were contrived in order to explore the range of system design trade-offs. The features of these schemes, which are similar to those of the DCT- and VQ-based systems of Tables 3.5 and 4.3 of the previous two chapters, are summarized in Table 5.6. In System 1 we employed the $R = 71/127 \approx 0.56$-rate BCH(127,71,9) code in both 16QAM subchannels. After BCH coding the 11.36 kbit/s QT-coded video sequence generated by Codec 1, the bitrate became $11.36 \times 127/71 = 20.32$ kbit/s. These system features are tabulated in Table 5.6. In System 2, on the other hand, we opted for the stronger BCH(127,50,13) code, which after FEC coding the 11 kbit/s stream produced by Codec 1-bis yielded a bitrate of 27.94 kbit/s. Since System 2 has a higher transmission rate, it will inevitably support a lower number of users than System 1. This comparison will allow us to decide whether it is worthwhile in terms of increased system robustness using a stronger BCH code at the cost of reducing the number of users supported.

The transmission packets used in System 1 are comprised by a Class One BCH(127,71,9) codeword conveyed over the C1 16QAM subchannel, plus a Class Two BCH(127,71,9) codeword transmitted over the C2 subchannel. A BCH(127,50,13) codeword having a stronger error correction capability is assigned to the packet header. The 381-bit packets are converted to 96 16QAM symbols, and 11 pilot symbols are inserted with a pilot spacing of $P = 10$. Lastly, four ramp symbols are concatenated and were invoked in order to allow smooth power amplifier on/off ramping, which mitigates spectral spillage into adjacent frequency bands. Eight 111-symbol packets are required to transmit an entire 1136-bit image frame; hence, the signaling rate becomes 111 symb/12.5 ms $\approx$ 9 kBd. This allows us to support 144 kBd/9 kBd = 16 video users, which is identical to the number of half-rate speech users accommodated by the GSM system [256] often alluded to throughout our system design examples. Hence eight combined video/speech users can be supported. The packet format

Table 5.6: Summary of QT-based Video System Features [253] ©IEEE, 1996, Streit, Hanzo

Feature	System 1	System 2	System 3	System 4
Video Codec	Codec 1	Codec 1-bis	Codec 1	Codec 1-bis
Video rate (kbps)	11.36	11	11.36	11
Frame Rate (fr/s)	10	10	10	10
C1 FEC	BCH(127,71,9)	BCH(127,50,13)	BCH(127,71,9)	BCH(127,50,13)
C2 FEC	BCH(127,71,9)	BCH(127,50,13)	BCH(127,71,9)	BCH(127,50,13)
Header FEC	BCH(127,50,13)	BCH(127,50,13)	BCH(127,50,13)	BCH(127,50,13)
FEC-coded Rate (kbps)	20.32	27.94	20.32	27.94
Modem	4/16-PSAQAM	4/16-PSAQAM	4/16-PSAQAM	4/16-PSAQAM
ARQ	No	No	Yes	Yes
User Signal. Rate (kBd)	18 or 9	12.21 or 24.75	18 or 9	12.21 or 24.75
System Signal. Rate (kBd)	144	144	144	144
System Bandwidth (kHz)	200	200	200	200
No. of Users	8 or 16	5 or 11	6 or 14	3 or 9
Eff. User Bandwidth (kHz)	25 or 12.5	40 or 18.2	33.3 or 14.3	66.7 or 22.2
Min. AWGN SNR (dB) 4/16QAM	7.5/13	7.5/12	6/12	6/11
Min. Rayleigh SNR (dB) 4/16QAM	20/20	15/18	8/14	8/14

5.7. QT-CODEC-BASED VIDEO TRANSCEIVERS

of System 2 is identical to that of System 1, but eleven packets are required to transmit an entire 1100-bit/100 ms frame. Thus, the signaling rate becomes 11×111 symb/100 ms = 12.21 kBd. Then the number of users supported by System 2 at the 144 kBd TDMA rate is INT[144 kBd/12.21] = 11, where INT[$\bullet$] represents the integer part of [$\bullet$].

When the channel signal-to-noise ratio (SNR) becomes insufficient for 16QAM communications, since, for example, the portable station (PS) moves away from the base station (BS), the BS reconfigures the HSDPA-style [198] transceivers as 4QAM schemes, which require twice as many timeslots but a lower SNR value. The 381-bit packets are now converted to 191 4QAM symbols and after inserting 30 pilot symbols and four ramp symbols, the packet length becomes 225 symb/12.5 ms, yielding a signaling rate of 18 kBd. Hence, the number of videophone users supported by System 1 is reduced to eight, as in the full-rate GSM speech channel used in our design examples, which is equivalent to four combined video/speech users. The signaling rate of the 4QAM mode of System 2 becomes 11×225 symb/100 ms = 24.75 kBd, and the number of users is $INT[144/24.75] = 5$.

The system also supports mixed-mode operation, where the more error-resilient 4QAM users must reserve two slots in each 12.5 ms TDMA frame while roaming near the fringes of the cell. In contrast, in the central section of the cell, 16QAM users will only require one slot per frame in order to maximize the number of users supported. Assuming an equal proportion of System 1 users in their 4QAM and 16QAM modes of operation, we see that the average number of users per carrier becomes 12. The equivalent user bandwidth of the 4QAM PSs is 200 kHz/8 = 25 kHz, while that of the 16QAM users is 200 kHz/16 = 12.5 kHz. The characteristics of the whole range of our candidate systems are highlighted in Table 5.6.

Systems 3 and 4 are identical to Systems 1 and 2, respectively, except for the fact that the former schemes can invoke Automatic Repeat Requests (ARQ), when the received video bits are erroneous. In the past the employment of ARQ schemes was limited to data communications [199–202], where longer delays are tolerable than in interactive videotelephony. In modern wireless systems, such as, for example, the 3G systems of [203], however, there is a full duplex control link between the BS and PS, which can be used for message acknowledgments, while the short TDMA frame length ensures a low packet delay. Hence, ARQ can be realistically invoked. Therefore, Systems 3 and 4 are expected to have a higher robustness than Systems 1 and 2, which dispense with ARQ assistance. The penalty must be paid in terms of a reduced number of users supported since the ARQ attempts occupy some of the timeslots.

It is unrealistic to expect the system to operate at such low channel SNR values, where the probability of packet corruption and hence the relative frequency of ARQ attempts is high, since then the number of users supported and the teletraffic carried would become very low. In order to compromise, the number of transmission attempts was limited to three in our system, which required two earmarked slots for ARQ. Furthermore, in any frame only one user was allowed to invoke ARQ, namely, the specific user whose packet was first corrupted within the frame. For the remaining users, no ARQ was allowed in the current TDMA frame. An attractive feature of this ARQ scheme is that if three copies of the transmitted packet are received, majority logic decisions can be invoked on a bit-by-bit basis, should all three received packets be corrupted. The basic features of Systems 3 and 4 are summarized in Table 5.6.

Similarly to our DCT- and VQ-based video transceivers, here we also opted for a twin-class scheme using the BCH(127,71,9) or the BCH(127,50,13) codecs in both the higher

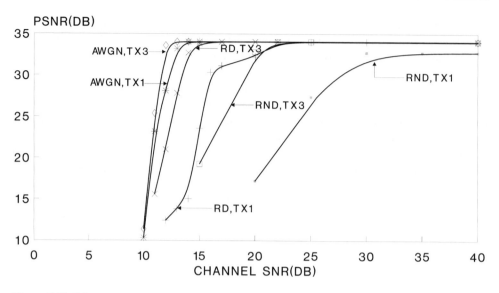

Figure 5.15: PSNR versus channel SNR performance of the 16QAM mode of operation of the QT codec-based video Systems 1 and 3 in Table 5.6 [253]. ©IEEE, 1996, Streit, Hanzo.

integrity C1 and lower integrity C2 16QAM subchannels for the transmission of the more sensitive and less sensitive video bits, respectively. In the case of Codec 1, there are $1136/2 = 568$ Class One and 568 Class Two bits, while for Codec 1-bis there are 550 Class One and Class Two bits and 550, respectively. Although the variable-length QT code bits nearly fill the capacity of the C1 subchannel, up to 68 of the high-sensitivity PFU and MV MSBs of Figure 5.14 can be directed to the higher integrity C1 subchannel.

5.8 QT-based Video-transceiver Performance

Let us now turn to the overall video system performance evaluated in terms of the PSNR versus channel SNR curves depicted in Figures 5.15–5.18 for Systems 1–4. In all the figures, six performance curves are displayed — three curves with ARQ using three transmissions (TX3) and three curves without ARQ, that is, using a single transmission attempt (TX1). Both the 4QAM and 16QAM mode of operation of Systems 1–4 are featured over AWGN channels, Rayleigh-fading channels with diversity (RD) and with no diversity (RND). Perceptually unimpaired video performance was typically achieved by these systems, when the PSNR degradation was less than about 1 dB. Therefore, in Table 5.6 we summarized the minimum required channel SNR values to ensure a PSNR degradation of less than 1 dB, as the systems' operating channel SNR.

The full-scale exploration of the various system design trade-offs using Figures 5.15–5.18 and Table 5.6 is left to the reader for the sake of compactness. We will restrict our comments to a description of some prevalent trends, which in general follow our expectations, rather than comparing each system to its counterparts. As anticipated, the most robust performance was achieved by all systems over AWGN channels, typically requiring a channel

5.8. QT-BASED VIDEO-TRANSCEIVER PERFORMANCE

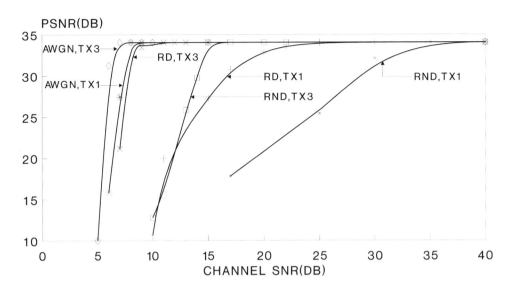

Figure 5.16: PSNR versus channel SNR performance of the 4QAM mode of operation of the QT codec-based video Systems 1 and 3 in Table 5.6 [253]. ©IEEE, 1996, Streit, Hanzo.

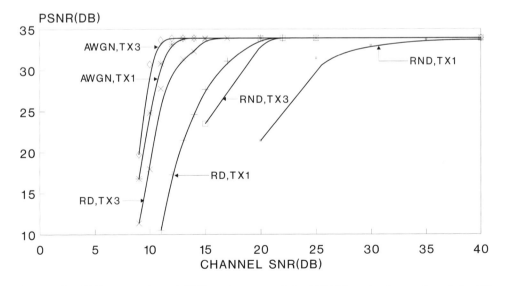

Figure 5.17: PSNR versus channel SNR performance of the 16QAM mode of operation of the QT codec-based video Systems 2 and 4 in Table 5.6 [253]. ©IEEE, 1996, Streit, Hanzo.

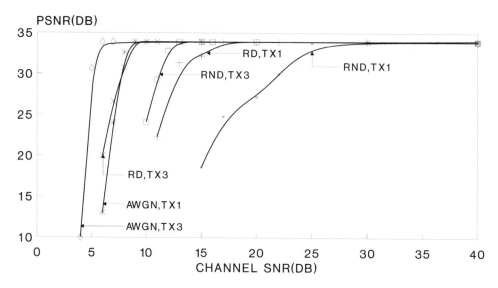

Figure 5.18: PSNR versus channel SNR performance of the 4QAM mode of operation of the QT codec-based video Systems 2 and 4 in Table 5.6 [253]. ©IEEE, 1996, Streit, Hanzo.

SNR of about 11–12 dB, when using the 16QAM mode of operation. In the 4QAM mode, the minimum required channel SNR was around 6–7 dB. The benefits of using ARQ over AWGN channels were limited to a channel SNR reduction of about 1 dB in almost all modes of operation while reducing the number of users supported by two. This was a consequence of the stationary, time-invariant channel statistics, which inflicted similar channel impairments during re-transmission attempts. In contrast, for Rayleigh channels, typically the error-free reception probability of a packet was substantially improved during the ARQ attempts, effectively experiencing time diversity. Hence during re-transmissions the PS often emerged from the fade by the time ARQ was invoked. The employment of the stronger BCH(127,50,13) code failed to substantially reduce the required channel SNRs over Gaussian channels.

Over Rayleigh channels, the effect of diversity was extremely powerful in the case of all systems, typically reducing the minimum channel SNR requirements by about 10 to 12 dB in the 4QAM mode and even more in the 16QAM modes of operation for both the weaker and stronger BCH codes. This was a ramification of removing the residual BER of the modems. In the 4QAM mode without diversity, the stronger BCH(127,50,13) code reduced the channel SNR requirement by about 8 dB and with diversity by 5 dB, when compared to the systems using the weaker BCH code. On the same note, in the 16QAM mode without diversity the BCH(127,50,13) code managed to eradicate the modem's residual BER. With diversity, its supremacy over the higher user-capacity, lower complexity BCH(127,71,9) coded system eroded substantially to about 2 dB. This indicated that in System 1 the weaker BCH code struggled to remove the modem's residual BER, but with diversity assistance the employment of the stronger code became less important.

Similarly dramatic improvements were observed when using ARQ, especially when no diversity was employed or the weaker BCH code was used. Again, this was due to removing the modem's residual BER. Specifically, in the 4QAM mode without diversity and with the BCH(127,71,9) code, an SNR reduction of about 16 dB was possible due to ARQ, which reduced to around 12 dB in case of the stronger code. Remaining in the 4QAM mode but invoking diversity reduced the ARQ gain to about 7 dB for the inherently more robust BCH(127,50,13) code, while the weaker code's ARQ gain in System 3 was only slightly less with diversity than without, amounting still to about 12 dB. Similar findings pertain to the 16QAM ARQ gains when comparing Systems 1 and 3 as well as 2 and 4, which were of the order of 10 dB without diversity and about 5 dB with diversity for both BCH codes.

Both diversity and ARQs mitigate the effects of fading, and in broad terms they achieve similar robustness. However, the teletraffic reduction due to the reserved timeslots is a severe disadvantage of the ARQ assistance. Diversity, on the other hand, can readily be employed in small handsets. When using a combination of diversity and ARQ, near-Gaussian performance was maintained over fading wireless channels effectively by all four systems but, as evidenced by Table 5.6, at a concomitant low carried teletraffic.

Consequently, the number of users supported by the different systems and operating modes varies over a wide range — between 3 and 16 — and the corresponding effective user bandwidth becomes 66.7 kHz and 12.5 kHz, respectively. The improved robustness of some of the ARQ-assisted schemes becomes less attractive in the light of their significantly reduced teletraffic capacity. It is interesting to compare in robustness terms the less complex 4QAM mode of System 1 supporting eight users with the more complex 16QAM mode of System 4, which can accommodate a similar number of users, namely nine. The 4QAM mode necessitates a channel SNR of about 7 dB over AWGN channels, while the more complex 16QAM requires about 11 dB, indicating a clear preference in favor of the former. Interestingly, the situation is reversed over the diversity-assisted Rayleigh channel, requiring SNR values of 20 and 14 dB, respectively. A range of further interesting system design aspects can be inferred from Figures 5.15–5.18 and Table 5.6 in terms of system complexity, robustness, and carried teletraffic.

5.9 Summary of QT-based Video Transceivers

This chapter has presented a range of HSDPA-style [198] reconfigurable QT-based, parametrically enhanced wireless videophone schemes. The various systems contrived are portrayed in Table 5.6, and their PSNR versus channel SNR performances are characterized by Figures 5.15–5.18. The QT codecs contrived have a programmable bitrate, which can vary over a wide range, as demonstrated by Figure 5.11, and hence they lend themselves to a variety of wireless multimedia applications. Because of this video quality/bitrate flexibility, these video codecs are ideal for employment in intelligent HSDPA-style [198] reconfigurable multimode terminals. The proposed video codec can be used for wireless videophony over fixed-rate mobile radio speech channels at a rate of 13 kbit/s, although preferably higher video rates should be maintained in order to ensure a sufficiently high video quality for high-dynamic video sequences. In this respect, the 3G systems [83] having an approximately 5MHz bandwidth and a chip-rate of 3.84 MChip/s are more flexible, since they are capable of supporting higher voice/video bitrates. Furthermore, depending on the spreading factor employed, they are capable of potentially accommodating a significantly higher number of users, although naturally, the video coding rate only depends on the video source codec.

Our future work is targeted at improving the complexity, video quality, and robustness of the video codec as well as the overall system performance by contriving adaptive system reconfiguration algorithms. Let us now briefly summarize our findings presented in the context of proprietary fixed-rate QCIF videophone codecs in this part of the book and compare them to some of the existing standard-based video codecs.

5.10 Summary of Low-rate Video Codecs and Transceivers

In Chapter 2, we studied several fractal codecs in terms of image quality, computational complexity, and compression ratio trade-offs. Two of the candidate codecs, a 0.28 bpp and a 1.1 bpp fractal codec, were also subjected to bit-sensitivity analysis. We noted a drawback of the fractal technique in terms of its incompatibility with motion compensation, which prohibits fractal codecs from achieving high compression ratios.

In Section 3.1, we reviewed a range of popular MC techniques and analyzed their efficiency in light of the bitrate investment required for the MVs. We found that the best performance was achieved when using cost/gain-controlled MV encoding. In Chapter 3, we used these results in order to contrive a fully cost–gain controlled DCT-based coding scheme, which exhibits a constant but programmable bitrate. Two different schemes — the more bandwidth-efficient but more error-sensitive Codec I and the higher bitrate but more robust Codec II — were proposed for various applications. In Chapter 4, we derived a set of VQ codecs, which guaranteed similar or better performances. Lastly, in Chapter 5 we presented a range of quad-tree-based schemes, which incorporated a parametric encoding enhancement option for the critical eye–mouth regions.

All codecs have been subjected to bit-sensitivity analysis in order to assess their performance under erroneous conditions and to assist in the design of source-matched FEC schemes for various channels. The associated transmission issues have been discussed in depth in [151, 195, 219, 253], respectively, for the various DCT-, VQ-, and QT-based video codecs. The performances of the various transceivers contrived were compared in Sections 3.13, 4.5, and 5.7 and in Tables 3.5, 4.3, and 3.5, respectively.

Let us finally compare the DCT-, VQ-, and QT-based codecs proposed to widely used standard codecs, namely, the MPEG-2 [257], H261 [258], and H263 [259] codecs. The latter standard schemes are variable-rate codecs, which make extensive use of variable-length compression techniques, such as RL-coding and entropy coding [10]. The H.261 and H.263 coding family will be detailed in the next part of this book. These standard codecs also require the transmission of intra-frame coded frames, which is carried out at selectable regular intervals. In contrast to our fixed-rate, distributed approach, however, these standard arrangements include a full intra-coded (I) frame, yielding a regular surge in the bitrate. This is unacceptable in fixed-rate mobile radio systems. In addition to the I frames, the MPEG-2 codec uses two more modes of operation, namely, predicted frames (P) and bidirectional (B) coding modes. Motion estimation is used in both frame types, while the P frames rely on additional differential coding strategies, invoked with reference to the surrounding I and B frames.

In the experiments we portrayed in Figure 5.19, we stipulated a fixed bitrate of 10 kbit/s for our three prototype codecs and adjusted the parameters of the H261, H263, and

5.10. SUMMARY OF LOW-RATE VIDEO CODECS AND TRANSCEIVERS 167

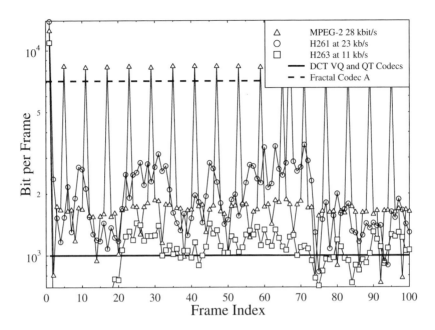

Figure 5.19: Bitrate comparison of the previously presented proprietary 10 kbit/s codecs and three standard codecs, namely, H.261, H.263, and MPEG-2.

MPEG-2[3] codecs, in order to provide a similar video quality associated with a similar average PSNR performance. The corresponding PSNR curves are displayed in Figure 5.20. Observe in Figure 5.19 that the number of bits/frame for our proposed codecs is always 1000 and averages about twice as high at 22 kbit/s for the H261 codec, exhibiting a random fluctuation for the H261 codec. The MPEG codec exhibits three different characteristic bitrates, corresponding to the I, B, and P frames in decreasing order from around 8000 bits/frame to about 1800 and 1300, respectively. The H263 standard superseded the H261 and shows drastic improvements in terms of bitrate at a similar PSNR performance.

A direct comparison of five codecs in Figures 5.19–5.22 reveals the following findings:

- Our programmable-rate proprietary codecs achieve a similar performance to the MPEG-2 codec at less than half the bitrate. The H-261 codec at 22 kbit/s outperforms our codecs by about 2 dB in terms of PSNR. Note furthermore that our fixed-rate DCT and VQ codecs require about 20 frames to reach their steady-state video quality owing to the fixed bitrate limitation, which is slightly prolonged for the QT codec. At the same average bitrate, the H263 standard shows a better PSNR performance than our codecs. The subjective difference seems to be less drastic than the objective measure indicates.

- The delay of our codecs and that of the H-261 codecs is in principle limited to one frame only. The delay of the MPEG-2 and H263 codecs may stretch to several

---
[3]The sequences may be viewed under the World Wide Web address
`http://www-mobile.ecs.soton.ac.uk/jurgen/vq/vq.html`

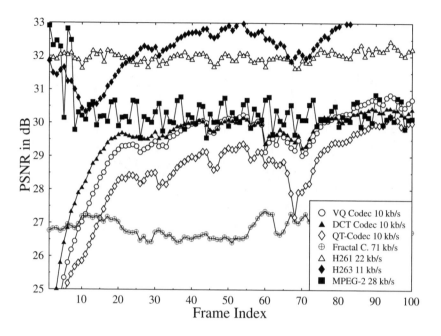

Figure 5.20: Performance of the previously presented proprietary 10 kbit/s codecs and three standard codecs, namely, H.261, H.263, and MPEG-2.

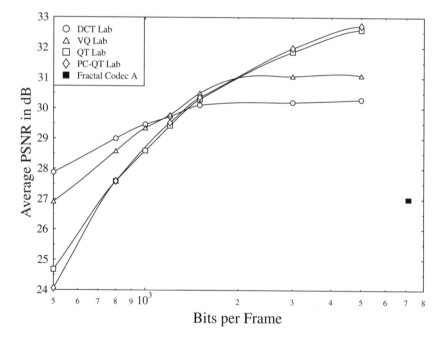

Figure 5.21: Performance versus bitrate comparison of the proposed programmable-rate codecs.

5.10. SUMMARY OF LOW-RATE VIDEO CODECS AND TRANSCEIVERS

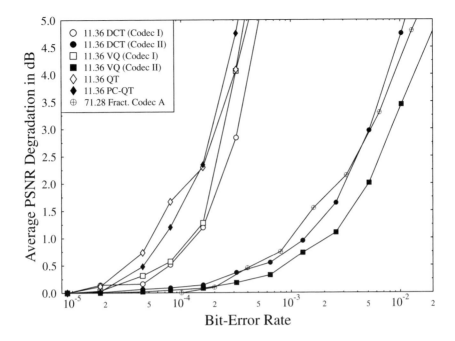

Figure 5.22: PSNR degradation versus BER for various fixed-rate codecs.

frames owing to the P frames. In order to smooth the teletraffic demand fluctuation of the MPEG-2 and H263 codecs, typically adaptive feedback-controlled output buffering is used, which further increases the delay by storing the bits that cannot be instantaneously transmitted in a buffer.

- The error resilience of those specific codecs, which use the active/passive table concept, is very limited, as is that of the standard codecs. These arrangements have to invoke ARQ assistance over error-prone channels or the packet dropping regime of Section 10.3.6. Hence, in these codecs single-bit errors can corrupt an entire frame or several frames in the case of the MPEG-2 codec. Error-resilient techniques such as strong Forward Error Correction (FEC) or Error-Resilient Positional Codes (ERPC) [188, 260] are required to protect such schemes. These problems are avoided by the less bandwidth efficient non-run-length encoded DCT, VQ, and QT schemes, which therefore exhibit a strongly improved error resilience.

In conclusion, compression ratios as high as 200 were achieved for ten frames/s QCIF sequences by the proposed programmable-rate proprietary codecs, while maintaining peak signal-to-noise ratios (PSNR) around 30 dB, at implementational complexities around 20 MFLOP. The fractal and quad-tree codecs were inferior to the DCT and VQ codecs, with the VQ exhibiting a somewhat improved high-frequency representation quality in comparison to the DCT codec. As stated before, the various system design trade-offs become explicit from Sections 3.13, 4.5, and 5.7 and Tables 3.5, 4.3, and 3.5, respectively. In the following part of the book, we will concentrate on the H261/H263 coding family.

Part II

High-resolution Video Coding

Chapter 6

Low-complexity Techniques

Having compared a wide range of QCIF-sized monochrome codecs designed for very low bitrates, we now embark on high-quality color coding of higher resolution pictures spanning the range from CIF to High Definition Television (HDTV) resolution. We will also consider the error resilience of the codecs and attempt to evaluate their suitability for mobile video transmission.

Later in this chapter, we will evaluate two low-complexity methods before we exploit our previous results concerning motion compensation and DCT in the high-quality domain in the context of Chapter 7.

6.1 Differential Pulse Code Modulation

Differential Pulse Code Modulation (DPCM) is one of the simplest techniques in image coding, and though known for a very long time, it still attracts some interest as recent publications [59, 60] demonstrate. It requires virtually no buffering of the incoming signal and only a single multiplication and subtraction step per pixel.

6.1.1 Basic Differential Pulse Code Modulation

DPCM is a predictive technique whose aim is to remove mutual redundancy between adjacent samples. The idea is to predict the current sample i_n from the previous sample i_{n-1}. Instead of transmitting the sample i_n, we transmit the error of the prediction $e_n = i_n - p(i) \cdot i_{n-1}$, where $p(i)$ is the prediction function. If the prediction is successful, the energy $E(e_n)$ of the signal e_n is lower than that of the incoming signal. Hence we can reduce the number of reconstruction levels of the quantizer. The simple schematic of forward and backward predictive coding is depicted in Figure 6.1, and their differences will become explicit during our further discussions. The estimates of the feedforward scheme are based on unquantized input samples, which — following a somewhat simplistic argument — would result in better predictions at the encoder than the estimates of the backward predictive scheme that are based

Video Compression and Communications Second Edition
L. Hanzo, P. J. Cherriman and J. Streit © 2007 John Wiley & Sons, Ltd

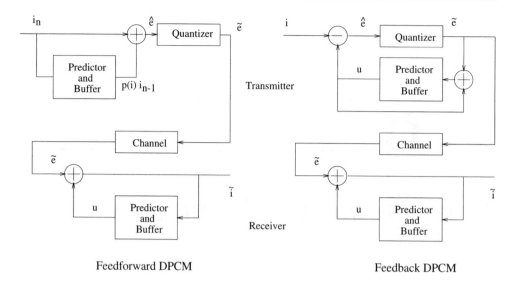

Figure 6.1: Feedforward (left) and feedback (right) DPCM schematic.

on quantized pixels. However, since unquantized pixels are not available at the decoder, the forward predictive scheme is typically outperformed by the backward predictive arrangement.

In video coding the prediction function $p(i)$ is often constituted by a constant a, which is referred to as a predictor coefficient. Hence the prediction error signal is defined as $e_n = i_n - a \cdot i_{n-1}$. If we aim to minimize the energy of e_n given by:

$$E(e_n) = E(i_n - a \cdot i_{n-1}) = \frac{1}{m} \sum_{n=1}^{m} (i_n - a \cdot i_{n-1})^2 \qquad (6.1)$$

we set the first derivative of Equation 6.1 with respect to a to zero, leading to:

$$\frac{\delta E(e_n)}{\delta a} = -2 \cdot \frac{1}{m} \sum_{n=1}^{m} (i_n - a \cdot i_{n-1}) i_{n-1}$$

$$= -2 \cdot \left[\frac{1}{m} \sum_{n=1}^{m} i_n i_{n-1} - a \cdot \sum_{n=1}^{m} i_{n-1}^2 \right] = 0. \qquad (6.2)$$

If we define the autocorrelation of i_n as $\rho_x(i_n)$, this finally leads to:

$$2 \cdot [\rho_1(i_n) - a\rho_0(i_n)] = 0 \implies a = \frac{\rho_1(i_n)}{\rho_0(i_n)}, \qquad (6.3)$$

which is the one-step correlation of the video signal. The prediction gain of the operation is obtained by evaluating Equation 6.1 for the optimum predictor coefficient a, which is given by:

$$gain[dB] = 10 \cdot \log_{10} \frac{E(i_n)}{E(e_n)} = 10 \cdot \log_{10} \frac{1}{2(1-a^2)} \qquad (6.4)$$

6.1. DIFFERENTIAL PULSE CODE MODULATION

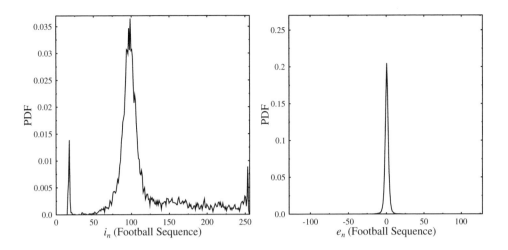

Figure 6.2: PDF of a typical frame (left) and the PDF of the prediction error signal (right).

Typical values for the adjacent-pixel correlation a in the case of images are in the range of $[0.9..0.99]$, resulting in a prediction gain ranging from 4.2 dB to 14 dB. The benefits of DPCM are exemplified in Figure 6.2, where we demonstrate how well this simple predictive technique compresses the widely spread PDF of the incoming video signal i_n shown at the left of the figure to values between -5 and 5 seen at the right of Figure 6.2. Note that the value of the parameter a affects the error resilience of the codec due to the predictive nature of the scheme. In the extreme case of $a = 1.0$, which is equivalent to using the previous pixel value to predict the current one, the reconstructed signal is fed back in the predictive loop without amplitude attenuation. Thus, eventual channel errors will be accumulated, which continuously affects the result. Practical values of a in video coding are in the range of $a \in [0.85..0.95]$, which assists in reducing the error propagation.

The disadvantage of feedforward prediction is that the quantization error results in a misalignment of the encoder's and decoder's reconstruction buffers because the decoder predicts the signal on the basis of the quantized signal. In order to overcome this problem, feedback prediction combined with local decoding is preferred in most applications, as seen at the right-hand side of Figure 6.1, unless the prediction residual can be noiselessly encoded. A simple possible implementation of such a forward predictive codec is based on the fact that the prediction residual signal is represented by a finite set of integer values, which can be noiselessly entropy coded.

6.1.2 Intra/Inter-frame Differential Pulse Code Modulation

In order to achieve a reasonable compressed bitrate below 4.5 bpp, we require coarse quantizers with eight or fewer reconstruction levels. Initially, we collected training data from our test sequences and trained a Max-Lloyd quantizer [10] for various quantizer resolutions.

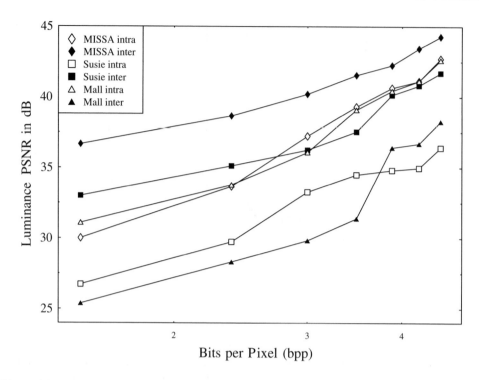

Figure 6.3: Intra-frame and inter-frame DPCM coding performance for various bitrates using symmetrical and asymmetrical quantizers for color images.

This process was carried out using both intra-frame and inter-frame training sequences. However, we found that extreme positive and negative samples of the training prediction error data resulted in inadequate quantizers, since the quantizer steps were too large. For a limited number of quantizers, it constituted no problem to find better quantizers by manually adjusting the reconstruction and decision levels, which led to the results depicted in Figure 6.3, where we optimized the quantizers for the "Miss America" sequence in case of both intra- and inter-frame coding. Figure 6.3 reveals that for intra-frame coding, quantizers with three to eight levels resulting in a bitrate of $1.5 \cdot \log_2(3) = 2.38$ to $1.5 \cdot \log_2(8) = 4.5$ bits per pixel are necessary to achieve a good quality, where the 1.5 multiplier is due to the 50% increase required by the subsampled color information. Furthermore, the video performance depends on the training sequence and the sequence to be coded. Observe, for example, that the performance results for the "Mall" sequence were significantly reduced in comparison to the "Miss America" based results.

We obtained similar results for inter-frame coded images. We opted to use simple frame-differencing as motion compensation and retrained the quantizers, since the statistics of the frame difference signal are different from those of the video signal itself. As demonstrated by Figure 6.3, the PSNR increases in excess of 5 dB are possible in case of the "Susie" and "Miss America" sequence, which are similar. Again we found that the quantizers derived were suboptimum for other sequences, such as the "Mall" sequence.

6.1.3 Adaptive Differential Pulse Code Modulation

The performance of DPCM can be improved if we use a more sophisticated predictor. Jayant [261] was the first to design quantizers, which are adaptive to the incoming samples, leading to the concept of adaptive DPCM or ADPCM. The idea is that the input data is observed, and if, for example, the quantizer is continuously overloaded, since the encoded samples are always assigned the highest reconstruction level, the quantizer step size will be scaled up by a certain factor. Conversely, the quantizer will be scaled down in case of the complementary event. The predictor adaptation is based on the reconstructed samples; hence, no additional side information has to be transmitted.

The prediction might be based on the previously transmitted sample [261] or, more often, on several surrounding pixels, as, for example, suggested by [262–264]. Below we illustrate the performance of the simplest case of predicting the quantizer step-size scaling on the basis of the previous sample. Remembering that a three-level quantizer results in a color-coded bitrate of more than 2.4 bpp when using bit packing or 3 bpp without, we attempted to investigate whether it was possible to improve the coding performance at this bitrate according to our required quality. Optimum step-size gain factors for Gaussian sources have been derived by Jayant [261], but because of the dissimilarity of standard distributions and that of images, an empirical solution is preferred. We forced the quantizer step size to be scaled up by a given scaling factor or to be scaled down by the reciprocal value. The results of the corresponding experiments show that a PSNR improvement in excess of 4 dB is possible, as evidenced by Figure 6.4. However, remembering the poor performance of the simple DPCM codec in Figure 6.3, we concluded that despite its implementational simplicity, DPCM and ADPCM are not suitable for high-quality, high-compression video coding. The possible PSNR improvements due to using more sophisticated two-dimensional predictors are limited to the range of about 1 dB [262] and so will not result in sufficient improvements necessary for this technique to become attractive for our purposes. Let us now concentrate on another low-complexity technique referred to as *block truncation coding*.

6.2 Block Truncation Coding

Block Truncation Coding (BTC) attracted continuous interest from researchers after it was proposed by Mitchell and Delp [265]. The BTC algorithm is very simple and requires limited memory, as only a few image scanlines have to be stored before encoding takes place. These are very useful features for real-time encoding of high-resolution image frames, such as HDTV sequences. BTC is also known for its error resilience, since no RL-encoding or similar variable-length coding methods are used.

6.2.1 The Block Truncation Algorithm

BTC is a block-based coding method based on the typical block sizes of 4×4 or 8×8, where every single pixel in the block is quantized with the same two-level quantizer. Hence, the attributes of BTC codecs are strongly dependent on the properties of this two-level quantizer. There are several approaches to designing the BTC quantizers.

According to a simple and plausible approach, the BTC arrangement's reconstruction levels are determined on the basis of assuming an identical mean and variance or first and second moment for the original and encoded blocks. This allows us to derive a set of two

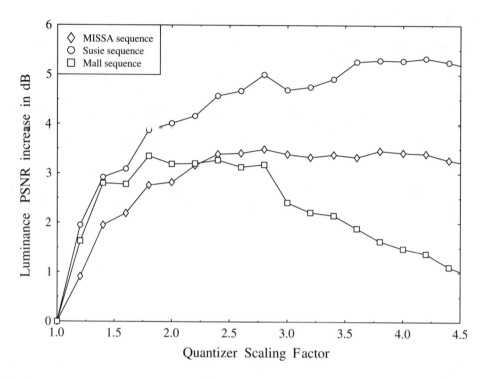

Figure 6.4: Intra-frame coded ADPCM PSNR performance improvement for various quantizer scaling factors.

simultaneous equations for finding the higher and lower reconstruction levels, respectively. Following this approach, we define the ith moment of the block as:

$$\overline{X^i} = \frac{1}{b^2} \sum_{k=1}^{b} \sum_{l=1}^{b} b^i_{(k,l)}, \qquad (6.5)$$

where the first moment is the mean $\overline{X}$ and the second moment is the energy $\overline{X^2}$ of the block to be encoded. Assuming that X_{th} is a given threshold and assigning each pixel to the low or high reconstruction levels q_L or q_H, we define each BTC encoded pixel $\hat{b}_{(k,l)}$ of the block as follows:

$$\hat{b}_{(k,l)} = \begin{cases} q_L & \text{if } b_{(k,l)} < X_{th} \\ q_H & \text{otherwise.} \end{cases} \qquad (6.6)$$

If we aim to preserve the first and second moment $\overline{X^1}$ and $\overline{X^2}$ of the original uncoded block, we obtain a set of two equations [266]:

$$\overline{X} = \frac{1}{b^2} \sum_{k=1}^{b} \sum_{l=1}^{b} \hat{b}_{(k,l)} = \frac{1}{b^2}((b^2 - n)q_L + nq_H) \qquad (6.7)$$

$$\overline{X^2} = \frac{1}{b^2} \sum_{k=1}^{b} \sum_{l=1}^{b} \hat{b^2}_{(k,l)} = \frac{1}{b^2}((b^2 - n)q_L^2 + nq_H^2), \qquad (6.8)$$

6.2. BLOCK TRUNCATION CODING

Table 6.1: PSNR Upper Bounds for BTC Using Unquantized q_L and q_H Values for the "Susie," "Football," and "Mall" Sequences

Sequence	"Susie"	"Football"	"Mall"
PSNR (dB) 4×4 BTC moment method	37.83	31.80	37.68
PSNR (dB) 4×4 BTC MAD method	38.17	32.14	38.17
PSNR (dB) 8×8 BTC moment method	34.40	28.92	33.70
PSNR (dB) 8×8 BTC MAD method	34.77	29.33	34.04

where n denotes the number of pixels exceeding the threshold X_{th}. Solving Equations 6.7 and 6.8 for q_L and q_H leads to:

$$q_L = \overline{X} - \sigma \sqrt{\frac{n}{b^2 - n}} \qquad (6.9)$$

$$q_H = \overline{X} + \sigma \sqrt{\frac{b^2 - n}{n}} \qquad (6.10)$$

$$\sigma = \sqrt{\overline{X^2} - \overline{X}^2},$$

where σ represents the standard deviation of pixel values in the block. The only unknown variable remains the threshold X_{th}. Delp argued [266] that setting X_{th} to $\overline{X}$ was a plausible choice. In fact, it is not too arduous a task to attempt setting the threshold to all possible levels and evaluate the associated coding performance. When we carried out this experiment, we found that there was only a negligible potential performance advantage due to optimizing the decision threshold.

Another possible technique of designing the BTC reconstruction levels would be to optimize x_L and x_H in the MSE or MAD sense. Accordingly, q_L is set to the median of all pixels below the threshold, and correspondingly, q_H is set to the median of all pixels exceeding the threshold X_{th} as seen in:

$$q_L = \frac{1}{b^2 - n} \sum_{k=1}^{b} \sum_{l=1}^{b} b_{(k,l)} \quad \forall b_{(k,l)} < X_{th} \qquad (6.11)$$

$$q_H = \frac{1}{n} \sum_{k=1}^{b} \sum_{l=1}^{b} b_{(k,l)} \quad \forall b_{(k,l)} \geq X_{th}. \qquad (6.12)$$

We examined the upper bound performance of BTC using both of these methods by leaving the q_L and q_H values unquantized. We obtained the results shown in Table 6.1. The upper bound values for the PSNR are around 38 dB in case of a block size of 4×4 and 34 dB for a block size of 8×8, which shows the limitations for this technique. This is not an impressive PSNR performance, since the binary quantized values q_L and q_H require a bitrate of one bit per pixel. Hence, the minimum bitrate is in excess of 1.5 bits per pixel for color blocks, where the chrominance information is stored at half the luminance resolution. Other optimization methods such as third-moment optimization [267, 268] exhibit similar performances.

Table 6.2: An eight-level (3-bit) Quantizer Reconstruction Table for 4×4 Sized Blocks, Trained for Inter- and Intra-coded Frames

Recon. Lv.	1	2	3	4	5	6	7	8
4×4 Intra q_H	25	71	95	106	125	157	186	222
4×4 Intra q_L	18	46	72	87	98	127	160	203
4×4 Inter q_H	−76	−39	−13	5	13	37	66	103
4×4 Inter q_L	−105	−68	105	−12	−3	14	37	70

6.2.2 Block Truncation Codec Implementations

The bitstream generated by the BTC codec contains two different types of data:

(1) the 1 bit per pixel bit map conveying the q_L and q_H information as to whether a given pixel is above or below the threshold; and

(2) the quantized values of either σ and $\overline{X}$ or q_L and q_H.

A large body of research has been devoted to determining the optimum selection of quantizers [266], and Delp found that allocating more than 8 bits for each quantized value results in no significant performance improvement. In a second step, Delp jointly quantized σ and $\overline{X}$ and achieved satisfying BTC performance results at around 10 bits per σ and $\overline{X}$ pair. This bit assignment results in a bitrate of

$$\frac{4 \times 4 \text{ bit} + 10 \text{ bit}}{4 \times 4 \text{ pixel}} = 1.625 \text{ bpp}$$

for monochrome 4×4 blocks and

$$\frac{8 \times 8 \text{ bit} + 10 \text{ bit}}{8 \times 8 \text{ pixel}} = 1.16 \text{ bpp}$$

for 8×8 monochrome blocks, respectively. In case of color images and subsampled Cr_u and Cr_v planes, these values increase by a factor of 1.5 to 2.44 bpp and 1.73 bpp, respectively.

6.2.3 Intra-frame Block Truncation Coding

In our further investigations, we initially focused on intra-frame BTC schemes and opted for a direct quantization of q_L and q_H. We collected sample data from three image sequences separately for our two different block sizes of 8×8 and 4×4 and used appropriately trained Max-Lloyd quantizers [269] for various resolutions ranging from 2 to 8 bits. The optimized Max-Lloyd quantizers for q_H and a block size of 4×4 are portrayed in Figure 6.5. As an example, the set of q_L and q_H reconstruction levels obtained for the 3-bit, eight-level quantizer is summarized in Table 6.2. Observe in the figure that the quantizers are very similar to simple linear quantizers in the range of $[0..255]$. Therefore, if the training

6.2. BLOCK TRUNCATION CODING

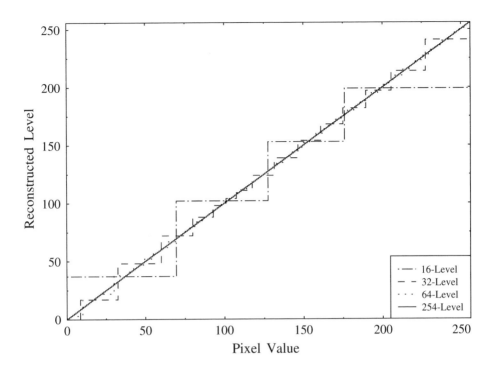

Figure 6.5: Trained Max-Lloyd quantizers characteristics for q_H at a block size of 4×4 for various quantizer resolutions.

sequence is large and statistically independent from the source sequences, the quantizers for BTC become almost uniform quantizers. The quantizers for q_L share the same characteristics with a negative x-axis offset of about one step size.

The results in terms of PSNR versus bitrate performance are shown in Figure 6.6. This figure suggests that the reconstructed quality saturates for quantizer resolutions in excess of 5 bits for q_L and q_H each. This coincides with the estimated quantizer resolution found in [266], where σ and $\overline{X}$ were jointly quantized using 10 bits. In contrast to the above jointly quantized scheme, in our implementation of the BTC codec the q_L and q_H values are obtained from a simple lookup table, which does not require the evaluation of Equations 6.9 and 6.10 and therefore exhibits a reduced complexity. Furthermore, the error resilience is improved because a bit error for jointly quantized values will affect the entire block, whereas in the latter case only n or $b^2 - n$ pixels are affected, depending on whether q_L or q_H was corrupted. The optimum trade-off in terms of quality and bitrate lies around 2.6 bpp for a block size of 4×4, resulting in a PSNR of up to 38 dB. For 8×8 blocks the optimum coding rate is around 1.8 bpp, and the associated maximum PSNR was around 34 dB. These values are very close to the upper bounds reflected in Table 6.1 and hence underline that our quantization method is quite efficient. On the other hand, we will see in Chapter 7 that the codec is readily outperformed by a DCT-based codec. Let us now consider the performance of inter-frame BTC schemes.

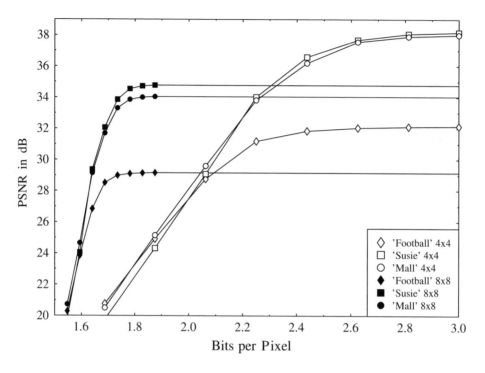

Figure 6.6: PSNR versus bitrate intra-frame coded BTC codec performance for block sizes of 4 and 8 × 8 using the 4CIF and 16CIF sequences of Figures 1.1 and 1.3.

6.2.4 Inter-frame Block Truncation Coding

BTC coding is one of the few coding techniques that does not inflict any spatial frequency limitations. Other methods, such as DCT or Subband Coding (SBC), capitalize on the fact that most of the energy of a given source tends to reside in the lower spatial frequency domain. Therefore higher resolution quantizers are allocated in the lower frequency band in subband coding or for the low-frequency DCT coefficients in DCT coding. BTC may be considered an amplitude-variant, frequency-invariant filter. This is explained by the fact that any two-level black-and-white pattern will be reconstructed as a two-level pattern, which differs from the original one only by an offset and a scaling factor. On the other hand, blocks that contain more than two different pixel values will contain an irreversible coding error, even if σ and $\overline{X}$ are not quantized. This is the reason for the relatively low PSNR upper bounds seen in Table 6.3.

In inter-frame coding, we are encoding a highly nonstationary MCER signal, and hence we have to retrain the quantizers on this basis. A typical 3-bit BTC reconstruction level quantizer is exemplified in Table 6.2 for both intra- and inter-frame coding. Surprisingly, the ranges of the quantizers for inter- and intra-frame coding are rather similar. One might expect that the quantizer's range is more limited owing to the lower variance and to the more limited-range PDF of the MCER. On the other hand, the PDF of a typical MCER sequence potentially may cover a dynamic range twice as wide as the PDF of a normal image.

6.3. SUBBAND CODING

Table 6.3: PSNR Upper Bounds for Inter-frame BTC Derived by Using the Unquantized q_L and q_H Values for the "Susie," the "Football," and the "Mall" Sequences

Sequence	"Susie"	"Football"	"Mall"
PSNR 4×4 BTC MAD in dB	38.18	31.93	38.00
PSNR 8×8 BTC MAD in dB	34.78	29.31	34.05

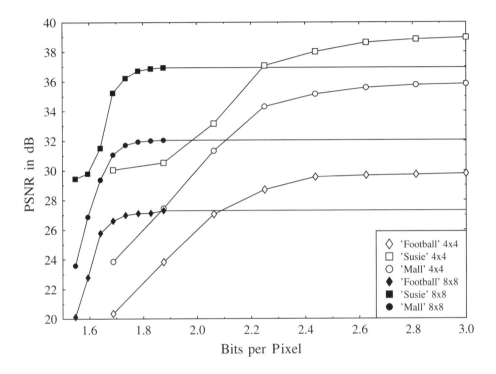

Figure 6.7: PSNR versus bitrate performance of the inter-frame coded BTC codec for block sizes of 4 and 8×8.

These effects seem to cancel each other, when applying a Max-Lloyd quantizer. The PSNR versus bitrate performance of the inter-frame BTC codec is depicted in Figure 6.7, which is slightly worse than that of the intra-frame codec, clearly indicating that the simple BTC algorithm is unsuitable for encoding the nonstationary MCER.

6.3 Subband Coding

Split-band or subband coding (SBC) techniques were first applied in the field of speech coding by Crochière [270, 271]. They became well established in this field before their potential for image coding was discovered [272–275]. Recently, the technique has attracted

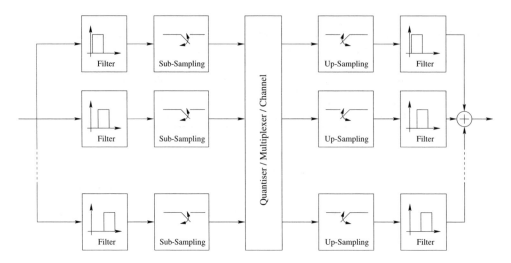

Figure 6.8: Schematic of a split-band codec.

the interest of many researchers in the field of HDTV video coding [276–278], since its complexity is lower than that of other techniques, such as VQ- and DCT-based arrangements.

The concept of subband coding, as depicted in Figure 6.8, is similar to that of transform coding [10, 180] (TC), which was the topic of Chapter 3 in the context of DCT-based compression. As we have seen, in TC-based compression the temporal-domain video signal is transformed to the spatial frequency domain using an orthogonal transformation such as the DCT, where the highest-energy coefficients are quantized and transmitted to the decoder. As seen in Figure 6.8, the subband codec invokes a simpler frequency-domain analysis in that the signal is split by a bank of filters in a number of different frequency bands [279]. These bands can then be quantized independently, according to their perceptual importance. Note that in its simplest implementation this filtering can be carried out after scanning the two-dimensional video signal on a line-by-line basis to a one-dimensional signal, where scanning is seamlessly continued at the end or beginning of each scanline by reading the pixel immediately below the last pixel scanned rather than returning always to the beginning of the next scanline in zig-zag fashion. This scanning pattern ensures that the adjacent pixels are always correlated.

Since the resulting filtered bandpass signals are of lower bandwidth than the original signal, they can be represented using a lower sampling rate than the full-band signal and so we can subsample them. This operation is also shown in Figure 6.8, which is followed by the quantization of the subband signals. This simple frequency-domain analysis allows the codec to identify which frequency band most of the video signal's energy resides in, facilitating the appropriate allocation of the encoding bits to the most important bands. If the full-band image contains low energy in a certain frequency band, the output of the corresponding filter also has low energy and vice versa. The subband signals can be losslessly encoded using entropy-coding schemes [280], which leads to relatively small compression ratios of up to 2.0. Higher compression ratios are possible if we apply quantizers optimized for the individual bands [281]. The original signal is recovered by superimposing the appropriately upsampled and filtered subband signals. The design of the band-splitting filters is crucial, since the

6.3. SUBBAND CODING

overlapping spectral lobes result in aliasing distortion, which cannot be retrospectively removed.

In this chapter, we briefly review the appropriate filtering techniques used to prevent spectral domain aliasing, which are referred to as quadrature mirror filters (QMF). This discussion is followed by the design of the required subband quantizers. Instead of attempting to provide an in-depth treatment of subband coding, here we endeavor to design a practical codec in order to be able to relate its characteristics to those of other similar benchmarkers studied. Hence readers who are more interested in the practical codec design aspects rather than in the mathematical background of QMFs may skip the next subsection.

6.3.1 Perfect Reconstruction Quadrature Mirror Filtering

6.3.1.1 Analysis Filtering

The success of QMF-based band-splitting hinges on the design of appropriate analysis and synthesis filters, which do not interfere with each other in their transition bands, that is, avoid introduction of the **aliasing distortion** induced by subband overlapping due to an insufficiently high sampling frequency, that is, due to unacceptable subsampling. If, on the other hand, the sampling frequency is too high, or for some other reason the filter bank employed generates a spectral gap, again the signal reproduction quality suffers. In simplistic approach this would imply employing filters having a zero-width transition band, associated with an infinite-steepness cutoff slope. Clearly, this would require an infinite filter order, which is impractical. As a practical alternative, Esteban and Galand [282] introduced an ingenious band-splitting structure referred to as a Quadrature Mirror Filter (QMF), and their construction will be detailed later. QMFs have a finite filter order and remove the spectral aliasing effects by cancellation in the overlapping frequency-domain transition bands.

As mentioned before, Quadrature Mirror Filters (QMF) were introduced by Esteban and Galand [282], while Johnston [283] designed a range of QMFs for a variety of applications. The principle of QMF analysis/synthesis filtering can be highlighted following their deliberations and considering the twin-channel scheme portrayed in Figure 6.9, where the subband signals are initially unquantized for the sake of simplicity. The corresponding spectral-domain operations can be viewed in Figure 6.10. Furthermore, for the sake of simplicity here we consider a one-dimensional band-splitting arrangement, assuming that the two-dimensional video signal was scanned in a one-dimensional sequence, as mentioned above.

If most of the energy of the signal is confined to the frequency $f_s/2$, it can be bandlimited to this range and sampled at $f_s = 1/T = \omega_s/2\pi$, to produce the QMF's input signal $x_{in}(n)$, which is input to the QMF analysis filter of Figure 6.9. As seen in the figure, this signal is filtered by the low-pass filter $H_1(z)$ and the high-pass filter $H_2(z)$ in order to yield the low-band signal $x_1(n)$ and the high-band signal $x_2(n)$, respectively. Since the energy of $x_1(n)$ and $x_2(n)$ is now confined to half of the original bandwidths bandwidth of $x(n)$, the sampling rate of the subbands can be halved by discarding every second sample to produce the *decimated signals* $y_1(n)$ and $y_2(n)$.

In the subband synthesis stage of Figure 6.9 the decimated signals $y_1(n)$ and $y_2(n)$ are *interpolated* by inserting a zero-valued sample between adjacent samples, in order to generate the up-sampled sequences $u_1(n)$ and $u_2(n)$. These are then filtered using the

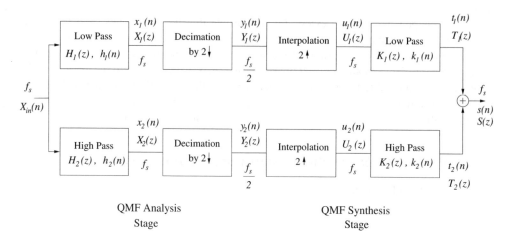

Figure 6.9: QMF analysis/synthesis arrangement.

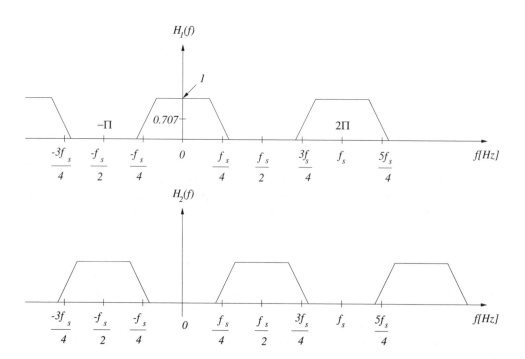

Figure 6.10: Stylized spectral domain transfer function of the lower- and higher-band QMFs.

6.3. SUBBAND CODING

z-domain transfer functions $K_1(z)$ and $K_2(z)$ in order to produce the discrete-time sequences $t_1(n)$ and $t_2(n)$, which now again have a sampling frequency of f_s and the filtering operation reintroduced nonzero samples in the positions of the previously injected zero in the process of interpolation. Finally, the $t_1(n)$ and $t_2(n)$ are superimposed on each other, delivering the recovered video signal $s(n)$.

Esteban and Galand [282] have shown that if the low-pass (LP) filter transfer functions $H_1(z), K_1(z)$ and their high-pass (HP) counterparts $H_2(z), K_2(z)$ satisfy certain conditions, perfect signal reconstruction is possible, provided the subband signals are unquantized. Let us assume that the transfer functions obey the following constraint:

$$\left|H_1\left(e^{j\omega T}\right)\right| = \left|H_2\left(e^{j\left(\frac{\omega_s}{2}-\omega\right)T}\right)\right| \tag{6.13}$$

where ω is the angular frequency, $2\pi = \omega_s$ and the imposed constraint implies a mirror-symmetric magnitude response around $f_s/4$, where the 3 dB-down frequency responses, corresponding to $|H(\omega)| = 0.5$ cross at $f_s/4$. This can be readily verified by the following argument referring to Figure 6.9. Observe that $H_1(\omega)$ is equal to the $\omega_s/2 = \pi$-shifted version of the mirror image $H_2(-\omega)$, which becomes explicit by shifting $H_2(-\omega)$ to the right by $\omega_s/2 = \pi$ at the bottom of the figure. Furthermore, it can also be verified in the figure that by shifting $H_1(\omega)$ to the left by $\omega_s/2 = \pi$ the following relationship holds:

$$\left|H_2\left(e^{j\omega T}\right)\right| = \left|H_1\left(e^{-j\left(\frac{\omega_s}{2}-\omega\right)T}\right)\right|. \tag{6.14}$$

Upon exploiting that:

$$\begin{aligned} e^{-j\left(\frac{\omega_s}{2}-\omega\right)T} &= e^{-j(\pi-\omega T)} \\ &= \cos(\pi - \omega T) - j\sin(\pi - \omega T) \\ &= -\cos(\omega T) - j\sin(\omega T) \\ &= -e^{j\omega T}. \end{aligned} \tag{6.15}$$

Equation 6.14 can also be written as:

$$\left|H_2\left(e^{j\omega T}\right)\right| = \left|H_1\left(-e^{j\omega T}\right)\right|, \tag{6.16}$$

and upon taking into account that $z = e^{j\omega T}$, in z-domain we have $H_1(z) = H_2(-z)$. Following a similar argument, it can also be easily shown that the corresponding HP filters $K_1(z)$ and $K_2(z)$ also satisfy Equation 6.16.

Let us now show how the original full-band signal can be reproduced using the required filters. The z-transform of the LP-filtered signal $x_1(n)$ can be expressed as:

$$X_1(z) = H_1(z)X(z) \tag{6.17}$$

or alternatively as:

$$X_1(z) = a_0 + a_1 z^{-1} + a_2 z^{-2} + a_3 z^{-3} + a_4 z^{-4} + \cdots \tag{6.18}$$

where $a_i, i = 1, 2, \ldots$, are the z-transform coefficients. Upon decimating $x_1(n)$, we arrive at $y_1(n)$, which can be written in z-domain as:

$$Y_1(z) = a_0 + a_2 z^{-1} + a_4 z^{-2} + \cdots, \tag{6.19}$$

where every other sample has been discarded and the previous even samples now become adjacent samples, which corresponds to halving the sampling rate. Equation 6.19 can also be decomposed to the following expression:

$$\begin{aligned} Y_1(z) &= \frac{1}{2}\left[a_0 + a_1 z^{-\frac{1}{2}} + a_2(z^{-\frac{1}{2}})^2 + a_3(z^{-\frac{1}{2}})^3 + a_4(z^{-\frac{1}{2}})^4 + \cdots\right] \\ &+ \frac{1}{2}\left[a_0 + a_1(-z^{-\frac{1}{2}}) + a_2(-z^{-\frac{1}{2}})^2 + a_3(-z^{-\frac{1}{2}})^3 + \cdots\right] \\ &= \frac{1}{2}\left[X_1(z^{\frac{1}{2}}) + X_1(-z^{\frac{1}{2}})\right], \end{aligned} \quad (6.20)$$

which represents the decimation operation in z-domain.

6.3.1.2 Synthesis Filtering

The original full-band signal is reconstructed by interpolating both the low-band and high-band signals, filtering them and adding them, as shown in Figure 6.9. Considering the low-band signal again, $y_1(n)$ is interpolated to give $u_1(n)$, whereby the injected new samples are assigned zero magnitude, yielding:

$$\begin{aligned} U_1(z) &= a_0 + 0 \cdot z^{-1} + a_2 z^{-2} + 0 \cdot z^{-3} + a_4 z^{-4} + \cdots \\ &= Y_1(z^2). \end{aligned} \quad (6.21)$$

From Figure 6.9, the reconstructed low-band signal is given by:

$$T_1(z) = K_1(z) U_1(z). \quad (6.22)$$

When using Equations 6.17 to 6.22, we arrive at:

$$\begin{aligned} T_1(z) &= K_1(z) U_1(z) \\ &= K_1(z) Y_1(z^2) \\ &= K_1(z) \frac{1}{2}[X_1(z) + X_1(-z)] \\ &= \frac{1}{2} K_1(z) [H_1(z) X(z) + H_1(-z) X(-z)]. \end{aligned} \quad (6.23)$$

Following similar arguments in the lower branch of Figure 6.9 as regards the high-band signal, we arrive at:

$$T_2(z) = \frac{1}{2} K_2(z) [H_2(z) X(z) + H_2(-z) X(-z)]. \quad (6.24)$$

Upon adding the low-band and high-band signals, we arrive at the reconstructed signal:

$$\begin{aligned} S(z) &= T_1(z) + T_2(z) \\ &= \frac{1}{2} K_1(z) [H_1(z) X(z) + H_1(-z) X(-z)] \\ &+ \frac{1}{2} K_2(z) [H_2(z) X(z) + H_2(-z) X(-z)]. \end{aligned}$$

6.3. SUBBAND CODING

This formula can be rearranged in order to reflect the partial system responses due to $X(z)$ and $X(-z)$:

$$S(z) = \frac{1}{2}\left[H_1(z)K_1(z) + H_2(z)K_2(z)\right]X(z)$$
$$+ \frac{1}{2}\left[H_1(-z)K_1(z) + H_2(-z)K_2(z)\right]X(-z), \quad (6.25)$$

where the second term reflects the aliasing effects due to decimation-induced spectral overlap around $f_s/4$, which can be eliminated following Esteban and Galand [282], if we satisfy for the following constraints:

$$K_1(z) = H_1(z) \quad (6.26)$$
$$K_2(z) = -H_1(-z) \quad (6.27)$$

and invoke Equation 6.16, satisfying the following relationship:

$$H_2(z) = H_1(-z). \quad (6.28)$$

Upon satisfying these conditions, Equation 6.25 can be written as:

$$S(z) = \frac{1}{2}\left[H_1(z)H_1(z) - H_1(-z)H_1(-z)\right]X(z)$$
$$+ \frac{1}{2}\left[H_1(-z)H_1(z) - H_1(z)H_1(-z)\right]X(-z),$$

simplifying the aliasing-free reconstructed signal's expression to:

$$S(z) = \frac{1}{2}\left[H_1^2(z) - H_1^2(-z)\right]X(z). \quad (6.29)$$

If we exploit that $z = e^{j\omega T}$, we arrive at:

$$S\left(e^{j\omega T}\right) = \frac{1}{2}\left[H_1^2\left(e^{j\omega T}\right) - H_1^2\left(-e^{j\omega T}\right)\right]X\left(e^{j\omega T}\right),$$

and from Equation 6.15 by symmetry we have:

$$-e^{-j\omega T} = e^{j\left(\frac{\omega_s}{2}+\omega\right)T}, \quad (6.30)$$

leading to:

$$S\left(e^{j\omega T}\right) = \frac{1}{2}\left[H_1^2\left(e^{j\omega T}\right) - H_1^2\left(e^{j\left(\frac{\omega_s}{2}+\omega\right)T}\right)\right]X\left(e^{j\omega T}\right). \quad (6.31)$$

6.3.1.3 Practical QMF Design Constraints

Having considered the analysis/synthesis filtering, we see that the elimination of aliasing becomes more explicit in this subsection. Let us now examine how the imposed filter design constraints can be satisfied. Esteban and Galand [282] have proposed an elegant solution in case of finite impulse response (FIR) filters, having a z-domain transfer function given by:

$$H_1(z) = \sum_{n=0}^{N-1} h_1(n)z^{-n}, \quad (6.32)$$

where N is the FIR filter order. Since $H_2(z)$ is the mirror-symmetric replica of $H_1(z)$, below we show that its impulse response can be derived by inverting every other tap of the filter impulse response $h_1(n)$. Explicitly, from Equation 6.28 we have

$$H_2(z) = H_1(-z)$$
$$= \sum_{n=0}^{N-1} h_1(n)(-z)^{-n}$$
$$= \sum_{n=0}^{N-1} h_1(n)(-1)^{-n}z^{-n}$$
$$= \sum_{n=0}^{N-1} h_1(n)(-1)^n z^{-n}, \qquad (6.33)$$

which obeys the above-stated symmetry relationship between the low-band and high-band impulse responses.

According to Esteban and Galand, the low-band transfer function $H_1(z)$, which is a symmetric FIR filter, can be expressed by its magnitude response $H_1(\omega)$ and a linear phase term, corresponding to the filter-delay $(N-1)$, as follows:

$$H_1\left(e^{j\omega T}\right) = H_1(\omega)e^{-j(N-1)\pi(\omega/\omega_s)}. \qquad (6.34)$$

Upon substituting this linear-phase expression in the reconstructed signal's expression in Equation 6.31 and taking into account that $2\pi/\omega_s = 2\pi/(2\pi f_s) = T$, we arrive at:

$$S\left(e^{j\omega T}\right) = \frac{1}{2}\left[H_1^2(\omega)e^{-j2(N-1)\pi(\omega/\omega_s)}\right.$$
$$\left. - H_1^2\left(\omega + \frac{\omega_s}{2}\right) e^{-j2(N-1)\pi\left(\frac{\omega}{\omega_s}+\frac{1}{2}\right)}\right] X\left(e^{j\omega T}\right)$$
$$S\left(e^{j\omega T}\right) = \frac{1}{2}\left[H_1^2(\omega) - H_1^2\left(\omega + \frac{\omega_s}{2}\right) e^{-j(N-1)\pi}\right]$$
$$e^{-j(N-1)2\pi(\omega/\omega_s)} X\left(e^{j\omega T}\right). \qquad (6.35)$$

As to whether the aliasing can be perfectly removed, we have to consider two different cases, depending on whether the filter order N is even or odd.

1. **The filter order N is even.**

    In this case we have:
    $$e^{-j(N-1)\pi} = -1, \qquad (6.36)$$
    since the expression is evaluated at odd multiples of π on the unit circle. Hence, the reconstructed signal's expression in Equation 6.35 can be formulated as:

    $$S\left(e^{j\omega T}\right) = \frac{1}{2}\left[H_1^2(\omega) + H_1^2\left(\omega + \frac{\omega_s}{2}\right)\right] e^{-j(N-1)\omega T} X\left(e^{j\omega T}\right). \qquad (6.37)$$

6.3. SUBBAND CODING

In order to satisfy the condition of a perfect all-pass system, we have maintained:

$$H_1^2(\omega) + H_1^2\left(\omega + \frac{\omega_s}{2}\right) = 1 \tag{6.38}$$

yielding:

$$S\left(e^{j\omega T}\right) = \frac{1}{2}e^{-j(N-1)\omega T} X\left(e^{j\omega T}\right), \tag{6.39}$$

which can be written in the time domain as:

$$s(n) = \frac{1}{2}x(n - N + 1). \tag{6.40}$$

In conclusion, if the FIR QMF filter order N is even, the reconstructed signal is an $N-1$-sample delayed and $\frac{1}{2}$-scaled replica of the input video signal, implying that all aliasing components have been removed.

2. **The filter order N is odd.**

   For an odd filter-order N, we have:

$$e^{-j(N-1)\pi} = 1, \tag{6.41}$$

since the exponential term is evaluated now at even multiples of π; hence, the reconstructed signal's expression is formulated as:

$$S\left(e^{j\omega T}\right) = \frac{1}{2}\left[H_1^2(\omega) - H_1^2\left(\omega + \frac{\omega_s}{2}\right)\right]e^{-j(N-1)\omega T} X\left(e^{j\omega T}\right). \tag{6.42}$$

Observe that due to the symmetry of $H_1(\omega)$ we have $H_1(\omega) = H_1(-\omega)$. Thus, the square-bracketed term becomes zero at $\omega = -\omega_s/4$, and the reconstructed signal $S\left(e^{j\omega T}\right)$ is different from the transmitted signal. As a consequence, perfect-reconstruction QMFs have to use even filter orders.

In conclusion, the conditions for perfect reconstruction QMFs are summarized in Table 6.4. Johnston [283] has proposed a set of perceptually optimized real QMF filter designs, which process real-time signals. A range of complex quadrature mirror filters (CQMFs) potentially halving the associated computational complexity have been suggested by Nussbaumer [284] and Galand [285].

6.3.2 Practical Quadrature Mirror Filters

When designing and implementing practical QMFs having a finite filter order and hence a nonzero width frequency-domain transition region between the frequency-domain passband and stopband, a certain amount of energy will always spill into the neighboring frequency regions, as illustrated in Figure 6.11. Unless special attention is devoted to this problem, it causes aliasing, manifesting itself in terms of video distortion. *Recall from the previous subsection that in case of even filter orders, while assuming infinite filter coefficient precision and unquantized subband signals, the QMF analysis/synthesis scheme of Figure 6.8 facilitates perfect signal reconstruction. However, in reality only the first condition*

Table 6.4: Conditions for Perfect Reconstruction QMF

$H_1(z)$ is a symmetric FIR filter of even order: $h_1(n) = h_1(N - 1 - n), n = 0 \ldots (N - 1)$

$H_2(z)$ is an antisymmetric FIR filter of even order: $h_2(n) = -h_2(N - 1 - n), n = 0 \ldots (N/2) - 1$

Mirror symmetry: $H_2(z) = H_1(-z)$ $h_2(n) = (-1)^n h_1(n)$ $n = 0 \ldots (N - 1)$

$K_1(z) = H_1(z)$

$K_2(z) = -H_2(z)$

All-pass criterion: $\left(H_1^2(\omega) + H_1^2\left(\omega + \dfrac{\omega_s}{2}\right)\right) = 1$

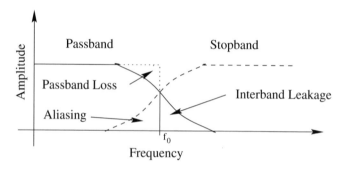

Figure 6.11: Example of two-band split using QMF filters.

is practically realizable, and hence practical subband video codecs often suffer from aliasing effects.

Unfortunately, the perfect reconstruction conditions can only be satisfied for $N = 2$ and $h_{low}(1) = \frac{1}{2}$ [187]. However, this two-tap filter has a very poor stopband rejection and so it often fails to adequately reject the signal energy leakage from the adjacent subband, when the subband signals are quantized, as suggested by the slowly decaying slope of the two-tap QMF in Figure 6.12. (These effects will be studied in the context of Figure 6.14, quantitatively.) The choice of near-perfect QMFs is usually based on a trade-off between a low interband leakage and a low ripple in the passband, a topic extensively studied in [283, 286–288].

QMFs split the band into a higher and lower band of equal bandwidth. They may be applied recursively in order to obtain further subbands. Therefore, they are directly suitable for one-dimensional data, such as speech signals. If only the lower subband is split further in consecutive steps, the resultant scheme is referred to as a tree-structured QMF [289]. In image coding, the source signal is two-dimensional, which has to be split into four components, namely, the vertical and horizontal low and high bands, respectively. In order to solve the problem using a one-dimensional filter set, we may scan the two-dimensional video signal on the basis of scanlines to a one-dimensional signal and then first apply the QMFs horizontally

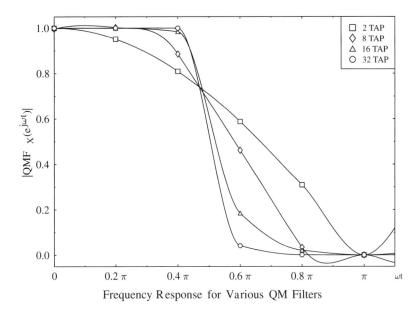

Figure 6.12: Frequency responses for various QM filters.

and obtain the low-horizontal and high-horizontal frequency bands. Each of these two bands is then vertically scanned and again QMF filtered, using the result in all four combinations of horizontal and vertical band combinations. The lowest frequency subband contains a vertically and horizontally low-pass filtered version of the original video frame and therefore contains most of the energy. The efficiency of the spectral analysis and hence that of subband coding may be increased with the aid of further consecutive band-split steps. These are commonly applied to the lowest band only [193], resulting in multiband schemes having 4, 7, 10, or more bands. Other design options are suggested in [290, 291].

A typical example of a 10-band split frame of the "Football" sequence is depicted in Figure 6.13. The benefit of more highly resolved band-splitting is that the codec can take advantage of the individual frequency-domain characteristics of each band. If the interband leakage due to using low-order QMFs is significant, each band will contain aliasing from other bands, which may defeat the benefits of frequency-selective encoding. The two-tap QMF filter is characterized by a weak stopband rejection and hence may not be adequate in certain applications. Higher-order, near-perfect QMFs have a far better stopband rejection, but since they do not satisfy the perfect-reconstruction QMF criteria specified in Table 6.4, they inflict impairments, even when the subband signal components are unquantized. This problem can be overcome, when using relatively long filters, as demonstrated by Figure 6.14 in case of unquantized subband signals for a range of different filter orders.

Specifically, for two-tap QMFs the reconstructed video signal's PSNR exceeds $20 \times \log(256/1) \approx 48$ dB, a PSNR value, which corresponds to the lowest possible error in the reconstructed signal, when using an 8-bit resolution or 256 possible integer values to represent the luminance component. In other words, the PSNR degradation due to this

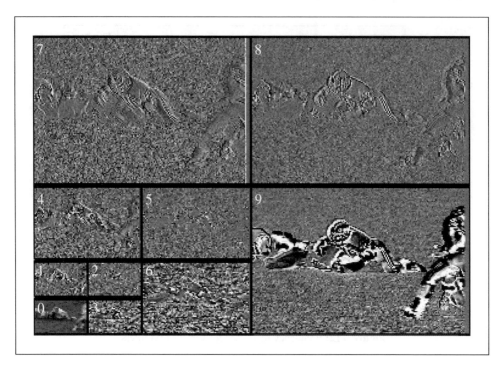

Figure 6.13: Ten-band split example of the 16CIF "American Football" sequence of Figure 1.3. Band 0 represents the lowest horizontal (H) and vertical (V) frequency bands, while band 8 indicates the highest H and V frequency bands. Similarly, band 9 portrays the highest horizontal frequency band associated with low vertical energy contents.

low-order filter in case of unquantized subband signals is very low. For longer near-perfect QMF filters and unquantized subband signals, the PSNR performance depends on the number of band splits, the actual filter length, and the accuracy of the arithmetic operations. Figure 6.14 is based on high-precision 64-bit arithmetics and reveals that a minimum QMF length of 16 taps is necessary for maintaining a relatively low PSNR penalty due to spectral spillage in case of nonperfect QMFs. The actual PSNR achieved is dependent on the number of subbands, but values in excess of about 35 dB are possible at a filter order of 8. For very high-quality coding we propose employing a 32-band filter, which almost reaches the reconstruction qualities of the ideal two-tap filter, achieving PSNRs in excess of about 55 dB. The cost of the improved stopband rejection is a 16-times more complex filter than the perfect-reconstruction two-tap scheme. The employment of long filter impulse responses has been proven to be important in the context of speech coding [283, 292, 293], but we found that in image coding the difference between a two-tap filter and higher-order filters having 16 taps or more may not be very significant. This difference may be attributed to the different sensitivity of the human eye and ear to errors in the frequency domain.

6.3. SUBBAND CODING

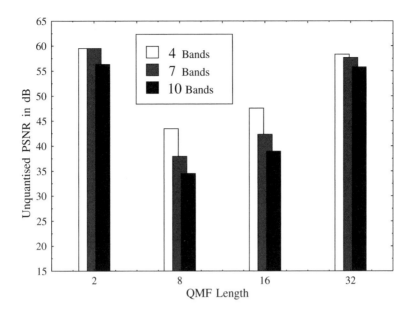

Figure 6.14: Reconstructed video quality versus QMF order without quantization for four-, seven- and ten-band subband intra-frame coding at QMF lengths of 2, 8, 16, and 32 coefficients for the "Susie" sequence using 64-bit number representation accuracy.

6.3.3 Run-length-based Intra-frame Subband Coding

Traditional subband coding as proposed by Gharavi [275] applies a combination of a linear subband quantizer with a dead zone and variable-length coding (VLC). The motivation behind this technique will be detailed later in this chapter. Figure 6.15 shows the PDF of subbands 1 to 9, which exhibit a highly peaked distribution in the range of $[-5..5]$, suggesting the presence of low-energy, noise-like components. Band 0 is not included in the figure, since it contains the lowest frequencies and the DC component. Thus, its PDF is spread across the whole dynamic range of $[0..255]$. Indeed, these noise-like components are typically due to camera-noise, and hence their exact encoding is actually detrimental with regard to the reconstructed video quality. Hence, the potential for compression lies in the fact that quantizers exhibiting a dead zone in this low-amplitude, high-probability region may be applied without significantly impairing the reconstructed video quality. The corresponding quantizer characteristic is shown in Figure 6.16.

Figure 6.17 characterizes the codec's PSNR performance, when experimenting with a variety of dead zone intervals. In this experiment, all values of the higher frequency bands 1 to 9 in the range [−dead zone ... dead zone] were set to zero, whereas all remaining values outside the dead zone were left unquantized. This figure confirms Gharavi's results [275] showing that a dead zone of ± 4 guarantees a good reconstructed video quality having PSNR values in excess of about 35 dB. For very high-quality reproduction, the dead zone has to be of a narrower width, around ± 2. Because of the high-probability subband PDF peaks near zero, a quantizer having such a dead-zone will set the majority of the subband video

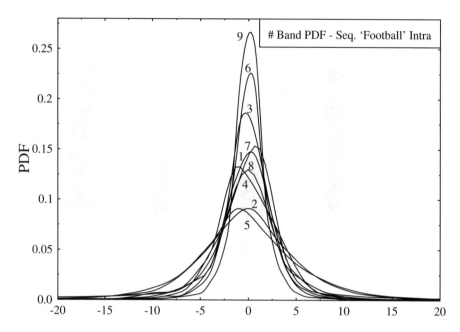

Figure 6.15: PDF of Band 1 to Band 9 of an intra-frame 10-band SBC for the 16CIF "Football" sequence of Figure 1.3.

samples to zero, introducing a grade of predictability concerning the zero sample, which is synonymous with redundancy. This redundancy may be exploited using variable-length codes or run-length codes, as we will show below. Suitable run-length codes are, for example, the B codes [246], which are described in Section 7.4.1.

Following a range of investigations using various numbers of quantization levels as shown in Table 6.5, we concluded that good results are achieved with a 17- or 33-level linear quantizer, characterized by a ± 4 dead zone and reconstruction range of ± 40. This quantizer may be used for all higher frequency bands [294]. The lowest frequency band containing the DC component is characterized by a widely spread PDF, covering the range of [0..255] and hence has been processed by invoking different compression methods, such as vector quantization [294, 295] or DPCM [275]. Because of its reduced size after subsampling, in a 10-band scheme, where there are three consecutive subsampling steps, the bitrate contribution of the lowest subband corresponds to an unquantized data rate of $\frac{1}{4} \cdot \frac{1}{4} \cdot \frac{1}{4} \cdot 8 \text{ bit} = 0.125$ bpp, which is negligible compared to the data rate of the other, less subsampled and hence larger-area bands. Furthermore, the high-quality reconstruction of the lowest band is vital, and hence the potential compression gains in this band seem to be limited. We opted for quantizing the lowest band with an 8-bit linear quantizer in the range of [0..255]. When applying such a scheme, we achieved good PSNR results. The 4CIF "Susie" and the CIF "Miss America" sequences of Figures 1.2 and 1.1 result in PSNR values just below 40 dB. The average entropy (see Chapter 1 of [161]) for all bands remains with 0.70 and 0.79 for the "Susie" and "Miss America" sequences below 1 bpp, similar to those reported earlier by other researchers, for example, [275, 294]. The corresponding PSNR values are around 40 dB

6.3. SUBBAND CODING

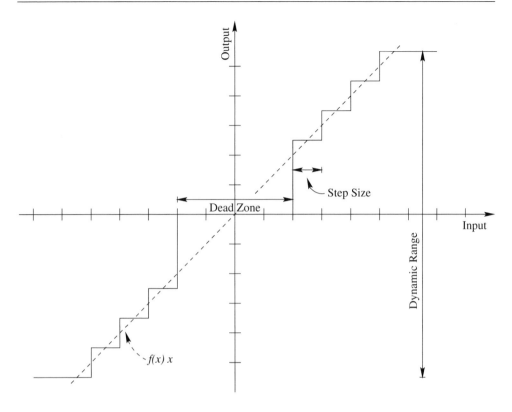

Figure 6.16: Dead-zone based quantizer characteristic.

Table 6.5: Intra-frame Subband Codec Performance for Various Quantizers with a Dead Zone of $[-4\ldots 4]$ and a Dynamic Range of ± 40, for the "Susie" Sequence

Reconstr. Levels	Entropy in bpp	Lumin. PSNR(dB)	Cr_u PSNR(dB)	Cr_v PSNR (dB)
3	0.59	33.00	42.79	42.85
5	0.60	33.85	42.75	42.82
9	0.63	37.29	43.83	43.95
17	0.66	38.90	43.99	44.09
33	0.70	39.46	44.02	44.11

for both the luminance and chrominance values for the sequences tested. These results are summarized along with the associated PSNR values for various CIF to CCIR 601 resolution video sequences in Table 6.6.

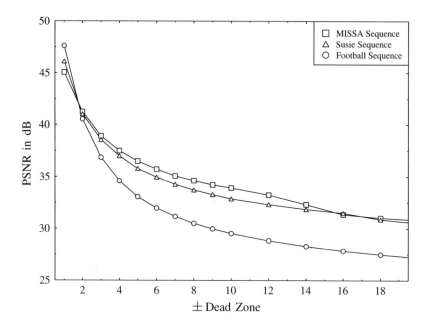

Figure 6.17: Dead zone based intra-frame subband codec PSNR versus dead zone width, setting all values of the higher frequency bands within the dead zone to zero for the 4CIF and 16CIF sequences of Figures 1.1 and 1.3.

Table 6.6: Subband Codec Performance with a 33-Level Quantizer

Sequence	Format	Luminance PSNR (dB)	Cr_u PSNR (dB)	Cr_v PSNR (dB)	% of "0" values	Entr. in bpp
"Miss America"	CIF	39.82	38.40	38.33	86.3	0.79
"Susie"	4CIF	39.46	44.02	44.11	88.4	0.70
"Football"	4CIF	36.23	40.89	40.16	74.5	1.29

6.3.4 Max-Lloyd-based Subband Coding

The drawback of the variable-length coding scheme is its error-sensitive nature, which we attempted to avoid throughout this chapter. Therefore, we endeavored to explore the potential of Max-Lloyd quantizers, individually trained, according to the significance and PDF of each subband. In other words, we attempted to exploit the fact that some bands are perceptually more important than others. We chose the resolution of the quantizer for each band based on a compromise in order to satisfy our bitrate and quality requirements. In order to achieve such a quantizer allocation scheme, we first derived Max-Lloyd quantizers for each band at all typical resolutions, that is, for $2, 4, 8 \ldots 256$ reconstruction levels, one set for each of the luminance and chrominance components.

Initially, we allocated a 1-bit quantizer for the most important band #0, whereas all other bands were quantized to 0. The overall bitrate generated by such a scheme is due to its small

6.3. SUBBAND CODING

Table 6.7: Number of Quantizer Levels per Subband for the 10-Band Scheme and PSNR as Well as Bit Rate in BPP (monochrome) for the 4CIF "Susie" Sequence

Band #	Number of quantizer levels per band										PSNR (dB)	Luminance BPP
	0	1	2	3	4	5	6	7	8	9		
1	2	0	0	0	0	0	0	0	0	0	17.067	0.0156
2	4	0	0	0	0	0	0	0	0	0	19.941	0.0313
3	8	0	0	0	0	0	0	0	0	0	22.642	0.0469
4	16	0	0	0	0	0	0	0	0	0	25.619	0.0625
5	32	0	0	0	0	0	0	0	0	0	26.540	0.0781
6	64	0	0	0	0	0	0	0	0	0	26.972	0.0938
7	64	2	0	0	0	0	0	0	0	0	27.196	0.1094
8	64	4	0	0	0	0	0	0	0	0	27.899	0.1250
9	64	8	0	0	0	0	0	0	0	0	28.323	0.1406
10	128	8	0	0	0	0	0	0	0	0	28.502	0.1563
11	128	8	2	0	0	0	0	0	0	0	28.737	0.1719
12	128	8	4	0	0	0	0	0	0	0	29.204	0.1875
13	128	8	8	0	0	0	0	0	0	0	29.511	0.2031
14	128	8	8	0	2	0	0	0	0	0	29.735	0.2656
15	128	8	8	0	4	0	0	0	0	0	30.177	0.3281
16	128	8	8	0	8	0	0	0	0	0	30.550	0.3906
17	128	8	8	0	8	2	0	0	0	0	30.768	0.4531
18	128	8	8	0	8	4	0	0	0	0	31.195	0.5156
19	128	8	8	0	8	8	0	0	0	0	31.492	0.5781
20	128	8	16	0	8	8	0	0	0	0	31.711	0.5938
21	128	8	16	0	8	8	0	2	0	0	31.930	0.8438
22	128	8	16	0	8	8	0	4	0	0	32.928	1.0938
23	128	8	16	0	8	8	0	8	0	0	33.444	1.3438
24	256	8	16	0	8	8	0	8	0	0	33.615	1.3594
25	256	8	16	0	8	16	0	8	0	0	33.886	1.4219
26	256	16	16	0	8	16	0	8	0	0	34.147	1.4375
27	256	16	16	0	8	16	0	8	2	0	34.433	1.6875
28	256	16	16	0	8	16	0	8	4	0	35.150	1.9375
29	256	16	16	0	8	16	0	8	8	0	35.733	2.1875
30	256	16	16	0	8	16	0	8	16	0	36.117	2.4375
⋮	⋮	⋮	⋮	⋮	⋮	⋮	⋮	⋮	⋮	⋮	⋮	⋮
40	256	16	16	8	16	32	4	32	16	2	40.015	3.4844
60	256	128	128	16	64	64	16	64	256	16	45.687	5.9063
70	256	128	128	64	128	256	32	256	256	64	47.003	7.1875
80	256	256	256	256	256	256	256	256	256	256	47.050	8.0000

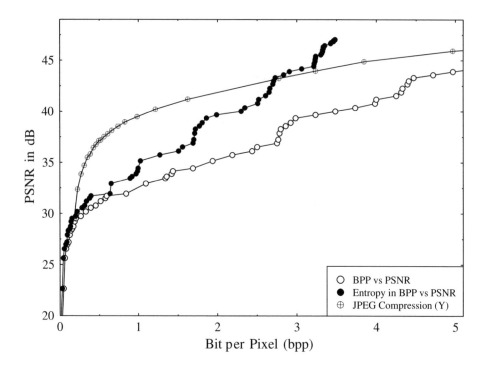

Figure 6.18: PSNR versus bits/pixel performance of the 10-band intra-frame SBC codec for the 4CIF "Susie" sequence of Figure 1.1 in comparison to the Joint Photographic Expert Group (JPEG) codec [296].

relative size of a $\frac{1}{64}$th of the original frame, yielding a bitrate of $\frac{1}{64} \cdot 1$ bit $= 0.0156$ bpp for the luminance information. As a consequence, the reconstructed PSNR of our example using the "Susie" sequence was fairly low at 17 dB, as seen in Table 6.7.

In order to increase the video quality at the lowest possible bitrate investment, we tentatively increased the quantizer resolution of the band 0 quantizer by 1 bit and assigned an additional two-level quantizer for Band 1, while noting the PSNR improvement and bitrate increase in both cases. The same relative figure of merit was then evaluated for each of the remaining nine bands, and the additional bit was finally allocated for the band, which resulted in the best figure of merit, increasing the bitrate and PSNR, as depicted in Figure 6.18 and Table 6.7. Continuing in the same fashion, we continuously increased the quantizer resolutions and derived a bit-allocation table up to 8 bpp for monochrome images or 12 bpp in case of color images, when assigning a 256-level quantizer to each of the luminance and chrominance bands. As an example, we refer to line 19 of Table 6.7, where assigning a 7-bit/128-level quantizer for band 0 and a 3-bit/8-level quantizer for each of the bands 1, 2, 4, and 5 results in a PSNR of 31.492 dB and a bitrate of 0.5781 bpp (0.87 bpp) for a monochrome (color) frame of the "Susie" sequence.

Although we used Max-Lloyd quantizers, the quantized signals still contained redundancy. We therefore determined the entropy of each of the quantized bands and determined the information contents of all bands together in terms of bpp, which is also shown as entropy

6.3. SUBBAND CODING

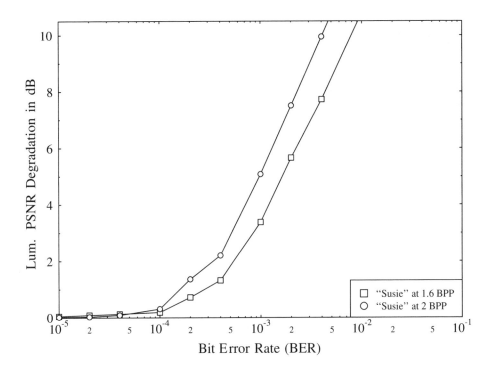

Figure 6.19: PSNR degradation versus BER for the 10-band subband codec, using the 4CIF "Susie" sequence of Figure 1.1, at 1.6 bpp and 2 bpp.

versus bitrate curve in Figure 6.18. This remaining redundancy is not sufficiently high to justify the introduction of further compression schemes, which would strongly increase the error sensitivity.

The full quantizer allocation table is summarized in Table 6.7. Good subjective video qualities are achievable for bitrates in excess of 2.1 bpp for color images, which is associated with PSNRs in excess of 34 dB. However, this PSNR performance is below that of the DCT-based standard JPEG codec seen in the figure, as well as below that of the DCT codec, which will be presented in the next chapter. The advantage of subband coding without RL compression is its error resilience. In case of RL-compression, the transmission errors inflicted propagate and are spread across a certain frequency range rather than being confined to a certain location of a subband.

In order to quantify the effect of transmission errors, we introduced random bit errors and evaluated the objective PSNR degradation measure as a function of the BER for two different bitrates, as seen in Figure 6.19. The BER at which the video quality becomes rather corrupted is around 10^{-3}, but the difference in comparison to block-based coding methods, such as the DCT, for example, is the variable size of the distorted areas, corresponding to the different subbands. For block-based methods using a constant block size, such as the schemes presented in Chapters 2–4, a single bit error inflicts video degradation in that particular fixed-size block. In contrast, in the case of subband coding, a single-frequency band at a certain location is affected. For our typical 10-band codec, bands 0 to 3 are down-sampled by a factor

Figure 6.20: Subband encoded frame of the "Susie" sequence at 2 bpp and subjected to a BER of 10^{-3}, yielding a reduced PSNR of 28.3 dB.

of 8, corresponding to a 64-fold reduction in size. Hence, in SBC a 1-bit error may impair a subband, corresponding to an area of 64 pels, which is equivalent to an 8×8 block. In the case of bands 4 to 6, the affected area shrinks to a size of 4×4 and a smaller area of 2×2 is affected, if the highest bands are corrupted. An example of a frame subjected to a BER of 10^{-3} is given in Figure 6.20.

The above scheme is not amenable to further compression due to motion compensation. Motion compensation would decrease the entropy of the original video signal, but it would not allow the use of more coarse quantizers, because of the unstationarity of the MCER. This problem is discussed in more detail in Chapter 7.

6.4 Summary and Conclusions

In this chapter we studied a number of low-complexity video compression methods. The simplest techniques, such as DPCM and ADPCM, do not achieve a satisfactory video quality at bitrates below 3 bpp for either intra- or inter-frame coded scenarios. DPCM and BTC were shown to achieve reasonable results at a bitrate of 2.5 bpp, reaching PSNRs in excess of 38 dB, which proved adequate for our purposes. These results will be compared with the performance of various other techniques of the following chapters, which have higher compression ratios at the cost of a higher computational demand and lower error resilience.

6.4. SUMMARY AND CONCLUSIONS

The simplicity of both methods facilitates their employment as part of a hybrid or hierarchical coding scheme. DPCM is often used in conjunction with run-length coding or subband coding [287]. In the latter case, it is used to efficiently encode each subband. BTC coding might be combined with quad-tree splitting [297] or vector quantization [298].

Subband coding is a convenient multiresolution technique that allowed us to invest the bitrate budget in those bands, where it was most beneficial in terms of PSNR improvements, as was suggested by Table 6.7. The associated PSNR improvements were characterized by Figure 6.18. Let us now focus our attention on high-quality DCT-based coding in the forthcoming chapter.

Chapter 7

High-resolution DCT Coding

7.1 Introduction

In this chapter, based on our findings from Chapter 3 we apply some of the principles in the context of high-quality image coding. Specifically, the next section explores the corresponding requirements as regards the MC algorithm, before designing a switched intra/inter-frame coded DCT codec.

At the time of writing HDTV coding is often implemented using the MPEG1 [299] and MPEG2 [300] standards, as well as the H263 [259] standard. All of these codecs employ similar compression methods for the quantization of the DCT coefficients, which is based on zig-zag scanning and run-length coding. This method guarantees a near-constant video quality at the cost of a widely fluctuating bitrate and a bitstream, which is vulnerable to bit corruption. This problem may be overcome only using strong error correction and/or error resilient entropy coding (EREC) [260].

Our aim in this chapter is to avoid run-length coding methods, where possible, which reduce the achievable compression ratio but improve the robustness against channel errors. Therefore we need bit allocation tables and appropriate quantizers, which can be derived by adopting an approach similar to that detailed in Section 3.3. For very low bitrates we considered bit-allocation schemes with less than 15 bits per 8×8 block and were able to determine the best allocation scheme by simple trial and error investigations. For the range of video qualities and video sequences considered in this section, we needed a more general approach leading to bit-allocation schemes requiring up to 3 bpp, in order to provide the required HDTV quality.

7.2 Intra-frame Quantizer Training

Determining the appropriate quantizers requires a range of different bit-allocation schemes and corresponding quantizers for various bitrates, up to 2–3 bpp. The aim is similar to that in Section 3.3, where we exploited the energy compaction characteristics of the DCT.

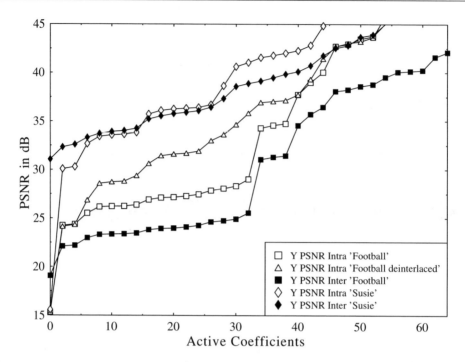

Figure 7.1: PSNR performance of the DCT versus the number of active DCT coefficients for various high-resolution video clips without motion compensation, when applying a zonal mask around the low spatial frequency top-left corner of the DCT block without quantization of the DCT coefficients.

In order to evaluate the potential of such schemes, in Figure 7.1 we examined the PSNR performance of the DCT versus the number of active coefficients per 8×8 block without coefficient quantization, when applying a simple technique called zonal masking. The zonal masking was realized by zig-zag scanning the coefficients according to Figure 7.2, followed by a cancellation or obliteration of all masked coefficients, which were set to zero.

Figure 7.1 underlines the fact that simple masking of the top-left corner [10] is not advantageous, since typically more than 30 out of 64 coefficients must be retained in order to be able to exceed a PSNR of 30 dB before quantization for the considered high-resolution video clips. We also endeavored to evaluate the effects of camera interlacing, since most state-of-the-art cameras record images in a two-phase, interlaced fashion. First, all even lines are recorded, and then in a second scan, the remaining odd lines are stored. This method was invented in order to artificially double the frame rate and to increase the viewing comfort. Since both half-frames are taken with a delay of half the frame scanning interval duration, significant motion may take place between these instants. This leads to a horizontal striped pattern at the edges of moving objects, which will increase the energy of the stored coefficients for horizontal frequencies, causing the distinct PSNR step-around coefficient 34 in Figure 7.1.

The question that arises is whether there are suitable fixed masking patterns, that will allow satisfactory quality. In order to evaluate the upper bound performance of a masking-

7.2. INTRA-FRAME QUANTIZER TRAINING

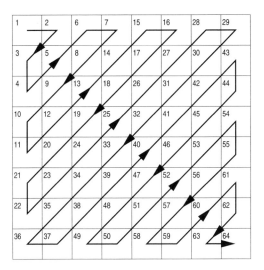

Figure 7.2: Zig-zag scanning of DCT coefficients.

type quantization scheme, we determined individual optimum masks for each block, which *retained the required number of highest energy coefficients*. Figure 7.3 reveals the PSNR versus number of retained coefficients performance, if an individual optimum mask is determined for each block before the masking takes place. The results indicate that for intra-frame-oriented DCT coding, about 10 to 15 coefficients are necessary to guarantee a good video reproduction quality. This explains why the scanning method used in the various standards, such as JPEG [296], MPEG-X [299, 300], and H.26X [258, 259], is so efficient.

An automated algorithm was necessary, which allowed us to determine all required bit-allocation tables. Starting from a 0 bpp quantizer, we used the pseudo-quantizer from Equation 3.22 on page 57 in order to emulate the distortion of the quantization process. In order to identify the DCT coefficient position, where the bit for a 1-bit per 8×8 block quantizer should be allocated for achieving the lowest distortion, the algorithm first allocated the bit at position (0,0) within the block. An entire picture was DCT transformed and quantized. The coefficient (0,0) was quantized with a two-level pseudo-quantizer, and all other coefficients were forced to zero. Finally, the inverse transformation took place, and the resulting PSNR was noted. The same DCT quantization, inverse DCT process was started again, with the bit allocated at the next possible position, for example (0,1), and so on. The bit was eventually allocated to the DCT coefficient position, which resulted in the highest PSNR, and at this stage we arrived at the one bit per block bit-allocation table. Because of the importance of the DC coefficient, the first bit was naturally allocated for this particular coefficient, namely, for the coefficient at position (0,0).

In the next step the process was repeated, and the PSNR benefit of a 1-bit increase at each of the 64 possible positions was tested. This implied that the number of levels of the pseudo-quantizer at a certain bit position was doubled, or, if no bit was allocated at that position before, the number of levels was set to two, yielding a 1-bit quantizer. Following this principle, we generated the bit allocation tables from 1 bit per block up to 256 bits per block. Example bit-allocations for four reconstruction qualities are depicted in Figure 7.4, which

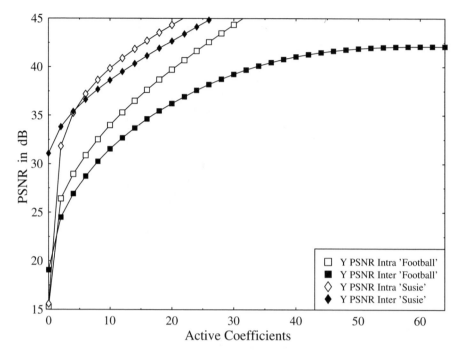

Figure 7.3: PSNR versus the number of active DCT coefficients per 8 × 8 block performance of the DCT, when applying a mask retaining the required number of highest energy coefficients in each block without quantization for the 4CIF sequences of Figure 1.1.

was trained on the "Football" sequence, but we found that the bit-allocation varied very little with different sequences.

It was expected that the resolution of quantizers would shrink with the distance of each coefficient from the top-left corner of the block. For Figure 7.4, we therefore zig-zag scanned the coefficients according to Figure 7.2 in order to determine how well the allocation table correlated with these expectations. As the figure confirms by comparing the active DCT coefficient positions with Figure 7.2, the allocation table follows the expected tendency except that further quantizers are allocated around positions 20–23 and 35–38. As seen from Figure 7.2, these coefficients are located at the very left of the block and are caused by high vertical frequencies, that is, horizontal stripes due to interlacing, as mentioned earlier.

Based on the above tentative quantizer design, appropriate final quantizers had to be trained. Hence, we gathered training data using the above initial quantizers and trained the required Max-Lloyd quantizers, as proposed by Noah [269]. The quantizers should be trained using coefficients, which contain a range of distinct patterns, since, for example, blocks from a flat background will detrimentally narrow the dynamic range of quantizers. This results in increased quantizer overload distortion and should be avoided. Therefore, during the training process we chose from every training sequence those blocks for quantizer training which had the highest variance. A total of more than 50,000 training entries guaranteed the statistical independence of the quantizers from the training sequences. The conventional Max-Lloyd

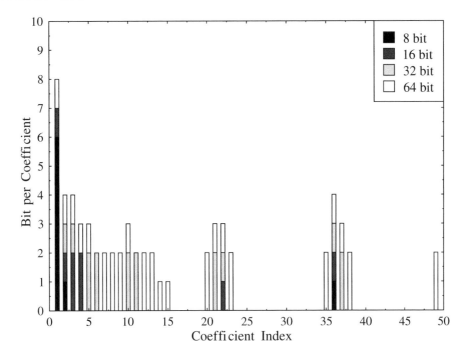

Figure 7.4: DCT coefficient quantizer bit-allocation for the zig-zag scanned DCT intra-frame coefficients for 8, 16, 32, and 64 bits per block schemes.

algorithm was slightly modified so that the generated quantizers were almost symmetrical to the y-axis and were of the mid-tread type [187].

Figure 7.5 reveals the performance of the quantizers. Good still-image quality was guaranteed for bitrates in excess of 0.75 bpp, where the luminance PSNR was around 27 dB, while the Cr_u and Cr_v components exhibited a PSNR close to 40 dB. Since the spatial frequency-domain spectrum of the Cr_u and Cr_v components is more compact, their reproduction quality expressed in terms of PSNR is superior to that of the Y component at the same bitrate associated with a certain quantizer resolution. Because of this property most codecs allocate a lower fraction of the bits to the chrominance components. The performance of the JPEG standard was also characterized in Figure 7.5 in order to provide a benchmarker. The JPEG codec achieves a better performance due to using different quantizers for different blocks and exploiting the benefits of run-length encoding. Further performance examples are given in Figures 7.6 and 7.7 using different video clips, which follow similar trends.

7.3 Motion Compensation for High-quality Images

In Chapter 3, we found that we were able to restrict the MC search to a window of 4×4 pixels. This was possible for relatively low-activity head-and-shoulders videophone sequences, where the image size was fixed to the comparatively small QCIF. Furthermore,

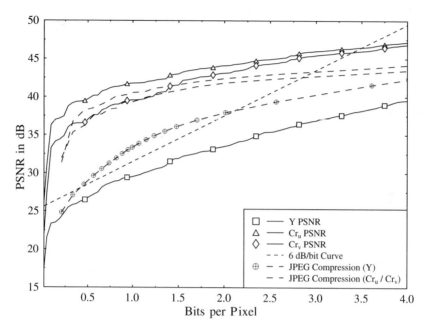

Figure 7.5: PSNR versus bitrate performance of the DCT quantizers for rates up to 4 bits per pixel in comparison to the JPEG codec for the 4CIF "Football" sequence of Figure 1.1.

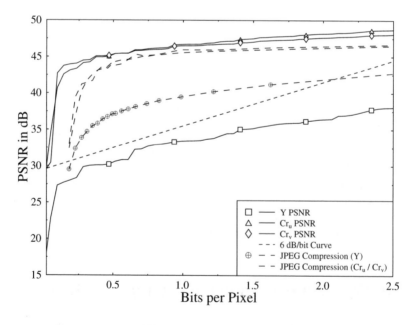

Figure 7.6: PSNR performance of the DCT quantizers for bitrates of up to 4 bit per pixel in comparison to the JPEG codec for the 4CIF "Susie" sequence of Figure 1.1.

7.3. MOTION COMPENSATION FOR HIGH-QUALITY IMAGES

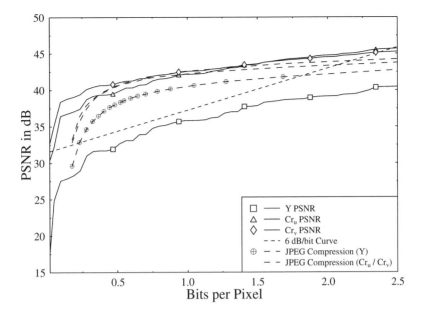

Figure 7.7: PSNR performance of the DCT quantizers for bitrates of up to 4 bits per pixel in comparison to the JPEG codec for the 16CIF "Mall" sequence of Figure 1.3.

we allowed the image quality to drop for short intervals due to the paucity of bits imposed by the constant bitrate requirement.

In high-resolution image coding, however, we have to provide an adequate image quality at all times. In addition, we have to cater for image sequences of varying natures, as, for example, a news-presenter or a football game clip. The motion becomes more complex, and hence, unlike in Part I of this book, we cannot assume that the camera is stationary. The camera might pan or zoom and cause global motion of the image, and so we have to accept that the movement of objects in terms of pixel-separation between consecutive video frames depends on the type of sequence as well as on the resolution. Therefore, it is important to know the statistical properties of the sequences processed so that the search window and the search algorithm are chosen accordingly.

Table 7.1 demonstrates the PSNR video degradation due to limiting the MC search scope to values between 0–32 pixels for two specific sequences. In the case of the 4CIF "Susie" sequence and frames No. 1 and No. 2, a relatively small search scope of 4 pixels in any direction exploits all the temporal redundancy, leading to only marginal PSNR improvements for further MC scope extensions. During later frames in the sequence, when "Susie" suddenly moves her head, the search scope needs to be extended to more than 8 pixels. For the higher activity "Football" sequence, however, a search scope of 64 is needed. This reflects the high-motion activity of the sequence, causing two problems. First, there are $(2 \times 64)^2 = 16\,384$ possible MVs, requiring a 14-bit MV identifier, and the computational demand for a CCIR 601-sized image exceeds 14 billion integer operations! At the time of writing, the implementation of such an algorithm in a real-time codec appears unrealistic. Employment of the full search (FS) algorithm at this search scope might be reserved for

Table 7.1: PSNR Degradation for the Full-search Algorithm, When Using Various MC Search Scopes in the Range of 0–32 Pixels Measured Relative to a Search Scope of 64 for the 4CIF "Susie" and "Football" Sequences of Figure 1.1

	0	1	2	4	8	16	32
Susie Frame #1/#2	6.05	5.32	2.02	0.17	0.13	0.09	0.03
Susie Frame #71/#72	10.55	10.09	9.81	7.27	2.68	0.15	0.08
Football	9.81	8.98	8.14	6.74	4.58	2.03	0.39

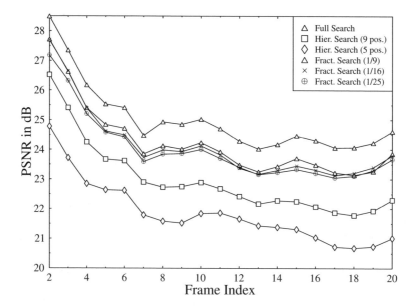

Figure 7.8: PSNR versus frame index performance comparison for motion compensation using various search techniques, a search scope of 32, 16CIF "Football" sequence, no MCER coding, assuming that the first frame was available at the decoder.

applications, such as video on demand, where the encoding process takes place independently from the decoding, as in a non-real-time scenario.

For real-time or quasi-real-time applications, the computational demand must be reduced. In Figures 7.8 and 7.9, we characterized the performance of various search strategies for the "Football" sequence. Figure 7.8 characterizes the PSNR video performance of a hypothetical codec for the first 20 frames of the "Football" sequence, when no MCER was transmitted. This uses three different MC search strategies, namely, the optimum full search (FS) and two suboptimum algorithms, the hierarchical search (HS) and the fractional search (FC). These methods were introduced, among others, in Section 3.2.2; for the sake of simplicity, we have concentrated here on these three common methods.

7.3. MOTION COMPENSATION FOR HIGH-QUALITY IMAGES

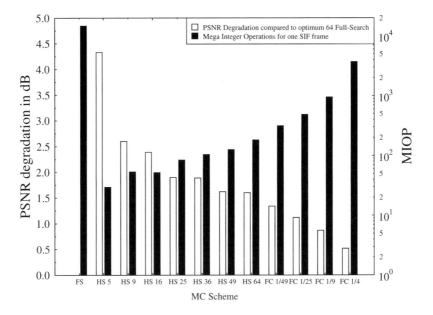

Figure 7.9: PSNR degradation and complexity for the first motion compensated frame, when applying various search methods. No MCER coding was used, and the search scope was 64 for the "Football" sequence.

Figure 7.8 reveals that the hierarchical search of Section 3.2.2 is unsuitable for CCIR 601-sized images exhibiting high-motion content. Its performance is up to 4.5 dB worse than that of the FS algorithm. This is because during the initial search step, the whole search window is evaluated for only 5 (HS 5) or 9 (HS 9) block matches. A search window of 32 offers $(2 \cdot 32)^2 = 4096$ search positions, and motion over this large area cannot be adequately estimated by just five or nine tentative motion compensation attempts. We enhanced the performance of the algorithm by increasing the number of block matches during each step to 16 (HS 16), 25 (HS 25), 36 (HS 36), 49 (HS 49), and even 64 (HS 46), as seen in Figure 7.9. This causes the PSNR penalty inflicted by suboptimum MC to drop from 4.5 dB for HS 5 to about 1.5 dB for HS 64. This is associated with an increase of the computational demand from 30 Mega Integer Operations (MIOP) to 180 MIOP. We also evaluated the potential of the fractional search technique at a quarter of the complexity (FC 1/4), a ninth (FC 1/9), a 25th (FC 1/25), and a 49th (FC 1/49) of the complexity of the exhaustive search. This leads to the range of search complexities and PSNR degradations depicted in Figure 7.9.

Figures 7.10–7.12 show the inter- and intra-frame correlation of motion vectors for the 4CIF "Football" and "Susie" sequences of Figure 1.1. The inter-frame MV correlation in this context is defined as the correlation of MVs for the corresponding blocks in consecutive frames. A high inter-frame correlation of MVs could be exploited by a telescopic MV scheme, where one transmits the difference of the MV of a certain block with respect to the previous MV at the block position under consideration. Such a scheme is an optional feature of the MPEG codecs [299, 300]. It is shown that the inter-frame correlation of MVs is below 20% for consecutive frames, even in the case of the high-motion "Football" sequence. The telescopic

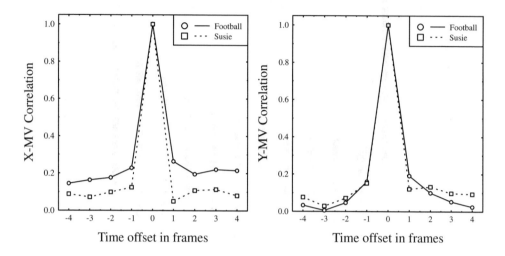

Figure 7.10: Inter-frame correlation of the MVs versus time offset for consecutive frames of the 4CIF "Football" and "Susie" sequences of Figure 1.1. The inter-frame MV correlation in this context is defined as the correlation of MVs for the corresponding blocks in consecutive frames.

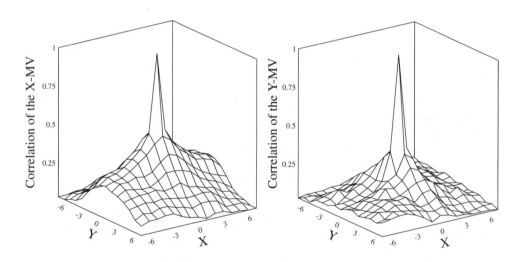

Figure 7.11: Intra-frame correlation of the MVs versus block offset for the 4CIF "Football" sequence of Figure 1.1.

7.4. INTER-FRAME DCT CODING

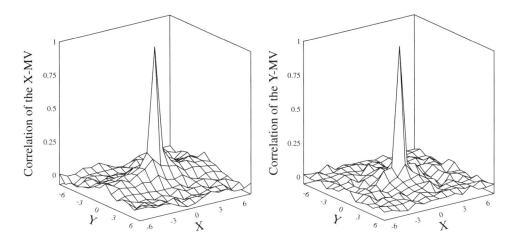

Figure 7.12: Intra-frame correlation of the MVs versus block offset for the 4CIF "Susie" sequence of Figure 1.1.

MV scheme has been abandoned for both the H.261 [258] and H.263 codecs [259], they both rely on exploiting the intra-frame MV correlation, as depicted in Figures 7.11 and 7.12.

The H.261 scheme transmits the difference vector with respect to the MV at the left of the current block, while the H.263 arrangement determines a prediction vector from three surrounding MVs before encoding the difference of the predicted MV and the MV under consideration, $MV_{(x,y)}$. These issues are discussed in more depth in Section 9.2.3. Briefly, the MV of the block to the left, $MV_{(x-1,y)}$, the MV of the corresponding block in the previous row $MV_{(x,y-1)}$, and that of the next block in the previous row $MV_{(x+1,y-1)}$ are used to derive the MV estimate. Although the exploitable correlation is below 40%, typically around 20%, since the H.261 codec uses error-sensitive variable-length coding (VLC) anyway, it is useful to exploit this potential redundancy. It does not dramatically impair the overall codec resilience. This is not true in the case of our proposed codecs refraining from using run-length and variable-length coding, since the increased error sensitivity would overshadow the potential bitrate savings. Having reviewed the potentially suitable MC techniques for high-resolution video compression, let us now concentrate on the high-resolution encoding of the MCER.

7.4 Inter-frame DCT Coding

7.4.1 Properties of the DCT Transformed MCER

The DCT is commonly used due to its so-called energy-compaction property, if the block exhibits spatial correlation in the time domain. The process of motion compensation removes most of the temporal redundancy inherent in the corresponding positions of consecutive frames and a substantial part of the spatial redundancy, which manifests itself in the form of the similarity of the adjacent luminance and chrominance pixels. In order to quantify how much of the spatial redundancy is removed, in Table 7.2 we summarized the first autocorrelation coefficient for certain frames of three sequences and their respective frame

Table 7.2: The First Autocorrelation Coefficients for Three Sample Images and for Typical Frame Differences of the Respective Sequences

SEQUENCE	Y-Cor.	U-Cor.	V-Cor.
Miss America #1	0.98	0.99	0.95
Miss America #1/#2 MCER	0.02	0.70	0.75
Susie #1	0.98	0.98	0.92
Susie #1/#2 MCER	0.56	0.14	−0.03
Football #1	0.96	0.93	0.96
Fb #1/#2 MCER	0.93	0.83	0.89

differences. We observed that in all three cases, the adjacent-pixel correlation of the sequences is higher than 0.95, and this is the reason why DCT is such a successful method in still-image compression. This is a consequence of the Wiener-Khintchine theorem, stating that the autocorrelation function (ACF) and the power spectral density (PSD) are Fourier Transform (FT) pairs. Since the DCT is closely related to the FT, a flat ACF is associated with a compact DCT spectrum located in the vicinity of the zero frequency. However, the autocorrelation properties of the MCER — which in this case is constituted simply by the frame difference of two consecutive frames — are rather different, since the correlation has been mainly removed, returning reduced or in some cases negative correlation values, as evidenced by the table. Clearly, MC may remove almost all adjacent-pixel autocorrelation.

Strobach [160] found that transform coding is not always efficient for encoding the MCER, since it sometimes results in energy expansion rather than energy compaction due to the temporal decorrelation process. In [160] he introduced an efficiency-of-transform test, which is based on the statistical moments of the MCER signal. His approach was as follows.

Equation 7.1 represents the energy preservation property of the DCT in time and frequency domains:

$$\sum_{i=0}^{b}\sum_{j=0}^{b} b_{time}^2(i,j) = \sum_{i=0}^{b}\sum_{j=0}^{b} b_{freq}^2(i,j), \tag{7.1}$$

where $b_{time}(i,j)$ and $b_{freq}(i,j)$ correspond to the time- and frequency-domain representations of a given block. However, the higher moments of these time- and frequency-domain coefficients are different in general, which is formulated as:

$$\sum_{i=0}^{b}\sum_{j=0}^{b} b_{time}^k(i,j) \neq \sum_{i=0}^{b}\sum_{j=0}^{b} b_{freq}^k(i,j) \text{ for } k > 2. \tag{7.2}$$

If the transformation applied results in significant energy compaction, concentrating most of the energy in a few low-frequency coefficients, then the frequency-domain higher-order statistical moments will become even more dominant after raising the high coefficient values to a high power. Therefore, Strobach argues that it is useful to define the higher-order energy

7.4. INTER-FRAME DCT CODING

compaction coefficients ρ_k as follows:

$$\rho_k = \frac{\sum_{i=0}^{b} \sum_{j=0}^{b} b_{freq}^k(i,j)}{\sum_{i=0}^{b} \sum_{j=0}^{b} b_{time}^k(i,j)} \quad \text{for } k > 2 \text{ (k: even)}, \tag{7.3}$$

where again $b_{time}(i,j)$ and $b_{freq}(i,j)$ represent a given block in the time and transform domain, respectively. The transformation is deemed efficient and compacts the bulk of the energy in a low number of high-energy coefficients, if $\rho_k > 1$; otherwise it results in energy expansion. In order to circumvent this problem, Strobach proposes time-domain MCER compression methods such as vector quantization or quad-tree decomposition.

In order to better understand the statistical properties of the MCER signal, we analyzed the blocks in both the time and transform domain and classified each pixel or coefficient into a high-energy and a low-energy class by a threshold decision given by:

$$thres(x) = \begin{cases} 1 & \text{if } x > x_0 \\ 0 & \text{otherwise} \end{cases} \quad x_0 \in [4..15], \tag{7.4}$$

where the threshold x_0 was in the range of [4...15] for both the time and frequency domain. For each block we counted the high-energy coefficients or pixels in each class and noted them as sum S_{high}:

$$S_{high} = \sum_{i=0}^{b} \sum_{j=0}^{b} thres|b_{time}(i,j)|. \tag{7.5}$$

This sum S_{high} is evaluated in both the time and frequency domain, and the difference ρ_{thres}:

$$\rho_{thres} = S_{high}(freq.) - S_{high}(time), \tag{7.6}$$

gives us an indication as to whether the DCT transform was efficient ($\rho_{thres} > 0$) or inefficient ($\rho_{thres} < 0$). For the evaluations below, we skipped those flat-blocks, which resulted in a time-domain count of four or less.

In Table 7.3 we listed the proportion of blocks exhibiting energy expansion on the basis of the previously defined ρ_{thres} efficiency measure and also summarized their total contribution to the energy of the MCER, when using a range of different MC search scopes. Observe in the table that a MC search scope of $\pm$ zero corresponds to no MC at all, implying that simple frame-differencing was applied. Surprisingly, about 5% to 9% of the blocks were classified as energy expanding, and their contribution to the overall energy of the MCER is up to 30.6%, which is quite significant. These results varied only insignificantly, when the energy classification threshold x_0 was kept within the range of [4..15]. The results shown here are based on an x_0 value of 5. The DCT's inability to compact the MCER energy increases with the increasing efficiency of the MC, since the more spatial and temporal redundancy is removed, the more unstationary and uncorrelated the MCER becomes.

In order to further explore the efficiency of the DCT, we analyzed the PDF of our test-of-efficiency criterion ρ_{thres} in Figures 7.13 and 7.14 for both the first frame and the first frame difference of the 4CIF "Susie" sequence. The marked areas to the left and right of the efficiency threshold in Figure 7.13 represent the blocks which exhibit energy expansion, or energy compaction in the latter case. The excellent compaction properties of intra-frame

Table 7.3: Percentage of Energy Expanding Blocks and their Relative Energy Contributions for the MCER of the 4CIF "Susie" and "Football" Sequences and the CIF "Miss America" Sequence

Seq.	MC-Scope	% of Energy Expanding Blks.	% of Energy in Energy Exp. Blks.
Susie	± 0	4.7	4.37
	± 2	7.2	14.12
	± 4	5.2	7.28
Football	± 0	6.8	1.28
	± 4	8.6	1.48
	± 16	8.0	7.17
	± 64	9.63	4.62
Miss America	± 0	23.1	12.3
	± 2	30.6	23.80
	± 4	28.59	22.61

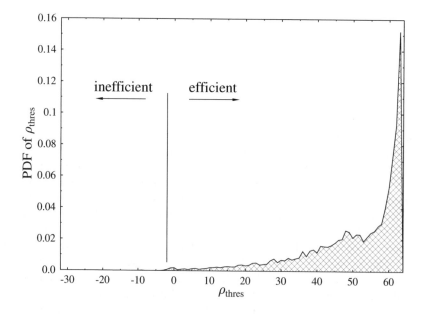

Figure 7.13: Energy compaction efficiency of the intra-frame DCT transform for frame 1 of the 4CIF "Susie" sequence of Figure 1.1, quantified in terms of the distribution of ρ_{thres}.

7.4. INTER-FRAME DCT CODING

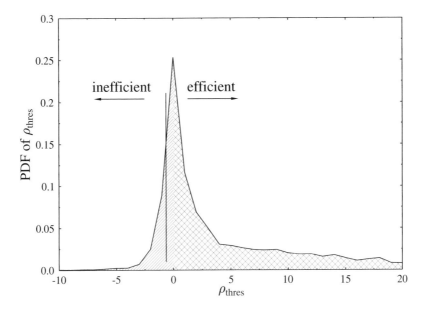

Figure 7.14: Energy compaction efficiency of the DCT transform for the frame difference of frame 1 and 2 of the 4CIF "Susie" sequence of Figure 1.1, quantified in terms of the distribution of ρ_{thres}.

DCT coding are clearly demonstrated by Figure 7.13. In this case most of the energy was compacted to less than ten coefficients for the majority of the blocks and typically positive ρ_{thres} values were recorded.

However, the energy compaction performance is drastically reduced, when the DCT is applied to a MCER frame, as evidenced by Figure 7.14, which shows that for about 20% of the blocks, the DCT fails to compact the energy. This expansion however is typically limited to an increase of five additional positions in total. In other words, the DCT of a MCER frame may result in energy expansion, but it does not result in a catastrophic spread of the energy across the block. This explains why the DCT coding is still favored for MCER coding in the MPEG and H.26x standards. On the other hand, for about 30% of the blocks the DCT compacts the energy by more than five coefficients, and hence the overall DCT compression of MCER images is still deemed advantageous.

In order to further elaborate on this energy expansion property of the DCT, when applied to the reduced-correlation MCER, we refer below to the performance of the H.261 standard. Using a H.261 codec, we compressed 20 frames of the "Miss America" sequence and analyzed the encoded bitstream. The first frame was intra-frame coded, and it was found that every block required an average coding rate of 0.5 bit per pixel. All active blocks of the remaining 19 frames were inter-frame coded, and they yielded an average bitrate of 1.1 bits per pixel, implying the surprising fact that the inter-frame coding in this particular scenario turned out to be less efficient than intra-frame coding and hence increased the required bitrate.

In order to further explore this fundamental problem, we tested the performance of a technique that we referred to as the scanning method. Accordingly, all blocks of a given

Table 7.4: Mapping Run-lengths and Colors to B1 Codewords (left) and Modified B1 Codewords (Right) (C Refers to the Run Color)

Run-length	B1 code	Run-length	Modified B1 code
1	C0	1	C0
2	C1	2	C1
3	C0C0	3	C0C0
4	C0C1	4	C0C1
5	C1C0	5	C1C0
6	C1C1	EOL,ESC	C1C1
7	C0C0C0	6	C0C0C0
8	C0C0C1	7	C0C0C1
9	C0C1C0	8	C0C1C0
⋮	⋮	⋮	⋮

frame were DCT transformed and quantized using the H.263 quantizers [259]. The H.263 standard offers 32 quantizers, each of linear type and a resolution of 8 bits. For each block, we chose that particular quantizer from the possible 32 characteristics, which gave a mean absolute error just below 4, resulting in a mean PSNR of $20 \cdot \log(256/4) = 36$ dB. The quantized coefficients were then scanned according to the standard's zig-zag scanning scheme in Figure 7.2, subsequently run-length encoded, and the resulting run-length was noted for each block. For the sake of easy implementation we chose the B codes as run-length codes [289].

As a brief excursion from our original topic, namely, the energy compaction-expansion problem of the DCT, Table 7.4 shows how each run-length and color maps to a unique B1 code, where C represents the run color, black or white. As a simple example, the bitstream ...0,0,0,0,0,1,1,1,0,0,1... would be coded as [black run = 5] [white run = 3] [black run = 2] [white run = 1], yielding the codes ... [0100][1010][01][10] ..., where the run color 0 is coded as 0 and color 1 as 1, respectively. Often the modified B codes are used where a certain run-length is used to signal a special event, such as end-of-line (EOL) and escape (ESC).

The run-length decoding algorithm is straightforward. Black and white run-lengths are read from the input bitstream. The run-lengths are used to reconstruct a scanline of, for example, the coded picture. After each line is coded, an EOL code is inserted into the bitstream; thus, once a full line has been decoded, an EOL is read. The redundancy introduced by the EOL code, coupled with the fact that the sum of run-lengths on a given scanline is constant, gives the coding technique some error detection capability.

Returning to the topic of the DCT's energy compaction-expansion property, we then evaluated the run-length probability density function (PDF), and its integral, the cumulative distribution function (CDF), was derived and depicted in Figure 7.15 — both with intra-frame and inter-frame coding for the "Football" and the "Susie" sequences.

Naturally, for high coding efficiency, short B-coded codewords were expected. Our first observation concerning Figure 7.15 is that the "Susie" sequence results in a substantially

7.4. INTER-FRAME DCT CODING

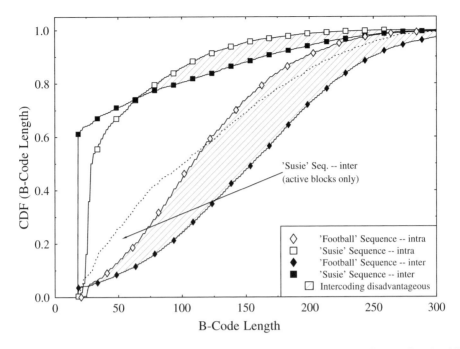

Figure 7.15: CDF of the B-code length of the scanned and quantized coefficients for the 4CIF "Football" and "Susie" sequences in both inter- and intra-frame coding.

higher proportion of short B-coded words and higher coding efficiency than the "Football" clip both in intra- and inter-coded modes. This is due to the "flat" background of the image and its lower motion activity. For any given run-length that particular technique of the intra- and inter-coding schemes is more successful, which results in a higher proportion of short B-coded words, that is, exhibiting a higher CDF value.

The important issue in Figure 7.15 is the B-code length ratio for inter- and intra-mode of operation. For the 4CIF "Susie" sequence, for example, 80% of the inter-frame blocks results in a B-code length of 100 or less, whereas 85% of the intra-coded blocks results in a B-code length of 100 or less. In other words, intra-coding would lead to a better compression ratio.

Note in Figure 7.15 that for the "Susie" sequence 60% of the inter-frame coded blocks resulted in the shortest possible run-length of 18. This happens, if all coefficients are set to zero (i.e., it was not necessary to encode the corresponding block to fulfill the quality criterion). Omitting those "passive" blocks, we generated the new CDF for the "Susie" sequence marked as a dotted line in Figure 7.15. Since this CDF is always below the CDF for the intra-frame mode, this experimental fact corroborates Strobach's claims. Similar results were also valid for the "Football" sequence.

The ITU Study Group 15, which contrived the H.263 standard [259], also produced a document referred to as the Test Near Model version 5, which is commonly termed TMN5 [301]. This document provides a skeleton outline of a codec, which meets the H.263 standard. In the TMN5 document, the study group proposed a solution to overcome the intra/inter coded decision problem using an inter/intra-coded decision for each macroblock b_{MB}, which is composed of four luminance blocks b_n. The chrominance information is not relevant as to the inter/intra-coded decision. This decision is based on:

$$SAD_n(x,y) = \sum_{i=1,j=1}^{b,b} |b_{\text{frame n}}(i,j) - b_{\text{frame n-1}}(i,j)|$$
$$\forall\, x,y \in [-15, 15], b = 8, 16 \tag{7.7}$$

$$SAD_{inter} = \min\left[SAD_{b_{MB}}, \sum_{n=1}^{4} SAD_{b_n}\right] \tag{7.8}$$

$$SAD_{intra} = \sum_{i=1,j=1}^{16,16} |b_{MB} - \overline{b_{MB}}| \tag{7.9}$$

$$\text{where } \overline{b_{MB}} = \sum_{i=1,j=1}^{16,16} |b_{MB}|$$

$$\text{Mode} = \begin{cases} \text{Inter} & \text{if } SAD_{intra} < SAD_{inter} - 500 \\ \text{Intra} & \text{otherwise} \end{cases}. \tag{7.10}$$

which are defined in the TMN5 document [301]. Specifically, Equation 7.7 defines the Sum of Absolute Difference (SAD) criterion for blocks in consecutive frames. The SAD is then evaluated for a whole macroblock and for the four luminance blocks of a macroblock according to Equation 7.8, where the SAD$_{\text{inter}}$ is the minimum of the SAD for the macroblock and the sum of the SAD for the four luminance blocks. The absolute difference of the intra-coded macroblock relative to its mean ($\overline{b_{MB}}$) is derived in Equation 7.9. Finally, the results of both the intra- and inter-coded SAD in Equations 7.8 and 7.9 are then compared in Equation 7.10. Basically, the smoothness of the potential inter- and intra-coded macro-block candidates decides upon the mode.

However, this intra- versus inter-frame classification constitutes a suboptimum approach, since the optimum technique would be to compare the generated bitrate and the resulting reconstructed quality in case of both inter- and intra-frame coding a block. In order to find an attractive solution to this problem, we explored the possibility of joint MC and DCT-based encoding, which is the topic of the following section.

7.4.2 Joint Motion Compensation and Residual Encoding

Inter-frame coding is based on a two-stage approach, exploiting temporal redundancy in the MC step and capitalizing on the spatial frequency-domain redundancy in the subsequent MCER encoding step. The first encoding step minimizes the energy of the MCER, and the second stage attempts to reconstruct the best possible replica of the MCER, using the DCT. Previous sections have shown that an increase in the efficiency of the MC results in a decreased efficiency with regard to the DCT encoding of the MCER, inspiring us to attempt finding the joint optimum. Rather than finding the optimum MVs by decreasing the MCER energy as much as possible, we evaluated the reconstruction error after DCT transformation for each legitimate MV. This process is associated with a huge complexity increase, as the number of required DCT transformations is multiplied by the number of pixels covered by the search window. Nonetheless, at this stage we were motivated by identifying only the order of magnitude of the potential video quality and/or bitrate benefits.

7.4. INTER-FRAME DCT CODING

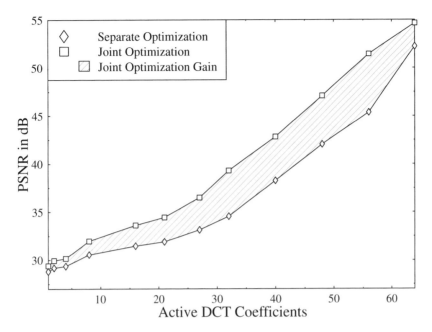

Figure 7.16: PSNR versus number of active DCT coefficients performance comparison of the joint MC-DCT coding and of the traditional two-stage approach using the 4CIF "Football" sequence of Figure 1.1 for frame 1, search scope 32, 1-64 active DCT coefficients, assuming that all blocks are active.

In a simple approach, we used DCT-coefficient masking without quantization of the unmasked coefficients, and Figure 7.16 portrays the results for both the jointly optimized and the traditional approaches. The potential PSNR gain is about 2 dB, if more than 30 DCT coefficients are used for a block-size of 8×8, although better results are achieved for the highest qualities, when aiming for PSNRs in excess of 40 dB. Here, we found that the jointly optimized approach was capable of identifying the MV, for which the best overall quality was achieved.

In order to obtain results, when applying quantization to the DCT coefficients, we generated bit-allocation tables, as described in Section 7.2. As was expected, we found that because of the uncorrelated nature of the MCER, we were unable to obtain satisfactory results when using pure inter-frame DCT coding. Even a multiclass quantization scheme with 16 different quantizer allocations for different frequency-domain energy distributions was inferior in PSNR terms to the intra-coding approach. A quantization scheme, in which the quantizers are allocated beforehand is in most cases simply not sufficiently flexible to encode the DCT cocfficients of a MCER block. Here the algorithm used by the various standard codecs, such as the JPEG, MPEG, and H.26x schemes, which identifies and scans the nonzero DCT coefficients for run-length coding is more suitable, although it exhibits a high error sensitivity.

In conclusion, the results of this and the previous section prompted us to employ a combined inter/intra-coded approach for our proposed high-quality codec, being able to switch between inter- and intra-coded modes of operation on a block-by-block basis.

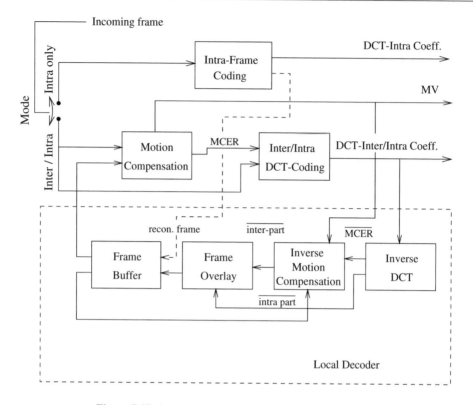

Figure 7.17: Schematic of the intra/intercoded DCT codec.

7.5 The Proposed Codec

An attractive codec is composed of the same blocks as our low-rate videophone codecs proposed in Part I of the book. The initial coarse intra-frame codec of Section 3.5, which relied on the block averages is replaced here by the intra-frame DCT codec from this chapter, which is seen at the top of Figure 7.17. Since no previous image information is available, every luminance and chrominance block is transformed and the quantized coefficients are transmitted to the receiver and also passed to the local decoder. The quantizer bit-allocation is selected from the range of quantizer configurations we derived earlier according to the bitrate and quality requirements. Hence, the codec generates a high initial bitrate according to the relationship of *Quantizer Index* $\times 1.5/64$, which typically lies around 0.75 bpp. This information is transmitted and loaded into the frame buffer of the local decoder.

In contrast to our previous low-rate, low-quality codecs, we used combined inter-/intraframe coding, as depicted in Figure 7.17. Applying this mode, the incoming frame is passed to the MC codec and the resulting MCER is passed to the inter/intra-frame coded DCT codec, along with the unmodified incoming frame. These operations are highlighted in simple terms in Sections 7.5.1 and 7.5.2 using Figures 7.17–7.18. Here we will only briefly allude to them. After the inter/intra-frame coding decision has been carried out, the actual DCT encoding and quantization takes place, and all relevant parameters are transmitted to the decoder. The

7.5. THE PROPOSED CODEC

parameters are also decoded locally and passed to the local decoder. Then the inverse DCT step is carried out, which results in the reconstructed MCER and the reconstructed intra-coded frame. Recall that two modes are possible, and the reconstructed intra-coded frame contains only the actually intra-coded blocks, while it contains blank blocks in the remaining positions. The same is valid for the reconstructed MCER, where only those blocks selected for inter-frame DCT coding are present. The MCER is passed through the inverse MC step. This reconstructed inter-coded frame is overlaid on the reconstructed intra-coded frame, giving the final reconstructed frame.

Further details of the encoding strategy will be described in the forthcoming Sections.

7.5.1 Motion Compensation

Let us first discuss the MC step, where we have to incorporate the results from Section 7.4.1. Previously, we attempted to combine MC and DCT using our transform efficiency measure, which may be disadvantageous for two reasons. First, the intra-frame coding block might lead to a better PSNR than the jointly optimized MC and DCT steps at the same bitrate, since a substantial bitrate contribution must be allocated to the encoding of the MVs, thereby disadvantaging the MCER coding due to the paucity of bits. Second, we unnecessarily increase the vulnerability of the encoded bitstream, since the MVs are sensitive to channel errors and result in error propagation through the reconstructed frame buffer. Specifically, a single MV, which can take 32 horizontal and vertical positions, for example, requires $5 + 5 = 10$ bits. Similarly, a high number of bits is required for the encoding of the block index, signaling the position of each block for which MVs are transmitted. For example, when using 1408×1152 pixel 16CIF resolution images, there is a total of $253\,448 \times 8$ blocks, requiring a 15-bit block index, unless specific coding techniques are employed to reduce this number. Hence a total of up to 25 bits/block may be necessary for the MC information of an $8 \times 8 = 64$ pixel block, which would severely limit the achievable compression ratio. The associated bitrate contribution would be $25/64 \approx 0.39$ bits/pixel. Therefore, we have to carefully consider the active MV selection criterion.

Thus, during the MC process, we first check whether our PSNR or MSE quality requirements are satisfied without MC or DCT coding, as seen in the flowchart of Figure 7.18. Hence, we defined a block quality criterion, which is a more sophisticated version of the criterion that is used by the H.263 codec. More specifically, this quality criterion is stated in Equation 7.13 using two thresholds, T_{mean}, for testing the maximum mean distortion, D_{mean}, in Equation 7.11 and T_{peak}, for checking the value of the maximum peak distortion D_{peak} in Equation 7.12. Then in the context of the blocks b_1 and b_2 we have:

$$D_{mean} = \frac{1}{b^2} \sum_{i=1,j=1}^{b,b} |b_1(i,j) - b_2(i,j)| \qquad (7.11)$$

$$D_{peak} = MAX(|b_1(i,j) - b_2(i,j)|) \quad \forall\, x, y \in [1, b] \qquad (7.12)$$

$$\text{Quality} = \begin{cases} \text{adequate} & \text{if } (D_{mean} < T_{mean}) \wedge (D_{peak} < T_{peak}) \\ \text{inadequate} & \text{otherwise} \end{cases}. \qquad (7.13)$$

The video quality is deemed adequate, if both distortion terms are below their respective thresholds and inadequate otherwise.

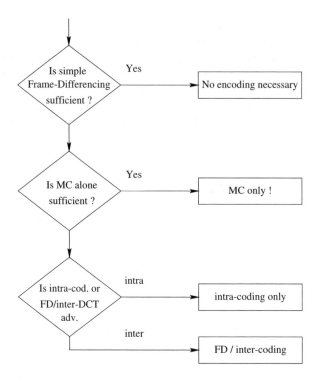

Figure 7.18: Flowchart of the block-encoding decisions.

Initially, the corresponding blocks in consecutive frames are subjected to our quality test expressed in terms of Equation 7.13. If the criterion is satisfied, since these blocks are sufficiently similar, it is not necessary to encode the corresponding block. This block is marked as passive, which is shown in Figure 7.18. If a block does not satisfy the criterion, a motion vector is determined and the criterion is applied again to the corresponding blocks in consecutive frames but after motion compensation. If after MC the quality criterion is met, it is signaled to the subsequent inter/intra-frame coded codec that no further encoding is necessary for the block concerned. If the MC step above fails to provide the required quality, either intra-frame DCT coding is invoked or frame-differencing (FD) combined with inter-frame-oriented DCT coding is employed, as seen at the bottom of Figure 7.18. This approach is adopted, when the additional bitrate requirement of the MVs along with MCER coding would be excessive.

7.5.2 The Inter/Intra-DCT Codec

The DCT codec of Figure 7.17 processes both the MCER and the incoming frame. Previous evaluations of the MC have shown which of the blocks are deemed to be motion passive or have been successfully subjected to MC according to the flowchart of Figure 7.18. Those blocks are ignored at this stage, reducing the computational demand. The remaining blocks failed our preset quality criterion of Equation 7.13 and must be encoded by the DCT codec.

In the previous section we stated that joint MC and inter-frame coding would unnecessarily inflate our bitrate. At this stage, we determined whether frame-differencing and inter-frame DCT coding, or simple intra-frame coding leads to a better reconstructed video quality. The outcome of this decision is signaled to the decoder using a one-bit flag.

7.5.3 Frame Alignment

Because of the high video quality requirements and in contrast to the codecs proposed in Part I of the book, we cannot maintain a constant bitrate per transmitted frame. Hence, we have to signal the length of the MV and DCT data segments to the decoder. This could be easily accomplished by a small header segment at the beginning of every frame, storing the number of MVs and DCT encoded blocks. This side-information would be very vulnerable to channel errors, since even a single bit error causes misalignment of the encoder and decoder, and forces the whole decoding process to collapse. Another, more error-resilient method is to insert frame alignment words [302] (FAW) constituted by unique bit patterns of a given length N, which are detected by correlation of the bitstream and the FAW at the decoder side. If the detection process is successful, the bitstream is separated in different substreams for the MV and the DCT data, and the correct number of MVs and DCT blocks is recovered for each video frame after identifying the length of each stream. It is important to ensure that the bitstream generated by the codec is not strongly correlated with the FAW, which could easily lead to frame synchronization problems in case of corrupting a few bits of the FAW. Therefore, the codec checks the encoded output and eventually rearranges the transmitted data, so that the Hamming distance of the generated data and the FAW remains sufficiently high. This can be achieved, for example, by inserting additional bits into the stream, which are then recognized and removed at the decoder. This is possible, because the proposed codec allows us to swap entire blocks of bits in the bitstream. In contrast to [303], we do not use bit stuffing, as for this method is vulnerable to channel errors.

In this section, we attempted to determine the minimum necessary length of the FAW, so that it becomes highly unlikely that the FAW is emulated by the video data. The longer the FAW, the less likely it will be emulated by the encoded data, reducing the false FAW detection probability and hence the probability of false frame synchronization. However, the longer the FAW, the more complex the correlation process required for its detection. This complexity problem is particularly grave if the decoder has no prior knowledge as to the likely position of the frame commencement, since then it is required to continuously search for it. A further problem is that the received FAW may contain a number of transmission errors, and it must not be erroneously rejected due to this phenomenon, forcing us to accept it, if a few bits are corrupted. However, this increases the probability of its emulation by video-coded data.

The bitstream under consideration contains two different types of data, namely, the FAW and the video-encoded data. Assuming an n-tuple FAW and a given number of arbitrary random error events, we quantify below the probability of emulating the FAW by the video data.

In the simplest transmission scenario, the FAW is not corrupted and the probability $p_0(n)$ that this error-free FAW of length n is emulated perfectly by random data is given by:

$$p_0(n) = \frac{1}{2^n}. \tag{7.14}$$

Since data corruption is unavoidable over practical channels, we have to allow for a certain number of errors in the FAW. The number of possible bit patterns, which emulate the FAW with exactly one bit error, is simply $C_1(n) = n$. Hence, the probability $p_1(n)$ corresponding to this event is given by:

$$p_1(n) = \frac{n}{2^n}. \tag{7.15}$$

The problem is slightly less obvious, when evaluating the probability $p_2(n)$ for two simultaneous bit errors, which can be solved with reference to the probability $p_1(n)$ given above. For the given FAW having n bits $(1, 2, 3, 4, \ldots, n)$, we assume that bit 1 is corrupted. The remaining $n - 1$ bits then contain exactly one error, leading to $C_1(n - 1) = n - 1$ possibilities as regards the error position. We can now assume that bit 2 is corrupted and bit 1 is assumed to be uncorrupted, since it was considered to be corrupted above. Hence, the second bit error is in one of the bit positions from 3 to n, which leads to $C_1(n - 2) = n - 2$ possible combinations. Following this principle, we see that the number of combinations for exactly two bit errors $C_2(n)$ is defined as:

$$C_2(n) = \sum_{i=1}^{n-1} C_1(i) = (i-1) + (i-2) + \cdots + 1 = \frac{n(n-1)}{2}. \tag{7.16}$$

This approach may be generalized for the case of $C_x(n)$ as follows:

$$C_x(n) = \sum_{i=1}^{n-1} C_{(x-1)}(i), \tag{7.17}$$

which is a recursive formula and makes it easy to evaluate all required probabilities $p_i(n)$. The probability $P_x(n)$, that a FAW of length n is emulated by data, when allowing up to x bit errors, is then defined by summing all possible error probabilities $p_i(n)$ for $i = 1, 2, \ldots x$:

$$P_x(n) = \sum_{i=1}^{n} p_i(n). \tag{7.18}$$

The probability of the FAW being emulated by random video data, while tolerating a certain number of errors, is depicted in Figure 7.19.

It is now possible for us to choose the required length of the FAW. At data rates around 1 bpp, an encoded HDTV frame is up to 2 Mbits long. Allowing a misalignment once every 10^7 fames ($\approx$ once every four days at 30 frames/s), we require a misalignment probability of less than $2 \cdot 10^{-13}$, which is satisfied by a FAW length of 48 bits and longer, as seen in Figure 7.19. When opting for an even longer, 64-bit word and allowing up to five errors in the detection process, the equivalent BER is about $7.8 \cdot 10^{-2}$. Since high-quality video communications is impossible at this BER, we are certain to have obtained a safe FAW length estimate. In our scheme two different FAWs are used in order to signal the beginning of the MV and DCT information segments. Therefore, we transmit 128 extra bits per frame, which results in a marginal increment of the data rate.

So far we have determined the probability of the FAW being emulated by random data. This is an upper bound estimate for the FAW being emulated, since the codec actually arranges the bitstream so that the correlation of the uncorrupted data with the FAW is low.

7.5. THE PROPOSED CODEC

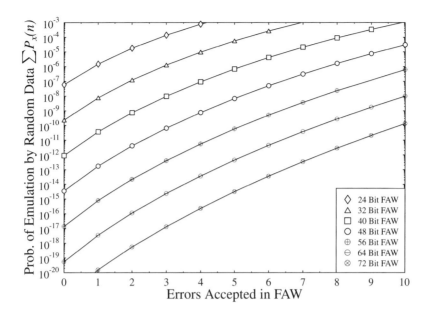

Figure 7.19: Probability of corrupted FAWs being emulated by random video data.

Now we have to choose the appropriate FAW so that FAW detection leads to the right frame alignment.

The choice of the appropriate FAW is based on its autocorrelation properties. Explicitly, the FAW must exhibit a high main peak to secondary peak ratio in order for the correlation process not to mistake a high secondary peak for a corrupted, and hence reduced, main peak, which would inevitably lead to a frame synchronization error. The set of best sequences in this sense for a certain length is then usually found by computer search, where all possible sequences of the required length are generated and the one exhibiting the highest main-to-side lobe ratio is deemed to be the best. We chose our FAW following Al-Subbagh's work [304], who listed the best patterns up to a word length of 32, and extended the optimum 32-bit FAW to 64 bits by simply repeating it. Their autocorrelation displayed in Figure 7.20 shows that the secondary peaks are substantially lower than the main peak. Explicitly, their difference amounts to 28. Therefore, up to 13 one-bit errors may be tolerated within the FAW without mistaking the secondary ACF peak for the main one.

Having considered the algorithmic details of the proposed HDTV codec, we can now proceed with the characterization of the bit-allocation strategy.

7.5.4 Bit-allocation

The resulting bitstream is composed of intra-frame coded DCT coefficient data, MVs, and combined inter/intra-DCT data as shown in Table 7.5. In the intra-frame coded mode of operation, we transmit the DCT coefficients of every single block at a rate of $32 - 64$ bits per 8×8 block, so that the initial frame's contribution is in the range of 0.75 to 1.5 bpp or 259 to 518 kbits for CCIR 601 images. In the inter-frame coded mode, the MVs, the

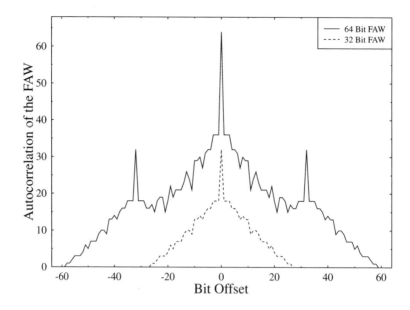

Figure 7.20: Autocorrelation function of the 32- and 64-bit FAWs used for the simulations.

Table 7.5: Bit-allocation Table for CCIR 601-sized Frames to HDTV-sized Frames

Mode	FAW	MV Index	MV	DCT Index	DCT Coeff.
Intra	64	–	–	–	$32 - 64$
Inter	2×64	$13 - 15$	$8 \ldots 12$	$14 - 16 + 1$ (inter/intra flag)	$32 - 64$

DCT coefficients, the inter/intra-coded flag and their respective indices are to be sent. For 720×480 pixel CCIR 601 images there are 5400 8×8 blocks and MVs, requiring a 13-bit block index. For a HDTV resolution of 1920×1080 pixels there are 32 400 blocks, requiring a 15-bit index. The number of DCT blocks is increased by 50% for the chrominance blocks; hence one extra bit is necessary for their indexing. The MV search scope must be restricted to less than 32 for computational complexity reasons, which leads to a MV coordinate of $6 + 6 = 12$ bits at most.

These information segments, composed of the MV and DCT data, are concatenated to form a single bitstream, separated by the FAW discussed above. At the decoder we detect the FAWs, which indicate the beginning, the end, and the type of the transmitted data segments.

7.5.5 The Codec Performance

The PSNR versus frame index performance figures for the 4CIF "Football" and "Susie" sequences of Figure 1.1 are given in Figures 7.21 and 7.22. For both simulations we selected quality thresholds of $T_{mean} = 4$ and $T_{peak} = 8$ for the luminance component, resulting

7.5. THE PROPOSED CODEC

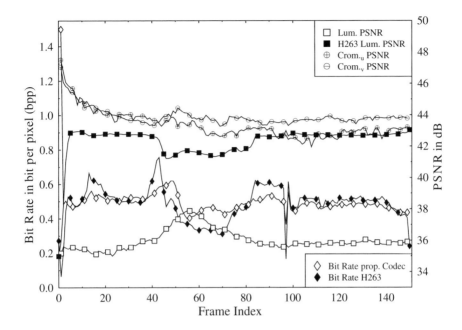

Figure 7.21: Performance of the proposed DCT codec for the 4CIF "Susie" sequence of Figure 1.1, using a MV search scope of 16.

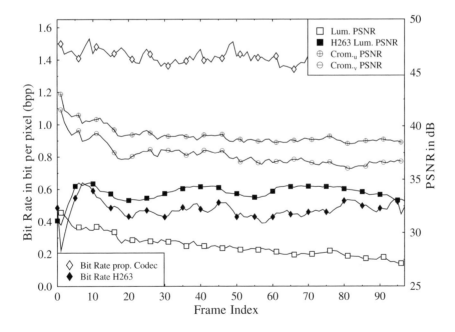

Figure 7.22: Performance of the proposed DCT codec for the 4CIF "Football" sequence of Figure 1.1, for a MV search scope of 32.

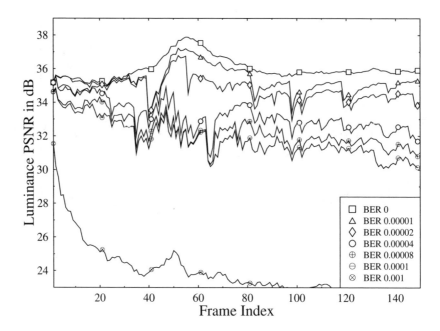

Figure 7.23: PSNR versus frame-index performance of the non-RL coded DCT codec under erroneous conditions, for a MV search scope of 16, and an average bitrate of 0.55 bpp at various BERs.

in a PSNR of about 36 dB. Initially, we applied the same criterion to the chrominance blocks, and consequently the chrominance PSNR dropped from the initial 47 dB PSNR of Figure 7.21 to about the same value as the luminance PSNR. We found that the subjective color reproduction suffered tremendously as a result. In our experiments we needed an enhanced chrominance quality of about 1.5, according to our quality criterion of Equation 7.13 compared to the luminance in order to achieve a subjectively similar luminance and chrominance reproduction. This resulted in a PSNR improvement of about 8 dB, which was rated as a subjectively pleasant quality. Apparently, the human eye seems to be more sensitive to luminance shifts than brightness shifts. We found that especially for the 4CIF "Susie" sequence a good color reproduction was crucial. Note furthermore, that the quality of the "Susie" sequence is about 6–8 dB better at a significantly lower bitrate than that of the high-activity 4CIF "Football" sequence. Comparing the proposed codec with the performance of the H.263 standard codec, we found that in the case of the 4CIF "Susie" sequence at almost the same bitrate the H.263 codec outperformed our more error resilient non-RL coded codec by about 6–7 dB in terms of the PSNR.

Results for the "Football" sequence demonstrate a worst-case scenario. Our codec detects high activity in about 80% of the blocks, forcing almost every block to be fully encoded. Accordingly, the bitrate increases to about 1.5 bpp. The H.263 codec performs better in terms of bitrate and PSNR values. The subjective image-quality difference was found to be less significant as the objective PSNR difference.

7.5. THE PROPOSED CODEC

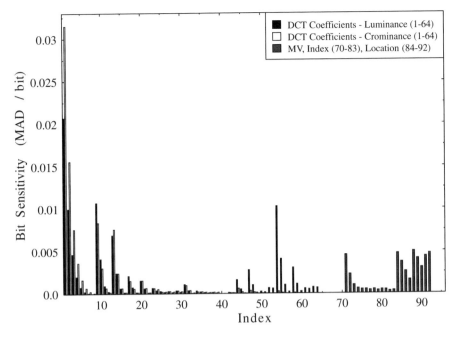

Figure 7.24: Bit sensitivities of the MVs, intra- and inter-frame-coded DCT bits, for the 4CIF "Susie" sequence, MV window size of 16 × 16, and for an average bitrate of 0.55 bpp.

7.5.6 Error Sensitivity and Complexity

The previous section characterized the performance of the non-RL coded DCT codec, which was outperformed by the H.263 standard codec. The advantage of our codec is its error resilience. Figure 7.23 characterizes the error sensitivity of the generated bitstream, which shows error resilience up to a BER of 10^{-4}. A more quantitative characterization of the bit sensitivities ensues from Figure 7.24 for the luminance and chrominance DCT coefficient encoding of bits of 1–64 of the 8 × 8 blocks. Bit positions 70–83 are the 14 MV bits of the blocks, where the MSB at position 70 inflicts the highest Mean Absolute Difference (MAD) degradation, while the 13 block location indices exhibit a more uniform MAD degradation, which was computed by consistently corrupting the bits concerned and evaluating the associated MAD. Examples of images after exposure to a BER of 10^{-3} and 10^{-4} are given in Figures 7.25 and 7.26.

Figure 7.25 shows an even distribution of the errors, whereas at the ten times higher BER of Figure 7.26 many of the errors are concentrated around the fringes of the frame. This is due to the varying encoding probability of the blocks in each frame and is explained as follows. Blocks of the background are less frequently coded than those of active locations of the image. The latter blocks are more frequently subjected to intra-frame coding and are therefore more often updated. In addition, if the index of a MV or DCT block is corrupted, the corrupted index points to a new, false location which might belong to those locations, which are less frequently updated. Apparently, this effect becomes more dominant at higher

Figure 7.25: Picture quality at a constant BER of 10^{-4} for frame 100 of the "Susie" sequence at 0.65 bpp.

Figure 7.26: Picture quality at a constant BER of 10^{-3} for frame 100 of the "Susie" sequence at 0.65 bpp.

BERs, and the problem might be solved by forcing the codec into intra-coded mode in order to encode every block after a preset number of frames, if it is used at such a high BER.

The complexity in terms of computational demand depends strongly on the MC parameters, search scope, and search algorithm, as well as on the activity within the frame and, of course, on the frame size. As shown in Section 7.3, the computational demand of the MVs easily reaches the region of 10^9 Flops, when applying Equation 3.10 at 5400 blocks per CCIR 601 frame. For real-time applications we propose small search windows at the size of 4×4 or the use of simple frame-differencing. This will slightly increase the bitrate, but the computational demand is reduced to that of the DCT transform. The optimized DCT transformation requires 768 Flop, which results in a computational demand of 768×8100 blocks = 6.2 MFlop per color CCIR 601 frame. In the worst case, this is carried out for both the inter- and intra-coded frame during each inter-frame coding step, thus doubling the computational demand. As the signals have to be locally decoded for the inter- and intra-coded decisions, the inverse DCT has to be executed, doubling the computational demand again. Hence, an upper bound complexity estimate for the inter/intra-frame codec for a MV search scope of zero is 25 Mflop. Since some of the operations may be executed in parallel, a real-time implementation in silicon is possible.

7.6 Summary and Conclusions

In this chapter the exploitation of the temporal redundancy was combined with DCT-based coding. We attempted to capitalize on our results from Part I of the book in the field of high-resolution image coding, where a persistent high quality is required. In the context of these codecs, we had to accept a variable bitrate, since the initial, fully intra-coded frames often impose a higher bitrate than the inter-coded frames. The proposed intra-frame subband codec achieved a PSNR of about 33.4 dB at a bitrate of about 2 bits/pixel. In Section 7.4.1 we explored the characteristics of the MCER and then contrived an intra/inter-coded DCT codec, achieving a PSNR of 36 dB at about 0.6 bit/pixel for color sequences.

Both Parts I and II of this book proposed methods for error-resilient image coding with a breakdown BER of the presented codecs in the range of 10^{-4} to 10^{-2}, which are more error resilient than existing standard video codecs, such as the H.26x and MPEG schemes. As more and more public services are provided over mobile channels, such codecs will become more important and stimulate further research. The codecs presented in Part I of this book were designed to allow direct replacement of mobile radio voice codecs in second-generation wireless systems, such as the Pan-European GSM, the American IS-54 and IS-95, as well as the Japanese systems, operating at 13, 8, 9.6, and 6.7 kbit/s, respectively. They are also amenable to transmission over third-generation wireless systems. The results in Part II of the book are applicable to HDTV and WLAN-type systems. As demand for these multimedia mobile services emerges, error-resilient schemes at all rates, qualities, and resolutions will be required in order to satisfy the demands of tomorrow's customers. In the next part of the book, commencing with the next chapter, we consider the standard H.261 and H.263 video codecs and design interactive wireless transmission systems.

Part III

H.261, H.263, H.264, MPEG2 and MPEG4 for HSDPA-style Wireless Video Telephony and DVB

H.261 for HSDPA-style Wireless Video Telephony

8.1 Introduction

This chapter investigates the properties of the H.261 video codec and designs a wireless system facilitating its employment in a mobile radio environment. Initially, a detailed investigation of the properties of the H.261 standard codec [258] was carried out.

Before designing a wireless transceiver facilitating the employment of the H.261 codec in a mobile environment, several investigations of particular coding aspects had to be made. A thorough investigation of the error effects on the H.261-encoded data stream was also undertaken, and different schemes for improving the error resilience of the H.261-based system were studied. Finally, an error-resilient H.261 system was designed and investigated in a wireless mobile scenario.

8.2 The H.261 Video Coding Standard

8.2.1 Overview

The International Telecommunications Union's (ITU) H.261 standard [258] specifies a video codec for audiovisual communications services. It was designed for data rates at multiples of 64 Kbps, thus it is well suited to Integrated Services Digital Network (ISDN) lines, which also have the low Bit Error Rate (BER) required for its operation. A so-called hybrid coding algorithm is employed and inter-frame prediction is used to reduce the inter-frame temporal redundancy, while transform coding is used to reduce the spatial redundancy exhibiting itself in the spatial frequency domain. The motion translation information extracted by motion compensation is represented by motion vectors (MV). The system was designed for bitrates in the range of approximately 40 Kbps to 2 Mbit/s. The block diagram of the H.261 codec is shown in Figure 8.1. Other sources of information about the H.261 video codec are,

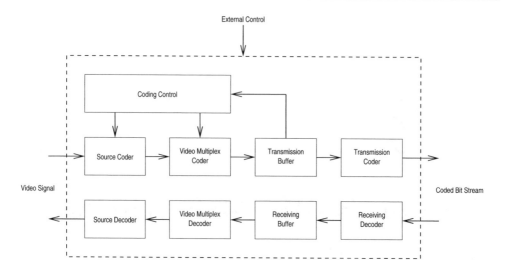

Figure 8.1: Block diagram of the H.261 scheme.

for example, T.Turletti's report on an H.261 video codec for use on the Internet [305] and British Telecom's audiovisual telecommunications book [306].

8.2.2 Source Encoder

The source encoder operates on noninterlaced pictures at approximately 30 frames/s. . These pictures are encoded as luminance and two color difference components (Y, C_B, C_R). Two picture resolution formats are specified, the 352×288-pixel Common Intermediate Format (CIF) and the 176×144 Quarter Common Intermediate Format (QCIF). The color differences are sampled at half the rate of the luminance, as shown in Figure 8.2. In order to be able to adjust the frame rate for a variety of applications having different quality and/or bitrate requirements, the H.261 scheme can disable the transmission of 0–3 frames between transmitted ones.

The video encoding algorithm consists of three main elements: motion prediction, block transformation, and quantization. A video frame can be encoded in one of two modes, either in the intra-frame coded mode, where the original video frame is encoded or in the inter-frame coded mode, where the motion-compensated frame difference related to the previous decoded frame is encoded. In each case, the frame is divided into blocks of 8×8 pixels. Four adjacent luminance blocks and two color difference blocks are then combined to form a so-called *macroblock*. Therefore a macroblock consists of 16×16 luminance pixels and 8×8 chrominance pixels, as shown in Figure 8.2. The block diagram of the H.261 source encoder is shown in Figure 8.3.

In the simplest implementation of the inter-frame coded mode, the previous frame can be used as the prediction for the next frame, a technique often referred to as frame-differencing. However, the inter-frame prediction can be improved by the optional use of full motion compensation relying on motion vectors and/or on spatial filtering. The motion vector-based motion compensation is optional in an H.261 encoder but required in all decoders. A distinct

8.2. THE H.261 VIDEO CODING STANDARD

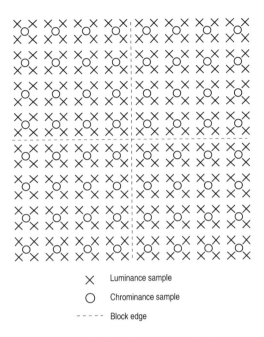

Figure 8.2: Positioning of luminance and chrominance samples in a macroblock containing four luminance blocks.

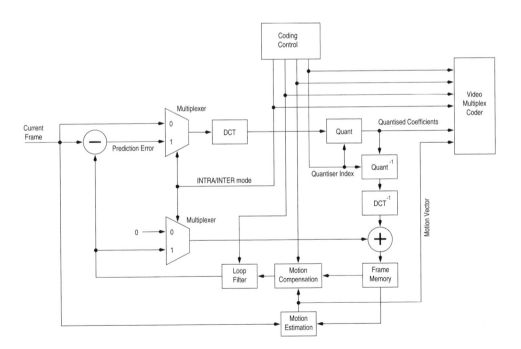

Figure 8.3: H.261 source encoder.

motion vector can be defined for each macroblock. The motion vector defines a translation in two dimensions, and therefore it has two coordinates for the x and y directions, respectively. Each motion vector can assume integer values in the range of ± 15, but they obey the restriction that the pixels referenced must be within the coded picture area. Accordingly, they can be encoded using 4 bits for each dimension, and therefore $(4+4) = 8$ bits are required for each motion vector. The prediction can also be modified by the use of a two-dimensional spatial filter [258] that operates on blocks of 8×8 pixels, mitigating the blockiness of the reconstructed video at low bitrates. The filter can only be used for all or none of the blocks in a macroblock.

The prediction error of the 8×8 pixel blocks is then transform coded. A separable two-dimensional discrete cosine transform (DCT) is used. Its output is clipped to the range of -255 to $+256$. The final element of the source coder is quantization of the DCT coefficients. There is a specific quantizer for the intra-frame coded zero-frequency or direct current (DC) component and 31 different quantizers for all other DCT coefficients. Selecting one of the 31 quantizers constitutes a convenient way of controlling the output bitrate. The quantizers are numbered 1 to 31 from the finest to the coarsest. All quantizers are linear, and apart from the DC-quantizer they all exhibit a central dead-zone around zero. The DC-quantizer has a step size of 8, while the higher-frequency or non-DC quantizers have even-valued step sizes in the range of 2–62. Generally, the coarser the quantizer the lower the bitrate and the transmitted image quality. The quantizer dead-zone is wider than the step size, so that camera noise does not contaminate the signal to be quantized.

The DCT coefficients of the motion prediction error are quantized. The encoder incorporates a local decoder in order to ensure that the motion compensation scheme operates using the reconstructed picture at both ends in its future motion estimation steps rather than the original frame at the encoder and the reconstructed frame at the remote decoder, which would result in misalignment between their operation. As seen in Figure 8.3, the quantized DCT coefficients are passed through an inverse quantizer in the local decoder. The combined transfer function of the quantizer and inverse quantizer is shown in Figure 8.4. The global shape of the transfer function is approximately linear. However, as the figure shows, there is a saturation limit, depending on the required bitrate and video quality, which translates into a quantizer index. The full dynamic range cannot be quantized, if a low quantizer index is required (*index* < 7), as it becomes clear from Figure 8.4, since the output is clipped to a maximum value for quantizers with an index less than 7. The figure also shows that the stepsize of the quantizer is $2n$ where n is the quantizer index. The dead-zone of the quantizer is defined as $6(n-1) + 3$, which is again introduced in order to remove low-level camera noise as well as to reduce the bitrate. Therefore the size of the dead-zone increases, as the step size and quantizer index increase. The coarsest quantizers have the largest dead-zones and step sizes. The finest quantizers have the smallest dead-zone and step size, but they have a restricted dynamic range.

The reconstructed video frames are clipped to the range of 0–255 in order to avoid arithmetic overflow in other parts of the encoder or decoder.

8.2.3 Coding Control

The coding control block adjusts the required performance of the codec by modifying the coding parameters in order to attain the required image quality versus bitrate compromise.

8.2. THE H.261 VIDEO CODING STANDARD

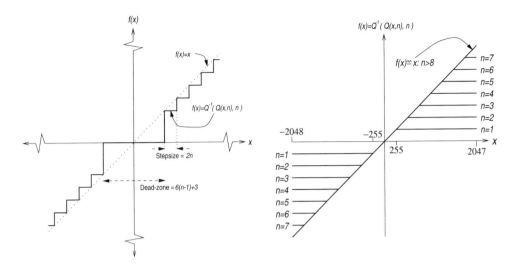

$Q(x,n)$ is the quantizing function.

$Q^{-1}(x,n)$ is the inverse quantizing function.

x denotes the DCT coefficients, n is the index of the quantizer used in the range 1-31.

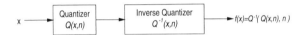

Figure 8.4: H.261 Quantizer transfer function.

Several parameters can be changed in order to modify the behavior of the codec. These can include pre-processing before source encoding, changing the quantizer, changing the block selection criterion, and temporal subsampling.

Specifically, not every block of a frame needs to be updated in each frame, because some have little or no differences in consecutive video frames. The block selection criterion evaluates for each block the difference between the current input block and the previous locally decoded frame. If it is above a certain threshold, the relevant parameters of the block are transmitted. By controlling the threshold, more or less blocks are transmitted, and hence the bitrate of the codec is modified. Temporal subsampling is performed by discarding whole frames, when there is little or no difference since the previous frame. Furthermore, the coding control block ensures that each macroblock is forcibly updated with an intra-frame coded macroblock at least once every 132 times it is transmitted.

8.2.4 Video Multiplex Coder

The video multiplex coder groups the source-coded data into groups and codes them into symbols to be transmitted. The video multiplex coder also attempts to further reduce the residual redundancy in the data stream and organizes the source-coded data into a hierarchical structure. This structure is shown in Figure 8.5. The data stream is then converted into

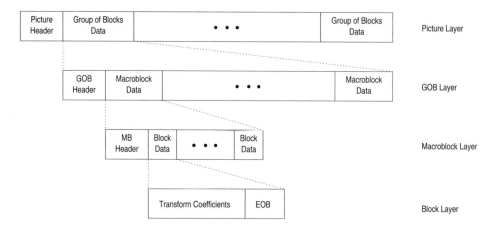

Figure 8.5: Hierarchical structure of a H.261 frame.

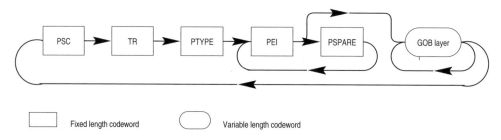

Figure 8.6: Picture layer structure for the H.261 codec.

a bitstream to be transmitted. Most of the transmitted bitstream comprises variable-length codewords, which are used to remove any remaining redundancy. Although these variable-length codes are vulnerable against channel errors, the transmission syntax was constructed to include unique words at various hierarchal levels, which facilitate resynchronization at various stages. Let us now consider how the video stream to be transmitted is constructed.

8.2.4.1 Picture Layer

As seen in Figure 8.5, this layer consists of a picture layer header followed by the Group of Block (GOB) data. The picture layer header is detailed in Figure 8.6, where the self-descriptive transmission packetization format becomes explicit. As seen in Figure 8.5, similarly to the picture layer, all other layers of the hierarchy commence with a self-contained header, describing the forthcoming information, followed by the encoded information of the next lower layer.

The picture-layer header information commences with the Picture Start Code (PSC), which is a specific 20-bit codeword facilitating the recognition of the start of a new frame. This unique word allows the codec to resynchronize, using a simple pattern recognition technique by identifying this code in the received bitstream after the loss of synchronization due to channel errors.

8.2. THE H.261 VIDEO CODING STANDARD

Table 8.1: Summary of the Picture Layer Information in the H.261 Codec

PSC	Picture Start code (20 bits) – fixed.
TR	Temporal reference (5 bits).
PTYPE	Information concerning the complete picture, (6bits) such as split screen, freeze, and CIF/QCIF.
PEI	Extra insertion information (1 bit) – set if PSPARE to follow.
PSPARE	Extra information (0/8/16... bits) – not used, always followed by PEI.

The Temporal Reference (TR) is a 5-bit counter labeling the video frames. If there are a number of nontransmitted frames between transmitted ones, it will be incremented by their number in order to reflect the index of the input frames rather than that of the transmitted ones. The temporal reference word is followed by the 6-bit Picture Type (PTYPE) code, which reflects a number of picture type options, such as whether the frame it refers to is a CIF or QCIF frame, whether the picture freeze option needs to be set or reset, and so on. As demonstrated by Figure 8.6, the next information field is the Extra Insertion (PEI) bit, which signals to the decoder whether further bits are to follow. This option allows the codec to incorporate future services by appending an arbitrary multiple number of 8 bits, as long as PEI is set to logical one. Following this field, the GOB layer information is appended, which will be considered in the next section. The role and the associated number of bits of the various segments are summarized in Table 8.1.

8.2.4.2 Group of Blocks Layer

At the next hierarchical level of Figure 8.5, the picture is divided into Groups of Blocks (GOB), each constituted by 176×48 luminance pels. A CIF and QCIF frame divided into GOBs is shown in Figure 8.7. The GOB layer consists of the GOB header seen in Figure 8.5, followed by the GOB information constituted by the macroblock information. The GOB header commences with the GOB Start Code (GBSC), followed by a number of other control words and macroblock information concerning all macroblocks. The GOB header is characterized in detail in Figure 8.8.

The GOB layer information commences with a 16-bit unique word, the so-called GOB Start Code (GBSC), again, similarly to the PSC word, in order to allow the decoder to resynchronize at the start of the next GOB word in case of loss of synchronization induced by channel errors. The position of any specific GOB in the CIF and QCIF frames is reflected by the Group Number (GN), which is a 4-bit word, allowing the encoding of 16 GOB positions, although only 12 indices are required in Figure 8.7. This segment is followed by a 5-bit field referred to as Group Quantizer (GQUANT), specifying the quantizer to be used in the current GOB. This 5-bit identifier allows the codec to encode up to 32 different quantizers. This quantizer is used until it is superseded by any other quantizer selection information, such as the Macroblock Quantizer (MQUANT) code of the macroblock layer. Similarly to the picture layer, the GOB layer also exhibits a spare capacity for future services, which is signaled to

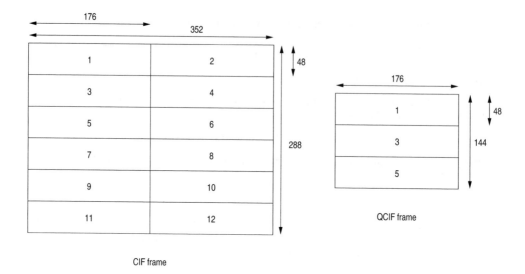

Figure 8.7: CIF and QCIF frames divided into H.261 GOBs.

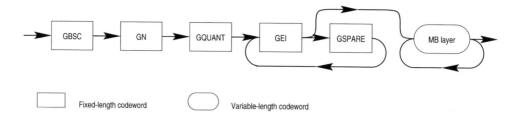

Figure 8.8: Group of blocks layer structure for the H.261 codec.

Table 8.2: Summary of GOB Layer Information in the H.261 Codec

GBSC	Group of blocks start code (16 bits) – fixed.
GN	Group number (4 bits) – binary representation of 1–12, 13–15 reserved. Indicates GOB position in Figure 8.7.
GQUANT	Group quantizer information (5 bits) – codeword to indicate quantizer used for group. Active until overridden by MQUANT.
GEI	Same function and size as PEI in Table 8.1.
GSPARE	Same function and size as PSPARE in Table 8.1.

the decoder using the Extra Insertion (GEI) flag followed by the $n \times 8$-bit GOB-layer spare bits, (GSPARE). These fields are summarized in Table 8.2, indicating also the number of bits assigned. From this transmission format, it is clear that the information stream has a self-contained descriptive structure, which is attractive in terms of compactness and flexibility but is also rather vulnerable to transmission errors.

8.2.4.3 Macroblock Layer

Each GOB is divided into 33 so-called Macroblocks (MB) in an 11×3 arrangement, each constituted by 16×16 luminance pels or four 8×8 blocks, as portrayed in Figure 8.2. Recall, however, that this will correspond to an 8×8 block in terms of the subsampled chrominance values. The MB layer header is described in Figure 8.9.

The MB address (MBA) of Figure 8.9 is a variable-length codeword specifying the position of each macroblock in a GOB. For the first MB of each GOB, the MB address is the absolute address given in Figure 8.7, while for the remaining ones it is a variable-length coded relative address. The lower the index, the shorter the codeword assigned to it. The macroblock type (MTYPE) segment conveys information concerning the nature of a specific MB, whether it is an intra- or inter-frame coded MB and whether full motion compensation was invoked or simple frame-differencing. The 5-bit optional macroblock quantizer (MQUANT) field of Figure 8.9 specifies a new quantizer to be invoked, which therefore supersedes the previous GQUANT/MQUANT code. If the previously included MTYPE code specified the presence of motion compensation (MC), the MQUANT segment is followed by the variable-length motion vector data (MVD). The MVD is encoded relative to the Motion Vector (MV) of the previous MB, where possible, which is again prone to errors, but reduces the bitrate, since MBs close to each other are likely to have similar MVs, yielding a difference close to zero. Therefore, the highest probability MVs' coordinates are assigned a short codeword, while rare vectors are allocated long codewords.

Observe furthermore in Figure 8.9 that the variable-length coded block pattern (CBP) is optional. It conveys a pattern to the decoder, which indicates the index of those blocks in the MB, for which at least one transform coefficient (TC) was encoded. Since for some blocks no TCs will be transmitted and some actively encoded block patterns are more frequent than others, this measure again results in some coding economy. The MB-layer codes and their roles are summarized in Table 8.3, along with the number of bits assigned. Let us now consider the lowest hierarchical layer constituted by the blocks, where block activity control and transform coding are carried out.

8.2.4.4 Block Layer

A macroblock is comprised of four luminance blocks and one each of the two subsampled color difference blocks. The transmitted data of a block consists of codewords for the transform coefficients (TCOEFF), followed by an end of block (EOB) marker. This structure is portrayed in Figure 8.10. The coding parameters are summarized in Table 8.4.

In the intra-frame coded mode, the TCOEFF field is always present for the four luminance and two color difference blocks. In the inter-frame coded modes the macroblock type, namely, MTYPE, and the coded block pattern, namely CBP, indicate for the decoder these blocks, that have nonzero coefficients. The transform coefficients of the active encoded blocks are

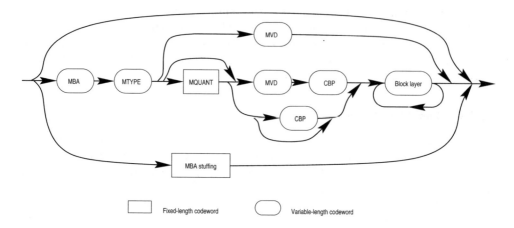

Figure 8.9: Macroblock layer structure for the H.261 codec.

Table 8.3: Summary of MB Layer Information in the H.261 Codec

MBA	Macroblock address (variable length) — Indicates position in GOB. For first transmitted MB, MBA is the absolute address, subsequent MBAs are the difference in address.
MTYPE	Type Information (variable length) — Intra/Inter, Motion Compensation, Filter, MQUANT flag.
MQUANT	Macroblock quantizer information (5 bits). Overrides GQUANT for this and future macroblocks.
MVD	Motion vector data (variable length) — included if flag set in MTYPE. MVD = vector for MB — vector for previous MB. The previous motion vector is zero if not available. Coded with variable-length code for horizontal, then vertical vector component.
CBP	Coded block pattern (variable length) — present if indicated by flag in MTYPE. Codeword specifies those blocks in macroblock, for which at least one transform coefficient is transmitted.

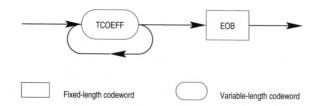

Figure 8.10: Block layer structure for the H.261 codec.

8.2. THE H.261 VIDEO CODING STANDARD

Table 8.4: Summary of Block-layer Codes in the H.261 Codec

TCOEFF	Transform coefficients – Data is present for all six blocks in a macroblock, if intra-mode. In other cases, MTYPE and CBP indicate whether data is transmitted. Coefficients in the 8 by 8 block are reordered in a zig-zag pattern from top-left to bottom-right. The most common combinations are transmitted with variable-length codes, the other combinations are encoded as a 20-bit word $(6 + 6 + 8)$.
EOB	A fixed-length code that represents the end of the coefficients in the block.

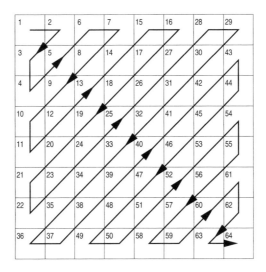

Figure 8.11: Run-length-coding block parameters along a zig-zag path.

scanned in a diagonal zig-zag pattern, commencing at the top left corner of the 8×8 transform coefficient matrix, as shown in Figure 8.11. This scanning procedure orders them in a more-or-less descending order of their magnitudes, compounding the highest values at the beginning of the scanned sequence. Hence at the beginning of this stream, the length of consecutive zero coefficient runs is likely to be zero, but toward the end of the stream there will be many nonzero coefficients, separated by long runs of zero coefficients. This property can be advantageously exploited to further reduce the required bitrate as follows. The most frequently encountered combination of successive zero run-lengths denoted by RUN and their associated magnitude denoted by LEVEL are assigned a variable-length code from a standardized coding table. Those combinations, which are infrequent and hence were not included in the standard table, are then subsequently encoded using a 20-bit code. This 20-bit code contains a 6-bit escape code to inform the variable-length decoder that a fixed-length code is being used. The 6-bit escape code is followed by 6 bits that indicate the run-length and 8 bits for the coefficient value following the specified length zero run.

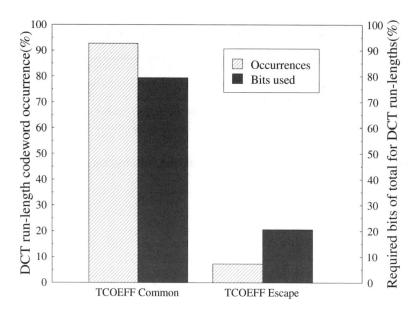

Figure 8.12: A comparison of the occurrence and bits required for coding parameters (codewords) containing run-length coded DCT coefficients.

In simulations, 93% of the run-lengths of the quantized DCT coefficients are in the set of common run-lengths which are coded with variable-length codewords. On average, these variable-length codes require approximately 6 bits per run-length. This is much less than the 20 bits required for the other run-lengths. This is why the other run-lengths, which occur in only 7% of the cases, require 20% of the bits needed to code all the quantized DCT run-lengths. This can be seen in the bar graph of Figure 8.12.

The transmission coder of Figure 8.1 provides a modest error correcting capability using a simple binary BCH codec, which is only suitable for low-BER ISDN lines. The transmission coder also carries out framing before transmission.

8.2.5 Simulated Coding Statistics

In order to find the relative importance of the different coding parameters or codewords in the H.261 bitstream, video-coding simulations were conducted. These results also assisted us in the design of the proposed error-resilient H.261 videophone system. In order to show the relative importance of all the coding parameters, two simulations were conducted:

- Coding with a fixed quantizer.

- Coding with a variable quantizer.

The default mode of the video codec was designed to use a fixed quantizer and motion vectors. The error-resilient videophone system to be described in Section 8.4 was used for the second type of simulation. This system used a variable quantizer to control the bitrate and did not use motion vectors. Both simulations used the same gray-scale "Miss America"

8.2. THE H.261 VIDEO CODING STANDARD

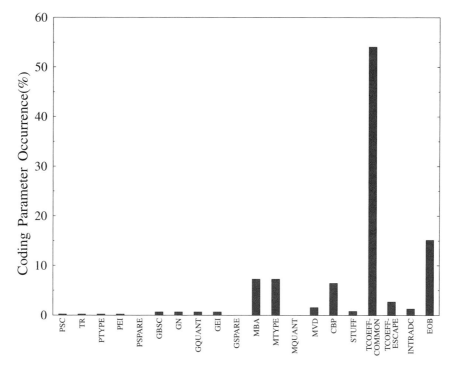

Figure 8.13: Probability of each type of coding parameter (codeword) in the bitstream, for a fixed-quantizer simulation.

video sequence, with just one initial intra-frame. The following sections describe the results for these simulations.

8.2.5.1 Fixed-quantizer Coding

The results of this section show the statistics of the coding parameters or codewords when a fixed quantizer is used. The bar graph in Figure 8.13 shows the probability that each type of coding parameter occurs in a representative coded video sequence. More than half of the coding parameters in a video sequence are the commonly occurring transform coefficient run-lengths (TCOEFF-COMMON) which use variable-length codes. Only 2.7% of the coding parameters used are the DCT coefficient run-lengths that need to be fixed-length coded (TCOEFF-ESCAPE).

The design of the error-resilient videophone system to be highlighted in Section 8.4 used the results of Figure 8.14 to estimate the probability that a random bit error would corrupt a particular coding parameter. As can be seen from the bar graph, over 60% of random bit errors would corrupt a TCOEFF-COMMON coding parameter. The effect of an error on such a parameter is severe due to the double effect of both the variable-length coding itself and the run-length coding used to represent the runs of zero-valued transform coefficients.

Also note that while the fixed-length coded DCT coefficient run-lengths constitute only 2.7% of the total coding parameters, they require over 10% of the bits used to code the video

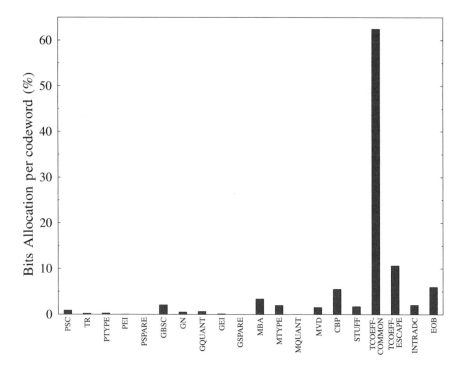

Figure 8.14: Probability of a bit being allocated to a particular coding parameter (codeword), for a fixed-quantizer simulation upon weighting the probabilities of Figure 8.13 by the number of bits allocated to the coding parameters.

sequence. The bar graph in Figure 8.15 shows the average number of bits required for each coding parameter, where some of these coding parameters are fixed length, but the majority are variable-length codewords.

8.2.5.2 Variable Quantizer Coding

The results of this section show the statistics of the coding parameters or codewords when a variable quantizer is used. The results were generated using the error-resilient videophone system to be described in Section 8.4. These simulations did not use motion vectors — only frame-differencing — but the MQUANT coding parameter was used to vary the quantizer and thereby to control the output bitrate of the video codec.

The bar graph in Figure 8.16 shows the relative quantities of various coding parameters that can be contrasted with the fixed-quantizer results of Figure 8.13. The coding parameter MQUANT was not used by the fixed-quantizer simulations. However, the bar graph shows that it is a relatively infrequently occurring coding parameter. The MQUANT coding parameter is used to change the quantizer employed for the DCT coefficients.

The only major difference between these variable-quantizer simulations and the fixed-quantizer simulations is that more TCOEFF-ESCAPE coding parameters are used. The TCOEFF-ESCAPE coding parameter is used to transmit run-length coded DCT coefficients,

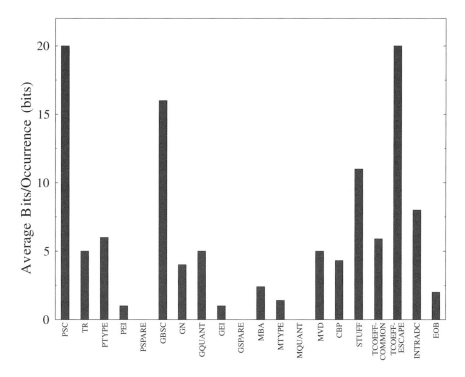

Figure 8.15: Average number of bits used by each coding parameter (codeword), for a fixed-quantizer simulation.

which are not very common. This implies that controlling the bitrate and packetizing the bitstream may cause more uncommon DCT run-lengths to occur.

The bar graph in Figure 8.17 shows the allocation of bits for the various coding parameters in the coded bitstream. Again, as in the fixed-quantizer simulation of Figure 8.14, a random bit error would have a 60% chance of corrupting a TCOEFF-COMMON coding parameter. The increase in the generation of TCOEFF-ESCAPE coding parameters in the variable-quantizer simulation caused the probability of a random bit error corrupting a TCOEFF-ESCAPE codeword to increase to 17% from 10% in the fixed-quantizer simulations.

Figure 8.18 is a bar graph of the average bits per coding parameter for the variable-quantizer simulation, which can be contrasted with the fixed-quantizer results of Figure 8.15. In the next section, we consider the problems associated with employment of the H.261 codec in wireless environments.

8.3 Effect of Transmission Errors on the H.261 Codec

8.3.1 Error Mechanisms

The H.261 standard was designed mainly for ISDN lines, with a channel capacity of at least 64 kbit/s and a low BER. For a mobile environment, the error rate is typically significantly higher, and according to the Shannon-Hartley law, the available channel capacity is dependent

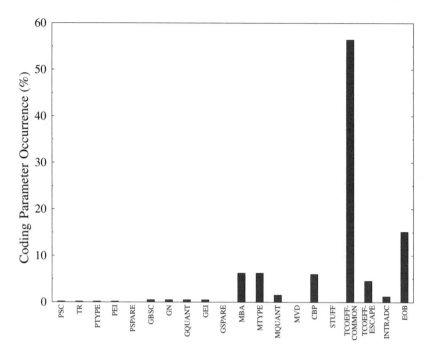

Figure 8.16: Probability of each type of coding parameter (codeword) in the bitstream, for a variable-quantizer simulation.

on the channel conditions. Furthermore, a 64 kbit/s radio channel would be very inefficient and costly in terms of bandwidth. Therefore, in order to enable the employment of the H.261 codec in a mobile environment, the bitrate has to be reduced and the error resilience has to be improved. The video quality versus bitrate balance of the codec can be controlled by the appropriate choice of the quantizers invoked. Improving the error resilience is more of a problem due to the extensive employment of bandwidth-efficient but vulnerable variable-rate coding techniques.

The transmission errors that occur in a H.261 bitstream can be classified into three classes, namely, detected, detected late, and undetected. A detected error is one that is recognized immediately, since the received bit pattern was not expected for the current codeword. Errors that are detected late are caused by the error corrupting a codeword to another valid codeword, with the result that within a few codewords an error is detected. The final type of error is an undetected error, which also corrupts a legitimate codeword to another valid codeword, but no error is detected later.

Detected errors cause little noticeable performance degradation, but errors that are not detected or detected late cause the bitstream to be decoded incorrectly. An incorrectly decoded bitstream usually corrupts one or more macroblocks, and the error spreads to larger areas within a few frames due to motion compensation. These precipitated errors are only corrected by the affected macroblocks being updated in intra-mode.

Generally, when an error is detected, the H.261 decoder will search through the incoming data stream for a picture start codeword (PSC) or group of blocks start code (GBSC) and when

8.3. EFFECT OF TRANSMISSION ERRORS ON THE H.261 CODEC

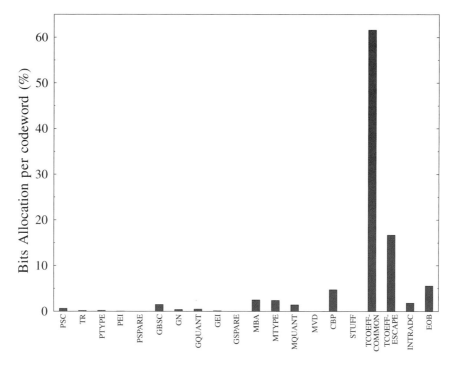

Figure 8.17: Probability of a bit being allocated to a particular coding parameter (codeword), for a variable-quantizer simulation.

this is found, the decoding process is restarted. This implies that all blocks following an error in a frame are lost. If the encoder is in its inter-frame coded mode, the local decoder of the encoder and the remote decoder will become misaligned, precipitating further degradation until the affected macroblocks are updated in intra-frame coded mode, or in other words, until an intra-coded frame arrives, facilitating resynchronization. In order to recover from transmission errors, the update of macroblocks in intra-mode is of paramount importance. However, if the intra-frame coded macroblocks are corrupted, the resulting degradation can be worse than that which the intra-frame coded macroblock was attempting to correct.

A further problem associated with employing the H.261 codec in a mobile environment is packet dropping, which can be inflicted, for example, by contention-based multiple access schemes. Dropping a H.261 transmission packet would be equivalent to a large number of bit errors. One solution to mitigate this problem is employing Automatic Repeat Request (ARQ) techniques. However, invoking ARQs regardless of the number of transmission attempts until a packet is received correctly is impractical.

8.3.2 Error Control Mechanisms

8.3.2.1 Background

In order to solve the problems associated with employment of the H.261 video codec in a mobile environment, the following issues have to be addressed:

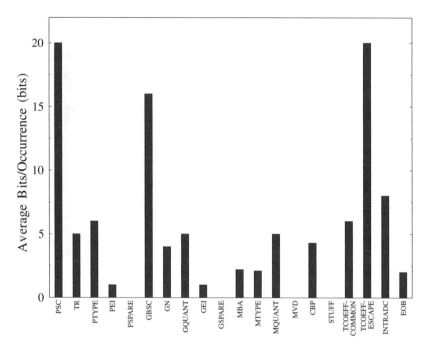

Figure 8.18: Average number of bits used by each coding parameter (codeword), for a variable-quantizer simulation.

- Reliable detection of errors in the bitstream in order to prevent incorrect decoding and error propagation effects.

- Increased resynchronization frequency in order to assist the decoder's recovery from error events, which would reduce the number of blocks that are lost, following an error in the bitstream.

- Identifying mechanisms for dropping segments of the bitstream without propogating errors.

- Contriving synchronization techniques for maintaining the synchronous relationship of the local and remote decoders even in the presence of errors.

A variety of solutions for addressing one or more of these problems are described below. Most radio systems make use of forward error correction (FEC) codecs and/or Cyclic Redundancy Check-sums (CRC) applied to the transmitted radio packets. These techniques can assist in ensuring reliable error detection and hence curtailing error propagation effects.

8.3.2.2 Intra-frame Coding

The H.261 standard relies on the intra-frame coded macroblocks in order to mitigate and ultimately to remove the effects of transmission errors that have occurred. Therefore, by increasing the number of intra-frame coded macroblocks in the coded bitstream, the error recovery is accelerated. Increasing the number to the ultimate limit of having all frames coded as intra-frame coded macroblocks curtails the duration of any error that occurred in the current frame to this particular frame. The disadvantage of this method is that the bitrate increases dramatically. Simulations have shown that for the same image quality, this bitrate increase is of the order of a factor of 10. Since bandwidth in a mobile scenario is always at premium, the image quality that could be maintained with this method under the constraint of a fixed bitrate budget is drastically reduced.

8.3.2.3 Automatic Repeat Request

Automatic Repeat Request (ARQ) techniques can be invoked in order to improve the BER of the channel, at the expense of increased delay. However, if the channel becomes extremely hostile, then either the data will never get through or the delay will increase to a level that is impractical for an interactive videophone system. If ARQ was combined with a regime of allowing packets in the data-stream to be dropped after a few transmission attempts, then the maximum delay could be bounded, while limiting the channel capacity reduction inflicted by ARQs. However, ARQs require a low-rate acknowledgment channel, and hence specific system-level measures are required in a mobile scenario in order to support their employment.

ARQ has been used for several video-coding systems, designed for mobile scenarios. Both [307] and [195] used ARQ and interleaving to improve the error resilience of video transmitted over mobile channels.

8.3.2.4 Reconfigurable Modulations Schemes

Employment of a reconfigurable modulation scheme lends the system a higher grade of flexibility for varying channel conditions. For example, a scheme using 4-level Quadrature Amplitude Modulation (4QAM, 16QAM, and 64QAM) can operate at the same signaling rate or symbol rate, while transmitting 2, 4, and 6 bits/symbol. This allows a 1:2:3 ratio of available channel capacities under improving channel conditions, which can benefit the video codec by allowing a concomitantly higher transmission rate and better video quality in a fixed bandwidth.

Explicitly, for a high-quality channel, the more vulnerable but higher capacity 64QAM mode of operation can be invoked, providing higher quality video due to the extra capacity. When the channel is poor, the more robust 4QAM scheme can be employed; due to the lower channel capacity, the image quality is lower. However, the 4QAM scheme is more robust to channel errors and hence ensures operation at a lower BER. Therefore, a system could switch between modulation schemes depending on channel conditions. The image quality that is dependent on the bitrate therefore increases under more benign channel conditions. The combination of ARQ with reconfigurable modulation schemes would allow for rapid response to time-variant channel conditions, invoking the more robust modulation schemes during ARQ attempts that would maintain a higher chance of getting through uncorrupted.

8.3.2.5 Combined Source/Channel Coding

When invoking ARQ or reconfigurable modulation schemes, the range of channel SNRs for which the H.261 bitstream can be transmitted without errors increases. However, this does not improve the ability of the decoder to cope with an erroneous bitstream.

In order to reduce the image-quality degradations due to transmission errors, it is beneficial to detect the transmission errors externally without relying on the variable-length codeword decoder's ability to detect them. This prevents any potential degradation caused by incorrect decoding of the data stream. If, however, there are uncorrectable transmission errors, the H.261 codec will attempt to resynchronize with the local decoder of the encoder by searching for a PSC or GBSC in the data stream. The decoding process can then be resumed from the PSC or GBSC. However, the data between the error and resynchronization point is lost. This loss of synchronization inflicts prolonged video quality degradation because the encoder's local decoder and the remote decoder become misaligned. This degradation lasts until the local decoder and remote decoder can be resynchronized by updating the affected macroblocks in intra-frame coded mode.

If the encoder's local decoder could identify the lost data stream segment, it could compensate for it by keeping the local decoder and remote decoder aligned. This would ensure that the image segment obliterated by the error could be kept constant in the forthcoming frame. There is a small image-quality degradation due to this obsolete, but error-free, image segment, but this is readily replenished within a few frames, and it is small compared with loss of decoder synchronization or incorrect decoding.

By using a low-rate feedback channel, which can be simply superimposed on the reverse-direction packet of a duplex system, the encoder can be notified of packet losses inflicted by the mobile network. By including this packet-loss channel feedback message in the encoder's coding control feedback loop, both decoders can take adequate account of it and maintain their synchronous relationship. The source-sensitivity matched joint optimization of the source encoder and channel encoder is extremely beneficial in a mobile scenarios. For the channel feedback information to be exploited efficiently, the round-trip delay must be short.

When the source encoder is informed that a packet has been lost, it can adjust the contents of the local decoder in order to reflect this condition. As a result, the decoder will use the corresponding segment of the previous frame for the specific part of the frame that was lost, which is nearly imperceptible in most cases. The lost segment of the frame will then be updated in future frames, and hence within a few frames the effects of the error will decay.

8.3.3 Error Recovery

In order to identify the most efficient error control mechanisms, it was necessary to investigate the effects of transmission errors on the H.261 data stream for its employment in a mobile environment. Therefore, the methods of recovering from these errors were investigated, and the effects of bit errors on different coding parameters in the H.261 data stream were studied.

Many of the codec parameters in the H.261 data stream are encoded using variable-length binary codes. This implies that when an error occurs, the decoder is oblivious of where the bitstream representing the next codec parameter starts. The decoder then has to search through the data stream for a symbol or coding parameter that can be easily and uniquely recognized. Two symbols were designed to be uniquely identifiable: the picture start code (PSC) and the group of blocks start code (GBSC). Both symbols begin with the same initial 16 bits so that the decoder can search for both simultaneously. The other variable-length codewords were designed so that a series of symbols could not be misinterpreted as a PSC or GBSC. Naturally, because of transmission errors, this situation can still be induced, but in case of long unique words the probability of this occurrence is sufficiently low.

The decoder can resume decoding at either of these symbols. However, if an error was not detected previously in the data stream, causing incorrect decoding to occur, it may not be sufficiently safe to resynchronize at the GBSC. For example, when bits in the PTYPE symbol are corrupted, the error cannot be detected by the decoder. Specifically, if, for example, the QCIF/CIF bit in the PTYPE code is corrupted, then the expected frame size is different, hosting three or twelve GOBs. Thus, resuming decoding at a GBSC may cause misplacement of the forthcoming GOBs, inflicting more severe video degradation than would have occurred upon waiting for the next PSC.

In a QCIF frame, for example, there are only four possible resynchronization points per frame, of which three are GBSC points and the fourth is the PSC. When an error occurs, a large segment of a frame can be lost before resynchronization can take place. Upon finding a mechanism of increasing the number of resynchronization points, the error effects are expected to decay more rapidly. If the decoder is restarted at a GBSC, the maximum loss is a GOB or a third of a QCIF frame. If the decoder waits for a PSC before restarting the decoding operations, then the maximum loss is a whole frame.

8.3.4 Effects of Errors

Since a single bit error can cause the loss of the remainder of the current GOB or frame, it could be argued that the effect of bit errors on different codec parameters in the H.261 data stream was more or less equal and in general very detrimental. The effect of a single error could be quantified on the basis of how much of the GOB or frame is lost.

However, if the H.261 data stream was rendered more resilient, the effect of errors on different symbols or codec parameters would vary more widely. It is relatively easy to prognose the worst-case effect of a single bit error on a particular codec parameter, but it is harder to quantify the Peak Signal to Noise Ratio (PSNR) degradation in terms of dB, which it would inflict. The only way to identify the quantitative effect of a single bit error on a symbol type or codec parameter is to simulate an error and evaluate the degradation caused. However, it is not always feasible to simulate a single transmission error and ensure that the resulting degradation is close to the worst case for that symbol or codec parameter. This is because the amount of degradation is dependent on the specific effects and mechanism of incorrectly decoding the bits of the variable-length codeword following or including the single transmission error. Let us now analyze the qualitative effects of the specific error events.

8.3.4.1 Qualitative Effect of Errors on H.261 Parameters

This section lists all the H.261 codec parameters or symbols and analyzes the typical and worst-case degradations if such a symbol is corrupted.

PSC The loss of a picture start code may in theory inflict the loss of a whole frame. However, the codec may be able to resynchronize at the beginning of the next GOB, upon receiving the next GBSC. Specifically, resynchronization becomes possible if the other parameters of the corrupted picture layer header, such as the picture type code PTYPE, are the same as in the previous frame. This allows the codec to reuse the corresponding parameters from the previous one.

TR The temporal reference is an indication of when the frame was coded with respect to the last one, and the effect of its corruption is not explicit in our codec implementation.

PTYPE The picture-type parameter contains specific bits that can cause significant video degradation, while other bits hosted by it may have little or no effect. The most vulnerable bit is the QCIF/CIF flag bit. If this bit is corrupted, the decoder could switch from QCIF mode to CIF and thus cause a loss of at most one frame before receiving the next PTYPE bit in the forthcoming frame.

PEI The extra insertion bit is a very vulnerable one. It was designed for future expansion of the standard, but corruption of this bit leads to the loss of a minimum of 8 bits, since it indicates whether any PSPARE bits will follow. If it is corrupted, the next 8 bits may be misinterpreted as PSPARE bits and hence are not decoded as the next GBSC sequence and vice versa. The maximum degradation yielded by a PEI error is a whole frame, since the PEI/PSPARE syntax can loop for many iterations. The maximum loss possible is a whole frame, but the typical corruption of this bit would cause the loss of the group of blocks (GOB) following the corrupted PEI bit.

PSPARE Corruption of the PSPARE bits has no effect in the current version of the H.261 standard, since these bits are not actively utilized but may be used in the future.

GBSC Loss of a group of blocks startcode will yield the loss of a whole GOB, but resynchronization is possible at the next GBSC.

GN The group number specifies the position of the GOB within the frame about to be decoded. If this value is corrupted to another valid GOB position, the effect of the error can be large because the whole of the GOB will be misplaced. The ramifications of this in an intra-coded frame would be very grave, leading to a substantial degradation of image quality. However, if the group number is corrupted to an invalid value, then the following GOB will be lost. Loss of a GOB is preferable to misplacement in terms of image degradation that could be caused.

GQUANT Corruption of the group quantizer parameter will result in employing the wrong quantizer for decoding at the remote decoder. However, the effect on image degradation will depend on how significant the change in quantizer value is due to the corruption. The amount of degradation is also dependent on the number of macroblocks to follow in the current GOB.

GEI Another very vulnerable bit, that can be used for future expansion of the standard, for applications unknown at the time of writing is the group of blocks layer extra insertion bit. For the bits of the GEI the error effects are the same as for those of the PEI. Corruption of this bit can also result in the corruption of the first macroblock address MBA in the given GOB, which inflicts further image degradation.

GSPARE Corruption of the group of block layer spare bits GSPARE has no effect at present, but it may affect future versions of the H.261 standard, as we have stated in the context of the PSPARE symbol.

8.3. EFFECT OF TRANSMISSION ERRORS ON THE H.261 CODEC

MBA The macroblock address (MBA) is the first of the variable-length coded parameters. The first MBA in a GOB indicates the absolute position of that macroblock within the GOB. Further MBAs in a GOB represent positions relative to the last MBA. Therefore, corruption of a MBA causes all further macroblocks in the GOB to be misplaced. Corruption of this symbol or codec parameter often leads to incorrect decoding owing to its variable-length coding, which can inflict further precipitated image-quality degradation. Furthermore, incorrect decoding of the MBA often leads to the decoder losing synchronization and having to resynchronize, thereby losing the rest of the GOB. Hence, the corruption of this symbol leads to one or more of the following phenomena: misplacements, incorrect decoding, or loss of the rest of the GOB.

MTYPE Corruption of the variable-length coded macroblock-type parameter, reflecting the intra/inter-frame coded, motion-compensated, filtered, nature of the macroblock can also lead to incorrect decoding and often to the loss of the rest of the GOB. However, it can also change the decoder from inter- or intra-frame coded modes, turn on or off motion vector compensation, and loop filtering. Incorrect decoding in one of these incorrect modes can lead to significant degradation of the image quality.

MQUANT The macroblock quantizer parameter is similar to the GQUANT parameter in terms of its error sensitivity and video quality degradation effects. Corruption of the macroblock quantizer identifier will result in employing a different quantizer and inverse quantizer at the encoder and decoder, respectively. The effect of its corruption on image degradation will depend on how different the choice of the quantizer and inverse quantizer is due to the corruption. The amount of degradation is also proportional to the number of macroblocks to follow, before another MQUANT or GQUANT parameter is encountered.

MVD The motion vector displacement parameter is a variable-length coded quantity. Therefore, the corruption of this symbol will typically result in incorrect decoding of the rest of the GOB, forcing the current macroblock to use the wrong macroblock-sized segment for prediction, which does not necessarily map to particular legitimate macroblock position. This type of error will then spread through the reconstruction buffer of the decoder to nearby macroblocks within a few consecutive frames. This can only be corrected by the affected macroblocks being updated in intra-frame coded mode.

CBP The coded block pattern parameter explicitly specifies for which of the blocks we can expect transform coefficient values in the block layer. Corruption of this variable-length coded parameter often leads to erroneous decoding and loss of the rest of the GOB.

intraDC The intraDC parameters are used to encode the DC component of the intra-coded frames. This is a fixed-length coded symbol, and hence the amount of image-quality degradation is dependent on the difference between the corrupted value and uncorrupted value. Because this is concerned with encoding the intra-frame coded macroblocks, any video degradation due to an error in this parameter will typically be prolonged until the affected macroblock is updated with another intra-frame coded macroblock.

TCOEFF The DCT transform coefficient parameters are split into two classes: a variable-length coded class for frequently encountered coefficient sequences and a fixed-length 20-bit class for those that are not included in the run-length/value coding tables. The variable-length class symbols will generally cause incorrect decoding and loss of the rest of the GOB when they are corrupted. The fixed-length 20-bit coded transform coefficients will inflict an error only within the specific block concerned, when they are corrupted.

EOB The end-of-block symbol signifies the end of the block layer and the start of another macroblock, GOB, or frame. When an EOB symbol is corrupted, it can at worst cause the loss of a frame but most often the loss of part or all of a GOB.

Having considered the qualitative effects of transmission errors as regards the typical and worst-case video degradations, let us now consider the quantitative effects of error events.

8.3.4.2 Quantitative Effect of Errors on a H.261 Data Stream

The only practical way of identifying the quantitative effect of errors is to corrupt the specific symbol concerned in the H.261 data stream and quantify the inflicted video degradation. The image degradation can be quantified by finding the difference between the decoded PSNR of the corrupted and uncorrupted H.261 data streams. However, it cannot be ensured that the inflicted one is the worst possible degradation. Nonetheless, this technique does give an estimate of the severity of such corruption compared to that of other parameters. These results can be invoked in order to classify symbols into sensitivity classes dependent on their vulnerability to errors.

In order to investigate the effect of errors on particular H.261 codec parameters, wide-ranging simulations were carried out using a 50-frame "Miss America" (MA) sequence. Simulations were conducted for the following four situations:

- Error in intra-coded frame (frame 1/50).

- Error in inter-coded frame (frame 2/50).

- Error in inter-coded frame using motion vectors (frame 2/50).

- Error in inter-coded frame at low quality (frame 2/50).

The following H.261 parameters were corrupted during these investigations: PEI, GN, GQUANT, MBA, MTYPE, MVD. These symbols were selected because they were the ones for which the effects of the corruption were difficult to estimate. The following H.261 symbols were not corrupted in these investigations because the effects of their corruption could be estimated reasonably well without simulations: PSC, TR, PTYPE, PSPARE. Furthermore, the MQUANT, GBSC, GEI, and GSPARE parameters were not corrupted because their effects would be similar to those of GQUANT, PSC, PEI, and PSPARE. The remaining H.261 parameters, namely, CBP, intraDC, TCOEFF, and EOB, were not corrupted, since their position was not readily identified in the data stream. Let us now highlight the quantitative error effects in intra-coded frames.

8.3. EFFECT OF TRANSMISSION ERRORS ON THE H.261 CODEC

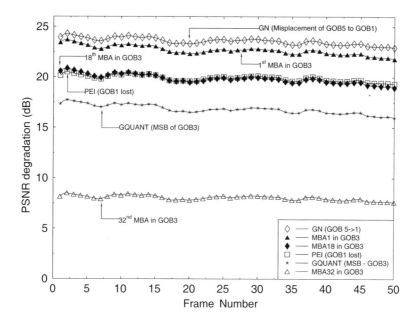

Figure 8.19: Image degradation due to corruption of H.261 coding parameters in an intra-coded frame.

8.3.4.2.1 Errors in an Intra-coded Frame These simulations were carried out in order to demonstrate the difference between the effect of errors in inter- and intra-coded frames. Let us begin our in-depth investigations in the intra-coded frame mode, concentrating on the following error events:

- Corrupted group number (GN) causing a GOB misplacement.

- Corrupted most significant bit of group quantizer (GQUANT).

- Corrupted picture extra information bit (PEI) causing loss of first GOB.

- Corrupted macroblock addresses (MBA) leading to the loss of the rest of the GOB, considering the 1st, 18th, and 32nd macroblock in the GOB.

The codec was configured to transmit an intra-coded frame every 50 frames. The graph of peak signal-to-noise ratio (PSNR) degradation versus frame index for the "Miss America" sequence is shown in Figure 8.19. The top curve represents the case when the GN parameter of the intra-coded frame was corrupted, misplacing the fifth group of blocks to the first position. These positions can be located with the aid of Figure 8.7 for our QCIF frame. The corresponding subjective effects can be inspected at the top right corner of Figure 8.20, which shows the misplaced GOB scenario. As Figure 8.19 shows, the PSNR degradation is very severe and persists until the arrival of the next intra-frame coded macroblocks.

When the most significant bit of the group quantizer (GQUANT) for the third group of blocks GOB3 is corrupted, the objective PSNR degradation in Figure 8.19 is also rather high — in excess of 15 dB and nondecaying. However, its subjective effects are less annoying,

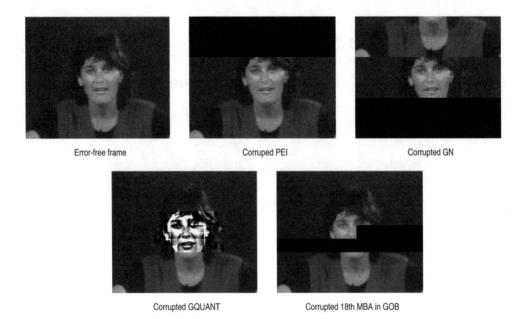

Figure 8.20: Image degradation due to corruption of H.261 coding parameters in an intra-coded frame.

manifesting themselves mainly in terms of a bright segment in the center of MA's face at the bottom left corner of Figure 8.20.

The corrupted PEI parameter of the picture layer header results in a misinterpreted GOB1 segment, which was deemed to be a PSPARE segment due to the corruption of the PEI bit. This is portrayed in the center of Figure 8.20. The associated PSNR degradation is around 20 dB and nondecaying.

Lastly, when the macroblock address MBA of the third GOB is corrupted, the extent of the video quality degradation is dependent on the position of the corrupted macro-block. If, for example, the MBA of the last but one MB, namely, that of the 32nd, is corrupted, the damage inflicted is limited to the (1/33)rd of the third GOB, which has a similar effect to that shown for the 18th macroblock in the center of the bottom right subfigure of Figure 8.20. In contrast, if the first MBA is corrupted, the rest of the GOB is lost and becomes black. When the 18th MBA is perturbed by a transmission error, about half of GOB3 is lost, as demonstrated at the bottom right corner of Figure 8.20. The corresponding PSNR degradations are shown in Figure 8.19.

All in all, as these graphs reveal, intra-frame errors result in image degradations that diminish very little with time. The effects of these errors will last until the affected macroblocks are retransmitted in intra-frame coded mode. The GOB misplacement caused by the GN error causes the worst degradation. As expected, the PSNR degradations due to MBA errors demonstrated that the position of the error in the GOB governs the amount of degradation when the remainder of the GOB is lost. Following this discourse on the effects on errors in intra-frame coded mode, let us now consider the inter-frame coded scenario.

8.3. EFFECT OF TRANSMISSION ERRORS ON THE H.261 CODEC

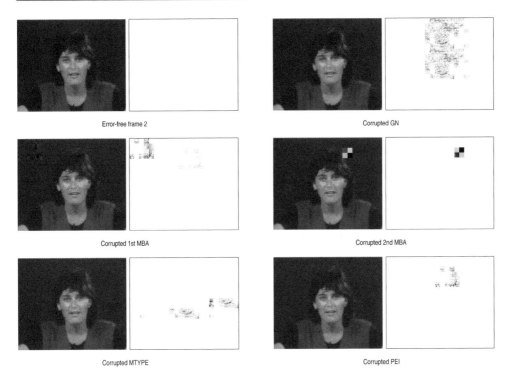

Figure 8.21: Subjective image degradation due to corruption of H.261 coding parameters in an inter-coded frame.

8.3.4.2.2 Errors in an Inter-coded Frame Most types of errors were investigated in more depth in inter-coded frames than in the case of intra-coded frames because inter-coded frames are more frequent. These investigations were conducted using the second frame of the "Miss America" sequence. The corrupted frames in Figure 8.21 are shown as pairs of pictures: at the left, the corrupted decoded frame, and at the right, the difference between the corrupted frame and the uncorrupted frame. In order to augment the visibility of errors and artifacts in the error frame, the largest luminance value in the error frame was always scaled up to 255, which was represented as a black pixel. Thus, the nature of real image degradation is best observed in the image frame itself, whereas the error images assist in observing the extent of error propagation and the size of the affected areas, but the shade of different error images cannot be compared with each other.

The difference image sometimes shows two areas of difference for a single transmission error. This is typically due to blocks being misplaced, where one of the corrupted areas is, where the blocks have moved from, and the second is where they moved to. The error events investigated for inter-coded frames were:

- Corrupted picture extra information bit (PEI) causing the loss of the first GOB.

- Corrupted group number (GN) causing a GOB misplacement.

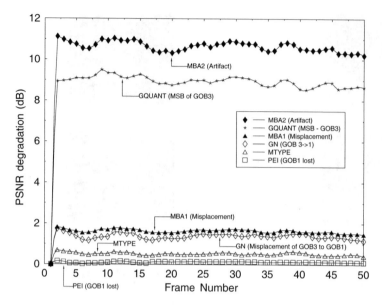

Figure 8.22: Image degradation in terms of PSNR due to corruption of H.261 coding parameters in an inter-coded frame.

- Corrupted first macroblock address (MBA) in a GOB causing all macroblocks in the GOB to be misplaced due to erroneous decoding.

- Corrupted second macroblock address (MBA) in a GOB. Here the decoder does not lose synchronization.

- Corrupted macroblock type (MTYPE) causes an error that affects all remaining macroblocks in the current GOB.

The corrupted frames are displayed in Figure 8.21 while the corresponding peak signal-to-noise ratio degradation versus video frame index is shown in Figure 8.22. The effect of the corrupted GQUANT parameter shown in Figure 8.22 was omitted from the corrupted frames in Figure 8.21, since the quantizers are investigated in greater depth at a later stage. At this stage, we note that these results analyzing the ramifications of corrupted parameters in inter-coded frames are not worst-case results. In fact, since in these studies the second MA frame was used, which follows immediately the intra-coded frame, only a very limited amount of new information was introduced during this inter-coded frame. Hence, if a GOB was corrupted, it did not result in subjectively annoying artifacts over the whole area of the GOB, since the MBs in the background region of the frame contained virtually no new information and therefore no transform coefficients. In general, the higher the number of macroblocks in the GOB where the error occurs, the worse the video degradation. During these investigations most of the errors occurred in the first GOB, which generally has fewer macroblocks per GOB than the other two GOBs in a typical head-and-shoulders QCIF frame.

Beginning with the corrupted PEI bit, it is expected that similarly to the intra-frame coded scenario, due to misinterpreting the forthcoming GOB data as PSPARE bits, the information of GOB1 is lost. However, both the subjective and objective effects of this event are relatively limited, as supported by Figures 8.21 and 8.22. Again, this was due to the virtually unchanged background area of the frame. Had the frame been more motion-active, there would have been more active MBs in the background, which would have resulted in more annoying artifacts, such as those that we have seen for the PEI bit in the case of the intra-coded frame in Figure 8.19.

When the group number GN of GOB3 is erroneously decoded as GOB1, the error effects are more severe, but not as annoying, as in the case of corrupting the same parameter of an intra-coded frame, which is demonstrated at the top right corner of Figure 8.21. Again, this leads to corruption in two GOBs, but its subjective effects are more mitigated than in Figure 8.19 owing to the previously mentioned sparsity of active MBs near the edges of the frame.

Similarly, the previously mentioned double-error pattern is observed in the left-center subfigure of Figure 8.21 due to the corrupted MBA1 parameter of GOB1, which is associated with misplacing all the MBs in GOB1. The corresponding PSNR degradation is around 2 dB in Figure 8.22. The objective PSNR reduction due to incorrectly decoding MBA2 of GOB1 is seen to be quite dramatic in Figure 8.22 — around 11 dB. This type of error also tends to spread to adjacent macroblocks within a few frames. The perceptual error can be clearly seen in Figure 8.21. The macroblock classifier MTYPE conveys the intra/inter-coded classifier, motion vector information, and so on, and its corruption affects all remaining MBs in GOB3. The subjective effects of the GQUANT MSB error in GOB3 are not demonstrated here, although the associated PSNR degradation of Figure 8.22 is rather severe.

In summary, the errors in an inter-coded frame do not affect the image quality as adversely as intra-coded frame errors, which is due mainly to the lower number of motion-active MBs. However, as with intra-coded frame errors, the effects diminish very little over time, persisting until the reception of the re-transmission of the affected macroblocks in intra-frame coded mode.

8.3.4.2.3 Errors in Quantizer Indices Since the quantizer indices are the most commonly used fixed-length coded symbols in a H.261 data stream, a deeper investigation of the effect of errors on these specific bits was carried out. These simulations were conducted using inter-coded frames.

In general, a quantizer selection error will have a graver effect when there are more macroblocks in a GOB. For the "Miss America" QCIF sequence, generally there are more macroblocks per GOB in the central GOB (GOB3) than in the bottom GOB (GOB5). It was also found that there were usually more macroblocks per GOB in the bottom GOB (GOB5) than in the top GOB (GOB1). Further simulations were necessary to evaluate whether the number of MBs in a group of blocks made any difference to the vulnerability of a specific bit in the quantizer selection parameter GQUANT. If it did, then the GQUANT parameters in GOB3 could be classified as more vulnerable than those in GOB1 in head-and-shoulders videophony, since we argued above that the most MBs are likely to be in GOB3.

The 5-bit GQUANT parameter has 32 values associated with 32 different quantizers with bit 0 assigned to the least significant bit (LSB) and bit 4 to the most significant GQUANT bit (MSB). The following GQUANT bit errors were injected:

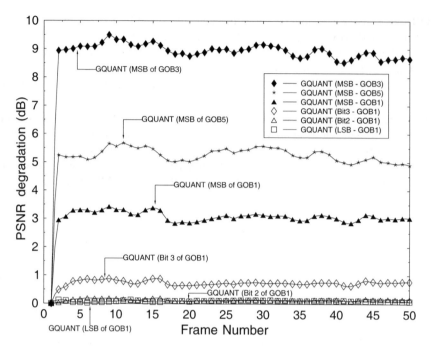

Figure 8.23: Image degradation in terms of PSNR due to corruption of the H.261 GQUANT parameter in an inter-coded frame.

- Corrupted MSB (Bit 4) of GOB1 (top GOB in QCIF frame).
- Corrupted MSB of GOB3 (middle GOB in QCIF frame).
- Corrupted MSB of GOB5 (bottom GOB in QCIF frame).
- Corrupted Bit 3 of GOB1.
- Corrupted Bit 2 of GOB1.
- Corrupted LSB (Bit 0) of GOB1.

The peak signal-to-noise ratio degradation versus frame index plot for these bits is shown in Figure 8.23. The corresponding corrupted frames are portrayed in Figure 8.24. The graph confirms that the amount of PSNR degradation increases with the significance of the bit corrupted. Furthermore, as we hypothesized before, the GQUANT parameter of GOBs hosting more MBs may be more significant. This can be confirmed by inspecting Figure 8.23, where the sensitivity order GOB3, GOB5, GOB1 is confirmed. This tendency is also explicit in Figure 8.24. The three LSBs of GQUANT were found to be relatively robust, as shown by Figure 8.23.

8.3.4.2.4 Errors in an Inter-coded Frame with Motion Vectors The H.261 codec optionally employs motion vectors in order to improve the efficiency of frame-differencing

8.3. EFFECT OF TRANSMISSION ERRORS ON THE H.261 CODEC

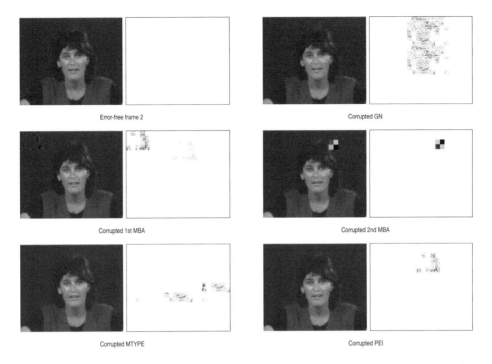

Figure 8.24: Subjective image degradation due to corruption of the H.261 GQUANT parameter in an inter-coded frame.

by invoking full correlative motion compensation. The effect of corrupting the motion vector displacement (MVD) parameter was investigated throughout the following simulation studies. Specifically, two sets of simulations were carried out in order to compare the sensitivity of a motion vector compensated H.261 data stream to an ordinary frame-differencing based H.261 data stream. The simulations were conducted by coding the video sequence in the same way as in the inter-coded frame simulations, except for the fact that the motion vectors were turned on. The following error events were simulated:

- The macroblock type (MTYPE) parameter was corrupted, so that a macroblock dispensing with using motion vectors was incorrectly decoded as a data stream that contains motion vectors. This prompted a macroblock to be predicted using a specific assumed MV, although this should not have happened.

- A motion vector displacement (MVD) was corrupted, displacing a block differently from what was used by the local decoder of the encoder, which may result in prolonged error propagation.

The resultant peak signal-to-noise ratio degradation versus frame index performance of the codec is shown in Figure 8.25 as regards both the MTYPE and MVD parameters. The associated corrupted frames in which the error occurred are shown in Figure 8.26.

When MVD errors occur, the use of motion vectors allows the effects of the errors to spread over the picture. This error-spreading phenomenon implies that after an error the

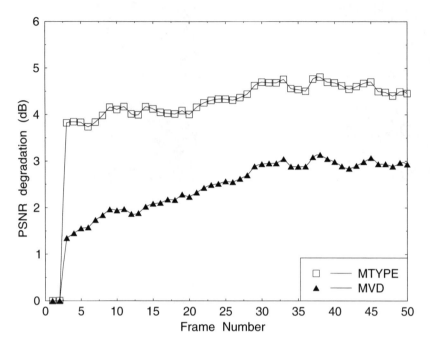

Figure 8.25: Image degradation in terms of PSNR due to corruption of H.261 motion vectors in an inter-coded frame.

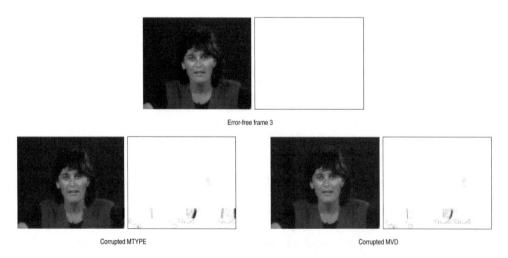

Figure 8.26: Subjective image degradation due to corruption of H.261 motion vectors in an inter-coded frame.

8.3. EFFECT OF TRANSMISSION ERRORS ON THE H.261 CODEC

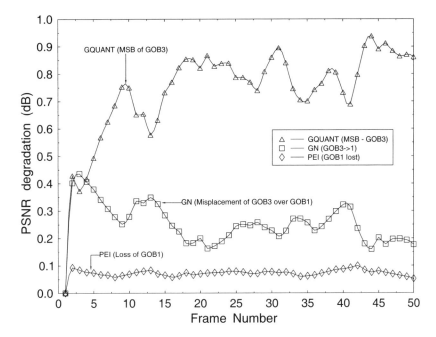

Figure 8.27: Image degradation in terms of PSNR due to corruption of H.261 codec parameters in an inter-coded frame at low bitrate.

degradation spreads further rather than being confined to the MB it was originally related to. This becomes explicit in Figure 8.25, where the PSNR degradation increases rather than decreases. Observe that the effect of the MTYPE error is in this case more detrimental than that of the MVD error in both Figures 8.25 and 8.26. In conclusion of our deliberations on the impairments introduced by motion vector corruption, we note, however, that the effect of a motion vector error can be much more aggravated than these illustrations suggest. In the next section, we discuss error effects in very low-rate coding.

8.3.4.2.5 Errors in an Inter-coded Frame at Low Rate In our next investigations the codec was constrained to operate under very low-rate conditions, using coarse quantizers. These simulations were conducted in order to ascertain whether the effect of errors was better or worse at low rates associated with lower image qualities. The error events simulated included the following:

- Corrupted MSB of GQUANT in GOB3.

- Corrupted Group number (GN) causing GOB misplacement.

- Corrupted picture extra information bit (PEI) resulting in the loss of GOB1.

The peak signal-to-noise ratio degradation versus frame index performance of the codec under these conditions is shown in Figure 8.27. The corresponding corrupted frames in which each error occurred are displayed in Figure 8.28. The effect of errors on the image quality

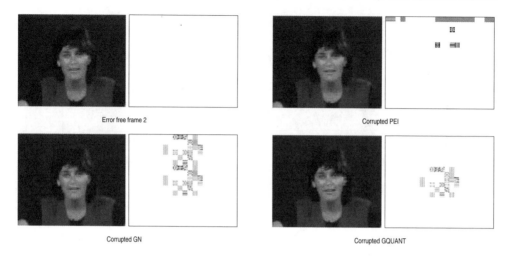

Figure 8.28: Subjective image degradation due to corruption of H.261 codec parameters in an intercoded frame at low bitrate.

appears to be more limited at this low quality than in the previous scenarios, which is plausible on the grounds of being more masked by the course quantization used. As we have seen before, these error effects also persist until the reception of the next intra-coded frame. The image degradation inflicted by the GQUANT error increases with future frames, which is likely to be due to the fact that in the case of low-quality coding the macroblocks are updated less often. This implies that typically it would take more frames than usual for the degradation to peak and then start to decay slightly. In conclusion, our low-rate simulations suggest that the effect of errors is less dramatic in terms of image degradation, since many error effects are masked by coarse quantization.

In this section we have studied the objective and subjective image-quality degradations caused by transmission errors corrupting various H.261 codec parameters. Errors contaminating the intra-coded frames are particularly detrimental, and both the intra- and intercoded frames have a persistent degradation due to errors, which typically does not decay until the retransmission of the affected macroblocks in intra-frame coded mode. With this experience, we then embarked on designing a wireless video system, which is the topic of our next section.

8.4 A Reconfigurable Wireless Videophone System

8.4.1 Introduction

In recent years, there has been increased research activity in the field of mobile videophony [191, 195, 307–312]. Some authors have investigated the employment of the H.261 video codec over mobile channels [307–312]. The other systems [191, 195] used proprietary video codecs. The H.261-based system developed by Redmill [311] used a technique called Error-Resilient Entropy Code (EREC) which was developed by Cheng [188, 310]. Matsumara [312] used an error-resilient syntax similar to EREC.

8.4.2 Objectives

The purpose of the error sensitivity investigations concerning the H.261 video codec was to use the knowledge gained in order to design an H.261-based videophone system for a mobile environment with the following aims:

- Better error resilience than that of the standard H.261 codec.

- A data stream that conforms to the H.261 standard or is amenable to conversion with a simple transcoder.

- A reconfigurable system so that it can adjust to different channel conditions.

In order to meet the above objectives, a flexible wireless system architecture was designed that exhibited the following features:

- A low-rate feedback channel was used for informing the encoder about the events of packet loss or corruption, for ensuring that the encoder can remain synchronized with the remote decoder. This is a requirement for any ARQ system.

- Each transmitted radio packet contained an integer number of macroblocks, which implies that the decoding process can resynchronize at the start of the next packet after an error.

- However, not allowing macroblock splitting between packets reduces the bit packing efficiency. This problem was mitigated by using a specific macroblock packing algorithm.

- FEC codecs were employed in order to decrease the channel BER and to balance the effects of the QAM subchannels' differing BERs.

- The system was rendered reconfigurable by being able to encode the video sequence for different bitrates. When the bitrate was reduced, the packet size was also reduced so that it could be transmitted using a more robust modulation scheme at the same signaling rate.

- Any corrupted packets were dropped in order to prevent erroneous decoding and error propagation, which would badly corrupt the received video sequence and result in error propagation effects.

8.4.3 Bitrate Reduction of the H.261 Codec

In order to facilitate the programmable-rate operation of the H.261 codec, its performance was studied using various quantizers. The bitrate of the H.261 codec may be reduced in a number of ways.

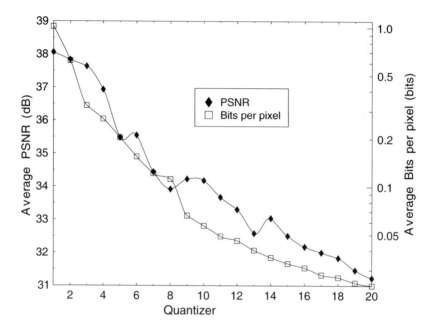

Figure 8.29: The effect of different quantizers on the H.261 coded image quality and bitrate.

- Opting for coarse quantization. Quantizers are numbered from fine (1) to coarse (31).
- Increasing the threshold associated with the active macroblock selection criterion, so that fewer macroblocks are transmitted. This implies that only the macroblocks with the most significant changes are transmitted.
- Reducing the frame rate, although a fixed frame rate is preferable.

Of these methods, changing the quantizer ensured the widest range of operation. Therefore, this technique was invoked in order to control the bitrate in the multiple bitrate coding algorithm. Figure 8.29 shows the image quality in terms of PSNR and bits per pixel versus different quantizers for 100 frames of the commonly used "Miss America" (MA) video sequence. The larger the quantizer index, the coarser the quantization of the DCT coefficients. For this low-activity sequence, a PSNR in excess of 30 dB can be maintained using an average of about 0.02 bits/pixel. However, as is true of many block-based video codes, the image quality is degraded at low bitrates, due mainly to "blocking" effects, where the macroblock boundaries become visible. Much work has been done to improve this by using pre- and post-filtering, particularly by Ngan and Chai [313, 314].

8.4.4 Investigation of Macroblock Size

At this stage of the system design, it is beneficial to evaluate the histogram of the number of bits generated by each macroblock. The packetization algorithm mapping the video information onto transmission packets aims to pack an integer number of macroblocks into a packet with the least amount of unused space in the packet. The graphs in Figure 8.30 show the histograms of the macroblock size in bits for various quantizers, intra-coded frame rates, both with and without motion vectors. In the latter case, simple frame-differencing is used. No macroblock length limitations or restrictions were applied. Therefore, if a packet contains one macroblock, the packet size is the macroblock size. At a later stage, we will evaluate the same histograms with packet-length restrictions, implying that some large macroblocks' bitstreams are shortened to fit them into the packets. All histograms were generated by coding a 100-frame "Miss America" video sequence.

All the histograms have a similar shape, where the leading edge rapidly rises to a peak and then tails off slowly for increasing macroblock sizes. Generally the coarser quantizers having a higher quantizer index generate a higher number of smaller macroblocks than the lower-index quantizers, moving the histogram peak toward smaller macroblock sizes. These intra-coded frames seem to generate several macroblocks with a size of 65 bits, which is associated with a distinct little spike in the histograms. When many intra-coded frames are encoded, this peak becomes more obvious. A further tendency is that when motion vectors are used, more small macroblocks are generated than in the case of a sequence-coded otherwise identically without motion vectors. This renders the motion vector-based histogram similar to one with a coarser quantizer but without MVs.

In order to design the packetization algorithm, it is necessary to study the effects of shortening the macroblock bitstreams in order to fit them into a transmission packet. This shortening can be achieved by increasing the quantizer index for the macroblock using the MQUANT or GQUANT coding parameter. All macroblocks are initially coded with the specified quantizer. If, however, the number of bits produced by the current macroblock is higher than the maximum packet size, then the quantizer index is continuously increased, until the number of bits produced for the current macroblock becomes smaller than the size limit imposed by the packet length. The effect of macroblock shortening is shown in the histograms of Figure 8.31, where the packet length was set to 128, 192, 384, and 384, while the initial quantizers were 3 and 5.

As seen in these figures, this technique changes the shape of the histograms. This is because many of the originally longer macroblocks are being shortened, causing the shortened macroblocks to have sizes just below the required packet-length limit. This procedure removes the low-probability histogram tails corresponding to the long macroblock bitstreams and increases the relative frequency of the macroblock sizes toward the required packet length. When a coarse quantizer is used, the size limit has little or no effect, since typically the generated macroblock bitstreams are relatively short. In contrast, when a fine quantizer is used in conjunction with a small packet-size limit, the histogram tends to a ramp-type shape, with the relative frequency of a given macroblock size expressed in terms of the number of bits increasing for larger macroblocks sizes, as seen in the top two graphs of Figure 8.31. In this context, the macroblock size is referring to the number of bits generated rather than the number of pixels covered by the macroblock, which is always 16×16.

8.4.5 Error Correction Coding

Forward error correction codes can reduce the number of transmission errors at the expense of an increased bitrate. Therefore the use of FEC can increase the range of operating channel

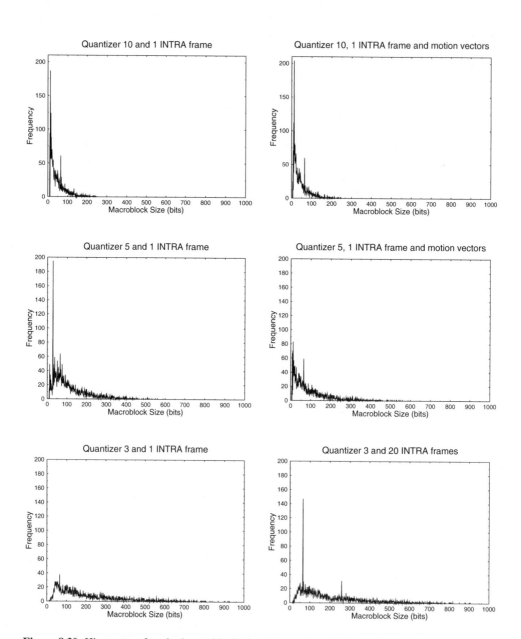

Figure 8.30: Histogram of packet/macroblock sizes for an "unrestricted" H.261 codec using various quantizers for encoding the 100-frame "Miss America" sequence using one or twenty intra-coded frames for generating the statistics, both with and without motion vectors.

8.4. A RECONFIGURABLE WIRELESS VIDEOPHONE SYSTEM

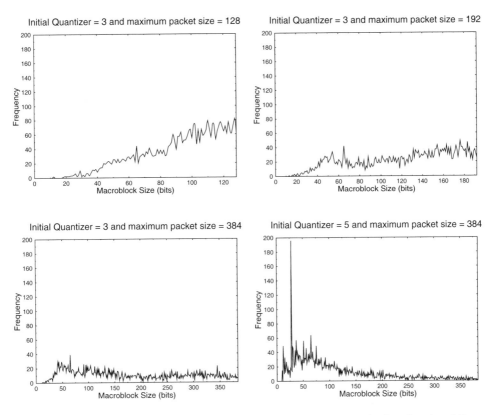

Figure 8.31: Histogram of macroblock sizes when imposing a maximum packet length, using different initial quantizers, without motion vectors.

SNR, for which transmissions are error-free. When the error rate becomes too high, a FEC is said to become overloaded, a condition that precipitates a very sharp increase in terms of BER.

When using Quadrature Amplitude Modulation (QAM) schemes, as argued in Chapter 5 of [196], certain bits in the nonbinary QAM transmission symbols have a higher probability of errors than others [315]. For example, 6 bits can be transmitted in a 64QAM symbol, and careful mapping of the bit patterns to the various 64QAM symbols allows some bits to be better protected than others. The symbols are grouped into classes according to the probability of an error. These different integrity bit classes are also often referred to as subchannels. By using different strength FEC codes on each QAM subchannel, it is possible to equalize the probability of errors on the subchannels. This means that all subchannels' FEC codes should break down at approximately the same channel SNR. This is desirable if all bits to be transmitted are equally important. Since the H.261 data stream is mainly variable-length coded, one error can cause a loss of synchronization. Therefore, in this respect most bits are equally important, and so equalization of the QAM subchannels' BER is desirable for the H.261 data stream. The specific FEC schemes used are listed in Table 8.7.

8.4.6 Packetization Algorithm

Our proposed packetization algorithm had the following objectives:

- It is important to pack an integer number of macroblocks into a packet with the least amount of wasted space.

- If the receiver asks for a packet to be re-transmitted with a smaller packet size, the previously transmitted macroblocks have to be re-encoded at a lower bitrate in order to fit into the shorter transmission packet. This enables the use of ARQ in the transmission system. ARQ issues are addressed later, especially in Section 10.5.1.

- If the transmitter is informed by the receiver acknowledgment that a packet was dropped, the encoder has to modify its parameters in order to ensure that the local decoder and remote decoder remain in synchronization.

The default action of the packetization algorithm was to encode each macroblock using a certain fixed quantizer, storing the encoded bitstream in a buffer. When the buffer was filled with video-encoded bits above the transmission packet size, the buffer contents up to the end of the previous completed macroblock would be transmitted. The remainder of the buffer contents would then be moved to the start of the buffer, before encoding the next macroblock. The fixed quantizer was set depending on the required packet size, and the employment of smaller packets inevitably required a coarser quantizer. The quantizers for different packet sizes were set so that a more robust, but lower capacity, modulation scheme in conjunction with a smaller packet size would require approximately the same number of packets per video frame as a less robust but a higher capacity modem scheme. This would allow the system to exploit the prevailing benign channel conditions by transmitting at a higher video quality, while dropping the number of modulation levels when the channel quality degrades. This would be accompanied by a concomitantly lower video quality.

In some ARQ transmission systems, the re-transmission attempts are carried out at the original bitrate. However, in our intelligent reconfigurable transceivers, the ARQ attempts are invoked at a lower bitrate but identical signaling rate, using a more robust modulation scheme. Hence, the re-transmission attempts have a higher success rate than the initial attempts. In order to enable the employment of ARQ and reconfigurable modulation schemes, the packetization algorithm had to be able to re-encode a packet at a lower target bitrate, in order to fit the video bits into a smaller packet size. For the sake of this re-encoding procedure, the previous history of various coding parameters had to be stored before each macroblock was encoded. This history had to be remembered by the transceiver for each macroblock until it had been successfully transmitted, or it was dropped.

8.4.6.1 Encoding History List

The history of coding parameters was implemented as a bidirectionally linked list, having a structure as shown in Figure 8.32. Each element of the list represents a macroblock and contains the coding parameters of the codec before the macroblock was coded.

Each new macroblock is added to the tail of the list, encoded, and the coded bitstream is appended to a buffer. When the buffer fullness exceeds the defined limit constituting a

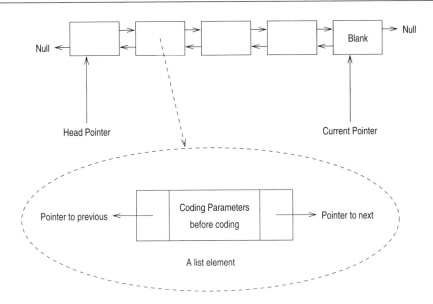

Figure 8.32: Structure of bidirectionally linked list for storing history of coding parameters.

transmission packet, the codec is reset to its state before encoding the current macroblock that resulted in exceeding the packet length. The buffer contents containing all but the current macroblock are transmitted. Then the linked list elements for the transmitted macroblocks are removed from the head of the list. The current, untransmitted macroblock then becomes the head of the linked list.

The algorithm has two pointers to the linked list: *head* points to the head of the list and *current* points to the next element/macroblock to be encoded. When a new macroblock is added to the list, it is placed in the blank list element at the tail of the list, and a new blank list element is created.

8.4.6.2 Macroblock Compounding

The initial packetization algorithm was not very efficient, since if one macroblock took the buffer occupancy just 1 bit over the packet size limit, then no attempt was made to fit it in by encoding the macroblock in such a way as to generate fewer bits. Thus, packets had much more wasted space than they needed to, which rendered the whole system less efficient and caused more packets to be transmitted.

In order to render the packetization more efficient, a method for shortening macroblocks was designed. Macroblocks were shortened by re-encoding them with a coarser quantizer. Nonetheless, this procedure sometimes forced the macroblock bitstream to become longer, since the encoded macroblock had to incorporate an MQUANT parameter if the previously valid quantizer had to be superceded. However, in the majority of cases, the macroblocks were encoded with fewer bits, when a coarser quantizer was used. When a macroblock was encoded which overfilled the buffer, there were two options: (1) to shorten the macroblocks in order to fit the new one in the packet buffer, or (2) to transmit the current packet buffer and

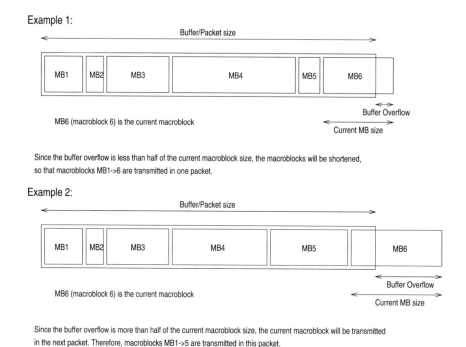

Figure 8.33: Examples of the decision threshold for compounding macroblocks.

leave the macroblock for the next packet. The threshold for deciding on which option to opt for was defined as follows. If the buffer overflow due to the current packet was more than half the size of the current macroblock, then the macroblock was transmitted in the next packet. Examples of the compounding decision threshold are shown in Figure 8.33.

When a macroblock is shortened by using a coarser quantizer, all the following macroblocks have to be re-encoded, since earlier those were encoded using the previously employed quantizer. Hence, without re-encoding them, the decoder would employ the most recently received MQUANT code in order to identify the inverse quantizer to be used for their decoding, which would be the correct one for the shortened last MB bitstream, but not for the others in the packet. An alternative would be to inject an MQUANT code after the shortened and re-encoded macroblock, which would reflect the index of the previously tentatively used quantizer for the remaining MB strings in the packet. This would, however, increase the length of the packet, which may result in packet-length overflow again. After some investigations, we found that re-encoding all the MB strings in the packet was the most attractive option to use. However, this extra computational burden makes an optimal macroblock compounding algorithm more computationally demanding.

In order to reduce the associated implementational complexity, the following sub-optimal algorithm was contrived for compounding the macroblocks, which proved to be fast in most cases. Initially, the fixed quantizer's index is increased by one step in order to render it coarser and generate a lower number of bits. The last macroblock is then re-encoded. If the buffer is still overflowing, the last two macroblocks are re-encoded, followed by the last three

8.4. A RECONFIGURABLE WIRELESS VIDEOPHONE SYSTEM

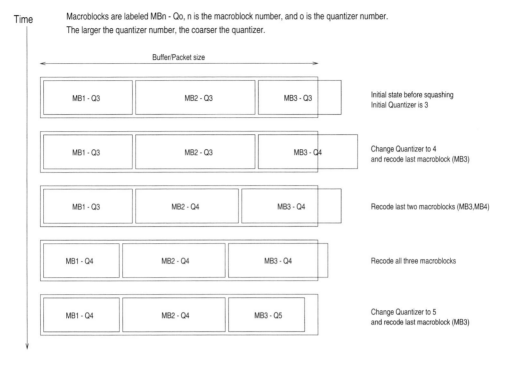

Figure 8.34: Example of macroblock compounding.

macroblocks, and so on. This procedure continues until all the macroblocks in the buffer have been re-encoded with the new coarser quantizer, or until the buffer is no longer overflowing. If all macroblocks in the buffer have been re-encoded, and the buffer is still overflowing, then the quantizer index is increased by another step and the re-encoding procedure resumes with the last macroblock in the buffer. Once the compounding algorithm has shortened the macroblocks sufficiently, such that the buffer is no longer overflowing, then the procedure is concluded. An example of macroblock compounding is shown in Figure 8.34. The figure is self-explanatory; hence we will refrain from detailing the individual re-encoding steps seen in the figure.

In a second phase, the algorithm will attempt to fill the small unused space in the buffer created by the compounding algorithm with a new macroblock. If this macroblock does not fit in the packet, further compounding may be attempted. This "squashing" algorithm improved the efficiency of the packetization algorithm.

8.4.6.3 End of Frame Effect

For implementational reasons, it is not possible to re-encode macroblocks from a previous video frame. Therefore, all buffers and the history list need to be reset at the end of a video frame. This means that the last packet of a frame may be very inefficiently filled with video bits. In order to overcome this problem and increase the packing efficiency at the end of a frame, the following solution was advocated. If the buffer of the last packet within the frame was not sufficiently full, its coded data could be left in the buffer and its entries removed from the history list. This meant that these bits could not be re-encoded or dropped. However, this was an acceptable compromise, since the end of this packet was then filled with the picture start code (PSC) of the next frame, and the picture start code was the only symbol that could not be dropped without catastrophic video quality degradation and hence had to be retransmitted.

8.4.6.4 Packet Transmission Feedback

When a packet is transmitted, the packetization algorithm receives feedback from the receiver concerning the success/failure of the packet. The algorithm is also aware of the required size of the next packet. Therefore, the transmitter can change packet size for different modulation schemes. The initial quantizer used by the packetization algorithm is set according to the requested packet size expressed in terms of bits. This is basically a "Stop and Wait ARQ" protocol [202, 316]. A more complex ARQ protocol, such as, for example, the "Selective Repeat ARQ" scheme [202, 317], would be much more complex to implement. Hence, in order to gauge the achievable performance gains, here the simpler "Stop and Wait ARQ" protocol is used. The transceiver uses three types of feedback messages.

- Packet received without error.

- Packet received with error, retransmission request.

- Packet received with error, dropping request.

If the packet is received without errors, the corresponding history list elements are removed from the list and the buffer is cleared. If the packet was received with errors and a retransmission is requested, the following actions are taken. First, the codec is reset to the state which it was in at the start of the history list. Then all the macroblocks are re-encoded. If the re-transmission is requested with a smaller packet size, then the initial quantizer will be coarser. When the recently cleared buffer becomes full, the new packet is transmitted.

If a packet is received with error and a request is made to drop it, then the following actions are taken. In case the packet contained a picture start code (PSC), then a retransmission is invoked. However, if the packet did not contain a PSC, then the codec is reset to the state stored in the head of the history list. All the macroblocks in the packet are removed from the history list.

8.4.6.5 Packet Truncation and Compounding Algorithms

In order to show the benefits of the proposed packetization algorithm, it is necessary to carry out comparisons with and without the algorithm. Recall that the graphs shown in Figure 8.31 portray the probability density function (PDF) of the various macroblock sizes occurring, when applying a simple macroblock size truncation algorithm. These graphs show that the packing efficiency is very low due to the large proportion of packets, which are well below the packet size limit. This is in contrast to an optimum packetization scheme, which would be characterized by a histogram exhibiting a peak at the required packet length and zero

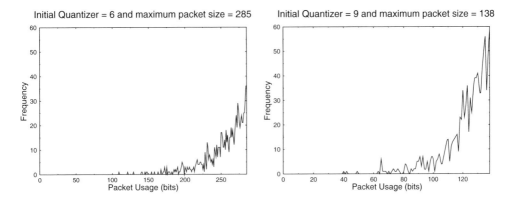

Figure 8.35: Histograms of packet usage using the compounding packetizing algorithm of Section 8.4.6.2.

relative frequencies otherwise. The corresponding relative frequency graphs, when using the proposed packetization algorithm, for example, for packet lengths of 285 and 138 bits, are displayed in Figure 8.35. Here we introduce the informal term *packet usage*, which we define as the number or percentage of useful video bits in each packet, where a full packet would have a packet usage of 100% or the same number of bits as the packet size. Upon returning to the figure, the associated curves exhibit a higher concentration of longer packets near the packet size limit, demonstrating the increased packing efficiency of the compounding algorithm over the simple truncation technique of Figure 8.31.

For the sake of better illustration, Figures 8.31 and 8.35 are combined and shown as cumulative density functions (CDF) in Figure 8.36. The CDF graphs show that the packet compounding algorithm guarantees a much more efficient packing than the simple truncation algorithm. This manifests itself explicitly in terms of the rapid early rise of the CDFs in case of the simple truncation algorithm, indicating a high probability of short packets associated with low packing efficiency. In contrast, the compounding packetization exhibits a very low probability tail for low packet usage, while the majority of the registered occurrences were in the vicinity of the maximum allowed packet length. Again, this is in harmony with what would be expected from an ideal packing algorithm, where all packets would be completely filled by useful data bits.

8.5 H.261-based Wireless Videophone System Performance

8.5.1 System Architecture

The performance of the proposed error-resilient H.261-based reconfigurable videophone system was evaluated in a wireless environment over the best-case stationary Gaussian and the worst-case Rayleigh-fading channels using the proposed reconfigurable modulation and ARQ schemes. The specific system parameters employed are summarized during our later discussions in Table 8.10 on page 292.

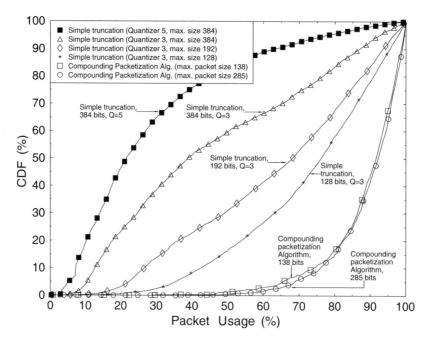

Figure 8.36: CDFs of the packet usage with simple truncation according to Figure 8.31 and with the packet compounding algorithm of Section 8.4.6.2.

This section briefly describes the specific system parameters employed during the performance evaluation of the H.261-based videophone system proposed. The reconfigurable modem invoked one of three modulation schemes, namely, 64-, 16-, and 4-QAM combined with a range of binary Bose-Chaudhuri-Hocquenghem (BCH) forward error correcting codes in order to increase the system's robustness. The system could reconfigure itself depending on the prevailing channel conditions, similar to a scheme proposed by Sampei [318] et al. An intelligent system could switch modulation scheme either under network control or, for example, by exploiting the inherent channel quality information in Time Division Duplex (TDD) systems upon evaluating the received signal level (RSSI) and assuming reciprocity of the channel [319–321]. A more reliable alternative to the use of RSSI as a measure of channel quality is the received BER. In interference-limited, dispersive environments, where reciprocity does not apply, the receiver attaches a channel-quality side-information message to the reverse-direction message transmitted. The reception of this channel quality information will however suffer from some latency. This reconfiguration could take place arbitrarily frequently, even on a packet-by-packet basis, depending on the channel conditions. The image quality is dependent on the bitrate and therefore increases with better channel conditions.

For the sake of mitigating the effect of fading, linearly interpolated pilot symbol assisted modulation (PSAM) was used. The number of bits per symbol and the previously mentioned number of associated modem subchannels are summarized for each modulation scheme in Table 8.5.

8.5. H.261-BASED WIRELESS VIDEOPHONE SYSTEM PERFORMANCE

Table 8.5: Basic Characterization of Modulation Schemes

Modulation Scheme	Bits/Symbol	Number of Subchannels
64QAM	6	3
16QAM	4	2
4QAM	2	1

Table 8.6: System Capacity in Terms of Bitrates and Bits/Pixel at 10 Frames/s for Each Modulation Scheme

Modulation Scheme	Bitrate (Kbit/s)	Bits/Pixel at 10 Frames/s
4QAM	11.76	0.0464
16QAM	23.52	0.0928
64QAM	35.60	0.1405

All results were generated assuming a single user and no co-channel base stations, which would be realistic in benign indoor picocells, where the partitioning walls and ceilings contribute to the required co-channel attenuation, resulting in a noise-, rather than interference-limited scenario. Simulations were done using a propagation frequency of 1.9 GHz, with mobile handsets moving at a vehicular speed of 13 Km/h. Second-order diversity was also used to mitigate the effects of the multipath fading. The pilot symbol spacing was $P = 10$. In order for the system to be able to use the proposed reconfigurable modulation schemes in conjunction with ARQ, it was necessary for all transmission packets to host the same number of symbols for each modulation scheme. This was achieved by using a higher number of bits per packet for 64QAM and a lower number of bits for 16- and 4QAM, where the ratio of the number of bits per transmission packet was determined by the number of bits per symbol for each modulation scheme. In our simulations, we used a conventional TDMA frame structure [322], having 8 slots per frame and a frame duration of 12.5 ms. The user baud rate was 11.84 Kbaud, and therefore the system baud-rate was 94.72 Kbaud. The system capacity in terms of bitrates for each modulation scheme is summarized in Table 8.6, and these values are further justified at a later stage in Table 8.10 on page 292. The table also shows the average bits/pixel required by the codec to achieve an average frame rate of 10 Hz.

An appropriate target quantizer was then chosen for each modulation scheme that produced approximately the required bitrate. The graph in Figure 8.37 shows how the bitrate varies for the quantizer used with each modulation scheme. Ideally, the target bitrate for each modulation scheme should approximate the average bitrate of the codec. This is not possible for all the modulation schemes because of the limited number of quantizers, and the fixed symbol rate. The ideal solution is to implement a rate control for the video codec. This is used in our H.263 videophone system described in Section 10.3. An alternative is to use the histogram of the video bitrate to predict the future bitrate, a method suggested by Schwartz [323] *et al.*

The packet sizes expressed in terms of the number of bits for each modulation scheme were found by initially setting the packet size for 4QAM. The packet sizes for 16- and

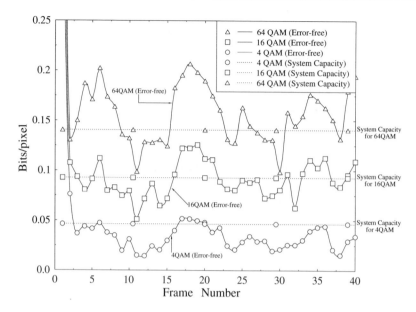

Figure 8.37: Bitrate in terms of bits/pixel versus frame index for each modulation scheme used in the H.261 videophone system.

64QAM were then set accordingly in order to keep the same number of transmission symbols per packet. The unrestricted macroblock length histograms of Figure 8.30 were invoked to determine the quantizer index chosen for 4QAM in order to set the appropriate packet size. Specifically, the transmission packet length in terms of bits was chosen so that the majority of macroblocks would fit into a single packet without shortening.

As mentioned earlier, the modem subchannels in 16- and 64QAM exhibit different BERs. This property can be exploited in order to provide more sensitive bits with increased protection from errors, if the source bits are mapped onto them using identical FEC codes. However, we have argued that owing to their inherent sensitivity to transmission errors, all the variable-length coded H.261 bits must be free from errors. Hence, the different subchannel error rates had to be equalized using different FEC codes. This was ensured by using stronger FEC codes over the higher BER modem subchannels. Taking also into account the practical issues of bit-packing requirements, the specific FEC codecs were selected on the basis of satisfying the BER equalization criterion as closely as possible. The different binary BCH FEC codes chosen for the various modulation modes are shown in Table 8.7, while Table 8.8 summarizes the packet sizes derived from the FEC codes and the initial quantizer indices used for each modulation scheme. Having summarized the system parameters of our H.261-based videophone, let us now characterize the system's performance.

8.5.2 System Performance

The ultimate aim of these performance investigations was to produce graphs of video PSNR versus channel SNR in a variety of system scenarios, both with and without ARQ over

8.5. H.261-BASED WIRELESS VIDEOPHONE SYSTEM PERFORMANCE

Table 8.7: FEC Codes Used for 4-, 16- and 64QAM in the H.261 Videophone System

Modulation Scheme	FEC Codes Used
4QAM	BCH(255,147,14)
16QAM	Class 1: BCH(255,179,10)
	Class 2: BCH(255,115,21)
64QAM	Class 1: BCH(255,199,7)
	Class 2: BCH(255,155,13)
	Class 3: BCH(255,91,25)

Table 8.8: H.261 Packet Size and Initial Quantizer Used for 4-, 16-, and 64QAM

Modulation Scheme	H.261 Packet Length (bits)	Initial Quantizer Index
4QAM	147	9
16QAM	294	6
64QAM	445	4

Table 8.9: Summary of H.261-based Videophone System Performance

Transmission Scenarios	Channel SNR for 1 dB PSNR Degradation	
	AWGN	Rayleigh
4QAM	5.78	7.8
16QAM	11.6	13.5
64QAM	17.46	20.09
ARQ: 64-16-4QAM	16.18	15.46
ARQ: 16-4-4QAM	9.06	8.81
ARQ: 4-4-4QAM	5.18	5.35

Gaussian and Rayleigh channels. The video sequences corresponding to the results in this section are available on the World Wide Web.[1] The "Miss America" video sequence was used, and the ARQ attempts are described as A-B-C, where A is the initial transmission, B is the first, and C is the second re-transmission attempt. The range of transmission scenarios used is summarized in Table 8.9.

The PSNR versus frame index curves of Figure 8.38 show the decoded video quality for the three different modes of operation over a perfect error-free channel. These curves demonstrate that the image quality improves when using the higher bitrates facilitated by

---

[1] http://www-mobile.ecs.soton.ac.uk/peter/robust-h261/robust.html

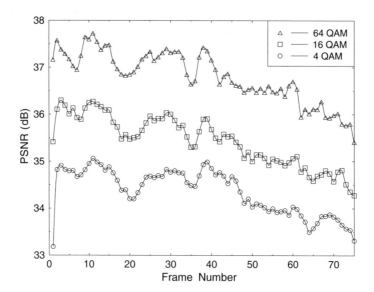

Figure 8.38: PSNR versus frame-index comparison between modulation schemes over error-free channels.

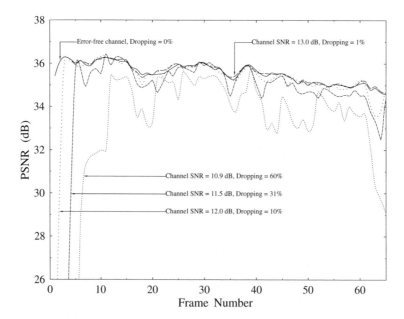

Figure 8.39: PSNR versus frame-index performance of the H.261-based videophone in its 16QAM modem mode, for various packet dropping rates over Gaussian channels.

8.5. H.261-BASED WIRELESS VIDEOPHONE SYSTEM PERFORMANCE

the higher-order modulation schemes. Observe also that the first frame, which is intra-frame coded, has a lower quality than the following frames due to the relative paucity of bits, since the packetization algorithm imposes a more stringent bitrate constraint on the comparatively long intra-frame coded macroblock bitstreams. An advantage of this, however, is that the inevitable bitrate increase for intra-coded frames is mitigated. Again, the video sequences of Figure 8.38 can be viewed on the WWW.[2] Packet loss concealment techniques were also investigated by Ghanbari and Seferidis [324] using a H.261 codec.

Let us now turn to the system's performance over the best-case Gaussian channel encountered in stationary wireless scenarios. Figure 8.39 demonstrates the effect of video packet dropping over Gaussian channels exhibiting different channel SNRs, when using the 16QAM mode of operation and channel SNRs of 20, 13, 12, 11.5, and 10.9 dB, which were associated with packet dropping rates of 0, 1, 10, 31, and 60%, respectively. The decoded video sequences of these results for different dropping rates can be viewed on the WWW.[3] When a packet is dropped, the macroblocks contained in the dropped packet will not be updated for at least one video frame. The dropped macroblocks will be transmitted in future video frames until the dropped macroblocks are received without errors. In case of adverse channel conditions and excessive packet dropping, some macroblocks may not be updated for several video frames. Observe in the figure that when a packet containing the initial intra-coded frame is lost, its effect on the objective PSNR and subjective video quality is more detrimental than in the case of inter-coded frame errors. This is because the decoder cannot use the corresponding video contents of the previous inter-coded frame to replenish the intra-frame coded dropped macroblocks. As seen in the figure, after the initial few frames, every macroblock has been transmitted at least once; hence, the PSNR gradually improves to around 30–40 dB. Figure 8.39 also demonstrates that for a 1% packet dropping probability, where 1 in every 100 packets is lost, the PSNR is almost identical to the error-free case, but the higher the dropping rate, the longer it takes to build the PSNR up to the normal operating range of around 35 dBs. However, even with 60% dropping, where only two in five packets are received, the PSNR varies within 4–5 dB of the error-free performance. At these high dropping rates, the delayed updating of the macroblocks can be observed, and its subjective effect is becoming objectionable.

In order to be able to observe the effect of various channel conditions on the different modes of operation of the proposed videophone system, the average PSNR versus channel SNR performance was evaluated. The performance of the 4QAM, 16QAM, and 64QAM modes dispensing with ARQ assistance is shown in Figure 8.40. These graphs demonstrate the sharp reduction of PSNR, as the channel quality degrades. As expected, the curves corresponding to Gaussian channel conditions have a much sharper reduction of PSNR than those associated with Rayleigh-fading channels, which is a consequence of the typically "average" rather than time-variant behavior of Gaussian channels, associated with the same average number of errors per transmission packet. In contrast, Rayleigh channels exhibit a more "bursty" error distribution, associated with occasional bad and good bimodal behavior. This is the reason for the more slowly decaying PSNR curves for Rayleigh channels. This graph may also be used to estimate the channel conditions, for which a decision to switch to a more robust modulation scheme becomes effective. The Gaussian

---

[2]http://www-mobile.ecs.soton.ac.uk/peter/robust-h261/robust.html#MODULATION
[3]http://www-mobile.ecs.soton.ac.uk/peter/robust-h261/robust.html#DROPPING

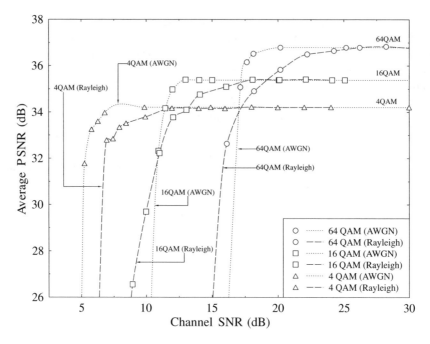

Figure 8.40: PSNR versus channel SNR performance of the H.261-based videophone for 4-, 16-, and 64QAM over both Gaussian and Rayleigh-fading channels.

performance curves also characterize the expected video performance over conventional telephone channels.

Let us now continue our system performance study by analyzing the benefits of ARQ assistance. Figure 8.41 portrays the system's PSNR versus channel SNR performance for 4QAM with and without ARQ, while the minimum channel SNR values for maintaining less than 1 dB PSNR degradation associated with near-unimpaired video quality are summarized for all investigated systems in Table 8.9. The operating channel SNR range over Rayleigh channels is extended to lower values more substantially than in case of Gaussian channels, when using ARQ. This is a consequence of the fact that over Gaussian channels each video packet experiences a similar dropping probability during each re-transmission attempt, while over fading channels the re-transmitted packet may have a higher probability of error-free reception due to the portable station emerging from a fade during the re-transmission attempt.

The performance of the more elaborate ARQ schemes, such as, for example, those using 64-, 16-, and 4QAM in successive ARQ attempts is analyzed in Figures 8.42 and 8.43 in contrast to the schemes without ARQ. Again, the corresponding minimum required channel SNRs for all systems studied are presented in Table 8.9. The graphs demonstrate that the PSNR performance of the ARQ schemes gracefully degrades in the medium SNR range, tending toward the corresponding performance curves of the simple nonaided arrangements, as the channel SNR decreases. This is because while during the first transmission attempt the highest order modulation is used, due to their failed attempts, many of the successful transmissions were actually carried out at the lower-rate, lower-quality mode. Nonetheless,

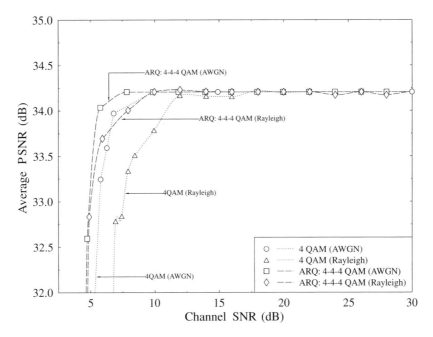

Figure 8.41: PSNR versus channel SNR performance of the H.261-based videophone for 4QAM with and without ARQ over both Gaussian and Rayleigh-fading channels.

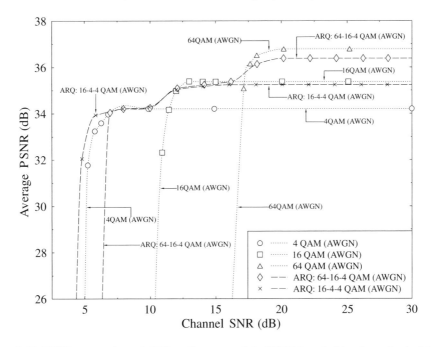

Figure 8.42: PSNR versus channel SNR performance of the H.261-based videophone for 4-, 16-, and 64QAM with and without ARQ over Gaussian channels.

Table 8.10: Summary of System Features for the H.261 Reconfigurable Wireless Videophone System

Features	H.261 Multirate System		
Modem	4QAM	16QAM	64QAM
Bits/Symbol	2	4	6
Number of subchannels	1	2	3
Pilot assisted modulation	yes	yes	yes
Video codec	H.261	H.261	H.261
Frame rate (Fr/s)	10	10	10
Video resolution	QCIF	QCIF	QCIF
Color	No	No	No
C1 FEC	BCH (255,147,14)	BCH (255,179,10)	BCH (255,199,7)
C2 FEC	BCH N/A	BCH (255,115,21)	BCH (255,155,13)
C3 FEC	BCH N/A	BCH N/A	BCH (255,91,25)
Data symbols/TDMA frame	128	128	128
Pilot symbols/TDMA frame	14	14	14
Ramp symbols/TDMA frame	4	4	4
Padding symbols/TDMA frame	2	2	2
Symbols/TDMA frame	148	148	148
TDMA frame length (ms)	12.5	12.5	12.5
User symbol rate (kBd)	11.84	11.84	11.84
Slots/frame	8	8	8
No. of users	8	8	8
System symbol rate (kBd)	94.72	94.72	94.72
System bandwidth (kHz)	200	200	200
Effective user bandwidth (kHz)	25	25	25
Coded bits/TDMA frame	147	294	445
Coded bitrate (kbit/s)	11.76	23.52	35.6
Vehicular speed (m/s)	13.4	13.4	13.4
Propagation frequency (GHz)	1.8	1.8	1.8
Normalized Doppler frequency	6.2696×10^{-4}	6.2696×10^{-4}	6.2696×10^{-4}
Pathloss model (Power Law)	3.5	3.5	3.5
Max. PSNR degradation (dB)	-1	-1	-1
Min. AWGN SNR (dB)	5.78	11.6	17.46
Min. AWGN BER (%)	28	25	24
Min. Rayleigh SNR (dB)	7.8	13.5	20.09
Min. Rayleigh BER (%)	29	26	17

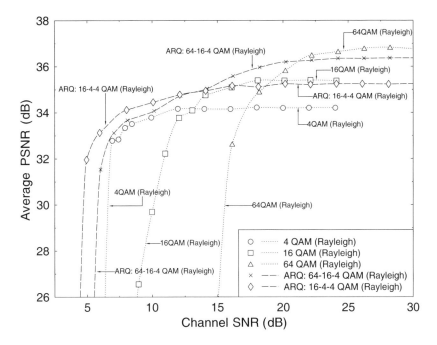

Figure 8.43: PSNR versus channel SNR performance of the H.261-based videophone for 4-, 16-, and 64QAM with and without ARQ over Rayleigh channels.

in subjective video quality terms it is less objectionable to view an uncorrupted, but initially lower quality, video sequence than an originally more pleasant but corrupted sequence. In other words, during the second and third transmission attempts, the high-level modulation schemes become unusable as the channel degrades, and when using the lower-order modem modes, the video quality inevitably degrades owing to the more stringent bitrate constraint imposed by the constant symbol-rate requirement. Furthermore, the ARQ-assisted PSNR performance is slightly below that of the corresponding schemes dispensing with ARQ assistance, but using the same modulation scheme, as during the first attempt of the ARQ-aided arrangements. This is because the packetization algorithm attempts to reduce the delay of the codec by minimizing the effect caused by the end of a frame, as discussed in Section 8.4.6.3. Specifically, the algorithm minimizes the end-of-frame effect by only allowing the bitstream to be left in the transmission buffer if it is shorter than the minimum packet size. The minimum packet size is governed by the lowest order modulation scheme used in the ARQ system. This effect occurs only when the ARQ scheme contains two or more different modulation modes. The above phenomenon can be observed, for example, in case of the 64-16-4 and 16-4-4 ARQ schemes in Figures 8.42 and 8.43.

8.6 Summary and Conclusions

In this chapter the H.261 [258] standard and its sensitivity to transmission errors were analyzed. An appropriate wireless videophone system was designed for the transmission of H.261-encoded video signals over mobile radio channels. Using error concealment and recovery techniques, the system was rendered amenable to operation under hostile channel conditions associated with high packet dropping rates. The proposed reconfigurable system allows the system to adapt to a range of channel conditions. The minimum required channel SNR values for near-unimpaired video communications are listed for the range of system scenarios studied in Table 8.9. A summary of features of the H.261 wireless reconfigurable videophone system is given in Table 8.10.

Comparative Study of the H.261 and H.263 Codecs

9.1 Introduction

In 1990 the International Telecommunications Union (ITU) published the H.261 video coding standard [258] designed to support audiovisual services over Integrated Services Digital Networks (ISDNs) at rates of multiples of 64 Kbit/s. Hence, it is sometimes referred to as a $p \times 64$ Kbit/s standard, where p is in the range of 1 to 30. More recent advances in video coding have been incorporated in the ITU-T H.263 Recommendation [259]. Following our previous chapter on the H.261 codec, in this chapter we describe the H.263 standard and highlight the differences between the H.261 and H.263 standards, while endeavoring to present a comparative study of their performance differences. Throughout this chapter familiarity with the H.261 standard is assumed.

As we have seen, the H.261 coding algorithm is a hybrid of inter-picture prediction, transform coding, and motion compensation. The coding algorithm was designed to support bitrates between 40 Kbits/s and 2 Mbits/s. The inter-frame prediction removes the temporal redundancy associated with consecutive video frames, while the consecutive transform coding of the MCER exploits the spatial frequency-domain redundancy of the video stream in order to reduce the required bitrate [10, 187]. Motion vectors are used to help the codec compensate for motion translation between adjacent video frames. In order to remove further redundancy from the transmitted bitstream, variable-length (VL) coding is used [10, 187]. Recall that the H.261 codec supports two different resolutions, namely, QCIF and CIF. Let us now consider the H.263 standard.

The H.263 ITU-T standard [259] is similar to the H.261 recommendation [258], but due to a number of measures introduced, it has a higher coding efficiency and it is capable of encoding a wider variety of input video formats, which will be discussed in Table 9.1. The main differences between the H.261 and H.263 coding algorithms are as follows. In the H.263 scheme, half-pixel precision [165] is used for motion compensation, whereas H.261 used full-pixel precision [165] and a loop filter. Some parts of the hierarchical structure of the

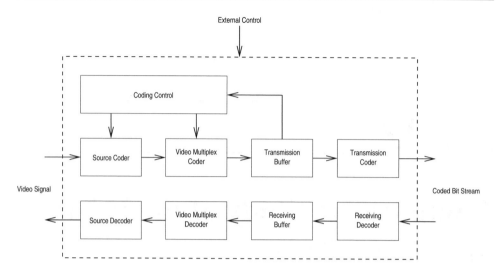

Figure 9.1: Block diagram of the video codecs.

Table 9.1: Picture Formats Supported by the H.261 and H.263 Codecs

Picture Format	Lumin. Cols.	Lumin. Lines	H.261 Support	H.263 Support	Uncompressed Bitrate (Mbit/s)			
					10 Frame/s		30 Frame/s	
					Gray	Color	Gray	Color
SQCIF	128	96	No	Yes	1.0	1.5	3.0	4.4
QCIF	176	144	Yes	Yes	2.0	3.0	6.1	9.1
CIF	352	288	Optional	Optional	8.1	12.2	24.3	36.5
4CIF	704	576	No	Optional	32.4	48.7	97.3	146.0
16CIF	1408	1152	No	Optional	129.8	194.6	389.3	583.9

data stream are now optional; hence, the codec can be configured for a lower data rate or better error recovery. Furthermore, in the H.263 arrangement four negotiable coding options are now included in order to improve the coding performance: Unrestricted Motion Vectors, Syntax-based Arithmetic Coding, Advanced Prediction, and Predicted and Bidirectional (P-B) frames. These options are described in further detail later in this chapter. The H.263 codec's schematic is identical to that of the H.261 scheme, which was shown in Figure 8.1 and is repeated in Figure 9.1 for convenience. As noted earlier, the H.263 scheme supports five different video resolutions, which are discussed in Table 9.1. In addition to QCIF and CIF which were supported by H.261, the H.263 scheme can also encode Sub-QCIF (SQCIF), 4CIF, and 16CIF video frames, which are summarized along with their associated number of pixels or resolutions in Table 9.1. As can be seen from the table, the SQCIF exhibits approximately half the resolution of QCIF, while 4CIF and 16CIF have four and sixteen times the resolution of CIF, respectively. The ability to support 4CIF and 16CIF resolutions implies that the H.263 codec could "compete" with other higher bitrate video coding schemes, such as the Motion Picture Expert Group (MPEG) standard codec, MPEG2, for example, in High-Definition Television (HDTV) applications. In contrast, for low-rate wireless videotelephony, the QCIF and SQCIF modes are realistic in bitrate terms.

9.2 The H.263 Coding Algorithms

9.2.1 Source Encoder

The H.263 codec outline is identical to that of the H.261 scheme, and is repeated for convenience in Figure 9.1. The source encoders of both the H.261 and H.263 codecs operate on noninterlaced pictures at approximately 30 frames/s. The pictures are encoded in terms of three components, a luminance component, and two color difference components, which are denoted by (Y, C_b, C_r). The color components are subsampled at a quarter of the spatial resolution of the luminance component.

Again, the picture formats supported by the codecs are summarized in Table 9.1 along with their uncompressed bitrate at frame scanning rates of both at 10 and 30 frames/s for both gray and color video. As mentioned, the H.261 standard defines two picture resolutions, QCIF and CIF, while the H.263 codec has the ability to support five different resolutions. All H.263 decoders must be able to operate in SQCIF and QCIF modes and optionally support CIF, 4CIF, and 16CIF.

Let us now concentrate on the main elements of the H.263 codecs, namely, inter-frame prediction, block transformation, and quantization.

9.2.1.1 Prediction

Both the H.261 and H.263 codecs employ inter-frame prediction based on frame-differencing, which may be improved with the aid of full motion compensation using motion vectors. When motion prediction is applied, the associated mode of operation is referred to as inter-frame coding. The coding mode is called intra-frame coding, when temporal motion prediction is not used. The H.263 coding standard has a negotiable coding option, referred to as P-B mode. This adds a new type of coding mode, where the B pictures can be partly bidirectionally predicted, which requires picture storage at the decoder, introducing latency, as will be discussed in Section 9.2.4.4. Let us now continue our discussions by considering the H.263 motion compensation scheme.

9.2.1.2 Motion Compensation and Transform Coding

Motion compensation is optional in the encoder, but it is required in all decoders in order to be able to decode the received sequence of any encoder. Both the H.261 and H.263 schemes generate one motion vector (MV) per macroblock. Some of the most important differences between the H.261 and H.263 source coding algorithms are concerned with motion compensation. The H.261 standard allows the optional use of a loop filter in order to improve the performance of the motion compensation. This can be turned on or off for each macroblock. In contrast, the H.263 scheme does not use a loop filter since it has more advanced motion compensation.

A further difference is that the H.263 arrangement also has an advanced prediction mode (which will be discussed in Section 9.2.4.3), allowing for one or four motion vectors per macroblock. When the H.263 P-B prediction mode (described at a later stage in Section 9.2.4.4) is used, an additional delta motion vector can be transmitted for each macroblock. Although the H.261 motion compensation scheme opts for using full pixel resolution, the H.263 codec employs a more accurate half-pixel precision for the motion vectors. This typically results in a reduced motion compensated prediction error residual. The motion vectors are usually restricted, so that all pixels referenced by them are within the coded picture area. This restriction can be removed when using the unrestricted motion vector mode to be detailed in Section 9.2.4.1 in case of the H.263 codec.

In order to remove the spatial frequency-domain redundancy, transform coding is used. Both the H.261 and H.263 schemes employ an 8×8 pixel discrete cosine transform (DCT). The DCT operates on either the intra-frame coded picture or in the inter-coded mode on the motion-compensated prediction error residual.

9.2.1.3 Quantization

After the picture or the MCER has been transform coded, the resulting coefficients are quantized. As we have seen for the H.261 codec, the bitrate can be reduced by using a coarser quantizer, which typically results in longer runs of zero-valued quantized coefficients. Longer runs of zero-valued quantized transform coefficients can be more efficiently coded with run-length coding.

The quantizers used by the H.261 and H.263 codecs are the same, which were briefly characterized in Figure 8.4. There is a specific quantizer for the DC coefficient of intra-frame coded blocks, and there are 31 different quantizers for all other transform coefficients. Accordingly, the GQUANT and MQUANT parameters are encoded by 5 bits, as we have seen for the H.261 scheme. The DC quantizer is a linear quantizer with no dead-zone. The other quantizers are nominally linear with a dead-zone near zero and with varying step sizes, as can be seen in Figure 8.4. The thirty one quantizers are numbered, and the smaller the quantizer index, the finer the quantizer. The fine quantizers cannot represent the full dynamic range of the transform coefficients, but they provide a high resolution in a more limited range.

In both codecs, the coding control block of Figure 9.1 regulates the performance of the codec by modifying the various codec parameters in order to attain goals such as a certain image quality or bitrate. The coding control algorithms are not defined in the standards, since they are dependent on the application in which the codec is used.

9.2.2 Video Multiplex Coder

Concentrating now on the video multiplex coder, we see that its main function is to convert the quantized transform coefficients and motion vectors into symbols to be transmitted. The multiplex coder also imposes a structure onto the data generated by the source encoder in order to aid its error recovery.

Each video frame is divided into blocks of 8×8 pixels. Similarly to the H.261 codec, in the H.263 scheme four adjacent luminance and two color difference blocks are grouped

9.2. THE H.263 CODING ALGORITHMS

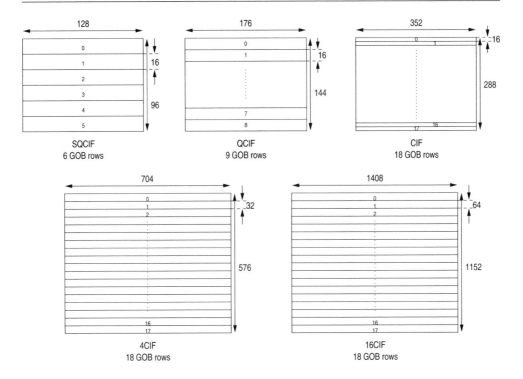

(All measurements are in Luminance pixels)

Figure 9.2: GOB structure for H.263 video codec, for frames of different resolutions.

together, which are referred to as macroblocks, as we portrayed it in the context of the H.261 codec in Figure 8.2. Each macroblock represents 16×16 pixels. A range of macroblocks are then grouped together in order to form a 176×48 pixel group of blocks (GOB), which were displayed in Figure 8.7 for the H.261 scheme in case of CIF and QCIF frames, defining a GOB as a group of 33 macroblocks in an 11×3 arrangement. Therefore, there are three GOBs in a QCIF frame. In contrast, the H.263 codec defines a GOB as a row of macroblocks for SQCIF, QCIF- and CIF, two rows of macroblocks for 4CIF, and four rows of macroblocks for 16CIF. The layout of the GOB in H.263 frames is shown in Figure 9.2, with a macroblock being 16×16 constituted by luminance pixels. These different resolutions are summarized in Table 9.1. The aim of grouping blocks of the frame into a hierarchical structure is to aid error-recovery.

The video multiplex schemes of the H.261 and H.263 codecs are arranged in an identical hierarchical structure based on four layers. From top to bottom of the hierarchy, these layers are: the Picture-, Group of Blocks- (GOB), Macroblock-, and Block-layers, which were shown in Figure 8.5.

We again emphasize that the remainder of this chapter is written assuming familiarity with the H.261 codec. Hence, we will concentrate on describing the differences between the H.263 and H.261 multiplex codecs.

9.2.2.1 Picture Layer

Some elements of the H.261 and H.263 picture layers are common in both standards. The information conveyed by these layers was summarized for the H.261 codec in Section 8.2.4.1, Figure 8.6, and Table 8.1, and their inspection is necessary in order to follow our deliberations. Briefly, the common H.261 and H.263 features are as follows:

The Picture Start Code(PSC) : a unique start code designed to aid resynchronization after a transmission error.

The Temporal Reference(TR) : a coding parameter expressed in terms of time as regards to when the current frame was coded relative to previous frames.

The Picture Type Information(PTYPE) : a parameter concerned with the format of the complete picture, specifying features, such as split screen, freeze, and CIF/QCIF operation.

In the H.263 standard, the picture start code (PSC) must be byte aligned, which is not required for H.261. The Temporal Reference is a 5-bit encoded value in H.261, which was increased to 8 bits in the H.263 scheme. The PTYPE symbol is constituted by 6 bits in the H.261 recommendation, which was increased to 13 bits in the H.263 scheme. This is required because of the range of extra options available in the H.263 standard. The only significant change to PTYPE in the H.263 codec is a Picture Coding Type flag, which is set to either intra- or inter-frame coded. A variety of other important changes to the picture layer in the H.263 scheme were concerned with the following parameters:

Quantizer (PQUANT) — The quantizer can now be changed in the picture layer, while in the H.261 scheme this could only be carried out in the GOB and macroblock layers.

Stuffing (STUF) — Bit Stuffing can now be invoked after a frame has been coded. This is useful for fitting the H.263 data into transmission packets. In the H.261 codec, bit stuffing was available only at the macroblock layer.

The remaining coding parameters of the H.263 picture layer are concerned with a variety of new options that the H.261 scheme did not support, such as the P-B prediction mode (see Section 9.2.4.4).

9.2.2.2 Group of Blocks Layer

This section relies on our previous discussions concerned with the H.261 GOB layer, which were summarized in Table 8.2 and Figure 8.8. As we have seen in the case of the picture layers, some of the coding parameters in the Group of Blocks layer are also common to the H.261 and H.263 codecs, which are as follows.

The Group of Block Start Code (GBSC) was incorporated in order to aid resynchronization after a transmission error.

The Group Number (GN) indicates which group of blocks is to follow.

9.2. THE H.263 CODING ALGORITHMS

The GOB Quantizer (GQUANT) conveys a request to change the quantizer commencing from the current GOB.

The changes concerning these symbols in the H.263 codec are minor. In the H.263 scheme, the group of blocks start code is byte aligned, like the picture start code, while as we noted before the H.261 codec's start codes need not be byte aligned. Byte-aligning the start code makes it easier to search for a start code after a transmission error, thereby potentially reducing hardware costs. The group number coding symbol has increased from 4 bits in the H.261 scheme to 5 bits in the H.263 arrangement. This was necessary because the GOBs were made smaller, and therefore they are more numerous, as shown in Figure 9.2. The GQUANT symbol is the same in both the H.261 and H.263 codecs.

The only other significant coding parameter in the H.263 group of blocks layer is the GOB frame ID (GFID). This symbol aids resynchronization after a transmission error in order to enable the decoder to identify whether the current GOB found is from the current frame or from a forthcoming frame. This reduces the likelihood of a whole GOB being decoded incorrectly, since this may cause catastrophic image degradation.

In the H.263 standard, the Group of Block layer header does not need to be coded for the first Group of Blocks, since a Picture layer header implies it. All other GOB headers in the frame can also be empty, depending on the encoding strategy. This implies that a video sequence can be coded with few GOB headers in order to reduce the bitrate or with more to increase the error-recovery capabilities. This renders the H.263 codec much more flexible than the H.261 scheme.

9.2.2.3 H.261 Macroblock Layer

In Section 8.2.4.3, Table 8.3, and Figure 8.9, we summarized the features of the macroblock layer in the H.261 codec. The H.263 schemes MB layer is very different. Although the picture and GOB layers in the H.263 scheme can be thought of as extensions of the corresponding H.261 layers, this is not true for the macroblock layer. The macroblock layers of both the H.261 and the H.263 codecs were designed to remove some redundancy from the data received from the source encoder by transmitting symbols containing the differential changes of parameters. However, the H.263 macroblock layer makes more intensive use of this differential coding. Recall from Table 8.3 that the H.261 macroblock consists of five different coding parameters, which are summarized now for convenience:

Macroblock Address (MBA) — used to indicate the position of the macroblock in the current group of blocks.

Macroblock Type (MTYPE) — signaling the coding mode, in which the macroblock was coded (inter/intra), and whether motion compensation or the loop filter was used for the current macroblock.

Quantizer (GQUANT/MQUANT) — employed to convey a change of the current quantizer for this and forthcoming macroblocks.

Motion Vector (MVD) — containing the values of the horizontal and vertical components of the motion vector for the current macroblock.

Coded Block Pattern (CBP) — transmitted in order to inform the decoder, blocks that are active in the current macroblock.

For the first macroblock in a group of blocks, the MBA symbol contains the absolute address of the macroblock. Subsequent macroblocks in the group of blocks contain the differential macroblock address in the MBA parameter. Specifically, the differential macroblock address is the difference between the current macroblock address and the address of the previously transmitted macroblock. The MBA parameter is coded with variable-length codewords, where the shorter codewords are assigned to the smallest values. This implies that consecutive adjacent macroblocks, which are most common, use fewer bits for the MBA than those that are far apart.

The MTYPE codec parameter in the H.261 standard is variable length coded like the MBA parameter, and the most common modes are assigned short codewords. The fixed-length encoded MQUANT parameter changes the absolute value of the quantizer. The motion vector displacement (MVD) transmits the difference in the motion vectors between the current and previous macroblock. If the previous macroblock did not contain a motion vector, then the previous motion vector is assumed to be zero. The MVD symbol is variable length coded, with the shorter codes given to smaller differences.

Lastly, the coded block pattern parameter (CBP) indicates which blocks are active within a macroblock. There are 64 different patterns, since each macroblock can contain up to four luminance blocks and up to two chrominance blocks. The active blocks are transmitted in sequential order after the macroblock header. Use of a CBP symbol removes the need for each active block to have a block address, which would be rather wasteful. The CBP is coded with variable-length codewords, where the common active block patterns are given the shorter codes.

9.2.2.4 H.263 Macroblock Layer

As mentioned before, the H.263 macroblock layer is different from the H.261 macroblock layer. As seen in Figure 9.3, eight different types of coding parameters can be transmitted in the H.263 macroblock header, as follows:

The coded macroblock indicator (COD) is a one-bit flag, which is set to 0 for all active macroblocks, while it is set to logical 1 in order to indicate that no further information is to be transmitted in the current macroblock. The COD bit is present in only those video frames for which the PTYPE codec parameter implies inter-frame coding.

The MB-type and chrominance block pattern (MCBPC) is a variable-length coded parameter, and it is always present in coded macroblocks, conveying information about the type of the current macroblock and as to which chrominance blocks are to be transmitted. Depending on whether the coding mode is intra (I) or motion predicted inter (P), two different variable-length coding tables can be used, which improves the coding efficiency by taking advantage of the different relative frequencies of the various coded block patterns. The assigned codewords also depend on one of five different types of MB, depending on the presence or absence of some of the MB layer's parameters. For details of the specific coding tables, the interested reader is referred to the H.263 Recommendation [259]. Without elaborating on the specific codeword

9.2. THE H.263 CODING ALGORITHMS

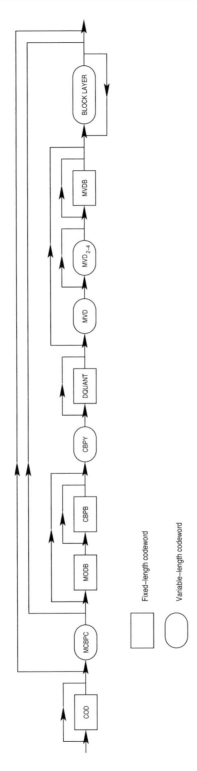

Figure 9.3: H.263 macroblock layer structure.

allocations for the Type 0-4 MBs, we note that there is a stuffing code for both the I and P frames, which indicates that the forthcoming section of the MB layer is skipped. This code is then naturally discarded by the decoder.

Macroblock mode for B blocks (MODB) represents the macroblock type for the bidirectionally predicted macroblocks or B blocks, which can occur in conjunction with any of the MB types 0–4 mentioned in the previous paragraph, but only when PTYPE implies that the P-B prediction mode is used. This variable-length coded parameter indicates the presence or absence of the forthcoming Coded Block Pattern for B blocks (CBPB) parameter and that of the Motion Vector Data for B macroblocks (MVDB), described later in this chapter.

Coded block pattern for B blocks (CBPB) indicates the specific pattern of the coded blocks for bidirectionally predicted macroblocks, which again occur only in P-B mode. As mentioned above, its presence is indicated by the preceding variable-length coded MODB parameter. Since CBPB is a 6-bit parameter, it can differentiate among 64 various coded patterns.

The coded block pattern for luminance (CBPY) parameter is variable-length encoded and conveys the coded block pattern of the active luminance blocks in the current macroblock, for which a minimum of one transform coefficient other than the DC component is transmitted. In both the intra- and inter-frame coded modes, there are four luminance blocks per MB, yielding a total of sixteen possible coded luminance patterns. Again, the most likely patterns are allocated shorter codewords, while the less likely ones are assigned longer ones. For details of the specific coding table, the reader is referred to the Recommendation [259].

The differential quantizer (DQUANT) codec parameter is used to differentially change the quantizer used for this and future macroblocks with respect to the previously specified PQUANT or GQUANT values of the picture layer or GOB layer. Since it is encoded by two bits, it can modify the previously stipulated quantizer index by ± 2. Furthermore, as the quantizer index is limited to the range [1–31], DQUANT values requiring quantizer indices outside this range are clipped.

Motion vector data (MVD) contains the variable-length coded horizontal and vertical components of the motion vector of the current inter-frame coded macroblock. Furthermore, for P-B frames, the MVD is also included in the intra-frame coded mode. The horizontal and vertical motion vector components are encoded independently, and they are represented using a half-pixel resolution. Again, the variable-length coding table assigns between 1 and 13 bits to the various vectors on the basis of their relative frequencies in order to maintain high coding efficiency. Specifically, instead of coding the motion vectors directly, a motion vector prediction process is invoked, and the difference between the actual and predicted vector is variable-length encoded. There are 128 possible motion translation vector differences in the coding table. However, there are only sixty-four codewords in the table. Each codeword can be interpreted as one of two differences. Since the new motion vector components are constituted by the sum of the predicted vector component plus the transmitted difference, where the sum

9.2. THE H.263 CODING ALGORITHMS

must be in the range $-16 \leq n \leq 15.5$, only one of the two possible differences for each codeword can be valid.

Motion vector data for B macroblock (MVDB) is a codec parameter describing the motion vector for bidirectionally predicted (B) macroblocks, which occur only when the P-B prediction mode is used. The MVDB parameter is included in the P-B frames only if its presence was previously indicated by the MODB parameter.

Here we curtail the description of the various MB layer parameters and note that the parameters MODB, CBPB, and MVDB will be discussed further in Section 9.2.4.4 concerning the P-B prediction mode. Recall that the codec's philosophy was to introduce the COD flag in order to mark the active MBs, for which then a coded block pattern was incorporated. This implies that each nonactive macroblock requires only a 1 bit macroblock header per macroblock.

In the H.263 codec, the macroblock type and the coded block pattern for chrominance have been combined into one parameter, namely, the MCBPC. The macroblock-type MCBPC informs the decoder as to what other macroblock layer parameters to expect for the current macroblock. The coded block pattern for luminance, namely, CBPY, is a separate codec parameter. The function of these symbols is similar to that of the corresponding coded block pattern parameters in the H.261 codec, where again, both are variable-length coded. The chrominance and luminance coded block patterns are separated because the luminance changes more frequently than the chrominance. Therefore, a combined coded block pattern like CBP in H.261 may be needlessly transmitting chrominance information every time the luminance changes.

An interesting feature of the H.263 codec is that the quantizer information at the macroblock layer is differentially coded. In the H.261 scheme, the quantizer could be set to an absolute value at the macroblock layer. The H.263 quantizer information at the macroblock level is conveyed using the DQUANT symbol. The quantizer index can only be changed in steps of $-2, -1, +1, +2$ with respect to its previous value in the H.263 arrangement, or it can be left unaltered by simply not transmitting the DQUANT parameter.

Recall that the motion vector coordinates in the H.263 codec are transmitted in the MVD codec parameter. We also remind the reader that in the H.261 codec the difference between the motion vector and the previous vector is transmitted in the MVD symbol. This can be thought of as using the previous motion vector as a prediction of the current vector and the MVD symbol transmitting the prediction error. In the H.263 scheme, a more complex motion vector predictor is used, which is based on up to three previous motion vectors. This new motion vector predictor is described in more detail in Section 9.2.3.1. The prediction error at the output of this predictor is transmitted in the MVD codec parameter. Having detailed the MB layer, let us now consider the lowest hierarchical level, the block layer.

9.2.2.5 Block Layer

In both the H.261 and H.263 coding schemes, a block contains the quantized transform coefficients of an 8×8 group of luminance or chrominance pixels. A macroblock can contain up to four luminance blocks and up to two chrominance blocks. In case of the H.261 scheme, the active blocks of each macroblock are transmitted in sequential order, separated by an end-of-block symbol. The encoded block pattern in the macroblock layer informs the decoder as to which blocks were deemed to be active. In the H.263 codec, the active blocks are transmitted in sequential order, as in the H.261 arrangement. However, there are no end-of-block symbols between the blocks. Instead, a flag bit referred to as LAST was introduced, which is set to logical 1 only in the last block of a macroblock; otherwise it is 0. Although this LAST flag bit is rather vulnerable, the bitrate contribution of the 2-bit end-of-block codec parameter of the H.261 scheme is higher than that of the 1-bit LAST flag of the H.263 codec.

The transform coefficients of the active blocks are encoded using the same method in both the H.261 and H.263 codecs. The first component in an intra-frame coded block is the DC component, which is encoded differently from the other coefficients, transmitting a fixed-length binary code representing its value.

For all other transform coefficients, run-length coding with variable-length codewords is used, employing a top-left to bottom-right oriented zig-zag scanning pattern in order to increase the length of the runs of zero-valued DCT coefficients. Following this scanning operation, the most common runs of zero coefficients followed by their specific nonzero values are assigned variable-length codewords on the basis of their relative frequency. Again, the length of the variable-length codewords is inversely proportional to their relative frequency. The remaining, less typical run-length and coefficient value combinations are encoded with an escape code followed by the run-length encountered and the transform coefficient value, which is encoded using fixed-length binary codewords.

In the case of the H.261 codec, there are 64 common run-length codewords, and other run-lengths are encoded using 20 bits, consisting of 6 bits escape, 6 bits run-length, and an 8-bit transform coefficient value. As regards the H.263 scheme, there are 44 common run-length codewords for the last block in a macroblock, where the LAST flag was set to logical 1 and a total of 58 common run-length codewords for the other blocks. In contrast to the 20-bit H.261 representation of uncommon run/value combinations, in the H.263 standard these combinations are encoded employing 22 bits. Specifically, the 7-bit escape code is followed by the 1-bit LAST flag in order to indicate whether it is the last block in a macroblock, while 6 bits are assigned to the run-length value and an 8-bit codeword to the transform coefficient value.

9.2.3 Motion Compensation

A common feature of both codecs is that motion compensation is used to produce the motion-compensated prediction error residual and hence to reduce the bitrate of the codec. Both codecs rely on motion vectors in order to achieve improved motion compensation in comparison to simple frame-differencing.

Both codecs specify in their respective recommendations how the motion vectors are conveyed. However, the motion vector search algorithm is not defined. Hence, significant research efforts have been devoted to finding motion vector search algorithms, which are fast, reduce the bitrate of the coders, and can be easily implemented in hardware, particularly, for example, by Chung *et al.* [325].

In the video multiplex layer, both the H.261 and H.263 schemes transmit a differential rather than absolute motion vector, encoded by the MVD codec parameter. The differential motion vector is the difference between the predicted and actual vector for a macroblock. The H.261 arrangement uses a simple predictor, where the predicted motion vector is given by the motion vector of the macroblock to the left of the current one. If the macroblock to the left

9.2. THE H.263 CODING ALGORITHMS

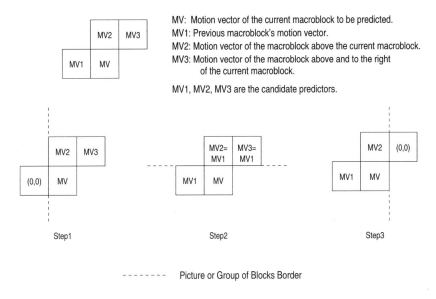

Figure 9.4: H.263 motion vector prediction scenarios.

of the current one does not exist or does not have a motion vector, then the predicted motion vector is 0.

9.2.3.1 H.263 Motion Vector Predictor

The H.263 motion compensation uses a more complex motion vector predictor. The prediction of the motion vector is a function of up to three previous motion vectors. The three candidate predictive motion vectors are the motion vectors of the macroblocks to the left, above, and above-right relative to the current MB. The arrangement of the motion vector and its candidate predictors is shown in Figure 9.4. The candidate predictors are modified when they are adjacent to a group of blocks boundary, a situation that is portrayed in Figure 9.4.

Step 1 The candidate predictive motion vector, MV1, is set to zero if it is outside the GOB or the picture frame containing the motion vector MV to be predicted.

Step 2 Then the candidate predictive motion vectors MV2 and MV3 are set to MV1, which may now be 0 if the macroblock being predicted is at the top edge of a GOB or picture.

Step 3 Then the candidate predictor MV3 is set to 0 if it is outside the GOB or picture frame containing the motion vector MV to be predicted.

Step 4 If any of the candidate predictive motion vectors' macroblocks were coded in intra-mode or were inactive and therefore are not encoded, then the candidate predictor would be 0.

The motion vectors have two components: horizontal and vertical. Each component of the predicted vector is found separately. The predicted motion vector is the median value of

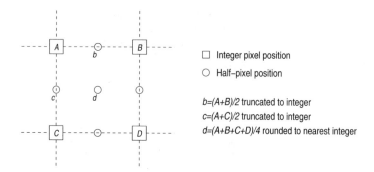

Figure 9.5: H.263 bilinear interpolation for subpixel prediction.

the three candidate predictors, as formally stated by Equation 9.1:

$$P_{Horiz} = Median(MV1_{Horiz}, MV2_{Horiz}, MV3_{Horiz})$$
$$P_{Vert} = Median(MV1_{Vert}, MV2_{Vert}, MV3_{Vert}). \quad (9.1)$$

The difference between the predicted and actual motion vector is encoded by the MVD codec parameter, as suggested by Equation 9.2:

$$MVD_{Horiz} = MV_{Horiz} - P_{Horiz}$$
$$MVD_{Vert} = MV_{Vert} - P_{Vert}. \quad (9.2)$$

This motion vector predictor is much more effective than the one used by the H.261 codec. It therefore reduces the bitrate because the differential motion vectors are typically smaller, requiring a lower number of bits for their variable-length encoding.

9.2.3.2 H.263 Subpixel Interpolation

As mentioned earlier, the H.263 codec uses half-pixel precision for its motion vectors, while the H.261 standard used full-pixel precision. Half-pixel precision allows the motion vector to more closely compensate for motion, thereby reducing the error residual and bitrate after inter-frame prediction. In order to predict the half-pixel values, the H.263 scheme invokes bilinear interpolation, which is portrayed in Figure 9.5.

Assuming that the pixel-spaced luminance values A, B, C, D are known, the half-pixel-spaced luminances are computed by averaging the pixel-spaced values surrounding the half-pixel-spaced positions. This bilinear interpolation effectively quadruples the original resolution and facilitates the "oversampled" high-resolution motion compensation.

Having described the various hierarchical layers of the H.263 codec and their relationship with the corresponding layers of the H.261 scheme, let us now concentrate on the H.263 codec's negotiable options.

9.2. THE H.263 CODING ALGORITHMS

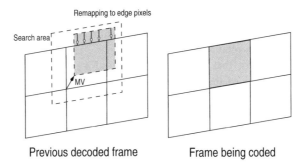

Figure 9.6: An example of the unrestricted motion vector pointing outside the coded picture area. Pixels outside the picture area are remapped to pixels on the edge of the picture area.

9.2.4 H.263 Negotiable Options

The H.263 standard specifies four negotiable coding options, which can be selected between the encoder and decoder when a connection is set up. The H.263 encoder and decoder can optionally support these configurations, and the negotiations can ascertain which options are supported by both parties. The negotiable options are additional coding modes, which can improve the system's performance at the expense of extra computation; therefore, they are more demanding and expensive in hardware terms. These options are defined below.

9.2.4.1 Unrestricted Motion Vector Mode

In the H.261 scheme and in the baseline H.263 coding modes, motion vectors are limited to reference pixels within the currently coded picture area, where the term "baseline" refers to an H.263 codec refraining from using the negotiable coding modes. In the unrestricted motion vector mode, this limitation is removed. This allows motion vectors to point outside the corresponding coded picture area in the previous decoded frame. When a block position referenced by a motion vector is outside the current picture area, a pixel on the edge of the picture area is used instead of the nonexisting pixel. This edge pixel is the last full pixel inside the picture area along the path of the motion vector. Edge pixels are found on a pixel by pixel basis and are calculated separately for each component of the motion vector. The mapping of pixels outside the coded picture area onto edge pixels is shown in Figure 9.6. In the figure, the shaded block in the currently coded frame was found to be best matched with the shaded block in the previously decoded frame; hence, the offset of this block is given by the motion vector MV.

For example, if the coordinates of the referenced pixel (which may not refer to a valid pixel position within the confines of the frame to be coded) are (x, y), then the coordinates are modified using Equation 9.3, into coordinates of the edge pixel (x', y'):

$$x' = \begin{cases} 0 & \text{if } x < 0 \\ 175 & \text{if } x > 175 \\ x & \text{otherwise,} \end{cases} \quad y' = \begin{cases} 0 & \text{if } y < 0 \\ 143 & \text{if } y > 143 \\ y & \text{otherwise.} \end{cases} \quad (9.3)$$

If pixel (x, y) exists, then $x = x'$ and $y = y'$. Clearly, Equation 9.3 is valid only for QCIF pictures of 176 by 144 pels resolution. However, the numbers 175 and 143 in the equation need to be changed for other picture sizes.

Experiments have shown that in this mode a significant gain is achieved if there is movement around the edges of the picture, especially for smaller picture resolutions. The results of Whybray et al. [326] show approximately an additional 9% bitrate reduction, when using the unrestricted motion vector mode in the context of the "Foreman" video sequence. The "Foreman" sequence contains a large amount of camera shake. The camera shake introduces movement at the edge of the picture, which the unrestricted motion vector mode handles well.

In addition to allowing motion vectors to point outside the coded picture area, the unrestricted motion vector mode allows the range of motion vectors to be expanded. In the default H.263 mode, motion vectors are restricted to the range $[-16, 15.5]$. However, in the unrestricted motion vector mode, the range of motion vectors is extended to $[-31.5, 31.5]$ with a few restrictions. If the predictor (P) of a motion vector (MV) is in the range $[-15.5, 16]$, the range of motion vectors is limited to $[-16, 15.5]$ around the predictor (P). If, however, the motion vector predictor (P) is outside the range $[-15.5, 16]$, then motion vectors in the range $[0, 31.5]$ or $[-31.5, 0]$ can be reached, depending on the sign of the predictor (P). These relationships are summarized in Equation 9.4:

$$\begin{array}{rcl rcrcl} -31.5 \leq & MV & \leq 0 & \text{if} & -31.5 \leq & P & \leq -16 \\ -16 + P \leq & MV & \leq 15.5 + P & \text{if} & -15.5 \leq & P & \leq 16 \\ 0 \leq & MV & \leq 31.5 & \text{if} & 16.5 \leq & P & \leq 31.5. \end{array} \quad (9.4)$$

The full range of motion vectors is split into these three ranges, so that the same coding table can be used, as in the baseline mode; only the equations to generate the motion vector from the predictor and motion vector difference are different. The extension of the motion vector range in this mode allows large motion vectors to be used, which is especially useful in the case of camera movement.

9.2.4.2 Syntax-based Arithmetic Coding Mode

When the syntax-based arithmetic coding (SAC) is used, the coding parameters are no longer coded with variable-length codewords but with syntax-based arithmetic coding [1]. The reconstructed frames will remain the same; however, the bitrate is reduced. The performance of the syntax-based arithmetic coding (SAC) mode was evaluated by Whybray et al. [326], who found bitrate savings between 1.6% and 6.3% for different video sequences.

SAC is a variant of arithmetic coding. It can be used to losslessly encode the video stream as an alternative to the widely used variable-length coding. The variable-length codewords are calculated from the entropy of the data. Since "conventional" codewords have to contain an integer number of bits, the rounding of the entropy values to fit a particular integer codeword length introduces inefficiency. Arithmetic coding almost completely eliminates this inefficiency by effectively allowing fractional bits per codeword. This is achieved by estimating the probability of a particular codeword in the data stream, where these estimates are referred to as the model.

The disadvantage of SAC is that it is difficult to implement. Furthermore, since each codeword is effectively constituted by a fractional number of bits, it is not possible to recognize particular symbols in the data stream. Another problem is that errors are not detected at all, unlike in the case of variable-length codewords, where errors are detected within a few codewords after an error. Lastly, if the above-mentioned model and the statistics of the actual data being coded are mismatched, the coding efficiency is reduced, which can lead to a high bitrate.

9.2.4.2.1 Arithmetic Coding [1] Is based on an abstract generic concept, that subdivides its number space incrementally into nonoverlapping intervals. This somewhat abstract concept will be "demystified" at the end of this section using a tangible practical example. Each interval corresponds to a particular event, and the size of the interval is proportional to the probability of the particular event occurring. An "event" in the context of syntax-based arithmetic coding is the encountering of a specific coding parameter, which has to be coded. The probability of encountering that particular coding parameter is proportional to the size of the subinterval for that event or coding parameter. Below we first introduce the abstract concept, which again will be made more explicit at a later stage using an example.

The arithmetic coder uses the sequence of events to incrementally subdivide the intervals. The coder then outputs as many bits as required to distinguish the subinterval or sequence of events that occurred. In general the arithmetic coders output these bits as soon as they become known. A comprehensive overview of arithmetic coding was given by Howard and Vitter [1]. The basic algorithm is as follows:

1. Each encoding stage starts with the "Current Interval", which is initially $[0, 1)$.

2. The current interval is subdivided into subintervals, one for each possible event that could occur next. The size of each event's subinterval is proportional to the model's estimated probability for that event at this particular instant.

3. The subinterval corresponding to the event that actually occurs next becomes the new "Current Interval".

4. The coder outputs a sufficient number of bits to distinguish the final current interval from all other possible final intervals.

An example of arithmetic coding is shown in Figure 9.7. In this example, the first event is either "a" or "b" with equal probability of $\frac{1}{2}$. For example, event "a" could mean that a COD coding parameter is being transmitted, and event "b" could mean that a MCBPC coding parameter is to be transmitted. In our example, event "a" was chosen since a COD coding parameter was to be transmitted. The subinterval "a" now becomes the current interval. The new current interval now has two possible next events, "c" (with probability $\frac{1}{3}$) and "d" (with probability $\frac{2}{3}$). At this stage event "d" was assumed to be encountered and became the new current interval. This new interval is $[\frac{1}{6}, \frac{1}{2}]$, which is further subdivided for events "e" (probability $\frac{1}{4}$) and "f" (probability $\frac{3}{4}$). If the final event encoded is event "f", then the interval for this event is $[\frac{1}{4}, \frac{1}{2}]$, as can be seen in Figure 9.7.

The arithmetic encoder has to transmit binary data to inform the decoder that the sequence of events was "a", "d", "f". This is accomplished by transmitting bits to define what the final interval was, which in this case was $[\frac{1}{4}, \frac{1}{2}]$. The interval is encoded by sending the value of the largest binary fraction that is fully contained inside the interval, which uniquely identifies

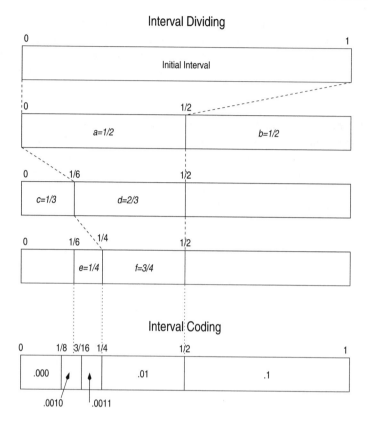

Figure 9.7: An example of arithmetic coding.

this interval using the lowest possible number of bits. The binary fractions seen at the bottom of Figure 9.7 are allocated as follows: .1 is $\frac{1}{2}$, .01 is $\frac{1}{4}$, .101 is $\frac{1}{2} + \frac{1}{8} = \frac{5}{8}$, and so on. The interval can therefore be encoded by recursively halving the number space, until a binary fraction interval is fully contained inside the interval to be coded. Hence, for our example in Figure 9.7 the sequence of events "a", "d", "f" has the interval $[\frac{1}{4}, \frac{1}{2}]$, which corresponds to the binary fraction .01. Therefore, the example sequence is encoded with the bits "01", implying that the final interval contains the binary fraction .01. If the sequence of events was "a", "d", "e", the final interval would be $[\frac{1}{6}, \frac{1}{4}]$. The largest binary fraction contained in that interval is .0011; hence, the encoded sequence of bits is "0011". For further details on the associated coding steps, the interested reader is referred to [1].

9.2.4.3 Advanced Prediction Mode

The advanced prediction mode is another negotiable option of the H.263 codec. This mode supports overlapped block motion compensation and the possibility of using four motion vectors per macroblock. In this mode, the motion vectors are allowed to cross picture boundaries, as in the unrestricted motion vector mode described in Section 9.2.4.1. The extended motion vector range feature is not part of the advanced prediction mode. It can,

9.2. THE H.263 CODING ALGORITHMS

however, be used with the advanced prediction mode if the unrestricted motion vector mode is also active. The overlapping motion compensation of the advanced prediction mode can only be used for the prediction of P-pictures, but not for B-pictures when the P-B mode (described in Section 9.2.4.4) is used.

The advanced prediction mode generally gives a considerable improvement by improving the subjective quality of the video. The results obtained by Whybray *et al.* [326] demonstrated that the bitrate increased for some video sequences when advanced prediction was used. However, it was noticed that the subjective quality of the video had improved, mainly because of the reduced blocking artifacts. This video quality increase is not clearly seen in the PSNR results, however. The bitrate generally was reduced only for inherently higher bitrate simulations. For example, the most dramatic bitrate reduction of 26% was obtained for the "Suzie" sequence. However, this technique may also increase the bitrate. For example, a 16% bitrate increase was experienced for the "Foreman" sequence.

Other techniques for reducing the blocking artifacts, which particularly occur at low bitrates, have been suggested by Ngan *et al.* [313, 314].

9.2.4.3.1 Four Motion Vectors per Macroblock
In the case of the H.261 scheme and the baseline H.263 coding modes, only one motion vector can be specified for each macroblock. In the advanced prediction mode, either one or four motion vectors can be transmitted for each macroblock. When one vector is transmitted, this is equivalent to four vectors of the same value. When four vectors per macroblock are transmitted, each motion vector references one luminance block and a quarter of the color blocks in the macroblock.

The use of four motion vectors per macroblock instead of one improves the motion prediction at the expense of an initially higher bitrate caused by the transmission of the extra motion vectors. This higher bitrate may be offset by a reduced-motion compensated error residual, which can be coded more efficiently by DCT coding.

The motion vectors are calculated and encoded in a similar fashion to the baseline coding mode. The only difference is that a different set of candidate predictive motion vectors are used. The new candidate predictive motion vectors are shown in Figure 9.8.

The candidate predictor MV3 for the top left block of the macroblock drawn in bold appears to be out of place. However, it cannot assume the obvious position adjacent to MV2 because if there was only one motion vector for the macroblock containing MV2, then MV2 and MV3 would be the same, and this would affect the quality of the motion vector prediction.

9.2.4.3.2 Overlapped Motion Compensation for Luminance
When the advanced prediction mode is used, each pixel in an 8×8 predicted luminance block is a weighted sum of three pixel values. These three pixels are found using three motion vectors, namely, the motion vector of the current luminance block and the motion vectors of two of the four adjacent blocks, termed "*remote vectors*". The corresponding four remote vectors are the motion vectors for the luminance blocks above, below, left, and right of the current luminance block. The two selected motion vectors are derived on a pixel by pixel basis for each pixel in the predicted block, using the motion vectors of the two nearest-neighbor blocks with respect to the position of the current pixel within the predicted block. For example, the 16 pixels in a 4×4 pixel group in the top-left of each predicted block use the motion vectors from the block above and to the left of the current block. As a further example, the pixel at the bottom-right

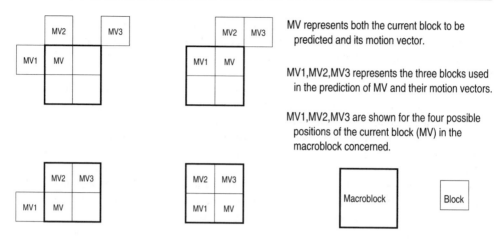

Figure 9.8: Redefinition of candidate predictive motion vectors MV1, MV2, and MV3 for advanced prediction mode using four motion vectors per macroblock.

of the predicted block is predicted from the motion vector of the current block, the motion vector of the block below, and the motion vector of the block to the right of the current block. If any of the chosen adjacent blocks do not exist, because the current block is at the border of the picture, then the motion vector of the current block is used instead. If the chosen adjacent block was not encoded or was encoded in intra-mode, then the remote vector is set to 0.

Having highlighted how the three motion vectors used for overlapped luminance motion compensation are derived, let us now explain how they are used in order to produce the predicted pixel values. Each pixel in the predicted block is derived from the weighted sum of three pixels. For example, the three pixels associated with the top-left pixel of the predicted block are the current pixel offset by the motion vector of the current block; the current pixel offset by the motion vector of the block above the current block; and the current pixel offset by the motion vector of the block to the left of the current one. If all three motion vectors used in the prediction are zero, then the three pixels used for predicting the current pixel are constituted by the pixel itself. Generally, however, the motion vectors are not zero, and therefore the predicted pixel can be generated by the weighted sum of up to three different pixels.

Specifically, the three pixels used for prediction are weighted using the associated weighting masks of the current and the four adjacent blocks shown in Figure 9.9. The predicted pixel value is then generated as the sum of the weighted pixels divided by 8. The five weighting masks portrayed in the figure actually overlap, and the sum of the weights for any pixel is 8. Therefore, if the three motion vectors used for prediction are 0, then the current pixel is weighted by three masks, summed and divided by 8, in order to produce the predicted pixel value, which in this case would be the same as the value of the current pixel.

Let us clarify the whole process with an example for the pixel to the left of the bottom-right pixel of the current block, where the corresponding weighting mask values are circled in Figure 9.9. The three motion vectors used for the prediction of this pixel are the motion vector associated with the current block, $[\Delta x_c, \Delta y_c]$, the motion vector of the block below the current block, $[\Delta x_b, \Delta y_b]$, and the motion vector of the block to the right of the current

9.2. THE H.263 CODING ALGORITHMS

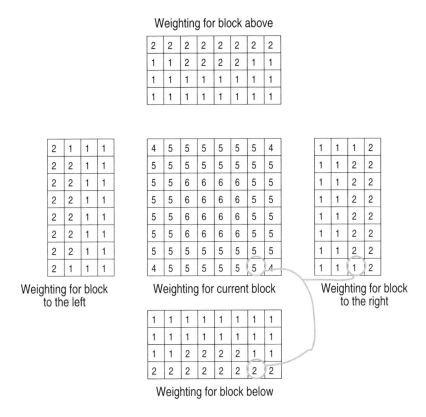

Figure 9.9: Overlapping motion compensation for luminance blocks using weighting masks.

block, $[\Delta x_r, \Delta y_r]$. Using this notation, the associated operations are summarized as follows:

$$L_{pred} = \frac{1}{8} \times \begin{bmatrix} [5 \times P(x + \Delta x_c, y + \Delta x_c)] + \\ [2 \times P(x + \Delta x_b, y + \Delta x_b)] + \\ [1 \times P(x + \Delta x_r, y + \Delta x_r)] \end{bmatrix} \quad (9.5)$$

where function $P(x, y)$ indicates the pixel luminance at the current coordinates, namely $[x, y]$.

9.2.4.4 P-B Frames Mode

The definition of the previously mentioned Predicted (P) and Bidirectional (B) frames borrowed from the Motion Pictures Expert Group (MPEG) standard [257] becomes explicit in Figure 9.10. A P-B frame is constituted by two pictures encoded as one unit, — namely by a P frame, which is predicted from the last decoded P frame, and a B picture, which is predicted both from the last decoded P picture and the P picture currently being decoded. The terminology B picture is justified because parts of it may be bidirectionally predicted from past and future P pictures, again as shown in Figure 9.10.

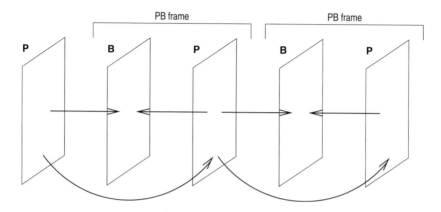

Figure 9.10: Prediction in P-B frames mode.

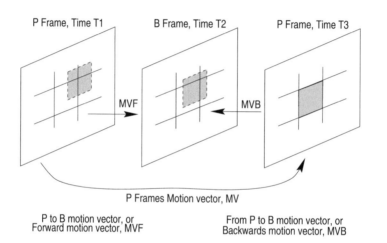

Figure 9.11: Motion vectors used in P-B prediction mode.

In the MPEG codec there can be several B pictures between P pictures, but in the H.263 standard only one B picture is allowed in order to reduce the decoding delay that P-B frames cause. In the P-B frame mode, a macroblock contains 12 blocks rather than just $4 + 2 = 6$ blocks. First, the six P blocks are transmitted as normal, followed by the six B blocks.

Motion vectors can be used to compensate for motion in B frames in a similar way to the usual P frames. The motion vectors for a P-B frame are shown in Figure 9.11.

The motion vector MV in the figure is the normal motion vector between two P frames, as in other H.263 modes. The forward (MVF) and backward (MVB) motion vectors are derived from MV but can be adjusted by the use of an optional delta motion vector. When a delta motion vector is not used, it is assumed that the motion vector changes linearly, between P frames. Therefore the forward and backward vectors can be found by interpolating the motion vector MV. If the motion vectors change linearly then $MVF - MVB = MV$. The forward and backward motion would be equal and opposite if the time between the three frames

9.2. THE H.263 CODING ALGORITHMS

(P,B,P) were the same. If the time between the three frames is different, then the forward and backward vector has to be adjusted, depending on the temporal references of the frame. The coding parameter TRB contains the number of untransmitted frames at 29.97 Hz between the B frame and the previous P frame ($T2 - T1$ in the figure). The equations for calculating the forward and backward motion vectors are shown in Equation 9.6:

$$MVF = \frac{TR_B \times MV}{TR_D} + MV_D$$
$$MVB = MVF - MV_D - MV, \qquad (9.6)$$

where, TR_B is the time between frames ($T1 - T2$), which is transmitted in the TRB coding parameter. TR_D is the time between the current and previous P frames ($T3 - T1$), while MV_D is the delta motion vector, which is optional. If it is not used, it is set to 0. Note that a positive delta motion vector increases the forward motion vector and reduces the backward vector.

The prediction for B blocks can be partly bidirectional. The part of each B block, which is bidirectionally encoded, is dependent on the backwards-oriented motion vector. The pixels in the B block referenced by the backward motion vector, which are inside the corresponding P macroblock, are bidirectionally predicted, while all other pixels use forward prediction only. If the whole B block were bidirectionally predicted, then the macroblock could not be decoded until all the macroblocks containing the pixels referenced by the backward motion vector were received, contributing additional latency to the decoding process. The bidirectionally predicted pixels use the average of the pixels referenced by the forward and backward motion vector, as the prediction value for the B-coded pixel. This prediction value is added to the decoded MCER in order to find the value of the B-coded pixel. The remaining pixels that use forward prediction employ the pixel values from the previous P frame as the prediction values for the pixels. An example showing the area of the B macroblocks that can be bidirectionally predicted is shown in Figure 9.12.

The P-B mode of the H.263 video codec can be utilized in two ways. First, it can be used to double the frame rate for a modest increase in bitrate. Increasing the frame rate of a coded video sequence gives a significant subjective improvement of quality. The second way to use P-B mode is to maintain the same frame rate but achieve greater compression by the use of the more efficient B frames. However, P-B mode cannot be used at low frame rates combined with a large amount of motion. This is because the process of interpolating the motion vector becomes inaccurate.

The investigations by Whybray *et al.* [326] demonstrated that the bitrate increased when P-B mode was used to double the frame rate. For video sequences with low to medium amounts of motions and a reasonable frame rate, it was found that the increase in bitrate became more moderate at higher bitrates. The increase in bitrate for these sequences ranged from 36% to 4.5%. For high-motion sequences at low frame rates, the bitrate increase was as much as 61%. It was noted that the PSNR of the P- and B- frames was very similar, but this may be due to the fixed quantizer that was used for these simulations. In general, the PSNR of the B-frame is typically less than that for the P-frames.

Following our comparative description of the H.261 and H.263 codecs, in the next section we provide a range of performance results for these schemes.

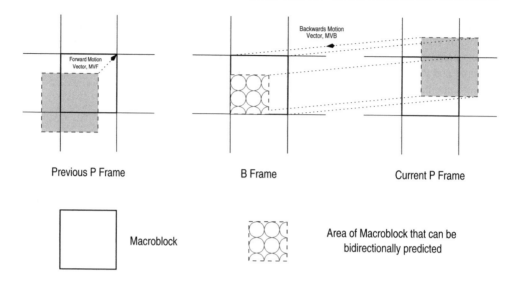

Figure 9.12: Motion vectors used in P-B prediction mode.

9.3 Performance Results

9.3.1 Introduction

The H.261 codec used in our investigations was a software implementation derived from the INRIA Videoconferencing System (IVS).[1] The H.263 codec was a modified version of a software implementation developed by Telenor Research and Development [327]. An enhanced version of the source code is now available from the University of British Columbia's World Wide Web site.[2]

Simulations were carried out with both the H.261 and H.263 codecs in order to comparatively study their performance. All simulations used well-known video sequences, such as the "Miss America", "Suzie", and "Carphone" clips and were performed with transmission frame rates of 10 and 30 frames/s.

The performance of both codecs depends very much on the quantizers used. The upper and lower limits of performance are found when the finest or coarsest quantizers are employed.

The H.261 simulations used gray-scale video sequences that were generated by discarding the color information, retaining just the luminance information. The H.263 simulations invoked in comparison with the H.261 codec used the same gray-scale video sequences. In this section, results are shown in graphical and tabular forms. To make the results comparable in tabular form, results were interpolated from the graphical curves in order to obtain results at the same bitrate or PSNR. Let us now turn to their performance comparison.

---

[1] http://www.inria.fr/rodeo/personnel/Thierry.Turletti/ivs.html
[2] http://www.ee.ubc.ca/spmg/research/motion/h263plus

9.3. PERFORMANCE RESULTS

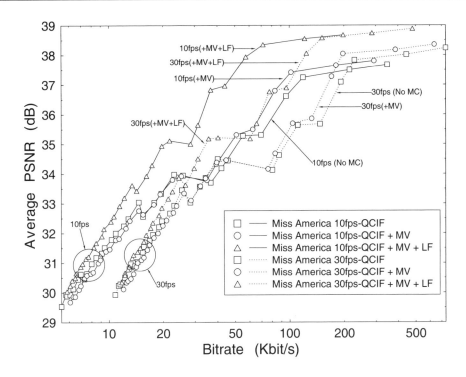

Figure 9.13: H.261 Image quality (PSNR) versus coded bitrate.

9.3.2 H.261 Performance

The H.261 simulations completed used the QCIF gray-scale "Miss America" video sequences at 10 and 30 frames/s. All simulations used inter-frame prediction with an intra-coded frame once every 132 frames. The intra-coded frame update rate of once every 132 frames was selected in order to meet the minimum macroblock forced update requirement of the codec. Simulations were carried out using the H.261 codec in one of the following three modes:

- No motion compensation.

- Motion compensation using motion vectors (MV).

- Motion compensation using motion vectors and loop filtering (MV+LF).

The simulations were performed at a wide range of bitrates. The results of these simulations are shown in Figures 9.13 and 9.14, where Figure 9.13 characterizes the average PSNR versus bitrate performance. The coded video sequences can also be seen on the WWW,[3] where readers can judge the associated perceptual video quality for themselves.

As can be seen in Figure 9.13 for a given bitrate, the simulations without motion compensation have typically the lowest image quality. This is a consequence of constraining the quantizers to maintain a given bitrate, while having to quantize the motion-compensated

---

[3] http://www-mobile.ecs.soton.ac.uk/peter/h261/h261.html#Examples

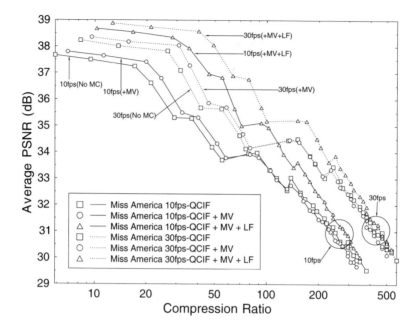

Figure 9.14: H.261 Image quality (PSNR) versus compression ratio.

prediction residual (MCPR) acquired by frame-differencing rather than by full motion compensation. Clearly, frame-differencing has a lower complexity than full motion compensation, and it can also dispense with transmitting the motion vectors, thereby potentially reducing the bitrate. However, the MCPR has a higher variance, which is thus less amenable to quantization, even when investing the extra bits, which would have been dedicated to the encoding of the MVs. Pairwise comparisons in Figure 9.13 reveal that the overall effect of MV-assisted full MC was found advantageous in terms of improved motion tracking, and it increased the image PSNR slightly. However, when motion vectors and loop filtering were used jointly, the image quality was substantially improved.

The image quality expressed in PSNR saturated at high bitrates, around 100 Kbit/s. However, in the bitrate range between 10 and 100 Kbit/s, the PSNR improved quasilinearly, as the bitrate was increased. The corresponding gradient expressed in terms of PSNR improvement versus bitrate increment was about 10 dB/decade. When maintaining a constant bitrate, the 30 frames/s curves typically exhibit a PSNR penalty due to the imposed bitrate constraint, since on the basis of simple logic at 30 frames/s a threefold bitrate increase is expected in comparison to 10 frames/s. This is a very coarse estimate, since due to the higher inter-frame correlation the MCPR has a typically reduced variance in comparison to 10 frames/s coding, which mitigates the above threefold bitrate increment estimate, but the bitrate penalty of 30 frames/s transmissions is certainly substantial. In many applications, it may be advantageous to vary the frame scanning rate under network and user control in order to arrive at the best scheme for a specific application.

Table 9.2 summarizes the decrease in bitrate achieved for the same PSNR at both 10 and 30 frames/s, with and without motion vectors (MV) and loop filtering (LF). The reductions are

9.3. PERFORMANCE RESULTS

Table 9.2: Relative Reduction (%) in Bitrate Required to Achieve Given PSNRs, Compared with Simulations at 10 Frames/s Using no Motion Compensation

Fixed PSNR (dB)	Percentage Bitrate Reduction (%)				
	10 frames/s		30 frames/s		
	MV	MV+LF	NoMC	MV	MV+LF
31	−8.28	10.96	−76.81	−78.81	−63.25
33	−23.55	19.53	−61.08	−67.05	−41.31
35	8.98	60.57	−85.16	−73.10	33.34
37	18.65	59.46	−72.33	−48.48	10.19

Table 9.3: Increase in PSNR (dB) Achieved at Given Bitrates, Compared with Simulations at 10 Frames/s Without Motion Compensation

Fixed Bitrate (Kbit/s)	PSNR Increase (dB)				
	10 frames/s		30 frames/s		
	MV	MV+LF	NoMC	MV	MV+LF
20	0.00	1.54	−0.98	1.15	−0.51
30	0.00	1.51	−0.57	−0.55	0.39
50	0.40	2.55	−0.44	−0.44	0.35
100	0.62	1.73	−1.56	−1.23	0.32
190	0.22	1.22	−0.41	0.40	1.18

relative to the 10 frames/s simulations without motion compensation (MC). The table shows the performance improvement provided by motion-vector-assisted full motion compensation combined with loop filtering. At high PSNRs, the simulations at 30 frame/s using motion vectors and loop filtering outperform the baseline simulation at 10 frames/s without motion compensation. For example, at a PSNR of 37 dB and a scanning rate of 30 frames/s with motion vectors and loop filtering, the required bitrate is 10.19% less than at 10 frames/s without motion compensation. In contrast, at 30 frames/s without motion compensation a 72% increase in bitrate is required. A range of further interesting trade-offs becomes explicit by scrutinizing Figure 9.13 and Table 9.2.

Viewing the codec's performance from a different angle, Table 9.3 shows the increase in image quality expressed in PSNR achieved at a range of fixed bitrates. The gains are relative to the 10 frames/s simulations without motion compensation. Again, the table characterizes the performance improvement provided by motion vector assisted full MC combined with loop filtering. For example, observe in Figure 9.13 and Table 9.3 that at a fixed bitrate of 50 Kbit/s, the 30 frames/s mode of operation without motion compensation has a 0.44 dB loss of PSNR compared with the 10 frames/s scenario without motion compensation. However, simulations conducted at 10 frames/s with motion-vector-assisted full motion compensation and loop filtering has a 2.55 dB increase in PSNR compared with the scenario at the same frame rate without motion compensation.

Figure 9.14 plots the results of the above investigations from a different perspective, as average PSNR versus the compression ratio. Again, the curves suggest a fairly linear relationship over a rather wide range, resulting in an approximately 10 dB/decade decay in PSNR terms as a result of increasing the compression ratio from around 50 to 500. As before, the improvement in performance achieved when using motion vectors and loop filtering is clearly shown in Figure 9.14. The graphs suggest that an approximately quadrupled compression ratio can be achieved for the same image quality, when using the more sophisticated MV-assisted loop-filtered full motion compensation. A rule of thumb was that at high bitrates, turning on motion vectors and loop filtering allowed the frame rate to be tripled, as well as reducing the bitrate. These results are compared with similar performance figures of the H.263 codec in Section 9.3.3.

9.3.3 H.261/H.263 Performance Comparison

In this section, we compare the H.261 performance figures derived in Section 9.3.2 with the corresponding characteristics of the H.263 codec using the same QCIF gray-scale "Miss America" video sequences. The H.263 experiments used none of the negotiable coding options. In this mode, the H.263 codec uses motion vectors and inter-frame prediction for motion compensation. The appropriate macroblock-forced update algorithm was invoked in order to ensure that each macroblock was coded in intra-mode, after a maximum of 132 inter-frame coded updates.

Similarly to our H.261 investigations, the H.263 experiments were conducted at 10 and 30 frames/s and at a wide range of bitrates. Figure 9.15 shows a comparison of the H.261 and H.263 video codecs in terms of image quality expressed in PSNR versus bitrate (Kbit/s). As can be seen from the graphs, the performance of the H.263 codec is significantly better than that of the H.261 scheme. Furthermore, the useful operating bitrate range of the H.261 codec was also extended by the H.263 scheme, ensuring an approximately 9 dB/decade PSNR improvement across the bitrate range of 5–500 Kbit/s.

Table 9.4 shows the decrease in terms of bitrate achieved for a range of specific fixed PSNR values when using the H.263 codec rather than the H.261 scheme. Specifically, the bitrate reductions are relative to the H.261 mode of operation at 10 frames/s without motion compensation. The table includes a column for the best H.261 performance, which uses motion-vector-assisted full MC and loop filtering. This table shows that the H.263 codec outperforms the H.261 scheme, even when the frame rate is three times that of the H.261 arrangement.

Viewing the higher performance of the H.263 codec from a different angle, Table 9.5 summarizes the increase in image quality measured in terms of PSNR achieved at a range of different fixed bitrate values. The gains are relative to the H.261 scenario at 10 frames/s without motion compensation. Again, it is clear from the table that the H.263 scheme outperforms the H.261 codec.

Figure 9.16 provides a comparison of the H.261 and H.263 video codecs in a different context, expressed as image PSNR versus compression ratio. Again, as can be seen from the graph, the H.263 codec exhibits a higher compression performance than the H.261 codec. As expected from the corresponding PSNR versus bitrate curves, the H.263 codec's

9.3. PERFORMANCE RESULTS

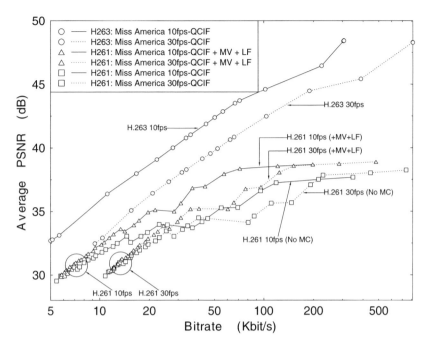

Figure 9.15: Image quality (PSNR) versus coded bitrate, for H.261 and H.263 simulations using grayscale QCIF "Miss America" video sequences at 10 and 30 frames/s.

Table 9.4: Relative Reduction (%) in Bitrate Required to Achieve Given PSNR Values, for the H.261 and H.263 Codecs Compared with the H.261 Performance at 10 Frames/s Using no Motion Compensation

Fixed PSNR (Kbit/s)	Percentage Bitrate Reduction (%)				
	10 frames/s		30 frames/s		
	H.261 (MV+LF)	H263	H.261 (NoMC)	H.261 (MV+LF)	H263
33	19.53	62.28	−61.08	−41.31	26.90
35	60.57	82.99	−85.16	33.34	70.30
37	59.46	87.86	−72.33	10.19	77.78

Table 9.5: Increase in PSNR (dB) Achieved at the Same Bitrate, for H.261 and H.263 Simulations Compared with the H.261 Simulation at 10 Frames/s Without Motion Compensation

Fixed Bitrate (Kbit/s)	PSNR Increase (dB)				
	10 frames/s		30 frames/s		
	H.261 (MV+LF)	H263	H.261 (NoMC)	H.261 (MV+LF)	H263
20	1.54	5.20	−0.98	−0.51	2.76
30	1.51	6.43	−0.57	0.39	3.98
50	2.55	7.50	−0.44	0.35	4.91
100	1.73	7.81	−1.56	0.32	5.59
190	1.22	8.49	−0.41	1.18	7.04

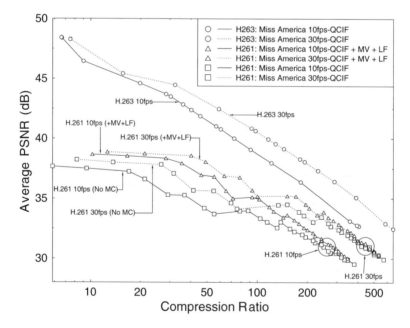

Figure 9.16: Image quality (PSNR) versus compression ratio performance for H.261 and H.263 simulations using gray-scale QCIF "Miss America" video sequences at 10 and 30 frames/s.

9.3. PERFORMANCE RESULTS

Table 9.6: Video Sequences Used for H.263 Simulations

Video Sequence	Size/s	Frame rates (frame/s)	Gray or Color
"Miss America"	QCIF	10, 30	Gray
"Miss America"	SQCIF, QCIF, CIF	10, 30	Color
"Carphone"	QCIF	10, 30	Color
"Suzie"	SQCIF, QCIF, 4CIF	10, 30	Color
"American Football"	4CIF	10, 30	Color
"Mall"	16CIF	10, 30	Color

compression ratio performance curves are near-linear on this log-log scaled graph, resulting in a predictable PSNR versus compression ratio relationship.

Recall that these investigations were conducted without the H.263 codec using the negotiable options that can be invoked in order to increase the image quality or to reduce the bitrate. Observe, furthermore, that the H.263 performance graphs in Figures 9.15 and 9.16 never intersect the H.261 curves, implying that the H.263 PSNR performance at 30 frames/s is better than the H.261 at 10 frames/s in the investigated bitrate range.

In addition to the results presented above, example video sequences can be viewed on the WWW for both H.261[4] and H.263[5] video codecs, where the associated subjective quality can be judged. Having contrasted the H.261 and H.263 PSNR versus bitrate performances, let us now study further aspects of the H.263 arrangement.

9.3.4 H.263 Codec Performance

In this section we report our findings on the trade-offs between image quality and bitrate for different image sizes, frame rates, and video sequences, when using the H.263 codec. All simulations were conducted using well-known video sequences at both 10 and 30 frames/s. The video sequences used are summarized in Table 9.6. The image sizes used were described in Table 9.1 on page 296. Video sequences at 10 frames/s were generated from the 30 frame/s versions. The SQCIF images were generated by re-sampling the QCIF sequences. The original "Mall" sequence was a High-Definition Television (HDTV) video sequence at a resolution of 2048×1048 pixels. In order to convert this to 16CIF format, a black border was added to the top and bottom of the 16CIF frame, and the left and right edges were cropped to fit the 16CIF frame format.

9.3.4.1 Gray-Scale versus Color Comparison

The previous H.261/H.263 comparison simulations were conducted using gray-scale video sequences. In this section we investigate the extra bitrate required to support color

---
[4] http://www-mobile.ecs.soton.ac.uk/peter/h261/h261.html#Examples
[5] http://www-mobile.ecs.soton.ac.uk/peter/h263/h263.html#Examples

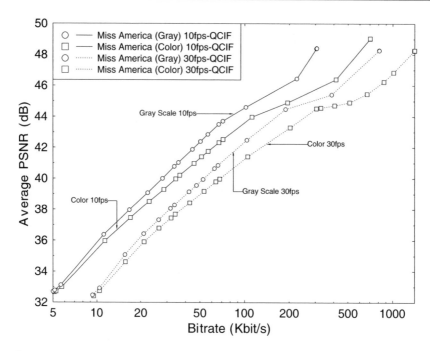

Figure 9.17: Image quality (PSNR) versus coded bitrate performance of the H.263 codec for the "Miss America" sequence in color and gray-scale at 10 and 30 frames/s.

video coding. The gray-scale video sequence was generated by extracting the luminance information from the color video sequence.

The results of our gray versus color investigations are presented in Figures 9.17 and 9.18. The corresponding graphs characterize the image quality expressed in terms of PSNR versus bitrate and compression ratio, respectively. These graphs suggest that the gray-scale and color PSNR performances are rather similar, but their difference becomes slightly more dominant with increasing bitrates. These tendencies imply that assuming a constant bitrate, only a small fraction of the overall bitrate has to be allocated to convey the color information. Therefore, the PSNR penalty due to reducing the bitrate and resolution of the luminance component is marginal. The previously observed near-linear PSNR versus bitrate and compression ratio relationship is also maintained for color communications.

Table 9.7 shows the relative reduction in bitrate for the same image quality when using gray-scale video rather than color. Observe in the table that the required bitrate difference is comparatively small for lower PSNR values, but this discrepancy increases for higher bitrates, where higher video quality is maintained. This confirms our previous observation as regards earmarking only a small fraction of the available bitrate to encoding the inherently lower resolution chrominance information.

Table 9.8 portrays the same tendency from a different perspective, expressed in terms of image quality gain (PSNR) for gray-scale video for the same bitrate as color video at the same frame rate. Again, this table suggests that the gray-scale PSNR performance is only marginally better for low bitrates, but the difference increases at higher bitrates.

9.3. PERFORMANCE RESULTS

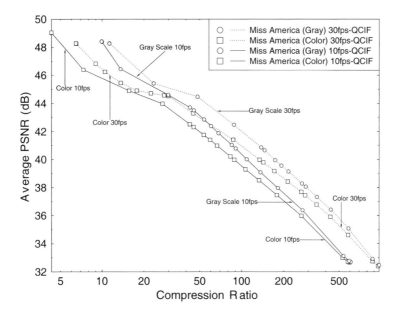

Figure 9.18: Image quality (PSNR) versus compression ratio performance of the H.263 codec for the "Miss America" sequence in color and gray-scale at 10 and 30 frames/s.

Table 9.7: Relative Reduction (%) in Bitrate Required to Achieve a Range of Fixed PSNR Values, Compared with the Simulations at the Same Frame Rate in Color

Fixed	Percentage Bitrate Reduction (%)	
PSNR (dB)	10 frames/s-Gray	30 frames/s-Gray
34	6.54	6.90
38	15.35	17.33
42	25.06	31.73
46	45.38	41.50
48	51.17	42.38

Table 9.8: Increase in PSNR(dB) Achieved at a Range of Fixed Bitrates, Compared with Simulations at the Same Frame Rate in Color

Fixed Bitrate (Kbit/s)	PSNR Increase (dB)	
	10 frames/s-Gray	30 frames/s-Gray
10	0.44	0.16
30	0.83	0.63
100	1.01	1.10
200	1.14	1.29

Overall, the perceived subjective video quality improvements due to using color, rather than gray-scale representation, are definitely attractive, unequivocally favoring the color mode of operation.

9.3.4.2 Comparison of QCIF Resolution Color Video

The investigations in this section were conducted in order to characterize the performance of the H.263 codec for a range of different video sequences. All simulations were carried out using color video sequences scanned at 10 and 30 frames/s. The video sequences used were "Miss America", "Suzie", and "Carphone". Some example QCIF video sequences can also be viewed on the World Wide Web.[6]

The results of these experiments are plotted in Figures 9.19 and 9.20. As expected on the basis of its low-motion activity, the graphs suggest that the "Miss America" sequence is the most amenable to compression, followed by "Suzie" and then the "Carphone" sequence. The achievable compression ratios span approximately one order of magnitude range, when comparing the lowest and highest activity sequences. The achievable relative bitrate reduction expressed in percentages is tabulated in Table 9.9 for the above video clips. Viewing these motion activity differences from a different angle, the PSNR reductions produced by the fixed bitrate constraints are summarized in Table 9.10. The negative PSNR gains reflect the above-stated tendencies and underline the importance of specifying the test sequence used in experimental studies. Let us now concentrate on analyzing the codec's performance in case of different video frame resolutions.

9.3.4.3 Coding Performance at Various Resolutions

The objective of this section is to evaluate the performance trade-offs for different video resolutions. Our investigations were conducted at both 10 and 30 frames/s, at resolutions of SQCIF, QCIF, and CIF. The SQCIF video sequences were generated by subsampling the QCIF video sequences. The QCIF and CIF video sequences originated from different sources and therefore are not identical.

The corresponding PSNR results are portrayed in Figures 9.21 and 9.22 as a function of the bitrate and compression ratio, respectively, when using the "Miss America" sequence.

---

[6]http://www-mobile.ecs.soton.ac.uk/peter/h263/h263.html#Examples

9.3. PERFORMANCE RESULTS

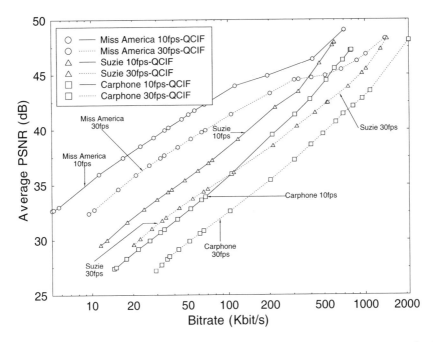

Figure 9.19: Image quality (PSNR) versus coded bitrate performance of the H.263 codec for QCIF resolution at 10 and 30 frames/s.

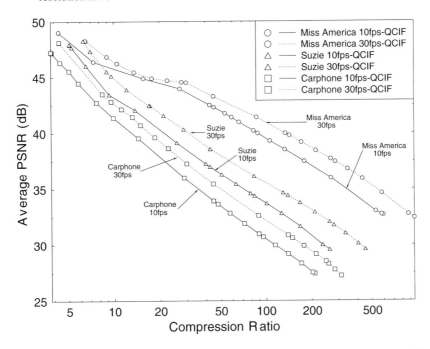

Figure 9.20: Image quality (PSNR) versus compression ratio performance of the H.263 codec for QCIF resolution at 10 and 30 frames/s.

Table 9.9: Relative Reduction (%) in Bitrate Required to Achieve Fixed PSNRs for QCIF Resolution Video for the H.263 Codec, Compared with the Corresponding PSNR for the "Miss America" Sequence at 10 Frames/s

Fixed PSNR (dB)	Percentage Bitrate Reduction (%)					
	Miss America	Suzie		Carphone		
	30 frames/s	10 frames/s	30 frames/s	10 frames/s	30 frames/s	
33	−93.8	−343.3	−678.2	−857.0	−1922.6	
34	−82.9	−331.9	−670.7	−805.8	−1893.8	
36	−89.6	−387.8	−823.8	−821.6	−1949.2	
38	−88.9	−363.6	−846.7	−724.4	−1683.7	
40	−98.3	−328.2	−760.3	−577.1	−1439.1	
42	−121.1	−257.3	−679.3	−466.5	−1187.5	
44	−129.0	−210.6	−570.2	−325.8	−945.7	
46	−127.1	−30.0	−206.4	−83.0	−358.8	

Table 9.10: Increase in PSNR (dB) Achieved at a Range of Fixed Bitrates for QCIF Resolution Video Coded with the H.263 Codec, Compared with the Simulation of "Miss America" at 10 Frames/s

Fixed Bitrate (Kbit/s)	PSNR Increase (dB)					
	Miss America	Suzie		Carphone		
	30 frames/s	10 frames/s	30 frames/s	10 frames/s	30 frames/s	
10	−2.67	—	—	—	—	
20	−2.37	−6.02	—	−9.23	—	
30	−2.25	−5.80	−8.08	−9.02	−12.13	
70	−2.39	−5.54	−7.82	−8.40	−11.38	
100	−2.32	−5.20	−7.76	−7.84	−11.04	
200	−1.69	−3.47	−6.68	−5.74	−9.60	
400	−1.59	−1.47	−5.06	−3.49	−7.68	
500	−2.29	−0.63	−4.97	−2.99	−7.47	

9.3. PERFORMANCE RESULTS

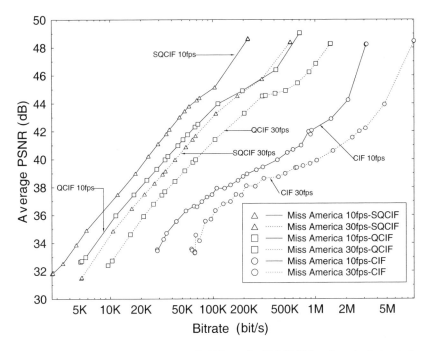

Figure 9.21: Image quality (PSNR) versus coded bitrate for H.263 "Miss America" simulations at 10 and 30 frames/s.

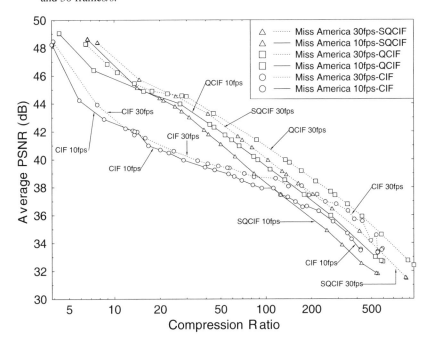

Figure 9.22: Image quality (PSNR) versus compression ratio, for H.263 "Miss America" simulations at 10 and 30 frames/s.

Table 9.11: Percentage Reduction in Bitrate Required to Achieved the Same PSNR, for "Miss America" Video Sequences Coded with H.263 Codec Compared with the QCIF Resolution Simulation at 10 Frames/s

Fixed PSNR (dB)	Percentage Bitrate Reduction (%)				
	SQCIF		QCIF	CIF	
	10 frames/s	30 frames/s	30 frames/s	10 frames/s	30 frames/s
33	32.98	−32.54	−93.83	—	—
34	37.29	−21.22	−82.94	−315.07	−780.63
36	26.76	−27.32	−89.57	−339.18	−772.36
38	30.60	−25.84	−88.94	−577.74	−945.79
40	34.88	−25.03	−98.29	−1159	−3042
42	39.86	−23.82	−121.12	−1389	−4259
44	44.12	−27.28	−129.01	−1614	−4110
46	62.74	7.71	−127.09	−613.04	−1808
48	66.12	10.01	−125.56	−419.78	−1388

The near-linear nature of the curves was maintained for the SQCIF and QCIF resolutions, but for the CIF resolution the coding performance curves became more nonlinear. The quadrupled number of pixels present in the CIF format resulted in an approximately fourfold increase of the bitrate. Furthermore, as in our previous investigations, the 30 frames/s scenarios typically required a factor of 2 higher bitrates in order to maintain a certain fixed PSNR. The same tendencies were observed in terms of the compression ratio in Figure 9.22. As before, we also summarized the relative bitrate reduction expressed in percentages with respect to the previously employed QCIF, 10 frames/s scenario, which is portrayed for the various PSNRs, resolutions, and frame rates in Table 9.11. The corresponding PSNR increment/reduction due to employing lower/higher resolutions, while maintaining certain fixed bitrates, was presented in Table 9.12 in order to help gauge the associated trade-offs.

Again, the CIF performance curves are more nonlinear than those of the lower resolution scenarios. A range of further interesting conclusions emerges from these graphs and tables. As an example, it is interesting to note in Figure 9.21 that by changing the resolution from QCIF to SQCIF, the frame rate can be tripled for just a 25% increase in bitrate, or a 0.5 dB loss of PSNR around a PSNR of 40 dB.

Let us now verify the above findings, which we arrived at on the basis of the "Miss America" sequence, also using the "Suzie" sequence, while replacing the CIF resolution with 4CIF representations. The remaining experimental conditions were unchanged. The corresponding results are plotted in Figures 9.23 and 9.24 in terms of PSNR versus the bitrate and compression ratio, respectively. Note, for example, in Figure 9.23 that the 30 frames/s SQCIF performance curve virtually overlaps with the 10 frames/s QCIF curve, indicating that the doubled number of pixels of the QCIF format requires a three times lower frame rate under the constraint of identical PSNR.

Similarly to the Miss America experiments, we also presented the relative bitrate reduction with respect to our reference scenario of QCIF resolution, 10 frames/s scanning

9.3. PERFORMANCE RESULTS

Table 9.12: Increase in PSNR (dB) Achieved at the Same Bitrate, for "Miss America" Video Sequences Coded with H.263 Codec Compared with the QCIF Resolution Simulation at 10 Frames/s

Fixed Bitrate (Kbit/s)	PSNR Increase (dB)				
	SQCIF		QCIF	CIF	
	10 frames/s	30 frames/s	30 frames/s	10 frames/s	30 frames/s
7	1.74	−1.05	—	—	—
10	1.43	−0.80	−2.67	—	—
20	1.47	−0.81	−2.37	—	—
50	1.90	−0.79	−2.37	−5.33	—
70	1.79	−0.69	−2.39	−5.85	−8.03
100	1.50	−0.55	−2.32	−5.97	−7.53
200	3.06	−0.16	−1.69	−6.14	−7.20
400	—	0.40	−1.59	−6.46	−7.59
600	—	—	−2.88	−7.41	−8.81

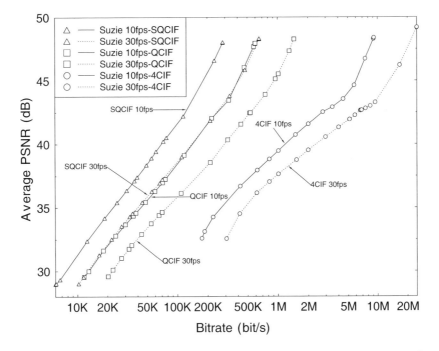

Figure 9.23: Image quality (PSNR) versus coded bitrate for H.263 "Suzie" simulations at 10 and 30 frames/s.

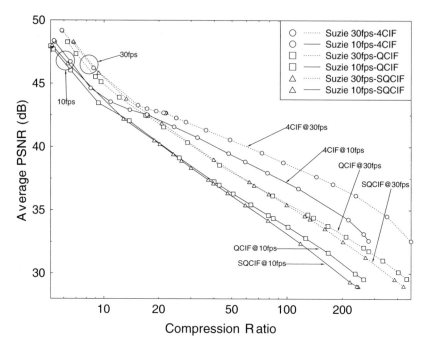

Figure 9.24: Image quality (PSNR) versus compression ratio for H.263 "Suzie" simulations at 10 and 30 frames/s.

Table 9.13: Percentage Reduction in Bitrate Required to Achieve the Same PSNR, for the Various Resolution "Suzie" Video Sequences Coded with H.263 Codec Compared with the QCIF Resolution Simulation at 10 Frames/s

Fixed PSNR (dB)	Percentage Bitrate Reduction (%)				
	SQCIF		QCIF	4CIF	
	10 frames/s	30 frames/s	30 frames/s	10 frames/s	30 frames/s
30	38.56	1.15	−70.55	—	—
32	41.43	2.03	−71.86	—	—
34	45.08	5.80	−78.44	−557	−1075
36	48.33	6.76	−89.38	−556	−969
38	50.74	4.36	−104	−585	−1170
40	52.94	0.86	−101	−720	−1558
42	49.46	−1.56	−118	−981	−2315
44	53.82	2.55	−116	−1310	−3055
46	52.87	−4.49	−136	−1385	−3381
47	53.14	−4.95	−135	−1361	−3448

9.4. SUMMARY AND CONCLUSIONS

Table 9.14: Increase in PSNR(dB) Achieved at the Same Bitrate for the Various Resolution "Suzie" Video Sequences Coded with H.263 Codec Compared with the QCIF Resolution Simulation at 10 Frames/s

Fixed Bitrate (Kbit/s)	PSNR Increase (dB)				
	SQCIF		QCIF	4CIF	
	10 frames/s	30 frames/s	30 frames/s	10 frames/s	30 frames/s
20	2.44	0.09	—	—	—
40	2.72	0.26	−2.17	—	—
70	3.12	0.25	−2.28	—	—
100	3.17	0.17	−2.56	—	—
200	3.86	0.02	−3.21	−7.93	—
300	—	0.16	−3.07	−7.93	—
500	—	−0.33	−4.34	−9.36	−11.33

rate, which is portrayed for the various fixed PSNRs, resolutions, and frame rates in Table 9.13. In contrast, Table 9.14 portrays the potential PSNR improvement or degradation at various fixed bitrates, resolutions, and frame rates. Again, comparing the SQCIF performance at 30 frames/s with the QCIF results at 10 frames/s produce interesting revelations. The results show that changing from QCIF to SQCIF resolution allows the frame rate to increase from 10 to 30 frames/s without any significant change in the bitrate requirements or loss of image quality.

In the final set of experiments of this chapter, compared the high-resolution H.263 modes of 4CIF and 16CIF representations. As before, our experiments were carried out at 10 and 30 frames/s. The previously employed 4CIF "Suzie" video sequence, a 4CIF "American Football" video sequence, and our cropped HDTV "Mall" video sequence were used for the 16CIF experiments. The associated performance curves are shown in Figures 9.25 and 9.26, where the 4CIF curves were repeated from Figures 9.23 and 9.24. Since the amount of image fine detail is quite different in the "Suzie", "American Football", and "Mall" sequences, the corresponding performance curves are not strictly comparable, but they give a coarse estimate of the expected bitrates in relative terms. As can be seen for the graphs in Figures 9.25 and 9.26, the 4CIF "American Football" sequence results in a similar performance to the 16CIF "Mall" sequence, primarily because of the large amount of motion in the "American Football" sequence. Before closing, let us finally summarize our experiences from this chapter.

9.4 Summary and Conclusions

In this chapter, we described the differences between the H.261 and H.263 video codecs, and a range of comparative performance curves were plotted. The differences between the two codecs in terms of source coding, multiplex coding, and motion compensation were described. From the range of algorithmic differences, motion compensation probably has

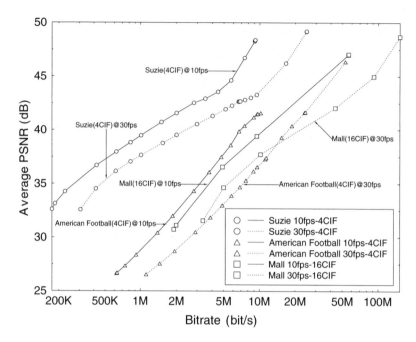

Figure 9.25: Image quality (PSNR) versus coded bitrate for H.263 high-resolution (4CIF and 16CIF) simulations at 10 and 30 frames/s.

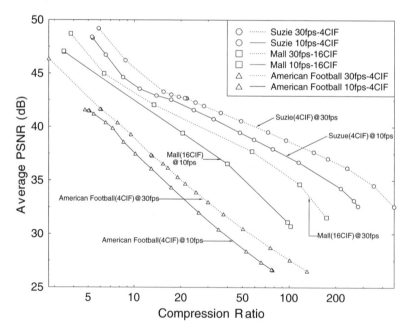

Figure 9.26: Image quality (PSNR) versus compression ratio for H.263 high-resolution (4CIF and 16CIF) simulations at 10 and 30 frames/s.

the greatest effect. Multiplex coding also contributes to the substantial performance gain of the H.263 codec over the H.261 scheme. The H.263 codec's negotiable options were also described briefly, and more details can be found in [259]. A variety of comparative performance curves were plotted, demonstrating how the performance of the H.263 codec varied after some of its parameters were altered.

Chapter 10

H.263 for HSDPA-style Wireless Video Telephony

10.1 Introduction

Having studied the differences and similarities between the H.261 and H.263 codecs, we designed a transceiver incorporating the ITU H.263 [259] video codec for visual communications in a mobile environment. We capitalized on our experience gained from our similar experiments using the H.261 [258] video codec. The software implementation of an H.263 codec was a modified version of the scheme developed by Telenor Research and Development [327]. The experimental conditions are summarized in Table 10.1. In this chapter, we review the H.263 codec, which was originally designed for benign Gaussian channels, and we design a system in order to facilitate its employment over hostile wireless channels.

10.2 H.263 in a Mobile Environment

10.2.1 Problems of Using H.263 in a Mobile Environment

The errors that occur in the H.263 bitstream can be divided into three classes: detected, detected-late, and undetected errors. A detected error is one that is recognized immediately, since the received bit was not expected for the current codeword. Errors that are detected late are caused by the error corrupting one codeword to another valid codeword. However, such errors are detected within the life span of a few codewords. Lastly, undetected errors also corrupt a valid codeword to another valid codeword, but no error is detected.

Detected errors result in picture quality degradation, but errors that are not detected or detected late cause the bitstream to be decoded incorrectly. An incorrectly decoded bitstream usually corrupts one or more macroblocks, and the error spreads to larger areas within a few frames. These types of errors can be corrected only by the retransmission of the affected macroblock in intra-frame coded mode.

Generally, when an error is detected, the H.263 decoder searches the incoming data stream for a picture start codeword. If this is found, the decoding process is resumed. This means that all blocks following an error in a video frame are lost. If the encoder is in inter-frame coded mode, the local decoder of the encoder and the remote decoder will become different, which we often refer to as being misaligned, causing further degradation until all the affected blocks are updated with intra-frame coded blocks. In order to recover from transmission errors, the updates using intra-frame coded blocks are very important. However, if intra-frame coded blocks are corrupted, the resulting degradation can become more perceptually objectionable than that which the intra-frame coded block was supposed to correct.

Another problem in using the H.263 codec in a mobile environment involves packet dropping. Packet dropping is common in hostile mobile scenarios, especially in conjunction with contention-based statistical multiple access schemes (PRMA). Dropping an H.263 packet would be equivalent to making a large number of bit errors. A feasible solution to this problem — especially in distributive video applications, where the latency is not a serious impediment — is using Automatic Repeat Request (ARQ) attempts. But to keep re-transmitting until a packet is received correctly is impractical, for it would require an excessive number of slots, inevitably disadvantaging other contending users.

10.2.2 Possible Solutions for Using H.263 in a Mobile Environment

The following problems have to be solved in order to use H.263 video codecs in a mobile environment:

- Reliably detecting errors in the bitstream in order to prevent incorrect decoding.

- Increasing the relative frequency of instances, where the decoding can re-start after an error. This reduces the number of blocks that are lost following an error event in the bitstream.

- Identifying ways of dropping segments of the bitstream without causing video decoding errors.

- Ensuring that the encoder's local decoder and remote decoder remain synchronized, even when there are transmission errors.

Various solutions to one or more of these problems are described here. Most radio systems make use of forward error correction (FEC) codecs and/or error detecting cyclic redundancy check (CRC) sums on the transmitted packets.

Significant research efforts have been devoted to the area of robust video coding [191, 195, 307–312, 328]. Some authors use proprietary video codecs [191, 195], while others employ the H.261 or other standard video codecs. A technique referred to as Error-Resilient Entropy Coding (EREC), developed by Cheng, Kingsbury, *et al.* [188, 310], was used by Redmill *et al.* [311] to design an H.261-based system. Matsumara *et al.* [312] used a similar technique to design a H.263-based system. EREC reduces the effects of errors but still requires low bit error rates in order to maintain perceptually unimpaired video quality. An EREC based technique is due to be used in a mobile version of the H.263 video coding standard, currently known as draft recommendation AV.26M.

10.2.2.1 Coding Video Sequences Using Exclusively Intra-coded Frames

As mentioned, the H.263 standard relies on the intra-frame coded macroblocks in order to remove the effects of any errors. Therefore, increasing the number of intra-frame coded macroblocks in the coded bitstream, enhances error recovery. Increasing the number of intra-frame coded macroblocks to the ultimate limit of having all frames coded as intra-frames has the advantage of resynchronizing the local decoder and the remote decoder at the start of every frame. Hence, any transmission errors can only persist for one frame duration.

The disadvantage of this method, however, is that the bitrate increases dramatically. Simulations have shown that in order to maintain a given image quality, the bitrate has to be increased by at least an order of magnitude in comparison to an inter-frame coded scenario. Since the bandwidth in mobile systems is at premium, the low bitrate requirement imposed drastically reduces the image quality maintained.

10.2.2.2 Automatic Repeat Requests

As a more attractive means of improving robustness, Automatic Repeat Request (ARQ) [202], can be used to improve the BER of the channel at the expense of increased delay and reduced useful teletraffic capacity. However, if the channel becomes very hostile, then either the video data will never get through or the delay and teletraffic penalty will increase to a level where communications become impractical for an interactive videophone system. If ARQ was combined with a regime of allowing packets in the data stream to be dropped after a few retransmission attempts, then the maximum delay could be bounded. ARQ has been used in several video coding systems designed for mobile environments [195, 307].

10.2.2.3 Multimode Modulation Schemes

The employment of reconfigurable modulation schemes renders the system more flexible in terms of coping with time-variant channel conditions. For example, a scheme using 4QAM, 16QAM, and 64QAM [222] can operate at the same Baud rate, for 2, 4, and 6 bits/symbol, respectively. This allows a 1:2:3 ratio of available channel bitrates for the various schemes.

For a high-quality, high-capacity channel 64QAM can be used, supporting higher quality video communications due to the increased channel capacity. When the channel is poor, the 4QAM scheme can be invoked, constraining the image quality to be lower due to the lower channel capacity. However, the 4QAM scheme is more robust to channel errors and so guarantees a lower BER. Therefore, an intelligent system could switch the modulation scheme either under network control, or, for example, by exploiting the inherent channel quality information in Time Division Duplex (TDD) systems upon evaluating the received signal level (RSSI) and assuming reciprocity of the channel [319–321]. A more reliable alternative to the use of RSSI as a measure of channel quality is the received BER. This reconfiguration could often take place arbitrarily, even on a packet-by-packet basis, depending on the channel conditions. The image quality is dependent on the bitrate and therefore increases with better channel conditions.

The combination of ARQ with multimode modulation schemes can facilitate a rapid system response to changing channel conditions. Re-transmission attempts can be carried out using the more robust, lower capacity modulation schemes that have a higher chance of delivering an inherently lower quality but uncorrupted packet.

10.2.2.4 Combined Source/Channel Coding

It follows from our previous arguments that using ARQs or multiple modulation schemes on their own increases the range of operating channel SNRs, for which the H.263 bitstream can be transmitted error-free. However, these measures do not improve the decoder ability is to cope with a corrupted bitstream.

In order to reduce the possible video degradations due to channel errors, it is more beneficial and reliable to detect channel errors externally to the video codec rather than relying on the variable-length codeword decoder's ability in the video codec. This can prevent any potential video degradations inflicted by incorrect decoding of the data stream. After a transmission error has been detected, either the codec will attempt to resynchronize with the data stream by searching for a picture start code (PSC) or group of blocks start code (GBSC) in the data stream. The decoding process can then restart from the correctly identified PSC or GBSC. However, the data between the transmission error and resynchronization point is lost. This data loss causes video degradation because the encoder's local decoder and the remote decoder become misaligned. Hence, they operate on the basis of different previous reconstructed frame buffers. This degradation lasts until the local decoder and remote decoder can be resynchronized by a re-transmission of the affected macroblocks in intra-frame coded mode.

If the encoder's local decoder could be informed when a segment of the data stream is lost, it could compensate for this loss by keeping the local decoder and remote decoder in alignment, simply by freezing their contents. This implies that the image segment obliterated by the transmission error can be updated in future frames. A small short-term image quality degradation occurs because part of the image is not being updated. This is mitigated within a few frames, and it is small compared with the loss of decoder synchronization or incorrect decoding.

When using transmission packet acknowledgment feedback information, the encoder can be notified of loss of packets in the mobile network. By including this channel feedback in the encoder's coding control feedback loop, the contents of transmitted packets do not need to be stored. The close coupling of the source decoder and channel decoder is possible in a mobile videophone application. For the channel feedback to operate effectively, the round-trip delay must be short. For microcellular mobile systems, the round-trip delay is negligible, and the algorithmic and packetization delay is also typically low. This packet acknowledgment feedback information can therefore be provided by the receiver.

When the source encoder is informed that a packet has been lost, it can infer the effect of this loss on the decoder's reconstructed frame buffer and adjust the contents of its local decoder in order to match this and maintain their synchronous relationship. Consequently, the decoder will use the corresponding segment of the previous frame for the areas of the frame that are lost. The effect of this is imperceptible in most cases. The lost section of the frame will then be updated in future frames. Thus, provided that no further errors occur, within a few video frames, the effects of the transmission error would have receded.

10.3 Design of an Error-resilient Reconfigurable Videophone System

10.3.1 Introduction

Having considered the behavior of the H.263 codec under erroneous conditions, let us now familiarize ourselves with the design principles of an error-resilient reconfigurable videophone transceiver. Our objective is to ensure better error resilience than that of the standard H.263 codec, which is achieved by adhering to the above suggested principles. Furthermore, the bitstream generated is expected to conform to the H.263 standard or to lend itself to low-complexity transcoding. The system's multimode reconfigurable operation guarantees near-optimum performance under time-variant propagation and teletraffic conditions. The system designed exhibits the following features:

- Transmission packet acknowledgment feedback is used in order to inform the encoder, when packets have been lost, so that it can adjust the contents of its local reconstructed frame buffer and keep synchronized with that of the decoder. This feature allows the packets to be dropped without ARQ attempts, thereby reducing any potential ARQ delay.

- Each transmitted packet contains a resynchronization pointer that points to the end of the last macroblock within the packet. This "address" can be used to resynchronize the encoder and decoder if the next packet is corrupted and the remainder of the partial macroblock is lost.

- FECs are used to decrease the channel BER and to balance the effects of the QAM subchannels' differing BERs [222].

- The system is rendered reconfigurable in order to be able to encode the video sequence for achieving different bitrates. When the affordable bitrate is reduced, the packet size is also reduced so that it can be transmitted using a more robust modulation scheme at the same Baud rate.

- Any received packets containing transmission errors are dropped in order to prevent erroneous decoding, which can badly corrupt the received video sequence.

- The transmitted bitstream conforms to the H.263 standard, and so there is no need for a transcoder.

10.3.2 Controlling the Bitrate

In this section, we highlight the concept of controlling the bitrate. As with most video codecs using variable-length coding techniques, the bitrate of the H.263 codec is inherently time variant. However, most existing mobile radio systems transmit at a fixed bitrate. Our proposed multimode system maintains a constant signaling rate, leading to a different constant bitrate for each modulation scheme invoked.

A straightforward bitrate control algorithm was used in order to maintain a fixed bitrate and fixed video frame rate. This algorithm modifies the quantizer in the video codec in order to maintain the target frame rate and bitrate. In these investigations, a fixed frame rate of 10 frames/s was employed. The target bitrate was set according to the past history of the modulation schemes used and to the packet dropping frequency experienced. The target bitrate was updated after every dropped packet or successful transmission according to Equation 10.1:

$$\text{Target Bitrate} = \frac{S_4 B_4 + S_{16} B_{16} + S_{64} B_{64}}{S_4 + S_{16} + S_{64} + D_4 + D_{16} + D_{64}} \tag{10.1}$$

where S_n is the number of successful packet transmissions, when using the n-QAM modulation scheme, D_n is the number of dropped packets in the n-QAM mode of operation, while B_n represents the various fixed bitrates for error-free transmission, again using the n-QAM modulation scheme.

The bitrate control algorithm constitutes a simple way of controlling the target bitrate. The rate-control algorithm in the video codec modifies the video codecs' parameters to meet the target bitrate requirement. This rate-control algorithm, combined with the equation to set the target bitrate, provided an adequate bitrate control algorithm, which adapted to varying channel conditions.

If the videophone system is operating in only a 4QAM mode, the target bitrate is simplified, as shown in Equation 10.2:

$$\text{Target Bitrate} = \frac{S_4 B_4}{S_4 + D_4}. \tag{10.2}$$

If the channel is nearly error-free, then D_4 will be virtually zero and the target bitrate will tend to the fixed bitrate for the 4QAM transmission mode (B_4). If, however, the channel causes a 50% frame error rate, then the number of dropped transmissions (D_4) will be similar to the number of successful transmissions (S_4). This will cause the target bitrate to reduce to approximately 50% of the fixed bitrate for the 4QAM scheme (B_4). Therefore, the target bitrate reduces in proportion to the channel frame error rate. If the video codec can meet the target bitrate requirements, the frame rate of the transmitted video should be nearly constant.

The videophone system can use modulation scheme switching or ARQ to improve the performance in the deep fades that occur in Rayleigh fading. When switching or ARQ is used, the target bitrate is proportional not only to the frame error rate, but also to the switching or ARQ changes. The fixed bitrate for each modulation scheme is weighted by the number of successful transmissions using that particular modulation mode. This system ensures that the target bitrate adjusts, so that the frame rate of the transmitted video remains as constant as possible.

The operation of this bitrate control algorithm is characterized by the normalized bitrate histograms of Figure 10.1 in the transceiver's different modes of operation using 64QAM, 16QAM, and 4QAM, respectively. As seen in this figure, in virtually all scenarios the instantaneous bitrate was within ±20% of the target mean bitrate, maintaining the target frame rate within this range. The corresponding frame rate versus time behavior of the system is characterized in Figure 10.2. Observe that after a slight initial delay, which is essentially due to the intra-frame coded bitrate surge of the first frame, the frame rate reaches the target of 10 frames/s, typically within less than 1s.

Much research has been conducted in the area of rate control for video codecs. Ding and Liu [329] gave a good introduction to rate-control algorithms and their application to the

10.3. DESIGN OF AN ERROR-RESILIENT RECONFIGURABLE VIDEOPHONE SYSTEM

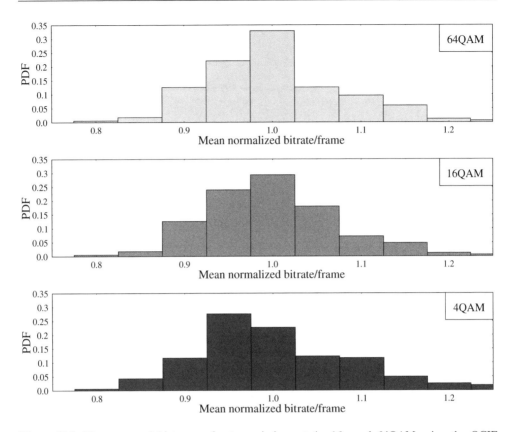

Figure 10.1: Bitrate control histogram for transmissions at 4-, 16-, and 64QAM using the QCIF resolution "Miss America" Sequence, in error-free channel conditions.

MPEG video codec. Schuster and Katsaggelos [330] introduced an optimal algorithm and showed its application to H.263. Martins, Ding, and Feig [331] introduced a variable bitrate, variable-frame-rate-control algorithm, and applied it to an H.263 video codec. The variable bitrate/frame rate is particularly suitable for very low bitrates. A rate-control algorithm designed specifically for H.263 was suggested by Wiegand *et al.* [332]. It optimized the codec parameters for each group of blocks (GOB) and found PSNR improvements of about 1 dB over more basic rate-control algorithms.

10.3.3 Employing FEC Codes in the Videophone System

Since forward error correction codes can reduce the number of transmission errors at the expense of an increased Baud rate, they also expand the useful channel SNR range, for which transmissions are error-free. As we have stated before, in Quadrature Amplitude Modulation (QAM) schemes, certain bits of a symbol have a higher probability of errors than others [315]. The bits of a symbol can be grouped into different integrity classes according to their probability of error, which are also often referred to as QAM subchannels. By using different-strength FEC codes on each QAM subchannel, it is possible to equalize

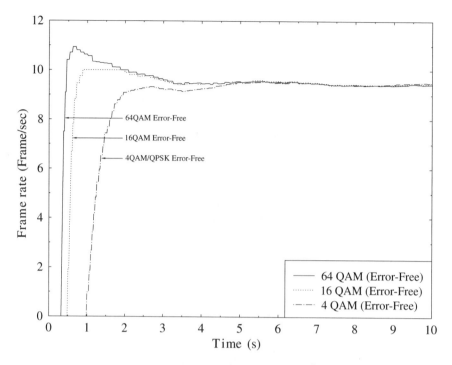

Figure 10.2: Frame rate versus time behavior of the frame-rate control algorithm, for transmissions at 4, 16, and 64QAM using QCIF resolution Miss America Sequence, in error-free channel conditions.

their probability of errors, implying that all subchannels' FEC codes become overloaded by transmission errors at approximately the same channel SNR. This is desirable if all bits to be transmitted are equally important. Since the H.263 data stream is mainly variable length coded, a single error can cause a loss of synchronization. Therefore, in this case most bits are equally important, and so the equalization of the QAM subchannels' BER is desirable for the H.263 data stream. Pelz [191] used a proprietary video codec and exploited the QAM subchannels to help classify the bitstream into classes.

10.3.4 Transmission Packet Structure

Our error-resilient H.263 codec generates video packets and conveys them to the error correction encoders. The structure of the packets generated by the H.263 video codec is shown in Figure 10.3.

As seen in the figure, the packets have two main constituent segments: the data and control information. The control information consists of 10 bits. As seen in Figure 10.3, 9 bits of the 10-bit control information are allocated to an index, which points to the end of the last whole macroblock in the packet. This packet may contain the bits representing a number of MBs plus a fraction of the most recently encoded MB, which was allocated to the current transmission packet in order not to waste channel capacity. The remainder of the

10.3. DESIGN OF AN ERROR-RESILIENT RECONFIGURABLE VIDEOPHONE SYSTEM

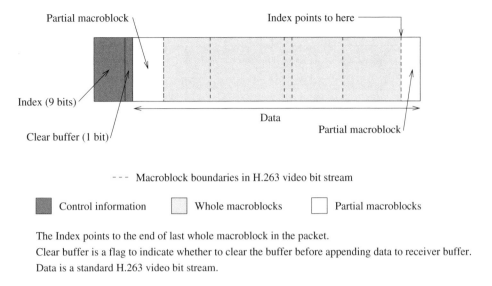

Figure 10.3: Structure of a transmission packet generated by the modified H.263 video codec.

partially transmitted MB is then transmitted at the beginning of the next transmission packet, as indicated in Figure 10.3. This pointer is used to ensure that the decoder only decodes whole macroblocks. The partial macroblock after the indexed point is buffered at the decoder, until the remainder of the partial macroblock is received. The control information segment also contains a 1 bit flag, which is used to inform the decoder if the decoder's received signal buffer containing an already error-freely received partial macroblock has to be cleared before appending the current packet. The encoder sets this flag in order to inform the decoder to drop the already error-freely received partial macroblock from its received signal buffer, when the remainder of the corresponding MB's information was lost due to packet dropping. Explicitly, by clearing the received signal buffer, this mechanism ensures that the first part of the error-freely received partial macroblock stored in the receiver's decoding buffer is removed, if the rest of the packet was corrupted.

10.3.5 Coding Parameter History List

In order to enable the re-encoding of the macroblocks for re-transmission or dropping, the history of various coding parameters has to be stored for every macroblock in the encoder's buffer. This is necessary, because the encoder has to remember, for example the unquantized DCT coefficients, the MB position, the current MB coding mode, and so on, until the error-free arrival of this specific MB's encoded information at the decoder. If the information is corrupted, for example, during its first transmission attempt, it may have to be re-encoded using more coarse quantizers, which would not be possible without this coding parameter history list. The history of coding parameters was implemented as a bidirectionally linked list with a structure as shown in Figure 10.4. Each element of the list represents a macroblock and contains the coding parameters of the codec before the macroblock was coded. Initially,

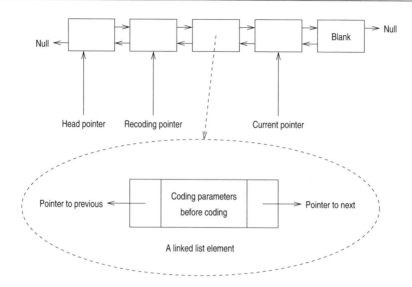

Figure 10.4: Structure of bidirectionally linked list for storing history of coding parameters.

then, a brief general description is given, which will depend substantially on the specific application scenarios to be considered in Section 10.3.6.1.

Each new macroblock is added to the tail of the list, then encoded, and the coded bitstream is appended to the transmission buffer. The packetization algorithm has three pointers to the linked list. *Head* points to the head of the list, *Recode* points to the first macroblock to be re-encoded for re-transmission, and *Current* points to the next element/macroblock to encode. Again, the role of these pointers will become clearer with reference to our example considered in Section 10.3.6.1. Here, only a few general comments concerning their employment are offered. For example, the *Recode* pointer points to the same macroblock as the *Head* pointer, when the decoder has no partial macroblocks in its buffer. When the decoder has a partial macroblock in its buffer, the *Recode* pointer points to the second macroblock in the list. This is because the partial transmitted macroblock, which is at the head of the list, can only be dropped or the remainder of the macroblock transmitted. When a new macroblock is added to the list, it is placed in the blank list element at the tail of the list and a new blank list element is created.

When the transmission buffer becomes filled above the defined limit corresponding to a transmission packet, its contents up to the limit is transmitted. If the transmission is successful, then the linked list elements for the fully transmitted macroblocks are removed from the head of the list. The partially transmitted macroblock then becomes the head of the linked list. The *Recode* and *Current* pointers point to the blank list element at the end of the list.

If the transmission fails, but a re-transmission is requested, then the codec is instructed to reset its state to what it was, before the macroblock pointed to by the *Recode* pointer was encoded. The remaining macroblocks are then re-encoded until the buffer becomes filled up to the new transmission packet size. This packet is then transmitted in the usual way.

10.3. DESIGN OF AN ERROR-RESILIENT RECONFIGURABLE VIDEOPHONE SYSTEM

If the transmission failed and re-transmission was not requested, then the codec is reset to the state before the macroblock pointed to by the *Head* pointer was encoded. The macroblocks in the history list are then re-encoded, as if they were empty. When this operation is complete, the encoding of new macroblocks continues as usual. However, when the next packet is transmitted, the clear buffer flag for the packet is set in order to clear the receiver's decoding buffer from the partially transmitted macroblock that may be in it.

The coding parameters saved in the history list are as follows:

- The unquantized DCT coefficients.

- Macroblock index.

- Buffer size before encoding the macroblock.

- Various quantizer identifiers used by the bitrate control algorithm.

- Current macroblock coding mode.

Now that the coding parameters' history list and the transmission packet structure have been discussed, the following section describes the packetization algorithm.

10.3.6 The Packetization Algorithm

The objectives of the packetization algorithm are to:

- Pack the H.263 bitstream into the data portion of the transmission packet, while setting the control information part of the packet so that the decoder can recover from packet losses.

- Exploit the feedback information in order to adjust the codec's bitrate to adapt to time-variant channel conditions.

- Generate variable-size packets, so that ARQ relying on different modulation schemes can be easily implemented.

- Ensure that the encoder's local decoder and the decoder are kept in synchronization after a packet is dropped.

The H.263 codec produces a bitstream, maintaining a fixed frame rate for the target bitrate, which is adjusted depending on the channel conditions. The packetizing algorithm operates at the macroblock layer. An example of operating scenarios for the packetization algorithm is given in the next section.

10.3.6.1 Operational Scenarios of the Packetizing Algorithm

In order to augment our discussions above, in this section we consider a few operational scenarios of the packetization algorithm, which are shown in Figure 10.5. At the commencement of communications, both the transmission buffer and the history list are empty, as suggested by the figure. When the video encoder starts generating macroblock bits, the transmission

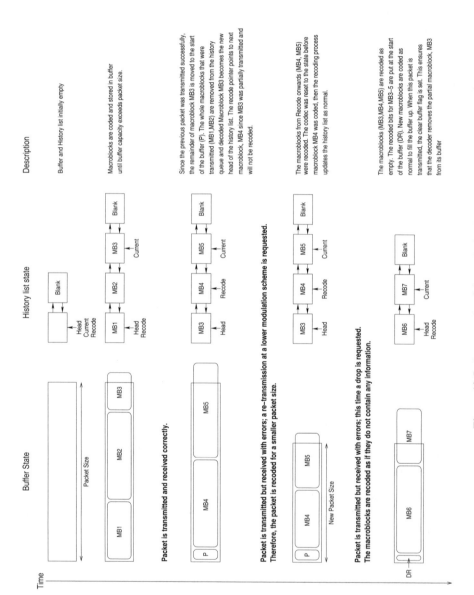

Figure 10.5: An example of the packetization algorithm.

10.3. DESIGN OF AN ERROR-RESILIENT RECONFIGURABLE VIDEOPHONE SYSTEM

buffer is filled by MB1-MB3, where MB3 is seen to "overfill" the buffer. The status of the pointers is also reflected in the second line of the figure. Assuming that this packet was transmitted successfully, the "overspilt" segment P of MB3 is transmitted at the beginning of the next packet, as displayed in the third line of Figure 10.5. The error-freely received and complete MB1 and MB2 can then be decoded, and their parameters can be removed from the history list. MB3 is now pointed to by the *Head* pointer, which implies that it constitutes the *Head* of the coding parameter history list. As mentioned earlier, the pointer *Recode* holds the position of the first MB to be re-encoded for a potential re-transmission. Lastly, the pointer *Current* identifies the MB to be encoded, which is in this case MB5. At this stage the second packet is filled up and ready for transmission, but MB5 "overfills" the packet, and hence the remaining segment P is assigned to the forthcoming packet, as seen in the fourth line of Figure 10.5.

Let us now consider an erroneous transmission scenario in which ARQ is invoked. Assume furthermore that the packet has to be re-transmitted using a more robust but lower transmission capacity modulation scheme, imposing a smaller packet size as shown in Figure 10.5. Since a substantial part of MB3 has already been received, error-free, MB3 is not re-encoded, but the length of MB4 and MB5 must be reduced. The position of the pointers is unaltered, as suggested by Figure 10.5. Hence the MBs starting at the position indicated by the pointer *Recode* are re-encoded using more coarse quantizers, which generate the required target bitrate. This operation is concerned with MB4 and MB5. Accordingly, the coding parameter history list is reset to its state, before MB4 was originally encoded, since the same unquantized DCT coefficients, MB index, MB coding mode, and so on, must be used, as during the first encoding operation. The re-encoded MB4 and MB5 now "overfill" the shortened transmission packet constrained by the lower-order modulation scheme.

Let us assume that this shortened packet transmitted using the most robust modulation scheme available is also corrupted by the channel, and the maximum number of re-transmission attempts has expired. In this case, the packet is dropped in order to prevent error propagation in the local reconstructed frame buffer. In the H.261 scheme, macroblocks could simply be dropped, since each macroblock contained an address. However, in the H.263 scheme, for every macroblock some information has to be transmitted, where inactive macroblocks are coded using a 1 bit codeword. Due to the packet drop request, the MBs are re-encoded as empty MBs conveying no information, requiring a 1 bit flag for each dropped macroblock. These bits are represented by the segment DR at the bottom of Figure 10.5. Then the forthcoming MBs, MB6 and MB7, are encoded as usual and assigned to the next transmission packet. When this packet is transmitted, the clear buffer flag is set in order to inform the decoder that the partially received MB3 has to be removed from the decoder's buffer.

This example demonstrates all possible transmission scenarios. It is worth noting that dropping macroblocks as described above, has the effect of some parts of the picture not being updated. However, in most cases the dropped macroblocks will be updated in the next frame. The effect of dropping is not noticeable, unless the frame contains a large amount of motion and the packet dropping probability is very high. These effects can be more easily appreciated by viewing the decoded video, which is available on the WWW.[1] Packet-loss concealment techniques were investigated by Ghanbari and Seferidis [324] using a H.261 codec.

---

[1] http://www-mobile.ecs.soton.ac.uk/peter/robust-h263/robust.html#DROPPING

Having described the video codec and the packetization algorithm, in the forthcoming section we concentrate our attention on the overall system's performance.

10.4 H.263-based Video System Performance

10.4.1 System Environment

The performance of the error-resilient H.263-based reconfigurable videophone system was evaluated in a mobile radio environment. These investigations identified the performance limits of the videophone system. Our experiments were carried out over Gaussian and Rayleigh-fading channels, using reconfigurable modulation schemes and ARQ. The simulation conditions were as follows: a propagation frequency of 1.8 GHz, vehicular speed of 13.4 m/s (or 30 mph), and a signaling rate of 94.72 kBaud. The corresponding normalized Doppler frequency was 6×10^{-4}.

Our reconfigurable modulation scheme was able to switch between 64-, 16-, and 4-level coherent, pilot-assisted QAM modulation [222], supported by a range of binary Bose-Chaudhuri-Hocquenghem (BCH) forward error correcting (FEC) codes [333], in order to increase the system's robustness. The number of transmitted bits/symbol and the associated number of subchannels are shown in Table 10.1 for each modulation scheme.

All our performance investigations were carried out for a single portable station (PS) and base station (BS) and for a pilot symbol separation of ten symbols. In order to be able to use our reconfigurable modulation schemes with ARQ assistance, all packets must have the same number of symbols for each modulation scheme. This is achieved by using a higher number of bits per packet for 64QAM and a lower number for 4QAM, where the ratio of the number of bits is determined by the number of bits/symbol for each modulation scheme.

Suitable values for the bitrates of the three modulation schemes were set. Specifically, the packet size, expressed in terms of the number of bits for each modulation scheme, was found by initially setting the packet size for 4QAM. Then the packet sizes for 16- and 64QAM expressed in terms of bits were set to keep the same number of symbols per packet, resulting in a bitrate ratio of 1:2:3 for the 4-, 16-, and 64-QAM arrangements.

The previously mentioned modulation subchannels in 16- and 64QAM exhibit different BER performances. This is usually exploited to provide more sensitive bits with increased protection against channel errors. Since all the video bits in the proposed H.263-based videophone system are sensitive to transmission errors, no source sensitivity matched FEC coding is employed. Hence, the subchannel error rates have to be equalized, which can be achieved using stronger FEC codes in the weaker subchannels. The FECs were chosen to equalize the different BERs of the specific modulation subchannels as closely as possible. The specific FEC codes chosen for the various modulation schemes are summarized in Table 10.1. The table also portrays the corresponding packet sizes derived on the basis of the FEC codes for each modulation scheme in the H.263-based system. All the simulation parameters and results are also summarized in Table 10.1, which we will frequently refer to throughout this chapter.

10.4. H.263-BASED VIDEO SYSTEM PERFORMANCE

Table 10.1: Summary of System Features for H.263 Wireless Reconfigurable Videophone System. Minimum Required SNR and BER Derived from Figure 10.9

	Modem-mode Specific System Parameters		
Features	H.263 Multirate System		
Modem	4-QAM	16-QAM	64-QAM
Bits/symbol	2	4	6
Number of subchannels	1	2	3
C1 FEC	BCH(255,147,14)	BCH(255,179,10)	BCH(255,199,7)
C2 FEC	N/A	BCH(255,115,21)	BCH(255,155,13)
C3 FEC	N/A	N/A	BCH(255,91,25)
Bits / TDMA frame	147	294	445
Transmission bitrate (kbit/s)	11.76	23.52	35.6
Packetization header (bits)	10	10	10
Video bits / TDMA frame	137	284	435
Useful video bitrate (kbit/s)	10.96	22.72	34.8
Min. PSNR threshold (dB)	-1	-1	-1
Min. AWGN SNR (dB)	7.0	12.7	18.6
Min. AWGN BER (%)	4.0	2.5	3.2
Min. Rayleigh SNR (dB)	11.5	17.3	23.2
Min. Rayleigh BER (%)	4.4	3.3	3.4

Modem-mode Independent System Parameters	
Features	General System
Pilot-Assisted Modulation	yes
Video Codec	H.263
Frame rate (Fr/s)	10
Video resolution	QCIF
Color	Yes
User data symbols / TDMA frame	128
User pilot symbols / TDMA frame	14
User ramping symbols / TDMA frame	4
User padding symbols / TDMA frame	2
User symbols / TDMA frame	148
TDMA frame length (ms)	12.5
User symbol rate (kBd)	11.84
Slots/frame	8
No. of users	8
System symbol rate (kBd)	94.72
System bandwidth (kHz)	200
Effective user bandwidth (kHz)	25
Vehicular speed (m/s)	13.4
Propagation frequency (GHz)	1.8
Normalized doppler frequency	6.2696×10^{-4}
Path-loss model	Power law exponent 3.5

10.4.2 Performance Results

10.4.2.1 Error-free Transmission Results

Having described the system environment, let us now focus our attention on the achievable system performance. Our ultimate aim was to produce graphs of video PSNR versus channel SNR. Our investigations were carried out both with and without ARQ in order to assess to what extent ARQ can improve system performance. The "Miss America" video sequence was H.263 encoded, packetized and transmitted over Gaussian and Rayleigh channels with and without ARQ assistance. Our ARQ-assisted investigations are described by the ARQ-regime A-B-C, where A characterizes the initial transmission, B the first retransmission attempt, and C the second. Our experiments entailed the following transmissions:

- 4QAM

- 16QAM

- 64QAM

- ARQ: 4-4-4QAM.

In order to establish the potential benefits of reconfigurable transmissions, Figure 10.6 characterizes the decoded video quality in terms of PSNR versus frame index for the three modulation schemes invoked under error-free channel conditions. Observe that the image quality improves with the higher bitrates achieved by the help of the higher channel capacity modulation schemes. This graph also shows how the image quality improves following the initial intra-coded frame, which requires a substantially higher number of bits than inter-coded frames. However, at the commencement of transmission, the intra-coded frame is encoded using a rather coarse quantizer at a comparatively low quality in order to reduce the associated bitrate peak. This bitrate peak is mitigated by the bitrate control and packetization algorithm at the cost of a concomitantly reduced image quality. An approximately 3 dB PSNR advantage is observed, when using 16QAM instead of 4QAM, which is further increased by another 1.5 dB, when opting for 64QAM facilitated by the prevailing friendly channel conditions.

10.4.2.2 Effect of Packet Dropping on Image Quality

Let us now consider the more realistic scenario of imperfect channel conditions. Figure 10.7 characterizes the effect of video packet dropping on image quality as a function of the video frame index for a range of packet dropping probabilities between 0 and 73%. When a packet is dropped, it implies that the macroblocks contained in that packet will not be updated in the encoder's and the decoder's reconstruction buffer. These MBs will be updated in the next frame as long as the corresponding packet is not dropped. In the event of excessive packet dropping, some macroblocks may not be updated for several frames. When a packet containing the initial intra-coded frame is lost, its effect on the PSNR is more obvious. This is because the decoder cannot use the previous frame for dropped macroblocks, since that is typically an inter-frame coded one. After the initial few frames, every macroblock has been transmitted at least once. The PSNR then improves to around 35 dB. For 2% packet

10.4. H.263-BASED VIDEO SYSTEM PERFORMANCE

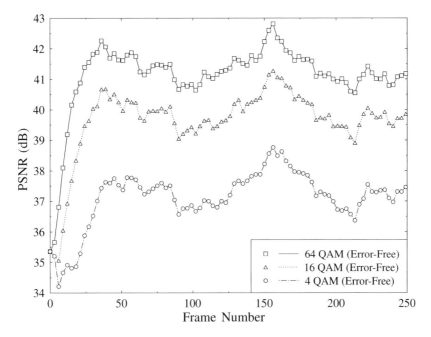

Figure 10.6: PSNR versus frame index comparison between modulation schemes over error-free Gaussian channels, using the system of Table 10.1.

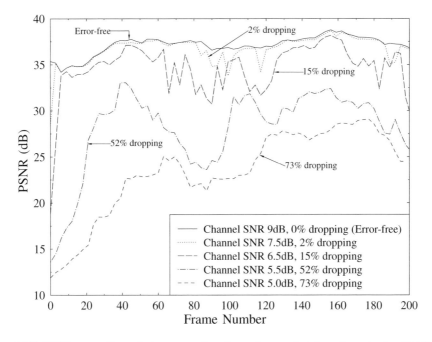

Figure 10.7: PSNR versus frame index comparison for various packet dropping rates for 4QAM over Gaussian channels, using the system of Table 10.1.

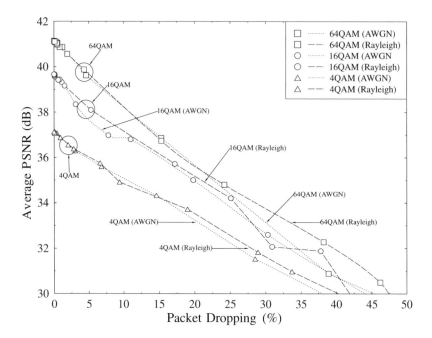

Figure 10.8: Average PSNR versus packet dropping rate for 4-, 16-, and 64QAM over Gaussian and Rayleigh channels, using the system of Table 10.1.

dropping, where 2 in every 100 packets are lost, the PSNR is almost identical to the error-free case. The higher the dropping rate, the longer it takes to build the PSNR up to the normal operating range of around 35 dB. However, even with 52% packet dropping, where only about half the packets are received, the PSNR averages around 28 dB. When the video contains a large amount of motion, the packet losses make the macroblocks that were not updated more perceptually obvious, which ultimately results in a lower PSNR. Toward high dropping rates, the delayed updating of the macroblocks became more objectionable, but the effect was not catastrophic. The decoded video sequences with packet dropping can be viewed on the WWW,[2] to help judge the subjective quality loss caused by packet dropping.

Figure 10.8 shows the average PSNR of the decoded video versus the packet dropping rate for 4-, 16-, and 64QAM over Gaussian and Rayleigh-fading channels. The PSNR degradation is approximately linear as a function of the increase in packet dropping probability.

10.4.2.3 Image Quality versus Channel Quality without ARQ

In order to show the effect of varying channel quality on the videophone system, the average PSNR versus channel SNR performance of the system was evaluated in Figure 10.9 for all three modulation modes over both Gaussian and Rayleigh channels.

This graph shows the sharp reduction of PSNR as the FEC schemes become overloaded by channel errors in the low SNR region. This effect is especially pronounced for Gaussian

---

[2]http://www-mobile.ecs.soton.ac.uk/peter/robust-h263/robust.html#DROPPING

10.4. H.263-BASED VIDEO SYSTEM PERFORMANCE

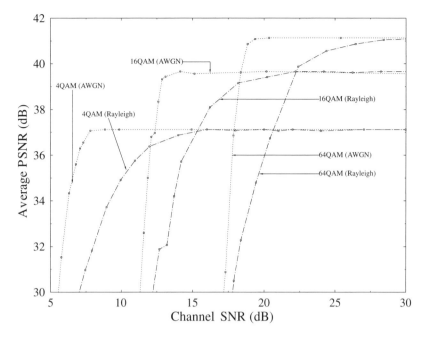

Figure 10.9: PSNR versus channel SNR for 4, 16, and 64QAM over both Gaussian and Rayleigh-fading channels, using the system of Table 10.1.

channels, regardless of the modulation mode used, since the channel errors are more uniformly distributed over time than in the case of Rayleigh-channels. Accordingly, this breakdown is less sharp for Rayleigh fading channels because of the bursty nature of the errors, overwhelming the FEC scheme only occasionally. This graph can be used to estimate the conditions for which a switch to a more robust modulation scheme becomes effective. These effects are shown from a different perspective, in terms of frame error rate versus channel SNR, in Figure 10.10.

10.4.2.4 Image Quality versus Channel Quality with ARQ

When using ARQ-assisted scenarios, the number of frame errors can be reduced, thereby improving the decoded image quality. The improvement of the frame error rate versus channel SNR performance due to ARQ is shown in Figure 10.11. It can be seen that the use of ARQ extends the range of channel SNRs, for which the system can be used, while maintaining a reasonable image quality. Observe in the figure that ARQ gives a greater improvement in Rayleigh channels than over Gaussian channels. This is because over Gaussian channels the packet typically faces similar channel conditions during every transmission attempt, whereas over Rayleigh channels the receiver may have emerged from a fade, by the time re-transmission takes place. This provides better reception chances for re-transmitted packets over fading channels.

The same tendencies are reinforced in Figure 10.12 in terms of average PSNR versus channel SNR with and without ARQ.

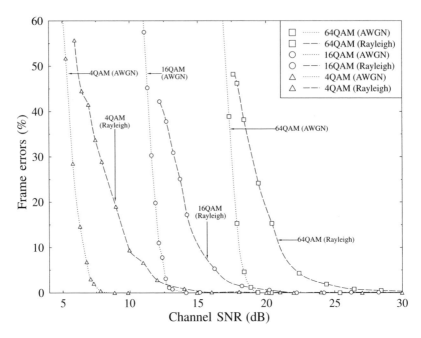

Figure 10.10: Frame errors versus channel SNR for 4-, 16-, and 64QAM over both Gaussian and Rayleigh fading channels, using the system of Table 10.1.

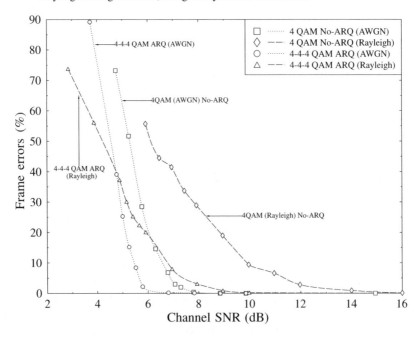

Figure 10.11: Frame errors versus channel SNR for 4QAM over both Gaussian and Rayleigh fading channels, with and without ARQ. The 4-4-4ARQ regime attempts up to three transmissions at 4QAM, using the system of Table 10.1.

10.4. H.263-BASED VIDEO SYSTEM PERFORMANCE

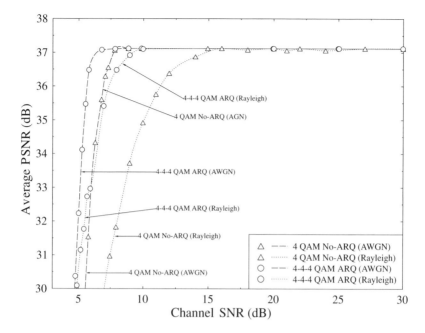

Figure 10.12: PSNR versus channel SNR for 4QAM over both Gaussian and Rayleigh fading channels, with and without ARQ. The 4-4-4ARQ regime invokes up to three transmissions at 4QAM, using the system of Table 10.1.

10.4.3 Comparison of H.263 and H.261-based Systems

In this section, we compare the proposed H.263-based videophone transceiver's performance to that of a similar H.261-based scheme. The previously described H.261 system of Section 8.4 was slightly less advanced and used the less efficient H.261 video coding standard. Figure 10.13 portrays the PSNR versus channel SNR performance of these schemes over Gaussian channels, and Figure 10.14 shows the corresponding results over Rayleigh-fading channels.

Observe in the figures that the H.263-based video system substantially outperforms the H.261 scheme in terms of decoded image quality. This is because of the improvements in video compression achieved by H.263 codec and the more efficient transmission of packets in the H.263 system. The PSNR advantage of the H.263 scheme is between 3 and 5 dB for the 4-, 16-, and 64QAM modulation modes.

Note, however, that at very low channel SNRs, the H.261 system has a slightly higher image quality. This is because H.263 makes more intensive use of motion vectors in order to increase compression. However, the employment of motion vectors is a problem at very high packet dropping rates because the image quality degrades more rapidly. Similar trends are also observed for Rayleigh-fading channels in Figure 10.14, although the video degradations become more apparent at higher channel SNRs than in the case of Gaussian channels. These degradations are more gracefully decaying, however.

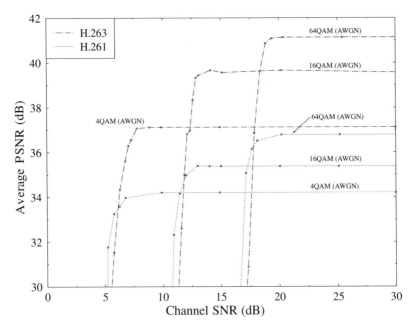

Figure 10.13: A comparison of mobile videophone systems based on the H.261 and H.263 video coding standards, over Gaussian radio channels, using the system of Table 10.1.

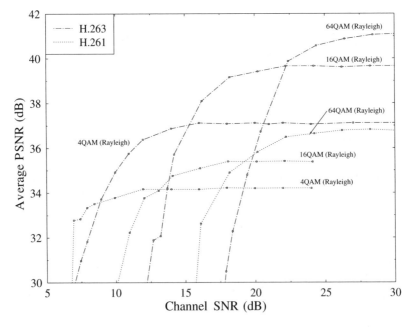

Figure 10.14: A comparison of mobile videophone systems based on the H.261 and H.263 video coding standards over Rayleigh-fading radio channels, using the system of Table 10.1.

10.4. H.263-BASED VIDEO SYSTEM PERFORMANCE

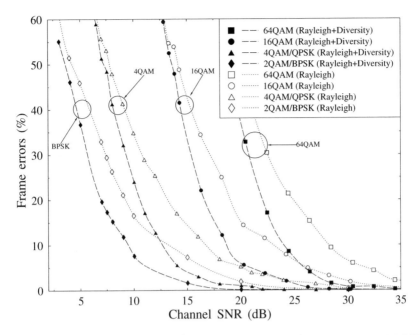

Figure 10.15: Frame errors versus channel SNR for 64QAM, 16QAM, 4QAM, and BPSK over Rayleigh-fading channels with and without dual antenna diversity, using the system of Table 10.1.

10.4.3.1 Performance with Antenna Diversity

A common technique used to improve reception in mobile radio systems is to employ antenna diversity. The technique is conceptually simple. The radio receiver has two or more aerials, which are separated by some spatial distance. Provided that the distance between the aerials is sufficiently high in order to ensure that their received signals are decorrelated, if the signal received by one aerial is deeply faded, the signal from the other aerials is likely not to be. The disadvantage of this technique is the expense and inconvenience of two aerials, in addition to the cost of extra circuitry. The receiver demodulates the signal from that antenna, which is deemed to result in the best reception. Several criteria [334] can be used to decide which antenna signal to demodulate; a few are as follows:

- maximum data symbol power
- maximum pilot symbol power
- minimum phase shift between pilots.

Alternatively, the received signals of all the antennas can be coherently combined for best performance. The best but most complex technique to optimally combine the received signals of the antennas is to employ maximum ratio combining [335].

The most common diversity technique is switched diversity, in which the antenna whose signal has the highest received signal power (maximum symbol power) is chosen by the

Table 10.2: Channel SNR Required for 5% Frame Error Rate with and without Dual Antenna Diversity, and SNR Margin Improvement with Diversity, for BPSK, 4QAM, 16QAM and 64QAM over Rayleigh Channels

Mode	Required SNR without Diversity (dB)	Required SNR with Diversity (dB)	SNR Margin Improvement (dB)
BPSK	17.1	12.2	4.9
4QAM/QPSK	20.2	14.4	5.8
16QAM	26.2	20.9	5.3
64QAM	31.6	26.0	5.6

receiver in order to demodulate the received signal. If pilot symbol-assisted modulation (PSAM) is used, then a better choice is the maximum pilot power, since the received pilot's signal strength can be used to estimate the current fading of the channel. Therefore, by demodulating the signal from the antenna with the highest received pilot symbol power, the least faded channel is chosen. An alternative antenna selection criterion in conjunction with PSAM is to choose the antenna that has the least channel-induced phase shift between consecutive pilots.

When using switched antenna diversity, the bit and frame error rates are reduced for the same SNR compared to the no-diversity scenario. Alternatively, the SNR required to maintain the target bit or frame error rate is lower, when using antenna diversity. The SNR gain becomes higher for marginally impaired channels, but the improvement is negligible for poor channels, where the frame error rate becomes very high. Figure 10.15 shows the frame error rate for all the modulation schemes of the multirate system, with and without diversity. The graph demonstrates that at any given SNR the frame error rate is lower with diversity. Furthermore, it can be seen that to maintain a desired frame error rate performance, the SNR can be consistently reduced when diversity is used. However, it can be observed that this SNR improvement reduces at high frame error rates. The SNR improvement margin for a 5% frame error rate is shown in Table 10.2 for various modulation modes.

The corresponding graph of the decoded video PSNR versus channel SNR is shown in Figure 10.16. The graphs demonstrate the improvement in video quality in terms of PSNR, when using dual antenna diversity.

We have shown results using the mobile radio systems whose parameters were summarized in Table 10.1 on page 353. In order to demonstrate the wide applicability of our error-resilient video system, we also investigated the performance of our scheme in the context of the Pan-European DECT cordless telephone system, an issue discussed in the next section.

10.4.3.2 Performance over DECT Channels

The Digital European Cordless Telecommunications (DECT) system [336, 337] was defined as a digital replacement for analog cordless telephones, which are commonly used in indoor residential and office scenarios, although its employment today is beyond these fields. The DECT system is TDMA/TDD-based with 12 channels per carrier, operating in the

10.4. H.263-BASED VIDEO SYSTEM PERFORMANCE

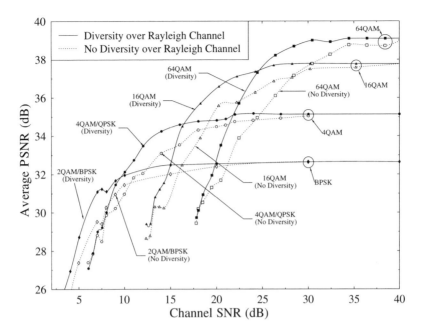

Figure 10.16: PSNR versus channel SNR for 64QAM, 16QAM, 4QAM, and BPSK over Rayleigh-fading channels, with and without dual antenna diversity, when using the system of Table 10.1.

1880–1900 MHz band. Systems based on Time Division Multiple Access (TDMA) multiplex users onto the same carrier frequency by dividing time into slots and allocating a slot every TDMA frame to each user. In the case of DECT, there are 12 timeslots for 12 users on each carrier frequency. Time Division Duplex (TDD) systems use the same carrier frequency for up-link and down-link, by dividing frames into up-link and down-link portions. These up-link and down-link sections of the TDD frame are usually further divided by using TDMA. The DECT standard employs Adaptive Differential Pulse Coded Modulation (ADPCM) at 32 Kbps for encoding the voice signal. Since the system was defined for indoor microcellular environments, it does not use channel coding or equalization. The TDMA/TDD frame structure is shown in Figure 10.17. The role of the control fields in Figure 10.17 becomes clear with reference to [336, 338]. As can be seen in the figure, the data channel's "payload" is 320 bits. Therefore the up-link/down-link data capacity is 320 bits every 10 ms, corresponding to a bitrate of 32 Kbps.

Using simulated DECT channel error profiles related to the scenarios described in [337], we investigated our H.263-based videophone system over these DECT channels. The corresponding DECT channel parameters are summarized in Table 10.3. The DECT system uses Gaussian Minimum Shift Keying (GMSK) modulation, as in the GSM system [256]. We simulated the videophone system with and without channel coding over these DECT channels, and the simulation parameters are shown in Table 10.4. For the uncoded simulations, the video bitrate is 32 Kbps. However, only a single-bit error is required to cause the loss of a video packet. For the FEC-coded simulations, the BCH codes were chosen to match the

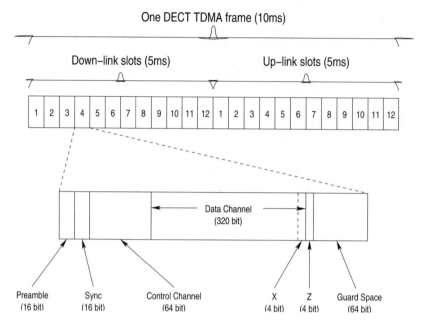

Figure 10.17: The Digital European Cordless Telecommunications (DECT) system's TDMA/TDD frame structure [336].

Table 10.3: DECT Channel Parameters, Channels Derived from [337]

Feature	DECT Channel 1	DECT Channel 2
Modulation	GMSK	GMSK
Carrier frequency	1.9 GHz	1.9 GHz
Wavelength	15.8 cm	15.8 cm
Doppler frequency	2.5 Hz	2.5 Hz
Vehicular speed	0.39 m/s	0.39 m/s
SNR	20 dB	30 dB
BER	2.11×10^{-2}	2.11×10^{-3}

amount of error correction used in previous simulations. This resulted in a reduced video bitrate of 21.6 Kbps for the FEC-coded simulations in order to arrive at a FEC-coded rate of 32 Kbps.

Using the DECT Channels 1 and 2 of Table 10.3 and the artificial error-free channel, the 32 Kbps videophone system was simulated with and without channel coding, assuming that video users are assigned an additional timeslot per frame. An attractive feature of the DECT system is that up to 23 timeslots can be allocated to a single user. This could support video rates up to 22×32 Kbps $= 704$ Kbps, while reserving one 32 Kbps timeslot for the speech information. In practice because of the system's internal overhead the maximum achievable single-user data rate is about 500 Kbps. The resultant video quality in terms of the average

10.4. H.263-BASED VIDEO SYSTEM PERFORMANCE

Table 10.4: DECT-like System Parameters

Feature	DECT-based System Parameters	
	Uncoded	Coded
Channel data rate	32 Kbps	32 Kbps
C1 FEC	n/a	BCH(255,171,11)
C2 FEC	n/a	BCH(63,45,3)
Net video bits/TDMA-frame	320	216
Net video rate	32 Kbps	21.6 Kbps

PSNR is plotted against the frame error rate in Figure 10.18. When there is no FEC coding, any bit error in a TDMA frame causes a frame error. When FEC coding is used, a transmission frame error occurs only if the number of bit errors overloads the error correcting capacity of the BCH codes used. The figure shows how the transmission frame error rate increases, as the channel changes from error-free to the 0.2% BER Channel(2), and then to the 2% BER Channel(1), where the three measurement points in Figure 10.18 correspond to these BERs. When FEC coding is used, the video quality is inherently reduced due to the reduced video bitrate because of accommodating the FEC parity bits. However, the proportion of erroneous transmission frames is also reduced at both BER = 0.2% and 2%. It was found that the effect of transmission frame errors was not objectionable up to an FER of about 10%, which is supported by the PSNR versus FER curves of Figure 10.18. This implies that dropping one in ten frames does not significantly impair the video quality as long as the acknowledge flag feedback keeps the encoder's and decoder's reconstruction frame buffer in synchronization. The implementation of this will be the subject of Section 10.5.

In subjective video-quality terms, it is often more advantageous to change to a more robust modulation or coding scheme, once the PSNR degradation exceeds about 1 dB. This is demonstrated by Figure 10.19. This figure shows the average decoded video quality in terms of PSNR (dB) versus channel SNR, for 16QAM, 4QAM, and BPSK modes of operation. The 5% FER switching levels are also shown in the figure, which shows that the switching occurs when the higher PSNR but corrupted video becomes subjectively more objectionable than the inherently lower PSNR but unimpaired video.

The DECT simulations show an improvement in frame error robustness over previous simulations, increasing the maximum tolerable FER from 5% to 10%. This is because the DECT TDMA frame length of 10 ms is half that of the previous simulations, which used a frame length of 20 ms. Therefore, each TDMA packet contains on average a smaller proportion of the video frame. Hence, the loss of a packet has a reduced effect on the PSNR and on the viewer's perception of the video quality.

Using FEC coding accordingly improves the range of channel SNRs, where the videophone system can operate. Figure 10.18 also shows that the videophone quality in the presence of transmission errors is fairly independent of the video sequence and resolution used, as evidenced by contrasting the more motion-active "Carphone" sequence and the CIF resolution "Miss America" sequence with the QCIF resolution Miss America clip. In the next

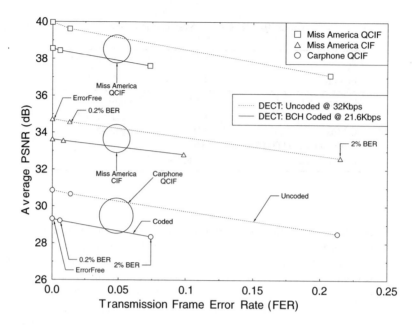

Figure 10.18: Simulation of the H.263 videophone system over DECT-like channels, with and without channel coding, and in error-free conditions. Using the "Miss America" video sequence at QCIF and CIF resolutions, and the "Carphone" sequence at QCIF resolution.

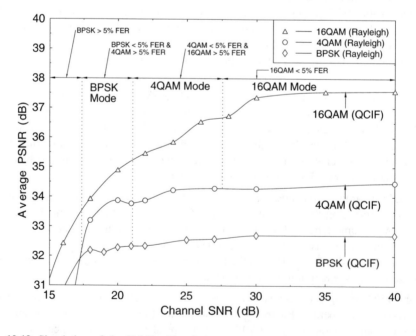

Figure 10.19: Simulation of the H.263 videophone system over Rayleigh-faded channels at QCIF resolution.

10.5. TRANSMISSION FEEDBACK

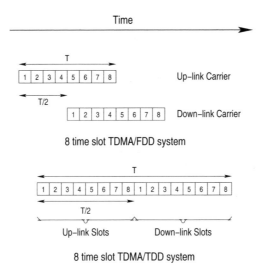

Figure 10.20: Up-link and down-link timeslots for TDMA/FDD and TDMA/TDD systems, where the TDMA frame length is T seconds, and the offset between the corresponding up-link and down-link timeslots is T/2 seconds.

section, we discuss the issues related to the transmission feedback mechanism that is used in our error-resilient mobile video system.

10.5 Transmission Feedback

The proposed wireless videophone system requires a feedback channel from the receiver to the transmitter in order to acknowledge the loss of packets and to keep the local and remote decoders in synchronization. This feedback channel can also be used to convey the ARQ requests, when ARQ is used. This feedback mechanism can be implemented in a number of ways, depending on the application. For two-way videotelephony the most efficient mechanism is to superimpose the up-link feedback information on the down-link data stream and vice versa. Normally, the up-link and down-link slots are offset by half a TDMA frame in time in both FDD and TDD systems. In Frequency Division Duplex (FDD), the up-link and down-link information is transmitted on different carriers. However, to reduce the complexity of the handset, the corresponding up-link and down-link slots are offset in time by half a TDMA frame in order to allow the handset time to process the information it receives and to reduce high-power transmitter leakage into the receiver, which is processing low-level received signals. In Time Division Duplex (TDD), the up-link and down-link information is transmitted on the same carrier; however, the carrier is divided in up-link and down-link timeslots. Again, these slots are normally offset by half a TDMA frame in time as in the DECT system shown in Figure 10.17. The up-link and down-link slot arrangements for a TDMA/FDD and a TDMA/TDD system are shown in Figure 10.20.

For videotelephony, where the feedback information is superimposed on the reverse channel, the delay before the feedback information is received is half the TDMA frame

Table 10.5: Summary of System Features for the Quadruple-mode Reconfigurable Mobile Radio System

Modem-mode Independent System Parameters	
Features	General System
User data symbols/TDMA frame	128
User pilot symbols/TDMA frame	14
User ramping symbols/TDMA frame	2
User padding symbols/TDMA frame	2
User symbols/TDMA frame	146
TDMA frame length (ms)	20
User symbol rate (kBd)	7.3
Slots/frame	18
No. of users	9
System symbol rate (kBd)	131.4
System bandwidth (kHz)	200
Eff. user bandwidth (kHz)	11.1
Vehicular speed	13.4 m/s or 30 mph
Propagation frequency (GHz)	1.8
Fast-fading normalized doppler frequency	6.2696×10^{-4}
Log-normal shadowing standard deviation (dB)	6
Path-loss model	Power law 3.5
Base station separation (km)	1

length. If the wireless video system is used for applications such as remote surveillance where video is only transmitted back to a central control room, the feedback information is conveniently implemented using a broadcast message from the base station. The proposed wireless video system is not suitable for broadcast video applications, such as television, because of the need of a reverse channel for feedback in this system. For broadcast video, the video stream has to be rendered more robust and able to recover from errors.

From now on, we will concentrate on interactive videotelephony applications, where there are both up-link and down-link video streams. In this case, the feedback signaling is superimposed on the reverse channel's video stream. We note that the system parameters used for investigating the transmission feedback effects are different from those used in the early part of this chapter, namely, in Table 10.1 on page 353.

Specifically, the transmission feedback aided investigations were conducted in the context of a similar HSDPA-style [198] quadruple-mode scheme, which also supported a BPSK mode. The channel parameters shown in Table 10.5. The error correction codes and the useful video bitrates of each of the four HSDPA-style [198] modes are shown in Table 10.6.

An example of this transmission feedback signaling and timing is shown in Figure 10.21 in the context of our system described in Tables 10.5 and 10.6. Upon receiving a transmission packet, the transceiver has two tasks to perform before the next transmission. First, the video data stream has to be demodulated and the FEC decoded in order to check whether

10.5. TRANSMISSION FEEDBACK

Table 10.6: Summary of System Features for the Reconfigurable Mobile Radio System, Including Transmission Feedback Implemented Using the Majority Logic Code ML(27,1,13)

Features	Multirate System			
	PSA-BPSK	4-PSAQAM	16-PSAQAM	64-PSAQAM
Modem	1	2	4	6
Bits/symbol	1	1	2	3
No. of modem sub-channels				
C1 FEC	BCH(127,85,6)	BCH(255,171,11)	BCH(255,191,8)	BCH(255,199,7)
C2 FEC	N/A	N/A	BCH(255,147,14)	BCH(255,163,12)
C3 FEC	N/A	N/A	N/A	BCH(255,131,18)
Padding bits	5	9	18	27
Total bits	132	264	528	792
Payload bits	90	180	356	520
Feedback bits	27	27	27	27
Video data bits	63	153	329	493
Video bitrate (Kbit/s)	3.15	7.65	16.45	24.65

370 CHAPTER 10. H.263 FOR HSDPA-STYLE WIRELESS VIDEO TELEPHONY

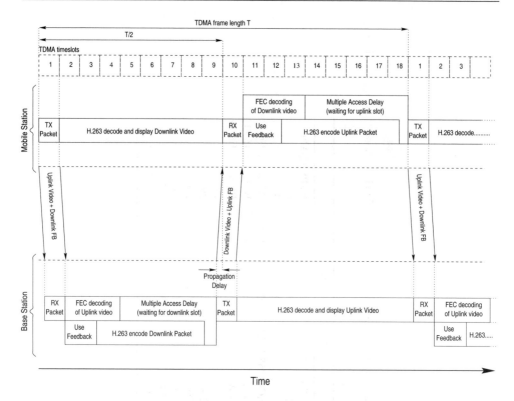

Figure 10.21: Transmission feedback timing, showing the feedback signaling superimposed on the reverse channel video data stream. The tasks that have to be performed in each time interval are shown for both the mobile station and the base station.

any error occurred since this event has to be signaled in the next transmission burst to the transmitter.

The second task is to produce the next video packet for the forthcoming transmission burst, as indicated in Figure 10.21. The next packet of video data cannot be encoded until the feedback signaling for the previous transmission is received. This is because the motion vectors used in the next video data packet may depend on blocks transmitted in the previous packet. Once the feedback acknowledgment flag of the previous packet's transmission is received, which was superimposed on the reverse channel video data, the effect of the loss or success of the previous packet can be taken into account in the local decoder of the H.263 codec. If the previous packet was lost, the effective changes made to the local decoder for this packet are discarded. If the previous packet was received without errors, then the local decoder changes pertinent to the previous packet are made permanent. If ARQ is used, then the transmission feedback signaling may have requested a re-encoding of the packet at the same or a lower bitrate. In this case the reconstruction buffer changes of the previous packet are discarded, and the codec parameters are modified to reduce the video bitrate temporarily. Once the feedback signaling was invoked in order to synchronize the reconstruction frame buffers of the local and remote H.263 codecs, the H.263 codec starts to encode the next video packet, as indicated in Figure 10.21.

The time between receiving a packet and transmitting a new packet is approximately half the TDMA frame length, T. Since we are considering microcells, the propagation delays are very low; for example, for a distance of 1 Km the delay is 3.3 μs. Within the time period T/2 between receiving and transmitting packets, the transceiver has to calculate the feedback signaling for the received packet and to generate a new video packet for transmission. These two tasks can be done in parallel or in series, but they must be completed within the T/2 time period. After a packet is transmitted, there is another T/2 time period before the next received packet arrives. This time period, as shown in Figure 10.21, is used to decode the H.263 video data received in the previous packet and to update the video display.

10.5.1 ARQ Issues

The employment use of ARQ results in several difficulties in conjunction with the use of transmission feedback signaling. If ARQ is used, a transmission feedback message is required for each transmission. In systems where a user is allocated one slot per TDMA frame, ARQ introduces latency and slot allocation problems, causing the video frame rate to fluctuate. Furthermore, use of additional slots for ARQ constrains the throughput bitrate of the system, reducing in video quality. Therefore, for systems with only one slot per TDMA frame, ARQ is not recommended because it reduces the video quality and causes a fluctuating frame rate. For these systems, it is better not to attempt to re-transmit lost packets with an ARQ request, but instead to invoke re-transmission at a higher level by transmitting the corresponding macroblocks in later packets. This policy is followed in our system.

The type of systems, in what ARQ can be used effectively are high-rate, short-frame duration wireless LAN-type systems. In these systems, a video user typically transmits less than one packet per TDMA frame. In other words, on average a video user is allocated a timeslot once every "n" TDMA frames. In such systems, the ARQ re-transmission can be implemented by requesting additional timeslots during the "n" TDMA frame period. Except in highly loaded systems, this will reduce the ARQ latency, yielding a more stable video frame rate. Hence upon using additional timeslots for the re-transmissions, the video bitrate for the user will remain fairly constant, resulting in reduced video-quality degradation.

For our simulations we have used "Stop and Wait"-type feedback systems, similar to "Stop and Wait" ARQ [202, 316]. It would be possible to invoke more complex types of feedback/ARQ systems, such as "selective repeat" [202, 317]. But since this would make the packet "dropping" more complex, particularly for the motion compensation, this was not implemented for this initial version of our system.

10.5.2 Implementation of Transmission Feedback

The transmission feedback messages need to be very robust to channel errors. If a transmission feedback message is received incorrectly, the local and remote decoders lose synchronization. A feedback message received incorrectly will cause the video picture to become corrupted, and this type of corruption can be corrected only by the affected macroblocks being updated in intra-frame coded mode. However, due to motion compensation, additional macroblocks surrounding the affected blocks usually need to be intra-updated as well.

Therefore, initially the best way to implement the feedback message was deemed to use a very strong block code to encode the feedback message. The following three short block codes from the BCH family were considered:

- BCH(7,4,1)
- BCH(15,5,3)
- BCH(31,6,7).

When there is a deep fade, as experienced in mobile radio channels, however, the bit error-rate within the TDMA transmission burst can be very high. (Henceforth we will refer to the bit-error ratio (BER) within a TDMA transmission burst as the In-Burst BER.) The BCH(15,5,3) code needs an In-Burst BER of higher than $\frac{3}{15}$ or 20% to become overloaded and cause the feedback message to be received in error. In our simulations we found that this BCH code would become overloaded quite often, when deep fades occurred in the mobile radio channel. A more robust coding method for the transmission of feedback messages is therefore needed, and we eventually decided on a majority logic code.

10.5.2.1 Majority Logic Coding

Majority logic codes use a fairly simple principle. The information message is transmitted several times, and at the receiver all the messages are demodulated. The correct message is decided by majority voting. For example, in order to transmit a 1-bit message, which happens to be a logical 1, this 1-bit message is transmitted, say, nine times. Therefore, the coding has caused the 1-bit message to be turned into a 9-bit encoded message. The receiver decides the intended 1-bit message based on whether it receives more logical 0s or more 1s. Therefore, the message is received correctly as long as no more than four logical 1s are corrupted by the channel and changed to logical 0s. Therefore, the code can be described as ML(9,1,4) in the same notation as the BCH codes. This means that a 1-bit message is encoded into 9 bits, and the code can cope with up to 4 bit errors. If five or more errors are caused by the channel, then majority voting will cause the decoded message to be incorrect. This code can cope with an In-Burst BER of less than $\frac{4}{9}$ or 44%, which is much better than the above BCH codes. However, these codes are unaware of when they become overloaded, unlike the BCH codes. By increasing the number of encoded bits, the maximum In-Burst BER the code can cope with can be brought closer to the 50% limit. For example, the ML(31,1,15) code can cope with an In-Burst BER of up to 48.4%. This is much better than the correction capability of the BCH(31,6,7) code. However, the majority logic code is unaware of the event, when it becomes overloaded.

Majority Logic Codes can be characterized mathematically using the binomial distribution. The probability of n bit errors in k bits is given by:

$$P_n^k = \binom{k}{n} p^n (1-p)^{(k-n)}, \qquad (10.3)$$

where p is the probability of a bit error or the bit-error rate and the binomial coefficient $\binom{k}{r}$ is defined as:

$$\binom{k}{r} = \left(\frac{k!}{r!\,(k-r)!}\right) \qquad (10.4)$$

10.5. TRANSMISSION FEEDBACK

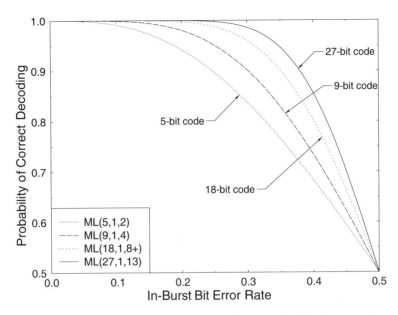

Figure 10.22: Numerical evaluation of the probability of correct decoding (P_{CD}) of majority logic codes.

which gives the number of combinations of r picks from k, not reusing picked values therefore, the probability of less than or equal to n errors in k is given by:

$$P^k_{(i \leq n)} = \sum_{i=0}^{i=n} (P^k_i) = \sum_{i=0}^{i=n} \left(\binom{k}{i} p^i (1-p)^{k-i} \right) \quad (10.5)$$

where p is again the bit-error rate. Hence, for a majority logic code of 9 bits, ML(9,1,4), the probability of four or fewer bit errors in 9 bits is $P^9_{(i \leq 4)}$. If the number of encoded bits is an even number, say 10, then the total probability of correct decoding is given by the probability of four or fewer bit errors in 10 bits ($P^{10}_{(i \leq 4)}$), plus half the probability of 5 bit errors in 10 (P^{10}_5). This additional term for even-length codes occurs, because the majority voting can result in a draw, and the outcome is then decided by chance. Therefore, for a 10 bit code there is a 50% chance of correct decoding, even when there are 5 bit errors. We describe this code as ML(10,1,4+). This means that the code can correct 4 bit errors and sometimes one more. The general formula for the correct decoding probability (P_{CD}) of majority logic codes is given by:

$$P_{CD} \text{ of ML}(k,1,?) = \begin{cases} P^k_{(i \leq (\frac{k-1}{2}))} & k \text{ is odd} \\ P^k_{(i \leq (\frac{k}{2}-1))} + \dfrac{P^k_{(k/2)}}{2} & k \text{ is even.} \end{cases} \quad (10.6)$$

The probability of correct decoding of the 5-, 9-, 18-, and 27-bit majority logic codes was evaluated for the range of bit-error rates 0% to 50% in Figure 10.22.

In order to find the length of majority logic code required for our system, the histogram of In-Burst bit error rates was calculated as described in the context of Tables 10.5 and 10.6.

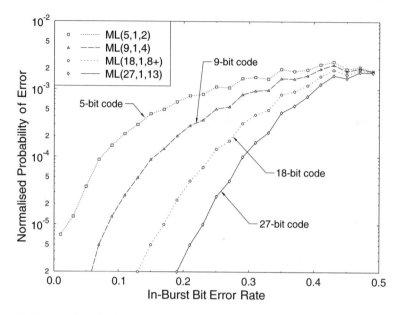

Figure 10.23: Numerical evaluation of probability of error in majority logic codes, normalized by the chance of occurrence of the given In-Burst BER.

This system experienced both Rayleigh fast fading and shadow fading. The combination of Rayleigh and shadow fading can cause very deep fades, generating very high In-Burst BER. The probability of a majority logic code in error versus the In-Burst BER was normalized by the probability of the occurrence of such an In-Burst BER, which is shown in Figure 10.23.

Figure 10.23 shows the chance of an error in a majority code for the system described in Tables 10.5 and 10.6. Even for the weakest 5-bit Majority Logic codes the probability of error is on average once in 400 packets for the worst-case situation. Our proposed system will change to a more robust modulation scheme, request a handover, or drop the call when the channel becomes very hostile. Hence, such high bit error rates will be avoided in all but a few cases.

The transmission feedback can then be implemented using a majority logic code on the reverse channel. The video data for the reverse link and the transmission feedback information majority logic code can be transmitted together in the same packet, as suggested above.

In order to evaluate the performance of this transmission feedback scheme in our mobile video system, we used the multirate scheme defined in Tables 10.5 and 10.6. We used a 27-bit majority logic code, ML(27,1,13), to encode the transmission feedback information bit. These 27 encoded bits were transmitted and exploited previously unused padding bits, where they were available. These padding bits were uncoded, while the remaining bits were transmitted within the FEC-encoded portion of the data packet. Because we are using systematic BCH coding, even if the error correction becomes overloaded, the majority logic bits can typically be recovered. The employment of the protected transmission feedback flag reduced the effective video source bitrate to the values shown in Table 10.6, which particularly affected the BPSK mode of operation. In this mode the video bitrate is limited to about 3 Kbit/s, which

10.5. TRANSMISSION FEEDBACK

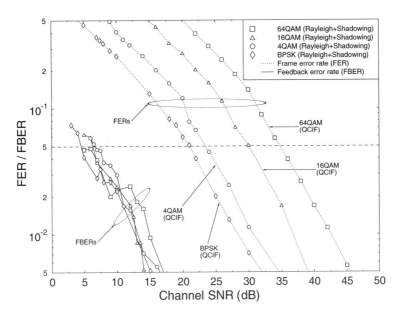

Figure 10.24: Frame error rate (FER) and feedback error rate (FBER) versus channel SNR, for the parameters defined in Table 10.6 and channel parameters of Tables 10.5 and 10.6.

means that the video will have to be of SQCIF resolution in order to maintain a reasonable video frame rate.

Figure 10.24 shows the Frame Error Rate (FER) versus channel SNR performance for the four modes of operation listed in Table 10.6. The figure also displays the feedback error rate versus channel SNR. In order to distinguish between them, the feedback error rate is referred to as FBER. Previously, we found that with the aid of our acknowledgment technique the video quality does not degrade significantly at FERs below 5%. For these simulations, the feedback error rate (FBER) was zero, since for the channel conditions where the FER is below 5%, we have FBER = 0. Therefore, if the system switches to a more robust modulation scheme or requests a handover before the 5% FER threshold is reached, then there are no feedback errors.

When feedback errors do occur, which is fairly infrequently, this effect can be removed during the next intra-frame transmission or during the intra-update of the macroblocks affected. The rate at which these intra-updates occur is dependent on the additional robustness required. However, the more frequent the intra-updates, the lower the video quality, since the intra-blocks consume a larger fraction of the available bitrate.

Figure 10.25 shows the PSNR degradation associated with different rates of feedback errors. The effect of the feedback errors is not removed by intra-frame coded replenishments in these simulations. In the majority of cases, the average PSNR was reduced to about 15 dB.

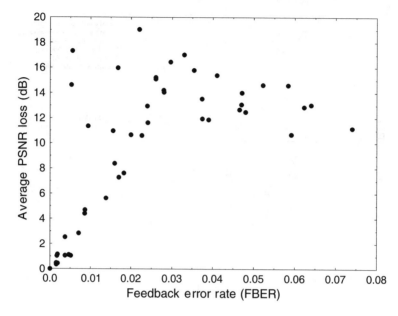

Figure 10.25: Average video-quality degradation in terms of PSNR (dB) versus feedback error rate (FBER).

10.6 Summary and Conclusions

This chapter described the design and performance of an H.263-based videophone system. The system performance was also compared to that of the previously characterized H.261 arrangement of section 8.4. The main system features were summarized in Table 10.1. In Section 10.2, we described the problems associated with using the H.263 video codec in a mobile environment. We discussed potential solutions to these problems, together with the advantages and disadvantages of each solution.

In Section 10.3, we described the design of our error-resilient reconfigurable videophone system, using transmission feedback in order to keep the local and remote decoders in synchronization. A packetization scheme is used to aid the resynchronization of the video data stream and to help conceal the loss of packets. Forward error correction (FEC) is used to decrease the channel BER and thereby to improve the error resilience. The system is designed to be reconfigurable in order to be able to adapt to time-variant channel conditions. This improves the video quality in periods of good channel conditions. A bitrate control algorithm is used to allow the video codec to vary its target bitrate based on the current channel conditions and to control the choice of the modulation scheme used by the reconfigurable system. The system can invoke ARQ in order to re-transmit lost packets. However, because of the system's reconfigurability, the re-transmissions can use a more robust modulation scheme, increasing the probability of success for the re-transmission. The system does not allow the standard H.263 decoder to decode erroneous data, and so errors cannot appear in the decoded video. The transmitted bitstream conforms to the H.263

10.6. SUMMARY AND CONCLUSIONS 377

standard, and no transcoder is necessitated. The receiver employs a standard H.263 decoder, which is fed by the de-packetization system.

Section 10.4.2 characterized the performance achieved by our error-resilient reconfigurable videophone system, which could cope with very high packet-loss rates. Specifically, it was found that an average packet-loss rate of 5% was the highest rate the system could comfortably cope with, before the effects of the packet loss became perceptually obvious. However, the system would recover from packet-loss conditions as soon as the channel quality improved, or the system was reconfigured to a more robust modulation scheme. We investigated how system performance improved with use of ARQ and antenna diversity. We also studied the error-resilient reconfigurable videophone system using the DECT cordless telephone system and found that the system performed in a similar manner. However, due to the higher TDMA frame rate in the DECT scheme, it was found that the maximum average packet-loss rate could be increased to 10%.

Finally, in Section 10.5 we discussed the issues concerned with the transmission feedback that was used in the error-resilient video system. We described how we protected this transmission feedback using majority logic-coding. We then presented performance results using the majority logic-based transmission feedback system. These results showed that the transmission feedback was error-free within the operational SNR range of the video system that is governed by the reconfigurable modulation scheme switching. In the next chapter, we investigate the effects of co-channel interference, which is the main limiting factor in radio spectrum reuse efficiency. The results from this chapter show the potential benefits of our reconfigurable modulation system.

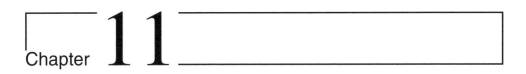

Chapter 11

MPEG-4 Video Compression

J-Y. Chung and L. Hanzo

11.1 Introduction

The "Moving Picture Experts Group" (MPEG) was established in 1988 [339], within the International Organization for Standardization (ISO) Steering Group (SG) 29, which was responsible for the encoding of moving pictures and audio. The MPEG commenced the development of the MPEG-1 standard in 1988, which it released in 1993 and embarked on the standardization of the MPEG-2 scheme in 1990 [340]. The MPEG-1 standard was mainly targeted at CD-ROM applications dedicated to recording video at bitrates of up to 1.5 Mbit/s [340, 341]. In contrast, the MPEG-2 standard was designed for substantially higher quality, namely for audiovisual applications such as today's home entertainment systems and digital broadcasting systems requiring video bitrates between 2 and 30 Mbit/s [342, 343].

The MPEG-4 standardization process was initiated in 1994 with the mandate of standardizing algorithms for audiovisual coding in multimedia applications, while allowing for interactivity, and supporting high compression as well as universal accessibility and portability of both the audio and video content [344].

The MPEG-4 Visual standard was developed by the ISO/IEC 14496-2,[1] and its Version 1 was released in 1998; additional tools and profiles were added in two amendments of the standard, culminating in Version 2 during late 2001. The operating bitrates targeted by the MPEG-4 video standard are between 5 and 64 kbit/s in the context of mobile or Public Switched Telephone Network (PSTN)-based video applications, spanning up to 4 Mbit/s for digital TV broadcast applications, and even to rates in excess of 100 Mbit/s in High Definition TV (HDTV) studio applications [345].

The MPEG-4 video coding standard is capable of supporting all functionalities already provided by MPEG-1 and MPEG-2. The MPEG-4 Visual standard improves on the popular MPEG-2 standard both in terms of the achievable compression efficiency, at a given visual quality, as well as in terms of the attainable flexibility that facilitates its employment in a

---

[1]This is the project's profile name for the ISO/IEC. For example, the profile index 15444 is for JPEG, 11172 is for MPEG-1, 13818 is for MPEG-2, 14496 is for MPEG-4, etc.

wide range of applications. It achieves these substantial advances by making use of more advanced compression algorithms and by providing an extensive set of "tools" for coding and manipulating digital media. The MPEG-4 Visual standard consists of a "core" video encoder/decoder model that invokes a number of additional coding tools. The core model is based on the well-established hybrid Differential Pulse Code Modulation/Discrete Cosine Transform (DPCM/DCT) coding algorithm and the basic function of the core is extended by "tools" supporting an enhanced compression efficiency and reliable transmission. Furthermore, MPEG-4 facilitates an efficient and novel coded representation of the audio and video data that can be "content based", which is a concept that is highlighted in Section 11.3.

To elaborate a little further, the MPEG-4 video standard compresses the video signal with the aid of a compression toolbox constituted by a set of encoding tools supporting several classes of functionalities. In short, the most important features supported by the MPEG-4 standard are *a high compression efficiency, content-based interactivity, and universal access*, which are summarized as follows [340, 346].

- *Achieving high compression efficiency* has been a core feature of both MPEG-1 and MPEG-2. The storage and transmission of audiovisual data requires a high compression efficiency, while reproducing a high-quality video sequence, hence enabling applications such as HDTV and Digital Video Disc (DVD) storage.

- *Content-based interactivity* represents video on an "object basis", rather than on a video "frame basis", which is one of the novel features offered by MPEG-4. The concept of content-based functionality is elaborated on in Section 11.3.

- *Universal access* allows audiovisual information to be transmitted and accessed in various network environments such as mobile networks as well as wire line-based systems.

This chapter provides a rudimentary overview of the MPEG-4 video standard. Following the overview of the standard, its approach and its features, the philosophy of the object-oriented coding scheme is discussed in Section 11.3. This is followed by a discussion on the so-called profiles defined for the coding of arbitrary-shaped objects and rectangular video frames in Section 11.3.3. Then the profiles defined for scalable coding of video objects are highlighted in Section 11.4 and subjective video quality measurement methods as well as our experimental results are discussed in Sections 11.5 and 11.6.

11.2 Overview of MPEG-4

11.2.1 MPEG-4 Profiles

The MPEG-4 standard aims at satisfying the requirements of various visual communications applications using a toolkit-based approach for encoding and decoding of visual information [25, 346]. In the following we describe some of the key features of the MPEG-4 video standard, which are superior in comparison to the previous video coding standards.

- The core compression tools are based on those of the International Telecommunications Union—Telecommunication Standardisation Sector (ITU-T) H.263 standard, which are more efficient than those of the MPEG-1 [347] and MPEG-2 [348] video

11.2. OVERVIEW OF MPEG-4

compression schemes. Efficient compression of progressive and interlaced video sequences as well as optional additional tools were introduced for the sake of further improving the attainable compression efficiency.

- Coding of video objects, having both rectangular-shaped and irregular-shaped objects. This is a new concept in the context of standard-based video coding and enables the independent encoding of both foreground and background objects in a video scene.

- Support for error-resilient transmission over hostile networks. A range of error resilience tools were included in the MPEG-4 codec for the sake of assisting the decoder in recovering from transmission errors and for maintaining a successful video connection in an error-prone network. Furthermore, the scalable coding tools are capable of supporting flexible transmission at a range of desired coded bitrates.

- Coding of still image within the same framework as full-motion video sequences.

- Coding of animated visual objects, such as 2D and 3D computer-generated polygonal meshes, animated objects, etc.

- Coding for specialist applications, such as very high "studio"-quality video. In this application maintaining a high visual quality is more important than attaining high compression.

Table 11.1 lists the MPEG-4 visual profiles invoked for coding video scenes. These profiles range from the so-called simple profile derived for the encoding of rectangular video frames through profiles designed for arbitrary-shaped and scalable object coding to profiles contrived for the encoding of studio-quality video.

11.2.2 MPEG-4 Features

Similarly to MPEG-1 [347] and MPEG-2 [348], the MPEG-4 specifications cover both the presentation and transmission of digital audio and video. However, in this book we only consider the specifics of the video coding standard. The block diagram of a basic videophone scheme is shown in Figure 11.1. Let us now consider some of the associated operations in slightly more detail.

Pre-processing and Video Encoding

According to the MPEG ISO standard [349], the MPEG-4 video codec only supports YUV 4:2:0[2] Quarter Common Intermediate Format (QCIF) or Common Intermediate Format (CIF) video representations in the context of compression [33]. Here, the pre-processing block of Figure 11.1 performs all necessary processing of the camera input in order to create the required 4:2:0 YUV QCIF- or CIF-based sequences. In order to encode the YUV video sequence into the MPEG-4 bitstream, the MPEG-4 encoder adopts the well-established

---

[2]A color encoding scheme [350] in which the luminance (Y) and the so-called color-difference signals U and V are represented separately. The human eye is less sensitive to color variations than to intensity variations. Hence, the YUV format allows the luminance (Y) information to be encoded at full resolution and the color-difference information to be encoded at a lower resolution.

Table 11.1: MPEG-4 Visual Profiles for Coding Natural Video [25, 346]

MPEG-4 Visual profile	Main features
☐ Simple	Low-complexity coding of rectangular video frames
☐ Advanced Simple	Coding rectangular frames with improved efficiency and support for interlaced video
☐ Advanced Real-Time Simple	Coding rectangular frames for real-time streaming
☐ Core	Basic coding of arbitrary-shaped video objects
☐ Main	Feature-rich coding of video objects
☐ Advanced Coding Efficiency	Highly efficient coding of video objects
☐ N-Bit	Coding of video objects with sample resolutions other than 8 bits
☐ Simple Scalable	Scalable coding of rectangular video frames
☐ Fine Granular Scalability	Advanced scalable coding of rectangular frames
☐ Core Scalable	Scalable coding of video objects
☐ Scalable Texture	Scalable still texture with improved efficiency and object-based features
☐ Advanced Scalable Texture	Scalable still texture with improved efficiency and object-based features
☐ Advanced Core	Combines the features of Simple, Core and Advanced Scalable Texture profiles
☐ Simple Studio	Object-based coding of high-quality video sequences
☐ Core Studio	Object-based coding of high-quality video with improved compression efficiency

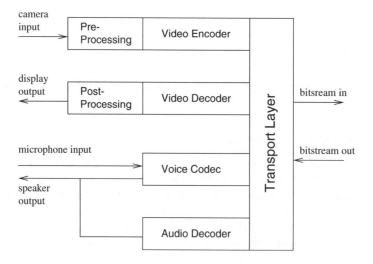

Figure 11.1: Simplified videophone schematic.

11.2. OVERVIEW OF MPEG-4

Motion Compensation (MC) and DCT-based structure shown in Figure 11.2. The block-by-block discussion of the MPEG-4 encoder's components is postponed until Sections 11.3.1 and 11.3.2.

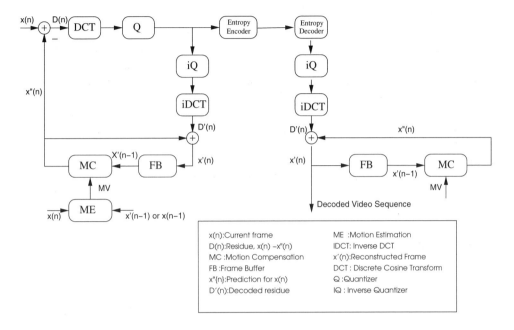

Figure 11.2: Block diagram of the MPEG-4 encoder and decoder.

Transport Layer

In MPEG-4, the transport of the video stream is divided into four layers, namely the Elementary Stream [351], the Synchronization Layer, the "Flexmuxed" [352] stream and the "Transmux" [353] stream. The MPEG-4 system's architecture and transport layer have been nicely documented in the literature, for example in [351, 354]. We briefly outline these transport layer characteristics as follows.

- The term *Elementary Streams* [351] refers to data that fully or partially contain the encoded representation of a single audio or video object, scene description information or control information.

- The *Synchronization Layer (SL)* [355] adds the identification of the information sources, such as audio or video sources, as well as time stamps.

- *Flexmuxed Streams* [352] convey groups of elementary streams according to a specific set of common attributes, such as quality requirements.

- *Transmux Streams* [353] are constituted by streams transmitted over the network using transport protocols, such as the Real-Time Protocol (RTP) [356] used for transmission over the Internet.

Video Decoder and Post-processing

Figure 11.2 portrays the simplified block diagram of the MPEG-4 video decoder. Observe that the structure of the decoding process is identical to that of the encoder's local decoder. Motion compensation, which has been comparatively studied for example in [5], is the most important process both in the video encoder and decoder in terms of achieving a high video compression. Motion Compensation (MC) generates the Motion Vectors (MVs) on the basis of identifying the most likely position within the previous video frame where the current 8×8-pixel video block has originated from, as it moved along a certain motion trajectory in the consecutive video frames. This MC process involves allocating a certain search area in the previous frame and then sliding the current block over this search area in an effort to find the position of highest correlation. Once this position has been identified, the Motion Compensated Prediction Residual (MCPR) [5] is formed by subtracting the two blocks from each other.

11.2.3 MPEG-4 Object-based Orientation

One of the functionalities defined by the MPEG-4 standard is the audiovisual "object-based" processing, which forms the "object-based" representation of the audio or video signal [342]. A video object can be exemplified by a person walking against the backdrop of mountains. Both the object and the background can be substituted by another object or by another backdrop and as expected certain encoding techniques perform better for certain objects. This representation supports *"content-based interactivity"*, which is discussed in more detail in Section 11.3.

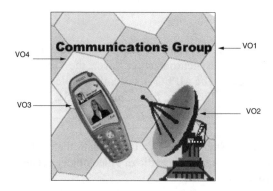

Figure 11.3: Original decoded video scene.

Figure 11.3 shows an example of a video frame extracted from a video scene, which consists of several objects namely text, an antenna, a mobile phone and the background scene.

11.2. OVERVIEW OF MPEG-4

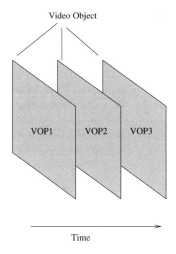

Figure 11.4: VOPs and VO (rectangular).

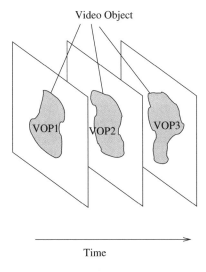

Figure 11.5: VOPs and VO (arbitrary shape).

Table 11.2: Different MPEG-4 Object-oriented Representations of Various Video Scenes

Name	Description
☐ Visual Object Sequence (VS)	The complete MPEG-4 scene, which may contain 2D or 3D natural as well as computer-generated objects.
☐ Video Object (VO)	A 2D video object. In the simplest case this may be a rectangular frame, or an arbitrarily shaped element of the video frame corresponding to an object or background of the scene.
☐ Video Object Layer (VOL)	Every VO can be encoded in a scalable fashion, i.e. at different bitrates, using a multi-layer representation constituted by the so-called base layer and enhancement layer. Alternatively, it may be encoded in a non-scalable (i.e. fixed-bitrate) form using a base layer but no enhancement layer, depending on the application. These layers are referred to as VOLs. The VOL facilitates scalable coding, where the video object can be encoded using spatial and/or temporal scalability.
☐ Video Object Plane (VOP)	A VOP is a time-domain sample of a video object. The VOPs can be encoded independently of each other, i.e. using intra-frame coding or inter-frame as well as bidirectional coding techniques employing MC.

Again, in MPEG-4-based coding [349] these objects are referred to as "Video Objects" (VO). The syntax of this representation may be written as VO1 (text), VO2 (the antenna), VO3 (mobile phone) and VO4 (background).

An MPEG-4 video scene may consist of one or more VOs. A VO is an area of the video scene that may occupy an arbitrarily shaped region and may be present for an arbitrary length of time. An instance of a VO at a particular point in time is a Video Object Plane (VOP). This definition encompasses the traditional approach of coding complete frames, in which each VOP is a single frame of video and a sequence of frames forms a VO. For example, Figure 11.4 shows a VO consisting of three rectangular VOPs, however in the MPEG-4 video standard, the introduction of the arbitrarily shaped VO concept allows for more flexibility. Figure 11.5 shows a VO that consists of three irregular-shaped VOPs, each one present within a frame and each encoded separately, hence leading to the concept of object-based coding, which is discussed in more detail in Section 11.3. The VO can be in binary shapes such as VO1, VO2, VO3 in Figure 11.6 or in rectangular shapes such as VO4 in Figure 11.6, which is equivalent to the dimensions of the entire video frame's size. For example, if a QCIF video format is used, the dimension would be 176×144 pixels.

In Table 11.2 we have summarized some of the important nomenclature, which is frequently used when referring to the MPEG-4 video coding syntax. Let us now consider content-based interactivity in the context of MPEG-4-based coding in more detail.

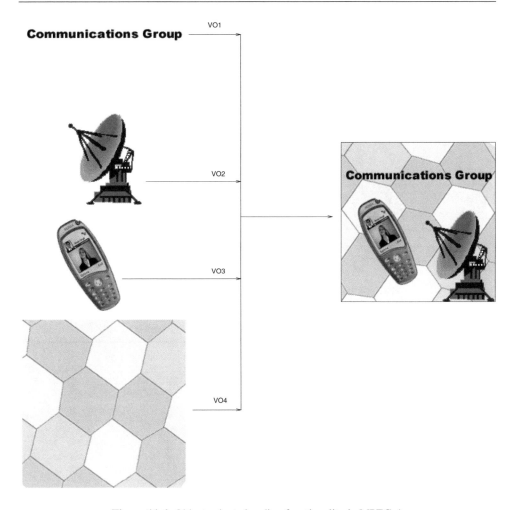

Figure 11.6: Object-oriented coding functionality in MPEG-4.

11.3 MPEG-4: Content-based Interactivity

"Content-based interactivity" attempts to encode an image scene in a way that will allow the separate decoding and reconstruction of the various objects as well as facilitating the manipulation of the original scene with the aid of simple operations carried out in the form of its bitstream representation [342, 357]. As mentioned in Table 11.2, the MPEG-4 video coding standard provides an "object-layered" bitstream referred to as a Video Object Layer (VOL) for supporting this function. Hence, at the encoder the bitstream will be object-layered and the shape, the grade of transparency of each object as well as the spatial coordinates and additional parameters describing object scaling, rotation, etc., are described by the bitstream of each VOL [342]. The received bitstream including all of the information bits is decoded and reconstructed by displaying the objects in their original size and at the original location, as depicted in Figure 11.7. Alternatively, it is possible to manipulate the image sequence according to the user's preference, allowing the scaling shifting and other transformations of the objects, as seen in Figure 11.8.

Figure 11.7: Original decoded video scene.

Figure 11.8: Decoded video scene according to the user's preference. The content-based approach adopted by the MPEG-4 video coding standard allows flexible decoding, representation, and manipulation of the VOs in a scene, where for example different-resolution video decoding is facilitated.

As illustrated in Figure 11.8, the mobile phone VO was not decoded, the satellite ground station was decoded and displayed using scaling or rotation. In addition, a new mobile videophone object defined by the user was included, which did not belong to the original scene. Since the bitstream of the sequence is organized in an "object-layered" form, the object manipulation is performed at the bitstream level by adding or deleting the appropriate object bitstreams [358, 359].

11.3. MPEG-4: CONTENT-BASED INTERACTIVITY

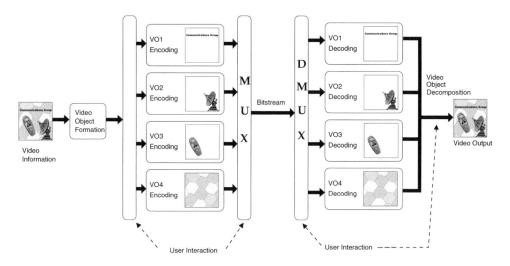

Figure 11.9: The content-based approach adopted by the MPEG-4 video coding standard allows flexible decoding, representation, and manipulation of VOs in a scene.

11.3.1 VOP-based Encoding

Before we can commence with the encoding of an object, it must be sampled. Most objects are sampled at regular time intervals corresponding to the frame-scanning rate, and each sample corresponding to the object's spatial representation at an instant in time is known as a VOP. Hence each object in the scene is represented by a series of VOPs. In more familiar terms, a camera views a scene and captures the information by sampling the scene (either by canning or by shuttering and scanning). The camera provides its output as a sequence of frames or, in MPEG-4 terminology, the texture part of a sequence of VOPs. A VOP contains texture data and either rectangular shape information or more complex shape data associated with the object. VOPs, like frames in earlier versions of the MPEG codec family [30, 31], may be encoded using intra-frame coding or by using MC.

The MPEG-4 standard introduces the concept of VOPs for supporting the employment of content-based interactive functionalities [340]. The associated concept is illustrated in Figure 11.9. The content of each video input frame is segmented into a number of arbitrarily shaped image regions, i.e. into VOs, and each VO is sampled in the time domain by extracting the corresponding area of the consecutive video frames. Each time domain sample of a VO which corresponds to its image in a consecutive video frame constitutes a VOP. The shape and location of each VO may vary from frame to frame, which can be visualized by considering the example shown in Figure 11.10. More explicitly, Figure 11.10 shows five consecutive frames of a particular VO, namely that of the paraboloid antenna, which is rotating from the left to the right during the time interval specified by the five consecutive frames spanning the interval of Frame 1 to Frame 5. Hence, in this case the paraboloid antenna constitutes a VO, while the successive frames portraying the VO constitute VOPs.

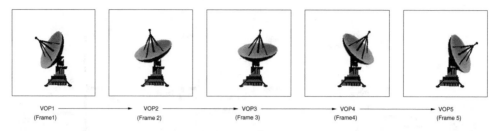

VOP1 (Frame1) → VOP2 (Frame 2) → VOP3 (Frame 3) → VOP4 (Frame4) → VOP5 (Frame 5)

Figure 11.10: The encoding of image sequences using MPEG-4 VOPs enables the employment of content-based functionalities, where the VOP contains five consecutive time domain samples of a VO extracted from five consecutive frames representing a video scene.

In contrast to the MPEG-1 [347] and MPEG-2 [348] standards, the VOP used in MPEG-4 is thus no longer considered to be a rectangular region. Again, the VOP extracted from a video scene contains motion parameters, shape information, and texture data. These are encoded using an arrangement similar to a macroblock coding scheme that is reminiscent of the corresponding schemes used the MPEG-1 and MPEG-2 standards, as well as in the ITU H.263 coding standard [32].

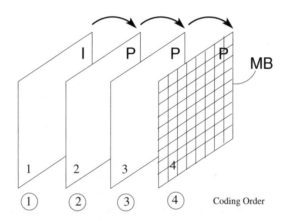

Figure 11.11: I-frames (I-VOP) and P-frame (P-VOP) in a video sequence. The P frames are encoded by using motion compensated prediction based on the nearest previous VOP.

11.3.2 Motion and Texture Encoding

As mentioned earlier, the MPEG-4 video encoding algorithm has a similar structure to that of the well-established MPEG-1/2 and H.263 coding algorithms. Schäfer and Sikora argued in [340] that the MPEG-4 coding algorithm encodes the first VOP in the intra-frame VOP coding mode (I-VOP), while each subsequent frame is encoded using inter-frame coded or predicted VOPs (P-VOP) and only data which accrues from the previous coded VOP frame is used for prediction. As can be seen from Figure 11.11, in this particular case the VO

11.3. MPEG-4: CONTENT-BASED INTERACTIVITY

is treated in a rectangular form. The first frame is encoded as an I-VOP frame, while the second frame (which is the first P-VOP frame) is encoded using the previous I-VOP frame as a reference for temporal coding. Each subsequent P-VOP frame uses the previous P-VOP frame as its reference. Let us now consider how motion and texture encoding is carried out in MPEG-4 [340]. Both the motion and texture encoding of a VOP sequence are block-based. A block in video coding is typically defined as a rectangular array of 8 × 8 pixels [344]. Since the chrominance of the video signal components is typically sampled at a spatial frequency, which is a factor of two lower than the spatial sampling frequency of the luminance (Y), each chrominance (C) block carries the color-difference related information corresponding to four luminance (Y) blocks. The set of these six 8 × 8-pixel blocks (4Y and 2C) is referred to as a Macroblock (MB), as shown in Figures 11.12 and 11.13. A MB is treated as a single encoded unit during the encoding process [360]. In addition, the MPEG-4 scheme uses Y, U, V coding and a 4:2:0 structure of color information [350]. This means that the luminance is encoded for every pixel, but the color difference information is filtered and decimated to half the luminance resolution, both horizontally and vertically. Thus an image area represented by a block of 16 × 16 luminance pixels requires only 8 × 8 values for U and 8 × 8 values for V. Since the standard uses 8 × 8-pixel blocks for the DCT [349], the MB consists of four blocks of luminance samples (Y) and one block U as well as V samples. Figure 11.13 shows the MB encoding order for four luminance (Y) and two chrominance (U, V) blocks whereas Figure 11.12 shows the spatial relationship between the luminance and color difference samples in YUV format.

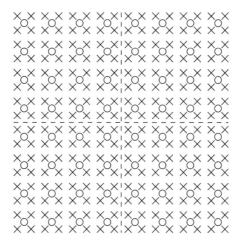

     × Luminance sample
     ○ Chrominance sample
  – – – Block edge

Figure 11.12: Positioning of luminance and chrominance samples in a MB containing four 8 × 8-pixel luminance blocks having a total area of 16 × 16 pixels in a video frame. Both color-difference signals separating the chrominance samples are processed at half the spatial resolution.

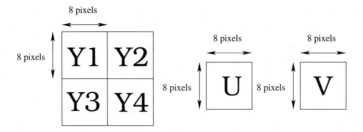

Encoding Order : Y1, Y2, Y3, Y4, U, V

Figure 11.13: Encoding order of blocks in a MB.

The block diagram of an idealized video encoder and decoder has been shown earlier in Figure 11.2 of Section 11.2. In this section, we discuss each individual block of the MPEG-4 video codec in more depth. The MPEG-4 video encoder and decoder are shown in Figures 11.14 and 11.15, respectively. The first frame in a video sequence is encoded in the intra-frame coded mode (I-VOP) without reference to any past or future frames. As seen in Figure 11.2, at the encoder the DCT is applied to each 8×8-pixel luminance and chrominance block. Then each of the 64 DCT coefficients is quantized (Q) in the block. After quantization, the lowest-frequency DCT coefficient, namely the DC coefficient, is treated differently from the remaining coefficients, which are also often referred to as the "Alternating Current" (AC) coefficients. The DC coefficient corresponds to the average luminance intensity of the block considered and it is encoded by using a differential DC component encoding method, employing the DC value of the previous frame as reference, when predicting and encoding the current one. The non-zero quantized values of the remaining DCT coefficients and their locations are "zig-zag"-scanned and run-length or entropy coded by means of Variable-Length Coding (VLC) tables similarly to the techniques shown from the MPEG-1 [30], MPEG-2 [31], and H.263 [28] codecs. More DCT algorithms are discussed in Section 12.4.3. Readers interested in the details of zig-zag scanning are referred to [5], for example.

When considering P-VOP coding, the previous I- or P-VOP frame, namely frame $n-1$, is stored in the reconstructed Frame Buffer (FB) of both the encoder and decoder for frame reference. MC is performed on a MB basis. Hence, only one MV is estimated for the frame VOP-n for a particular MB to be encoded. These motion vectors are encoded and transmitted to the receiver. The motion-compensated prediction error or block residue $D(n)$ seen in Figure 11.14 is calculated by subtracting each pixel in a MB from its motion-shifted counterpart in the previous VOP frame, namely in VOP-$(n-1)$. Then an 8×8-dimensional DCT is applied to each of the 8×8 blocks contained in the MB, followed first by quantization of the DCT coefficients and then by run-length coding and entropy coding, both of which constitute VLC techniques.

The decoder of Figure 11.15 uses the "inverse" routine of the encoder for reproducing a MB of the Nth VOP frame at the receiver. After decoding the variable-length words contained in the video decoder's buffer, the pixel values of the motion prediction error or block residue $D(n)$ are reconstructed with the aid of the Inverse Quantizer (iQ) and inverse DCT blocks of

11.3. MPEG-4: CONTENT-BASED INTERACTIVITY

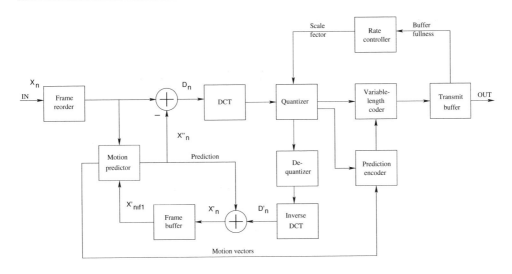

Figure 11.14: Block diagram of the MPEG-4 encoder.

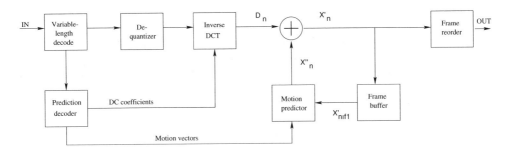

Figure 11.15: Block diagram of the MPEG-4 decoder.

Figure 11.15. The motion-compensated pixels of the previous VOP frame, namely those of VOP-$(n-1)$, are contained in the VOP frame buffer of the decoder, which are added to the motion prediction error $D(n)$ after appropriately positioning them according to the MVs, as seen in Figure 11.15 in order to recover the particular MB of VOP-n.

11.3.3 Shape Coding

A VO can be rectangular or of arbitrary shape. For a rectangular VOP, the encoding process is similar to that of the MPEG-1 [347] and MPEG-2 [348] standard. However, if a VO is of arbitrary shape, a further coding step is necessary prior to motion and texture coding, as illustrated in Figure 11.16 [361]. Specifically, in the MPEG-4 visual standard two types of shape information are considered as inherent characteristics of a VO, binary and gray-scale shape information [362].

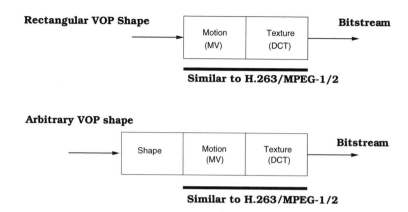

Figure 11.16: MPEG-4 shape coding structures.

11.3.3.1 VOP Shape Encoding

In the MPEG-4 video compression algorithm, the shape of every VOP is coded along with its other parameters, such as its texture and motion vectors. A binary alpha plane defines which pixels within the boundary box belong to the VO at a given instant of time [363]. The VOP shape information, or in this case the binary alpha plane is most commonly represented by a matrix having the same size as the VOP, where each element of the matrix may assume one of two possible values, namely 255 or 0, depending on whether the pixel is inside or outside the video object [362]. If the corresponding pixel belongs to the object, then the element is set to 255, otherwise it is set to 0. This matrix is referred to as a *binary mask* or as a *bitmap*. Figure 11.17 shows the binary alpha plane of the "Miss America" and "Akiyo" VOP.

Before encoding, the binary alpha plane is then partitioned into binary alpha blocks (BABs) of size 16×16 pixels. Each BAB is encoded separately. It is not surprising that a BAB may contain identical values, which are either all 0, in which case the BAB is referred to as a transparent blocks, or all 255, when the BAB is said to be an opaque block. The main MPEG-4 tools used for encoding BABs are the Context-based Arithmetic Encoding (CAE) algorithm and the MC scheme[3] [362]. Inter-frame CAE (InterCAE) and intra-frame CAE (IntraCAE) are the two variants of the CAE algorithm used in conjunction with P- or I-VOPs, respectively. The InterCAE scenario involves motion vectors, which are based on finding and encoding the best-matching position of the previous VOP frame, whereas IntraCAE is used without MC and is referred to as IntraCAE. A BAB of the current VOP may be encoded in one of the following seven possible modes [364].

1. If the entire BAB is flagged as transparent, no shape encoding is necessary at all and hence texture information is not encoded for this BAB.

2. If the entire BAB is flagged as opaque, no shape encoding is necessary at all, but the texture information is encoded for the VOP.

---

[3]Any video sequence can be viewed as a set of consecutive snapshots of still images of a scene. Therefore, the consecutive snapshots are correlated. It is this form of predictability or redundancy that the motion prediction mechanism is exploiting. In a basic form, we can simply use the previous frame for predicting the current frame.

11.3. MPEG-4: CONTENT-BASED INTERACTIVITY

Figure 11.17: MPEG-4 binary plane of the "Miss America" and "Akiyo" frames.

3. The BAB is encoded using IntraCAE without any reference to previous frames and motion compensation.

4. The block is not updated with respect to the same block of the previous frame if we have zero Motion Vector Difference (MVD) for the block concerned between the previous and current frame.

5. Even if the MVD is zero, the content of the block may be updated. In this case, InterCAE is used for encoding the block update.

6. The MVD is non-zero and no update is necessary, thus neither the texture nor the shape of the block is encoded.

7. The MVD is non-zero and the block has to be updated. In this case, InterCAE is used for encoding both the texture and the shape of the block.

Modes 1 and 2 require no shape coding. For mode 3, shape is encoded using IntraCAE. For modes 4-7, motion estimation and MC are employed. The MVD is the difference between the shape MV and its predicted value (MVP). This predicted value is estimated from either the neighboring shape MVs or from the co-located texture MVs. When the mode indicates that no update is required, then the MV is simply used to copy an appropriately displaced 16×16-pixel block from the reference binary alpha plane to the current BAB. If, however, the mode indicates that an update is required, then the update is coded using InterCAE.

11.3.3.2 Gray-scale Shape Coding

Instead of having only 0 and 255 as possible values for the shape encoding matrix, the shape encoding may assume a range of values spanning from 0 to 255, which represent the degree of transparency for each pixel, where 0 corresponds to a transparent pixel and 255 represents an opaque pixel. As regards to the values between 0 and 255, the smaller the value, the more transparent the pixel. Similarly, the larger the value, the more opaque the pixel [363]. This scenario is referred to as gray-scale shape encoding, rather than binary shape encoding. These values may also be stored in a matrix form for representing the shape of VOP. The gray-scale shape information is also encoded using a block-based DCT similar to the conventional approach used in texture coding.

11.4 Scalability of Video Objects

In terms of the scalability of the text, images, and video to be encoded, the MPEG-4 standard provides a procedure for the supporting complexity-based, spatial, temporal, and quality scalability [364], which is the most frequently used parlance for indicating that the MPEG-4 codec may be configured in a plethora of different coding modes for the sake of striking different trade-offs in terms of the achievable implementation complexity, spatial and temporal resolution, video quality, etc. More specifically, the trade-offs are summarized as follows.

- Complexity-based scalability of the encoder and decoder facilitates the encoding of images or video at different levels of algorithmic complexity, where the complexity affects the quality of the reconstructed object.

- Spatial scalability makes it possible for the bitstream to be decoded in subsets so that the spatial resolution of the objects would be improved upon decoding each consecutive subset. A maximum of three specific scalability levels are supported for video objects and 11 different levels for still images as well as for text.

- The philosophy of temporal scalability is similar to that of spatial scalability, except that the video is displayed at a reduced temporal resolution, rather than reduced spatial resolution, for example at a lower frame-rate.

- Quality-motivated scalability implies that a bitstream could be separated into several layers, corresponding to different bitrates. When the layers are decoded, the quality is determined by the number of layers that was used.

Important considerations for video coding schemes to be used within future wireless networks are the achievable compression efficiency, the attainable robustness against packet loss, the ability to adapt to different available bandwidths, different amounts of memory, and computational power for different mobile clients, etc. Scalable video coding schemes have been proposed in the literature [73, 109, 119–121, 125], which are capable of producing bitstreams decodable at different bitrates, requiring different computational power and channel bitrate.

An example associated with two layers of scalability coding is shown in Figure 11.18. The enhancement layers are encoded by a motion-compensated hybrid codec, where the DCT

11.4. SCALABILITY OF VIDEO OBJECTS

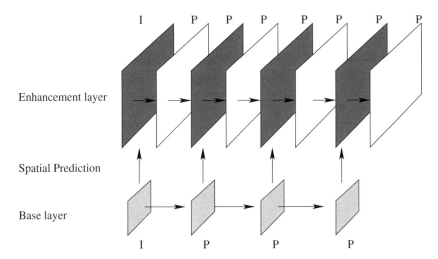

Figure 11.18: Example of a possible configuration of the spatio-temporal resolution pyramid of the scalable video codec. One base layer relying on an intra-frame and three interpolated inter-coded frames are shown, supplemented by an enhancement layer having twice the spatial and temporal resolution. The horizontal and vertical arrows denote temporal and spatial prediction, respectively.

has been replaced by lattice vector quantization of the Motion Compensated Error Residual (MCER). This approach leads to a coding efficiency attained by the layered coding scheme, which is comparable to that of single-layer codecs [344]. Since the ability to decode an enhancement layer depends on the reception of the base layer and lower enhancement layers, an efficient transmission scheme is expected to ensure that these layers are transmitted such that the associated packet loss is kept as low as possible even for high overall packet loss rates. In addition to the ability to adapt to different clients, we can also ensure a sufficiently graceful degradation of the associated video quality in the case of packet loss in this scenario.

As mentioned earlier, MPEG-4 provides both spatial and temporal scalability at the object level [25, 364]. In both of these scenarios this technique is invoked for the sake of generating a *base layer*, representing the lowest quality to be supported by the bitstream, and one or more *enhancement layers*. These layers may all be produced in a single encoding step. The scaling can be implemented in two different ways. When there are known bandwidth limitations, the different-rate versions of the bitstream may be used that include only the base layer, or the base layer plus lower-order enhancement layers. Alternatively, all layers may be transmitted and the scaling decision may be left for the decoder's discission. If the display device at the receiver side has a low resolution, or if the available computational resources are insufficient, the enhancement layers may be ignored.

Figure 11.19 shows the concept of an encoder exploiting spatial scalability; in this case at just two levels. The input VOP is down-converted to a lower resolution, resulting in the base layer. This layer is encoded and then a decoder reconstructs the base-layer VOP, as it will appear at the decoder's display. This VOP is then up-converted to the same resolution as the input, and a subtraction operation generates the difference in comparison to the original image. These are separately encoded in an enhancement-layer encoder. Note that each stream of the encoded VOPs forms a VOL. The base-layer VOL uses both intra- and inter-frame coding, but the enhancement layer uses only predictive coding. The base-layer VOPs are used as references, as shown in Figure 11.18

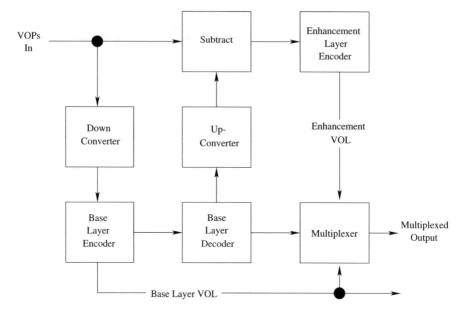

Figure 11.19: Spatially scalable encoder for a single enhancement layer.

11.5 Video Quality Measures

In this section, the objective video quality measure used during our investigations of the various wireless video transceivers is defined. Quantifying the video quality is a challenging task, because numerous factors may affect the results. Video quality is inherently *subjective* and our human perception is influenced by many factors.

11.5.1 Subjective Video Quality Evaluation

Several test procedures designed for subjective video quality evaluation were defined in the ITU-R Recommendation BT.500-11 [365]. A commonly-used procedure outlined in the standard is the so-called Double Stimulus Continuous Quality Scale (DSCQS) method, in which an assessor is presented with a pair of images or short video sequences A and B, one after the other, and is asked to assign both A and B a quality score on a continuous scale having five intervals ranging from "Excellent" to "Bad". In a typical test session, the assessor is shown a series of pairs of sequences and is asked to grade each pair. Within each pair of

11.5. VIDEO QUALITY MEASURES

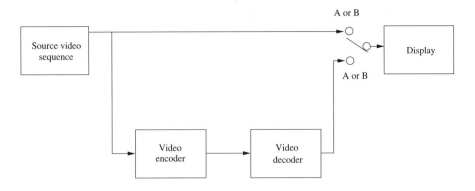

Figure 11.20: DSCQS testing system.

sequences, one sequence is an unimpaired "reference" sequence and the other is the same sequence, modified by a system or process under test. Figure 11.20 shows an experimental setup for the testing of a video codec, where the original sequence is compared with the same sequence after encoding and decoding. In addition, the order in which the sequence "A" and "B" are presented is random. The evaluation of subjective measures, such as the DSCQS measure, is time consuming and expensive. Hence, contriving an objective video quality measure that is capable of reliably predicting the subjective quality is desirable.

11.5.2 Objective Video Quality

The designers and developers of video compression and processing systems rely heavily on objective quality measures. The most widely used measure is the Peak Signal-to-Noise Ratio (PSNR). The PSNR is measured on a logarithmic scale and depends on the normalized Mean Squared Error (MSE) between the original and the reconstructed as well as potentially channel-impaired image or video frame, relative to $(2^n - 1)^2$, namely normalized by the square of the highest possible pixel value in the image, where n is the number of bits per image sample, yielding:

$$\text{PSNR} = 10 \log_{10} \frac{(2^n - 1)^2}{\text{MSE}}. \tag{11.1}$$

The PSNR may be conveniently calculated. However, its evaluation requires the availability of the unimpaired original image or video signal for comparison, which may not be available.

Unfortunately the PSNR defined above does not always constitute a reliable image quality measure. For example, if the received image is shifted by one pixel compared with the original image, the human eye would hardly notice any difference, while the PSNR objective measures would indicate a more substantial degradation in quality.

Nonetheless, owing to its appealing implicity, in this book, the PSNR is used as the predominant image quality measure.

11.6 Effect of Coding Parameters

The previous sections described the MPEG-4 encoding and decoding process with the aid of block diagrams shown in Figures 11.14 and 11.15. In video compression, some of the encoding parameters seen in the context of Figures 11.14 and 11.15 may directly affect the resultant reconstructed video quality. Therefore, in this section we briefly demonstrate how these parameters may affect the encoded/decoded video quality.

The MPEG-4 source code used in our simulations was a modified version of the software implementation provided by the Mobile Multimedia Systems (MoMuSys) Pan-European project. Simulations were carried out for the sake of characterizing the achievable performance of the MPEG-4 codec. Most of our simulations used the popular "Akiyo", "Miss America", "Suzi", or "Foreman" 144×176 pixel resolution QCIF video sequence at a transmission frame scanning rate of 10 or 30 frames/s. In addition, the length of the video sequence used in our simulations was 100 frames and, apart from the first frame, no intra-frame coded update was used, unless otherwise stated. Our results are shown below in graphical form in Figures 11.25–Figure 11.29.

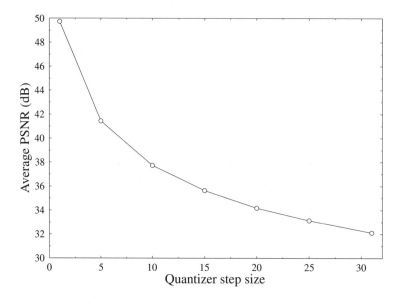

Figure 11.21: MPEG-4 average video quality (PSNR) versus quantizer step size ranging from 1 to 31 using the 144×176-pixel QCIF "Miss America" sequence.

Our investigations commenced by demonstrating the effect of the quantizer index employed in the MPEG-4 encoder. In video compression, the quantization operation is typically constituted by a "forward quantizer" in the encoder and an "inverse quantizer" in

11.6. EFFECT OF CODING PARAMETERS

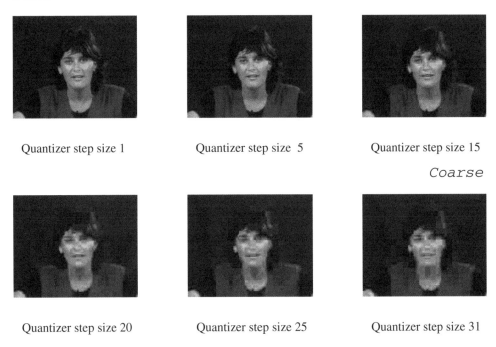

Figure 11.22: The effect of quantization step size on the MPEG-4 encoder in the second frame of the 144 × 176-pixel QCIF "Miss America" sequence.

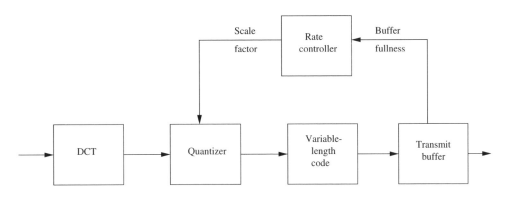

Figure 11.23: Rate control in MPEG-4.

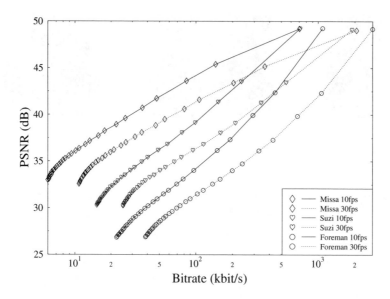

Figure 11.24: Image quality (PSNR) versus coded bitrate performance of the MPEG-4 codec for various QCIF-resolution video sequences scanned at 10 and 30 frames/s.

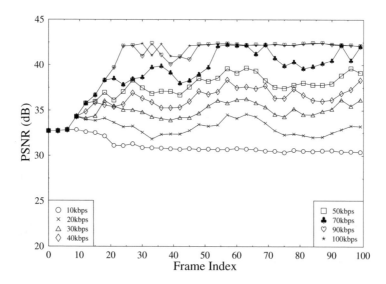

Figure 11.25: MPEG-4 video quality (PSNR) versus frame index performance for the bitrate range spanning from 10 to 100 kbit/s using the 144 × 176-pixel QCIF "Akiyo" sequence.

11.6. EFFECT OF CODING PARAMETERS

the decoder. A critical parameter is the quantizer step size. If the step size is large, a highly compressed bitstream is typically generated. However, the reconstructed values provide a crude approximation of the original signal. If the step size is small, the reconstructed values match the original signal more closely at the cost of a reduced compression efficiency. The effect of different quantizer step sizes is demonstrated in Figure 11.21 in terms of the video PSNR, while the associated subjective video quality is illustrated in Figure 11.22.

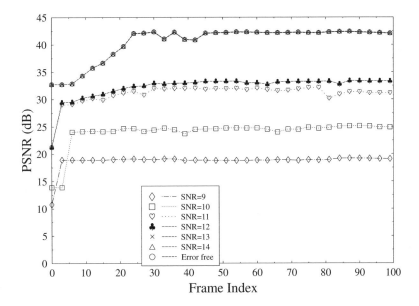

Figure 11.26: Decoded video quality (PSNR) versus frame index when transmitting over an AWGN channel using a BPSK modem. These results were recorded for the "Akiyo" video sequence at a resolution of 144 × 176 pixels and 100 kbit/s video bitrate. The 81st decoded video frame was shown in Figure 11.28.

Let us now consider the effect of different bitrates, although the effects of the video bitrate and quantizer step size are highly correlated. In order to achieve a given target bitrate, the encoded video frame buffer is used for adjusting the quantizer step size. More explicitly, the encoder produces a variable-rate stream and the buffer may be emptied at a near-constant transmission rate. If the encoder produces bits at a low rate, the buffer may become empty. In contrast, if the encoded bitrate is too high, the buffer may overflow and the data may become irretrievably lost. For the sake of avoiding these two problems, we typically feed a measure of buffer fullness to the rate controller, which reacts by appropriately adjusting the quantization step size of the DCT coefficients of the video blocks that have not yet been encoded. Figure 11.23 shows the rate control mechanism of the basic MPEG-4 encoder. The actions of the rate-controller are characterized in Figures 11.24 and 11.25. More specifically, Figure 11.24 characterizes the average PSNR versus bitrate performance, while Figure 11.25 shows the PSNR versus the video frame index. The bitrate range used in these simulations spanned between 10 kbit/s and 1000 kbit/s. As expected, it can be seen from the graphs that

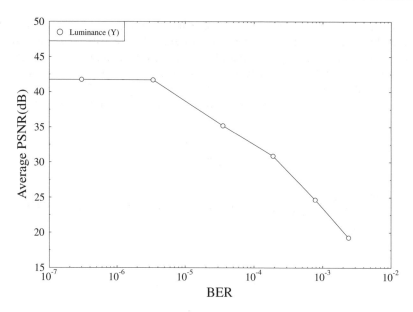

Figure 11.27: Average video quality (PSNR) versus BER for transmission over an AWGN channel using BPSK modulation and the 144 × 176-pixel QCIF "Akiyo" sequence.

the video quality increases upon increasing the video bitrate. Figure 11.25 plots the PSNR versus frame index performance for the "Akiyo" sequence scanning at the video frame rate of 10 frames/s in bitrate range spanning from 10 to 100 kbit/s.

So far, our experiments were carried out in an error-free environment for various encoding parameters. The next experiments were carried out by transmitting the MPEG-4 bitstream over an Additive White Gaussian Noise (AWGN) channels using a simple Binary Phase Shift Keying (BPSK) modem. Note that the system used in this experiment employed no channel coding. Figure 11.26 portrays the PSNR versus video frame index performance for transmission over the AWGN channel for the channel Signal-to-Noise Ratio (SNR) range spanning from 9 to 14 dB. Figure 11.27 shows the average luminance (Y) PSNR versus Bit Error Ratio (BER). Observe in Figures 11.26 and 11.27 that as expected, the decoded picture quality increases as the channel SNR increases. At low channel SNRs, the channel inflicts a high BER, hence the decoded video frame sequence will be highly degraded. This is demonstrated in Figure 11.28. Viewing the codec's performance from a different perspective, Figure 11.29 shows the average PSNR degradation versus BER.

11.7 Summary and Conclusion

In Section 11.1, we commenced our discourse with a brief historical perspective on the development of the MPEG-4 visual standard, the MPEG-4 visual profiles and features. These include the diverse set of coding tools described in the standard that are capable of supporting a wide range of applications, such as efficient coding of video frames, video coding for

11.7. SUMMARY AND CONCLUSION

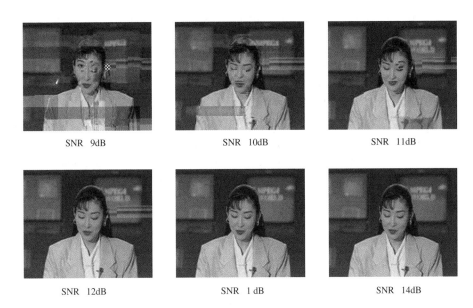

Figure 11.28: Image degradation due to corruption of the MPEG-4 bitstream over AWGN channels using a BPSK modem in the 81st frame of the 144×176-pixel QCIF "Akiyo" sequence. The BER within this frame at the above SNRs were between 10^{-7} and 10^{-3}, while the PSNR ranged from 41.74 to 19.27 dB.

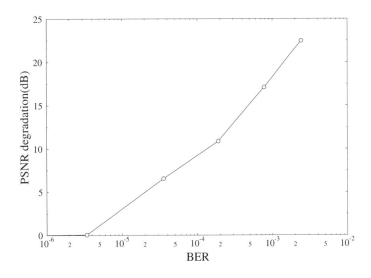

Figure 11.29: Decoded video quality degradation in terms of PSNR versus BER for transmission over an AWGN channel using BPSK modulation due to corruption of the MPEG-4 bitstream during transmission. The results were recorded for the QCIF resolution "Akiyo" video sequence.

error-prone transmission networks, object-based coding and manipulation, as well as the interactive visual applications.

In Section 11.3, we introduced the MPEG-4 visual encoding standard that supports an object-based representation of the video sequence. This allows convenient access to and manipulation of arbitrarily shaped regions in the frames of the video sequence.

Scalable coding is another feature supported by the MPEG-4 codec. Scalability is supported in terms of generating several layers of information. In addition to the base layer, enhancement layers may be decoded, which will improve the resultant image quality either in terms of the achievable temporal or spatial resolution.

In Section 11.5 the PSNR objective measure was introduced for quantifying the video quality. Finally, the effect of certain important video encoding parameters were discussed in Section 11.6, commencing with an emphasis on the *quantizer step size* parameter, which also affects the resultant video *target bitrate*.

The specifications of MPEG-4 standard continue to evolve with the addition of new tools, such as the recently introduced profile supporting video streaming [25]. However, amongst developers and manufacturers, the most popular elements of the MPEG-4 Visual standard to date have been the simple and the advanced simple profile tools, which were summarized in Table 11.1. Having studied the MPEG-4 codec, let us know focus our attention on one of its relatives, namely the H.264 [27] codec, in the next chapter.

Comparative Study of the MPEG-4 and H.264 Codecs

J-Y. Chung and L. Hanzo

12.1 Introduction

In this chapter, we provide a comparative study between the MPEG-4 and H.264 codecs and briefly outline the role of the International Organization for Standardization (ISO) Moving Picture Experts Group (MPEG) and International Telecommunications Union (ITU) Video Coding Experts Group (VCEG) groups in developing these standards. Creating, maintaining and updating the ISO/IEC 14496 (MPEG-4) standards is the responsibility of the MPEG under the auspices of the ISO. The H.264 Recommendation (also known as MPEG-4 Part 10, "Advanced Video Coding" and formerly known as H.26L [20]) emerged as a joint effort of the MPEG and the VCEG, another study group of the ITU.

Prior to these activities, MPEG developed the highly successful MPEG-1 and MPEG-2 standards for encoding of both video and audio, which are now widely used for the transmission and storage of digital video, as well as the MPEG-4, MPEG-7, and MPEG-21 standards. In contrast, VCEG was responsible for the first widely-used videotelephony standard (H.261) and its successors, the H.263, H.263+, and H.263++ and schemes as well as for the early development of the H.264 codec. The two groups set up the collaborative Joint Video Team (JVT) for the sake of finalizing the H.264 proposal and for creating an international standard (H.264/MPEG-4 Part 10) published by both ISO/International Electrotechnical Commission (IEC) and International Telecommunications Union—Telecommunications Standardization Sector (ITU-T).

12.2 The ITU-T H.264 Project

The H.264 standard is the result of a recent joint research initiative of the ITU-T VCEG and the ISO/IEC MPEG standardization committee. The H.264 codec offers substantially improved coding efficiency at the same video quality as the MPEG-4 or H.263

schemes [20]. The main goal of this new ITU-T H.264 standardization effort was that of enhancing the achievable compression ratio, while providing a "network-friendly" packet-based video representation addressing both real-time "conversational" videotelephony and "non-conversational", i.e. storage, broadcast or streaming type applications [20].

One of the fundamental concepts of H.264 is the separation of the design into two distinct layers, namely the video coding layer and the network adaptation layer. The video coding layer is responsible for efficiently representing the video content, while the network adaptation layer is responsible for packaging the coded data in an appropriate manner for transmission over the network [366]. In this chapter we focus our attention mainly on the video coding layer.

12.3 H.264 Video Coding Techniques

The coding techniques employed by the H.264 codec are to adopt similar to the schemes that have been successfully employed in earlier video coding standards. Hence, the same basic functional elements such as the prediction, transform, quantization, and entropy coding stages of the previous standards such as the MPEG-1 [30], MPEG-2 [31], MPEG-4 [25], H.261 [29], and H.263 [28] codecs may be readily identified. Moreover, there are some important changes in each functional element of the H.264 scheme.

The schematic diagram of the H.264 encoder is depicted in Figure 12.1. The encoder consists of two data flow paths, a "forward" path oriented from left to right and a "reconstruction" path evolving from right to left. The H.264 corresponding decoder is shown in Figure 12.2, which mimics the structure of the H.264 encoder's local decoder.

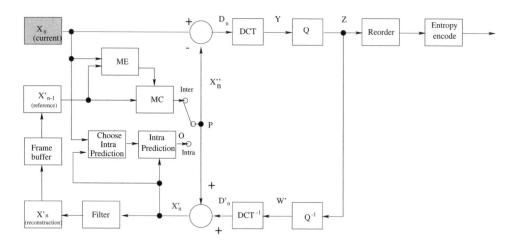

Figure 12.1: Schematic of the H.264 encoder.

12.3.1 H.264 Encoder

An input frame X_n is presented for encoding in Figure 12.1. The frame is processed in units of a Macroblock (MB) corresponding to 16×16 pixels in the original image. As in the previous standards, each MB is encoded in either the intra- or inter-frame coding mode. In either case, a predicted MB **P** is formed based on a reconstructed frame. In intra-frame mode, **O** is formed from samples in the current frame, namely frame n that has been previously encoded, decoded and reconstructed, resulting in the reconstructed frame X'_n in Figure 12.1. In contrast, in the inter-frame coding mode, **P** is formed by motion-compensation-aided prediction from one or more reference frame(s); in Figure 12.1, the reference frame is shown as the previous encoded frame X'_{n-1}. Note that the prediction of each MB may be formed by a single or up to five past or future frames that have already been encoded, reconstructed, and stored in the reconstruction frame buffer.

As seen in Figure 12.1, the predicted MB **P** is subtracted from the current MB for the sake of producing a Motion Compensated Error Residual (MCER) or difference MB D_n, which is then discrete cosine transformed and quantized to produce a set of quantized transform coefficients (Tcoeff) **Z**. These Tcoeffs are re-ordered using zig-zag scanning [5] and entropy encoded. The entropy encoded Discrete Cosine Transform (DCT) coefficients, together with the accompanying side information such as the MB prediction mode, quantizer step size, motion vector information required for decoding the MB form the compressed bitstream. After concatenating the control headers the encoded information is then transmitted via the channel in a compressed bitstream format.

In the reconstruction path, the quantized MB coefficients **Z** are decoded in order to reconstruct a frame, which may then be used for the encoding of further MBs. The coefficients **Z** are inverse quantized in the block Q^{-1} of Figure 12.2 and inverse transformed in the block T^{-1} for producing the reconstructed MCER MB D'_n. Note that this is not identical to the original difference MB D_n; the quantization process introduces granular effects and therefore D'_n is an approximate replica of D_n. The predicted MB **P** is then added to D'_n for creating an approximate replica of the original MB, namely the reconstructed MB X'_n. Finally, a smoothing filter is applied to reduce blocking effects and the reconstructed reference frame is created from a series of MBs X'_n, depending on how many of the previous MBs were involved for creating the MCER.

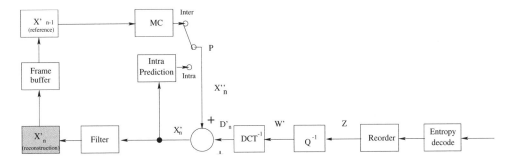

Figure 12.2: Schematic of the H.264 decoder.

12.3.2 H.264 Decoder

The decoder receives a compressed bitstream from the channel. The bitstreams are entropy decoded and appropriately ordered for the sake of producing a set of DCT coefficients **Z**. These are then inverse quantized and also inverse transformed for the sake of producing the decoded MCER D'_n. In the absence of transmission errors this quantity is identical to D'_n generated by the encoder of Figure 12.1. In the decoder, the first group of information to be decoded is the header information section of the received bitstream. The decoder then creates a predicted MB **P**, which is again, in the absence of transmission errors, identical to the original predicted frame **P** formed in the encoder. This predicted frame **P** is then added to D'_n for the sake of producing X'_n, which is then smoothed by a filter for the sake of creating the decoded MB X'_n.

12.4 H.264 Specific Coding Algorithm

Having described the encoding and decoding process of the H.264 encoder and decoder, in this section we investigate some additional features of the H.264 standard. Some examples of the novel techniques first introduced into standard codecs are constituted, for example, by spatial prediction [101] in intra-frame coding, motion compensation using an adaptive block size [98], 4×4-pixel integer DCT [367], Universal Variable Length Coding (UVLC) [368], Context-based Adaptive Binary Arithmetic Coding (CABAC) [369], and de-blocking filtering [66], etc.

12.4.1 Intra-frame Prediction

Intra-frame coded pictures are typically encoded by directly applying the DCT to the different MBs in the frame. The intra-frame encoded pictures typically generate a high number of bits, since no temporal redundancy removal is used as part of the encoding process. In order to increase the achievable efficiency of the intra-frame coding process in the context of the H.264 codec, the spatial correlation between adjacent MBs of a given frame is exploited. The associated philosophy is based on the observation that statistically speaking the adjacent MBs tend to have similar luminance and color difference signals. Therefore, as a first step in the H.264 encoding process of a given MB, one may predict the MB of interest from the surrounding MBs of the intra-frame coded picture, which are the MBs located at the top and left of the given MB, because those MBs would have already been encoded.

For the luminance of the corresponding MBs intra-frame prediction may be performed for each 4×4-pixel sub-block or for a 16×16-pixel MB. The H.264 codec offers a total of eight different modes for the prediction of the 4×4-pixel luminance blocks. The predicted block is calculated based on the samples labeled A–Q[1] in Figure 12.3, where the samples A–Q are at the edge of the neighboring sub-blocks. Specifically, we may observe in Figure 12.3 that samples A–H are to the top and samples I–J are to the left of the "current" MB which have

---

[1]Note that each letter A to Q in Figure 12.3 represents a single pixel in the MB, where A to H are the horizontal neighboring pixels at the top of the current MB. Furthermore, I to P are the vertical neighboring pixels that are to the left of the current MB, while Q belongs to the bottom right pixel of another MB positioned at the top left corner of the current MB.

12.4. H.264 SPECIFIC CODING ALGORITHM

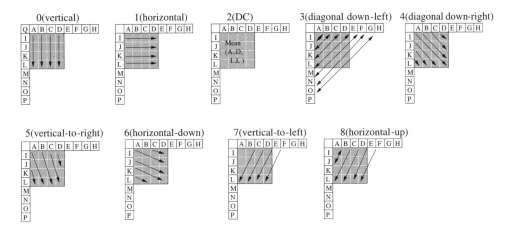

Figure 12.3: 4 × 4-pixel intra-frame prediction modes.

previously been encoded and reconstructed. The intra-frame prediction process is illustrated in Table 12.1 and Figure 12.3. The arrows in Figure 12.3 indicate the direction of prediction in each mode.

Table 12.1: 4 × 4-pixel Sub-block Intra-frame Prediction Modes

Modes	Description
Mode 0 (Vertical)	The pixels A,B,C,D are used for vertical prediction.
Mode 1 (Horizontal)	The left pixels I,J,K,L are used for horizontal prediction.
Mode 2 (DC)	All pixels are predicted by the mean of A–D and I–L.
Mode 3 (Diagonal Down-left)	The pixels are diagonally predicted at a angle of $45°$ between the lower-left and upper-right corners of the block.
Mode 4 (Diagonal Down-right)	The pixels are at an angle of $45°$ downwards with an orientation of left to right.
Mode 5 (Vertical Left-to-right)	Prediction at an angle of $22.5°$ with respect to the vertical axis from left to right.
Mode 6 (Horizontal Down)	Prediction at an angle of $22.5°$ horizontal from left to right.
Mode 7 (Vertical Right-to-left)	Prediction at an angle of $22.5°$ to the left of vertical.
Mode 8 (Horizontal Up)	Extrapolation at an angle of $22.5°$ above horizontal.

For picture regions exhibiting less spatial detail, H.264 also supports intra-frame coding based on 16×16-pixel blocks. As an alternative to the 4×4-pixel intra-frame prediction modes described above, the entire 16 × 16-pixel luminance component of a MB can be predicted in one of four possible prediction modes, as shown in Table 12.2 and Figure 12.4.

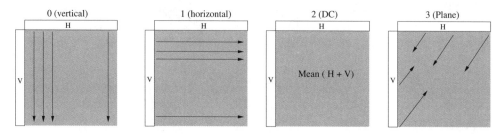

Figure 12.4: 16 × 16-pixel intra-frame prediction modes.

Table 12.2: 16 × 16-Pixel Intra-frame Prediction Modes

Modes	Description
Mode 0 (Vertical)	Prediction from the upper pixels (H).
Mode 1 (Horizontal)	Prediction from the left pixels (V).
Mode 2 (DC)	All pixels are predicted from the mean of the left and upper samples.
Mode 3 (Plane)	A linearly sloping "plane" function is fitted to the upper and left-hand side pixels H and V. This prediction mode performs well in areas of smoothly varying luminance.

Finally, the prediction mode of each block is encoded by assigning shorter prediction-mode signaling symbols to more likely modes, where the probability of each mode is determined based on the modes used for encoding the surrounding blocks.

12.4.2 Inter-frame Prediction

In inter-frame coding motion estimation and compensation is employed for the sake of exploiting the temporal redundancies that exist between successive frames, hence providing an efficient encoding of video sequences. The motion estimation regime of the H.264 codec supports most of the features found in earlier video standards [366], however its efficiency was further improved. The following sections describe in detail the four main motion estimation modes that are used in H.264, which are (1) the employment of various shapes and block sizes, (2) the use of high-precision sub-pixel motion vectors, (3) the utilization of multiple reference frames, and (4) the employment of de-blocking or smoothing filters in the prediction loop.

12.4.2.1 Block Sizes

Motion compensation of each of the 16 × 16-pixel MB can be performed using a number of different block or sub-block sizes and shapes. These are depicted in Figure 12.5. As shown in the figure, seven different block sizes are supported in H.264, namely 16 × 16-, 8 × 16-,

12.4. H.264 SPECIFIC CODING ALGORITHM

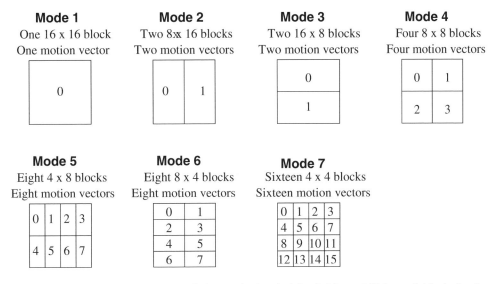

Figure 12.5: Different inter-frame prediction modes invoked for dividing a MB into sub-blocks for the sake of improving the accuracy of motion estimation in H.264.

16×8-, 8×8-, 4×8-, 8×4-, and 4×4-pixel blocks [366]. Individual motion vectors can be transmitted for blocks as small as 4×4, therefore up to a total of 16 motion vectors may be transmitted for a single MB. The advantage of having smaller motion compensation blocks is that it has the potential of improving the quality of prediction. In particular, the employment of small blocks allows the model to handle fine motion details and hence typically results in an improved subjective video quality, because it prevents the occurrence of blocking artifacts. Table 12.3 shows the luminance Peak Signal-to-Noise Ratio (PSNR) results of the various motion compensation search modes. Experiments were conducted for three different Quarter Common Intermediate Format (QCIF) resolution video sequences, namely for the "Foreman", "Suzi", and "Miss America" sequences. Motion compensation using a small 4×4-pixel block yields a better picture quality in terms of PSNR, than the larger 16×16-pixel block size.

12.4.2.2 Motion Estimation Accuracy

The prediction capability of the motion compensation algorithm used in the H.264 codec may be further improved by allowing motion vectors to be determined with a higher spatial accuracy than in the existing hybrid coding standards. Quarter-pixel accurate motion compensation is currently the lowest-accuracy form of motion compensation in H.264, in contrast with prior standards based primarily on half-pixel accuracy, with quarter-pixel accuracy only available elsewhere in the newest version of MPEG-4, while eighth-pixel accuracy is being adopted as a feature that will likely be useful for increased coding efficiency at high bitrates and high video resolutions.

Table 12.3: Luminance PSNR Results for the Various Motion Compensation Search Modes

	Y-PSNR (dB)		
Block size	"Foreman"	"Suzi"	"Miss America"
16 × 16	35.39	37.12	40.01
16 × 8	35.55	37.20	40.14
8 × 16	35.59	37.22	40.22
8 × 8	35.85	37.34	40.35
8 × 4	35.98	37.41	40.42
4 × 8	36.00	37.43	40.44
4 × 4	36.01	37.45	40.45

12.4.2.3 Multiple Reference Frame Selection for Motion Compensation

The H.264 standard also offers the advanced option of having multiple reference frames in inter-frame picture coding. Up to five different reference frames could be selected, resulting in an improved subjective video quality and more efficient encoding of the video frame under consideration. Moreover, using multiple reference frames might assist in rendering the H.264 coded bitstream more error-resilient. However, from an implementation point of view, an additional processing delay, increased implementational complexity, and higher memory requirements are imposed at both the encoder and decoder. Table 12.4 shows the achievable PSNR results for different numbers of reference pictures. As can be seen in Table 12.4, the video quality of the codec relying on an increased number of reference frames for motion compensation is increased.

Table 12.4: Luminance PSNR for Different Numbers of Frames Invoked During Motion Compensation

Number of reference frames	Y-PSNR (dB)		
	"Foreman"	"Suzi"	"Miss America"
1	36.06	37.34	40.32
2	36.15	37.42	40.47
3	36.18	37.44	40.48
4	36.23	37.44	40.48
5	36.24	37.44	40.48

12.4.2.4 De-blocking Filter

The H.264 codec specifies the employment of a deblocking or smoothing filter that mitigates the visual affects of the horizontal and vertical block edges which may appear as a consequence of truncating the 2D video frame at the block edges before the DCT is invoked. More explicitly, this truncation operation imposes 2D Gibbs oscillation in the

12.4. H.264 SPECIFIC CODING ALGORITHM

frequency domain. The filtering operation is generally based on 4×4-pixel block boundaries, in which two pixels on either side of the boundary may be updated using a three-tap smoothing filter. The rules of applying the de-blocking filter are intricate and quite complex. Therefore, substantial research efforts have been dedicated to reducing the complexity of the de-blocking filter, which is expected to further halve before the H.264 standard is finalized.

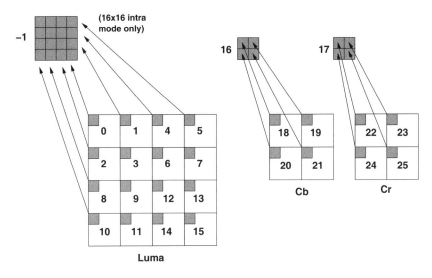

Figure 12.6: Scanning order of residual blocks within a MB.

12.4.3 Integer Transform

Following the prediction process, each Motion Compensated Error Residual (MCER) MB is transformed, quantized, and entropy coded. Earlier standards such as MPEG-1, MPEG-2, MPEG-4, and H.263 used the well known 8×8 pixel DCT [14, 370] as their basic transform. The H26L codec is unique in that it employs a purely integer valued spatial transform, which is an approximation of the DCT. The transform process operates on 4×4-pixel blocks of MCER data. The H.264 codec uses three different transforms, depending on the type of MCER data as detailed in the following.

(1) Transformation of 4×4-pixel luminance DC coefficients in intra-frame MBs. The order of the transformation steps within a MB of the H.264 codec is shown in Figure 12.6. The numerical values indicated in the figure represent the order of encoding for each block in a MB. More explicitly, if the MB is coded in the 16×16-pixel intra-frame coded mode, the block labeled as "-1" in Figure 12.6 is transformed first. This block contains the DC coefficient of each 4×4 luminance (luma) block (otherwise known as a Y block). Then the luma residual blocks labeled 0–15 are transformed with their DC coefficients set to zero in a 16×16-pixel intra-frame coded MB.

(2) Transformation of a 2×2 array of chroma DC coefficients. Blocks 16 and 17 consist of a 2×2 array of DC coefficients representative of the C_b and C_r chroma components, respectively, as seen in Figure 12.6.

(3) Transformation of all other 4 × 4 blocks (both luma and chroma) in the residual data. Finally, the chroma residual blocks 18–25 having zero DC coefficients are transformed.

The 4 × 4 block size assists in reducing blocking and ringing artifacts, while the employment integer-valued operations eliminates any mismatch between the encoder and decoder in the inverse transform. The 4 × 4-pixel integer transform is based on the DCT, although there are some fundamental differences as follows.

1. It is an integer-valued transform where all operations can be carried out using an integer-valued arithmetic, which results in no loss of accuracy.

2. The inverse transform is fully specified in the H.264 standard [366] and provided that this specification is followed, no mismatch occurs between the encoder and decoder.

The entire process of integer transformation and quantization may be carried out using a 16-bit integer-valued arithmetic and only a single multiplication per coefficient is required without any loss of accuracy.

12.4.3.1 Development of the 4 × 4-pixel Integer DCT

The DCT was developed by Ahmed and Rao in 1974 [371]. There are four slightly different versions of the DCT [350] and the one commonly used for video coding is referred to as DCT type II or simply as DCT-II. The 2D DCT-II of an $N \times N$ block of pixels is given by [14, 350]

$$F(U,V) = C(u)C(v) \sum_{x=0}^{N-1} \sum_{y=0}^{N-1} f(x,y) \cos\left(\frac{(2x+1)u\pi}{2N}\right) \cos\left(\frac{(2y+1)v\pi}{2N}\right), \quad (12.1)$$

where $f(x,y)$ is the pixel value at location (x,y) within the block, $F(U,V)$ is the corresponding transform coefficient, where we have $0 \leq u, v, x, y \leq N-1$, and

$$C(u) = C(v) = \begin{cases} \sqrt{\dfrac{1}{N}} & u,v = 0, \\ \sqrt{\dfrac{2}{N}} & \text{otherwise.} \end{cases} \quad (12.2)$$

The DCT operation of Equation 12.1 can be expressed in terms of matrix multiplications as

$$\mathbf{Y} = \mathbf{AXA}^{\mathrm{T}}, \quad (12.3)$$

where $\mathbf{X}$ represents the original image block and $\mathbf{Y}$ represents the resultant DCT coefficients. The elements of $\mathbf{A}$ are defined for an $M \times N$-pixel image block as follows:

$$A_{mn} = k_n \cos\left[\frac{(2m+1)n\pi}{2N}\right], \quad m = 0, 1, \ldots, M-1, \ n = 0, 1, \ldots, N-1, \quad (12.4)$$

12.4. H.264 SPECIFIC CODING ALGORITHM

where

$$k_n = \begin{cases} \sqrt{\dfrac{1}{N}} & n = 0, \\ \sqrt{\dfrac{2}{N}} & n = 1, 2, \ldots, N-1. \end{cases} \quad (12.5)$$

For example, for a 4×4-pixel image block size, Equation 12.4 becomes

$$A_{mn} = k_n \cos\left[\frac{(2m+1)n\pi}{8}\right], \quad m = 0, 1, 2, 3, \; n = 0, 1, 2, 3, \quad (12.6)$$

where we have

$$k_n = \begin{cases} \dfrac{1}{2\sqrt{2}} & n = 0, \\ \dfrac{1}{2} & n = 1, 2, 3. \end{cases} \quad (12.7)$$

Let us now describe the DCT of a 4×4-pixel array $\mathbf{X}$. From Equation 12.3 we arrive at

$$\mathbf{Y} = \mathbf{A}\mathbf{X}\mathbf{A}^{\mathrm{T}}$$

$$= \begin{bmatrix} a & a & a & a \\ b & c & -c & -b \\ a & -a & -a & a \\ c & -b & b & -c \end{bmatrix} \begin{bmatrix} x_{00} & x_{01} & x_{02} & x_{03} \\ x_{10} & x_{11} & x_{12} & x_{13} \\ x_{20} & x_{21} & x_{22} & x_{23} \\ x_{30} & x_{31} & x_{32} & x_{33} \end{bmatrix} \begin{bmatrix} a & b & a & c \\ a & c & -a & -b \\ a & -c & -a & b \\ a & -b & a & -c \end{bmatrix}, \quad (12.8)$$

where according to Equation 12.6 we have

$$a = \frac{1}{2},$$

$$b = \sqrt{\frac{1}{2}} \cos\left(\frac{\pi}{8}\right),$$

$$c = \sqrt{\frac{1}{2}} \cos\left(\frac{3\pi}{8}\right).$$

It can be readily shown that the matrix multiplication of Equation 12.8 can be factorized [366] according to the following equivalent form:

$$\mathbf{Y} = (\mathbf{C}\mathbf{X}\mathbf{C}^{\mathrm{T}}) \otimes \mathbf{E}$$

$$= \left(\begin{bmatrix} 1 & 1 & 1 & 1 \\ 1 & d & -d & -1 \\ 1 & -1 & -1 & 1 \\ d & -1 & 1 & -d \end{bmatrix} \begin{bmatrix} x_{00} & x_{01} & x_{02} & x_{03} \\ x_{10} & x_{11} & x_{12} & x_{13} \\ x_{20} & x_{21} & x_{22} & x_{23} \\ x_{30} & x_{31} & x_{32} & x_{33} \end{bmatrix} \begin{bmatrix} 1 & 1 & 1 & d \\ 1 & d & -1 & -1 \\ 1 & -d & -1 & 1 \\ 1 & -1 & 1 & -d \end{bmatrix} \right)$$

$$\otimes \begin{bmatrix} a^2 & ab & a^2 & ab \\ ab & b^2 & ab & b^2 \\ a^2 & ab & a^2 & ab \\ ab & b^2 & ab & b^2 \end{bmatrix}, \quad (12.9)$$

where $\mathbf{CXC}^T$ is the "core" 2D transform in the integer transform. Furthermore $\mathbf{E}$ is a matrix of scaling factors and the symbol $\otimes$ indicates that each element of $(\mathbf{CXC}^T)$ is multiplied by the corresponding scaling factor appearing in the same position in the matrix $\mathbf{E}$, which is a scalar multiplication, rather than matrix multiplication. The constants a and b are the same as before, while d is $c/b \approx 0.414$.

For the sake of simplifying the implementation of the transform, d is approximated by 0.5. Furthermore, in order to ensure that the transform remains orthogonal, b also has to be modified so that [366]

$$a = \frac{1}{2},$$

$$b = \sqrt{\frac{2}{5}},$$

$$d = \sqrt{\frac{1}{2}}.$$

The second and fourth rows of the matrix $\mathbf{C}$ and the second and fourth columns of matrix $\mathbf{C}^T$ are scaled by a factor of 2 and the matrix $\mathbf{E}$ in Equation 12.9 is scaled down by the same factor in the appropriate positions in order to compensate. For these adjustments, this avoids multiplications by $\frac{1}{2}$ in the "core" transform $\mathbf{CXC}^T$, which would result in a loss of accuracy, when using an integer arithmetic. The final forward transform becomes

$$\mathbf{Y} = (\mathbf{C_f X C_f^T}) \otimes \mathbf{E_f}$$

$$= \left(\begin{bmatrix} 1 & 1 & 1 & 1 \\ 2 & 1 & -1 & -2 \\ 1 & -1 & -1 & 1 \\ d & -2 & 2 & -1 \end{bmatrix} \begin{bmatrix} x_{00} & x_{01} & x_{02} & x_{03} \\ x_{10} & x_{11} & x_{12} & x_{13} \\ x_{20} & x_{21} & x_{22} & x_{23} \\ x_{30} & x_{31} & x_{32} & x_{33} \end{bmatrix} \begin{bmatrix} 1 & 2 & 1 & 1 \\ 1 & 1 & -1 & -2 \\ 1 & -1 & -1 & 2 \\ 1 & -2 & 1 & -1 \end{bmatrix} \right)$$

$$\otimes \begin{bmatrix} a^2 & ab/2 & a^2 & ab/2 \\ ab/2 & b^2/4 & ab/2 & b^2/4 \\ a^2 & ab/2 & a^2 & ab/2 \\ ab/2 & b^2/4 & ab/2 & b^2/4 \end{bmatrix}. \quad (12.10)$$

This transform constitutes an approximation of the 4×4 pixel DCT. As a consequence of changing the factors d and b, the output of the new transform will not be identical to that of the 4×4 DCT.

The inverse transform defined in [372] is given by

$$\mathbf{X'} = \mathbf{C_i^T}(\mathbf{Y} \otimes \mathbf{E_i})\mathbf{C_i}$$

$$= \left(\begin{bmatrix} 1 & 1 & 1 & 1/2 \\ 1 & 1/2 & -1 & -1 \\ 1 & -1/2 & -1 & 1 \\ 1 & -1 & 1 & -1/2 \end{bmatrix} \begin{bmatrix} y_{00} & y_{01} & y_{02} & y_{03} \\ y_{10} & y_{11} & y_{12} & y_{13} \\ y_{20} & y_{21} & y_{22} & y_{23} \\ y_{30} & y_{31} & y_{32} & y_{33} \end{bmatrix} \begin{bmatrix} a^2 & ab & a^2 & ab \\ ab & b^2 & ab & b^2 \\ a^2 & ab & a^2 & ab \\ ab & b^2 & ab & b^2 \end{bmatrix} \right)$$

$$\otimes \begin{bmatrix} 1 & 1 & 1 & 1 \\ 1 & 1/2 & -1/2 & -1 \\ 1 & -1 & -1 & 1 \\ 1/2 & -1 & 1 & -1/2 \end{bmatrix}. \quad (12.11)$$

12.4. H.264 SPECIFIC CODING ALGORITHM

This time, $\mathbf{Y}$ is pre-scaled upon multiplying each of its coefficients by the appropriate weighting factor found in the corresponding position of matrix $\mathbf{E_i}$. The factors $\pm\frac{1}{2}$ seen in the matrices $\mathbf{C}$ and $\mathbf{C^T}$, which can be implemented by a right-shift without a significant loss of accuracy, because the coefficients of $\mathbf{Y}$ are pre-scaled.

12.4.3.2 Quantization

The H.264 codec uses a scalar DCT coefficient quantizer. The quantizer's implementation is complicated by the requirements of (1) avoiding division and/or floating point arithmetic operations and (2) by incorporating the post- and pre-scaling matrices $\mathbf{E_f}$ and $\mathbf{E_i}$ of Equations 12.10 and 12.11, as described in Section 12.4.3.1.

The basic DCT-coefficient quantization operation is as follows:

$$Z_{ij} = \text{round}\left(\frac{Y_{ij}}{Q_{\text{step}}}\right), \quad (12.12)$$

where Y_{ij} is a DCT coefficient generated by the transform described in Section 12.4.3.1, Q_{step} is a quantizer step size, and Z_{ij} is a quantized DCT coefficient. A total of 31 step-size values of Q_{step} are supported by the standard [372] and these are indexed by the quantization parameter, QP. The values of Q_{step} corresponding to each QP are shown in Table 12.5. Note that Q_{step} doubles in size for every increment of 6 in QP, Q_{step} increases by 12.5% for each increment of 1 in QP.

Table 12.5: Quantization Step Sizes in the H.264 Codec

QP	0	1	2	3	4	5	6	7	...	10	...	20	...	31
Q_{step}	0.625	0.6875	0.8125	0.875	1	1.125	1.25	1.375	...	2	...	6.5	...	22

The post-scaling factors a^2, $ab/2$, or $b^2/4$ of $\mathbf{E_f}$ are incorporated into the forward quantizer. First, the video input block $\mathbf{X}$ is transformed to give a block of unscaled coefficients $\mathbf{W} = \mathbf{CXC^T}$. Then, each DCT coefficient W_{ij} is quantized and scaled in a single operation according to

$$Z_{ij} = \text{round}\left(W_{ij} \cdot \frac{PF}{Q_{\text{step}}}\right), \quad (12.13)$$

where PF is a^2, $ab/2$ or $b^2/4$ depending on their position (i,j). The scaling operation of (PF/Q_{step}) is implemented in the H.264 reference model software [372] as a multiplication by the multiplication factor of *MF* and a right-shift,[2] thus avoiding any division operations:

$$Z_{ij} = \text{round}\left(W_{ij} \cdot \frac{MF}{2^{qbits}}\right), \quad (12.14)$$

---
[2]In integer arithmetic, Equation 12.14 can be implemented as follows:

$$|Zij| = (|W_{ij} \cdot MF + f) \gg qbits$$
$$\text{sign}(Zij) = \text{sign}(W_{ij})$$

where $\gg$ indicates a binary right-shift.

where

$$\frac{MF}{2^{qbits}} = \frac{PF}{Q_{\text{step}}} \tag{12.15}$$

$$qbits = 15 + floor\left(\frac{QP}{6}\right). \tag{12.16}$$

Let us now consider the inverse quantization process. The basic inverse quantizer or "rescale" operation is carried out as

$$Y'_{ij} = Z_{ij} \cdot Q_{\text{step}}. \tag{12.17}$$

The pre-scaling factor used for the inverse transform of Equation 12.11 implemented with the aid of the matrix $\mathbf{E}_i$, containing values of a^2, ab, and b^2 depending on the coefficient position is incorporated in this operation, together with a further constant scaling factor of 64, in order to avoid any rounding errors [372] according to

$$W'_{ij} = Z_{ij} \cdot Q_{\text{step}} \cdot PF \cdot 64. \tag{12.18}$$

In Equation 12.18, W'_{ij} is a scaled DCT coefficient, which is then transformed by the "core" inverse transform, namely by $\mathbf{C}_i^T \mathbf{W} \mathbf{C}_i$. The decoded pixel values generated by the inverse transform are divided by 64 in order to remove the scaling factor of 64, which may be implemented by shift operations. Section 12.4.3.3 summarizes the complete DCT transform, quantization, rescaling, and inverse transform process, while Section 12.4.3.4 provides some examples for characterizing the transformation process.

12.4.3.3 The Combined Transform, Quantization, Rescaling, and Inverse Transform Process

The entire process of generating the output residual block $\mathbf{X}'$ from input residual block $\mathbf{X}$ is described below in terms of a number of processing steps and illustrated in Figure 12.7 where the first three steps correspond to the encoding, while the remaining steps are decoding.

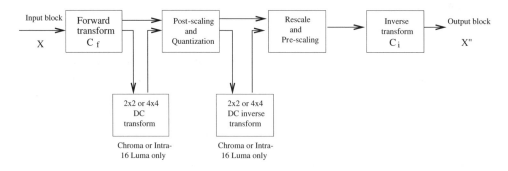

Figure 12.7: Transformation, quantization, rescale, and inverse transform flow diagram.

12.4. H.264 SPECIFIC CODING ALGORITHM

1. The input signal is constituted by the 4×4 MCER samples contained in $\mathbf{X}$.

2. Forward "core" transform: $\mathbf{W} = \mathbf{C_f X C_f}^T$.

3. Post-scaling and quantization: $\mathbf{Z} = \mathbf{W} \cdot (PF/Q_{\text{step}} \cdot 2^{qbits})$.

4. Re-scaling, incorporating inverse transform's pre-scaling operation: $\mathbf{W}' = \mathbf{Z} \cdot Q_{\text{step}} \cdot PF \cdot 64$.

5. Inverse "core" transform: $\mathbf{X}' = \mathbf{C_i^T W' C_i}$.

6. Post-scaling: $\mathbf{X}'' = round(\mathbf{X}'/64)$.

7. Output: 4×4-pixel block of MCER samples: $\mathbf{X}''$.

12.4.3.4 Integer Transform Example

Figure 12.8 shows a specific example of a 4×4-pixel block extracted from the "Foreman" video sequence having QCIF resolution.

Input block $\mathbf{X}$:

	$j=0$	1	2	3
$i=0$	46	49	55	64
1	38	39	48	49
2	44	38	39	42
3	63	58	56	54

Since the input MCER block appears to be fairly "flat" the DCT output is expected to be concentrated in the top-left corner, corresponding to the DC component. Output $\mathbf{W}$ of "core" transform:

	$j=0$	1	2	3
$i=0$	776	-44	24	-2
1	-29	-152	3	-31
2	114	0	-6	0
3	-27	-6	9	7

Before quantization and the post-scaling process, we first need to determine the value of multiplication factor *MF*. Using Equation 12.15, $MF = PF \times (2^{qbits}/Q_{\text{step}})$. In this example, $QP = 5$, hence $Q_{\text{step}} = 1.125$. Since *MF* is a multiplication factor, therefore some of the elements in a 4×4 matrix position are identical. Hence *MF*:

	$j=0$	1	2	3
$i=0$	7281	4605	7281	4605
1	4605	2912	4605	2912
2	7281	4605	7281	4605
3	4605	2912	4605	2912

Next, from Equation 12.14, the "core" transform $\mathbf{W}$ is multiplied by the *MF* producing the quantized transform coefficients $\mathbf{Z}$. Again, $\mathbf{Z}$ is expected to be concentrated in the top-left corner, furthermore, an increasing number of "zero" elements for the AC coefficient is expected. Output of forward quantizer $\mathbf{Z}$:

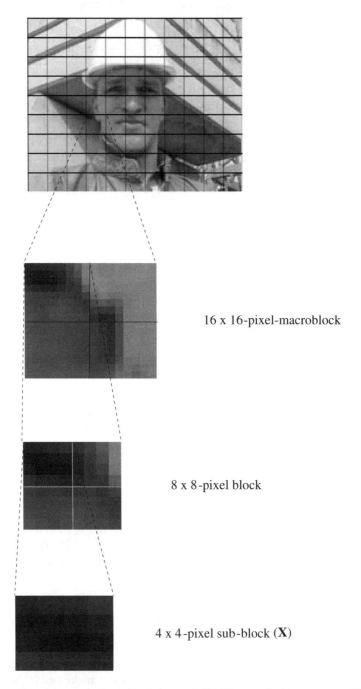

Figure 12.8: The selected 4×4-pixel sub-block from a QCIF "Foreman" video sequence. The 4×4-pixel sub-block **X** was used as a demonstration example in this section.

12.4. H.264 SPECIFIC CODING ALGORITHM

	$j=0$	1	2	3
$i=0$	172	−6	5	0
1	−4	−14	0	−3
2	25	0	−1	0
3	−4	−1	1	1

Next is the inverse quantization process: the receive quantized coefficients from the encoder are de-quantized or, in other words, "re-scaled"; the de-quantized coefficient before the main inverse transformation process are again has the same characteristic as the DCT output, where the DC value is much higher than AC values. Output of rescale $\mathbf{W}'$:

	$j=0$	1	2	3
$i=0$	3096	−136	90	0
1	−91	−403	0	−86
2	450	0	−18	0
3	−91	−28	22	22

For the inverse transformation process, the inverse transform coefficients are appear to be "flat". Output of $\mathbf{X}'$:

	$j=0$	1	2	3
$i=0$	2902	3120	3531	4083
1	2431	2523	2689	3122
2	2803	2415	2523	2659
3	4060	3690	3553	3426

The final step is to reconstruct, so that the reconstructed block is similar to the original block $\mathbf{X}$. Output of $\mathbf{X}''$:

	$j=0$	1	2	3
$i=0$	45	49	55	64
1	38	39	42	49
2	44	38	39	42
3	63	58	56	54

Finally, Figure 12.9 depicts the original and the reconstructed 4×4-pixel block of the selected "Foreman" video sequence.

12.4.4 Entropy Coding

After all of the transform coefficients have been quantized, the last step in the video coding process of Figure 12.1 is entropy coding. The H.264 codec has adapted two different approaches for entropy coding. The first approach is based on the use of UVLCs [368] and the second invokes CABAC [373]. Note that the type of entropy coding employed is selected by the user during the encoding process. When the picture parameter set flag "entropy_coding_mode" is set to "0", the block of quantized coefficients of either 4×4 or 16×16 are coded using the UVLC scheme, whereas when the picture parameter set flag "entropy_coding_mode" is set to "1", the CABAC scheme is employed.

Figure 12.9: Comparison of the 4 × 4-pixel sub-block extracted from the "Foreman" sequence, where (a) is the original 4 × 4-pixel sub-block before transformation and quantization, while (b) is the reconstructed 4 × 4-pixel sub-block after it has undergone the H.264 standard's integer transformation and quantization process.

12.4.4.1 Universal Variable Length Coding

Entropy coding [20] based on Variable Length Coding (VLC) is the most widely used method of further compressing the quantized transform coefficients, motion vectors, and other encoding parameters. Variable length codes are based on assigning shorter codewords to symbols having higher probability of occurrence and longer codewords to symbols with less frequent occurrences. The symbols and the associated codewords are organized in look-up tables, referred to as VLC tables, which are stored at both the encoder and decoder [373].

In some of the video coding standards such as H.263 [374] and MPEG-4 [349], a number of VLC tables are used, where each of the parameter types such as quantized DCT coefficients, motion vectors, etc., have their respective VLC tables. The H.264 codec employs a single universal VLC table that is to be used in entropy coding of all symbols in the encoder, regardless of the type of data those symbols represent.

12.4.4.2 Context-based Adaptive Binary Arithmetic Coding

Arithmetic coding makes use of a probability model at both the encoder and decoder for all the syntax elements such as transform coefficients and motion vectors. To increase the coding efficiency of arithmetic coding, the underlying probability model is adapted to the changing statistics with a video frame, through a process called context modeling. CABAC has three distinct advantages:

1. the context model provides an estimation of conditional probabilities of the coding symbols;

2. the arithmetic code permits a non-integer number if bits to be assigned to each symbol;

3. the adaptive arithmetic code permits the entropy coder to adapt itself to non-stationary symbol statistics.

12.4.4.3 H.264 Conclusion

The H.264 standard provides video coding mechanisms that are optimized for compression efficiency and aim at supporting practical multimedia communication applications. The range of available coding tools available is more restricted than in the MPEG-4 Visual standard owing to its narrower focus of applications, but there are still numerous possible choices for the coding parameters and strategies. The success of a practical H.264 implementation depends on the careful design of the codec and on the effective choices of the coding parameters. In the next section, we investigate the achievable performance of both codecs, while in Chapter 14 we propose a range of videophony schemes, where the H.264 video codec is employed.

12.5 Comparative Study of the MPEG-4 and H.264 Codecs

12.5.1 Introduction

Following the previous chapter introducing the basic features of the MPEG-4 codec and the above section on H.264 codec emerging from the ITU-T VCEG project, in this section we highlight the differences between the MPEG-4 and H.264 standards. Throughout this section, familiarity with the MPEG-4 standard is assumed.

The ISO/IEC standard MPEG-4 video coding standard [62] and the ITU-T VCEG H.264 codec [62, 366] exhibit some similarities in that they both define so-called block-based hybrid video codecs. Each video frame is divided into fixed size MBs of 16×16 pixels, which can be encoded in several coding modes. They both distinguish between intra-frame coding and inter-frame coding or predictive modes. In intra-frame mode, a MB is encoded without referring to other frames in the entire video sequence, while in the inter-frame coded or predictive mode, previously encoded images are used as the reference for forming the MCER signal. The resultant MCER signal is encoded using transform coding, where a MB is subdivided into a number of fixed size blocks. Prior to MCER encoding, each of these blocks is transformed using a block transform, and the transform coefficients are quantized and transmitted using entropy coding methods, as highlighted in Sections 12.3 and 12.4 [366].

Although both the MPEG-4 Visual layer and the H.264 codec define similar coding algorithms, they contain features and enhancements that render them different. These differences mainly concern the formation of the MCER signal, as well as the block sizes used for the transform coding and entropy coding methods. Following this brief introduction, let us now compare the features of these codecs in a little more detail in Sections 12.5.2–12.5.6.

12.5.2 Intra-frame Coding and Prediction

The MPEG-4 Visual layer and the H.263 codec have much in common [374], since the H.263 ITU-T Recommendation was the starting point of the MPEG-4 project and most additional work was carried out in the area of object-based coding, where each of the objects in a video frame was referred to as a Video Object (VO). Thus, the MPEG-4 coding standard supports three different picture types during the encoding of a so-called Video Object Plane (VOP), namely I-VOPs, P-VOPs, and B-VOPs. In I-VOPs each MB is encoded in intra-frame mode, and there is only one intra-frame mode in the MPEG-4 standard which uses 8×8-pixel sub-blocks. By comparison in the H.264 codec, sub-blocks of 4×4 samples are used for transform coding, and thus a sub-block consists of 16 luminance and 8 chrominance pixels, as shown in Figure 12.6. The conventional picture types known as I, P, and B pictures are still supported in the standard. Unlike in MPEG-4, in H.264 a MB can always be encoded in one of several intra-frames modes. There are two classes of intra-frame coding modes, which are denoted as intra 16×16 and intra 4×4 in the following. Moreover, in contrast to MPEG-4, where only some of the DCT coefficients can be predicted from neighboring intra-frame blocks, in H.264 intra-frame pixel prediction is always utilized in the spatial domain by referring to the neighboring pixels of already encoded blocks. When using the intra-frame coded 4×4-pixel mode, each 4×4-pixel block of the luminance component utilizes six prediction modes. The chosen modes are encoded and transmitted as side information. In the intra-frame coded 16×16 pixel mode, a uniform prediction is performed for the whole luminance component of a MB. Four different luminance modes are supported, as described in Section 12.4.1. Let us now consider the issue of intra-frame prediction in the next section.

12.5.3 Inter-frame Prediction and Motion Compensation

Both the MPEG-4 and the H.264 codec employ inter-frame prediction based on frame differencing as well as motion compensation using motion vectors. In MPEG-4, three different predictive coding modes are provided for P-VOPs, namely the inter 16×16, inter 8×8 and the Skip mode. For the Skip mode, a single-bit flag is required for signalling to the decoder that all samples of the entire MB are repeated from the reference frame. In contrast, the H.264 standard provides seven motion-compensated coding modes for MBs encoded in the inter-frame mode. Each motion-compensated mode corresponds to a specific partition of the MB into fixed size blocks used for motion description. Blocks with sizes of 16×16, $16 \times 8, 8 \times 16, 8 \times 8, 8 \times 4, 4 \times 8$, and 4×4 pixels are supported by the H.264 syntax [366] and thus up to 16 motion vectors can be transmitted for a MB.

Furthermore, H.264 supports the employment of multi-frame motion compensated prediction, where more than one previously encoded pictures can be used as reference for motion compensation. For the current version of the H.264 codec, up to five reference frames are available for motion prediction. In contrast, in the MPEG-4 standard only one reference frame is used for motion compensation.

Half- or quarter-pixel resolution motion compensation based on the previously coded I- or P-VOP is applied in MPEG-4 for forming the MCER signal in the inter 16×16 and inter 8×8 modes. In the inter 16×16 mode, the motion trajectory of an entire MB is specified by a single motion vector, whereas the inter 8×8 mode uses four motion vectors rounded to values associated with half-pixel accuracy. The horizontal and vertical components of each motion vector are encoded differentially by using median prediction based on three neighboring blocks, which have already been encoded.

The H.264 syntax supports both quarter- and eighth-pixel resolution motion compensation. The motion vector components are differentially encoded using either median or directional prediction relying on the motion vectors of the neighboring blocks. The selection of a specific prediction mode depends on the block's shape and on its position within the MB. Let us now focus our attention on the encoding of the MCER in the next section.

12.5.4 Transform Coding and Quantization

In MPEG-4 the DCT is employed for encoding of an intra- or inter-frame coded MB, but no arithmetic procedure is specified for computing the inverse transform. H.264 codec is basically similar to the MPEG-4 scheme and to other previous coding standards in that it utilizes transform coding of the prediction error signal. However, in the H.264 codec the transformation is applied to 4×4-pixel blocks and instead of the DCT, the H.264 codec uses the separable integer transform of Section 12.4.3, which has similar properties to the 4×4-pixel DCT. Since the inverse DCT of Section 12.4.3 is defined by exact integer operations, inverse-transform mismatches will never be encountered. An additional 2×2-pixel transform is applied to the four DC coefficients of each chrominance component. If the intra 16×16 mode is in use, a similar operation extending the length of the transforms basis functions is performed for the 4×4 DC-coefficients of the luminance signal.

The H.264 codec uses scalar quantization for encoding the DCT coefficients, but without the quantizer dead-zone around zero of the MPEG-4 codec. One of the 32 quantizers is selected for each MB by the quantization parameter QP. The quantizer levels are arranged such that there is an approximately 12.5% quantizer step-size increase from one QP to the next in the list of 32 quantizers. The quantized transform coefficients are scanned in a zig-zag fashion [5] and converted to compressed symbols by Run-Length Coding (RLC). In the MPEG-4 standard the quantized DCT coefficients of an 8×8-pixel block are zig-zag scanned, whereas in the H.264 standard zig-zag scanning is carried out in 4×4-pixel blocks. In the next section we consider the entropy coding schemes used.

12.5.5 Entropy Coding

Two different methods of entropy coding are supported by H.264. The first method is UVLC [368], which relies on potentially infinite-length codeword sets. Instead of designing a different VLC table for each syntax element, only the mapping to the single UVLC table is customized according to the statistics of the data to be encoded. The efficiency of entropy coding can be improved if CABAC [373] is used. On the one hand, the employment of arithmetic coding [366] allows the assignment of non-integer number of bits to each symbol of an alphabet, which is particularly beneficial in conjunction with symbol probabilities higher than 0.5. On the other hand, the use of adaptive codes supports adaptation to non-stationary symbol statistics. Another important property of the CABAC scheme [373] is its context modeling capability [366], where the statistics of already encoded syntax elements can be used for estimating the conditional probabilities of the symbols to be encoded. The inter-symbol dependencies may be exploited by switching amongst several estimated probability models according to the already coded symbols in the neighborhood of the symbol to be encoded.

12.5.6 De-blocking Filter

For the sake of mitigating the detrimental visual effects of block-edge artifacts, the H.264 codec design includes a deblocking filter. In contrast to the MPEG-4 codec, where the deblocking filter is an optional part of the recommendation, in the H.264 codec the filter is applied within the motion prediction loop and hence forms an integral part of the recommendation. The cut-off frequency of the filtering is adaptively controlled by the values of several coding syntax elements [366]. Following our rudimentary comparison of the elementary of the H.264 and MPEG4 codecs, let us now focus our attention on the comparison of their performance.

12.6 Performance Results

12.6.1 Introduction

Exclusive simulations were carried out using both the MPEG-4 and H.264 codecs in order to comparatively study their performance. In our experiments, well-known video sequences such as the "Miss America", "Suzi", and "Foreman" clips were used, which were scanned at video frame rates of 10 and 30 frames/s. The performance of both codecs depends highly on the quantizer step sizes used in the encoder. The upper and lower limits of video quality performance can be recorded when the finest or coarsest quantizers are employed.

Note that in our MPEG-4 and H.264 simulations we used identical YUV format input video sequences. Since both codecs support 4:2:0 format color YUV video sequences, our experiments were carried out using QCIF format color video sequences. Let us now turn our attention to their performance comparison.

12.6.2 MPEG-4 Performance

As stated above, in our MPEG-4 simulations we used the QCIF-format color "Miss America", "Suzi" and "Foreman" sequences scanned at 10 and 30 frames/s. Simulations were carried out using the MPEG-4 codec under the following conditions.

- Only one intra-frame coded picture was transmitted at the beginning of the sequence.

- Two B-pictures have been inserted between two successive P-pictures. More specifically, the frame-sequence had the following syntax: I,P,B,B,P,B,B,P....

- The video sequence length of the "Miss America" clip was 150 frames.

- The video sequence length of the "Suzi" clip was 150 frames.

- The video sequence length of the "Foreman" clip was 300 frames.

- The object-based coding option was deactivated, thus we treated the entire frame as one video object.

The simulations were performed at a wide range of bitrates. The results of these simulations are shown in Figures 12.10 and 12.11. (Figure 11.24 of Section 11.6 portrayed the PSNR video quality versus coded bitrate performance of the MPEG-4 codec for various QCIF-resolution video sequences.) These results are repeated here for convenience in Figure 12.10, while Figure 12.11 plots the results from a different perspective, namely as the average PSNR versus the compression ratio. As can be seen from Figure 12.10, for a given fixed bitrate, the simulations for the "Foreman" sequence resulted in the lowest image quality, followed by the "Suzi" sequence and finally by the "Miss America" sequence. This is because the "Miss America" clip is a pure head and shoulder's type video sequence, which

12.6. PERFORMANCE RESULTS

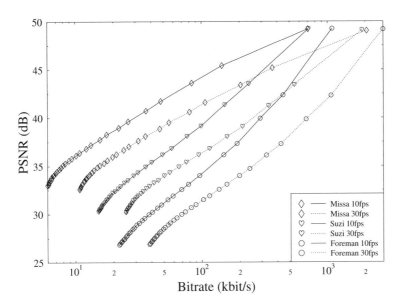

Figure 12.10: Image quality (PSNR) versus coded bitrate performance of the MPEG-4 codec for various QCIF-resolution video sequences scanned at 10 and 30 frames/s.

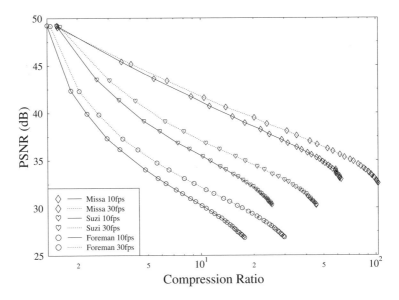

Figure 12.11: Image quality (PSNR) versus compression ratio performance of the MPEG-4 codec for various QCIF-resolution video sequences scanned at 10 and 30 frames/s.

is associated with the least motion activity, hence less motion vector information has to be transmitted. Therefore, the Skip mode is used more often during the encoding process which generates a single bit flag. Hence, a reduced number of bits are necessary for coding the entire sequence. In contrast, more motion takes place in the "Suzi" sequence. Hence, more motion vector information was required by the motion compensation process. In addition, less Skip-mode flags were used by the encoding process, which results in a higher coded bitrate compared with the "Miss America" video sequence. Needless to say, the "Foreman" sequence exhibits an even higher motion activity than the other two sequences used in our experiments, therefore it requires the highest number of bits during the encoding process, in particular for the motion vector parameters.

The image quality expressed in PSNR improved almost linearly as the bitrate was increased. When maintaining a constant bitrate, the 30 frames/s curves typically exhibit a PSNR penalty due to the fixed bitrate constraint imposed, because on the basis of simple logic at 30 frames/s a threefold bitrate increase is expected in comparison to 10 frames/s. Moreover, due to the higher inter-frame correlation of the 30 frames/s clip the MCER has a typically reduced variance in comparison to 10 frames/s coding, which mitigates the above threefold bitrate increment estimate to an approximate ratio of two.

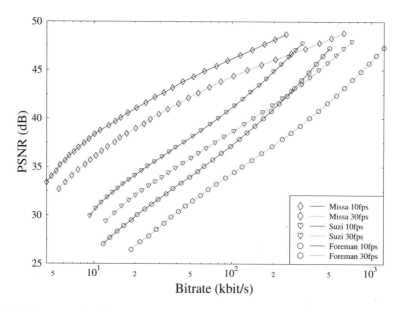

Figure 12.12: Image quality (PSNR) versus coded bitrate performance of the H.264 codec for various QCIF-resolution video sequences scanned at 10 and 30 frames/s.

12.6.3 H.264 Performance

In this section, we report our findings on the ITU-T standard H.264 codec using simulation conditions similar to those employed in the MPEG-4 experiments described in the previous section. Again, all simulations were conducted using the three well-known video sequences

12.6. PERFORMANCE RESULTS

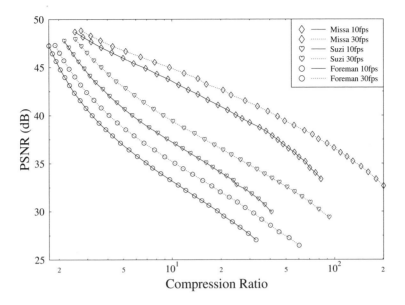

Figure 12.13: Image quality (PSNR) versus compression ratio performance of the H.264 codec for various QCIF-resolution video sequences scanned at 10 and 30 frames/s.

"Miss America", "Suzi", and "Foreman" at both 10 and 30 frames/s. The additional coding options used in the H.264 simulations are listed as follows.

- Entropy coding was performed using the CABAC technique [373].

- Five reference frames were used for inter-frame prediction.

- Eighth-pixel resolution motion compensation was used.

Figure 12.12 portrays the image quality (PSNR) versus coded bitrate performance of the H.264 codec for the "Miss America", "Suzi", and "Foreman" sequences scanned at 10 and 30 frames/s. As can be seen from the figure, it exhibits trends to those of the MPEG-4 scheme previously seen in Figure 12.10. This was expected, because both the MPEG-4 and H.264 scheme have similar coding algorithms, as discussed earlier in Sections 12.5.

As in MPEG-4, for a given bitrate, the "Miss America" sequence had the highest image quality (PSNR) followed by the "Suzi" then the "Foreman" clips. This is in line with the amount of motion content in the sequences. As shown in Figure 12.12, the 30 frames/s scenarios typically required a factor of two higher bitrates for maintaining a certain fixed PSNR. The same tendencies were observed in terms of the compression ratio in Figure 12.13. As expected from the corresponding PSNR versus bitrate curves, the H.264 codec's compression ratio performance curves are near-linear on this log–log scale graph, resulting in a predictable PSNR versus compression ratio relationship.

432   **CHAPTER 12. COMPARATIVE STUDY OF THE MPEG-4 AND H.264 CODECS**

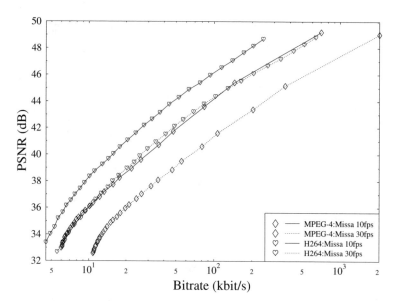

Figure 12.14: Image quality (PSNR) versus coded bitrate, for the MPEG-4 and the H.264 codecs for the "Miss America" color QCIF-resolution video sequence scanned at 10 and 30 frames/s.

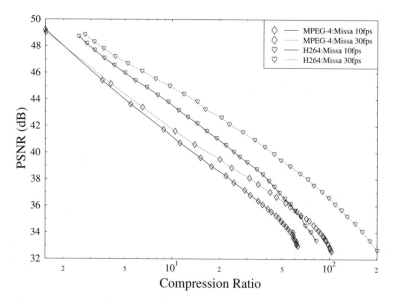

Figure 12.15: Image quality (PSNR) versus compression ratio, for the MPEG-4 and the H.264 codecs for the "Miss America" color QCIF-resolution video sequence scanned at 10 and 30 frames/s.

12.6. PERFORMANCE RESULTS

12.6.4 Comparative Study

Our investigations presented in this section were conducted in order to compare the MPEG-4 codec's performance with the characteristics of the H.264 codec. Both the MPEG-4 and the H.264 experiments were conducted at 10 and 30 frames/s and at a wide range of bitrates. We varied the quantizer step sizes for the sake of attaining a given target bitrate in our experiments. Thus, the quantizers having indices ranging from 0 to 31 were set approximately in each of the simulations. Figure 12.14 shows a comparison of the MPEG-4 and H.264 video codecs for the "Miss America" video sequence in terms of the achievable image quality expressed in PSNR versus bitrate (kbit/s). As can be seen from the graphs, the performance of the H.264 codec is significantly better than that of the MPEG-4 arrangement. Comparing the MPEG-4 codec's performance at 30 frames/s to that of the H.264 scheme at a frame rate of 10 frames/s produces interesting revelations. The results show that by replacing the MPEG-4 codec by the H.264 scheme allows the frame rate to increase from 10 to 30 frames/s without any significant change in the bitrate requirements or without loss of image quality. This was achieved by invoking more sophisticated and hence higher-complexity signal processing, resulting in a better image quality perception.

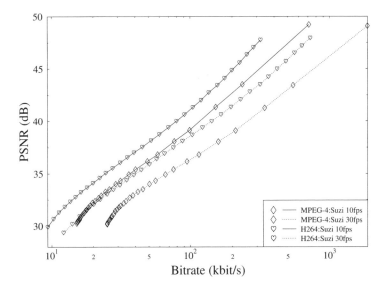

Figure 12.16: Image quality (PSNR) versus coded bitrate, for the MPEG-4 and the H.264 codecs for the "Suzi" color QCIF-resolution video sequence scanned at 10 and 30 frames/s.

Figure 12.15 provides a comparison of the MPEG-4 and H.264 video codecs in a different context, expressed as image PSNR versus comparison ratio. Again, as can be seen from the corresponding graphs, the H.264 codec exhibits a higher compression performance than the MPEG-4 codec.

Let us now verify the above findings, which we arrived at on the basis of the "Miss America" sequence, also using the "Suzi" sequence. The experimental conditions were unchanged, we simply replaced the "Miss America" video sequence with the "Suzi" sequence. The corresponding results are plotted in Figures 12.16 and 12.17 in terms of PSNR

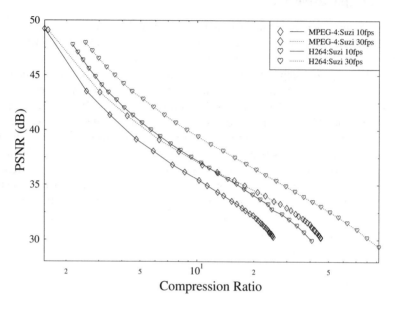

Figure 12.17: Image quality (PSNR) versus compression ratio, for the MPEG-4 and the H.264 codecs for the "Suzi" color QCIF-resolution video sequence scanned at 10 and 30 frames/s.

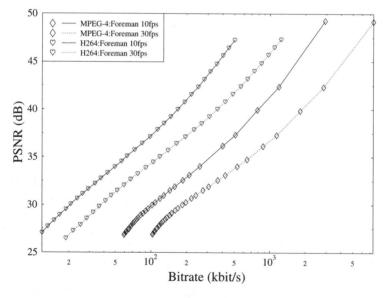

Figure 12.18: Image quality (PSNR) versus coded bitrate, for the MPEG-4 and the H.264 codecs for the "Foreman" color QCIF-resolution video sequence scanned at 10 and 30 frames/s.

12.6. PERFORMANCE RESULTS

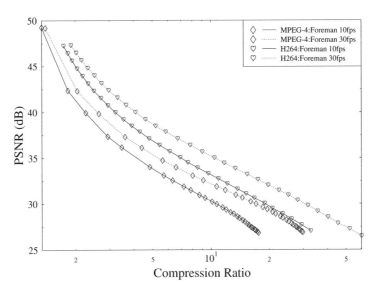

Figure 12.19: Image quality (PSNR) versus compression ratio, for the MPEG-4 and the H.264 codecs for the "Foreman" color QCIF-resolution video sequence scanned at 10 and 30 frames/s.

versus bitrate and versus compression ratio, respectively. As expected, the results are similar to those of the Miss America experiments. However, there is less similarity between the MPEG-4 curves recorded at 10 frames/s and the H.264 curves generated at 30 frames/s. More specifically, the H.264 codec operating at 30 frames/s outperforms the MPEG-4 scheme transmitting at 10 frames/s, at higher bitrates.

The final set of experiments provided in this chapter was based on replacing the "Miss America" video sequence with the "Foreman" video sequence. As before, our experiments were carried out at both 10 and 30 frames/s. The corresponding performance curves are shown in Figures 12.18 and 12.19. Since the amount of image fine detail is quite different for the "Miss America" and "Foreman" video sequences, the image quality (PSNR) performance was slightly degraded at a given coded bitrate. Observe that the curves seen in Figure 12.18 do not cross each other, which implies that the H.264 codec in operating at both 10 and 30 frames/s achieves a the higher picture quality than the MPEG-4 codec. Similar conclusions are valid also for the compression ratio.

12.6.5 Summary and Conclusions

In this chapter, we have described the differences between the MPEG-4 and H.264 video codecs, and a range of comparative performance curves have been plotted. The differences between the two codecs in terms of standards, functionalities, and coding algorithm were described. From the range of algorithmic differences, prediction methods (spatial prediction and motion compensation), transform coding, and entropy coding have the greatest effect on the coding efficiency. A variety of comparative performance curves with various video sequences were plotted, demonstrating how the performance of the H.264 codec varied after some of its parameters were altered.

Chapter 13

MPEG-4 Bitstream and Bit-sensitivity Study

J-Y. Chung and L. Hanzo

13.1 Motivation

The MPEG-4 video coding standard was briefly introduced in Chapter 11. As seen in Figure 13.1, the encoded video bitstream can be represented in terms of a number of hierarchical layers, where each higher-order layer contains lower-order layers, which in turn contain further lower-order layers. This figure will be discussed in more detail during our further discourse. We note, however, that it is rather difficult to visualize the corresponding bitstream in a well-presented, highly organized manner, because the MPEG-4 format is far more complicated than the MPEG-1, MPEG-2 as well as the International Telecommunications Union—Telecommunications Standardization Sector (ITU-T) H.263 bitstream formats. Nonetheless, in order to ease this task, the MPEG-4 bitstream structure has been partitioned into several categories.

In Chapter 11, we outlined the basic foundations for the MPEG-4 Visual standard's coding profiles and algorithms. In Chapter 12, we provided a comparative study in comparison to H.264. This chapter attempts to outline the MPEG-4-encoded hierarchical bitstream structure, which forms the basis of our error-resilience study in Section 13.7.

13.2 Structure of Coded Visual Data

The MPEG-4 encoded visual information may assume several different types, such as video information, still texture information, 2D mesh information or facial animation parameter information [364]. Synthetic objects such as video data, still image data, mesh data and animated data and their attributions are structured in a hierarchical manner for the sake of supporting both bitstream scalability and object scalability. The system-oriented part of the MPEG-4 specifications outlines the philosophy of spatial–temporal scene composition,

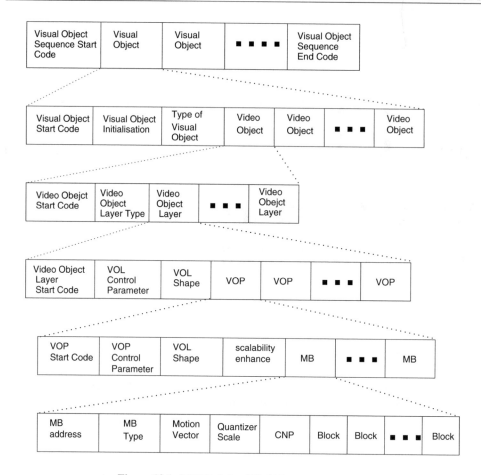

Figure 13.1: MPEG-4 simplified bitstream structure.

including normative 2D/3D scene graph nodes and their composition supported by the Binary Interchange Format Specification (BIFS) [364,375]. At this level, synthetic and natural object composition relies on systems using subsequent (non-normative) rendering performed by the application to generate specific pixel-oriented views of the models.

13.2.1 Video Data

The MPEG-4 codec facilitates the employment of the so-called embedded encoding model. Specifically, a low-quality low-rate mode referred to as the base-rate mode may be used by the decoder, if not all of the encoded bits reach the remote receiver. For the sake of ensuring that both the encoder's local decoder and the remote decoder use the same reconstructed video buffer contents during the process of motion compensation, the encoder also refrains from using the so-called enhancement bits during this process. However, if the remote decoder does receive the enhancement bits, they allow the improvement of decoded video quality.

13.2. STRUCTURE OF CODED VISUAL DATA

If there is only a single encoded bitstream, the coded video data is referred to as a non-scalable video bitstream. In contrast, if there is both a base and an enhancement bitstream, the encoded video bit pattern is termed a scalable video bitstream. In MPEG-4, video object layers are often differentiated as base layer and enhancement Layer.

13.2.2 Still Texture Data

Visual texture coding was designed for maintaining a high perceptual quality while maintaining a low-rate transmission rate and for rendering texture under widely varied viewing conditions, in particular, when encoding and for manipulating 2D/3D synthetic scenes. Still texture encoding facilitates the multi-layer representation of luminance, color and shape. This supports progressive transmission of the texture for image quality improvements, when more information is received by a terminal.

13.2.3 Mesh Data

Coded mesh data was introduced in [349] and it was well presented by Tekalp and Ostermann in [376]. In Chapter 11, we discussed that the MPEG-4 is an object-based multimedia compression standard, which allows for the encoding of different audiovisual objects. The mesh data in this category are some of the synthetic visual objects supported by MPEG-4, namely animated faces and animated arbitrary 2D uniform and Delaunay meshes [376]; the mesh data consists of a single non-scalable bitstream. This bitstream defines the structure and motion trajectory of the 2D mesh. The texture of the mesh has to be encoded as a separate video object.

13.2.4 Face Animation Parameter Data

The coded face animation parameter (FAP) data also consists of a single non-scalable bitstream. It defines the animation of the face model of the decoder. Face animation data is structured in a specific format outlined by the standard, which invoked downloadable face models and their animation controls, as well as a signal layer of compressed FAPs used for remote manipulation of the face model. The face constitutes a node in a scene graph that includes the face geometry ready for rendering. The shape, texture and facial expressions of a synthetic face are generally controlled by the bitstream containing Facial Definition Parameter (FDP) sets [349] or FAP sets [349].

However, in this book the focus of research is related to the encoding of video data. Therefore, still texture data, 2D mesh data and FAP data will not be discussed further in this discourse.

As mentioned above, the structure of the bitstream is organized into separate video parameter layers. We have seen in Figure 13.1 that it consists of visual sequence information, visual objects, Video Objects (VOs), the Video Object Layer (VOL), Video Object Plane (VOP), Macroblock (MB) and video block data.

13.3 Visual Bitstream Syntax

13.3.1 Start Codes

Start codes are constituted by a specific bit patterns that may be mimicked only with a low probability by the video stream. Each start code seen at the various layers of Figure 13.1 consists of a start code prefix followed by a start code value. The start code prefix is a string of 23 bits of zero value followed by a single bit of value one. The start code prefix is thus the bit string "00000000000000000000001". The start code value is an 8-bit integer, which identifies the type of the start code concerned. In the MPEG-4 bitstream syntax, the majority of the start codes has a single legitimate start code value. However, the video object start code and the VOL start code may be represented by numerous different start code values, where each VO start code marks the beginning of a new VOL hosting several VOPs, as seen in Figure 13.1.

All start codes are byte-aligned. This is achieved by inserting padding bits, namely a bit having a zero value and then, if necessary, by inserting bits having a logical value of one before the start code prefix, such that the first bits of the start code prefix becomes the first (most significant) bit of the byte concerned [364]. Table 13.1 explicitly defines the start code values for all start codes used in the video bitstream.

Table 13.1: Start Code Values Expressed in Hexadecimal Format

Start code type	Start code value (hexadecimal)
Video object start code	00 to 1F
Video object layer start code	20 to 2F
Visual object sequence start code	B0
Visual object sequence end code	B1
User data start code	B2
Group of VOP start code	B3
Video session error code	B4
Visual object start code	B5
VOP start code	B6
Face object start code	BA
Face object plane start code	BB
Mesh object start code	BC
Mesh object plane start code	BD
Still texture object start code	BE
Texture spatial layer start code	BF
Texture SNR layer start code	C0

- The Visual Object Sequence is the highest-order syntactic structure of the coded visual bitstream in Figure 13.1. A VO sequence commences with a VO start code, which is followed by a one or more VOs. The VO sequence is terminated by a VO sequence end code.

- A VO sequence commences with the VO start code of Table 13.1, which is followed by profile and level identification, and a visual object identifier and by a VO, a still texture object, a mesh object or a face object, which were introduced in Section 13.2.

- A VO commences with a VO start code, that is followed by one or more VOLs, as seen in Figure 13.1

13.4 Introduction to Error-resilient Video Encoding

MPEG-4 aims at achieving error resilience in order to access image or video information remotely, while communicating over hostile transmission media [377]. In particular, it is important to support access to audio and video information via wireless networks at low bitrates.

The so-called error-resilience tools developed for the MPEG-4 codec can be divided into three classes, namely into resynchronization, data recovery and error concealment algorithms. It should be noted that these categories are not unique to MPEG-4, they have been used by numerous researchers working in the area of error resilience for video [378].

13.5 Error-resilient Video Coding in MPEG-4

This section discusses the error-resilient encoding measures employed in the MPEG-4 standard and describes the different encoding tools adopted by the standard for enabling robust video communication over noisy wireless channels. When compressed video data is transmitted over a wireless communication channel, it is subjected to channel errors in the form of either single-bit errors or burst of errors or both [379]. In order to render the video codec more resilient to channel-induced quality degradations, various Forward Error Correction (FEC) codes may be employed by the encoder for protecting the bitstream prior to transmission to the decoder [380, 381]. At the decoder, these FEC codes are then used for correcting the transmission errors which employs the following processing stages [364]:

- error detection and localization;

- resynchronization;

- data recovery;

- error concealment.

Before any error-concealment technique can be applied at the decoder, it is necessary to ascertain whether a transmission error has occurred. FEC techniques can also be used for error-detection [380]. In the context of a typical block-based video compression technique that uses motion compensation and DCT, a number of error-detection techniques may be invoked, such as the motion vectors are out of range, an invalid Variable Length Coding (VLC) table entry is encountered, the number of DCT coefficients in a block exceeds 64. When one or more of these conditions is detected in the process of video decoding, the video decoder flags an error and activates the corresponding error-handling procedure. Due to the specific nature of the MPEG-4 video compression algorithm, which relies on VLC, the location where the decoder detects an error is not the same as where the error has actually occurred. As shown in Figure 13.2, these events may be an undetermined distance away from each other. Naturally, an error event may also induce the loss of synchronization with the encoder.

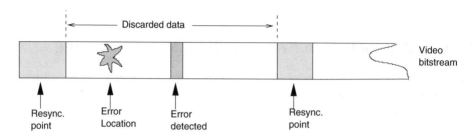

Figure 13.2: Stylized illustration of error event and its detection. It is usually impossible to detect the exact location of the occurrence of an error. Hence, all of the data between the two corresponding resynchronization points may have to be discarded.

While constructing the encoded bitstream, the encoder inserts unique resynchronization markers into the bitstream at approximately equally spaced intervals. Upon detecting an error, the decoder searches through the incoming bitstream, until the next resynchronization maker is found. Once this resynchronization marker is found, the decoder re-establishes synchronization with the encoder. Hence, all data that corresponds to the MBs between the resynchronization points has to be discarded, because the perceptual effects of displaying an image reconstructed from erroneous data are typically highly annoying.

After synchronization has been re-established, the MPEG-4 codec's data recovery tools attempt to recover the data that otherwise would be lost. More specifically, data recovery techniques such as reversible decoding [382] enable the decoder to salvage some of the data between the two resynchronization points by decoding the bitstream in both the forward and reverse direction. The philosophy of this method is described in greater detail in Section 13.6.

Error concealment is a further important component of any error-resilient video codec [383]. Similarly to the error-resilience tools discussed above, the efficiency of an error-concealment strategy is highly dependent on the performance of the resynchronization scheme used. More specifically, if the resynchronization method is capable of accurately localizing the error, then the error concealment becomes more efficient. In low-bitrate, low-delay applications the above-mentioned resynchronization scheme provides acceptable results in conjunction with a simple concealment strategy, such as the method of copying blocks from the corresponding video frame fraction of the previous frame. Upon recognizing the need to provide enhanced error-concealment capabilities, the MPEG-4 standardization body has developed an additional error-resilient mode that further improves the ability of the decoder to localize transmission error. Specifically, this approach utilizes data partitioning by separating the motion and texture related parts of the bitstream. This approach requires that a second resynchronization marker be inserted between the motion and texture information related bits. If the texture information is lost, this approach utilizes the motion information for concealing these errors. More explicitly, the corrupted texture information is discarded, but the motion vectors still can be used for motion compensation on the basis of the previous decoded VOP instead of the current one.

13.6 Error-resilience Tools in MPEG-4

A number of measures referred to in MPEG-4 parlance as "tools" have been incorporated into the MPEG-4 video encoder for the sake of rendering it more resilient to error. These tools include various techniques such as:

- Video packet resynchronization [355];

- Data Partitioning (DP) [384];

- Reversible VLCs (RVLCs) [382];

- Header Extension Codes (HECs) [355],

which are discussed below in a little more detail.

13.6.1 Resynchronization

As mentioned in Section 13.5, an erroneous video bitstream is expected to arrive at the decoder, when compressed video data is transmitted over error-prone communication channels. The decoder might lead to the loss of synchronization with the encoder, when an error-infested bitstream is received. Resynchronization tools, as the terminology implies, attempt to re-establish synchronization between the encoder and the decoder, when transmission errors have been detected. Generally, the data between the synchronization point prior to the error and the first point where synchronization is re-established, is discarded. If the resynchronization approach is successful in determining the amount of data discarded by the decoder, then the ability of other types of tools which recover data and/or conceal the effects of errors is greatly enhanced.

The resynchronization approach adopted by MPEG-4, referred to above as the video packet resynchronization approach, is similar to the Group of Blocks (GOBs) structure exploited by the ITU-T standards of H.261 and H.263 [63] codecs for re-establishing synchronization after encountering transmission errors. Let us hence briefly highlight the resynchronization procedure used by these codecs [385]. In these standards a GOB is defined as one or more rows of MBs within a specific video frame. At the commencement of a new GOB a GOB header is incorporated into the bitstream. This header information contains a GOB start code, which is different from a picture start code, and allows the decoder to locate this GOB. Furthermore, the GOB header contains information which allows the decoding process to be restarted by resynchronizing the decoder with the encoder, while resetting all predictively coded data. Figure 13.3 shows the GOB numbering scheme for H.263 for QCIF video format.

Again, while constructing the MPEG-4 bitstream, the encoder inserts unique resynchronization markers into the bitstream at approximately equally spaced intervals as seen in Figure 13.4 (b). Upon detecting an error, the decoder searches through the received bitstream, until the next resynchronization marker is found. Once this resynchronization marker is

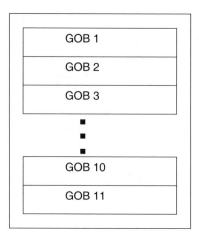

Figure 13.3: H.263 GOB numbering for a QCIF-resolution 176 × 144-pixel image.

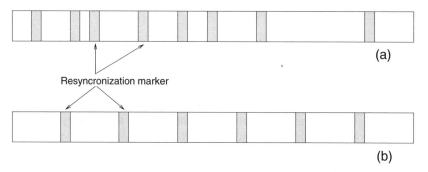

Figure 13.4: (a) Position of the synchronization markers in the bitstream for a baseline H.263 encoder showing the GOB headers. (b) Position of the resynchronization markers in the bitstream of an MPEG-4 encoder generating equal-length video packets.

found, the decoder re-establishes synchronization with the encoder. Hence, all data that corresponds to the MBs between the resynchronization points has to be discarded, because the effects of displaying an image reconstructed from erroneous bits may inflict highly annoying visual artifacts.

After synchronization has been re-established, the MPEG-4 data recovery tools attempt to recover data that in general would be lost. As mentioned earlier, data recovery techniques, such as reversible decoding [386] enable the decoder to salvage some of the data between the two resynchronization points, because the received bitstream can be decoded in both the forward and reverse direction, as will be described in Section 13.6.4.

The GOB start code aided resynchronization is based on spatial resynchronization. More explicitly, once a particular MB location is reached during the encoding process, a resynchronization marker is inserted into the bitstream. A potential problem associated with this approach is that because the encoding process results in a variable rate, these resynchronization markers will most likely be unevenly spaced throughout the bitstream.

13.6. ERROR-RESILIENCE TOOLS IN MPEG-4

Therefore, certain portions of the scene, such as highly motion-active areas, will be more susceptible to errors and these will also be more difficult to conceal.

MPEG-4 provides a similar method of resynchronization to that of the H.263 codec with one important difference: the MPEG-4 encoder is not restricted to inserting the resynchronization markers only at the beginning of each row of MBs. Instead, The encoder has the option of dividing the image into video packets, as seen in Figure 13.4, each constituted by an integer number of consecutive MBs. Hence, as portrayed in Figure 13.4, it forms periodic resynchronization markers throughout the bitstream. If the number of bits contained in the current video packet exceeds a predetermined threshold, then a new video packet is created at the start of the next MB. Thus, in the presence of a short burst of transmission errors, the decoder has the ability to promptly localize the error within a few MBs.

A resynchronization marker is used for distinguishing the start of a new video packet, as seen in Table 13.1 This marker is distinguishable from all possible VLC codewords as well as from the VOP start code. Header information is also provided at the start of a video packet as seen in Table 13.1. The information necessary for restarting the decoding process is contained in this header and includes the MB index of the first MB contained in this packet as well as the Quantization Parameter (QP) necessary for decoding the first MB. The MB index provides the necessary spatial resynchronization, while the knowledge of the QP allows the differential decoding process to be resynchronized.

13.6.2 Data Partitioning

Figure 13.5 depicts the organization of the video data within a packet for the MPEG-4 video compression scheme using no data partitioning. The video packet commences with a resynchronization marker, followed by the MB index and the QP and finally by the combined motion and Discrete Cosine Transform (DCT) data. Note that the combined motion and DCT data in general contains a lot more data than the three preceding header fields.

To elaborate a little further, Figure 13.6 shows the syntactic elements of each MB in the case of the MPEG-4/H.263 video encoder. This data is repeated for all MBs contained in the packet. The subscripts indicate the MB index. The parameter COD is a 1-bit field used for indicating whether a certain MB is encoded or skipped. In contrast MCBPC is a variable-length field used for indicating, first of all, the encoding mode of a MB, namely the `intra, inter, inter4V`[1] and `intra+Q`[2] coded modes. Secondly, MCBPC also signals which of the two chrominance blocks of the MB is coded. Furthermore, DQUANT is an optimal 2-bit fixed-length field used for indicating the incremental modification related to the quantizer index with respect to the previous MB's quantizer index. The parameter coded block pattern for luminance (CBPY) is a variable length code that indicates which of the four blocks of the MB are encoded. The motion vectors are transmitted in the form of the motion vector differences by a variable length code and, finally, the DCT data comprises the 64 DCT coefficients actually encoded, which are transmitted after zig-zag scanning and run-length encoding, using the VLC table of the MPEG-4 standard [349].

---

[1]The acronym implies the transmission of four motion vectors per MB. Each luminance block in the MB is allowed to have its own motion vector. This allows higher flexibility in obtaining the best match for the MB.

[2]The QP is modified for this MB with respect to the previous MB.

Resync. marker	MB index	QP	Combined motion and DCT data

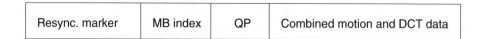

Figure 13.5: Organization of the data within an MPEG-4 video packet.

COD_1	$MCBPC_1$	$CBPY_1$	$DQUANT_1$	Encoded $MV(s)_1$	DCT $data_1$	■ ■ ■

Figure 13.6: Bitstream components for each MB within the MPEG-4 video packet.

Resync. marker	MB index	QP	Motion data	MBM	DCT data

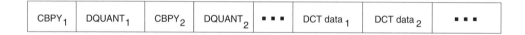

Figure 13.7: MPEG-4 bitstream organization using data partitioning for separating the motion and DCT data.

COD_1	$MCBPC_1$	Encoded $MV(s)_1$	COD_2	$MCNPC_2$	Encoded $MV(s)_2$	■ ■ ■ ■

Figure 13.8: Bitstream components of the motion data.

$CBPY_1$	$DQUANT_1$	$CBPY_2$	$DQUANT_2$	■ ■ ■	DCT $data_1$	DCT $data_2$	■ ■ ■

Figure 13.9: Bitstream components of the DCT data.

In MPEG-4, the data partitioning mode separates the data of a video packet into a motion-selected part and a texture-selected part separated by a unique motion boundary marker (MBM), as shown in Figure 13.7. Compared with Figure 13.5, the motion- and DCT-related parts are now separated by an MBM. All of the syntactic elements that have motion-related information are placed in the motion partition, and all of those related to DCT data are placed in the DCT data partition. Figure 13.8 shows the bitstream structure after reorganization of the motion part, while Figure 13.9 shows the bitstream structure of the DCT part. Note that the elements in both Figures 13.8 and 13.9 have been defined in the context of Figure 13.5.

When a transmission error is detected in the motion section of Figure 13.8, the decoder declares an error and replaces all of the MBs in the current packet with skipped blocks, until the next resynchronization marker is found. When a transmission error is detected in the texture section, but no errors are detected in the motion section, the corresponding MB's motion vectors are used during motion compensation for reconstructing the transmitted

MBs using the block DCT data, because the texture part is discarded and the decoder resynchronizes when detecting the next resynchronization marker. If no error is detected in the motion and texture sections, but the resynchronization marker is not found at the end of decoding, an error is declared. In this case, only the texture part of the MBs contained in the current packet is discarded. As reported in [387], data partitioning provides improved video quality under different error conditions.

13.6.3 Reversible Variable-length Codes

RVLCs [386] constitute a specific class of variable length codes that can be uniquely decoded in both directions. This advantageous property is a consequence of the fact that no codeword may constitute a prefix or postfix of another codeword, which allows their unambiguous decoding in both forward and backward directions. When the decoder detects an error, while decoding in the forward direction, it will immediately look for the next resynchronization marker and start decoding in the reverse direction, until an error is found. Based on the position of the detected errors in the forward and backward directions, the decoder is likely to be able to locate the error within a more limited region in the bitstream and hence recover additional data. As illustrated in Figure 13.10, only the data contained in the shaded region is discarded. It is important to mention that if RVLCs were not invoked, all of the data between the two consecutive resynchronization markers seen in Figure 13.10 would have to be discarded. In MPEG-4, RVLCs have been shown to provide a significant gain in subjective video quality terms in the presence of channel errors by enabling more data to be recovered [382, 386]. In MPEG-4, when the data partitioning mode is employed, the DCT coefficient information is encoded by means of RVLCs.

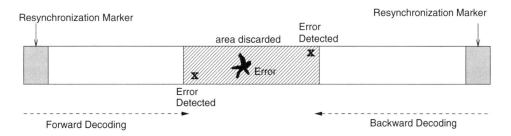

Figure 13.10: RVLCs enable the recovery of data by allowing decoding to take place both in the forward and reverse directions.

13.6.4 Header Extension Code

In video compression, the HEC is the most important information required by the decoder for decoding video bitstream. The header extension code allows the introduction of duplicate copies of important picture header information sequences in the video packets. The HEC is a fixed single bit, which indicates the presence of additional resynchronization information, including information such as the spatial resolution of the frame, its temporal location and the encoding mode of the frame considered. This technique has the potential to reduce the number of discarded video frames significantly in error-prone environments. This additional information is made available for the decoder, when the VOP header has been corrupted. This mode is supported in the MPEG-4 codec, and a similar tool is included also in the Real Time Payload (RTP) [356] specification of the H.263 version 2 standard, as defined in Request for the Comments 2429 [388].

In conclusion, wireless communication networks do not always guarantee error-free transmission. Hence, we have discussed a range of error-resilient coding methods that facilitate reliable video communications in error-prone environments. Many of these error-resilient methods have now been incorporated by the various video standards, such as H.263 and MPEG-4.

13.7 MPEG-4 Bit-sensitivity Study

13.7.1 Objectives

The purpose of the error-sensitivity investigations in this section is to use the knowledge gained in order to design an MPEG-4 based videophone system for a mobile environment with the following aims:

- better error resilience than the original MPEG-4 standard;

- a data stream that conforms to the MPEG-4 standard or is amenable to conversion with a simple transcoder;

- a reconfigurable system so that it can adjust to different channel conditions.

13.7.2 Introduction

In this section, we embark on presenting the results of our MPEG-4 bit-sensitivity study. Over the years a number of different techniques were used in the literature for quantifying the bit error-sensitivity of video codecs. The outcome of the investigations to be conducted is highly dependent on a number of factors, such as the video material used, the target bitrate of the video codec, and on the averaging procedure employed. Our experiments were carried out using the MPEG-4 encoder and decoder with the error-resilient mode disabled. This implies that whenever a bit was corrupted, the MPEG-4 codec did not attempt to correct or conceal the error effects. We purposely selected this mode in our experiments so that when a bit was corrupted, the corresponding video artifacts and video degradation were not obfuscated.

Here, we propose a simplified objective video quality measure-based bit-sensitivity evaluation procedure, which attempts to examine all of the major factors influencing the sensitivity of the MPEG-4 encoded bits. Specifically, the proposed procedure takes into account the MPEG-4 parameters contained in the entire bitstream, which also has the effect of losing synchronization, as well as error propagation effects.

At this stage, we assume that readers are familiar with the MPEG-4 standard at an outline of depth, which was provided in Chapter 11. The aim of our MPEG-4 bit-sensitivity study

13.7. MPEG-4 BIT-SENSITIVITY STUDY

was to quantify the average Peak Signal-to-Noise Ratio (PSNR) degradation caused by each erroneously decoded video codec parameter in the bitstream, so that appropriate channel coding-based error protection can be assigned to each parameter.

13.7.3 Simulated Coding Statistics

Video-coding simulations were conducted in order to find the relative importance of the different coding parameters or codewords in the MPEG-4 bitstream.

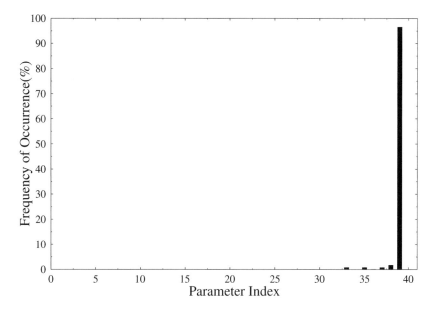

Figure 13.11: Probability of occurrence for the various MPEG-4 parameters for the "Miss America" QCIF video sequence encoded at 30 frame/s and 1.15 Mbit/s using the parameter indices listed in Table 13.2.

The results presented in this section quantify the relative frequency of occurrence for various coding parameters using the MPEG-4 codec. Specifically, Figures 13.11–13.14 show the probability of occurrence for the various MPEG-4 parameters in the context of the VOs, VOL, VOP and MB representations. The statistics shown in Figures 13.11–13.14 were derived for the same video clip, namely, for the "Miss America" Quarter Common Intermediate Formal (QCIF) video sequence, but the results are presented on different vertical scales, because some of the coding parameters have a relatively small probability of occurrence compared with others. Therefore results are presented on various vertical scales and the duplications of results is avoided in Figures 13.11–13.14. In addition, Table 13.2 shows the parameter name corresponding to the parameter indexes in Figures 13.11–13.14.

We now discuss the results shown in Figure 13.14, which shows the probability of occurrences confined to the range below 0.00015%. It can be seen that all parameters having indices of 1–20 in Table 13.2 except for the markers_bit have the same probability of

450   CHAPTER 13. MPEG-4 BITSTREAM AND BIT-SENSITIVITY STUDY

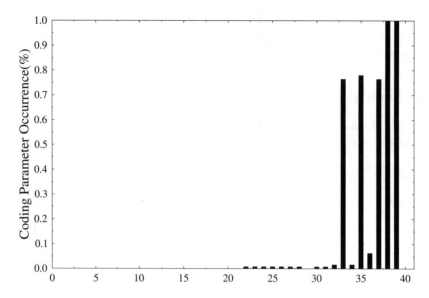

Figure 13.12: Probability of occurrence for the various MPEG-4 parameters for the "Miss America" QCIF video sequence encoded at 30 frame/s and 1.15 Mbit/s using the parameter indices listed in Table 13.2. This graph illustrates the probability of occurrence for the various types of MPEG-4 parameters in MBs.

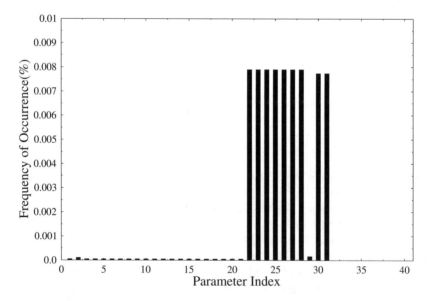

Figure 13.13: Probability of occurrence for the various MPEG-4 parameters for the "Miss America" QCIF video sequence encoded at 30 frame/s and 1.15 Mbit/s using the parameter indices listed in Table 13.2. This graph illustrates the probability of occurrence for the various types of MPEG-4 parameters in the VOP header.

13.7. MPEG-4 BIT-SENSITIVITY STUDY

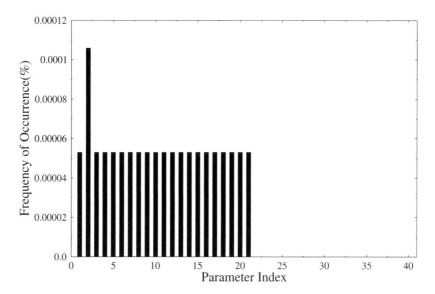

Figure 13.14: Probability of occurrence for the various MPEG-4 parameters for the "Miss America" QCIF video sequence encoded at 30 frame/s and 1.15 Mbit/s using the parameter indices listed in Table 13.2. This graph illustrates the probability of occurrence of the various types of MPEG-4 parameters in both the VOs and VOL headers.

occurrence. The parameter marker_bit exhibits a doubled probability of occurrence in comparison with the other VO and VOL header parameters, because it appeared twice in the VOL header. This is a 1-bit parameter that is set to logical 1 and assists in preventing the emulation of start codes by the video-encoded bits.

Figure 13.13 shows the probability of occurrence for the bits mainly belonging to the VOP seen between the indices of 22 and 31. It is obvious that the probability of occurrence for the parameters associated with the VOP is significantly higher compared with the parameters of the VO and associated VOL layer, because the VO and VOL header seen at indices of 1–18 only appear once at the very beginning of the entire video sequence. Likewise, in the MPEG-4 VOP coding, each VOP is associated with a video frame, which can be either intra- or inter-frame coded. We used a total of 150 video frames of the "Miss America" sequence in our experiments. Therefore, there were 150 VOP and hence 150 VOP headers for the entire video bitstream. As a result, the probability of encoded bits being allocated to the various VOP header parameters is higher than that of the parameters in VO and VOL.

Hence, it is plausible that all VOP header parameters, except for the VOP_Intra_Quantizer, VOP_Inter_Quantizer and VOP_FCode_Forward parameters, we have the same probability of occurrence, because they appear once for every coded video frame. The VOP_Inter_Quantizer and VOP_FCode_Forward parameters have a slightly lower probability of occurrence, than most of the remaining VOP header parameters seen at indices because these two parameters are only generated in inter-frame coding mode, while the VOP_Intra_Quantizer parameter has a significantly low probability of occurrence

Table 13.2: Reference to the MPEG-4 Coding Parameter Index as Shown in Figures 13.11–13.14

Index	Descriptions	Index	Descriptions
1	VO_start_code	21	VOL_Scalability
2	markers_bit	22	VOP_START_CODE
3	VO_id	23	VOP_Prediction_Type
4	VOL_START_CODE	24	VOP_timemovulo
5	VOL_Id	25	VOP_timeinc
6	VOL_RandomAccessible	26	Width_buffer
7	VOL_IsObjectLayerIdentifier	27	VOP_IntraDC_Vlc
8	VOL_ControlParameters	28	VOP_Interlaced
9	VOL_Shape	29	VOP_Intra_Quantizer
10	VOL_TimeIncrement_Resolution	30	VOP_Inter_Quantizer
11	VOL_FixedVopRate	31	VOP_FCodeFor
12	VOL_Width	32	PutMCBPC_Intra
13	VOL_Height	33	PutMCBPC_Inter
14	VOL_OBMCDisable	34	ACpred_flag
15	VOL_SpriteUsage	35	PutCBPY_index
16	VOL_Not_8_Bit	36	PutDCsize_lum
17	VOL_QuantType	37	PutDCsize_chrom
18	VOL_CompleEstimatDisable	38	PutMV
19	resyn_marker_flag	39	ALL_TCOEFF
20	VOL_DataPart_Enable		

because it is only required for intra-frame coding. There are only three intra-frames in our experiments, hence the low probability seen in Figure 13.13.

Let us now compare the relative frequency of occurrence for the parameters incorporated by the MB layer. In the MB layer, there are even more high probability of occurrence parameters, because there were 99 MBs per video frame for the QCIF "Miss America" video sequence we used in these experiments. The DCT parameter (ALL_TCOEFF) has the highest probability of occurrence, exceeding 90%. In our experiments, the video sequence was encoded at a high bitrate of 1.15 Mbit/s, hence at this relatively high bitrate it is not surprising to see that the DCT coefficients have the highest probability of occurrence. Let us now focus our attention on the effects of transmission errors in the next section.

13.7.4 Effects of Errors

In order to investigate the effects of transmission errors on the MPEG-4 parameters, a wide range of simulations were carried out using a 150-frame duration, QCIF-resolution "Miss America" sequence encoded at 1.15 Mbit/s. Simulations were conducted mainly for two situations, namely for encountering errors in intra-coded error as well as in inter-coded frames. Almost all of the MPEG-4 parameters from the VO, VOL, VOP and MB domains of Table 13.2 were corrupted during our investigations. The PSNR degradations were recorded for each of the parameters in Table 13.2 under the following simulation conditions:

13.7. MPEG-4 BIT-SENSITIVITY STUDY

- error in intra-coded frame (frame 1, VOP1);

- error in inter-coded frame (frame 2, VOP2).

The effects of corruption on the MPEG-4's VO, VOL and VOP header parameters were somewhat arduous to estimate because these parameters contain important information related to the entire video sequence, to a particular video frame or to a relatively small MB.

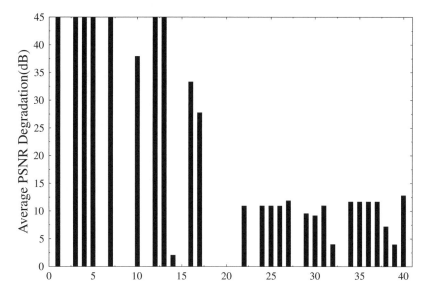

Figure 13.15: Average PSNR degradation for the various MPEG-4 parameters in the VO, VOL, VOP and MB domains for the "Miss America" QCIF video sequence encoded at 30 frame/s and 1.15 Mbit/s using the parameter indices of Table 13.2.

Figure 13.15 shows the average PSNR degradations inflicted by the various MPEG-4 coding parameters, while their relative frequencies were characterized in Figures 13.11–13.14. It can be seen from Figure 13.15 that most of the corrupted parameters from the VO and VOL domains of Table 13.2 resulted in a high PSNR degradation. Furthermore, the video sequence cannot be decoded at all when parameters such as the VO_start_code, VO_id, VOL_START_CODE, VOL_Id, VOL_IsObjectLayerIdentifier, VOL_Width and VOL_Height are corrupted. The decoder cannot even commence the decoding operations, when those parameters are corrupted. This situation is indicated in Figure 13.15 by setting the PSNR degradation to the average PSNR of the video sequence concerned, which is seen to be 45 dB and this implies the assumption of encountering a PSNR of 0 dB.

In contrast, some of the parameters, such as markers_bit, VOL_Random Accessible, VOL_ControlParameters, VOL_Shape, VOL_FixedVopRate, VOL_SpriteUsage, VOP_ComplexEstimatDisable, resyn_marker_flag, VOL_Data Part_Enable, VOL_Scalability, VOP_Interlaced and PutMCBPC_Inter are more error resilient and, hence, even when these parameters were corrupted, they did not dramatically affect the picture decoding process. Thus, no significant PSNR degradation

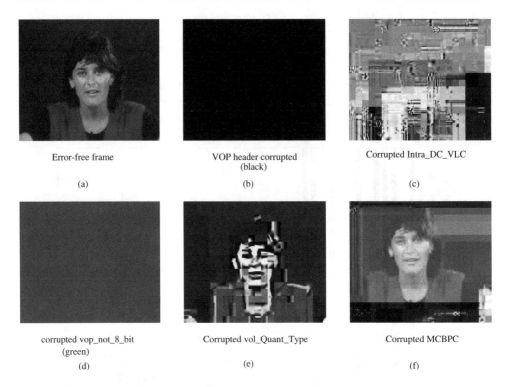

Figure 13.16: Image degradation due to the corruption of the various MPEG-4 coding parameters in an intra-coded frame.

was recorded. Since advanced functionalities such as binary shape coding [343] and sprite coding [349] have not been included in our MPEG-4 studies, the related parameters of Table 13.2 have been disabled during the encoding and decoding process.

When the parameter VOL_TimeIncrement_Resolution was corrupted, the PSNR degradation shown in Figure 13.15 became high, around 37 dB. As suggested by the acronym, when this parameter was corrupted, it imposed the wrong timing information during the decoding process. Specifically, the problems caused by the corruption of these parameters resulted in decoding the video frames in an incorrect order, while some video frames might have been skipped as a consequence of using an incorrect time increment.

The flag VOL_Not_8_bit contained at index 16 in the VOL header of Table 13.2 is set when the video pixel amplitude resolution is not 8 bits per pixel. Specifically, when this single bit parameter is set to binary "0" the video pixel amplitude resolution is 8 bits per pixel. In contrast, when this single bit parameter is set to binary "1" the video pixel amplitude resolution is not 8 bits per pixel. Hence, the video decoder will search through the bitstream for another parameter referred to as "bit_per_pixel". This "bit_per_pixel" parameter specifies the video pixel amplitude resolution in bits per pixel between 4 and 12. As we used the 8 bits per pixel video pixel amplitude resolution in our experiments, the parameter "bit_per_pixel" is not necessary, hence it is absent in the bitstream. However, when the VOL_Not_8_bit parameter is corrupted, the decoder is unable to find

13.7. MPEG-4 BIT-SENSITIVITY STUDY

the "bit_per_pixel" parameter, therefore the video decoder does not know the video pixel amplitude resolution and as a result, the decoded video frame appeared as a green frame when viewed in color and this frame was printed in Figure 13.16(d) as a dark-gray frame.

When the VOL_Quant_Type parameter of Table 13.2 is corrupted, the entire video frame appears to be "blocky", as illustrated in Figure 13.16(e). As expected, the PSNR degradation also becomes rather high.

When any of the parameters associated with the VOP header seen at indices 22–31 of Table 13.2 is corrupted, it results in a lower PSNR degradation compared with those in the VO and VOL layer, as seen in Figure 13.15. The parameters VOP_START_CODE, VOP_timemovulo, VOP_timeinc, VOP_Width_buffer, VOP_IntraDCVlcThr, VOP_Intra_Quantizer result in similar PSNR degradations. The parameters appearing at the beginning of every video frame bitstream may be considered vital in the VOP header. Again, as the VO and VOL header found at indices of Table 13.2, the integrity of the VOP header is also important, although not for the entire video sequence, but for a complete video frame. When these bits were corrupted, the entire video frame was skipped, but the decoder remained capable of decoding the rest of the video sequence. In our experiments, we corrupted the first intra VOP header found at indices 22–31 of Table 13.2, and hence as expected, the decoder skipped the first frame, because there was no previous information, therefore it appeared to be a black frame, as demonstrated in Figure 13.16(b).

The parameter VOP_IntraDCVlcThr allows the codec to switch between two sets of VLC which can be used to encode the intra-DC coefficients. When this parameter is corrupted, decoder uses the wrong VLC index for decoding the intra-frame DC coefficients. The associated error effects are severe and because most of the MPEG-4 bitstream is encoded using VLCs, even a single bit-error will detrimentally affect the achievable video quality. Figure 13.16(c) portrays an example when the VOP_IntraDCVlcThr parameter is corrupted and consequently the rest of the DC coefficients in the entire video frame are affected.

Another parameter that has to be characterized here is the so-called Macroblock type and Coded Block Pattern for Chrominance (MCBPC) parameter, which is also a variable-length code that is used for signalling the MB type and the coded block pattern for chrominance. This parameter is always included for every encoded MB. When it is corrupted, it may result in another "new" VLC, which is not included in the VLC index table. This would also affect the rest of the VLC positions in the bitstream or, even more seriously, could result in annoying artifacts as mentioned before. Figure 13.16(f) shows the effects of the corrupted MCBPC of MB1, which is the first MB at the top-left corner of the picture. Obviously, the affected part was concentrated in the area of MB1, but this parameter is variable-length coded, it may also gradually contaminate the rest of the MB in the picture.

To elaborate a little further, Figure 13.17 illustrates the effects of a bit error occurred in the MB83, where we observe that the picture is less contaminated compared with that illustrated in Figure 13.16(f). Annoying artifacts only start to emerge from MB83 onwards and there are total of 99 MBs in a QCIF-resolution video frame.

During our inter-frame coded bit-error-sensitivity study, investigations were conducted using the second frame of the "Miss America" sequence and the bitrate was the same as above. The corrupted frames are displayed in Figure 13.18, while the corresponding PSNR degradation is shown at index of Figure 13.15. Most of the video artifacts caused by errors

Figure 13.17: Image degradation due to the corruption of DCT coefficient in frame 1, MB83 for the "Miss America" QCIF video sequence encoded at 30 frame/s and 1.15 Mbit/s.

        (a)   (b)   (c)
Corrupted F_code_Forward   Corrupted DCT_coefficient   Error-free frame

Figure 13.18: Image degradation due to the corruption of the MPEG-4 coding parameters in an inter-frame coded sequence.

that occurred in the VOP header of inter-frame pictures are similar to those recorded in the intra-frame coded scenario as described above.

The VOP_fcode_forward parameter of Table 13.2 is only required in the inter-frame coding mode. This is used for decoding of the motion vectors in the forward direction. When this parameter is corrupted, it results in wrong forward prediction for the motion vectors and hence some of the MBs, which also have associated motion vectors might be lost or not decoded properly as shown in Figure 13.18(a). Figure 13.18(b) also displays the video degradation inflicted by errors in the inter-frame DCT coefficients. As can be seen in Figure 13.18, the video degradation is not as severe as in the intra-frame coding of Figure 13.16, and much less information is used to encode an inter-frame, which uses predicting information from the reference frame. So if the inter-DCT coefficient is corrupted, the corrupted information can be concealed by other DCT information from the predicted video frame.

In summary, as these graphs and figures reveal, the intra-frame errors result in image degradations that diminish very little with time; in addition errors will propagate to the subsequent inter-frames until the next intra-frame commences. In contrast, the errors in an inter-coded frame do not affect the image quality as adversely as the intra-coded frame errors.

13.8 Chapter Conclusions

In this chapter we have presented the error-resilience aspects of the MPEG-4 video standard. A number of error-resilience tools have been adapted into the MPEG-4 video standard, which facilitates the robust transmission of compressed video over error-prone communication channels, such as wireless links. In Section 13.6, we described these tools in detail and highlighted their relative merits.

Having discussed the MPEG-4 bitstream and error-resilience tools, in Section 13.5, we investigated the hierarchial MPEG-4 bitstream organization, more specifically the VO layer, VOL layer, VOP layer and MB layer. We also quantified the relative frequency of certain parameters. We have shown that the transform coefficient (TCOEFF) parameter has the highest occurrence of probability, representing in excess of 90% of the total bits at a bitrate of 1 Mbit/s using the QCIF "Miss America" video sequence. Then a series of MPEG-4 bit sensitivity studies were carried out, demonstrating that the header information of the VO, VOL, VOP, and MB hierarchies are extremely sensitive, because if the header's parameters are corrupted, the entire encoded bitstream may become undecodable. In contrast, in the presence of errors in the TCOFF, the bitstream can still be decoded, although the reconstructed video texture may be perceptually degraded, as seen in Figures 13.17 and 13.18.

The perceptual effects of some of the erroneous or corrupted bits can be concealed to a certain degree with the aid of the error resilience tools introduced in the MPEG-4 standard. However, we also proposed a number of other novel methods that further improve the performance of the wireless video codec, which were not incorporated the standard. If the encoder and decoder are aware of the limitations imposed by the communication channel, they are capable of further improving the resilience of the codec. This chapter was mainly dedicated to the error-resilience aspects of the video codec. Hence, in the forthcoming chapters we will focus our attention on the system architecture of a range of attractive wireless videophone systems.

Chapter 14

HSDPA-like and Turbo-style Adaptive Single- and Multi-carrier Video Systems

P.J. Cherriman, J-Y. Chung, L. Hanzo, T. Keller, E-L. Kuan, C.S. Lee, R. Maunder, S.X. Ng, S. Vlahoyiannatos, J. Wang and B.L. Yeap

14.1 Turbo-equalized H.263-based Videophony for GSM/GPRS[1,2]

14.1.1 Motivation and Background

The operational second-generation (2G) wireless systems [389] constitute a mature technology. In the context of 2G systems, and, with the advent of videotelephony, attractive value-added services can be offered to a plethora of existing users. Although the 2G systems have not been designed with video communications in mind, with the aid of the specially designed error-resilient, fixed-rate video codecs proposed in Chapters 3–5 it is nonetheless realistic to provide videophone services over these low-rate schemes. Specifically, in this chapter we designed a suite of fixed-rate, proprietary video codecs capable of operating at a video scanning or refresh rate of 10 frames/s over an additional speech channel of the 2G systems. These video codecs were capable of maintaining sufficiently low bitrates for the provision of videophony over an additional speech channel in the context of the operational 2G wireless systems [389] provided that low-dynamic head-and-shoulders video

---

[1] This section is based on **P. Cherriman, B. L. Yeap, and L. Hanzo**: Turbo-Equalized H.263-based videotelephony for GSM/GPRS; submitted to *IEEE Transactions on Circuits and Systems for Video Technology*, 2000.

[2] ©2000 IEEE. Personal use of this material is permitted. However, permission to reprint/republish this material for advertising or promotional purposes or for creating new collective works for resale or redistribution to servers or lists, or to reuse any copyrighted component of this work in other works, must be obtained from IEEE.

Table 14.1: System Parameters

Simulation Parameters	
Channel model	COST-207 hilly terrain
Carrier frequency	900 MHz
Vehicular speed	30 mph
Doppler frequency	40.3 Hz
Modulation	GMSK, $B_n = 0.3$
Channel coding	Convol.$(n, k, K) = (2,1,5)$
Octal generator polynomials	23, 33
Channel interleavers	Random (232, 928)
Turbo-coding interleavers	Random (116, 464)
Max turbo-equalizer iterations	10
No. of TDMA frame per packet	2
No. of slots per TDMA frame	1, 4
Convolutional decoder algorithm	LOG-MAP
Equalizer algorithm	LOG-MAP

sequences of the 176×144-pixel Quarter Common Intermediate Format (QCIF) or 128×96-pixel sub-QCIF video resolution are employed. We note, however, that for high-dynamic sequences the 32 kbit/s typical speech bitrate of the cordless telephone systems [389], such as the Japanese PHS, the Digital European Cordless Telephone (DECT), or the British CT2 system, is more adequate in terms of video quality. Furthermore, the proposed programmable video codecs are capable of multirate operation in the third generation (3G) Universal Mobile Telecommunications System (UMTS) or in the IMT2000 and cdma2000 systems, which were summarized in [83].

Chapters 3–5 used constant video rate proprietary video codecs and reconfigurable Quadrature Amplitude Modulation (QAM)-based [390] transceivers. In this chapter we advocate constant-envelope Gaussian Minimum Shift Keying (GMSK) [389]. Specifically, we investigated the feasibility of H.263-based videotelephony in the context of an enhanced turbo-equalized GSM-like system, which can rely on power-efficient class-C amplification. The associated speech compression and transmission aspects are beyond the scope of this book [391].

The outline of this section is as follows. Section 14.1.2 summarizes the associated system parameters and system's schematic, while Section 14.1.3 provides a brief overview of turbo equalization. Section 14.1.4 characterizes the system both in terms of turbo equalization performance and video performance. Lastly, Section 14.1.5 provides our conclusions. Let us now consider the outline of the system.

14.1.2 System Parameters

The associated video system parameters for the GSM system are summarized in Table 14.1, and the system's schematic is portrayed in Figure 14.1. An advanced feature of the system is its employment of joint channel decoding and channel equalization, which is referred to as

14.1. TURBO-EQUALIZED H.263-BASED VIDEOPHONY FOR GSM/GPRS

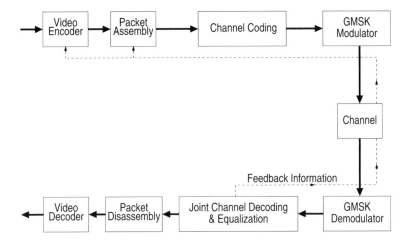

Figure 14.1: System schematic for turbo-equalized video system.

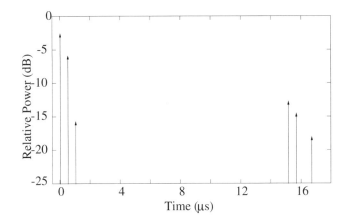

Figure 14.2: COST207-Hilly terrain channel model.

turbo equalization. The fundamental principles and motivation for using turbo equalization are described in Section 14.1.3.

The system uses the GSM frame structure of Chapter 8 in [389] and the COST-207 Hilly Terrain (HT) channel model, whose impulse response is shown in Figure 14.2. Each transmitted packet is interleaved over two GSM TDMA frames in order to disperse bursty errors.

The GPRS system allows the employment of multiple timeslots per user. We studied both a GSM-like system using 1 slot per TDMA frame and a GPRS-like arrangement with four slots per TDMA frame. In this scenario, the user is assigned half the maximum capacity of an eight-slot GPRS/GSM carrier. The bitrates associated with one and four slots per TDMA frame are shown in Table 14.2.

The effective video bitrates that can be obtained in conjunction with half-rate convolutional coding are 10 and 47.5 Kbit/s for the one and four slots per TDMA frame

Table 14.2: Summary of System-specific Bitrates

	Bitrates, etc.	
Slots/TDMA frame	1	4
Coded bits/TDMA slot	116	116
Data bits/TDMA slot	58	58
Data bits/TDMA frame	58	232
TDMA frame/packet	2	2
Data bits/packet	116	464
Packet header (bits)	8	10
CRC (bits)	16	16
Video bits/packet	92	438
TDMA frame length	4.615 ms	4.615 ms
TDMA frames/s	216.68	216.68
Video packets per sec	108.34	108.34
Video bitrate (kbit/s)	10.0	47.5
Video frame rate (frames/s)	10	10

scenario, respectively. Again, the system's schematic is shown in Figure 14.1. The basic prerequisite for the duplex video system's operation (as seen in Figure 14.1) is use of the channel decoder's output is used for assessing whether the received packet contains any transmission errors. If it does, the remote transmitter is instructed (by superimposing a strongly protected packet acknowledgment flag on the reverse-direction message) to drop the corresponding video packet following the philosophy of [392]. This prevents the local and remote video reconstructed frame buffers from being contaminated by channel errors.

14.1.3 Turbo Equalization

Turbo equalization [393] was proposed by Douillard, Picart, Jézéquel, Didier, Berrou, and Glavieux in 1995 for a serially concatenated rate $R = \frac{1}{2}$, convolutional-coded Binary Phase Shift Keying (BPSK) system. Specifically, Douillard *et al.* demonstrated that the turbo equalizer could mitigate the effects of intersymbol interference (ISI), provided that the channel impulse response (CIR) is known. Instead of performing the equalization and error correction decoding independently, better performance can be achieved by considering the channel's memory, when performing joint equalization and decoding iteratively. Gertsman and Lodge [394] then showed that the iterative process of turbo equalizers can compensate for the degradations caused by imperfect channel estimation. In the context of noncoherent detection, Marsland *et al.* [395] demonstrated that turbo equalization offered better performance than Dai's and Shwedyk's noncoherent, hard-decision-based receiver using a bank of Kalman filters [396]. Different iteration termination criteria [397], such as cross-entropy [398], were also investigated in order to minimize the number of iteration steps for the turbo equalizer. A turbo-equalization scheme for the Global System of Mobile Communications (GSM) was also proposed by Bauch and Franz [399], who investigated different approaches for overcoming the dispersion of the *a priori* information due to the

14.1. TURBO-EQUALIZED H.263-BASED VIDEOPHONY FOR GSM/GPRS

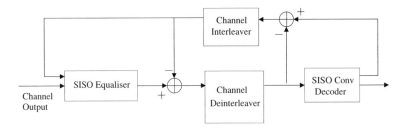

Figure 14.3: Structure of original turbo equalizer introduced by Douillard *et al.* [393].

interburst interleaving scheme used in GSM. Further research into combined turbo coding using convolutional constituent codes and turbo equalization has been conducted by Raphaeli and Zarai [400].

The basic philosophy of the original turbo-equalization technique derives from the iterative turbo-decoding algorithm consisting of two Soft-In/Soft-Out (SISO) decoders. This structure was proposed by Berrou *et al.* [401, 402]. Before proceeding with our in-depth discussion, let us briefly define the terms *a priori*, *a posteriori*, and extrinsic information, which we employ throughout this section.

A priori The *a priori* information associated with a bit v_m is the information known before equalization or decoding commences, from a source other than the received sequence or the code constraints. *A priori* information is also often referred to as intrinsic information to contrast it with extrinsic information.

Extrinsic The extrinsic information associated with a bit v_m is the information provided by the equalizer or decoder based on the received sequence and on the *a priori* information of all bits with the exception of the received and *a priori* information explicitly related to that particular bit v_m.

A posteriori The *a posteriori* information associated with a bit is the information that the SISO algorithm provides taking into account all available sources of information about the bit u_k.

The turbo equalizer of Figure 14.3 consists of a SISO equalizer and a SISO decoder. The SISO equalizer in the figure generates the *a posteriori* probability upon receiving the corrupted transmitted signal sequence and the *a priori* probability provided by the SISO decoder. However, at the initial iteration stages (i.e., at the first turbo-equalization iteration) no *a priori* information is supplied by the channel decoder. Therefore, the *a priori* probability is set to $\frac{1}{2}$, since the transmitted bits are assumed to be equiprobable. Before passing the *a posteriori* information generated by the SISO equalizer to the SISO decoder of Figure 14.3, the contribution of the decoder in the form of — *a priori* information — accruing from the previous iteration must be removed in order to yield the combined channel and extrinsic information. This also minimizes the correlation between the *a priori* information supplied by the decoder and the *a posteriori* information generated by the equalizer. The term "combined channel and extrinsic information" indicates that they are inherently linked. In fact, they are typically induced by mechanisms, which exhibit memory. Hence, they cannot be separated. The removal of the *a priori* information is necessary, to prevent the decoder from "reprocessing" its own information, which would result in the positive feedback phenomenon, overwhelming the decoder's current reliability-estimation of the coded bits, that is, the extrinsic information.

The combined channel and extrinsic information is channel-deinterleaved and directed to the SISO decoder, as depicted in Figure 14.3. Subsequently, the SISO decoder computes the *a posteriori* probability of the coded bits. Note that the latter steps are different from those in turbo decoding, which only produces the *a posteriori* probability of the source bits rather than those of all channel-coded bits. The combined deinterleaved channel and extrinsic information are then removed from the *a posteriori* information provided by the decoder in Figure 14.3 before channel interleaving in order to yield the extrinsic information. This approach prevents the channel equalizer from receiving information based on its own decisions, which was generated in the previous turbo-equalization iteration. The extrinsic information computed is then employed as the *a priori* input information of the equalizer in the next channel equalization process. This constitutes the first turbo-equalization iteration. The iterative process is repeated until the required termination criteria are met [397]. At this stage, the *a posteriori* information of the source bits, which has been generated by the decoder, is utilized to estimate the transmitted bits.

Recent work by Narayanan and Stüber [403] demonstrates the advantage of employing turbo equalization in the context of coded systems invoking recursive modulators, such as Differential Phase Shift Keying (DPSK). Narayanan and Stüber emphasized the importance of a recursive modulator and show that high-iteration gains can be achieved, even when there is no ISI in the channel (i.e., for transmission over the nondispersive Gaussian channel). The advantages of turbo equalization as well as the importance of a recursive modulator motivated our research on turbo equalization of coded partial response GMSK systems [389], since GMSK is also recursive in its nature. In our investigations, we have employed convolutional coding for the proposed turbo-equalized GSM-like video system as it has been shown in [404] that convolutional-coded GMSK systems are capable of providing large iteration gains — that is, gains in SNR performance with respect to the first iteration — with successive turbo-equalization iterations. We also observed that the convolutional-coded GMSK system employing turbo equalization outperformed the convolutional-coding based turbo-coded GMSK scheme, as demonstrated by Figure 14.4. Specifically, over a nondispersive Gaussian channel the convolutional-coded GMSK system had an approximately 0.8 dB better E_b/N_o performance than the corresponding convolutional-coding based turbo-coded GMSK scheme at BER $= 10^{-4}$. Although not explicitly shown here, similar findings were valid for dispersive Rayleigh-fading channels, where the advantage of the convolutional-coded system over the convolutional-coding-based turbo-coded scheme was approximately 1.0 dB.

These results were surprising since the more complex turbo-coded system was expected to form a more powerful encoded system compared to the convolutional-coded scheme. Below, we offer an interpretation of this phenomenon by considering the performance of both codes after the first turbo-equalization iteration. Specifically, over the nondispersive Gaussian channel in Figure 14.4, the convolutional-coded scheme yielded a lower BER than that of the turbo-coded system at E_b/N_o values below 4.5 dB, indicating that the *a posteriori* LLRs of the bits produced by the convolutional decoder had a higher reliability than that of the corresponding turbo-coded scheme. Consequently, upon receiving the higher-confidence

14.1. TURBO-EQUALIZED H.263-BASED VIDEOPHONY FOR GSM/GPRS

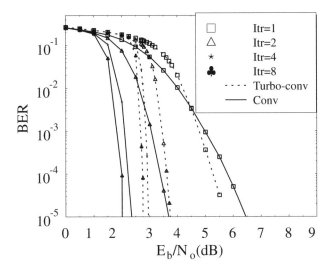

Figure 14.4: Comparison of the $R = 0.5$ convolutional-coded GMSK system with $R = 0.5$ convolutional-coding-based turbo-coded GMSK scheme, transmitting over the nondispersive Gaussian channel employing convolutional and turbo-coding-based turbo equalization, which performs eight turbo equalization iterations at the receiver.

LLR values from the decoder, the equalizer in the convolutional-coded scheme was capable of producing significantly more reliable LLR values in the subsequent turbo-equalization iteration, when compared to the turbo-coded system. After receiving these more reliable LLR values, the decoder of the convolutional-coded system will generate even more reliable LLR values. Hence, the convolutional-coded system outperformed the turbo-coded scheme after performing eight turbo-equalization iterations. Motivated by these trends — which were also confirmed in the context of dispersive Rayleigh-fading channels [404] — we opted for a convolutional-coded rather than turbo-coded GSM-like videophone system, which employs turbo equalization in order to enhance the video performance of the system.

14.1.4 Turbo-equalization Performance

Let us now characterize the performance of our video system. Figure 14.5 shows the bit error ratio (BER) versus channel SNR for the one- and four-slot scenarios, after one, two and ten iterations of the turbo equalizer. The figure shows the BER performance improvement upon each iteration of the turbo equalizer, although there is only a limited extra performance improvement after five iterations. The figure also shows that the four-slot scenario has a lower bit error ratio than the one-slot scenario. This is because the four-slot scenario has a longer interleaver, which renders the turbo-equalization process more effective due to its increased time diversity.

Let us now consider the associated packet-loss ratio (PLR) versus channel SNR performance in Figure 14.6. The PLR is a more pertinent measure of the expected video performance, since — based on the philosophy of [392] and on the previous two chapters — our video scheme discards all video packets, which are not error-free. Hence, our goal is

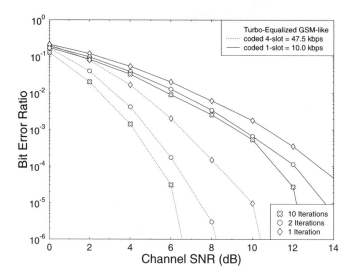

Figure 14.5: BER versus channel SNR for one and four slots per TDMA frame, and for 1, 2, and 10 turbo-equalizer iterations, over the channel of Figure 14.2.

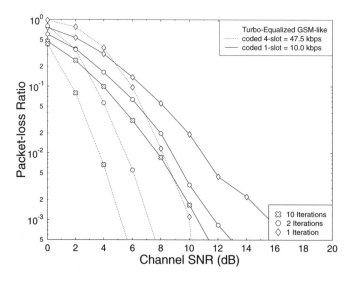

Figure 14.6: Video packet-loss ratio versus channel SNR for one and four slots per TDMA frame, and for 1, 2, and 10 turbo-equalizer iterations, over the channel of Figure 14.2 using convolutional coding.

14.1. TURBO-EQUALIZED H.263-BASED VIDEOPHONY FOR GSM/GPRS

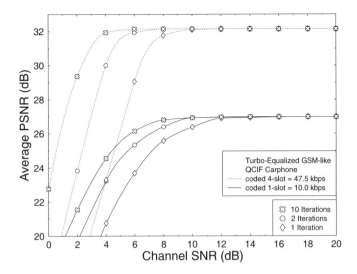

Figure 14.7: Video quality in PSNR (dB) versus channel SNR for one and four slots per TDMA frame, and for 1, 2, and 10 iterations of the turbo-equalizer upon using the highly motion active "Carphone" video sequence over the channel of Figure 14.2 using convolutional coding.

to maintain as low a PLR as possible. Observe in Figure 14.6 that the associated iteration gains are more pronounced in terms of the packet-loss ratio than in bit error ratio.

It should also be noted that for low SNRs the packet-loss performance of the four-slot system is inferior to that of the one-slot system, while the bit error ratio is similar or better at the same SNRs. This is because the probability of having a single-bit error in the four-slot video packet is higher due to its quadruple length. This phenomenon is further detailed in Section 14.1.4.2.

14.1.4.1 Video Performance

The PLR performance is directly related to the video quality of our video system. Figure 14.7 shows the associated average PSNR versus channel SNR performance, demonstrating that an improved video quality can be maintained at lower SNRs as the number of iterations increases. In addition, the higher bitrate of the four-slot system corresponds to a higher overall video quality. Up to 6 dB SNR-gain can be achieved after 10 turbo-equalization iterations, as seen in Figure 14.7.

Figure 14.7 characterizes the performance of the highly motion-active "Carphone" sequence. However, the performance improvements are similar for the low-activity "Miss America" video sequence, as seen in Figure 14.8. Observe that the lower-activity "Miss America" video sequence is represented at a higher video quality at the same video bitrate. A deeper insight into the achievable video-quality improvement in conjunction with turbo equalization can be provided by plotting the video quality measured in PSNR (dB) versus time, as seen in Figure 14.9 for the "Miss America" video sequence using the four-slot system at a channel SNR of 6 dB for one, two, and ten iterations of the turbo equalizer.

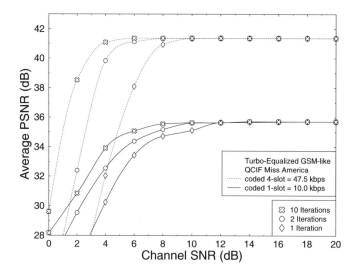

Figure 14.8: Video quality in PSNR (dB) versus channel SNR for one and four slots per TDMA frame, and for 1, 2, and 10 turbo-equalizer iterations, using the low-activity "Miss America" video sequence over the channel of Figure 14.2 using convolutional coding.

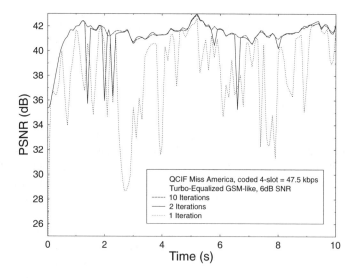

Figure 14.9: Video quality in PSNR (dB) versus time using four slots per TDMA frame, and for 1, 2, and 10 iterations of the turbo-equalizer for the low-activity "Miss America" video sequence using convolutional coding.

14.1. TURBO-EQUALIZED H.263-BASED VIDEOPHONY FOR GSM/GPRS

Table 14.3: Minimum Required Operating Channel SNR for the QCIF "Carphone" Sequence over the Channel of Figure 14.2

	Channel SNR for 1 dB Loss of PSNR	
Slots/TDMA Frame	1	4
1 Iteration	9.0 dB	7.5 dB
2 Iterations	7.2 dB	5.2 dB
3 Iterations	6.44 dB	3.9 dB
4 Iterations	6.37 dB	3.7 dB
10 Iterations	5.8 dB	3.4 dB

Specifically, the bottom-trace of the figure shows how the video quality varies in the one-iteration scenario, which is equivalent to conventional equalization. The sudden reductions in video quality are caused by packet-loss events, which result in parts of the picture being "frozen" for one or possibly several consecutive video frames. The sudden increases in video quality are achieved when the system updates the "frozen" part of the video picture in subsequent video frames. The packet-loss ratio for this scenario was 10% at the stipulated SNR of 6dB.

The video quality improved significantly with the aid of two turbo-equalizer iterations, while the packet-loss ratio was reduced from 10% to 0.7%. In the time interval shown in Figure 14.9, there are eight lost video packets, six of which can be seen as sudden reductions in video quality. However, in each case the video quality recovered with the update of the "frozen" picture areas in the next video frame.

We have found that the maximum acceptable PSNR video-quality degradation with respect to the perfect-channel scenario was about 1 dB, which was associated with nearly unimpaired video quality. In Table 14.3 we tabulated the corresponding minimum required channel SNRs that the system can operate at for a variety of scenarios, extracted from Figure 14.7. As the table shows, the minimum operating channel SNR for the one- and four-slot system using one iteration is 9 dB and 7.5 dB, respectively. This corresponds to a system using conventional equalization. A system using two turbo-equalization iterations can reduce these operating SNRs to 7.2 dB and 5.2 dB, respectively. The minimum operating SNRs can be reduced to as low as 5.8 dB and 3.4 dB for the one- and four-slot systems, respectively, when invoking ten iterations.

14.1.4.2 Bit Error Statistics

In order to demonstrate the benefits of turbo equalization more explicitly, we investigated the mechanism of how turbo equalization reduces the bit error and packet-loss ratios. We found that the distribution of the bit errors in video packets after each iteration provided interesting insights. Hence, the CDF of the number of bit errors per video packet was evaluated. In order to allow a fair comparison between the one- and four-slot system, we normalized the number of bit errors per packet to the video packet size, thereby producing the CDF of "in-packet" BER.

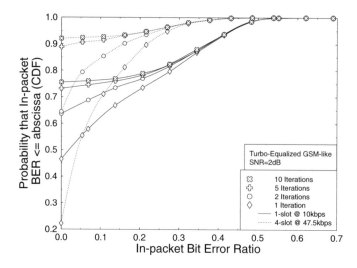

Figure 14.10: CDF of the "in-packet" BER at a channel SNR of 2 dB over the channel of Figure 14.2, and for various numbers of iterations for the turbo-equalized one- and four-slot systems using convolutional coding.

Figure 14.10 shows the CDF of the "in-packet" BER for a channel SNR of 2 dB and for one, two, five, and ten iterations for both the one- and four-slot systems. The value of the CDF for an "in-packet" BER of zero is the probability that a packet is error-free and so can be interpreted as the packet success ratio (PSR). The packet-loss ratio is equal to 1 minus the PSR. For example, the four-slot system in Figure 14.10 at one iteration has a packet success ratio of 0.22, which corresponds to a packet-loss ratio of 78%.

Both the one- and four-slot systems increase the PSR as the number of iterations increases. For example, the four-slot system increases the PSR from 22% to 92% as the number of iterations is increased from one to ten. This corresponds to a reduction in the packet-loss ratio from 78% to 8%. However, the CDF of "in-packet" BER can provide further insight into the system's operation. It can be seen in Figure 14.10 that the turbo-equalizer iterations reduce the number of packets having "in-packet" BERs of less than 30%. However, the probability of a packet having an "in-packet" BER higher than 35% is hardly affected by the number of iterations, since the number of bit errors is excessive, hence overwhelming even the powerful turbo equalization.

Figures 14.5 and 14.6, show that the four-slot system always has a lower BER than the one-slot system, although at low SNRs the PLR is higher for the four-slot system. The CDF in Figure 14.10 can assist in interpreting this further. The CDF shows that the PSR improves more significantly for the four-slot system than for the one-slot system as the number of iterations increases. This is because the four-slot system allows the employment of a longer interleaver, thereby improving the efficiency of the turbo equalizer. However, the CDF also underlines a reason for the lower BER of the four-slot system across the whole range of SNRs, demonstrating that the probability of packets having a high "in-packet" bit error rate is lower for the four-slot system. Since packets having a high "in-packet" BER have a more grave

14.1. TURBO-EQUALIZED H.263-BASED VIDEOPHONY FOR GSM/GPRS

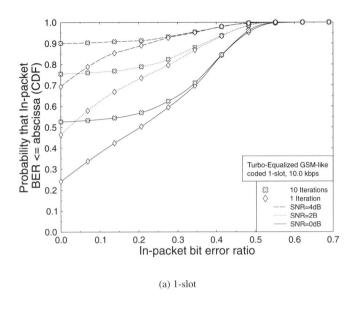

(a) 1-slot

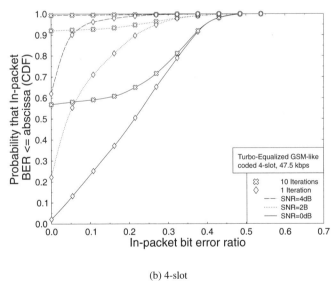

(b) 4-slot

Figure 14.11: CDF of "in-packet" BER performance over the channel of Figure 14.2 for the turbo-equalized one- and four-slot system for channel SNRs of 0 dB, 2 dB, and 4 dB and 1 and 10 iterations using convolutional coding.

effect on the overall BER than those packets having a low "in-packet" BER, this explains the inferior overall BER performance of the one-slot system.

Figure 14.11 shows the CDF of "in-packet" BER for conventional equalization and with the aid of ten turbo-equalizer iterations for 0 dB, 2 dB, and 4 dB channel SNRs. Figure 14.11(a) represents a one-slot system, and Figure 14.11(b) a four-slot system. The figures also show the packet-loss ratio performance improvement with the aid of turbo equalization.

14.1.5 Summary and Conclusions

In this section the performance of turbo-equalized GSM/GPRS-like videophone transceivers was studied over dispersive fading channels as a function of the number of turbo-equalization iterations. Iteration gains in excess of 4 dB were attained, although the highest per iteration gain was achieved for iteration indices below 5. As expected, the longer the associated interleaver, the better the BER and PLR performance. In contrast to our expectations, the turbo-coded, turbo-equalized system was outperformed by the less complex convolutional coded turbo-equalized system. In conclusion, GPRS/GSM are amenable to videotelephony, and turbo equalization is a powerful means of improving the system's performance. Our future work will improve the system's performance invoking the forthcoming MPEG4 video codec, using space–time coding and burst-by-burst adaptive turbo equalization.

14.2 HSDPA-style Burst-by-burst Adaptive CDMA Videophony: Turbo-coded Burst-by-burst Adaptive Joint Detection CDMA and H.263-based Videophony[3,4]

14.2.1 Motivation and Video Transceiver Overview

While the third-generation wireless communications standards are still evolving, they have become sufficiently mature for the equipment designers and manufacturers to complete the design of prototype equipment. One of the most important services tested in the field trials of virtually all dominant players in the field is interactive videotelephony at various bitrates and video qualities. Motivated by these events, the goal of this section is to quantify the expected video performance of a UMTS-like videophone scheme, while also providing an outlook on the more powerful burst-by-burst adaptive transceivers of the near future.

In this study, we transmitted 176×144 pixel Quarter Common Intermediate Format (QCIF) and 128×96 pixel Sub-QCIF (SQCIF) video sequences at 10 frames/s using a reconfigurable Time Division Multiple Access/Code Division Multiple Access (TDMA/CDMA) transceiver, which can be configured as a 1-, 2-, or 4-bit/symbol scheme. The H.263 video

---

[3]This section is based on P. Cherriman, E.L. Kuan, and L. Hanzo: Burst-by-burst Adaptive Joint-detection CDMA/H.263 Based Video Telephony, submitted to *IEEE Transactions on Circuits and Systems for Video Technology*, 1999.

[4]©1999 IEEE. Personal use of this material is permitted. However, permission to reprint/republish this material for advertising or promotional purposes or for creating new collective works for resale or redistribution to servers or lists, or to reuse any copyrighted component of this work in other works, must be obtained from IEEE.

Table 14.4: Generic System Parameters Using the Frames Spread Speech/Data Mode 2 Proposal [405]

Parameter	
Multiple access	TDMA/CDMA
Channel type	COST 207 Bad Urban
Number of paths in channel	7
Normalized Doppler frequency	3.7×10^{-5}
CDMA spreading factor	16
Spreading sequence	Random
Frame duration	4.615 ms
Burst duration	577 μs
Joint-detection CDMA receiver	Whitening matched filter (WMF) or minimum mean square error block decision feedback equalizer (MMSE-BDFE)
No. of slots/frame	8
TDMA frame length	4.615 ms
TDMA slot length	577 μs
TDMA slots/video packet	3
Chip periods/TDMA slot	1250
Data symbols/TDMA slot	68
User data symbol rate (kBd)	14.7
System data symbol rate (kBd)	117.9

codec [259] exhibits an impressive compression ratio, although this is achieved at the cost of a high vulnerability to transmission errors, since a run-length coded stream is rendered undecodable by a single-bit error. In order to mitigate this problem, when the channel codec protecting the video stream is overwhelmed by the transmission errors, we refrain from decoding the corrupted video packet in order to prevent error propagation through the reconstructed video frame buffer [392]. We found that it was more beneficial in video-quality terms, if these corrupted video packets were dropped and the reconstructed frame buffer was not updated, until the next video packet replenishing the specific video frame area was received. The associated video performance degradation was found to be perceptually unobjectionable for packet-dropping or transmission frame error rates (FER) below about 5%. These packet dropping events were signaled to the remote decoder by superimposing a strongly protected 1-bit packet acknowledgment flag on the reverse-direction packet, as outlined in [392]. Bose–Chaudhuri–Hocquenghem (BCH) [322] and turbo error correction codes [401] were used, and again, the CDMA transceiver was capable of transmitting 1, 2, and 4 bits per symbol, where each symbol was spread using a low spreading factor (SF) of 16, as seen in Table 14.4. The associated parameters will be addressed in more depth during our later discourse. Employing a low spreading factor of 16 allowed us to improve the system's multi-user performance with the aid of joint-detection techniques [406]. We also note that the implementation of the joint-detection receivers is independent of the number of bits per symbol associated with the modulation mode used, since the receiver simply inverts the associated system matrix and invokes a decision concerning the received symbol, regardless

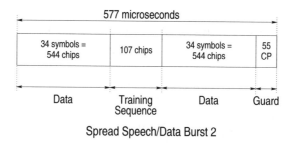

Figure 14.12: Transmission burst structure of the FMA1 spread speech/data mode 2 of the FRAMES proposal [405].

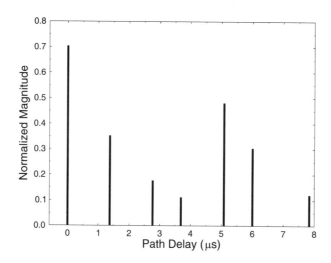

Figure 14.13: Normalized channel impulse response for the COST 207 [407] seven-path Bad Urban channel.

of how many bits per symbol were used. *Therefore, joint-detection receivers are amenable to amalgamation with the above 1-, 2-, and 4-bit/symbol modem, since they do not have to be reconfigured each time the modulation mode is switched.*

In this performance study, we used the Pan-European FRAMES proposal [405] as the basis for our CDMA system. The associated transmission frame structure is shown in Figure 14.12, while a range of generic system parameters is summarized in Table 14.4. In our performance studies, we used the COST207 [407] seven-path bad urban (BU) channel model, whose impulse response is portrayed in Figure 14.13.

Our initial experiments compared the performance of a whitening matched filter (WMF) for single-user detection and the minimum mean square error block decision feedback equalizer (MMSE-BDFE) for joint multi-user detection. These simulations were performed using four-level Quadrature Amplitude Modulation (4QAM), invoking both binary BCH [322] and turbo-coded [401] video packets. The associated bitrates are summarized in Table 14.5. The transmission bitrate of the 4QAM modem mode was 29.5 Kbps, which

14.2. HSDPA-STYLE BURST-BY-BURST ADAPTIVE CDMA VIDEOPHONY

Table 14.5: FEC-protected and Unprotected BCH and Turbo-coded Bit Rates for the 4QAM Transceiver Mode

Features	BCH coding	Turbo Coding
Modulation	4QAM	
Transmission bitrate (kbit/s)	29.5	
Video rate (kbit/s)	13.7	11.1
Video frame rate (Hz)	10	

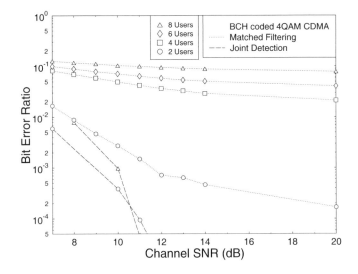

Figure 14.14: BER versus channel SNR 4QAM performance using BCH-coded, 13.7 Kbps video, comparing the performance of matched filtering and joint detection for two to eight users.

was reduced due to the approximately half-rate BCH or turbo coding, plus the associated video packet acknowledgment feedback flag error control [161] and video packetization overhead to produce effective video bitrates of 13.7 Kbps and 11.1 Kbps, respectively. A more detailed discussion on the video packet acknowledgment feedback error control and video packetization overhead will be provided in Section 14.2.2 with reference to the convolutionally coded multimode investigations.

Figure 14.14 portrays the bit error ratio (BER) performance of the BCH coded video transceiver using both matched filtering and joint detection for two to eight users. The bit error ratio is shown to increase as the number of users increases, even upon employing the MMSE-BDFE multi-user detector (MUD). However, while the matched filtering receiver exhibits an unacceptably high BER for supporting perceptually unimpaired video communications, the MUD exhibits a far superior BER performance.

When the BCH codec was replaced by the turbo-codec, the bit error ratio performance of both matched filtering and the MUD receiver improved, as shown in Figure 14.15. However, as expected, matched filtering was still outperformed by the joint-detection scheme for the

476　CHAPTER 14. HSDPA-LIKE AND TURBO-STYLE ADAPTIVE VIDEO SYSTEMS

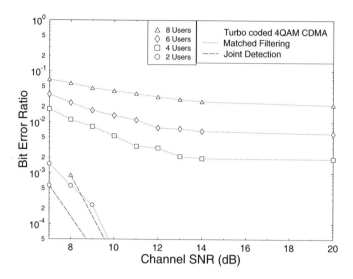

Figure 14.15: BER versus channel SNR 4QAM performance using turbo-coded 11.1 Kbps video, comparing the performance of matched filtering and joint detection for two to eight users.

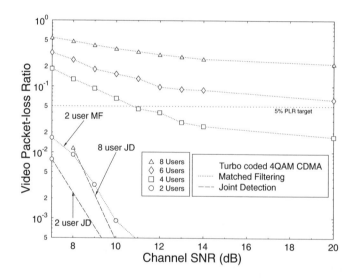

Figure 14.16: Video packet-loss ratio versus channel SNR for the turbo-coded 11.1 Kbps video stream, comparing the performance of matched filtering and joint detection for two to eight users.

same number of users. Furthermore, the matched filtering performance degraded rapidly for more than two users.

Figure 14.16 shows the video packet-loss ratio (PLR) for the turbo-coded video stream using matched filtering and joint detection for two to eight users. The figure clearly shows that the matched filter was only capable of meeting the target packet-loss ratio of 5% for up to four users, when the channel SNR was in excess of 11 dB. However, the joint detection

14.2. HSDPA-STYLE BURST-BY-BURST ADAPTIVE CDMA VIDEOPHONY

Table 14.6: Operational-mode Specific Transceiver Parameters for the Proposed Multimode System

Features	Multirate System		
Mode	BPSK	4QAM	16QAM
Bits/symbol	1	2	4
FEC	Convolutional Coding		
Transmitted bits/packet	204	408	816
Total bitrate (kbit/s)	14.7	29.5	58.9
FEC-coded bits/packet	102	204	408
Assigned to FEC-coding (kbit/s)	7.4	14.7	29.5
Error detection per packet	16 bit CRC		
Feedback bits/packet	9		
Video packet size	77	179	383
Packet header bits	8	9	10
Video bits/packet	69	170	373
Unprotected video rate (kbit/s)	5.0	12.3	26.9
Video frame rate (Hz)	10		

algorithm guaranteed the required video packet loss ratio performance for two to eight users in the entire range of channel SNRs shown. Furthermore, the two-user matched-filtered PLR performance was close to the eight-user MUD PLR.

14.2.2 Multimode Video System Performance

Having shown that joint detection can substantially improve our system's performance, we investigated the performance of a multimode convolutionally coded video system employing joint detection, while supporting two users. The associated convolutional codec parameters are summarized in Table 14.6.

We now detail the video packetization method employed. The reader is reminded that the number of symbols per TDMA frame was 68 according to Table 14.4. In the 4QAM mode this would give 136 bits per TDMA frame. However, if we transmitted one video packet per TDMA frame, then the packetization overhead would absorb a large percentage of the available bitrate. Hence we assembled larger video packets, thereby reducing the packetization overhead and arranged for transmitting the contents of a video packet over three consecutive TDMA frames, as indicated in Table 14.4. Therefore, each protected video packet consists of $68 \times 3 = 204$ modulation symbols, yielding a transmission bitrate of between 14.7 and 38.9 Kbps for BPSK and 16QAM, respectively. However, in order to protect the video data, we employed half-rate, constraint-length nine convolutional coding, using octal generator polynomials of 561 and 753. The useful video bitrate was further reduced due to the 16-bit Cyclic Redundancy Checking (CRC) used for error detection and the 9-bit repetition-coded feedback error flag for the reverse link. This results in video packet sizes of 77, 179, and 383 bits for each of the three modulation modes. The useful video capacity was reduced further by the video packet header of between 8 and 10 bits, resulting in useful

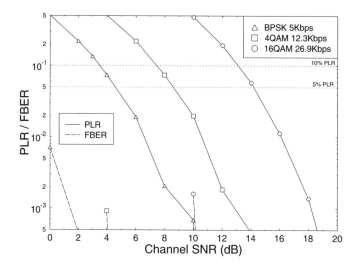

Figure 14.17: Video packet-loss ratio (PLR) and feedback error ratio (FBER) versus channel SNR for the three modulation schemes of the two-user multimode system using joint detection.

or effective video bitrates ranging from 5 to 26.9 Kbps in the BPSK and 16QAM modes, respectively.

The proposed multimode system can switch among the 1-, 2-, and 4-bit/symbol modulation schemes under network control, based on the prevailing channel conditions. As seen in Table 14.6, when the channel is benign, the unprotected video bitrate will be approximately 26.9 Kbps in the 16QAM mode. However, as the channel quality degrades, the modem will switch to the BPSK mode of operation, where the video bitrate drops to 5 Kbps, and for maintaining a reasonable video quality, the video resolution has to be reduced to SQCIF (128×96 pels).

Figure 14.17 portrays the packet loss ratio for the multimode system in each of its modulation modes for a range of channel SNRs. The figure shows that above a channel SNR of 14 dB the 16QAM mode offers an acceptable packet-loss ratio of less than 5%, while providing an unprotected video rate of about 26.9 Kbps. If the channel SNR drops below 14 dB, the multimode system is switched to 4QAM and eventually to BPSK, when the channel SNR is below 9 dB, in order to maintain the required quality of service, which is dictated by the packet-loss ratio. The figure also shows the acknowledgment feedback error ratio (FBER) for a range of channel SNRs, which has to be substantially lower than the video PLR itself. This requirement is satisfied in the figure, since the feedback errors only occur at extremely low-channel SNRs, where the packet-loss ratio is approximately 50%. It is therefore assumed that the multimode system would have switched to a more robust modulation mode, before the feedback acknowledgment flag can become corrupted.

The video quality is commonly measured in terms of the peak signal-to-noise ratio (PSNR). Figure 14.18 shows the video quality in terms of the PSNR versus the channel SNRs for each of the modulation modes. As expected, the higher throughput bitrate of the 16QAM mode provides a better video quality. However, as the channel quality degrades,

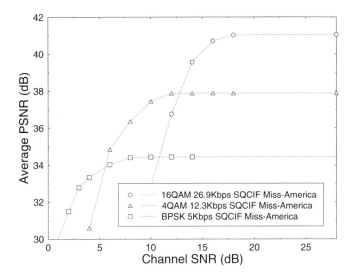

Figure 14.18: Decoded video quality (PSNR) versus channel SNR for the modulation modes of BPSK, 4QAM, and 16QAM supporting two users with the aid of joint detection. These results were recorded for the "Miss America" video sequence at SQCIF resolution (128×96 pels).

the video quality of the 16QAM mode is reduced. Hence, it becomes beneficial to switch from the 16QAM mode to 4QAM at an SNR of about 14 dB, as suggested by the packet-loss ratio performance of Figure 14.17. Although the video quality expressed in terms of PSNR is superior for the 16QAM mode in comparison to the 4QAM mode at channel SNRs in excess of 12 dB, because of the excessive PLR the perceived video quality appears inferior to that of the 4QAM mode, even though the 16QAM PSNR is higher for channel SNRs in the range of 12–14 dB. More specifically, we found that it was beneficial to switch to a more robust modulation scheme when the PSNR was reduced by about 1 dB with respect to its unimpaired PSNR value. This ensured that the packet losses did not become subjectively obvious, resulting in a higher perceived video quality and smoother degradation, as the channel quality deteriorated.

The effect of packet losses on the video quality quantified in terms of PSNR is portrayed in Figure 14.19. The figure shows that the video quality degrades as the PLR increases. In order to ensure a seamless degradation of video quality as the channel SNR is reduced, it is best to switch to a more robust modulation scheme when the PLR exceeded 5%. The figure shows that a 5% packet-loss ratio results in a loss of PSNR when switching to a more robust modulation scheme. However, if the system did not switch until the PSNR of the more robust modulation mode was similar, the perceived video quality associated with the originally higher rate, but channel-impaired, stream became inferior.

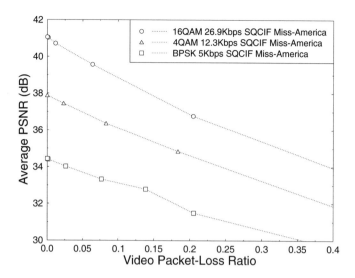

Figure 14.19: Decoded video quality (PSNR) versus video packet-loss ratio for the modulation modes of BPSK, 4QAM, and 16QAM, supporting two users with the aid of joint detection. The results were recorded for the "Miss America" video sequence at SQCIF resolution (128×96 pels).

14.2.3 Burst-by-burst Adaptive Videophone System

A burst-by-burst adaptive modem maximizes the system's throughput by using the most appropriate modulation mode for the current instantaneous channel conditions. Figure 14.20 exemplifies how a burst-by-burst adaptive modem changes its modulation modes based on the fluctuating channel conditions. The adaptive modem uses the SINR estimate at the output of the joint detector to estimate the instantaneous channel quality and therefore to set the modulation mode.

The probability of the adaptive modem using each modulation mode for a particular channel SNRs is portrayed in Figure 14.21. At high-channel SNRs, the modem mainly uses the 16QAM modulation mode, while at low-channel SNRs the BPSK mode is most prevalent.

The advantage of dynamically reconfigured burst-by-burst adaptive modem over the statically switched multimode system previously described is that the video quality is smoothly degraded as the channel conditions deteriorate. The switched multimode system results in more sudden reductions in video quality, when the modem switches to a more robust modulation mode. Figure 14.22 shows the throughput bitrate of the dynamically reconfigured burst-by-burst adaptive modem, compared to the three modes of the statically switched multimode system. The reduction of the fixed modem modes' effective throughput at low SNRs is due to the fact that under such channel conditions an increased fraction of the transmitted packets have to be dropped, reducing the effective throughput. The figure shows the smooth reduction of the throughput bitrate, as the channel quality deteriorates. The burst-by-burst modem matches the BPSK mode's bitrate at low-channel SNRs and the 16QAM mode's bitrate at high SNRs. The dynamically reconfigured HSDPA-style

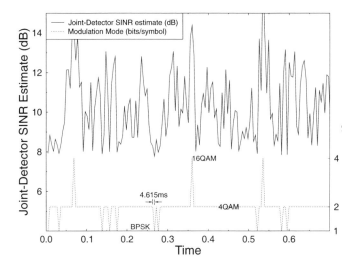

Figure 14.20: Example of modem mode switching in a dynamically reconfigured burst-by-burst modem in operation, where the modulation mode switching is based on the SINR estimate at the output of the joint detector.

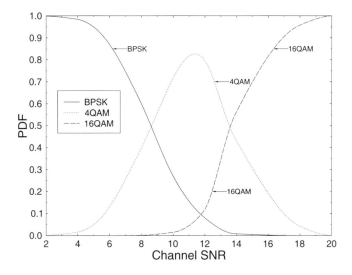

Figure 14.21: PDF of the various adaptive modem modes versus channel SNR.

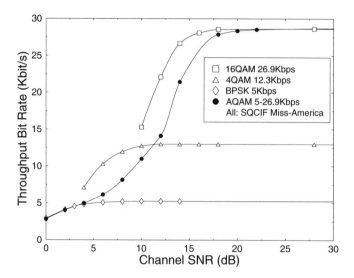

Figure 14.22: Throughput bitrate versus channel SNR comparison of the three fixed modulation modes (BPSK, 4QAM, 16QAM) and the adaptive burst-by-burst modem (AQAM), both supporting two users with the aid of joint detection.

burst-by-burst adaptive modem characterized in the figure perfectly estimates the prevalent channel conditions, although in practice the estimate of channel quality is imperfect and it is inherently delayed, which results in a slightly reduced performance when compared to perfect channel estimation [83, 85].

The smoothly varying throughput bitrate of the burst-by-burst adaptive modem translates into a smoothly varying video quality as the channel conditions change. The video quality measured in terms of the average peak signal-to-noise ratio (PSNR) is shown versus the channel SNR in Figure 14.23 in contrast to that of the individual modem modes. The figure demonstrates that the burst-by-burst adaptive modem provides equal or better video quality over a large proportion of the SNR range shown than the individual modes. However, even at channel SNRs, where the adaptive modem has a slightly reduced PSNR, the perceived video quality of the adaptive modem is better since the video packet-loss rate is far lower than that of the fixed modem modes.

Figure 14.24 shows the video packet-loss ratio versus channel SNR for the three fixed modulation modes and the burst-by-burst adaptive modem with perfect channel estimation. Again, the figure demonstrates that the video packet-loss ratio of the adaptive modem is similar to that of the fixed BPSK modem mode. However, the adaptive modem has a far higher bitrate throughput, as the channel SNR increases. The burst-by-burst adaptive modem gives an error performance similar to that of the BPSK mode, but with the flexibility to increase the bitrate throughput of the modem, when the channel conditions improve. If imperfect channel estimation is used, the bitrate throughput of the adaptive modem is reduced slightly. Furthermore, the video packet-loss ratio seen in Figure 14.24 is slightly higher for the AQAM scheme due to invoking higher-order modem modes, as the channel quality increases. However, we have shown in the context of wideband video transmission [161] that it is

14.2. HSDPA-STYLE BURST-BY-BURST ADAPTIVE CDMA VIDEOPHONY

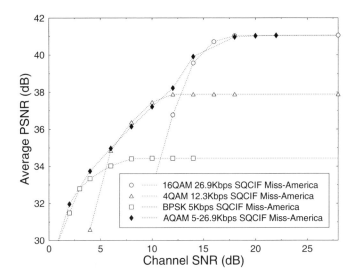

Figure 14.23: Average decoded video quality (PSNR) versus channel SNR comparision of the fixed modulation modes of BPSK, 4QAM and 16QAM, and the burst-by-burst adaptive modem — both supporting two users with the aid of joint detection. These results were recorded for the "Miss America" video sequence at SQCIF resolution (128 × 96 pels).

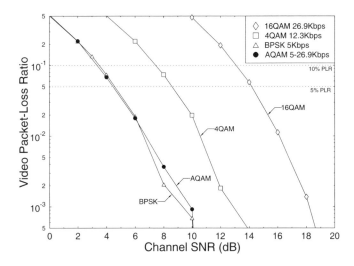

Figure 14.24: Video packet-loss ratio (PLR) versus channel SNR for the three modulation schemes of the multimode system, compared to the burst-by-burst adaptive modem. Both systems substain two users using joint detection.

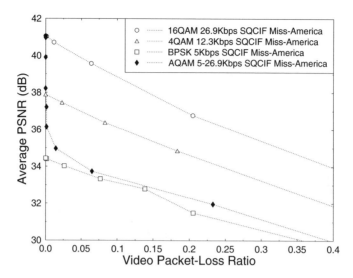

Figure 14.25: Decoded video quality (PSNR) versus video packet-loss ratio comparison of the fixed modulation modes of BPSK, 4QAM, and 16QAM, and the burst-by-burst adaptive modem. Both supporting two users with the aid of joint detection. These results were recorded for the "Miss America" video sequence at SQCIF resolution (128×96 pels).

possible to maintain the video packet-loss ratio within tolerable limits for the range of channel SNRs considered.

The interaction between the video quality measured in terms of PSNR and the video packet loss ratio can be seen more clearly in Figure 14.25. The figure shows that the adaptive modem slowly degrades the decoded video quality from that of the error-free 16QAM fixed modulation mode, as the channel conditions deteriorate. The video quality degrades from the error-free 41 dB PSNR, while maintaining a near-zero video packet-loss ratio, until the PSNR drops below about 36 dB PSNR. At this point, the further reduced channel quality inflicts an increased video packet-loss rate, and the video quality degrades more slowly. The PSNR versus packet-loss ratio performance then tends toward that achieved by the fixed BPSK modulation mode. However, the adaptive modem achieved better video quality than the fixed BPSK modem even at high packet-loss rates.

14.2.4 Summary and Conclusions

In conclusion, the proposed joint-detection assisted burst-by-burst adaptive CDMA-based video transceiver substantially outperformed the matched-filtering based transceiver. The transceiver guaranteed a near-unimpaired video quality for channel SNRs in excess of about 5 dB over the COST207 dispersive Rayleigh-faded channel. The benefits of the multimode video transceiver clearly manifest themselves in terms of supporting unimpaired video quality under time-variant channel conditions, where a single-mode transceiver's quality would become severely degraded by channel effects. The dynamically reconfigured burst-by-burst adaptive modem gave better perceived video quality due to its more graceful reduction in video quality, as the channel conditions degraded, than a statically switched multimode system.

Following our discussions on joint-detection assisted CDMA-based burst-by-burst adaptive interactive videotelephony, in the next two sections we concentrate on a range of multicarrier modems. The last section of the chapter considers distributive broadcast video transmission, based also on multicarrier modems.

14.3 Subband-adaptive Turbo-coded OFDM-based Interactive Videotelephony[5,6,7]

14.3.1 Motivation and Background

In Section 14.2 a CDMA-based video system was proposed, while the previous section considered the transmission of interactive video using OFDM transceivers in various propagation environments. In this section, burst-by-burst adaptive OFDM is proposed and investigated in the context of interactive videotelephony.

As mentioned earlier, burst-by-burst adaptive quadrature amplitude modulation [222] (AQAM) was devised by Steele and Webb [222, 321] in order to enable the transceiver to cope with the time-variant channel quality of narrowband fading channels. Further related research was conducted at the University of Osaka by Sampei and his colleagues, investigating variable coding rate concatenated coded schemes [408]; at the University of Stanford by Goldsmith and her team, studying the effects of variable-rate, variable-power arrangements [409]; and at the University of Southampton in the United Kingdom, investigating a variety of practical aspects of AQAM [410, 411]. The channel's quality is estimated on a burst-by-burst basis, and the most appropriate modulation mode is selected in order to maintain the required target bit error rate (BER) performance, while maximizing the system's Bit Per Symbol (BPS) throughput. Though use of this reconfiguration regime, the distribution of channel errors becomes typically less bursty, than in conjunction with nonadaptive modems, which potentially increases the channel coding gains [412]. Furthermore, the soft-decision channel codec metrics can also be invoked in estimating the instantaneous channel quality [412], regardless of the type of channel impairments.

A range of coded AQAM schemes was analyzed by Matsuoka *et al.* [408], Lau *et al.* [413] and Goldsmith *et al.* [414]. For data transmission systems, which do not necessarily require a low transmission delay, variable-throughput adaptive schemes can be devised, which operate efficiently in conjunction with powerful error correction codecs, such as long block length turbo codes [401]. However, the acceptable turbo interleaving delay is rather low in the context of low-delay interactive speech. Video communications systems typically require a higher bitrate than speech systems, and hence they can afford a higher interleaving delay.

---

[5]This section is based on P.J. Cherriman, T. Keller, and L. Hanzo: Subband-adaptive Turbo-coded OFDM-based Interactive Video Telephony, submitted to *IEEE Transactions on Circuits and Systems for Video Technology*, July 1999.

[6]Acknowledgment: The financial support of the Mobile VCE, EPSRC, UK and that of the European Commission is gratefully acknowledged.

[7]©1999 IEEE. Personal use of this material is permitted. However, permission to reprint/republish this material for advertising or promotional purposes or for creating new collective works for resale or redistribution to servers or lists, or to reuse any copyrighted component of this work in other works must be obtained from IEEE.

The above principles — which were typically investigated in the context of narrowband modems — were further advanced in conjunction with wideband modems, employing powerful block turbo-coded, wideband Decision Feedback Equalizer (DFE) assisted AQAM transceivers [412, 415]. A neural-network Radial Basis Function (RBF) DFE-based AQAM modem design was proposed in [416], where the RBF DFE provided the channel-quality estimates for the modem mode switching regime. This modem was capable of removing the residual BER of conventional DFEs, when linearly nonseparable received phasor constellations were encountered.

These burst-by-burst adaptive principles can also be extended to Adaptive Orthogonal Frequency Division Multiplexing (AOFDM) schemes [417] and to adaptive joint-detection-based Code Division Multiple Access (JD-ACDMA) arrangements [418]. The associated AQAM principles were invoked in the context of parallel AOFDM modems by Czylwik *et al.* [419], Fischer [420], and Chow *et al.* [421]. Adaptive subcarrier selection has also been advocated by Rohling *et al.* [422] in order to achieve BER performance improvements. Due to lack of space without completeness, further significant advances over benign, slowly varying dispersive Gaussian fixed links — rather than over hostile wireless links — are due to Chow, Cioffi, and Bingham [421] from the United States, rendering OFDM the dominant solution for asymmetric digital subscriber loop (ADSL) applications, potentially up to bitrates of 54 Mbit/s. In Europe OFDM has been favored for both Digital Audio Broadcasting (DAB) and Digital Video Broadcasting [423, 424] (DVB) as well as for high-rate Wireless Asynchronous Transfer Mode (WATM) systems due to its ability to combat the effects of highly dispersive channels [425]. The idea of "water-filling" — as allocating different modem modes to different subcarriers was referred to — was proposed for OFDM by Kalet [426] and later further advanced by Chow *et al.* [421]. This approach was adapted later in the context of time-variant mobile channels for duplex wireless links, for example, in [417]. Finally, various OFDM-based speech and video systems were proposed in [427, 428], while the co-channel interference sensitivity of OFDM can be mitigated with the aid of adaptive beam-forming [429, 430] in multi-user scenarios.

The remainder of this section is structured as follows. Section 14.3.1 outlines the architecture of the proposed video transceiver, while Section 14.3.5 quantifies the performance benefits of AOFDM transceivers in comparison to conventional fixed transceivers. Section 14.3.6 endeavors to highlight the effects of more "aggressive" loading of the subcarriers in both BER and video quality terms, while Section 14.3.7 proposed time-variant rather than constant rate AOFDM as a means of more accurately matching the transceiver to the time-variant channel quality fluctuations, before concluding in Section 14.3.8.

14.3.2 AOFDM Modem Mode Adaptation and Signaling

The proposed duplex AOFDM scheme operates on the following basis:

- *Channel quality estimation* is invoked upon receiving an AOFDM symbol in order to select the modem mode allocation of the next AOFDM symbol.

- *The decision concerning the modem modes for the next AOFDM symbol* is based on the prediction of the expected channel conditions. Then the transmitter has to select the appropriate modem modes for the groups or subbands of OFDM subcarriers, where the subcarriers were grouped into subbands of identical modem modes in order to reduce the required number of signaling bits.

- *Explicit signaling or blind detection of the modem modes* is used to inform the receiver as to what type of demodulation to invoke.

If the channel quality of the up-link and down-link can be considered similar, then the channel-quality estimate for the up-link can be extracted from the down-link and vice versa. We refer to this regime as open-loop adaptation. In this case, the transmitter has to convey the modem modes to the receiver, or the receiver can attempt blind detection of the transmission parameters employed. In contrast, if the channel cannot be considered reciprocal, then the channel-quality estimation has to be performed at the receiver, and the receiver has to instruct the transmitter as to what modem modes have to be used at the transmitter, in order to satisfy the target integrity requirements of the receiver. We refer to this mode as closed-loop adaptation. Blind modem mode recognition was invoked, for example, in [417] — a technique that results in bitrate savings due to refraining from dedicating bits to explicit modem mode signaling at the cost of increased complexity. Let us address the issues of channel quality estimation on a subband-by-subband basis in the next subsection.

14.3.3 AOFDM Subband BER Estimation

A reliable channel-quality metric can be devised by calculating the expected overall bit error probability for all available modulation schemes M_n in each subband, which is denoted by $\bar{p}_e(n) = 1/N_s \sum_j p_e(\gamma_j, M_n)$. For each AOFDM subband, the modem mode having the highest throughput, while exhibiting an estimated BER below the target value is then chosen. Although the adaptation granularity is limited to the subband width, the channel-quality estimation is quite reliable, even in interference-impaired environments.

Against this background in our forthcoming discussions, the design trade-offs of turbo-coded Adaptive Orthogonal Frequency Division Multiplex (AOFDM) wideband video transceivers are presented. We will demonstrate that AOFDM provides a convenient framework for adjusting the required target integrity and throughput both with and without turbo channel coding and lends itself to attractive video system construction, provided that a near-instantaneously programmable rate video codec — such as the H.263 scheme highlighted in the next section — can be invoked.

14.3.4 Video Compression and Transmission Aspects

In this study we investigate the transmission of 704 x 576 pixel Four-times Common Intermediate Format (4CIF) high-resolution video sequences at 30 frames/s using subband-adaptive turbo-coded Orthogonal Frequency Division Multiplex (AOFDM) transceivers. The transceiver can modulate 1, 2, or 4 bits onto each AOFDM subcarrier, or simply disable transmissions for subcarriers that exhibit a high attenuation or phase distortion due to channel effects.

The H.263 video codec [161] exhibits an impressive compression ratio, although this is achieved at the cost of a high vulnerability to transmission errors, since a run-length coded bitstream is rendered undecodable by a single bit error. In order to mitigate this problem, when the channel codec protecting the video stream is overwhelmed by the transmission

Table 14.7: System Parameters for the Fixed QPSK and BPSK Transceivers, as well as for the Corresponding Subband-adaptive OFDM (AOFDM) Transceivers for Wireless Local Area Networks (WLANs)

	BPSK mode	QPSK mode
Packet rate	4687.5 packets/s	
FFT length	512	
OFDM symbols/packet	3	
OFDM symbol duration	2.6667 μs	
OFDM time frame	80 timeslots = 213 μs	
Normalized Doppler frequency, f'_d	1.235×10^{-4}	
OFDM symbol normalized Doppler frequency, F_D	7.41×10^{-2}	
FEC coded bits/packet	1536	3072
FEC-coded video bitrate	7.2 Mbit/s	14.4 Mbit/s
Unprotected bits/packet	766	1534
Unprotected bitrate	3.6 Mbit/s	7.2 Mbit/s
Error detection CRC (bits)	16	16
Feedback error flag bits	9	9
Packet header bits/packet	11	12
Effective video bits/packet	730	1497
Effective video bitrate	3.4 Mbit/s	7.0 Mbit/s

errors, we refrain from decoding the corrupted video packet in order to prevent error propagation through the reconstructed video frame buffer [392]. We found that it was more beneficial in video-quality terms if these corrupted video packets were dropped and the reconstructed frame buffer was not updated, until the next video packet replenishing the specific video frame area was received. The associated video performance degradation was found perceptually unobjectionable for packet dropping- or transmission frame error rates (FER) below about 5%. These packet dropping events were signaled to the remote video decoder by superimposing a strongly protected one-bit packet acknowledgment flag on the reverse-direction packet, as outlined in [392]. Turbo error correction codes [401] were used.

14.3.5 Comparison of Subband-adaptive OFDM and Fixed Mode OFDM Transceivers

In order to show the benefits of the proposed subband-adaptive OFDM transceiver, we compare its performance to that of a fixed modulation mode transceiver under identical propagation conditions, while having the same transmission bitrate. The subband-adaptive modem is capable of achieving a low bit error ratio (BER), since it can disable transmissions over low-quality subcarriers and compensate for the lost throughput by invoking a higher modulation mode than that of the fixed-mode transceiver over the high-quality subcarriers.

Table 14.7 shows the system parameters for the fixed BPSK and QPSK transceivers, as well as for the corresponding subband-adaptive OFDM (AOFDM) transceivers. The system employs constraint length three, half-rate turbo coding, using octal generator polynomials of 5

14.3. ADAPTIVE TURBO-CODED OFDM-BASED VIDEOTELEPHONY

and 7 as well as random turbo interleavers. Therefore, the unprotected bitrate is approximately half the channel-coded bitrate. The protected to unprotected video bitrate ratio is not exactly half, since two tailing bits are required to reset the convolutional encoders' memory to their default state in each transmission burst. In both modes, a 16-bit Cyclic Redundancy Checking (CRC) is used for error detection, and 9 bits are used to encode the reverse link feedback acknowledgment information by simple repetition coding. The feedback flag decoding ensues using majority logic decisions. The packetization requires a small amount of header information added to each transmitted packet, which is 11 and 12 bits per packet for BPSK and QPSK, respectively. The effective or useful video bitrates for the BPSK and QPSK modes are then 3.4 and 7.0 Mbit/s.

The fixed-mode BPSK and QPSK transceivers are limited to 1 and 2 bits per symbol, respectively. In contrast, the proposed AOFDM transceivers operate at the same bitrate, as their corresponding fixed modem mode counterparts, although they can vary their modulation mode on a subcarrier- by-subcarrier basis between 0, 1, 2, and 4 bits per symbol. Zero bits per symbol implies that transmissions are disabled for the subcarrier concerned.

The "micro-adaptive" nature of the subband-adaptive modem is characterized by Figure 14.26, portraying at the top a contour plot of the channel signal-to-noise ratio (SNR) for each subcarrier versus time. At the center and bottom of the figure, the modulation mode chosen for each 32-subcarrier subband is shown versus time for the 3.4 and 7.0 Mbit/s target-rate subband-adaptive modems, respectively. The channel SNR variation versus both time and frequency is also shown in three-dimensional form in Figure 14.27, which may be more convenient to visualize. This was recorded for the channel impulse response of Figure 14.28. It can be seen that when the channel is of high quality — as for example, at about frame 1080 — the subband-adaptive modem used the same modulation mode, as the equivalent fixed-rate modem in all subcarriers. When the channel is hostile — for example, around frame 1060 — the subband-adaptive modem used a lower-order modulation mode in some subbands than the equivalent fixed-mode scheme, or in extreme cases disabled transmission for that subband. In order to compensate for the loss of throughput in this subband, a higher-order modulation mode was used in the higher quality subbands.

One video packet is transmitted per OFDM symbol; therefore, the video packet-loss ratio is the same as the OFDM symbol error ratio. The video packet-loss ratio is plotted against the channel SNR in Figure 14.29. It is shown in the graph that the subband-adaptive transceivers — or synonymously termed as microscopic-adaptive (μAOFDM), in contrast to OFDM symbol-by-symbol adaptive transceivers — have a lower packet-loss ratio (PLR) at the same SNR compared to the fixed modulation mode transceiver. Note in Figure 14.29 that the subband-adaptive transceivers can operate at lower channel SNRs than the fixed modem mode transceivers, while maintaining the same required video packet-loss ratio. Again, the figure labels the subband-adaptive OFDM transceivers as μAOFDM, implying that the adaption is not noticeable from the upper layers of the system. A macro-adaption could be applied in addition to the microscopic adaption by switching between different target bitrates, as the longer-term channel quality improves and degrades. This issue is the subject of Section 14.3.7.

Having shown that the subband-adaptive OFDM transceiver achieved a reduced video packet loss in comparison to fixed modulation mode transceivers under identical channel conditions, we now compare the effective throughput bitrate of the fixed and adaptive OFDM

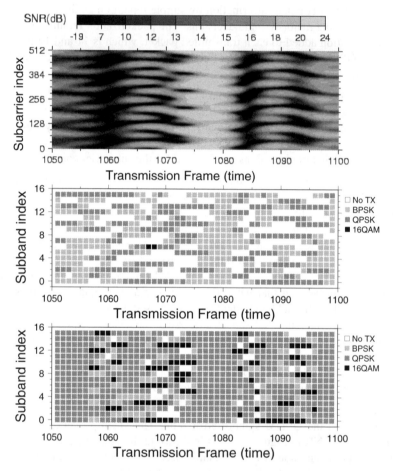

Figure 14.26: The micro-adaptive nature of the subband-adaptive OFDM modem. The top graph is a contour plot of the channel SNR for all 512 subcarriers versus time. The bottom two graphs show the modulation modes chosen for all 16 32-subcarrier subbands for the same period of time. The middle graph shows the performance of the 3.4 Mbit/s subband-adaptive modem, which operates at the same bitrate as a fixed BPSK modem. The bottom graph represents the 7.0 Mbit/s subband-adaptive modem, which operated at the same bitrate as a fixed QPSK modem. The average channel SNR was 16 dB.

14.3. ADAPTIVE TURBO-CODED OFDM-BASED VIDEOTELEPHONY

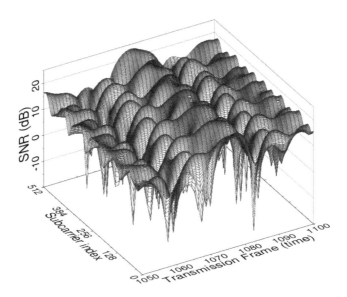

Figure 14.27: Instantaneous channel SNR for all 512 subcarriers versus time, for an average channel SNR of 16 dB over the channel characterized by the channel impulse response (CIR) of Figure 14.28.

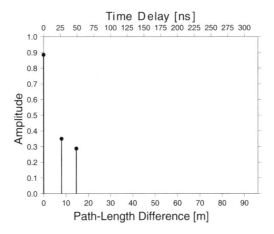

Figure 14.28: Indoor three-path WATM channel impulse response.

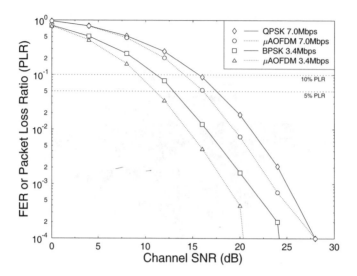

Figure 14.29: Frame Error Rate (FER) or video packet-loss ratio (PLR) versus channel SNR for the BPSK and QPSK fixed modulation mode OFDM transceivers and for the corresponding subband-adaptive μAOFDM transceiver, operating at identical effective video bitrates, namely, at 3.4 and 7.0 Mbit/s, over the channel model of Figure 14.28 at a normalized Doppler frequency of $F_D = 7.41 \times 10^{-2}$.

transceivers in Figure 14.30. The figure shows that when the channel quality is high, the throughput bitrates of the fixed and adaptive transceivers are identical. However, as the channel degrades, the loss of packets results in a lower throughput bitrate. The lower packet-loss ratio of the subband-adaptive transceiver results in a higher throughput bitrate than that of the fixed modulation mode transceiver.

The throughput bitrate performance results translate to the decoded video-quality performance results evaluated in terms of PSNR in Figure 14.31. Again, for high-channel SNRs the performance of the fixed and adaptive OFDM transceivers is identical. However, as the channel quality degrades, the video quality of the subband-adaptive transceiver degrades less dramatically than that of the corresponding fixed modulation mode transceiver.

14.3.6 Subband-adaptive OFDM Transceivers Having Different Target Bitrates

As mentioned earlier, the subband-adaptive modems employ different modulation modes for different subcarriers in order to meet the target bitrate requirement at the lowest possible channel SNR. This is achieved by using a more robust modulation mode or eventually by disabling transmissions over subcarriers having a low channel quality. In contrast, the adaptive system can invoke less robust, but higher throughput, modulation modes over subcarriers exhibiting a high-channel quality. In the examples we have previously considered, we chose the AOFDM target bitrate to be identical to that of a fixed modulation mode

14.3. ADAPTIVE TURBO-CODED OFDM-BASED VIDEOTELEPHONY

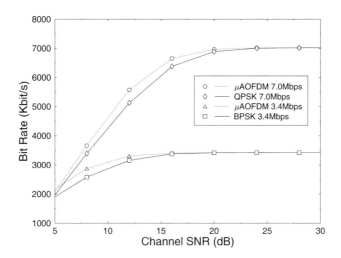

Figure 14.30: Effective throughput bitrate versus channel SNR for the BPSK and QPSK fixed modulation mode OFDM transceivers and that of the corresponding subband-adaptive or μAOFDM transceiver operating at identical effective video bitrates of 3.4 and 7.0 Mbit/s, over the channel of Figure 14.28 at a normalized Doppler frequency of $F_D = 7.41 \times 10^{-2}$.

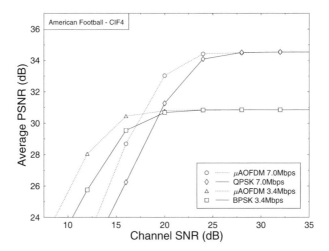

Figure 14.31: Average video quality expressed in PSNR versus channel SNR for the BPSK and QPSK fixed modulation mode OFDM transceivers and for the corresponding μAOFDM transceiver operating at identical channel SNRs over the channel model of Figure 14.28 at a normalized Doppler frequency of $F_D = 7.41 \times 10^{-2}$.

Table 14.8: System Parameters for the Four Different Target Bitrates of the Various Subband-adaptive OFDM (μAOFDM) Transceivers

Packet rate	\multicolumn{4}{c}{4687.5 packets/s}			
FFT length		512		
OFDM symbols/packet		3		
OFDM symbol duration		2.6667 μs		
OFDM time frame		80 timeslots = 213 μs		
Normalized Doppler frequency, f'_d		1.235×10^{-4}		
OFDM symbol normalized Doppler frequency, F_D		7.41×10^{-2}		
FEC-coded bits/packet	858	1536	3072	4272
FEC-coded video bitrate	4.0 Mbit/s	7.2 Mbit/s	14.4 Mbit/s	20.0 Mbit/s
No. of unprotected bits/packet	427	766	1534	2134
Unprotected bitrate	2.0 Mbit/s	3.6 Mbit/s	7.2 Mbit/s	10.0 Mbit/s
No. of CRC bits	16	16	16	16
No. of feedback error flag bits	9	9	9	9
No. of packet header bits/packet	10	11	12	13
Effective video bits/packet	392	730	1497	2096
Effective video bitrate	1.8 Mbit/s	3.4 Mbit/s	7.0 Mbit/s	9.8 Mbit/s
Equivalent modulation mode		BPSK	QPSK	
Minimum channel SNR for 5% PLR (dB)	8.8	11.0	16.1	19.2
Minimum channel SNR for 10% PLR (dB)	7.1	9.2	14.1	17.3

transceiver. In this section, we comparatively study the performance of various μAOFDM systems having different target bitrates.

The previously described μAOFDM transceiver of Table 14.7 exhibited a FEC-coded bitrate of 7.2 Mbit/s, which provided an effective video bitrate of 3.4 Mbit/s. If the video target bitrate is lower than 3.4 Mbit/s, then the system can disable transmission in more of the subcarriers, where the channel quality is low. Such a transceiver would have a lower bit error rate than the previous BPSK-equivalent μAOFDM transceiver and therefore could be used at lower average channel SNRs, while maintaining the same bit error ratio target. In contrast, as the target bitrate is increased, the system has to employ higher-order modulation modes in more subcarriers at the cost of an increased bit error ratio. Therefore, high target bitrate μAOFDM transceivers can only perform within the required bit error ratio constraints at high-channel SNRs, while low target bitrate μAOFDM systems can operate at low-channel SNRs without causing excessive BERs. Therefore, a system that can adjust its target bitrate as the channel SNR changes would operate over a wide range of channel SNRs, providing the maximum possible average throughput bitrate, while maintaining the required bit error ratio.

Hence, below we provide a performance comparison of various μAOFDM transceivers that have four different target bitrates, of which two are equivalent to that of the BPSK and QPSK fixed modulation mode transceivers of Table 14.7. The system parameters for all four different bitrate modes are summarized in Table 14.8. The modes having effective video bitrates of 3.4 and 7.0 Mbit/s are equivalent to the bitrates of a fixed BPSK and QPSK mode transceiver, respectively.

14.3. ADAPTIVE TURBO-CODED OFDM-BASED VIDEOTELEPHONY

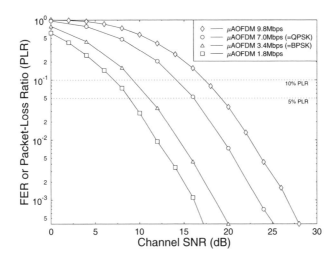

Figure 14.32: FER or video packet-loss ratio (PLR) versus channel SNR for the subband adaptive OFDM transceivers of Table 14.8 operating at four different target bitrates, over the channel model of Figure 14.28 at a normalized Doppler frequency of $F_D = 7.41 \times 10^{-2}$.

Figure 14.32 shows the Frame Error Rate (FER) or video packet-loss ratio (PLR) performance versus channel SNR for the four different target bitrates of Table 14.8, demonstrating, as expected, that the higher target bitrate modes require higher channel SNRs in order to operate within given PLR constraints. For example, the mode having an effective video bitrate of 9.8 Mbit/s can only operate for channel SNRs in excess of 19 dB under the constraint of a maximum PLR of 5%. However, the mode that has an effective video bitrate of 3.4 Mbit/s can operate at channel SNRs of 11 dB and above, while maintaining the same 5% PLR constraint, albeit at about half the throughput bitrate and so at a lower video quality.

The trade-offs between video quality and channel SNR for the various target bitrates can be judged from Figure 14.33, suggesting, as expected, that the higher target bitrates result in a higher video quality, provided that channel conditions are favorable. However, as the channel quality degrades, the video packet-loss ratio increases, thereby reducing the throughput bitrate and hence the associated video quality. The lower target bitrate transceivers operate at an inherently lower video quality, but they are more robust to the prevailing channel conditions and so can operate at lower channel SNRs, while guaranteeing a video quality, which is essentially unaffected by channel errors. It was found that the perceived video quality became impaired for packet-loss ratios in excess of about 5%.

The trade-offs between video quality, packet-loss ratio, and target bitrate are further augmented with reference to Figure 14.34. The figure shows the video quality measured in PSNR versus video frame index at a channel SNR of 16 dB as well as for an error-free situation. At the bottom of each graph, the packet-loss ratio per video frame is shown. The three figures indicate the trade-offs to be made in choosing the target bitrate for the specific channel conditions experienced — in this specific example for a channel SNR of 16dB. Note that under error-free conditions the video quality improved upon increasing the bitrate.

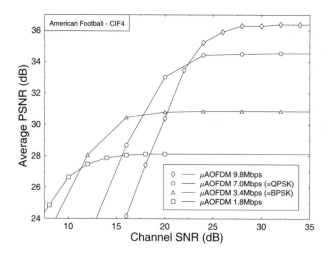

Figure 14.33: Average video quality expressed in PSNR versus channel SNR for the subband-adaptive OFDM transceivers of Table 14.8, operating at four different target bitrates, over the channel model of Figure 14.28 at a normalized Doppler frequency of $F_\text{D} = 7.41 \times 10^{-2}$.

Specifically, video PSNRs of about 40, 41.5, and 43 dB were observed for the effective video bitrates of 1.8, 3.4, and 7.0 Mbit/s. The figure shows that for the target bitrate of 1.8 Mbit/s, the system has a high grade of freedom in choosing which subcarriers to invoke. Therefore, it is capable of reducing the number of packets that are lost. The packet-loss ratio remains low, and the video quality remains similar to that of the error-free situation. The two instances where the PSNR is significantly different from the error-free performance correspond to video frames in which video packets were lost. However, in both instances the system recovers in the following video frame.

As the target bitrate of the subband-adaptive OFDM transceiver is increased to 3.4 Mbit/s, the subband modulation mode selection process has to be more "aggressive", resulting in increased video packet loss. Observe in the figure that the transceiver having an effective video bitrate of 3.4 Mbit/s exhibits increased packet loss. In one frame as much as 5% of the packets transmitted for that video frame were lost, although the average PLR was only 0.4%. Because of the increased packet loss, the video PSNR curve diverges from the error-free performance curve more often. However, in almost all cases the effects of the packet losses are masked in the next video frame, indicated by the re-merging PSNR curves in the figure, maintaining a close to error-free PSNR. The subjective effect of this level of packet loss is almost imperceivable.

When the target bitrate is further increased to 7.0 Mbit/s, the average PLR is about 5% under the same channel conditions, and the effects of this packet-loss ratio are becoming objectionable in perceived video-quality terms. At this target bitrate, there are several video frames where at least 10% of the video packets have been lost. The video quality measured in PSNR terms rarely reaches its error-free level, because every video frame contains at least one lost packet. The perceived video quality remains virtually unimpaired until the head

14.3. ADAPTIVE TURBO-CODED OFDM-BASED VIDEOTELEPHONY

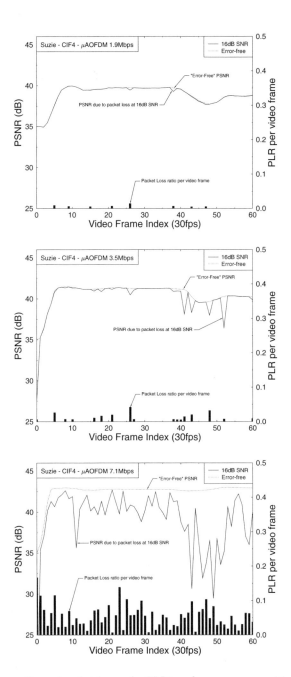

Figure 14.34: Video-quality and packet-loss ratio (PLR) performance versus video-frame index (time) comparison of subband-adaptive OFDM transceivers having target bitrates of 1.8, 3.4, and 7.0 Mbit/s, under the same channel conditions, at 16 dB SNR over the channel of Figure 14.28 at a normalized Doppler frequency of $F_D = 7.41 \times 10^{-2}$.

movement in the "Suzie" video sequence around frames 40–50, where the effect of lost packets becomes obvious, and the PSNR drops to about 30 dB.

14.3.7 Time-variant Target Bitrate OFDM Transceivers

By using a high target bitrate, when the channel quality is high, and employing a reduced target bitrate, when the channel quality is poor, an adaptive system is capable of maximizing the average throughput bitrate over a wide range of channel SNRs, while satisfying a given quality constraint. This quality constraint for our video system could be a maximum packet-loss ratio.

Because a substantial processing delay is associated with evaluating the packet-loss information, modem mode switching based on this metric is less efficient due to this latency. Therefore, we decided to invoke an estimate of the bit error ratio (BER) for mode switching, as follows. Since the noise energy in each subcarrier is independent of the channel's frequency domain transfer function H_n, the local signal-to-noise ratio (SNR) in subcarrier n can be expressed as

$$\gamma_n = |H_n|^2 \cdot \gamma, \tag{14.1}$$

where γ is the overall SNR. If no signal degradation due to Inter–Subcarrier Interference (ISI) or interference from other sources appears, then the value of γ_n determines the bit error probability for the transmission of data symbols over the subcarrier n. Given γ_j across the N_s subcarriers in the jth subband, the expected overall BER for all available modulation schemes M_n in each subband can be estimated, which is denoted by $\bar{p}_e(n) = 1/N_s \sum_j p_e(\gamma_j, M_n)$. For each subband, the scheme with the highest throughput, whose estimated BER is lower than a given threshold, is then chosen.

We decided to use a quadruple-mode switched subband-adaptive modem using the four target bitrates of Table 14.8. The channel estimator can then estimate the expected bit error ratio of the four possible modem modes. Our switching scheme opted for the modem mode, whose estimated BER was below the required threshold. This threshold could be varied in order to tune the behavior of the switched subband-adaptive modem for a high or a low throughput. The advantage of a higher throughput was a higher error-free video quality at the expense of increased video packet losses, which could reduce the perceived video quality.

Figure 14.35 demonstrates how the switching algorithm operates for a 1% estimated BER threshold. Specifically, the figure portrays the estimate of the bit error ratio for the four possible modem modes versus time. The large square and the dotted line indicate the mode chosen for each time interval by the mode switching algorithm. The algorithm attempts to use the highest bitrate mode, whose BER estimate is less than the target threshold, namely, 1% in this case. However, if all the four modes' estimate of the BER is above the 1% threshold, then the lowest bitrate mode is chosen, since this will be the most robust to channel errors. An example of this is shown around frames 1035–1040. At the bottom of the graph a bar chart specifies the bitrate of the switched subband adaptive modem versus time in order to emphasize when the switching occurs.

An example of the algorithm, when switching among the target bitrates of 1.8, 3.4, 7, and 9.8 Mbit/s, is shown in Figure 14.36. The upper part of the figure portrays the contour plot of the channel SNR for each subcarrier versus time. The lower part of the figure displays

14.3. ADAPTIVE TURBO-CODED OFDM-BASED VIDEOTELEPHONY

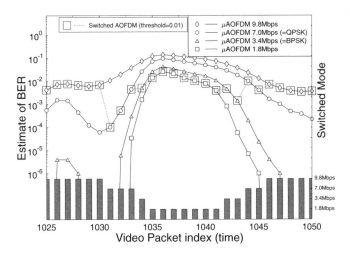

Figure 14.35: Illustration of mode switching for the switched subband adaptive modem. The figure shows the estimate of the bit error ratio for the four possible modes. The large square and the dotted line indicate the modem mode chosen for each time interval by the mode switching algorithm. At the bottom of the graph, the bar chart specifies the bitrate of the switched subband adaptive modem on the right-hand axis versus time when using the channel model of Figure 14.28 at a normalized Doppler frequency of $F_D = 7.41 \times 10^{-2}$.

the modulation mode chosen for each 32-subcarrier subband versus time for the time-variant target bitrate (TVTBR) subband adaptive modem. It can be seen at frames 1051–1055 that all the subbands employ QPSK modulation. Therefore, the TVTBR-AOFDM modem has an instantaneous target bitrate of 7 Mbit/s. As the channel degrades around frame 1060, the modem has switched to the more robust 1.8 Mbit/s mode. When the channel quality is high around frames 1074–1081, the highest bitrate 9.8 Mbit/s mode is used. This demonstrates that the TVTBR-AOFDM modem can reduce the number of lost video packets by using reduced bitrate but more robust modulation modes, when the channel quality is poor. However, this is at the expense of a slightly reduced average throughput bitrate. Usually, a higher throughput bitrate results in a higher video quality. However, a high bitrate is also associated with a high packet-loss ratio, which is usually less attractive in terms of perceived video quality than a lower bitrate, lower packet-loss ratio mode.

Having highlighted how the time-domain mode switching algorithm operates, we will now characterize its performance for a range of different BER switching thresholds. A low BER switching threshold implies that the switching algorithm is cautious about switching to the higher bitrate modes. Therefore the system performance is characterized by a low video packet-loss ratio and a low throughput bitrate. A high BER switching threshold results in the switching algorithm attempting to use the highest bitrate modes in all but the worst channel conditions. This results in a higher video packet-loss ratio. However, if the packet-loss ratio is not excessively high, a higher video throughput is achieved.

Figure 14.37 portrays the video packet-loss ratio or FER performance of the TVTBR-AOFDM modem for a variety of BER thresholds, compared to the minimum and maximum

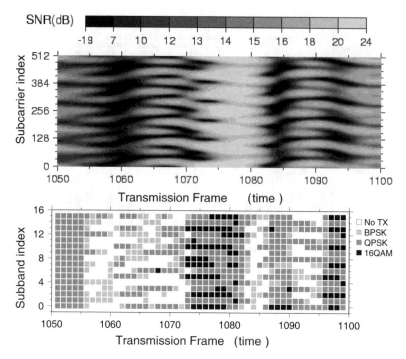

Figure 14.36: The micro-adaptive nature of the time-variant target bitrate subband adaptive (TVTBR-AOFDM) modem. The top graph is a contour plot of the channel SNR for all 512 subcarriers versus time. The bottom graph shows the modulation mode chosen for all 16 subbands for the same period of time. Each subband is composed of 32 subcarriers. The TVTBR AOFDM modem switches between target bitrates of 2, 3.4, 7, and 9.8 Mbit/s, while attempting to maintain an estimated BER of 0.1% before channel coding. Average Channel SNR is 16 dB over the channel of Figure 14.28 at a normalized Doppler frequency of $F_D = 7.41 \times 10^{-2}$.

rate unswitched modes. For a conservative BER switching threshold of 0.1%, the time-variant target bitrate subband adaptive (TVTBR-AOFDM) modem has a similar packet-loss ratio performance to that of the 1.8 Mbit/s nonswitched or constant target bitrate (CTBR) subband adaptive modem. However, as we will show, the throughput of the switched modem is always better than or equal to that of the unswitched modem and becomes far superior, as the channel quality improves. Observe in the figure that the "aggressive" switching threshold of 10% has a similar packet-loss ratio performance to that of the 9.8 Mbit/s CTBR-AOFDM modem. We found that in order to maintain a packet-loss ratio of below 5%, the BER switching thresholds of 2 and 3% offered the best overall performance, since the packet-loss ratio was fairly low, while the throughput bitrate was higher than that of an unswitched CTBR-AOFDM modem.

A high BER switching threshold results in the switched subband adaptive modem transmitting at a high average bitrate. However, we have shown in Figure 14.37 how the packet-loss ratio increases as the BER switching threshold increases. Therefore, the overall useful or effective throughput bitrate — that is, the bitrate excluding lost packets — may in fact be reduced in conjunction with high BER switching thresholds. Figure 14.38

14.3. ADAPTIVE TURBO-CODED OFDM-BASED VIDEOTELEPHONY

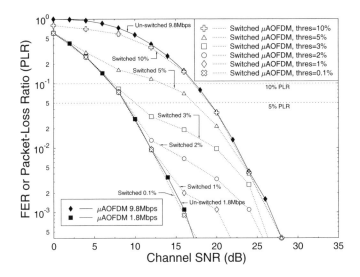

Figure 14.37: FER or video packet-loss ratio versus channel SNR for the TVTBR-AOFDM modem for a variety of BER switching thresholds. The switched modem uses four modes, with target bitrates of 1.8, 3.4, 7, and 9.8 Mbit/s. The unswitched 1.8 and 9.8 Mbit/s results are also shown in the graph as solid markers using the channel model of Figure 14.28 at a normalized Doppler frequency of $F_D = 7.41 \times 10^{-2}$.

demonstrates how the transmitted bitrate of the switched TVTBR-AOFDM modem increases with higher BER switching thresholds. However, when this is compared to the effective throughput bitrate, where the effects of packet loss are taken into account, the trade-off between the BER switching threshold and the effective bitrate is less obvious. Figure 14.39 portrays the corresponding effective throughput bitrate versus channel SNR for a range of BER switching thresholds. The figure demonstrates that for a BER switching threshold of 10% the effective throughput bitrate performance was reduced in comparison to some of the lower BER switching threshold scenarios. Therefore, the BER = 10% switching threshold is obviously too aggressive, resulting in a high packet-loss ratio and a reduced effective throughput bitrate. For the switching thresholds considered, the BER = 5% threshold achieved the highest effective throughput bitrate. However, even though the BER = 5% switching threshold produces the highest effective throughput bitrate, this is at the expense of a relatively high video packet-loss ratio, which, as we will show, has a detrimental effect on the perceived video quality.

We will now demonstrate the effects associated with different BER switching thresholds on the video quality represented by the peak signal-to-noise ratio (PSNR). Figure 14.40 portrays the PSNR and packet-loss performance versus time for a range of BER switching thresholds. The top graph in the figure indicates that for a BER switching threshold of 1% the PSNR performance is very similar to the corresponding error-free video quality. However, the PSNR performance diverges from the error-free curve when video packets are lost, although the highest PSNR degradation is limited to 2 dB. Furthermore, the PSNR curve typically reverts to the error-free PSNR performance curve in the next frame. In this example, about

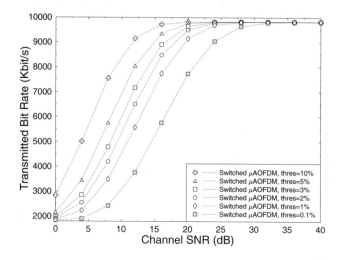

Figure 14.38: Transmitted bitrate of the switched TVTBR-AOFDM modem for a variety of BER switching thresholds. The switched modem uses four modes, having target bitrates of 1.8, 3.4, 7, and 9.8 Mbit/s, over the channel model of Figure 14.28 at a normalized Doppler frequency of $F_D = 7.41 \times 10^{-2}$.

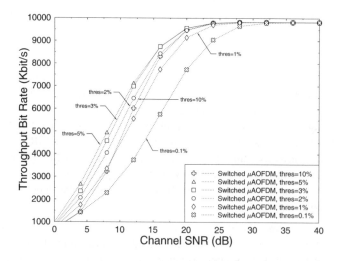

Figure 14.39: Effective throughput bitrate of the switched TVTBR-AOFDM modem for a variety of BER switching thresholds. The switched modem uses four modes, with target bitrates of 1.8, 3.4, 7, and 9.8 Mbit/s. The channel model of Figure 14.28 is used at a normalized Doppler frequency of $F_D = 7.41 \times 10^{-2}$.

14.3. ADAPTIVE TURBO-CODED OFDM-BASED VIDEOTELEPHONY

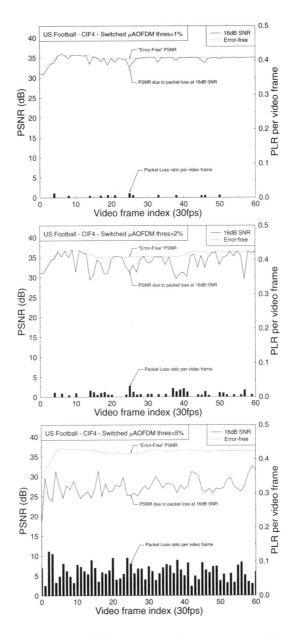

Figure 14.40: Video-quality and packet-loss ratio performance versus video-frame index (time) comparison between switched TVTBR-AOFDM transceivers with different BER switching thresholds, at an average of 16 dB SNR, using the channel model of Figure 14.28 at a normalized Doppler frequency of $F_D = 7.41 \times 10^{-2}$.

80% of the video frames have no video packet loss. When the BER switching threshold is increased to 2%, as shown in the center graph of Figure 14.40, the video-packet loss ratio has increased, such that now only 41% of video frames have no packet loss. The result of the increased packet loss is a PSNR curve, which diverges from the error-free PSNR performance curve more regularly, with PSNR degradations of up to 7 dB. When there are video frames with no packet losses, the PSNR typically recovers, achieving a similar PSNR performance to the error-free case. When the BER switching threshold was further increased to 3% — which is not shown in the figure — the maximum PSNR degradation increased to 10.5 dB, and the number of video frames without packet losses was reduced to 6%.

The bottom graph of Figure 14.40 depicts the PSNR and packet loss performance for a BER switching threshold of 5%. The PSNR degradation in this case ranges from 1.8 to 13 dB and all video frames contain at least one lost video packet. Even though the BER = 5% switching threshold provides the highest effective throughput bitrate, the associated video quality is poor. The PSNR degradation in most video frames is about 10 dB. Clearly, the highest effective throughput bitrate does not guarantee the best video quality. We will now demonstrate that the switching threshold of BER = 1% provides the best video quality, when using the average PSNR as our performance metric.

Figure 14.41(a) compares the average PSNR versus channel SNR performance for a range of switched (TVTBR) and unswitched (CTBR) AOFDM modems. The figure compares the four unswitched (i.e., CTBR subband adaptive modems) with switching (i.e., TVTBR subband adaptive modems), which switch between the four fixed-rate modes, depending on the BER switching threshold. The figure indicates that the switched TVTBR subband adaptive modem having a switching threshold of BER = 10% results in similar PSNR performance to the unswitched CTBR 9.8 Mbit/s subband adaptive modem. When the switching threshold is reduced to BER = 3%, the switched TVTBR AOFDM modem outperforms all of the unswitched CTBR AOFDM modems. A switching threshold of BER = 5% achieves a PSNR performance, which is better than the unswitched 9.8 Mbit/s CTBR AOFDM modem, but worse than that of the unswitched 7.0 Mbit/s modem, at low- and medium-channel SNRs.

A comparison of the switched TVTBR AOFDM modem employing all six switching thresholds that we have used previously is shown in Figure 14.41(b). This figure suggests that switching thresholds of BER = 0.1, 1, and 2% perform better than the BER = 3% threshold, which outperformed all of the unswitched CTBR subband adaptive modems. The best average PSNR performance was achieved by a switching threshold of BER = 1%. The more conservative BER = 0.1% switching threshold results in a lower PSNR performance, since its throughput bitrate was significantly reduced. Therefore, the best trade-off in terms of PSNR, throughput bitrate, and video packet-loss ratio was achieved with a switching threshold of about BER = 1%.

14.3.8 Summary and Conclusions

A range of AOFDM video transceivers has been proposed for robust, flexible, and low-delay interactive videotelephony. In order to minimize the amount of signaling required, we divided the OFDM subcarriers into subbands and controlled the modulation modes on a subband-by-subband basis. The proposed constant target bitrate AOFDM modems provided a lower BER than the corresponding conventional OFDM modems. The slightly more complex switched

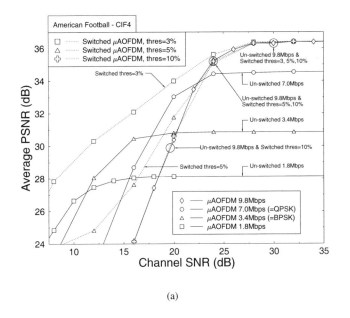

(a)

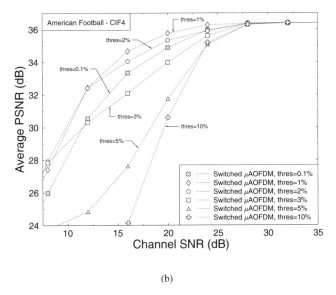

(b)

Figure 14.41: Average PSNR versus channel SNR performance for switched and unswitched subband adaptive modems. Figure (a) compares the four unswitched CTBR subband adaptive modems with switched TVTBR subband adaptive modems (using the same four modem modes) for switching thresholds of BER = 3, 5, and 10%. Figure (b) compares the switched TVTBR subband adaptive modems for switching thresholds of BER = 0.1, 1, 2, 3, 5, and 10%.

TVTBR-AOFDM modems can provide a balanced video-quality performance, across a wider range of channel SNRs than the other schemes investigated.

14.4 Burst-by-burst Adaptive Decision Feedback Equalized TCM, TTCM, and BICM for H.263-assisted Wireless Videotelephony[8]

14.4.1 Introduction

$\mathcal{M}$-ary Coded Modulation (CM) schemes such as Trellis Coded Modulation (TCM) [431] and Bit-interleaved Coded Modulation (BICM) [432, 433] constitute powerful and bandwidth-efficient forward error-correction schemes, which combine the functions of coding and modulation. It was found in [432, 433] that BICM is superior to TCM when communicating over narrowband Rayleigh fading channels, but inferior to TCM in Gaussian channels. In 1993, power efficient binary Turbo Convolutional Codes (TCCs) were introduced in [401], which are capable of achieving a low bit error rate at low Signal-to-Noise Ratios (SNRs). However, TCCs typically operate at a fixed coding rate of 1/2 and they were originally designed for Binary Phase Shift Keying (BPSK) modulation, hence they require the bandwidth to be doubled. In order to lend TCCs a higher spectral efficiency, BICM using TCCs was first proposed in [434], where it was also referred to as Turbo Coded Modulation (TuCM). As another design alternative, Turbo Trellis Coded Modulation (TTCM) was proposed in [435], which has a structure similar to that of the family of TCCs, but employs TCM codes as component codes. It was shown in [435] that TTCM performs better than TCM and TuCM at a comparable complexity. Many other bandwidth efficient schemes using turbo codes, such as multilevel coding employing turbo codes [436], have been proposed in the literature [437], but here we focus our study on the family of TCM, BICM, and TTCM schemes in the context of a wireless videotelephony system.

In general, fixed-mode transceivers fail to adequately accommodate and counteract the time varying nature of the mobile radio channel. Hence, their error distribution becomes bursty and this would degrade the performance of most channel coding schemes, unless long-delay channel interleavers are invoked. However, the disadvantage of long-delay interleavers is that owing to their increased latency they impair "lip-synchronization" between the voice and video signals. In contrast, in Burst-by-Burst (BbB) Adaptive Quadrature Amplitude (or Phase Shift Keying) Modulation (AQAM) schemes [410, 411, 414, 438–447] a higher-order modulation mode is employed when the instantaneous estimated channel quality is high for the sake of increasing the number of Bits Per Symbol (BPS) transmitted; conversely, a more robust but lower-throughput modulation mode is used, when the instantaneous channel quality is low, in order to improve the mean Bit Error Ratio (BER) performance. Uncoded AQAM schemes [410,411,414,439] and channel coded AQAM schemes [440–446] have been lavishly investigated in the context of narrowband fading channels. In particular, adaptive trellis-coded $\mathcal{M}$-ary Phase Shift Keying (PSK) was considered in [440] and coset codes were applied to adaptive trellis-coded $\mathcal{M}$-ary Quadrature Amplitude Modulation (QAM) in [442]. *However, these contributions were based on a number of ideal assumptions,*

---
[8]Ng, Chung, Cherriman and Hanzo: *IEEE Transactions on Circuits and Systems for Video Technology.*

such as perfect channel estimation and zero modulation mode feedback delay. Hence, in [443], adaptive TCM using more realistic outdated fading estimates was investigated. Recently, the performance of adaptive TCM based on realistic practical considerations such as imperfect channel estimation, modem mode signaling errors, and modem mode feedback delay was evaluated in [444], where the adaptive TCM scheme was found to be robust in most practical situations when communicating over narrowband fading channels. In an effort to increase the so-called time-diversity order of the TCM codes, adaptive BICM schemes were proposed in [445], although their employment was still limited to communications over narrowband fading channels.

On the other hand, for communications over wideband fading channels, a BbB adaptive transceiver employing separate channel coding and modulation schemes was proposed in [448]. The main advantage of this wideband BbB adaptive scheme is that regardless of the prevailing channel conditions, the transceiver achieves always the best possible source-signal representation quality such as video, speech, or audio quality by automatically adjusting the achievable bitrate and the associated multimedia source-signal representation quality in order to match the channel quality experienced. Specifically, this wideband BbB adaptive scheme employs the adaptive video rate control and packetization algorithm of [392], which generates exactly the required number of video bits for the channel-quality-controlled BbB adaptive transceiver, depending on the instantaneous modem-mode-dependent payload of the current packet, as determined by the current modem mode. Hence, a channel-quality-dependent variable-sized video packet is transmitted in each Time Division Multiple Access (TDMA) frame constituted by a fixed number of AQAM symbols and hence the best possible source-signal representation quality is achieved on a near-instantaneous basis under given propagation conditions in order to cater for the effects of path-loss, fast-fading, slow-fading, dispersion, co-channel interference, etc. More explicitly, a half-rate Bose–Chaudhuri–Hocquenghem (BCH) block code and a half-rate TCC were employed and the modulation modes were adapted according to the channel conditions. However, owing to the fixed coding rate of the system, the range of the effective video bitrates was limited. Hence, in this section our objective is to further develop the wireless videophone system of [448] by increasing its bandwidth efficiency up to a factor of two upon rendering not only the modulation-mode selection, but also the choice of the channel coding rate near instantaneously adaptive with the advent of the aforementioned bandwidth-efficient coded modulation schemes. As a second objective, we extend this adaptive coded modulation philosophy to multiuser scenarios in the context of a UMTS Terrestrial Radio Access (UTRA) [161, 449] system.

This section is organized as follows. In Section 14.4.2 the system's architecture is outlined. In Section 14.4.3, the performance of various fixed-mode CM schemes is characterized, while in Section 14.4.4 the performance of adaptive CM schemes is evaluated. In Section 14.4.5 the performance of the adaptive TTCM-based video system is studied in the UTRA Code Division Multiple Access (CDMA) environment. Finally, we will conclude in Section 14.4.6.

14.4.2 System Overview

The simplified block diagram of the BbB adaptive CM scheme is shown in Figure 14.42, where channel interleaving spanning one transmission burst is used. The length of the CM codeword is one transmission burst.

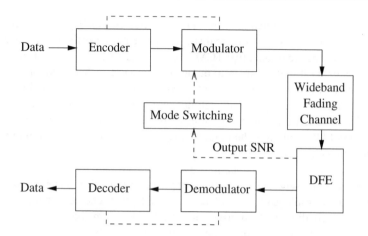

Figure 14.42: The block diagram of the BbB adaptive CM scheme.

We invoke four CM encoders for our quadruple-mode adaptive coded modulation scheme, each attaching one parity bit to each information symbol generated by the video encoder, yielding a channel coding rate of 1/2 in conjunction with the modulation modes of 4QAM, a rate of 2/3 for 8PSK, 3/4 for 16QAM and 5/6 for 64QAM. The complexity of the CM schemes is compared in terms of the number of decoding states and the number of decoding iterations. For a TCM or BICM code of memory M, the corresponding complexity is proportional to the number of decoding states $S = 2^M$. As TTCM schemes invoke two component TCM codes, a TTCM code employing t iterations and using an S-state component code exhibits a complexity proportional to $2 \cdot t \cdot S$ or $t \cdot 2^{M+1}$.

Over wideband fading channels, the employed Minimum Mean Squared Error (MMSE)-based Decision Feedback Equalizer (DFE) eliminates most of the channel-induced InterSymbol Interference (ISI). Consequently, the mean-squared error at the output of the DFE can be calculated and used as the channel quality metric invoked for switching the modulation modes. More explicitly, the residual signal deviation from the error-free transmitted phasors at the DFE's output reflects the instantaneous channel quality of the time varying wideband fading channel. Hence, given a certain instantaneous channel quality, the most appropriate modulation mode can be chosen according to this residual signal deviation from the error-free transmitted phasors at the DFE's output. Specifically, the SNR at the output of the DFE, γ_{DFE}, can be computed as [439]

$$\gamma_{\text{DFE}} = \frac{\text{Required signal power}}{\text{Residual ISI power} + \text{Effective noise power}}$$

$$= \frac{E\left[|s_k \sum_{m=0}^{N_f} C_m h_m|^2\right]}{\sum_{q=-(N_f-1)}^{-1} E\left[|\sum_{m=0}^{N_f-1} C_m h_{m+q} s_{k-q}|^2\right] + N_0 \sum_{m=0}^{N_f} |C_m|^2}, \quad (14.2)$$

where C_m and h_m denote the DFE's feed-forward coefficients and the Channel Impulse Response (CIR), respectively. The transmitted signal is represented by s_k and N_0 denotes the noise spectral density. Finally, the number of DFE feed-forward coefficients is denoted by N_f. The equalizer's output SNR, γ_{DFE}, in Equation 14.2, is then compared against a set of

14.4. BURST-BY-BURST ADAPTIVE TCM, TTCM, AND BICM FOR H.263 VIDEOTELEPHONY 509

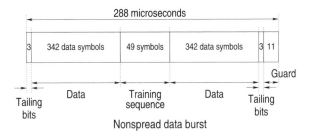

Figure 14.43: Transmission burst structure of the FMA1 non-spread data as specified in the FRAMES proposal [405].

adaptive modem mode switching thresholds f_n, and subsequently the appropriate modulation mode is selected [439].

In the adaptive transmission schemes the outdated channel quality estimates arriving after a feedback delay inevitably inflict performance degradations, which can be mitigated using powerful channel quality prediction techniques [450, 451]. However, in our proposed adaptive video system we adopted the practical approach of employing outdated, rather than perfect channel quality estimates. Specifically, a practical modem mode switching regime adapted to the specific requirements of wireless videotelephony is employed, where a suitable modulation mode is chosen at the receiver on a BbB basis and it is then communicated to the transmitter by superimposing the requested modulation mode identifier code onto the terminal's reverse-direction transmission burst. Hence, the actual channel condition is outdated by one TDMA/Time-division Duplex (TDD) frame duration.

At the receiver, the DFE's symbol estimate $\hat{s}_k$ is passed to the channel decoder and the log-domain branch metric is computed for the sake of maximum-likelihood decoding at time instant k as

$$m_k(s_i) = -\frac{|\hat{s}_k - s_i|^2}{2\sigma^2}, \quad i = \{0, \ldots, \mathcal{M} - 1\}, \tag{14.3}$$

where s_i is the ith legitimate symbol of the $\mathcal{M}$-ary modulation scheme and σ^2 is the variance of the Additive White Gaussian Noise (AWGN). Note that the equalizer output $\hat{s}_k$ is near-Gaussian, because the channel has been equalised [439]. In other words, the DFE has 'converted' the dispersive Rayleigh fading channels into an 'AWGN-like' channel. Hence, the TCM and TTCM codes that have been designed for AWGN channels will outperform the BICM scheme, as demonstrated in Figure 14.44.

The following assumptions are stipulated. First, we assume that the equalizer is capable of estimating the CIR perfectly with the aid of the equalizer training sequence hosted by the transmission burst of Figure 14.43. Secondly, the CIR is time-invariant for the duration of a transmission burst, but varies from burst to burst according to the Doppler frequency, which corresponds to assuming that the CIR is slowly varying.

14.4.2.1 System Parameters and Channel Model

For the sake of direct comparisons, we used the same H.263 video codec as in [448]. Hence, we refer the interested readers to [161] for a detailed description of the H.263 video codec.

Table 14.9: Operational-mode Specific Transceiver Parameters for TTCM

Features	Multi-rate System			
Mode	4QAM	8PSK	16QAM	64QAM
Transmission symbols/TDMA slot		684		
Bits/symbol	2	3	4	6
Transmission bits/TDMA slot	1368	2052	2736	4104
Packet rate		216.7/s		
Transmission bitrate (kbit/s)	296.4	444.6	592.8	889.3
Code termination symbols		6		
Data symbols/TDMA slot		678		
Coding rate	1/2	2/3	3/4	5/6
Information bits/symbol	1	2	3	5
Unprotected bits/TDMA slot	678	1356	2034	3390
Unprotected bitrate (kbit/s)	146.9	293.8	440.7	734.6
Video packet CRC (bits)		16		
Feedback protection (bits)		9		
Video packet header (bits)	11	12	12	13
Video bits/packet	642	1319	1997	3352
Effective video bitrate (kbit/s)	139.1	285.8	432.7	726.3
Video framerate (Hz)		30		

The transmitted bitrate of all four modes of operation is shown in Table 14.9 for the TTCM coding scheme. The associated bitrates are similar for the other CM schemes. The slight difference is caused by using different numbers of code termination symbols. The unprotected bitrate before channel coding is also shown in the table. The actual useful bitrate available for video encoding is slightly lower than the unprotected bitrate owing to the useful bitrate reduction required by the transmission of the strongly protected packet acknowledgement information and packetization overhead information. The effective video bitrate is also shown in the table, which varies from 139 to 726 kbit/s. We have investigated the video system concerned using a wide range of video sequences having different resolutions. However, for conciseness we only show results for the Common Intermediate Format (CIF) resolution (352 × 288 pixels) "Salesman" sequence at 30 frames/s.

Table 14.10 shows the modulation and channel parameters employed. Again, similarly to [448], the COST 207 [407] channel models, which are widely used in the community, were employed. Specifically, a four-path Typical Urban COST 207 channel [407] was used. The multi-path channel model is characterized by its discretized symbol-spaced CIR, where each path is faded independently according to a Rayleigh distribution. The non-spread data transmission burst structure FMA1 specified in the FRAMES proposal [405] was used, which is shown in Figure 14.43. Nyquist signaling was employed and the remaining system parameters are shown in Table 14.11.

A component TCM having a code memory of $M = 3$ was used for the TTCM scheme. The number of iterations for TTCM was fixed to $t = 4$ and hence the iterative scheme exhibited a similar decoding complexity to that of the TCM having a code memory $M = 6$ in terms of the number of coding states. The fixed-mode CM schemes that we invoked in our BbB AQAM schemes are Ungerböck's TCM [86, 431], Robertson's TTCM [86, 435],

Table 14.10: Modulation and Channel Parameters

Parameter	Value
Carrier Frequency	1.9 GHz
Vehicular Speed	30 mph
Doppler frequency	85 Hz
Normalized Doppler frequency	3.3×10^{-5}
Channel type	COST 207 Typical Urban [407]
Number of paths in channel	4
Data modulation	Adaptive coded modulation (4QAM, 8PSK, 16QAM, 64QAM)
Receiver type	Decision Feedback Equalizer Number of Forward Filter Taps = 35 Number of Backward Filter Taps = 7

Table 14.11: Generic System Features of the Reconfigurable Multi-mode Video Transceiver, using the Non-spread Data Burst Mode of the FRAMES Proposal [405] Shown in Figure 14.43

Features	Value
Multiple access	TDMA
Number of slots/Frame	16
TDMA frame length	4.615 ms
TDMA slot length	288 μs
Data symbols/TDMA slot	684
User data symbol rate (kBd)	148.2
System data symbol rate (MBd)	2.37
Symbols/TDMA slot	750
User symbol rate (kBd)	162.5
System symbol rate (MBd)	2.6
System bandwidth (MHz)	3.9
Effective user bandwidth (kHz)	244

and Zehavi's BICM [86, 432]. Soft decision trellis decoding utilizing the log-maximum *a posteriori* algorithm [452] was invoked for decoding.

Note that the parameters of the DFE employed, which are the same as those in [448], as well as that of the Joint Detection (JD) receiver in Section 14.4.5 were adjusted such that they achieve their best attainable performance in terms of removing the effect of channel induced multipath interference as well as the multiuser interference, respectively. Hence, opting for a more complex DFE or JD design would not improve the overall achievable performance. Furthermore, because the video quality expressed in terms of the luminance Peak Signal-to-Noise Ratio (PSNR) is a direct function of the system's effective throughput, we will use the PSNR value as the ultimate evaluation metric of the system's performance. As the same video system and channel model are employed, the complexity difference between the various CM-assisted video schemes is directly dependent on the decoding complexity of the CM schemes.

14.4.3 Employing Fixed Modulation Modes

Initial simulations of the videophone transceiver were performed with the transceiver configured in one of the four fixed modulation modes of Table 14.9. We commence by comparing the performance of TTCM in conjunction with a code memory of $M = 3$ and using $t = 4$ iterations with that of non-iterative TCM along with a code memory of $M = 6$, because the associated computational complexity is similar. We then also compare these results with those of TCM using a code memory of $M = 3$ and to BICM employing a code memory $M = 3$. Again, one video packet is transmitted in each TDMA frame and the receiver checks whether the received packet has any bit errors using the associated Cyclic Redundancy Check (CRC). If the received packet has been corrupted, a negative acknowledgement flag is transmitted to the video encoder in order to prevent it from using the packet just transmitted to update the encoder's reconstruction frame buffer. This allows the video encoder's and decoder's reconstruction frame buffer to use the same contents for motion compensation. This acknowledgement message is strongly protected using repetition codes and superimposed on the reverse link transmission. In these investigations a transmission frame error resulted in a video packet error. We characterize the relative frequency of these packet corruption events by the Packet Loss Ratio (PLR). The PLR of the CM schemes is shown in Figure 14.44. We emphasize again that the video packets are either error-free or discarded. Hence, the PLR is a more meaningful modem performance metric than the BER in this scenario.

From Figure 14.44, it is found that the BICM scheme has the worst PLR performance and the TCM6 scheme using a code memory $M = 6$ has a significant PLR performance advantage over the TCM3 scheme employing a code memory of $M = 3$. Furthermore, the TTCM scheme provides the best PLR performance, requiring approximately 2.5 dB lower channel SNR than the BICM scheme. This is because turbo decoding of the TTCM scheme is very effective in reducing the number of bit errors to zero in all of the received packets exhibiting a moderate or low number of bit errors before channel decoding. In contrast, the gravely error-infected received packets are simply dropped and the corresponding video frame segment is replaced by the same segment of the previous frame. The performance of BICM is worse than that of TCM due to the associated limited channel interleaving depth [432, 433] of the BICM scheme in our slow-fading wideband channels.

Figure 14.45 shows the error-free decoded video quality, measured in terms of the PSNR versus time for the CIF-resolution "Salesman" sequence for each of the four fixed modulation modes using the TTCM scheme. The figure demonstrates that the higher-order modulation modes, which have a higher associated bitrate, provide a better video quality. However, in an error-impaired situation a high error-free PSNR does not always guarantee a better subjective video quality. In order to reduce the detrimental effects of channel-induced errors on the video quality, we refrain from decoding the error-infested video packets and hence avoid error propagation through the reconstructed video frame buffer [392, 448]. Instead, these originally high-bitrate and high-quality but error-infested video packets are dropped and hence the reconstructed video frame buffer will not be updated until the next packet replenishing the specific video frame area arrives. As a result, the associated video performance degradation

14.4. BURST-BY-BURST ADAPTIVE TCM, TTCM, AND BICM FOR H.263 VIDEOTELEPHONY

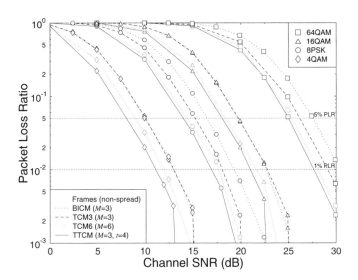

Figure 14.44: PLR versus channel SNR for the four fixed modem modes, using the four joint coding/modulation schemes considered, namely BICM, TCM3, TCM6, and TTCM, when communicating over the COST 207 channel [407].

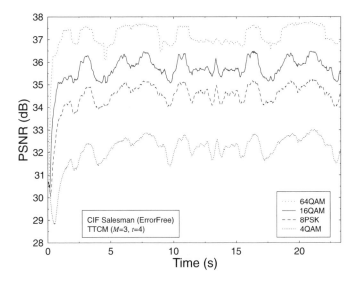

Figure 14.45: PSNR (video quality) versus time for the four fixed modulation modes, under error-free channel conditions using the CIF resolution "Salesman" video sequence at 30 frame/s. TTCM scheme using a code memory of $M = 3$ and $t = 4$ iterations was employed.

Table 14.12: Switching Thresholds According to Equation 14.4 at the Output of the Equalizer Required for each Modulation Mode of the BbB Adaptive Modem. The Threshold Types are Normal (N), Conservative (C) and Aggressive (A). M Denotes the Code Memory of the Encoder

				Thresholds (dB)		
No.	Scheme	M	Type	f_1	f_2	f_3
1.	TTCM	3	N	13.07	17.48	24.77
2.	TTCM	3	C	16.00	20.00	27.25
3.	TTCM	3	A	10.05	12.92	20.67
4.	TCM	3	N	14.48	18.86	26.24
5.	TCM	6	N	13.98	17.59	25.37
6.	BICM	3	N	15.29	18.88	26.49

becomes fairly minor for PLR values below 5% [448], which is significantly lower than in the case of replenishing the corresponding video frame area with the error-infested video packet.

14.4.4 Employing Adaptive Modulation

The BbB AQAM mode switching mechanism is characterized by a set of switching thresholds, by the corresponding random TTCM symbol interleavers, and by the component codes, as follows:

$$\text{Modulation mode} = \begin{cases} \text{4QAM}, I_0 = I_s, R_0 = 1/2 & \text{if } \gamma_{\text{DFE}} \leq f_1 \\ \text{8PSK}, I_1 = 2I_s, R_1 = 2/3 & \text{if } f_1 < \gamma_{\text{DFE}} \leq f_2 \\ \text{16QAM}, I_2 = 3I_s, R_2 = 3/4 & \text{if } f_2 < \gamma_{\text{DFE}} \leq f_3 \\ \text{64QAM}, I_3 = 5I_s, R_3 = 5/6 & \text{if } \gamma_{\text{DFE}} > f_3, \end{cases} \quad (14.4)$$

where $f_n, n = 1, \ldots, 3$, are the AQAM switching thresholds, which were set according to the target PLR requirements, while $I_s = 684$ is the number of data symbols in a transmission burst and I_n represents the random TTCM symbol-interleaver sized expressed in terms of the number of bits, which is not used for the TCM and BICM schemes.

The video encoder/decoder pair discards all corrupted video packets in an effort to avoid error-propagation effects in the video decoder. Therefore, in this section the AQAM switching thresholds f_n were chosen using an experimental procedure in order to maintain the required target PLR, rather than the BER. More specifically, we defined a set of "normal" thresholds for each adaptive CM scheme depending on their fixed modem mode's performance in terms of PLR versus average equalizer SNR. Explicitly, the "normal" threshold was set to maintain a PLR of around 3%. We also defined a "conservative" threshold, for a PLR of around 0.5%, and an "aggressive" threshold, for a PLR of around 20%, for the adaptive TTCM scheme. These modem mode switching thresholds are listed in Table 14.12.

The probability of each of the modulation modes versus channel SNR is shown in Figure 14.46 for the BbB adaptive TTCM scheme using the "normal" switching thresholds.

14.4. BURST-BY-BURST ADAPTIVE TCM, TTCM, AND BICM FOR H.263 VIDEOTELEPHONY

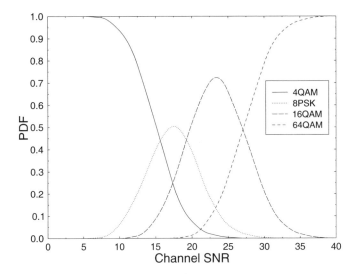

Figure 14.46: PDF of the various active mode versus channel SNR for the quadruple-mode BbB adaptive TTCM scheme using the "normal" thresholds employing a code memory of $M = 3$ and $t = 4$ iterations when communicating over the COST 207 channel [407].

The graph shows that the 4QAM mode is the most probable at low channel SNRs, and the 64QAM mode is predominant at high channel SNRs. In addition, for example at 20 dB, the 4QAM mode is being used about 8% of the time and 64QAM only 1% of the time, because most of the time the AQAM modem is operating in its 8PSK or 16QAM mode with an associated probability of 40% and 51%, respectively.

14.4.4.1 Performance of TTCM AQAM

In this section we compare the performance of the four fixed modulation modes with that of the quadruple-mode TTCM AQAM scheme using the "normal" switching thresholds of Table 14.12. The code memory is $M = 3$ and the number of turbo iterations is $t = 4$.

Specifically, in Figure 14.47 we compare the PLR performance of the four fixed modulation modes with that of the quadruple-mode TTCM AQAM arrangement using the "normal" switching thresholds. The figure shows that the performance at low channel SNRs is similar to that of the fixed TTCM 4QAM mode, while at high channel SNRs the performance is similar to that of the fixed 64QAM mode. At medium SNR values the PLR performance is near-constant, ranging from 1% to 5%. More explicitly, the TTCM AQAM modem maintains this near-constant PLR, which is higher than that of the fixed 4QAM modem, while achieving a higher throughput bitrate than the 2 bit/symbol rate of 4QAM. As our BbB AQAM videophone system's video performance is closely related to the PLR, the TTCM AQAM scheme provides a near-constant video performance across a wide range of channel SNRs. In addition, the BbB TTCM AQAM modem allows the throughput bitrate to increase, as the channel SNR increases, thereby supporting an improved video quality as the channel SNR improves, which is explicitly shown in Figure 14.48.

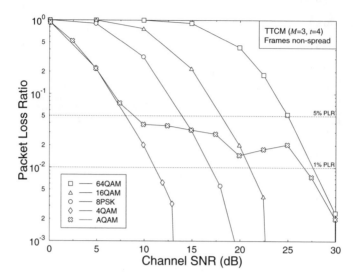

Figure 14.47: Packet loss ratio versus channel SNR for the four fixed TTCM modes and for the quadruple-mode AQAM scheme using the "normal" thresholds of Table 14.12, employing a code memory of $M = 3$ and $t = 4$ iterations when communicating over the COST 207 channel [407].

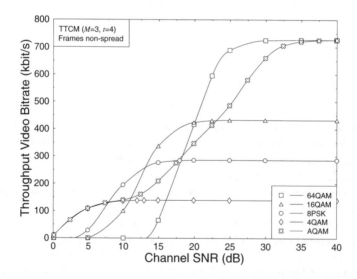

Figure 14.48: Throughput video bitrate versus channel SNR for the four fixed modes and for the quadruple-mode TTCM AQAM scheme using the "normal" thresholds of Table 14.12, employing a code memory of $M = 3$ and $t = 4$ iterations when communicating over the COST 207 channel [407].

14.4. BURST-BY-BURST ADAPTIVE TCM, TTCM, AND BICM FOR H.263 VIDEOTELEPHONY

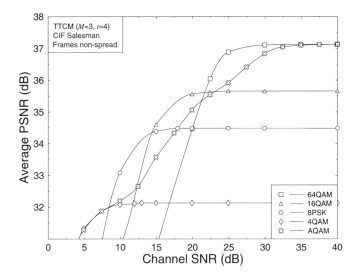

Figure 14.49: Average PSNR versus channel SNR for the four fixed TTCM modes and for the quadruple-mode TTCM AQAM scheme, using the "normal" thresholds of Table 14.12, and the CIF "Salesman" video sequence at 30 frame/s. A code memory of $M = 3$ and $t = 4$ iterations were invoked when communicating over the COST 207 channel [407].

The effective throughput bitrate of the fixed-mode modems drops rapidly due to the increased PLR, which is a consequence of discarding the "payload" of the corrupted video packets, as the channel SNR reduces. The TTCM AQAM throughput bitrate matches that achieved by the fixed modem modes at both low and high channel SNRs. At a channel SNR of 25 dB the fixed-mode 64QAM modem achieves an approximately 700 kbit/s throughput, while the TTCM AQAM modem transmits at approximately 500 kbit/s. However, by referring to Figure 14.48 it can be seen that the 700 kbit/s video throughput bitrate is achieved by 64QAM at a concomitant PLR of slightly over 5%, while the AQAM modem experiences a reduced PLR of 2%. As mentioned earlier, the associated video performance degradation becomes noticeable at a PLR in excess of 5%. Hence, the fixed-mode 64QAM modem results in a subjectively inferior video quality in comparison with the TTCM AQAM scheme at a channel SNR of 25 dB. The video quality expressed in terms of the average PSNR is closely related to the effective video throughput bitrate. Hence, the trends observed in terms of PSNR in Figure 14.49 are similar to those seen in Figure 14.48.

Note that the channel-quality related AQAM mode feedback is outdated by one TDMA transmission burst of 4.615 ms for the sake of providing realistic results. As shown in [448], in the idealiztic zero-delay-feedback AQAM scheme, the resultant PSNR curve follows the envelope of the fixed-mode schemes. However, a suboptimum modulation mode may be chosen due to the one-frame-feedback delay, which may inflict an increased video packet loss. As the effective throughput is quantified in terms of the average bitrate provided by all of the successful transmitted video packets but excluding the erroneous and hence dropped packets, the throughput and hence the PSNR of the realistic AQAM scheme using outdated channel quality estimates is lower than that of the ideal AQAM scheme. Furthermore, the AQAM system's effective throughput is slightly reduced by allocating some of the useful payload to the AQAM mode signaling information, which is protected by strong repetition coding. Therefore, the throughput or PSNR curves of the AQAM scheme plotted in Figures 14.48 and 14.49 do not strictly follow the throughput or PSNR envelopes of the fixed-mode schemes. However, at lower vehicular speeds the switching latency is less crucial and the practical one-frame-delay AQAM can achieve a performance that is closer to that of the ideal zero-delay AQAM.

Figure 14.49 portrays that the AQAM modem's video performance degrades gracefully, as the channel SNR degrades, while the fixed-mode modems' video performance degrades more rapidly, when the channel SNR becomes insufficient for the reliable operation of the specific modem mode concerned. As we have shown in Figure 14.45, the higher-order modulation modes, which have a higher associated bitrate, provide a higher PSNR in an error-free scenario. However, in the presence of channel errors, the channel-induced PSNR drops and the associated perceptual video quality degradations imposed by transmission errors are more dramatic when higher-order modulation modes are employed. Therefore, annoying video artifacts and a low subjective video perception will be observed when employing higher-order modulation modes such as 64QAM during instances of low channel quality. *However, these subjective video quality degradations and the associated artifacts are eliminated by the advocated AQAM/TTCM regime, because it operates as a "safety-net" mechanism, which drops the instantaneous AQAM/TTCM throughput in an effort to avoid the dramatic PSNR degradations of the fixed modes during instances of low channel quality, instead of dropping the entire received packet.*

14.4.4.2 Performance of AQAM Using TTCM, TCC, TCM, and BICM

Let us now compare the video performance of the TTCM-aided AQAM video system with that of the TCC-assisted AQAM video system of [448]. The lowest information throughput of the TTCM AQAM scheme was 1 BPS in the 4QAM mode here, while that of the TCC AQAM scheme of [448] was 0.5 BPS in the BPSK mode. Furthermore, the highest information throughput of the TTCM AQAM scheme was 5 BPS in the 64QAM mode here, while that of the TCC AQAM scheme of [448] was 3 BPS on the 64QAM mode. Hence, we can see from Figure 14.48 that TTCM has a video bitrate of about 100 kbit/s at SNR = 5 dB and 726 kbit/s at SNR = 35 dB. In contrast, the TCC in [448, Figure 12] has only a video throughput of 50 kbit/s at SNR = 5 dB and 409 kbit/s at SNR = 35 dB. Hence, we may conclude that at the symbol rate considered the TTCM scheme has succeeded in substantially increasing the achievable effective video bitrate of the TCC scheme of [448] which had an identical symbol rate. This increased effective video bitrate allows the TTCM scheme to transmit CIF video frames which are four times larger than the QCIF video frames transmitted by the TCC scheme of [448].

Let us now compare the video performance of the TTCM-aided AQAM video system with that of the TCM- and BICM-assisted AQAM video system. As was shown in Figure 14.44, the TTCM scheme achieved the best PLR performance in a fixed modulation scenario. Therefore, according to Table 14.12 the AQAM switching thresholds of the TTCM scheme can then be set lower, while still achieving the required PLR performance. The lower thresholds imply that higher-order modulation modes can be used at lower channel SNRs, and hence a higher video transmission bitrate is achieved with respect to the other joint coding and

14.4. BURST-BY-BURST ADAPTIVE TCM, TTCM, AND BICM FOR H.263 VIDEOTELEPHONY

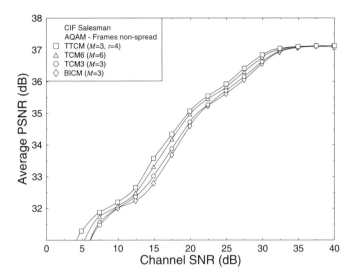

Figure 14.50: Average PSNR versus channel SNR for the quadruple-mode AQAM modems using the four joint coding/modulation schemes considered, namely BICM, TCM3, TCM6, and TTCM, when communicating over the COST 207 channel [407].

modulation schemes. The PSNR video quality is closely related to the video bitrate. As shown in Figure 14.50, the TTCM-based AQAM modem exhibits the highest PSNR video quality followed by the TCM6 scheme having a code memory of $M = 6$, the TCM3 scheme having a code memory of $M = 3$, and, finally, the BICM scheme having a code memory of $M = 3$.

As seen in Figure 14.50, the PSNR video performance difference of the BICM scheme compared with that of the TTCM scheme is only about 2 dB, because the video quality improvement of TTCM is limited by the moderate turbo interleaver length of our BbB AQAM scheme. Therefore, the low-complexity TCM3 scheme provides the best compromise in terms of the PSNR video performance and decoding complexity, because the BICM and TCM3 schemes are of similar complexity, but the TCM3 scheme exhibits a better video performance.

14.4.4.3 The Effect of Various AQAM Thresholds

In the previous sections we have studied the performance of AQAM using the switching thresholds of the "normal" scenario. In contrast, in this section we study the performance of the TTCM-aided AQAM scheme using the switching thresholds of the "conservative", "normal", and "aggressive" scenarios, which were characterized in Table 14.12. Again, the "conservative", "normal", and "aggressive" thresholds sets result in a target PLRs of 0.5%, 3%, and 20%, respectively.

The three sets of thresholds have allowed us to demonstrate how the performance of the AQAM modem was affected when the modem used the radio channel more "aggressively" or more "conservatively". This translated into a more- or less-frequent employment of the higher-throughput, but more error-prone AQAM modes, respectively. Explicitly, the more aggressive switching thresholds provide a higher effective throughput at a cost of a higher

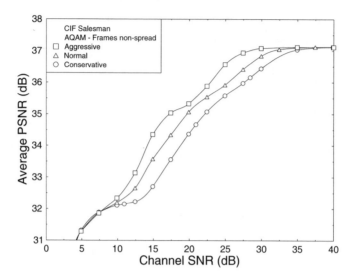

Figure 14.51: Average PSNR video quality versus channel SNR for the quadruple-mode TTCM AQAM scheme using the three different sets of switching thresholds from Table 14.12. TTCM scheme using a code of $M = 3$ and $t = 4$ iterations was used when communicating over the COST 207 channel [407].

PLR. The PSNR versus channel SNR performance of the three AQAM modem switching thresholds is depicted in Figure 14.51, where it is shown that the average PSNR of the AQAM modem employing "aggressive" and "normal" switching thresholds is about 4 and 2 dB better than that employing "conservative" switching thresholds, respectively, for channel SNRs ranging from 12 to 30 dB. However, in perceptual video quality terms, the best compromise was associated with the "normal" switching threshold set. This was because the excessive PLR of the "aggressive" set inflicted noticeable channel-induced video degradations, despite the favorable video quality of the unimpaired high-throughput video packets.

14.4.5 TTCM AQAM in a CDMA system

We have demonstrated that the adaptive video system aided by TTCM performs better than schemes aided by TCC [448], TCM, and BICM in a single-user scenario. We now study the performance of the most promising TTCM AQAM scheme in the context of Direct Sequence Code Division Multiple Access (DS-CDMA) when communicating over the UTRA wideband vehicular Rayleigh fading channels [449] and supporting multiple users[9].

In order to mitigate the effects of Multiple Access Interference (MAI) and ISI, while at the same time improving the system's performance by benefiting from the multipath diversity effects of the channels, we employ the MMSE-based Block Decision Feedback Equalizer (MMSE-BDFE) for JD [203] in our system. The multiuser JD-MMSE-BDFE receivers are derivatives of the single-user MMSE-DFE [203, 438], where the MAI is equalised as if it was

---
[9]The significance of this research is that although independent adaptive coding and modulation was standardized for employment in the 3G High Speed Data Packet Access mode, this was only proposed for high-latency data transmission. In contrast, here we employ joint adaptive CM in a low-latency, real-time video context.

14.4. BURST-BY-BURST ADAPTIVE TCM, TTCM, AND BICM FOR H.263 VIDEOTELEPHONY

another source of ISI. Interested readers are referred to [203, 438] for a detailed description of the JD-MMSE-BDFE scheme.

In JD systems the Signal-to-Interference-plus-Noise Ratio (SINR) of each user recorded at the output of the JD-MMSE-BDFE can be calculated by using the channel estimates and the spreading sequences of all of the users. By assuming that the transmitted data symbols and the noise samples are uncorrelated, the expression used for calculating the SINR γ_o of the nth symbol transmitted by the kth user was given by Klein *et al.* [453] as

$$\gamma_o(j) = \frac{\text{Required signal power}}{\text{Residual MAI and ISI power + Effective noise power}}$$
$$= g_j^2 [\mathbf{D}]_{j,j}^2 - 1, \quad \text{for } j = n + N(k-1), \tag{14.5}$$

where SINR is the ratio of the wanted signal power to the residual MAI and ISI power plus the effective noise power. The number of users in the system is K and each user transmits N symbols per transmission burst. The matrix $\mathbf{D}$ is a diagonal matrix that is obtained with the aid of the Cholesky decomposition [454] of the matrix used for linear MMSE equalization of the CDMA system [203, 453]. The notation $[\mathbf{D}]_{j,j}^2$ represents the element in the jth row and jth column of the matrix $\mathbf{D}$ and the value g_j is the amplitude of the jth symbol. The AQAM mode switching mechanism used for the JD-MMSE-BDFE of user k is the same as that of Equation 14.4, where γ_{DFE} of user k was used as the modulation switching metric, which can be computed from

$$\gamma_{\text{DFE}}(k) = \frac{1}{N} \sum_{n=1}^{N} \gamma_o(j), \quad j = n + N(k-1). \tag{14.6}$$

Again, the log-domain branch metric is computed for the CM trellis decoding scheme using the MMSE-BDFE's symbol estimate by invoking Equation 14.3.

The UTRA channel model and system parameters of the AQAM CDMA scheme are outlined as follows. Table 14.13 shows the modulation and channel parameters employed. The multi-path channel model is characterized by its discretized chip-spaced UTRA vehicular channel A [449]. The transmission burst structure of the modified UTRA Burst 1 [161] using a spreading factor of eight is shown in Figure 14.52. The number of data symbols per JD block is 20, hence the original UTRA Burst 1 was modified to host a burst of 240 data symbols, which is a multiple of 20.

The remaining system parameters are shown in Table 14.14, where there are 15 timeslots in one UTRA frame and we assign one slot for one group of CDMA users. More specifically, each CDMA user group employed a similar system configuration, but communicated with the base station employing another one of the 15 different timeslots.

In general, the total number of users supportable by the uplink CDMA system can be increased by using a higher spreading factor at the cost of a reduced throughput, because the system's chip rate was fixed at 3.84×10^6 chips/s, as shown in Table 14.13. Another option for increasing the number of users supported is by assigning more uplink timeslots for new groups of users. In our study, we investigate the proposed system using one timeslot only. Hence, the data symbol rate per slot per user is 24 kBd for a spreading factor of eight. Finally, Table 14.15 shows the operational-mode specific video transceiver parameters for the TTCM AQAM video system, where the effective video bitrate of each user ranges from 19.8

Table 14.13: Modulation and Channel Parameters for the CDMA System

Parameter	Value
Carrier frequency	1.9 GHz
Vehicular speed	30 mph
Doppler frequency	85 Hz
System Baud rate	3.84 MBd
Normalized Doppler frequency	$85/(3.84 \times 10^6) = 2.21 \times 10^{-5}$
Channel type	UMTS Vehicular Channel A [449]
Number of paths in channel	6
Data modulation	ACM (4QAM, 8PSK, 16QAM, 64QAM)
Receiver type	JD-MMSE-BDFE
Number of symbols per JD block	20

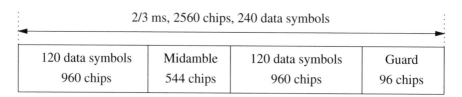

Figure 14.52: A modified UTRA Burst 1 [161] with a spreading factor of 8. The original UTRA burst has 244 data symbols.

to 113.8 kbit/s. As the video bitrate is relatively low as a consequence of CDMA spreading, we transmitted 176×144-pixel QCIF resolution video sequences at 30 frames/s based on the H.263 video codec [161].

14.4.5.1 Performance of TTCM AQAM in a CDMA system

The PLR and video bitrate performance of the TTCM AQAM CDMA scheme designed for a target PLR of 5% and for supporting $K = 2$ and 4 users is shown in Figure 14.53. The PLR was below the target value of 5% and the video bitrate improved as the channel SNR increased. As we employed switching thresholds which are constant over the SNR range, in the region of SNR $= 10$ dB the PLR followed the trend of 4QAM. Similarly, for SNRs between 17 and 20 dB the PLR-trend of 16QAM was predominantly obeyed. In both of these SNR regions a significantly lower PLR was achieved than the target value. We note, however, that it is possible to increase the PLR to the value of the target PLR in this SNR region for the sake of attaining extra throughput gains by employing a set of switching thresholds, where the thresholds are varied as a function of the SNR [455], but this design option was set aside for further research.

From Figure 14.53 we also note that the performance difference between the $K = 2$ and 4 user scenarios is only marginal with the advent of the powerful JD-MMSE-BDFE scheme. Specifically, there was only about 1 dB SNR penalty when the number of users

14.4. BURST-BY-BURST ADAPTIVE TCM, TTCM, AND BICM FOR H.263 VIDEOTELEPHONY

Table 14.14: Generic System Features of the Reconfigurable Multi-mode Video Transceiver, using the Spread Data Burst 1 of UTRA [161, 449] Shown in Figure 14.52

Features	Value
Multiple access	CDMA, TDD
Number of slots/frame	15
Spreading factor, Q	8
Frame length	10 ms
Slot length	2/3 ms
Data symbols/slot/user	240
Number of slot/user group	1
User data symbol rate (kBd)	$240/10 = 24$
System data symbol rate (kBd)	$24 \times 15 = 360$
Chips/slot	2560
Chips/frame	$2560 \times 15 = 38400$
User chip rate (kBd)	$2560/10 = 256$
System chip rate (MBd)	$38.4/10 = 3.84$
System bandwidth (MHz)	$3.84 \times 3/2 = 5.76$
Effective user bandwidth (kHz)	$5760/15 = 384$

Table 14.15: Operational-mode Specific Transceiver Parameters for TTCM in a CDMA System

Features	Multi-rate system			
Mode	4QAM	8PSK	16QAM	64QAM
Transmission symbols			240	
Bits/symbol	2	3	4	6
Transmission bits	480	720	960	1440
Packet rate			100/s	
Transmission bitrate (kbit/s)	48	72	96	144
Code termination symbols			6	
Data symbols			234	
Coding rate	1/2	2/3	3/4	5/6
Information bits/symbol	1	2	3	5
Unprotected bits	234	468	708	1170
Unprotected bitrate (kbit/s)	23.4	46.8	70.8	117.0
Video packet CRC (bits)			16	
Feedback protection (bits)			9	
Video packet header (bits)	11	12	12	13
Video bits/packet	198	431	671	1138
Effective video bitrate (kbit/s)	19.8	43.1	67.1	113.8
Video framerate (Hz)			30	

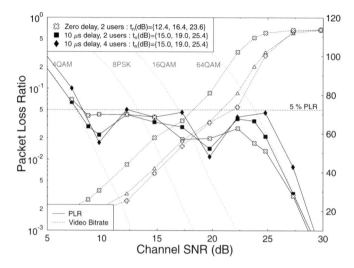

Figure 14.53: PLR and video bitrate versus channel SNR for the four fixed TTCM modes and for the quadruple-mode TTCM AQAM CDMA scheme supporting two and four users transmitting the QCIF video sequence at 30 frame/s, when communicating over the UTRA vehicular channel A [449].

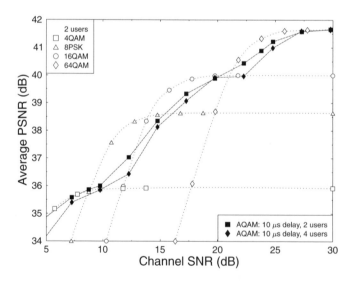

Figure 14.54: Average PSNR versus channel SNR for the four fixed TTCM modes and for the quadruple-mode TTCM AQAM CDMA scheme supporting two and four users transmitting the QCIF video sequence at 30 frame/s, when communicating over the UTRA vehicular channel A [449].

increased from two to four for the 4QAM and 64QAM modes at both low and high SNRs, respectively. The PLR of the system supporting $K = 4$ users was still below the target PLR when the switching thresholds derived for $K = 2$ users were employed. In terms of the video bitrate performance, the SNR penalty is less than 1 dB when the number of users supported is increased from $K = 2$ to 4 users. Note that the delay spread of the chip-spaced UTRA vehicular channel A [449] is 2.51 μs corresponding to a chip duration of $2.51 \times 3.84 \approx 10$ for the 3.84 MBd Baud rate of our system, as seen in Table 14.13. Hence, the delay spread is longer than the spreading code length ($Q = 8$ chips) used in our system and therefore the resultant ISI in the system is significantly higher than that of a system employing a higher spreading factor, such as $Q > 10$ chips. These findings illustrate the efficiency of the JD-MMSE-BDFE scheme in combating the high ISI and MAI of the system. More importantly, the employment of the JD-MMSE-BDFE scheme in our system allowed us to generalize our results recorded for the $K = 2$ users scenario to that of a higher number of users, because the performance penalty associated with supporting more users was found to be marginal.

We also investigated the effect of mode signaling delay on the performance of the TTCM AQAM CDMA scheme in Figure 14.53. The performance of the ideal scheme, where the channel quality estimation is perfect without any signaling delay, is compared with that of the proposed practical scheme, where the channel quality estimation is imperfect and outdated by the delay of one frame duration of 10 μs. For a target PLR of 5%, the ideal scheme exhibited a higher video bitrate than the practical scheme. More specifically, at a target PLR of 5%, about 2.5 dB SNR gain is achieved by the ideal scheme in the SNR region spanning from 8 to 27 dB. A channel quality signaling delay of one frame duration certainly represents the worst-case scenario. In general, the shorter the signaling delay the better the performance of the adaptive system. Hence, the performance of the zero-delay and one-frame-delay schemes represent the lower-bound and upper-bound performance, respectively, for practical adaptive systems, although employing the channel-quality prediction schemes of [450, 451] would allow us to approximate the perfect channel estimation scenario.

We now evaluate the PSNR performance of the proposed practical TTCM AQAM CDMA scheme. For maintaining a target PLR of 5% in conjunction with an adaptive mode signaling delay of one UTRA frame length of 10 ms, the average PSNR versus channel SNR performance of the quadruple-mode TTCM AQAM CDMA scheme is shown in Figure 14.54 together with that of the four fixed TTCM modes. As shown in Figure 14.54, the video performance of the TTCM AQAM CDMA scheme supporting $K = 4$ users degrades gracefully from the average PSNR of 41.5 to 34.2 dB, as the channel SNR degrades from 30 to 5 dB. Hence, an attractive subjective video performance can be obtained in the UTRA environment. Again, the PSNR performance difference between the $K = 2$ and 4 scenarios is only marginal with the advent of the powerful JD-MMSE-BDFE scheme invoked [203].

14.4.6 Conclusions

In this section various BbB AQAM TCM, TTCM, and BICM based video transceivers have been studied. The near-instantaneously adaptive transceiver is capable of operating in four different modulation modes, namely 4QAM, 8PSK, 16QAM, and 64QAM.

The advantage of using CM schemes in our near-instantaneously adaptive transceivers is that when invoking higher-order modulation modes, in the case of encountering a higher channel quality, the coding rate approaches unity. This allows us to maintain as high a throughput as possible. We found that the TTCM scheme provided the best overall video performance due to its superior PLR performance. However, the lower complexity TCM3-assisted scheme provides the best compromise in terms of the PSNR performance and complexity in comparison with the TTCM-, TCM6-, and BICM-assisted schemes.

The BbB AQAM modem guaranteed the same video performance as the lowest- and highest-order fixed-mode modulation schemes at extremely low and high channel SNRs with a minimal latency. Furthermore, in between these extreme SNRs the effective video bitrate smoothly increased as the channel SNR increased, whilst maintaining a near-constant PLR. By controlling the AQAM switching thresholds, a near-constant PLR can be maintained. We have also compared the performance of the proposed TTCM AQAM scheme with that of the TCC AQAM arrangement characterized in [448] under similar conditions and the TTCM scheme was found to have more advantages than the TCC scheme of [448] in the context of the adaptive H.263 video system. The best TTCM AQAM arrangement was also studied in the context of a DS-CDMA system by utilizing a JD-MMSE-BDFE scheme and promising results were obtained when communicating in the UTRA environment. It was also apparent from this study that, as long as the DFE or the JD-MMSE-BDFE scheme is capable of transforming the wideband Rayleigh fading channel's error statistics into "AWGN-like" error statistics, the performance trends of the CM schemes observed in AWGN channels will be preserved in the wideband Rayleigh fading channels. Specifically, the TTCM assisted scheme, which is the best performer when communicating over AWGN channels, is also the best performer when communicating over the wideband Rayleigh fading channels, when compared with the schemes assisted by TCC, TCM, and BICM. Hence, it is shown that by adapting the video rate, channel coding rate, and the modulation mode together according to the channel quality, the best possible source-signal representation quality is achieved efficiently in terms of bandwidth and power, on a near-instantaneous basis.

Our future research will focus on employing a similar BbB AQAM philosophy in the context of CDMA systems in conjunction with low-density parity check CM and the MPEG4 video codec.

14.5 Turbo-detected MPEG-4 Video Using Multi-level Coding, TCM and STTC[10]

14.5.1 Motivation and Background

TCM [86, 431] constitutes a bandwidth-efficient joint channel coding and modulation scheme, which was originally designed for transmission over AWGN channels. In contrast, BICM [432] employing parallel bit-based interleavers was designed for communicating over uncorrelated Rayleigh fading channels. Therefore, TCM outperforms BICM when communicating over AWGN channels because TCM exhibits a higher Euclidean distance.

---

[10]Ng, Chung and Hanzo, ©IEEE 2005. *IEE Proceedings Communications*, Vol. 152, No. 6, December 2005, pp. 1116–1124.

14.5. TURBO-DETECTED MPEG-4 VIDEO USING MULTI-LEVEL CODING, TCM AND STTC

The opposite is true when communicating over uncorrelated Rayleigh fading channels, because BICM exhibits a higher Hamming distance. Space–time Trellis Coding (STTC) schemes [456, 457], which employ multiple transmit and receive antennas, are capable of providing both spatial diversity gain and coding gain. Note that when the spatial diversity order is sufficiently high, the channel's Rayleigh fading envelope is transformed into a Gaussian-like near-constant envelope. Hence, the benefits of a TCM scheme designed for AWGN channels will be efficiently exploited when TCM is concatenated with STTC in comparison with BICM [458].

Multi-level Coding (MLC) schemes [459] have been widely designed for providing unequal error-protection capabilities [460]. In this section, we design a twin-class unequal protection MLC scheme by employing two different code-rate maximal minimum distance Non-Systematic Convolutional codes (NSCs) [317, p. 331] or Recursive Systematic Convolutional codes (RSCs) [86,431] as the constituent codes. More specifically, a stronger NSC/RSC is used for protecting the more-sensitive video bits, while a weaker NSC/RSC is employed for protecting the less-sensitive video bits. Note that TCM employs a Set Partitioning (SP)-based bit mapper [431], where the signal set is split into a number of subsets, such that the minimum Euclidean distance of the signal points in the new subset is increased at every partitioning step. Hence, the NSC/RSC encoded bits which are based on the more-sensitive video bits are also mapped to the constellation subsets having the highest minimum Euclidean distance for the sake of further enhanced protection. The TCM and STTC encoders may be viewed as a "coded mapper" for the unequal protected MLC scheme.

The MPEG-4 standard [25, 461] offers a framework for a whole range of multimedia applications, such as teleshopping, interactive TV, Internet games, or iterative mobile videotelephony. *The novel contribution of this section is that the state-of-the-art MPEG-4 video codec is amalgamated with a systematically designed sophisticated turbo transceiver using MLC for providing unequal error protection, TCM for maintaining bandwidth efficiency, and STTC for attaining spatial diversity. Extrinsic information was exchanged across three serially concatenated decoder stages and the decoding convergence was studied using novel 3D non-binary Extrinsic Information Transfer (EXIT) charts [462]. We refer to this unequal-protection joint MPEG-4 source-coding, channel-coding, modulation, and spatial diversity aided turbo-transceiver as the STTC-TCM-2NSC or STTC-TCM-2RSC scheme. We also investigate the STTC-BICM-2RSC scheme, where BICM is employed as the inner code for the sake of studying the performance difference between BICM and TCM as the inner code in the STTC-based unequal-protection turbo transceiver. It is shown that significant iteration gains are attained with the aid of an efficient iterative decoding mechanism.*

The rest of the section is structured as follows. In Section 14.5.2 we describe the proposed system's architecture and highlight the interactions of its constituent elements. The achievable channel capacity of the scheme is studied in Section 14.5.3, while the convergence of the iterative receiver is investigated in Section 14.5.4. We elaborate further by characterizing the achievable system performance in Section 14.5.5 and conclude with a range of system design guidelines in Section 14.5.6.

14.5.2 The Turbo Transceiver

A schematic of the serially concatenated STTC-TCM/BICM-2NSC/2RSC turbo scheme using a STTC, a TCM/BICM, and two NSCs/RSCs as its constituent codes is depicted in

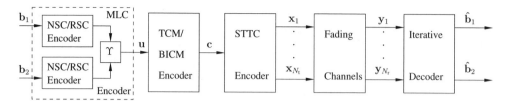

Figure 14.55: Block diagram of the serially concatenated STTC-TCM/BICM-2NSC/2RSC scheme. The notation b_i, $\hat{b}_i$, u, c, x_j, and y_k denotes the vectors of the class-i video bits, the estimates of the class-i video bits, the MLC coded symbols, the TCM coded symbols (or BICM coded bits), the STTC coded symbols for transmitter j, and the received symbols at receiver k, respectively. Furthermore, Υ is a bit-to-symbol converter, while N_t and N_r denote the number of transmitters and receivers, respectively. The symbol-based (or bit-based) channel interleaver between the STTC and TCM (or BICM) schemes as well as the two bit-based interleavers at the output of NSC/RSC encoders are not shown for simplicity. The iterative decoder seen on the right is detailed in Figure 14.56.

Figure 14.55. The MPEG-4 codec operated at $R_f = 30$ frames/s using the 176×144-pixel Quarter Common Intermediate Format (QCIF) "Miss America" video sequence, encoded at a near-constant bitrate of $R_b = 69$ kbit/s. Hence, we have $R_b/R_f = 2300$ bits per video frame. We partition the video bits into two unequal-protection classes. Specifically, class-1 and class-2 consist of 25% (which is 575 bits) and 75% (which is 1725 bits) of the total number of video bits. The more-sensitive video bits constituted mainly by the MPEG-4 framing and synchronization bits are in class-1 and they are protected by a stronger binary NSC/RSC having a coding rate of $R_1 = k_1/n_1 = 1/2$ and a code memory of $L_1 = 3$. The less-sensitive video bits (predominantly signaling the MPEG-4 Discrete Cosine Transform (DCT) coefficients and motion vectors) are in class-2 and are protected by a weaker non-binary NSC/RSC having a coding rate of $R_2 = k_2/n_2 = 3/4$ and a code memory of $L_2 = 3$. Hence, the effective code rate for the MLC scheme employing the $R_1 = 1/2$ and $R_2 = 3/4$ NSCs/RSCs is given by $R_{\text{MLC}} = (k_1 + k_2)/(n_1 + n_2) = 2/3$. Note that the number of MPEG-4 framing and synchronization bits is only about 10% of the total number of video bits. Hence, about $25\% - 10\% = 15\%$ of the class-1 bits are constituted by the video bits signaling the most-sensitive MPEG-4 DCT coefficients. We invoke code termination bits in both NSCs/RSCs and hence the number of coded bits emerging from the $R_1 = 1/2$ binary NSC/RSC encoder is $(575 + k_1 L_1)/R_1 = 1156$ bits, while that generated by the $R_2 = 3/4$ non-binary NSC/RSC encoder is $(1725 + k_2 L_2)/R_1 = 2312$ bits.

The class-1 and class-2 NSC/RSC coded bit sequences seen in Figure 14.55 are interleaved by two separate bit interleavers of length 1156 and 2312 bits, respectively. The two interleaved bit sequences are then concatenated to form a bit sequence of $1156 + 2312 = 3468$ bits. This bit sequence is then fed to the TCM/BICM encoder of Figure 14.55 having a coding rate of $R_3 = k_3/n_3 = 3/4$ and a code memory of $L_4 = 3$. When the SP-based TCM is employed, the Most Significant Bit (MSB) of the three-bit input symbol in the rate-3/4 TCM encoder has a higher protection. Therefore, we map the interleaved bit sequence of the class-1 NSC/RSC encoder to the MSB of the TCM scheme's three-bit input symbol for the sake of further protecting the class-1 video bits. Hence, the MLC encoder of Figure 14.55, which consists of two NSC/RSC encoders, can be viewed as a non-binary outer encoder providing 3-bit MLC symbols, denoted as **u** in Figure 14.55, for the rate-3/4 TCM/BICM encoder. We employ code termination also in the TCM/BICM scheme and hence at the TCM/BICM encoder's output we have $(3468 + k_3 L_3)/R_3 = 4636$ bits or $4636/4 = 1159$ symbols. The TCM symbol sequence (or BICM bit sequence) is then symbol-interleaved (or bit-interleaved) in Figure 14.55 and fed to the STTC encoder. We invoke a 16-state STTC scheme having a code memory of $L_4 = 4$ and $N_t = 2$ transmit antennas, employing $M = 16$-level Quadrature Amplitude Modulation (16QAM). We terminate the STTC code by a 4-bit 16QAM symbol, because we have $N_t = 2$. Therefore, at each transmit antenna we have $1159 + 1 = 1160$ 16QAM symbols or $4 \times 1160 = 4640$ bits in a transmission frame. The overall coding rate is given by $R = 2300/4640 = 0.496$ and the effective throughput of the system is $\log_2(M)R = 1.98$ BPS. The STTC decoder employed $N_r = 2$ receive antennas and the received signals are fed to the iterative decoders for the sake of estimating the video bit sequences in both class-1 and class-2, as seen in Figure 14.55.

14.5.2.1 Turbo Decoding

The STTC-TCM-2RSC scheme's turbo decoder structure is illustrated in Figure 14.56, where there are four constituent decoders, each labeled with a round-bracketed index. Symbol-based and bit-based Maximum *A Posteriori* (MAP) algorithms [86] operating in the logarithmic-domain are employed by the TCM as well as the rate $R_2 = 3/4$ RSC decoders and by the $R_1 = 1/2$ RSC decoder, respectively. The notation $P(.)$ and $L(.)$ in Figure 14.56 denotes the logarithmic-domain symbol probabilities and the Logarithmic-Likelihood Ratio (LLR) of the bit probabilities, respectively. The notation c, u and b_i in the round brackets $(.)$ in Figure 14.56 denote TCM coded symbols, TCM information symbols, and the class-i video bits, respectively. The specific nature of the probabilities and LLRs is represented by the subscripts a, p, e and i, which denote in Figure 14.56 *a priori, a posteriori, extrinsic,* and *intrinsic* information, respectively. The probabilities and LLRs associated with one of the four constituent decoders having a label of $\{1, 2, 3a, 3b\}$ are differentiated by the identical superscripts of $\{1, 2, 3a, 3b\}$. Note that the superscript 3 is used for representing the MLC decoder of Figure 14.56 which invokes the RSC decoders of $3a$ and $3b$.

As we can observe from Figure 14.56, the STTC decoder of block (1) benefits from the *a priori* information provided by the TCM decoder of block (2), namely from $P_a^1(c) = P_i^2(c)$ regarding the 2^{m+1}-ary TCM coded symbols, where m is the number of information bits per TCM coded symbol. More specifically, $P_i^2(c)$ is referred to here as the intrinsic probability of the 2^{m+1}-ary TCM coded symbols, because it contains the inseparable extrinsic information provided by the TCM decoder itself as well as the *a priori* information regarding the uncoded 2^m-ary TCM input information symbols emerging from the RSC decoders of block (3), namely $P_a^2(u) = P_e^3(u)$. Hence, the STTC decoder indirectly also benefits from the *a priori* information $P_a^2(u) = P_e^3(u)$ provided by the RSC decoders of block (3), potentially enhanced by the TCM decoder of block (2). Similarly, the intrinsic probability of $P_i^2(u)$ provided by the TCM decoder for the sake of the RSC decoders' benefit consists of the inseparable extrinsic information generated by the TCM decoder itself as well as of the systematic information of the STTC decoder, namely $P_a^2(c) = P_e^1(c)$. Note that after the symbol probability-to-LLR conversion, $P_i^2(u)$ becomes $L_i^2(u)$. Therefore, the RSC decoders of block (3) benefit directly from the *a priori* information provided by the TCM decoder of block (2), namely from $L_a^3(u) = L_i^2(u)$ as well as indirectly from the *a priori* information

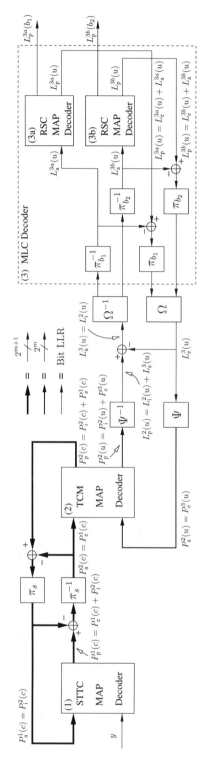

Figure 14.56: Block diagram of the STTC-TCM-2RSC turbo detection scheme. The notation $\pi_{(s,b_i)}$ and $\pi_{(s,b_i)}^{-1}$ denotes the interleaver and deinterleaver, while the subscript s denotes the symbol-based interleaver of TCM and the subscript b_i denotes the bit-based interleaver for class-i RSC. Furthermore, Ψ and Ψ^{-1} denote LLR-to-symbol probability and symbol probability-to-LLR conversion, while Ω and Ω^{-1} denote the parallel-to-serial and serial-to-parallel converter, respectively. The notation m denotes the number of information bits per TCM coded symbol. The thickness of the connecting lines indicates the number of non-binary symbol probabilities spanning from a single LLR per bit to 2^m and 2^{m+1} probabilities.

provided by the STTC decoder of block (1), namely from $P_a^2(c) = P_e^1(c)$. On the other hand, the TCM decoder benefits directly from the STTC and RSC decoders through the *a priori* information of $P_a^2(c) = P_e^1(c)$ and $P_a^2(u) = P_e^3(u)$, respectively, as shown in Figure 14.56.

14.5.2.2 Turbo Benchmark Scheme

For the sake of benchmarking the scheme advocated, we created a powerful benchmark scheme by replacing the TCM/BICM and NSC/RSC encoders of Figure 14.55 by a single NSC codec having a coding rate of $R_0 = k_0/n_0 = 1/2$ and a code memory of $L_0 = 6$. We will refer to this benchmark scheme as the STTC-NSC arrangement. All video bits are equally protected in the benchmark scheme by a single NSC encoder and a STTC encoder. A bit-based channel interleaver is inserted between the NSC encoder and STTC encoder. Taking into account the bits required for code termination, the number of output bits of the NSC encoder is $(2300 + k_0 L_0)/R_0 = 4612$, which corresponds to 1153 16QAM symbols. Again, a 16-state STTC scheme having $N_t = 2$ transmit antennas is employed. After code termination, we have $1153 + 1 = 1154$ 16QAM symbols or $4(1154) = 4616$ bits in a transmission frame at each transmit antenna. Similar to the STTC-TCM/BICM-2NSC/2RSC scheme, the overall coding rate is given by $R = 2300/4616 = 0.498$ and the effective throughput is $\log_2(16)R = 1.99$ BPS, both of which are close to the corresponding values of the proposed scheme.

Let us define a single decoding iteration for the proposed STTC-TCM/BICM-2NSC/2RSC scheme as a combination of a STTC decoding, a TCM/BICM decoding, a class-1 NSC/RSC decoding, and a class-2 NSC/RSC decoding step. Similarly, a decoding iteration of the STTC-NSC benchmark scheme is comprised of a STTC decoding and a NSC decoding step. We quantify the decoding complexity of the proposed STTC-TCM/BICM-2NSC/2RSC scheme and that of the benchmark scheme using the number of decoding trellis states. The total number of decoding trellis states per iteration of the proposed scheme employing two NSC/RSC decoders having a code memory of $L_1 = L_2 = 3$, TCM/BICM having $L_3 = 3$, and STTC having $L_4 = 4$, is $S = 2^{L_1} + 2^{L_2} + 2^{L_3} + 2^{L_4} = 40$. In contrast, the total number of decoding trellis states per iteration for the benchmark scheme having a code memory of $L_0 = 6$ and STTC having $L_4 = 4$, is given by $S = 2^{L_0} + 2^{L_4} = 80$. Therefore, the complexity of the STTC-TCM-2NSC/2RSC scheme having two iterations is equivalent to that of the benchmark scheme having a single iteration, which corresponds to 80 decoding states.

14.5.3 MIMO Channel Capacity

Let us consider a Multi-input Multi-output (MIMO) system employing N_t transmit antennas and N_r receive antennas. When two-dimensional L-ary PSK/QAM is employed at each transmit antenna, the received signal vector of the MIMO system is given by

$$\vec{y} = \mathbf{H}\vec{x} + \vec{n}, \tag{14.7}$$

where $\vec{y} = (\mathbf{y}_1, \ldots, \mathbf{y}_{N_r})^T$ is an N_r-element vector of the received signals, $\mathbf{H}$ is an $(N_r \times N_t)$-dimensional channel matrix, $\vec{x} = (\mathbf{x}_1, \ldots, \mathbf{x}_{N_t})^T$ is an N_t-element vector of the transmitted signals and $\vec{n} = (\mathbf{n}_1, \ldots, \mathbf{n}_{N_r})^T$ is an N_r-element noise vector, where each element in $\vec{n}$ is an AWGN process having a zero mean and a variance of $N_0/2$ per dimension. Note that in a MIMO system there are $M = L^{N_t}$ possible L-ary PSK/QAM phasor combinations in the transmitted signal vector $\vec{x}$. The STTC scheme of [456] designed for attaining transmit diversity may in fact be viewed as a rate-$1/N_t$ channel code, where there are only L^1 legitimate space–time codewords out of the L^{N_t} possible phasor combinations during each transmission period. In contrast, Bell Lab's Layered Space–Time (BLAST) scheme [463] designed for attaining multiplexing gain may be viewed as a rate-1 channel code, where all L^{N_t} phasor combinations are legitimate during each transmission period. Despite having different code rates, both the STTC and BLAST schemes have the same channel capacity.

The conditional probability of receiving a signal vector $\vec{y}$ given that an $M = L^{N_t}$-ary signal vector $\vec{x}_m$, $m \in \{1, \ldots, M\}$, was transmitted over Rayleigh fading channels is determined by the Probability Density Function (PDF) of the noise, yielding

$$p(\vec{y}|\vec{x}_m) = \frac{1}{\pi N_0} \exp\left(\frac{-\|\vec{y} - \mathbf{H}\vec{x}_m\|^2}{N_0}\right). \quad (14.8)$$

In the context of discrete-amplitude QAM [196] and PSK [464] signals, we encounter a Discrete-Input Continuous-Output Memoryless Channel (DCMC) [464]. We derived the channel capacity for a MIMO system, which uses two-dimensional M-ary signaling over the DCMC, from that of the Discrete Memoryless Channel (DMC) [465] as

$$C_{\text{DCMC}} = \max_{p(\vec{x}_1)\ldots p(\vec{x}_M)} \sum_{m=1}^{M} \int_{-\infty}^{\infty} p(\vec{y}|\vec{x}_m) p(\vec{x}_m)$$

$$\times \log_2 \left(\frac{p(\vec{y}|\vec{x}_m)}{\sum_{n=1}^{M} p(\vec{y}|\vec{x}_n) p(\vec{x}_n)}\right) d\vec{y} \quad [\text{bits/symbol}], \quad (14.9)$$

where $p(\mathbf{x}_m)$ is the probability of occurrence for the transmitted signal $\mathbf{x}_m$. It was shown in [465, p. 94] that for a symmetric DMC, the full capacity may only be achieved by using equiprobable inputs. Hence, the right-hand side of Equation 14.9, which represents the mutual information between $\vec{x}$ and $\vec{y}$, is maximized, when the transmitted symbols are equiprobably distributed, i.e. when we have $p(\vec{x}_m) = 1/M$ for $m \in \{1, \ldots, M\}$. Hence, by using $p(\vec{x}_m) = 1/M$ and after a range of mathematical manipulations, Equation 14.9 can be simplified to

$$C_{\text{DCMC}} = \log_2(M) - \frac{1}{M} \sum_{m=1}^{M} E\left[\log_2 \sum_{n=1}^{M} \exp(\Phi_{m,n}) \bigg| \vec{x}_m\right] \quad [\text{bits/symbol}], \quad (14.10)$$

where

$$E[f(\vec{y}|\vec{x}_m)|\vec{x}_m] = \int_{-\infty}^{\infty} f(\vec{y}|\vec{x}_m) p(\vec{y}|\vec{x}_m) \, d\vec{y}$$

is the expectation of the function $f(\vec{y}|\vec{x}_m)$ conditioned on $\vec{x}_m$. The expectation in Equation 14.10 is taken over $\mathbf{H}$ and $\vec{n}$, while $\Phi_{m,n}$ is given by

$$\Phi_{m,n} = \frac{-\|\mathbf{H}(\vec{x}_m - \vec{x}_n) + \vec{n}\|^2 + \|\vec{n}\|^2}{N_0},$$

$$= \sum_{i=1}^{N_r} \frac{-\left|\vec{h}_i(\vec{x}_m - \vec{x}_n) + \mathbf{n}_i\right|^2 + |\mathbf{n}_i|^2}{N_0}, \quad (14.11)$$

14.5. TURBO-DETECTED MPEG-4 VIDEO USING MULTI-LEVEL CODING, TCM AND STTC

where $\vec{\mathbf{h}}_i$ is the ith row of $\mathbf{H}$ and $\mathbf{n}_i$ is the AWGN at the ith receiver.

When the channel input is a continuous-amplitude, discrete-time Gaussian-distributed signal, we encounter a Continuous-Input Continuous-Output Memoryless Channel (CCMC) [464], where the capacity is only restricted by either the signaling energy or the bandwidth. It was shown in [466, 467] that the MIMO capacity of the CCMC can be expressed as

$$C_{\text{CCMC}} = E\left[WT \sum_{i=1}^{r} \log_2\left(1 + \lambda_i \frac{\text{SNR}}{N_t}\right) \right], \tag{14.12}$$

where W is the bandwidth, T is the signaling period of the finite-energy signaling waveform, and r is the rank of $\mathbf{Q}$, which is defined as $\mathbf{Q} = \mathbf{H}^H\mathbf{H}$ for $N_r \geq N_t$ or $\mathbf{Q} = \mathbf{H}\mathbf{H}^H$ for $N_r < N_t$. Furthermore, λ_i is the ith eigenvalue of the matrix $\mathbf{Q}$.

However, for the special case of an orthogonal MIMO transmission system, such as the orthogonal Space–Time Block Coding (STBC) scheme of [468, 469], the received signal in Equation 14.7 can be transformed into [470]

$$\mathbf{y}_i = \sum_{j=1}^{N_t} |\mathbf{h}_{i,j}|^2 \mathbf{x} + \mathbf{\Omega}_i = \chi^2_{2N_t,i} \mathbf{x} + \mathbf{\Omega}_i, \quad i = \{1, \ldots, N_r\}, \tag{14.13}$$

where $\mathbf{y}_i$ is the received signal at receiver i in the received signal vector $\vec{\mathbf{y}}$, $\mathbf{x}$ is the complex-valued (two-dimensional) transmitted signal, $\mathbf{h}_{i,j}$ is the complex-valued Rayleigh fading coefficient between transmitter j and receiver i, $\chi^2_{2N_t,i} = \sum_{j=1}^{N_t} |\mathbf{h}_{i,j}|^2$ represents a chi-squared distributed random variable having $2N_t$ degrees of freedom at receiver i, and $\mathbf{\Omega}_i$ is the ith receiver's complex-valued AWGN after transformation, which has a zero mean and a variance of $\chi^2_{2N_t,i} N_0/2$ per dimension. Due to orthogonal transmissions, the MIMO channel was transformed into a Single-Input Multi-Output (SIMO) channel, where the equivalent Rayleigh fading coefficient between the transmitter and the ith receiver is given by $\chi^2_{2N_t,i}$ and the equivalent noise at the ith receiver is given by $\mathbf{\Omega}_i$. As the MIMO channel has now been transformed into a SIMO channel, we have $M = L^1 = L$, because there is only a single transmit antenna in a SIMO scheme. The channel capacity of STBC can be shown to be

$$C_{\text{DCMC}}^{\text{STBC}} = \log_2(M) - \frac{1}{M} \sum_{m=1}^{M} E\left[\log_2 \sum_{n=1}^{M} \exp(\Phi_{m,n}^{\text{STBC}}) \,\bigg|\, \mathbf{x}_m \right] \text{ [bits/symbol]}, \tag{14.14}$$

where the expectation in Equation 14.14 is taken over $\chi^2_{2N_t,i}$ and $\mathbf{\Omega}_i$, while $\Phi_{m,n}^{\text{STBC}}$ is given by

$$\Phi_{m,n}^{\text{STBC}} = \sum_{i=1}^{N_r} \frac{-|\chi^2_{2N_t,i}(\mathbf{x}_m - \mathbf{x}_n) + \mathbf{\Omega}_i|^2 + |\mathbf{\Omega}_i|^2}{\chi^2_{2N_t,i} N_0}. \tag{14.15}$$

Furthermore, the CCMC capacity for STBC can be shown to be

$$C_{\text{CCMC}}^{\text{STBC}} = E\left[WT \log_2\left(1 + \sum_{i=1}^{N_r} \chi^2_{2N_t,i} \frac{\text{SNR}}{N_t}\right) \right] \text{ [bits/symbol]}. \tag{14.16}$$

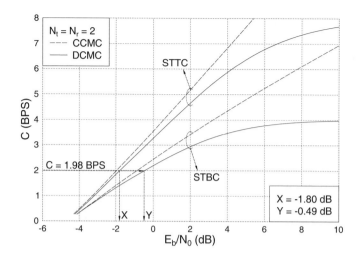

Figure 14.57: The MIMO channel capacity limit for STTC and STBC schemes employing 16QAM and $N_t = N_r = 2$.

Figure 14.57 shows the MIMO channel capacity limit of STTC and STBC schemes employing 16QAM and $N_t = N_r = 2$.

As we can see from Figure 14.57, the channel capacity of STBC is lower than that of STTC due to employing orthogonal transmissions. Note that STBC achieves only diversity gain but no coding gain. However, the coding gain of STTC is achieved at the cost of a higher trellis-based decoding complexity. The MIMO channel capacity limit of STTC determined from Equation 14.10 at a throughput of 1.98 BPS and 1.99 BPS is $E_b/N_0 = -1.80$ dB and -1.79 dB, respectively. The corresponding channel capacity limit of STBC evaluated from Equation 14.14 is $E_b/N_0 = -0.49$ dB and -0.47 dB, respectively.

14.5.4 Convergence Analysis

EXIT charts [471] have been widely used for analyzing the convergence behavior of iterative decoding aided concatenated coding schemes. A specific novelty of this section is that we employ the technique proposed in [462] for computing the *non-binary EXIT functions*, where the multidimensional histogram computation of [472, 473] is replaced by the lower-complexity averaging of the extrinsic symbol probabilities of the MAP decoders. Let us study the convergence of the proposed three-component STTC-TCM-2RSC scheme using 3D EXIT charts [474], when communicating over MIMO Rayleigh fading channels. As we can see from Figure 14.56, the TCM decoder receives inputs from and provides outputs for both the STTC and the MLC decoders of Figure 14.56. Hence, we have to compute two EXIT planes, the first corresponding to the TCM decoder's intrinsic probabilities $P_i^2(c)$ provided for the STTC decoder and the second corresponding to $P_i^2(u)$ supplied for the MLC decoders, as shown in Figure 14.56. In contrast, the STTC decoder has a single EXIT plane characterizing its extrinsic probability $P_e^1(c)$ forwarded to the TCM decoder in Figure 14.56. Similarly, the MLC decoder has one EXIT plane characterizing its extrinsic probability $P_e^3(u)$ forwarded to the TCM decoder in Figure 14.56.

14.5. TURBO-DETECTED MPEG-4 VIDEO USING MULTI-LEVEL CODING, TCM AND STTC

Let us denote the average *a priori* information and the average extrinsic (or intrinsic for TCM) information as I_A and I_E, respectively. The I_A (probabilities or LLRs) and I_E (probabilities or LLRs) quantities of TCM corresponding to the links with the STTC and MLC schemes are differentiated by the subscripts 1 and 2, respectively. Similar to computing the conventional EXIT curve in a 2D EXIT chart, we have to model/provide the *a priori* information I_A for each of the inputs of a constituent decoder in order to compute the EXIT plane of that constituent decoder in a 3D EXIT chart. When a long bit interleaver is used between two non-binary constituent decoders, I_A can indeed be sufficiently accurately modeled based on the assumption of having independent bits within the non-binary symbol [471]. More explicitly, for the bit-interleaved decoder, I_A can be computed based on the average mutual information obtained, when BPSK modulated signals are transmitted across AWGN channels. To expound further, because the MLC coded bits are bit-interleaved before feeding them to the TCM encoder, we model both the average *a priori* information provided for the TCM decoder, namely $I_{A2}(\text{TCM}) = f(P_a^2(u))$ corresponding to the non-binary TCM input symbol, as well as the average *a priori* information generated for the MLC decoder, namely $I_A(\text{MLC}) = f(P_i^2(c))$ corresponding to the non-binary MLC coded symbol, where $f(.)$ represents the EXIT function, by assuming that the MLC coded bits in a non-binary MLC coded symbol are independent of each other.

In contrast, the bits in a non-binary coded symbol of a symbol-interleaved concatenated coding scheme are not independent. Hence, for the symbol-interleaved links between the 16QAM-based TCM and STTC scheme, the distribution of the bits in a non-binary coded symbol cannot be sufficiently accurately modeled using independent BPSK modulation. In this scenario, we found that when the average *a priori* information generated for these 4-bit TCM coded symbols is modeled based on the average mutual information obtained when 16QAM modulated signals are transmitted across AWGN channels, a good EXIT plane approximation can be obtained. Note that the average mutual information of 16QAM in AWGN channels is given by the AWGN channel's capacity computed for 16QAM (i.e. the DCMC capacity [196]) provided that all of the 16QAM symbols are equiprobable. Therefore, the average *a priori* information provided for the TCM decoder, namely $I_{A1}(\text{TCM}) = f(P_a^2(c))$, and the average *a priori* information for STTC decoder, namely $I_A(\text{STTC}) = f(P_a^1(c))$, are generated based on the AWGN channel's capacity determined for 16QAM.

Figures 14.58 and 14.59 illustrate the 3D EXIT charts as well as the iteration trajectories for the proposed STTC-TCM-2RSC scheme at $E_b/N_0 = -0.5$ dB, when an interleaver block length of 10000 16QAM symbols is employed, where E_b/N_0 is the SNR per information bit.

Specifically, the EXIT plane marked with triangles in Figure 14.58 was computed based on the STTC decoder's output $P_e^1(c)$ at the given $I_E(\text{MLC})$ and $I_A(\text{STTC})$ abscissa values, while the EXIT plane drawn using lines in Figure 14.58 was computed based on the TCM decoder's output $P_i^2(c)$ at the given $I_{A1}(\text{TCM})$ and $I_{A2}(\text{TCM})$ value. Similarly, the EXIT plane of Figure 14.59 spanning from the vertical line $[I_E(\text{MLC}) = 0, I_A(\text{MLC}) = 0, I_E(\text{STTC}) = \{0 \rightarrow 4\}]$ to the vertical line $[I_E(\text{MLC}) = 3, I_A(\text{MLC}) = 3, I_E(\text{STTC}) = \{0 \rightarrow 4\}]$ was computed based on the MLC decoder's output $P_e^3(c)$ at the given $I_E(\text{STTC})$ and $I_A(\text{MLC})$. The other EXIT plane of Figure 14.59 spanning from the horizontal line $[I_{A2}(\text{TCM}) = \{0 \rightarrow 3\}, I_{E2}(\text{TCM}) = 0, I_{A1}(\text{TCM}) = 0]$ to the horizontal line $[I_{A2}(\text{TCM})$

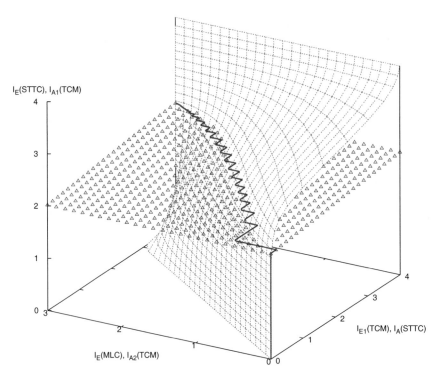

Figure 14.58: The 3D EXIT chart and the iteration trajectory between the STTC and TCM decoders at $E_b/N_0 = -0.5$ dB, when a block length of 10000 16QAM symbols is used. The EXIT plane marked with triangles was computed based on the STTC decoder's output $P_e^1(c)$ and the EXIT plane drawn using lines was computed based on the TCM decoder's output $P_i^2(c)$.

$= \{0 \rightarrow 3\}$, $I_{E2}(\text{TCM}) = 3$, $I_{A1}(\text{TCM}) = 4]$ was computed based on the TCM decoder's output $P_i^2(u)$ at the given $I_{A1}(\text{TCM})$ and $I_{A2}(\text{TCM})$ values.

As we can see from Figure 14.58, the iteration trajectory computed based on the average intrinsic information of the TCM decoder's output, namely $I_{E1}(\text{TCM}) = f(P_i^2(c))$, is under the STTC-EXIT plane marked with triangles and above the TCM-EXIT plane drawn using lines. Note that the approximated EXIT-planes in Figure 14.58 failed to mimic the exact distribution of the TCM coded symbols, and hence resulted in some overshooting mismatches between the EXIT-planes and the trajectory. However, as seen from Figure 14.59, the mismatch between the EXIT-planes and the trajectory computed based on the average intrinsic information of the TCM decoder's output, namely $I_{E2}(\text{TCM})=f(P_i^2(u))$, is minimal. Explicitly, the trajectory seen in Figure 14.59 is located on the right of the MLC-EXIT plane spanning two vertical lines and on the left of the TCM-EXIT plane spanning two horizontal lines. Note from Figure 14.59 that $I_{E2}(\text{TCM})$ is not strictly monotonically increasing with $I_E(\text{STTC})$, which is in contrast to the bit-interleaved system of [474]. Hence, we cannot combine Figures 14.58 and 14.59 into a single 3D EXIT chart, as in [474].

As we can see from Figure 14.58, the STTC-based EXIT plane spans from the horizontal line $[I_E(\text{MLC}) = \{0 \rightarrow 3\}$, $I_{E1}(\text{TCM}) = 0$, $I_E(\text{STTC}) = 2.0148]$ to the horizontal line

14.5. TURBO-DETECTED MPEG-4 VIDEO USING MULTI-LEVEL CODING, TCM AND STTC

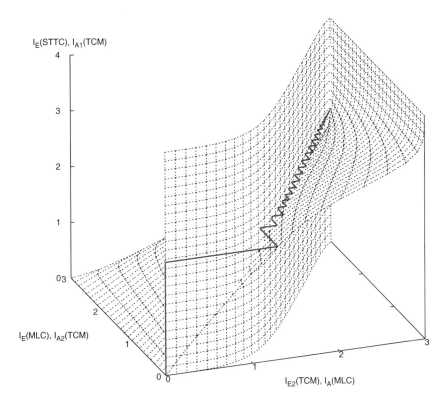

Figure 14.59: The 3D EXIT chart and the iteration trajectory between the TCM and MLC decoders at $E_b/N_0 = -0.5$ dB, when a block length of 10000 16QAM symbols is used. The EXIT plane spanning from the vertical line $[I_E(\text{MLC}) = 0, I_A(\text{MLC}) = 0, I_E(\text{STTC}) = \{0 \to 4\}]$ to the vertical line $[I_E(\text{MLC}) = 3, I_A(\text{MLC}) = 3, I_E(\text{STTC}) = \{0 \to 4\}]$ was computed based on the MLC decoder's output $P_e^3(c)$ and the other EXIT plane was computed based on the TCM decoder's output $P_i^2(u)$.

$[I_E(\text{MLC}) = \{0 \to 3\}, I_{E1}(\text{TCM}) = 4, I_E(\text{STTC}) = 2.3493]$. As the STTC decoder was unable to converge to the $I_E(\text{STTC}) = 4$ position, a two-stage concatenated scheme based on STTC, such as the STTC-NSC benchmark scheme, would fail to reach an error-free performance at $E_b/N_0 = -0.5$ dB. However, as we can see from Figures 14.58 and 14.59, the TCM decoder's output trajectories converged to the $[I_E(\text{MLC}) = 3, I_{E1}(\text{TCM}) = 4]$ and $[I_E(\text{MLC}) = 3, I_{E2}(\text{TCM}) = 3]$ positions, respectively. This indicates that an error-free performance can be attained by the three-stage concatenated STTC-TCM-2RSC scheme at $E_b/N_0 = -0.5$ dB, despite employing a poorly converging STTC scheme. As we can observe from Figure 14.59, the intersection of the EXIT planes includes the vertical line at $[I_E(\text{MLC}) = 3, I_{E2}(\text{TCM}) = 3, I_E(\text{STTC}) = \{1.9 \to 4\}]$, hence the recursive TCM decoder has in fact aided the non-recursive STTC decoder in achieving an early convergence at $I_E(\text{STTC}) = 1.9$, rather than only at $I_E(\text{STTC}) = 4$, when the STTC-TCM scheme is iteratively exchanging extrinsic information with the MLC decoder. This indicates that when a non-recursive STTC is employed, a three-stage concatenated coding scheme is necessary

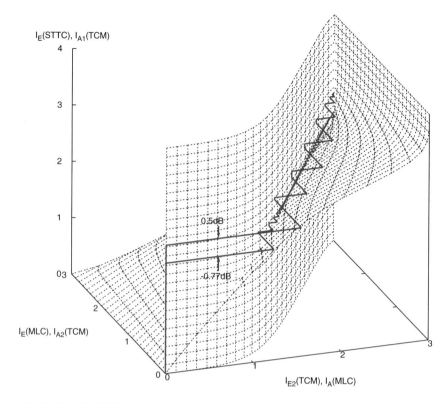

Figure 14.60: The 3D EXIT chart and the iteration trajectory between the TCM and MLC decoders at $E_b/N_0 = -0.77$ dB, when a block length of 100000 16QAM symbols is used, as well as at $E_b/N_0 = 0.5$ dB, when a block length of 1160 16QAM symbols is used. The EXIT plane spanning from the vertical line $[I_E(\text{MLC}) = 0, I_A(\text{MLC}) = 0, I_E(\text{STTC}) = \{0 \to 4\}]$ to the vertical line $[I_E(\text{MLC}) = 3, I_A(\text{MLC}) = 3, I_E(\text{STTC}) = \{0 \to 4\}]$ was computed based on the MLC decoder's output $P_e^3(c)$ and the other EXIT plane was computed based on the TCM decoder's output $P_i^2(u)$.

for approaching the MIMO channel's capacity. Better constituent codes may be designed for the three-stage concatenated coding scheme based on the 3D EXIT chart of Figure 14.59. More explicitly, good constituent codes would result in two EXIT planes that intersect at as low an $I_E(\text{STTC})$ value in Figure 14.59 as possible. It should be noted, however, that such schemes may require a high number of iterations, because they may operate between the cut-off rate and the capacity, which typically imposes a high delay and high complexity.

Figure 14.60 shows that convergence can be achieved at a low SNR value of $E_b/N_0 = -0.77$ dB when a longer interleaver block length of 100000 16QAM symbols is employed. In contrast, convergence is only achieved at a higher SNR value of $E_b/N_0 = 0.5$ dB, when a shorter interleaver block length of 1160 16QAM symbols is used. Hence, the lower-delay STTC-TCM-2RSC scheme of Section 14.5.2 employing an interleaver length of 1160 16QAM symbols is approximately 2.3 dB away from the STTC channel capacity of -1.80 dB and 0.99 dB from the STBC channel capacity of -0.49 dB at a throughput of 1.98 BPS,

14.5. TURBO-DETECTED MPEG-4 VIDEO USING MULTI-LEVEL CODING, TCM AND STTC

according to Figure 14.57. When a longer interleaver delay of 100000 16QAM symbols can be tolerated, the effective throughput becomes approximately 2.00 BPS, because the code rate loss due to termination symbols/bits has been slightly reduced. In this case, the STTC-TCM-2RSC scheme which converged at $E_b/N_0 = -0.77$ dB is only about 1 dB away from the STTC channel capacity of -1.77 dB and it performs 0.32 dB better than the STBC channel capacity of -0.45 dB at a throughput of 2.00 BPS.

14.5.5 Simulation Results

We continue our discourse by characterizing the attainable performance of the proposed MPEG-4 based videophone schemes using both the BER and the average video PSNR [5].

Figures 14.61 and 14.62 depict the class-1 and class-2 BER versus E_b/N_0 performance of the 16QAM-based STTC-TCM-2NSC and STTC-TCM-2RSC schemes, respectively, when communicating over uncorrelated Rayleigh fading channels.

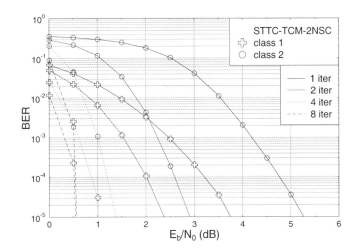

Figure 14.61: BER versus E_b/N_0 performance of the 16QAM-based STTC-TCM-2NSC assisted MPEG-4 scheme, when communicating over uncorrelated Rayleigh fading channels. The effective throughput was 1.98 BPS.

Specifically, the class-1 bits benefit from more than an order of magnitude lower BER at a given SNR, than the class-2 bits. Figure 14.63 compares the overall BER performance of the STTC-TCM-2NSC and STTC-TCM-2RSC schemes. More explicitly, the STTC-TCM-2RSC scheme is outperformed by the STTC-TCM-2NSC arrangement, when the number of iterations is lower than eight. At BER $= 10^{-4}$ an approximately 4 and 6 dB iteration gain was attained by the STTC-TCM-2NSC and STTC-TCM-2RSC schemes, respectively, when the number of iterations was increased from one to eight. Note in Figures 14.62 and 14.63 that the STTC-TCM-2RSC scheme suffers from an error floor, despite having a high iteration gain, which is due to the employment of RSC outer codes instead of the NSC outer codes.

The BER performance curves of STTC-BICM-2RSC and STTC-NSC are shown in Figure 14.64. Note that if we reduce the code memory of the NSC constituent code of the

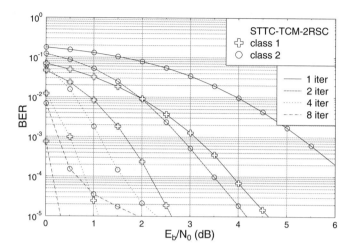

Figure 14.62: BER versus E_b/N_0 performance of the 16QAM-based STTC-TCM-2RSC assisted MPEG-4 scheme, when communicating over uncorrelated Rayleigh fading channels. The effective throughput was 1.98 BPS.

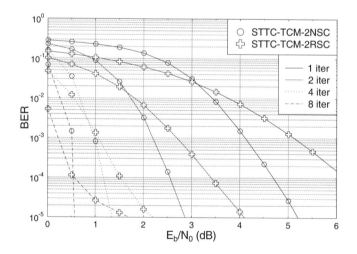

Figure 14.63: BER versus E_b/N_0 performance of the 16QAM-based STTC-TCM-2NSC and STTC-TCM-2RSC assisted MPEG-4 schemes, when communicating over uncorrelated Rayleigh fading channels. The effective throughput was 1.98 BPS.

14.5. TURBO-DETECTED MPEG-4 VIDEO USING MULTI-LEVEL CODING, TCM AND STTC

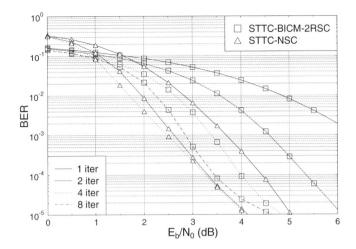

Figure 14.64: BER versus E_b/N_0 performance of the 16QAM-based STTC-BICM-2RSC and STTC-NSC assisted MPEG-4 schemes, when communicating over uncorrelated Rayleigh fading channels. The effective throughput of STTC-BICM-2RSC and STTC-NSC was 1.98 and 1.99 BPS, respectively.

STTC-NSC benchmark scheme from $L_0 = 6$ to 3, the best possible performance degrades. If we increased L_0 from 6 to 7 (or higher), the decoding complexity would be significantly increased, while the attainable best possible performance is only marginally improved. Hence, the STTC-NSC scheme having $L_0 = 6$ constitutes a powerful benchmark scheme in terms of its performance versus complexity trade-offs. As observed in Figure 14.64, the performance of the STTC-BICM-2RSC scheme is even worse than that of the STTC-NSC benchmark scheme. More explicitly, STTC-BICM-2RSC employing eight iterations cannot outperform the STTC-NSC arrangement employing two iterations. By changing the outer code to NSC, i.e. using the STTC-BICM-2NSC scheme, the attainable performance cannot be further improved. The complexity of the STTC-TCM-2NSC/2RSC arrangement having four (or eight) iterations corresponds to 160 (or 320) trellis states, which is similar to that of the STTC-NSC scheme having two (or four) iterations. Hence at a complexity of 160 (or 320) trellis states, the E_b/N_0 performance of the STTC-TCM-2NSC (or STTC-TCM-2RSC) scheme is approximately 2 dB (or 2.8 dB) better than that of the STTC-NSC benchmark scheme at BER $= 10^{-4}$.

Let us now consider the PSNR versus E_b/N_0 performance of the systems characterized in Figures 14.65 and 14.66. The PSNR performance trends are similar to our observations made in the context of the achievable BER results. The maximum attainable PSNR is 39.7 dB. Observe in Figure 14.65 that the BER floor of the STTC-TCM-2RSC scheme resulted in a slightly lower maximum attainable PSNR value when we had $E_b/N_0 < 6$ dB. Furthermore, when employing eight iterations at $E_b/N_0 = 0.5$ dB, the PSNR of STTC-TCM-2RSC was found to be slightly lower than that of the STTC-TCM-2NSC arrangement, although the BER of STTC-TCM-2RSC is significantly lower than that of the STTC-TCM-2NSC scheme, as it is evidenced in Figure 14.63. This is because STTC-TCM-2RSC suffers from a higher transmission frame error ratio, despite having a lower BER, in comparison with the STTC-TCM-2NSC scheme at $E_b/N_0 = 0.5$ dB.

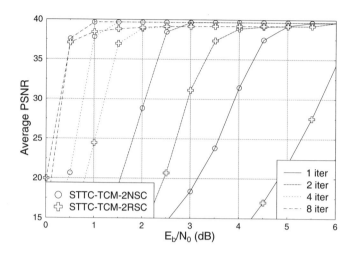

Figure 14.65: Average PSNR versus E_b/N_0 performance of the proposed 16QAM-based STTC-TCM-2NSC and STTC-TCM-2RSC assisted MPEG-4 schemes, when communicating over uncorrelated Rayleigh fading channels. The effective throughput was 1.98 BPS.

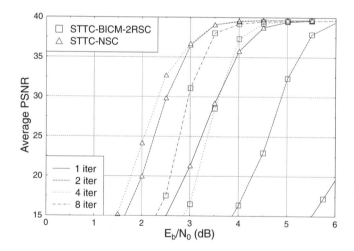

Figure 14.66: Average PSNR versus E_b/N_0 performance of the 16QAM-based STTC-BICM-2RSC and STTC-NSC assisted MPEG-4 benchmark schemes, when communicating over uncorrelated Rayleigh fading channels. The effective throughput of STTC-BICM-2RSC and STTC-NSC was 1.98 and 1.99 BPS, respectively.

14.5.6 Conclusions

In conclusion, a jointly optimized source-coding, outer channel-coding, inner CM and spatial diversity aided turbo transceiver was studied and proposed for MPEG-4 wireless videotelephony. With the aid of an MLC scheme that consists of two different-rate NSCs/RSCs the video bits were unequally protected according to their sensitivity. The employment of TCM improved the bandwidth efficiency of the system and spatial diversity was attained by utilizing STTC. The performance of the proposed STTC-TCM-2NSC/STTC-TCM-2RSC scheme was enhanced with the advent of an efficient iterative decoding structure exchanging extrinsic information across three consecutive blocks. Explicitly, it was shown in Section 14.5.4 that when a non-recursive STTC decoder is employed, a three-stage concatenated iterative decoding scheme is required for approaching the MIMO channel's capacity. It was shown in Figures 14.63 and 14.65 that the STTC-TCM-2RSC scheme required $E_b/N_0 = 0.5$ dB in order to attain BER $= 10^{-4}$ and PSNR > 37 dB, which is 2.3 dB away from the corresponding MIMO channel's capacity. However, if the proposed STTC-TCM-2RSC scheme is used for broadcasting MPEG-4 encoded video, where a longer delay can be tolerated, the required E_b/N_0 value is only 1 dB away from the MIMO channel's capacity, as evidenced by comparing Figures 14.57 and 14.60.

14.6 Near-capacity Irregular Variable Length Codes[11]

14.6.1 Introduction

Irregular Convolutional Coding (IrCC) [475] has been proposed for employment as an outer channel codec in iteratively decoded schemes. This amalgamates a number of component Convolutional Codes (CC) having different coding rates, each of which encodes an appropriately selected fraction of the input bitstream. More specifically, the appropriate fractions may be selected with the aid of EXIT chart analysis [476] in order to shape the inverted EXIT curve of the composite IrCC to ensure that it does not cross the EXIT curve of the inner codec. In this way, an open EXIT chart tunnel may be created at low E_b/N_0 values, which implies approaching the channel's capacity bound [477]. This was demonstrated, for example, for the serial concatenation of IrCCs combined with precoded equalization in [478].

Similarly to binary BCH codecs [479], the constituent binary CCs [479] of an IrCC codec are unable to exploit the unequal source symbol occurrence probabilities that are typically associated with audio, speech, image, and video sources [161, 391]. As the exploitation of all available redundancy is required for near-capacity operation [480], typically a Huffman source encoder [481] is employed to remove this redundancy before channel encoding commences. However, Huffman decoding is very sensitive to bit errors, requiring the low BER reconstruction of the Huffman encoded bits so that a low Symbol Error Ratio (SER) may be achieved [482].

This motivates the application of the Variable Length Error Correction (VLEC) class of variable length codes [482] as an alternative to the Huffman and BCH or CC coding of sequences of source symbols having values with unequal probabilities of occurrence.

---

[11]R. G. Maunder, J. Wang, S. X. Ng, L-L. Yang and L. Hanzo: On the Performance and Complexity of Irregular Variable Length Codes for Near-Capacity Joint Source and Channel Coding, submitted to *IEEE Transactions on Wireless Communications.*

In VLEC coding, the source symbols are represented by binary codewords of varying lengths. Typically, the more frequently that a particular source symbol value occurs, the shorter its VLEC codeword is, resulting in a reduced average codeword length L. In order that each valid VLEC codeword sequence may be uniquely decoded, a lower bound equal to the source entropy E is imposed on the average codeword length L. Any discrepancy between L and E is quantified by the coding rate $R = E/L \in [0, 1]$ and may be attributed to the intentional introduction of redundancy into the VLEC codewords. Naturally, this intentionally introduced redundancy imposes code constraints that limit the set of legitimate sequences of VLEC-encoded bits. Like the code constraints of CC and BCH coding [479], the VLC code constraints may be exploited for providing an additional error-correcting capability during VLEC decoding [482]. Furthermore, *unlike* in CC and BCH decoding [479], any redundancy owing to the unequal occurrence probabilities of the source symbol values may be additionally exploited during VLEC decoding [482].

Depending on the VLEC coding rate, the associated code constraints render VLEC decoding substantially less sensitive to bit errors than Huffman decoding. Hence, a coding gain of 1 dB at an SER of 10^{-5} has been observed by employing VLEC coding having a particular coding rate instead of a concatenated Huffman and BCH coding scheme having the same coding rate [482].

This motivates the application of the irregular coding concept to VLEC coding for employment in the near-capacity joint source and channel coding of sequences of source symbols with values with unequal occurrence probabilities. More specifically, we employ a novel Irregular Variable Length Coding (IrVLC) scheme as our outer source codec, which we serially concatenate [483, 484] with an inner channel codec for the sake of exchanging extrinsic information. In analogy to IrCC, the proposed IrVLC scheme employs a number of component VLEC codebooks having different coding rates, which are used for encoding appropriately selected fractions of the input source symbol stream. In this way, the resultant composite inverted EXIT curve may be shaped to ensure that it does not cross the EXIT curve of the inner channel codec.

The rest of this section is outlined as follows. In Section 14.6.2, we propose iteratively decoded schemes, in which we opt for serially concatenating IrVLC with TCM. Furthermore, Section 14.6.2 additionally introduces our benchmark schemes, where IrVLC is replaced by regular VLC having the same coding rate. The design and EXIT chart aided characterization of these schemes is detailed in Section 14.6.3. In Section 14.6.4, we quantify the attainable performance improvements offered by the proposed IrVLC arrangements compared with the regular VLC benchmarker schemes, as provided in [485]. Furthermore in Section 14.6.4 of this contribution, we additionally consider a Huffman coding and IrCC based benchmarker, as well as presenting significant new results pertaining to the computational complexity of the considered schemes. More specifically, we quantify the computational complexity required for achieving a range of source sample reconstruction qualities at a range of Rayleigh fading channel E_b/N_0 values. Finally, we offer our conclusions in Section 14.6.5.

14.6.2 Overview of the Proposed Schemes

In this section we provide an overview of a number of serially concatenated and iteratively decoded joint source and channel coding schemes. Whilst the novel IrVLC scheme introduced in this section may be tailored for operating in conjunction with any inner

14.6. NEAR-CAPACITY IRREGULAR VARIABLE LENGTH CODES

channel codec, we opt for employing TCM [486] in each of our considered schemes. This provides error protection without any bandwidth expansion or effective bitrate reduction by accommodating the additional redundancy by transmitting more bits per channel symbol. The choice of TCM is further justified, since *A Posteriori* Probability (APP) TCM Soft-In Soft-Out (SISO) decoding, similarly to APP SISO IrVLC decoding, operates on the basis of Add–Compare–Select (ACS) operations within a trellis structure. Hence, the APP SISO IrVLC and TCM decoders can share resources in systolic-array-based chips, facilitating a cost-effective implementation. Furthermore, we show that TCM exhibits attractive EXIT characteristics in the proposed IrVLC context even without requiring TTCM- or BICM-style internal iterative decoding [479].

Our considered schemes differ in the choice of the outer source codec. Specifically, we consider a novel IrVLC codec and an equivalent regular VLC-based benchmarker in this role. In both cases we employ both Symbol-Based (SB) [487, 488] and Bit-Based (BB) [489] VLC decoding, resulting in a total of four different configurations. We refer to these four schemes as the SBIrVLC-, BBIrVLC-, SBVLC-, and BBVLC-TCM arrangements, as appropriate. A schematic that is common to each of these four considered schemes is provided in Figure 14.67.

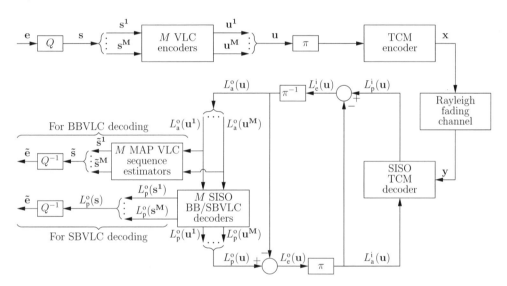

Figure 14.67: Schematic of the SBIrVLC-, BBIrVLC-, SBVLC-, and BBVLC-TCM schemes. In the IrVLC schemes, the M number of VLC encoders, APP SISO decoders and MAP sequence estimators are each based upon one of N number of VLC codebooks. In contrast, in the VLC benchmarkers, all of the M number of VLC encoders, decoders, and sequence estimators are based on the same VLC codebook.

14.6.2.1 Joint Source and Channel Coding

The schemes considered are designed to facilitate the near-capacity detection of source samples received over an uncorrelated narrowband Rayleigh fading channel. We consider the case of independent and identically distributed (i.i.d.) source samples, which may represent the prediction residual error that remains following the predictive coding of audio, speech, image, or video information [161, 391]. A Gaussian source sample distribution is assumed, because this has widespread applications owing to the wide applicability of the central limit theorem. Note, however, that with the aid of suitable adaptation, the techniques proposed in this book may be just as readily applied to arbitrary source sample distributions.

In the transmitter of Figure 14.67, the real-valued source samples are quantized to K number of quantization levels in the block Q. The resultant frame of quantized source samples is synonymously referred to as the frame of source symbols s here. Each source symbol in this frame indexes the particular quantization level $\tilde{e}^k$, $k \in [1\ldots K]$, that represents the corresponding source sample in the frame e with the minimum squared error. Owing to the lossy nature of quantization, distortion is imposed on the reconstructed source sample frame $\tilde{\mathbf{e}}$ that is obtained by the receiver of Figure 14.67, following inverse quantization in the block Q^{-1}. The total distortion expected depends on both the original source sample distribution as well as on the number of quantization levels K. This distortion may be minimized by employing Lloyd–Max quantization [490, 491]. Here, a $K = 16$-level Lloyd–Max quantization scheme is employed, which achieves an expected Signal-to-Quantization Noise Ratio (SQNR) of about 20 dB for a Gaussian source [490]. Note, however, that with suitable adaptation, the techniques advocated in this book may be just as readily applied to arbitrary quantizers. Also note that Lloyd–Max quantization results in a large variation in the occurrence probabilities of the resultant source symbol values. These occurrence probabilities are given by integrating the source PDF between each pair of adjacent decision boundaries. In the case of our $K = 16$-level quantizer, the source symbol values' occurrence probabilities vary by more than an order of magnitude between 0.0082 and 0.1019. These probabilities correspond to symbol informations spanning between 3.29 and 6.93 bits/symbol, motivating the application of VLC and giving a source entropy of $E = 3.77$ bits/symbol.

In the transmitter of the proposed scheme, the Lloyd–Max quantized source symbol frame s is decomposed into $M = 300$ sub-frames $\{\mathbf{s^m}\}_{m=1}^{M}$, as shown in Figure 14.67. In the case of the SBIrVLC- and SBVLC-TCM schemes, this decomposition is necessary for the sake of limiting the computational complexity of VLC decoding, because the number of transitions in the symbol-based VLC trellis is inversely proportional to the number of sub-frames in this case [487]. We opt to employ the same decomposition of the source symbol frames into sub-frames in the case of the BBIrVLC- and BBVLC-TCM schemes for the sake of ensuring that we make a fair comparison with the SBIrVLC- and SBVLC-TCM schemes. This is justified, because the decomposition considered benefits the performance of the BBIrVLC- and BBVLC-TCM schemes, as will be detailed below. Each source symbol sub-frame $\mathbf{s^m}$ comprises $J = 100$ source symbols. Hence, the total number of source symbols in a source symbol frame becomes $M \cdot J = 30000$. As described above, each Lloyd–Max quantized source symbol in the sub-frame $\mathbf{s^m}$ has a K-ary value $s_j^m \in [1\ldots K]$, where we have $j \in [1\ldots J]$.

As described in Section 14.6.1, we employ N number of VLC codebooks to encode the source symbols, where we opted for $N = 15$ for the SBIrVLC and BBIrVLC schemes and $N = 1$ for the regular SBVLC and BBVLC schemes. Each Lloyd–Max quantized source symbol sub-frame $\mathbf{s^m}$ is VLC-encoded using a single VLC codebook $\mathbf{VLC}^n$, where we have $n \in [1\ldots N]$. In the case of the SBIrVLC and BBIrVLC schemes, the particular fraction

14.6. NEAR-CAPACITY IRREGULAR VARIABLE LENGTH CODES

C^n of the set of source symbol sub-frames that is encoded by the specific VLC codebook **VLC**^{n} is fixed and will be derived in Section 14.6.3. The specific Lloyd–Max quantized source symbols having the value of $k \in [1 \ldots K]$ and encoded by the specific VLC codebook **VLC**^{n} are represented by the codeword **VLC**^{n,k}, which has a length of $I^{n,k}$ bits. The $J = 100$ VLC codewords that represent the $J = 100$ Lloyd–Max quantized source symbols in each source symbol sub-frame $\mathbf{s}^m$ are concatenated to provide the transmission sub-frame $\mathbf{u}^m = \{\mathbf{VLC}^{n,s_j^m}\}_{j=1}^J$.

Owing to the variable lengths of the VLC codewords, each of the $M = 300$ transmission sub-frames typically comprises a different number of bits. In order to facilitate the VLC decoding of each transmission sub-frame $\mathbf{u}^m$, it is necessary to explicitly convey its length $I^m = \sum_{j=1}^J I^{n,s_j^m}$ to the receiver. Furthermore, this highly error-sensitive side information must be reliably protected against transmission errors. This may be achieved using a low-rate block code, for example. For the sake of avoiding obfuscation, this is not shown in Figure 14.67. Note that the choice of the specific number of sub-frames M in each frame constitutes a trade-off between the computational complexity of SBVLC decoding or the performance of BBVLC decoding and the amount of side information that must be conveyed. In Section 14.6.3, we comment on the amount of side information that is required for reliably conveying the specific number of bits in each transmission sub-frame to the decoder.

In the scheme's transmitter, the $M = 300$ number of transmission sub-frames $\{\mathbf{u}^m\}_{m=1}^M$ are concatenated. As shown in Figure 14.67, the resultant transmission frame $\mathbf{u}$ has a length of $\sum_{m=1}^M I^m$ bits.

In the proposed scheme, the VLC codec is protected by a serially concatenated TCM codec. Following VLC encoding, the bits of the transmission frame $\mathbf{u}$ are interleaved in the block π of Figure 14.67 and TCM encoded in order to obtain the channel's input symbols $\mathbf{x}$, as shown in Figure 14.67. These are transmitted over an uncorrelated narrowband Rayleigh fading channel and are received as the channel's output symbols $\mathbf{y}$, as seen in Figure 14.67.

14.6.2.2 Iterative Decoding

In the receiver, APP SISO TCM- and VLC-decoding are performed iteratively, as shown in Figure 14.67. Both of these decoders invoke the Bahl–Cocke–Jelinek–Raviv (BCJR) algorithm [492] on the basis of their trellises. Symbol-based trellises are employed in the case of TCM [486], SBIrVLC, and SBVLC [487, 488] decoding, whilst BBIrVLC and BBVLC decoding rely on bit-based trellises [489]. All BCJR calculations are performed in the logarithmic probability domain and using an eight-entry lookup table for correcting the Jacobian approximation in the Log-MAP algorithm [479]. The proposed approach requires only ACS computational operations during iterative decoding, which will be used as our complexity measure, because it is characteristic of the complexity/area/speed trade-offs in systolic-array-based chips.

As usual, extrinsic soft information, represented in the form of Logarithmic Likelihood Ratios (LLRs) [493], is iteratively exchanged between the TCM and VLC decoding stages for the sake of assisting each other's operation, as detailed in [483] and [484]. In Figure 14.67, $L(\cdot)$ denotes the LLRs of the bits concerned (or the log-APPs of the specific symbols as appropriate), where the superscript i indicates inner TCM decoding, while o corresponds to outer VLC decoding. In addition, a subscript denotes the dedicated role of the LLRs (or log-APPs), with a, p and e indicating *a priori*, *a posteriori*, and extrinsic information, respectively.

Just as $M = 300$ separate VLC encoding processes are employed in the proposed scheme's transmitter, $M = 300$ separate VLC decoding processes are employed in its receiver. In parallel to the composition of the bit-based transmission frame $\mathbf{u}$ from its $M = 300$ sub-frames, the *a priori* LLRs $L_a^o(\mathbf{u})$ are decomposed into $M = 300$ sub-frames, as shown in Figure 14.67. This is achieved with the aid of the explicit side information that conveys the number of bits I^m in each transmission sub-frame $\mathbf{u}^m$. Each of the $M = 300$ VLC decoding processes is provided with the *a priori* LLR sub-frame $L_a^o(\mathbf{u}^m)$ and in response it generates the *a posteriori* LLR sub-frame $L_p^o(\mathbf{u}^m)$, $m \in [1 \ldots M]$. These *a posteriori* LLR sub-frames are concatenated in order to provide the *a posteriori* LLR frame $L_p^o(\mathbf{u})$, as shown in Figure 14.67.

In the case of SBIrVLC and SBVLC decoding, each of the $M = 300$ VLC decoding processes additionally provides log-APPs pertaining to the corresponding source symbol sub-frame $L_p^o(\mathbf{s}^m)$. This comprises a set of K number of log-APPs for each source symbol s_j^m in the sub-frame $\mathbf{s}^m$, where $j \in [1 \ldots J]$. Each of these log-APPs provides the logarithmic probability that the corresponding source symbol s_j^m has the particular value $k \in [1 \ldots K]$. In the receiver of Figure 14.67, the source symbols' log-APP sub-frames are concatenated to provide the source symbol log-APP frame $L_p^o(\mathbf{s})$. By inverse-quantizing this soft information in the block Q^{-1}, a frame of MMSE source sample estimates $\tilde{\mathbf{e}}$ can be obtained. More specifically, each reconstructed source sample is obtained by using the corresponding set of K source symbol value probabilities to find the weighted average of the K number of quantization levels $\{\tilde{e}^k\}_{k=1}^K$.

Conversely, in the case of BBIrVLC and BBVLC decoding, no symbol-based *a posteriori* output is available. In this case, each source symbol sub-frame $\mathbf{s}^m$ is estimated from the corresponding *a priori* LLR sub-frame $L_a^o(\mathbf{u}^m)$. This may be achieved by employing MAP sequence estimation operating on a bit-based trellis structure, as shown in Figure 14.67. Unlike in APP SISO SBIrVLC and SBVLC decoding, bit-based MAP sequence estimation cannot exploit the knowledge that each sub-frame $\mathbf{s}^m$ comprises $J = 100$ source symbols. For this reason, the resultant hard decision estimate $\tilde{\mathbf{s}}^m$ of each source symbol sub-frame $\mathbf{s}^m$ may or may not contain $J = 100$ source symbols. In order to prevent the loss of synchronization that this would imply, source symbol estimates are removed from, or appended to, the end of each source symbol sub-frame estimate $\tilde{\mathbf{s}}^m$ to ensure that they each comprise exactly $J = 100$ source symbol estimates. Note that it is the decomposition of the source symbol frame $\mathbf{s}$ into sub-frames that provides this opportunity to mitigate the loss of synchronization that is associated with bit-based MAP VLC sequence estimation. Hence, the decomposition of the source symbol frame $\mathbf{s}$ into sub-frames benefits the performance of the BBIrVLC- and BBVLC-TCM schemes, as mentioned above.

Following MAP sequence estimation, the adjusted source symbol sub-frame estimates $\tilde{\mathbf{s}}^m$ are concatenated for the sake of obtaining the source symbol frame estimate $\tilde{\mathbf{s}}$. This may be inverse-quantized in order to obtain the source sample frame estimate $\tilde{\mathbf{e}}$. Note that for the reconstruction of a source sample frame estimate $\tilde{\mathbf{e}}$ from a given *a priori* LLR frame $L_a^o(\mathbf{u})$, a higher level of source distortion may be expected in the BBIrVLC- and BBVLC-TCM schemes than in the corresponding SBIrVLC- and SBVLC-TCM schemes. This is due to the BBIrVLC- and BBVLC-TCM schemes' reliance on hard decisions as opposed to the soft decisions of the SBIrVLC- and SBVLC-TCM schemes. However, this reduced performance substantially benefits us in terms of a reduced complexity, because the bit-based VLC decoding trellis employed during APP SISO BBIrVLC and BBVLC decoding and MAP sequence estimation contains significantly less transitions than the symbol-based VLC decoding trellis of APP SISO SBIrVLC and SBVLC decoding.

In the next section we detail the design of our IrVLC scheme and characterize each of the SBIrVLC-, BBIrVLC-, SBVLC-, and BBVLC-TCM schemes with the aid of EXIT chart analysis.

14.6.3 Parameter Design for the Proposed Schemes

14.6.3.1 Scheme Hypothesis and Parameters

As described in Section 14.6.1, the SBIrVLC and BBIrVLC schemes may be constructed by employing a number of VLC codebooks having different coding rates, each of which encodes an appropriately chosen fraction of the input source symbols. We opted for using $N = 15$ VLC codebooks $\mathbf{VLC}^n$, $n \in [1 \ldots N]$, that were specifically designed for encoding $K = 16$-level Lloyd–Max quantized Gaussian i.i.d. source samples. These 15 VLC codebooks were selected from a large number of candidate codebooks in order to provide a suite of similarly-spaced EXIT curves. More specifically, the $N = 15$ VLC codebooks comprised 13 VLEC designs having various so-called minimum block, convergence, and divergence distances as defined in [494], complemented by a symmetric and an asymmetric Reversible Variable Length Coding (RVLC) design [494]. In all codebooks, a free distance of at least $d_f = 2$ was employed, because this supports convergence to an infinitesimally low BER [495]. The resultant average VLC codeword lengths of $L^n = \sum_{k=1}^{K} P(k) \cdot I^{n,k}$, $n \in [1 \ldots N]$, were found to range from 3.94 to 12.18 bits/symbol. When compared with the source symbol entropy of $E = -\sum_{k=1}^{K} P(k) \cdot \log_2(P(k)) = 3.77$ bits/symbol, these correspond to coding rates of $R^n = E/L^n$ spanning the range of 0.31 to 0.96.

As detailed below, our SBIrVLC and BBIrVLC schemes were designed under the constraint that they have an overall coding rate of $R = 0.52$. This value was chosen, because it is the coding rate of the VLC codebook $\mathbf{VLC}^{10}$, which we employ in our SBVLC and BBVLC benchmarkers using $N = 1$ codebook. This coding rate results in an average interleaver length of $M \cdot J \cdot E/R = 217500$ bits for all of the schemes considered.

In-phase Quadrature-phase (IQ)-interleaved TCM having eight trellis states per symbol along with 3/4-rate coded 16QAM is employed, because this is appropriate for transmission over uncorrelated narrowband Rayleigh fading channels. Ignoring the modest bitrate contribution of conveying the side information, the bandwidth efficiency of the schemes considered is therefore $\eta = 0.52 \times 0.75 \times \log_2(16) = 1.56$ bit/s/Hz, assuming ideal Nyquist filtering having a zero excess bandwidth. This value corresponds to the channel capacity of the uncorrelated narrowband Rayleigh fading channel at an E_b/N_0 value of 2.6 dB [496]. Given this point on the corresponding channel capacity curve, we will be able to quantify how closely the proposed schemes may approach this ultimate limit.

Recall from Section 14.6.2 that it is necessary to convey the length of each transmission sub-frame $\mathbf{u}^m$ to the receiver in order to facilitate its VLC decoding. It was found for all considered schemes that a single 10-bit fixed-length codeword of side information is sufficient for conveying the length of each of the $M = 150$ transmission sub-frames $\mathbf{u}^m$ in each transmission frame $\mathbf{u}$. As suggested in Section 14.6.2, this error-sensitive side information may be protected by a low-rate block code in order to ensure its reliable transmission. Note that because even a repetition code having a rate as low as $1/7$ encodes the side information with only about 1% of the expected transmission frame length, we consider the overhead of conveying side information to be acceptable.

14.6.3.2 EXIT Chart Analysis and Optimization

We now consider the EXIT characteristics of the various components of our various schemes. In all cases, EXIT curves were generated using uncorrelated Gaussian distributed *a priori* LLRs with the assumption that the transmission frame's bits have equiprobable values. This is justified, because we employ a long interleaver and because the entropy of the VLC encoded bits was found to be at least 0.99 for all considered VLC codebooks. All mutual information measurements were made using the histogram-based approximation of the LLR PDFs [476].

In Figure 14.68, we provide the EXIT curves $I_e^i(I_a^i, E_b/N_0)$ of the TCM scheme for a number of E_b/N_0 values above the channel capacity bound of 2.6 dB. The inverted EXIT curves $I_a^{o,n}(I_e^o)$ plotted for the $N = 15$ VLC codebooks, together with their coding rates R^n, are also given in Figure 14.68. Note that these curves were obtained using bit-based VLC decoding, but very similar curves may be obtained for symbol-based decoding.

The inverted EXIT curve of an IrVLC scheme $I_a^o(I_e^o)$ can be obtained as the appropriately weighted superposition of the $N = 15$ component VLC codebooks' EXIT curves,

$$I_a^o(I_e^o) = \sum_{n=1}^{N} \alpha^n I_a^{o,n}(I_e^o), \qquad (14.17)$$

where α^n is the fraction of the transmission frame **u** that is generated by the specific component codebook $\mathbf{VLC}^n$. Note that the values of α^n are subject to the constraints

$$\sum_{n=1}^{N} \alpha^n = 1, \quad \alpha^n \geq 0 \ \forall n \in [1 \ldots N]. \qquad (14.18)$$

The specific fraction of source symbol sub-frames $\mathbf{s}^m$ that should be encoded by the specific component codebook $\mathbf{VLC}^n$ in order that it generates a fraction α^n of the transmission frame **u**, is given by

$$C^n = \alpha^n \cdot R^n / R, \qquad (14.19)$$

where $R = 0.52$ is the desired overall coding rate. Again, the specific values of C^n are subject to the constraints

$$\sum_{n=1}^{N} C^n = \sum_{n=1}^{N} \alpha^n \cdot R^n / R = 1, \quad C^n \geq 0 \ \forall n \in [1 \ldots N]. \qquad (14.20)$$

Beneficial values of C^n may be chosen by ensuring that there is an open EXIT tunnel between the inverted IrVLC EXIT curve and the EXIT curve of TCM at an E_b/N_0 value that is close to the channel capacity bound. This may be achieved using the iterative EXIT-chart matching process of [475] to adjust the values of $\{C^n\}_{n=1}^{N}$ under the constraints of Equations

14.6. NEAR-CAPACITY IRREGULAR VARIABLE LENGTH CODES

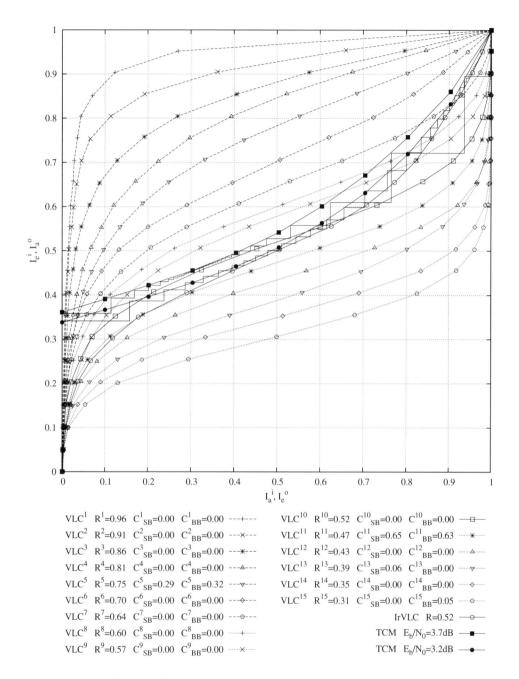

Figure 14.68: Inverted VLC EXIT curves and TCM EXIT curves.

14.18 and 14.20 for the sake of minimizing the error function

$$\{C^n\}_{n=1}^{N} = \underset{\{C^n\}_{n=1}^{N}}{\operatorname{argmin}} \left(\int_0^1 e(I)^2 \, dI \right), \tag{14.21}$$

where

$$e(I) = I_e^i(I, E_b/N_0) - I_a^o(I) \tag{14.22}$$

is the difference between the inverted IrVLC EXIT curve and the EXIT curve of TCM at a particular target E_b/N_0 value. Note that in order to ensure that the design results in an open EXIT tunnel, we must impose the additional constraint of

$$e(I) > 0 \quad \forall I \in [0, 1]. \tag{14.23}$$

Open EXIT tunnels were found to be achievable for both the SBIrVLC- and the BBIrVLC-TCM schemes at an E_b/N_0 of 3.2 dB, which is just 0.6 dB from the channel capacity bound of 2.6 dB. The inverted BBIrVLC EXIT curve is shown in Figure 14.68, which is similar to the SBIrVLC EXIT curve not shown here for reasons of space-economy. The corresponding values of C^n are provided for both the SBIrVLC- and the BBIrVLC-TCM schemes in Figure 14.68.

In contrast, the corresponding EXIT-tunnel only becomes open for the SBVLC- and BBVLC-TCM benchmarkers for E_b/N_0 values in excess of 3.7 dB, which is 1.1 dB from the channel capacity bound of 2.6 dB. We can therefore expect our SBIrVLC- and BBIrVLC-TCM schemes to be capable of operating at nearly half the distance from the channel capacity bound in comparison to our benchmarkers, achieving a gain of about 0.5 dB.

14.6.4 Results

In this section, we discuss our findings when communicating over an uncorrelated narrowband Rayleigh fading channel having a range of E_b/N_0 values above the channel capacity bound of 2.6 dB.

In addition to the proposed SBIrVLC-, BBIrVLC-, SBVLC-, and BBVLC-TCM schemes, in this section we also consider the operation of an additional benchmarker which we refer to as the Huffman–IrCC-TCM scheme. In contrast to the SBIrVLC-, BBIrVLC-, SBVLC- and BBVLC-TCM schemes, in the Huffman–IrCC-TCM scheme the transmission frame **u** of Figure 14.67 is generated by both Huffman and concatenated IrCC encoding the source symbol frame **s**, rather than by invoking VLC encoding. More specifically, Huffman coding is employed on a sub-frame-by-sub-frame basis, as described in Section 14.6.2. The resultant frame of Huffman encoded bits **v** is protected by the memory-4 17-component IrCC scheme of [497]. This was tailored to have an overall coding rate of 0.52 and an inverted EXIT curve that does not cross the TCM EXIT curve at an E_b/N_0 of 3.2 dB, just like the SBIrVLC and BBIrVLC designs detailed in Section 14.6.3. In the Huffman–IrCC-TCM receiver, iterative APP SISO IrCC and TCM decoding proceeds, as described in Section 14.6.2. Note that in addition to the *a posteriori* LLR frame $L_p^o(\mathbf{u})$ pertaining to the transmission frame **u**, the APP SISO IrCC decoder can additionally provide the *a posteriori* LLR frame $L_p^o(\mathbf{v})$ pertaining to the frame of Huffman encoded bits **v**. It is on the basis of this that bit-based MAP Huffman sequence estimation may be invoked on a sub-frame-by-sub-frame basis in order to obtain the source symbol frame estimate $\tilde{\mathbf{s}}$.

14.6. NEAR-CAPACITY IRREGULAR VARIABLE LENGTH CODES

14.6.4.1 Asymptotic Performance Following Iterative Decoding Convergence

For each of our schemes and for each value of E_b/N_0, we consider the reconstructed source sample frame $\tilde{e}$ and evaluate the SNR associated with the ratio of the source signal's energy and the reconstruction error energy that may be achieved following iterative decoding convergence. This relationship is plotted for each of the SBIrVLC-, BBIrVLC-, SBVLC-, and BBVLC-TCM schemes, as well as for the Huffman–IrCC-TCM scheme, in Figure 14.69.

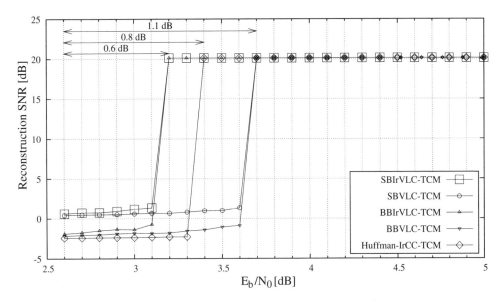

Figure 14.69: Reconstruction SNR versus E_b/N_0 for a Gaussian source using $K = 16$-level Lloyd–Max quantization for the SBIrVLC-, BBIrVLC-, SBVLC-, and BBVLC-TCM schemes, as well as for the Huffman–IrCC-TCM scheme, communicating over an uncorrelated narrowband Rayleigh fading channel following iterative decoding convergence.

At sufficiently high E_b/N_0 values, all considered schemes are capable of achieving source sample reconstruction SNRs of up to 20 dB, which represents the error-free case, where only quantization noise is present, while all channel-induced errors are absent. As shown in Figure 14.69, this may be achieved by the SBIrVLC- and BBIrVLC-TCM schemes at E_b/N_0 values above 3.2 dB, which is just 0.6 dB from the channel capacity bound of 2.6 dB. This represents a 0.5 dB gain over the SBVLC- and BBVLC-TCM schemes, which require E_b/N_0 values in excess of 3.7 dB, a value that is 1.1 dB from the channel capacity bound. In addition, the SBIrVLC- and BBIrVLC-TCM schemes can be seen to offer a 0.2 dB gain over the Huffman–IrCC-TCM scheme, which is incapable of operating within 0.8 dB of the channel capacity bound.

Note that our findings recorded in Figure 14.69 for the SBIrVLC-, BBIrVLC-, SBVLC- and BBVLC-TCM schemes confirm the EXIT chart predictions of Section 14.6.3. Figure 14.68 provides decoding trajectories for the BBIrVLC-TCM and BBVLC-TCM schemes at E_b/N_0 values of 3.2 and 3.7 dB, respectively. Note that owing to the sufficiently long interleaver length of $M \times J \times E/R = 217500$ bits, correlation within the *a priori* LLR frames $L_a^i(\mathbf{u})$ and $L_a^o(\mathbf{u})$ is mitigated and the recorded trajectories exhibit a close match with the corresponding TCM and inverted IrVLC/VLC EXIT curves. In both cases, the corresponding trajectory can be seen to converge to the $(1, 1)$ mutual information point of the EXIT chart after a number of decoding iterations. Note that with each decoding iteration, a greater extrinsic mutual information is achieved, which corresponds to a greater source sample reconstruction SNR.

At low E_b/N_0 values, the corresponding TCM EXIT curves cross the inverted IrVLC or VLC EXIT curves and the open EXIT chart tunnel disappears. In these cases, iterative decoding convergence to unity mutual information cannot be achieved, resulting in the poor reconstruction quality that may be observed at low values of E_b/N_0 in Figure 14.69.

In the case of the Huffman–IrCC–TCM scheme, however, poor reconstruction qualities were obtained despite the presence of an open EXIT chart tunnel for E_b/N_0 values between 3.2 and 3.4 dB in Figure 14.69. This surprising result may be attributed to the APP SISO IrCC decoder's relatively high sensitivity to any residual correlation within the *a priori* LLR frame $L_a^o(\mathbf{u})$ that could not be mitigated by the interleaver having an average length of 217500 bits. As a result of this, the Huffman–IrCC–TCM EXIT trajectory does not approach the inverted IrCC EXIT curve very closely and a wide EXIT chart tunnel is required for iterative decoding convergence to the $(1, 1)$ mutual information point of the EXIT chart.

The relatively high sensitivity of APP SISO CC decoding to any correlation within the *a priori* LLR frame $L_a^o(\mathbf{u})$ as compared with VLC decoding may be explained as follows. During APP SISO CC decoding using the BCJR algorithm, it is assumed that all *a priori* LLRs that correspond to bits within each set of L consecutive codewords are uncorrelated, where L is the constraint length of the CC [479]. In contrast, during APP SISO VLC decoding using the BCJR algorithm, it is assumed that all *a priori* LLRs that correspond to bits within each *single* codeword are uncorrelated [489]. Hence, the BCJR-based APP SISO decoding of a CC scheme can be expected to be more sensitive to correlation within the *a priori* LLR frame $L_a^o(\mathbf{u})$ than that of a VLC scheme having similar codeword lengths. As a result, a longer interleaver and, hence, a higher latency would be required for facilitating near-capacity CC operation.

14.6.4.2 Performance During Iterative Decoding

The achievement of iterative decoding convergence requires the completion of a sufficiently high number of decoding iterations. Clearly, each decoding iteration undertaken is associated with a particular computational complexity, the sum of which represents the total computational complexity of the iterative decoding process. Hence, the completion of a sufficiently high number of decoding iterations in order to achieve iterative decoding convergence may be associated with a high computational complexity. In order to quantify how this computational complexity scales as iterative decoding proceeds, we recorded the total number of ACS operations performed per source sample during APP SISO decoding and MAP sequence estimation.

Furthermore, the performance of the considered schemes was also assessed *during* the iterative decoding process, not only after its completion once convergence has been achieved. This was achieved by evaluating the source sample reconstruction SNR following the completion of *each* decoding iteration. The total computational complexity associated with this SNR was calculated as the sum of the computational complexities associated with

14.6. NEAR-CAPACITY IRREGULAR VARIABLE LENGTH CODES

all decoding iterations completed so far during the iterative decoding process. Clearly, as more and more decoding iterations are completed, the resultant source sample reconstruction SNR can be expected to increase until iterative decoding convergence is achieved. However, the associated total computational complexity will also increase as more and more decoding iterations are completed. Hence, this approach allows the characterization of the trade-off between reconstruction quality and computational complexity.

For each considered Rayleigh channel E_b/N_0 value, a set of source sample reconstruction SNRs and their corresponding computational complexities was obtained, as described above. Note that the size of these sets was equal to the number of decoding iterations required to achieve iterative decoding convergence at the particular E_b/N_0 value. It would therefore be possible to display the source sample reconstruction SNR versus both the E_b/N_0 and the computational complexity in a three-dimensional surface plot, for each of the SBIrVLC-, BBIrVLC-, SBVLC-, and BBVLC-TCM schemes. For clarity, however, these surfaces are projected in the direction of the source sample reconstruction SNR axis into two dimensions in Figure 14.70. We employ contours of constant source sample reconstruction SNR, namely 15 and 20 dB, to parameterize the relationship between the Rayleigh fading channel's E_b/N_0 value and the associated computational complexity. Note that the plot of Figure 14.69 may be thought of as a cross section through the surfaces represented by Figure 14.70, perpedicular to the computational complexity axis at 1×10^7 ACS operations per source sample. Note that this particular value of computational complexity is sufficiently high to achieve iterative decoding convergence at all values of E_b/N_0, in each of the considered schemes.

Note that the SBIrVLC and SBVLC decoders have a computational complexity per source sample that depends on the number of symbols in each source symbol sub-frame s^m, namely J. This is because the number of transitions in their symbol-based trellises is proportional to J^2 [487]. Hence, the results provided in Figure 14.70 for the SBIrVLC- and SBVLC-TCM schemes are specific to the $J = 100$ scenario. In contrast, the TCM, BBIrVLC, BBVLC, and IrCC decoders have a computational complexity per source sample that is independent of the number of symbols in each source symbol sub-frame s^m, namely J. This is because the number of transitions in their trellises is proportional to J [479, 486, 498]. Hence, the results for the BBIrVLC- and BBVLC-TCM schemes, as well as for the Huffman–IrCC-TCM scheme, provided in Figure 14.70 are *not* specific for the $J = 100$ case.

As shown in Figure 14.70, source sample reconstruction SNRs of up to 20 dB can be achieved within 0.6 dB of the channel's capacity bound of 2.6 dB for the SBIrVLC- and BBIrVLC-TCM schemes, within 1.1 dB for the SBVLC- and BBVLC-TCM schemes and within 0.8 dB for the Huffman–IrCC-TCM scheme. Note that these findings agree with those of the EXIT chart analysis and the asymptotic performance analysis.

14.6.4.3 Complexity Analysis

We now comment on the computational complexities of the considered schemes and select our preferred arrangement.

In all considered schemes and at all values of E_b/N_0, a source sample reconstruction SNR of 15 dB can be achieved at a lower computational complexity than an SNR of 20 dB can, as shown in Figure 14.70. This is because a reduced number of decoding iterations is required for achieving the extrinsic mutual information value associated with a lower reconstruction quality, as stated above. However, for all considered schemes operating at

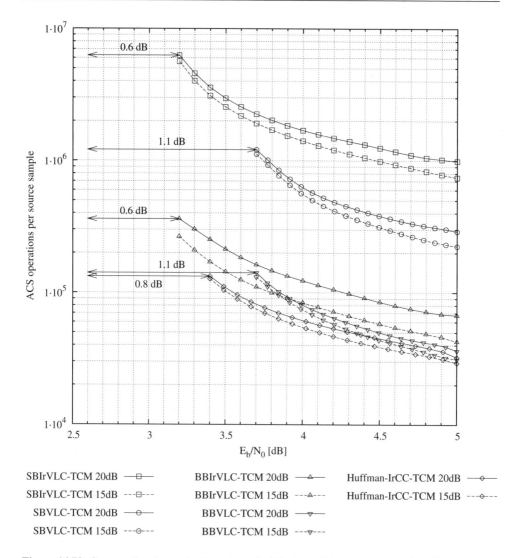

Figure 14.70: Computational complexity versus E_b/N_0 for a Gaussian source using $K = 16$-level Lloyd–Max quantization for the SBIrVLC-, BBIrVLC-, SBVLC- and BBVLC-TCM schemes, as well as for the Huffman–IrCC-TCM scheme, communicating over an uncorrelated narrowband Rayleigh fading channel, parameterized with the source sample reconstruction SNR.

high values of E_b/N_0, this significant 5 dB reduction in source sample reconstruction SNR facilitates only a relatively modest reduction of the associated computational complexity, which was between 9% in the case of the Huffman–IrCC-TCM scheme and 36% for the BBIrVLC-TCM scheme. Hence we may conclude that the continuation of iterative decoding until near-perfect convergence is achieved can be justified at all values of E_b/N_0.

14.6. NEAR-CAPACITY IRREGULAR VARIABLE LENGTH CODES

In addition, it may be seen that a given source sample reconstruction SNR may be achieved at a reduced computational complexity for all considered schemes as the E_b/N_0 value increases. This may be explained by the widening of the EXIT chart tunnel, as the E_b/N_0 value increases. As a result, less decoding iterations are required for reaching the extrinsic mutual information that is associated with a specific source sample reconstruction SNR considered.

In each of the considered schemes it was found that VLC and CC decoding is associated with a higher contribution to the total computational complexity than TCM decoding. Indeed, in the case of the SBIrVLC- and SBVLC-TCM schemes, it was found that VLC decoding accounts for about 97% of the numbers of ACS operations per source sample, having a complexity that is about 32.3 times higher than that of TCM decoding. In contrast, in the BBIrVLC- and BBVLC-TCM schemes, VLC decoding accounts for only 70% of the operations, having a complexity that is about 2.3 times that of TCM decoding. Similarly, CC decoding accounts for only 60% of the ACS operations in the Huffman–IrCC-TCM scheme, having a complexity that is about 1.4 times that of TCM decoding.

The high complexity of the SBIrVLC and SBVLC decoders may be attributed to the specific structure of their trellises, which contain significantly more transitions than those of the BBIrVLC, BBVLC, and IrCC decoders [487]. As a result, the SBIrVLC- and SBVLC-TCM schemes have a complexity that is about an order of magnitude higher than that of the BBIrVLC- and BBVLC-TCM schemes, as well as the Huffman–IrCC-TCM scheme, as shown in Figure 14.70. In light of this, the employment of the SBIrVLC- and SBVLC-TCM schemes cannot be readily justified.

Observe in Figure 14.70 that at high E_b/N_0 values, the SBIrVLC- and BBIrVLC-TCM schemes have a higher computational complexity than the corresponding SBVLC- or BBVLC-TCM scheme. This is due to the influence of their low-rate VLC codebook components. These codebooks comprise codewords with many different lengths, which introduce many transitions, when represented in a trellis structure. The observed computational complexity discrepancy is particularly high in the case of the schemes that employ the symbol-based VLC trellis, owing to its particular nature. For this reason, the SBIrVLC-TCM scheme has a computational complexity that is 240% higher than that of the SBVLC-TCM scheme.

In contrast, we note that at high values of E_b/N_0 the BBIrVLC-TCM scheme has only about a 60% higher computational complexity than the BBVLC-TCM scheme. Similarly, the BBIrVLC-TCM scheme has only twice the computational complexity of the Huffman–IrCC-TCM scheme. Coupled with the BBIrVLC-TCM scheme's ability to operate within 0.6 dB of the Rayleigh fading channel's capacity bound, we are able to identify this as our preferred arrangement.

14.6.5 Conclusions

In this section we have introduced a novel IrVLC design for near-capacity joint source and channel coding. In analogy to IrCC, IrVLC employs a number of component VLC codebooks having different coding rates in appropriate proportions. More specifically, with the aid of EXIT chart analysis, the appropriate fractions of the input source symbols may be chosen for directly ensuring that the EXIT curve of the IrVLC codec may be matched to that of a serially concatenated channel codec. In this way, an open EXIT chart tunnel facilitating near-capacity high-quality source sample reconstruction may be achieved.

We have detailed the construction of an IrVLC scheme that is suitable for the encoding of 16-level Lloyd–Max quantized Gaussian i.i.d. source samples and for use with IQ-interleaved TCM and 16QAM over uncorrelated narrowband Rayleigh fading channels. For the purposes of comparison, we also selected a regular VLC benchmarker, having a coding rate equal to that of our IrVLC scheme. Serially concatenated and iteratively decoded SBIrVLC-, BBIrVLC-, SBVLC-, and BBVLC-TCM schemes were characterized with the aid of EXIT chart analysis. These schemes have a bandwidth efficiency of 1.56 bits per channel symbol, which corresponds to a Rayleigh fading channel capacity bound of 2.6 dB. Using an average interleaver length of 217500 bits, the SBIrVLC- and BBIrVLC-TCM schemes were found to offer high-quality source sample reconstruction at an E_b/N_0 value of 3.2 dB, which is just 0.6 dB from the capacity bound. This compares favorably with the SBVLC- and BBVLC-TCM benchmarkers, which require an E_b/N_0 value of 3.7 dB. This also compares favorably with a Huffman–IrCC-TCM benchmarker, which requires an E_b/N_0 value of 3.4 dB owing to its slightly eroded performance when operating with the considered interleaver length. Owing to the higher computational complexity of the SBIrVLC-TCM scheme, the BBIrVLC-TCM arrangement was identified as our preferred scheme.

14.7 Digital Terrestrial Video Broadcasting for Mobile Receivers[12]

14.7.1 Background and Motivation

Following the standardization of the pan-European digital video broadcasting (DVB) systems, we have begun to witness the arrival of digital television services to the home. However, for a high proportion of business and leisure travelers, it is desirable to have access to DVB services while on the move. Although it is feasible to access these services with the aid of dedicated DVB receivers, these receivers may also find their way into the laptop computers of the near future. These intelligent laptops may also become the portable DVB receivers of wireless in-home networks.

In recent years three DVB standards have emerged in Europe for terrestrial [499], cable-based [500], and satellite-oriented [501] delivery of DVB signals. The more hostile propagation environment of the terrestrial system requires concatenated Reed–Solomon (RS) [389, 502] and rate-compatible punctured convolutional coding (RCPCC) [389,502] combined with orthogonal frequency division multiplexing (OFDM)-based modulation [222]. In contrast, the more benign cable and satellite-based media facilitate the employment of multilevel modems using up to 256-level quadrature amplitude modulation (QAM) [222]. These schemes are capable of delivering high-definition video at bitrates of up to 20 Mbit/s in stationary broadcast-mode distributive wireless scenarios.

Recently, a range of DVB system performance studies has been presented in the literature [503–506]. Against this background, in this contribution we have proposed turbo-coding improvements to the terrestrial DVB system [499] and investigated its performance

---

[12]This section is based on C. S. Lee, T. Keller, and L. Hanzo, OFDM-based turbo-coded hierarchical and non-hierarchical terrestrial mobile digital video broadcasting, *IEEE Transactions on Broadcasting*, March 2000, pp. 1–22, ©2000 IEEE. Personal use of this material is permitted. However, permission to reprint/republish this material for advertising or promotional purposes or for creating new collective works for resale or redistribution to servers or lists, or to reuse any copyrighted component of this work in other works must, be obtained from the IEEE.

14.7. DVB-T FOR MOBILE RECEIVERS

under hostile mobile channel conditions. We have also studied various video bitstream partitioning and channel coding schemes both in the hierarchical and nonhierarchical transceiver modes and compared their performance.

The rest of this section is divided as follows. In Section 14.7.2 the bit error sensitivity of the MPEG-2 coding parameters [507] is characterized. A brief overview of the enhanced turbo-coded and standard DVB terrestrial scheme is presented in Section 14.7.3, while the channel model is described in Section 14.7.4. Following this, in Section 14.7.5 the reader is introduced to the MPEG-2 data partitioning scheme [508] used to split the input MPEG-2 video bitstream into two error protection classes, which can then be protected either equally or unequally. These two video bit protection classes can then be broadcast to the receivers using the DVB terrestrial hierarchical transmission format [499]. The performance of the data partitioning scheme is investigated by corrupting either the high- or low-sensitivity video bits using randomly distributed errors for a range of system configurations in Section 14.7.6, and their effects on the overall reconstructed video quality are evaluated. Following this, the performance of the improved DVB terrestrial system employing the nonhierarchical and hierarchical format [499] is examined in a mobile environment in Sections 14.7.7 and 14.7.8, before our conclusions and future work areas are presented in Section 14.7.9. We note furthermore that readers interested mainly in the overall system performance may opt to proceed directly to Section 14.7.3. Let us commence our discourse in the next section by describing an objective method of quantifying the sensitivity of the MPEG-2 video parameters.

14.7.2 MPEG-2 Bit Error Sensitivity

At this stage we again note that a number of different techniques can be used to quantify the bit error sensitivity of the MPEG-2 bits. The outcome of these investigations will depend to a degree on the video material used, the output bitrate of the video codec, the objective video-quality measures used, and the averaging algorithm employed. Perceptually motivated, subjective quality-based sensitivity testing becomes infeasible due to the large number of associated test scenarios. Hence, in this section we propose a simplified objective video-quality measure based bit-sensitivity evaluation procedure, which attempts to examine all the major factors influencing the sensitivity of MPEG-2 bits. Specifically, the proposed procedure takes into account the position and the relative frequency of the MPEG-2 parameters in the bitstream, the number of the associated coding bits for each MPEG-2 parameter, the video bitrate, and the effect of loss of synchronization or error propagation due to corrupted bits. Nonetheless, we note that a range of similar bit-sensitivity estimation techniques exhibiting different strengths and weaknesses can be devised. No doubt future research will produce a variety of similarly motivated techniques.

In this section, we assume familiarity with the MPEG-2 standard [507, 508]. The aim of our MPEG-2 error resilience study was to quantify the average PSNR degradation caused by each erroneously decoded video codec parameter in the bitstream, so that appropriate protection can be assigned to each parameter. First, we define three measures, namely, the peak signal-to-noise ratio (PSNR), the PSNR degradation, and the average PSNR degradation, which are to be used in our subsequent discussions. The PSNR is defined

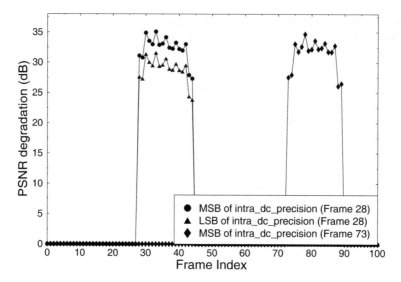

Figure 14.71: PSNR degradation profile for the different bits used to encode the intra_dc_precision parameter [507] in different corrupted video frames for the "Miss America" QCIF video sequence encoded at 30 frame/s and 1.15 Mbit/s.

as follows:

$$\text{PSNR} = 10 \log_{10} \frac{\sum_{n=0}^{N} \sum_{m=0}^{M} 255^2}{\sum_{n=0}^{N} \sum_{m=0}^{M} \Delta^2}, \quad (14.24)$$

where Δ is the difference between the uncoded pixel value and the reconstructed pixel value, while the variables M and N refer to the dimension of the image. The maximum possible 8-bit pixel luminance value of 255 was used in Equation 14.24 in order to mitigate the PSNR's dependence on the video material used. The PSNR degradation is the difference between the PSNR of the decoder's reconstructed image in the event of erroneous decoding and successful decoding. The average PSNR degradation is then the mean of the PSNR degradation values computed for all the image frames of the video test sequence.

Most MPEG-2 parameters are encoded by several bits, and they may occur in different positions in the video sequence. In these different positions, they typically affect the video quality differently, since corrupting a specific parameter of a frame close to the commencement of a new picture start code results in a lesser degradation than corrupting an equivalent parameter further from the resynchronization point. Hence the sensitivity of the MPEG-2 parameters is position-dependent. Furthermore, different encoded bits of the same specific MPEG-2 parameter may exhibit different sensitivity to channel errors. Figure 14.71 shows such an example for the parameter known as intra_dc_precision [507], which is coded under the picture coding extension [508]. In this example, the PSNR degradation profiles due to bit errors being inflicted on the parameter intra_dc_precision of frame 28 showed that the degradation is dependent on the significance of the bit considered. Specifically, errors in the most significant bit (MSB) caused an approximately 3 dB higher PSNR degradation than the least significant bit (LSB) errors. Furthermore, the PSNR degradation due to an MSB error of the intra_dc_precision parameter in frame 73 is similar to the PSNR

14.7. DVB-T FOR MOBILE RECEIVERS

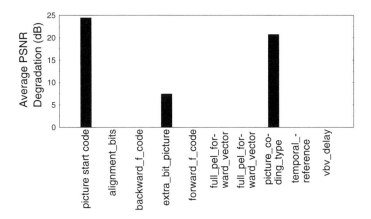

Figure 14.72: Average PSNR degradation for the various MPEG-2 parameters in picture header information for the "Miss America" QCIF video sequence encoded at 30 frame/s and 1.15 Mbit/s.

degradation profile for the MSB of the intra_dc_precision parameter of frame 28. Due to the variation of the PSNR degradation profile for the bits of different significance of a particular parameter, as well as for the same parameter at its different occurrences in the bitstream, it is necessary to determine the *average* PSNR degradation for each parameter in the MPEG-2 bitstream.

Our approach in obtaining the average PSNR degradation was similar to that suggested in the literature [195, 509]. Specifically, the average measure used here takes into account the significance of the bits corresponding to the MPEG-2 parameter concerned, as well as the occurrence of the same parameter at different locations in the encoded video bitstream. In order to find the average PSNR degradation for each MPEG-2 bitstream parameter, the different bits encoding a specific parameter, as well as the bits of the same parameter but occurring at different locations in the MPEG-2 bitstream, were corrupted and the associated PSNR degradation profile versus frame index was registered. The observed PSNR degradation profile generated for different locations of a specific parameter was then used to compute the average PSNR degradation. As an example, we will use the PSNR degradation profile shown in Figure 14.71. This figure presents three degradation profiles. The average PSNR degradation for each profile is first computed in order to produce three average PSNR degradation values corresponding to the three respective profiles. The mean of these three PSNR averages will then form the final average PSNR degradation for the intra_dc_precision parameter. The same process is repeated for all MPEG-2 parameters from the picture layer up to the block layer. The difference with respect to the approach adopted in [195, 509] was that while in [195, 509] the average PSNR degradation was acquired for each bit of the output bitstream, we have adopted a simpler approach in this contribution due to the large number of different parameters within the MPEG-2 bitstream. Figures 14.72 to 14.74 show the typical average PSNR degradations of the various MPEG-2 parameters of the picture header information, picture coding extension, slice layer macroblock layer and block layer [508], which were obtained using the 176×144 quarter common intermediate format (QCIF) "Miss America" (MA) video sequence at 30 frames/s and a high average bitrate of 1.15 Mbit/s.

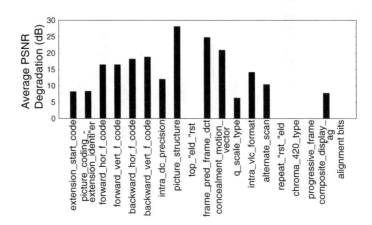

Figure 14.73: Average PSNR degradation for the various MPEG-2 parameters in picture coding extension for the "Miss America" QCIF video sequence encoded at 30 frame/s and 1.15 Mbit/s.

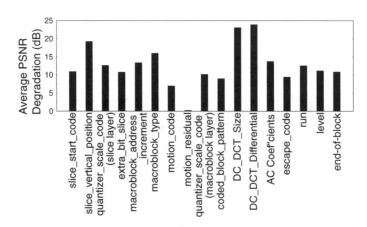

Figure 14.74: Average PSNR degradation for the various MPEG-2 parameters in slice, macroblock, and block layers for the "Miss America" QCIF video sequence encoded at 30 frame/s and 1.15 Mbit/s.

14.7. DVB-T FOR MOBILE RECEIVERS

The different MPEG-2 parameters or code words occur with different probabilities, and they are allocated different numbers of bits. Therefore, the average PSNR degradation registered in Figures 14.72 to 14.74 for each MPEG-2 parameter was multiplied, with the long-term probability of this MPEG-2 parameter occurring in the MPEG-2 bitstream and with the relative probability of bits being allocated to that MPEG-2 parameter. Figures 14.75 and 14.76 show the occurrence probability of the various MPEG-2 parameters characterized in Figures 14.72 to 14.74 and the probability of bits allocated to the parameters in the picture header information, picture coding extension, as well as in the slice, macroblock, and block layers [508], respectively, for the QCIF MA video sequence encoded at 1.15 Mbit/s.

We will concentrate first on Figure 14.75(a). It is observed that all parameters — except for the full_pel_forward_vector, forward_f_code, full_pel_backward_vector, and backward_f_code — have the same probability of occurrence, since they appear once for every coded video frame. The parameters full_pel_forward_vector and forward_f_code have a higher probability of occurrence than full_pel_backward_vector and backward_f_code, since the former two appear in both P frames and B frames, while the latter two only occur in B frames. For our experiments, the MPEG-2 encoder was configured so that for every encoded P frame, there were two encoded B frames. However, when compared with the parameters from the slice layer, macroblock layer, and block layer, which are characterized by the bar chart of Figure 14.75(b), the parameters of the picture header information and picture coding extension appeared significantly less frequently.

If we compare the occurrence frequency of the parameters in the slice layer with those in the macroblock and block layers, the former appeared less often, since there were 11 macroblocks and 44 blocks per video frame slice for the QCIF "Miss America" video sequence were considered in our experiments. The AC discrete cosine transform (DCT) [178] coefficient parameter had the highest probability of occurrence, exceeding 80%.

Figure 14.76 shows the probability of bits being allocated to the various MPEG-2 parameters in the picture header information, picture coding extension, slice, macroblock and block layers [508]. Figure 14.77 was included to more explicitly illustrate the probability of bit allocation seen in Figure 14.76(b), with the probability of allocation of bits to the AC DCT coefficients being omitted from the bar chart. Considering Figure 14.76(a), the two dominant parameters, with the highest number of encoding bits requirement are the picture start code (PSC) and the picture coding extension start code (PCESC). However, comparing these probabilities with the probability of bits being allocated to the various parameters in the slice, macroblock, and block layers, we see that the percentage of bits allocated can still be considered minimal due to their infrequent occurrence. In the block layer, the AC DCT coefficients require in excess of 85% of the bits available for the whole video sequence. However, at bitrates lower than 1.15 Mbit/s the proportion of AC coefficient encoding bits was significantly reduced, as illustrated by Figure 14.78. Specifically, at 30 frame/s and 1.15 Mbit/s, the average number of bits per video frame is about 38,000 and a given proportion of these bits is allocated to the MPEG-2 control header information, motion information, and the DCT coefficients. Upon reducing the total bitrate budget — since the number of control header bits is more or less independent of the target bitrate — the proportion of bits allocated to the DCT coefficients is substantially reduced. This is explicitly demonstrated in Figure 14.78 for bitrates of 1.15 Mbit/s and 240 kbit/s for the "Miss America" QCIF video sequence.

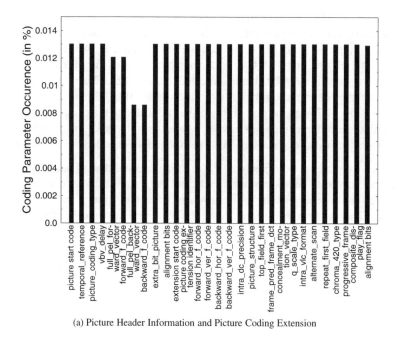

(a) Picture Header Information and Picture Coding Extension

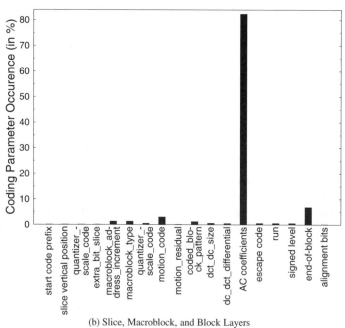

(b) Slice, Macroblock, and Block Layers

Figure 14.75: Occurrence probability for the various MPEG-2 parameters characterized in Figure 14.72 to Figure 14.74. (a) Picture header information and picture coding extension. (b) Slice, macroblock, and block layers for the "Miss America" QCIF video sequence encoded at 30 frame/s and 1.15 Mbit/s.

14.7. DVB-T FOR MOBILE RECEIVERS

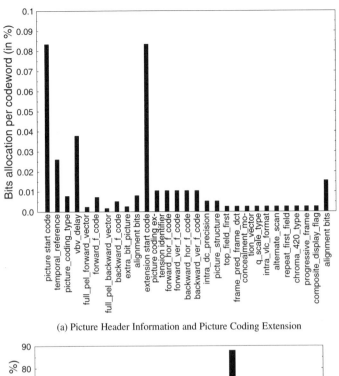

(a) Picture Header Information and Picture Coding Extension

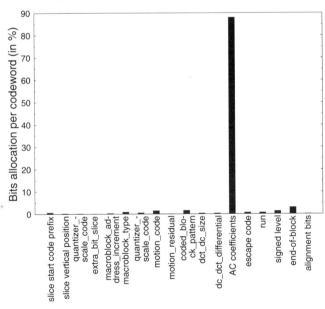

(b) Slice, Macroblock, and Block Layers

Figure 14.76: Probability of bits being allocated to parameters in (a) picture header information and picture coding extension; and (b) Slice, macroblock, and block layers for the "Miss America" QCIF video sequence encoded at 30 frame/s and 1.15 Mbit/s.

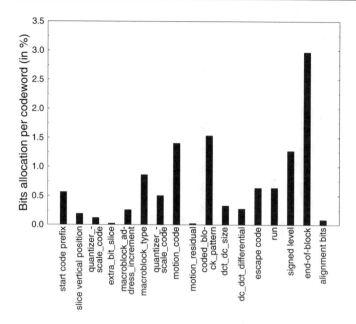

Figure 14.77: Probability of bits being allocated to the various MPEG-2 slice, macroblock, and block layer parameters, as seen in Figure 14.76(b), where the probability of bits allocated to the AC DCT coefficients was omitted in order to show the allocation of bits to the other parameters more clearly. This probability of bits allocation to the various MPEG-2 parameters is associated with the "Miss America" QCIF video sequence encoded at 30 frame/s and 1.15 Mbit/s.

The next process, as discussed earlier, was to normalize the measured average PSNR degradation according to the occurrence probability of the respective MPEG-2 parameters in the bitstream and the probability of bits being allocated to this parameter. The normalized average PSNR degradation caused by corrupting the parameters of the picture header information and picture coding extension [508] is portrayed in Figure 14.79(a). Similarly, the normalized average PSNR degradation for the parameters of the slice, macroblock, and block layers is shown in Figure 14.79(b). In order to visually enhance Figure 14.79(b), the normalized average PSNR degradation for the AC DCT coefficients was omitted in the bar chart shown in Figure 14.80.

The highest PSNR degradation was inflicted by the AC DCT coefficients, since these parameters occur most frequently and hence are allocated the highest number of bits. When a bit error occurs in the bitstream, the AC DCT coefficients have a high probability of being corrupted. The other parameters, such as DC_DCT_size and DC_DCT_differential, exhibited high average PSNR degradations when corrupted, but registered low normalized average PSNR degradations since their occurrence in the bitstream is confined to the infrequent intra-coded frames.

The end-of-block MPEG-2 parameter exhibited the second highest normalized average PSNR degradation in this study. Although the average number of bits used for the end-of-block is only approximately 2.17 bits, the probability of occurrence and the probability of bits being allocated to it are higher than for other parameters, with the exception of the

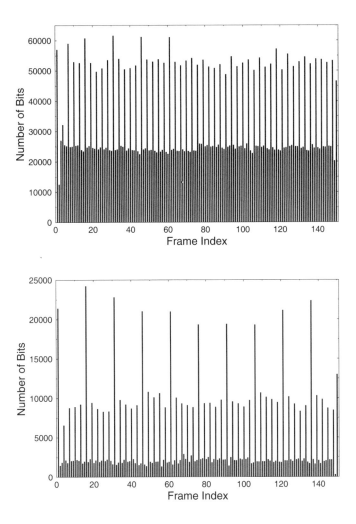

Figure 14.78: Profile of bits allocated to the DCT coefficients, when the 30 frame/s QCIF "Miss America" video sequence is coded at (a) 1.15 Mbit/s (top) and (b) 240 kbit/s (bottom). The sequence of frames is in the order I B B, P B B, P B B, P B B, and so on.

AC DCT coefficients. Furthermore, in general, the parameters of the slice, macroblock, and block layers exhibit higher average normalized PSNR degradations due to their more frequent occurrence in the bitstream compared to the parameters that belong to the picture header information and to the picture coding extension. This also implies that the percentage of bits allocated to these parameters is higher.

Comparing the normalized average PSNR degradations of the parameters in the picture header information and picture coding extension, we observe that the picture start code (PSC) exhibits the highest normalized average PSNR degradation. Although most of the parameters here occur with equal probability as seen in Figure 14.74(a), the picture start code

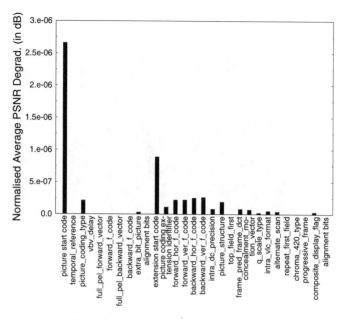

(a) Picture Header Information and Picture Coding Extension

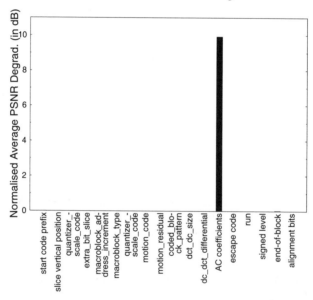

(b) Slice, Macroblock, and Block Layers

Figure 14.79: Normalized average PSNR degradation for the various parameters in (a) picture header information and picture coding extension (b) slice, macroblock, and block layers, normalized to the occurrence probability of the respective parameters in the bitstream and the probability of bits being allocated to the parameter for the "Miss America" QCIF video sequence encoded at 30 frame/s and 1.15 Mbit/s.

14.7. DVB-T FOR MOBILE RECEIVERS

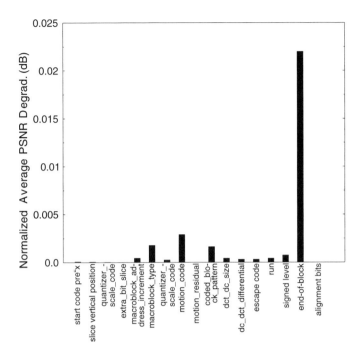

Figure 14.80: This bar chart is the same as Figure 14.79(b), although the normalized average PSNR degradation for the AC DCT coefficients was omitted in order to show the average PSNR degradation of the other parameters. This bar chart is presented for the "Miss America" QCIF video sequence encoded at 30 frame/s and 1.15 Mbit/s case.

requires a higher portion of the bits compared to the other parameters, with the exception of the extension start code. Despite having the same probability of occurrence and the same allocation of bits, the extension start code exhibits a lower normalized PSNR degradation than the picture start code, since its average unnormalized degradation is lower, as shown in Figure 14.72 to Figure 14.74.

In Figures 14.79 and 14.80, we observed that the video PSNR degradation was dominated by the erroneous decoding of the AC DCT coefficients, which appeared in the MPEG-2 video bitstream in the form of variable-length codewords. This suggests that unequal error protection techniques be used to protect the MPEG-2 parameters during transmission. In a low-complexity implementation, two protection classes may be envisaged. The higher priority class would contain all the important header information and some of the more important low-frequency variable-length coded DCT coefficients. The lower priority class would then contain the remaining less important, higher frequency variable-length coded DCT coefficients. This partitioning process will be detailed in Section 14.7.5 together with its associated performance in the context of the hierarchical DVB [499] transmission scheme in Section 14.7.8. Let us, however, first consider the architecture of the investigated DVB system in the next section.

14.7.3 DVB Terrestrial Scheme

The block diagram of the DVB terrestrial (DVB-T) transmitter [499] is shown in Figure 14.81, which consists of an MPEG-2 video encoder, channel-coding modules, and an orthogonal frequency division multiplex (OFDM) modem [222,510]. The bitstream generated by the MPEG-2 encoder is packetized into frames 188 bytes long. The video data in each packet is then randomized by the scrambler of Figure 14.81. The specific details concerning the scrambler have not been included in this chapter since they may be obtained from the DVB-T standard [499].

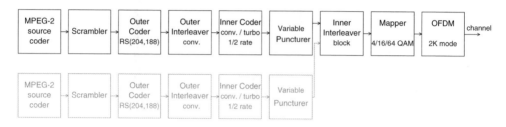

Figure 14.81: Schematic of the DVB terrestrial transmitter functions.

Because of the poor error resilience of the MPEG-2 video codec, powerful concatenated channel coding is employed. The concatenated channel codec of Figure 14.81 comprises a shortened Reed–Solomon (RS) outer code and an inner convolutional encoder. The 188-byte MPEG-2 video packet is extended by the Reed–Solomon encoder [389, 502], with parity information to facilitate error recovery in order to form a 204-byte packet. The Reed–Solomon decoder can then correct up to 8 erroneous bytes for each 204-byte packet. Following this, the RS-coded packet is interleaved by a convolutional interleaver and further protected by a half-rate inner convolutional encoder using a constraint length of 7 [389, 502].

Furthermore, the overall code rate of the concatenated coding scheme can be adapted by variable puncturing that supports code rates of $1/2$ (no puncturing) as well as $2/3, 3/4, 5/6$, and $7/8$. The parameters of the convolutional encoder are summarized in Table 14.16.

Table 14.16: Parameters of the $CC(n, k, K)$ Convolutional Inner Encoder of the DVB-T Modem

Convolutional Coder Parameters	
Code rate	1/2
Constraint length	7
n	2
k	1
Generator polynomials (octal format)	171, 133

If only one of the two branches of the transmitter in Figure 14.81 is utilized, the DVB-T modem is said to be operating in its nonhierarchical mode. In this mode, the modem can have a choice of QPSK, 16QAM, or 64QAM modulation constellations [222].

14.7. DVB-T FOR MOBILE RECEIVERS

Table 14.17: Parameters of the OFDM Module Used in the DVB-T Modem [499]

OFDM Parameters	
Total number of subcarriers	2048 (2 K mode)
Number of effective subcarriers	1705
OFDM symbol duration T_s	224 μs
Guard interval	$T_s/4 = 56$ μs
Total symbol duration (inc. guard interval)	280 μs
Consecutive subcarrier spacing $1/T_s$	4464 Hz
DVB channel spacing	7.61 MHz
QPSK and QAM symbol period	7/64 μs

A second video bitstream can also be multiplexed with the first one by the inner interleaver, when the DVB modem is in its hierarchical mode [499]. The choice of modulation constellations in this mode is between 16QAM and 64QAM. We employ this transmission mode when the data partitioning scheme of Section 14.7.5 is used to split the incoming MPEG-2 video bitstream into two video bit-protection classes, with one class having a higher grade of protection or priority than the other one. The higher priority video bits will be mapped to the MSBs of the modulation constellation points and the lower priority video bits to the LSBs of the QAM-constellation [222]. For 16QAM and 64QAM, the two MSBs of each 4-bit or 6-bit QAM symbol will contain the more important video data. The lower priority video bits will then be mapped to the lower significance 2 bits and 4 bits of 16QAM and 64QAM, respectively [222].

These QPSK, 16QAM, or 64QAM symbols are then distributed over the OFDM carriers [222]. The parameters of the OFDM system are presented in Table 14.17.

Besides implementing the standard DVB-T system as a benchmark, we have improved the system by replacing the convolutional coder with a turbo codec [401, 402]. The turbo codec's parameters used in our investigations are displayed in Table 14.18. The block diagram of the turbo encoder is shown in Figure 14.82. The turbo encoder is constructed of two component encoders. Each component encoder is a half-rate convolutional encoder whose parameters are listed in Table 14.18. The two-component encoders are used to encode the same input bits, although the input bits of the second component encoder are interleaved before encoding. The output bits of the two-component codes are punctured and multiplexed in order to form a single-output bitstream. The component encoder used here is known as a half-rate recursive systematic convolutional encoder (RSC) [511]. It generates one parity bit and one systematic output bit for every input bit. In order to provide an overall coding rate of $R = 1/2$, half the output bits from the two encoders must be punctured. The puncturing arrangement used in our work is to transmit all the systematic bits from the first encoder and every other parity bit from both encoders [512]. We note here that one iteration of the turbo decoder involves two Logarithmic Maximum A-Posteriori (log-MAP) [452] decoding operations, which we repeated for the eight iterations. Hence, the total turbo decoding complexity is about 16 times higher than a constraint length $K = 3$ constituent convolutional decoding. Therefore, the turbo decoder exhibits a similar complexity to the $K = 7$ convolutional decoder.

Table 14.18: Parameters of the Inner Turbo Encoder Used to Replace the DVB-T System's $K = 7$ Convolutional Encoder of Table 14.16 (RSC: recursive systematic code)

Turbo Coder Parameters	
Turbo code rate	1/2
Input block length	17, 952 bits
Interleaver type	Random
Number of turbo-decoder iterations	8
Turbo Encoder Component Code Parameters	
Component code encoder type	Recursive Systematic Convolutional (RSC)
Component code decoder type	log-MAP [452]
Constraint length	3
n	2
k	1
Generator polynomials (octal format)	7, 5

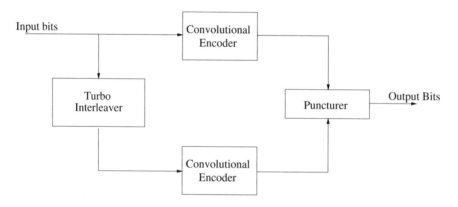

Figure 14.82: Block diagram of turbo encoder

In this section, we have given an overview of the standard and enhanced DVB-T system, which we have used in our experiments. Readers interested in further details of the DVB-T system are referred to the DVB-T standard [499]. The performance of the standard DVB-T system and the turbo-coded system is characterized in Sections 14.7.7 and 14.7.8 for nonhierarchical and hierarchical transmissions, respectively. Let us now briefly consider the multipath channel model used in our investigations.

14.7.4 Terrestrial Broadcast Channel Model

The channel model employed in this study was the 12-path COST 207 [513] hilly terrain (HT) type impulse response, with a maximal relative path delay of 19.9 μs. This channel

14.7. DVB-T FOR MOBILE RECEIVERS

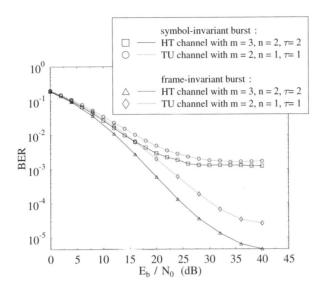

Figure 14.83: COST 207 hilly terrain (HT) type impulse response.

was selected in order to provide a worst-case propagation scenario for the DVB-T system employed in our study.

In the system described here, we have used a carrier frequency of 500 MHz and a sampling rate of 7/64 μs. Each of the channel paths was faded independently obeying a Rayleigh-fading distribution, according to a normalized Doppler frequency of 10^{-5} [389]. This corresponds to a worst-case vehicular velocity of about 200 km/h. The unfaded impulse response is depicted in Figure 14.83. For the sake of completeness we note that the standard COST 207 channel model was defined in order to facilitate the comparison of different GSM implementations [389] under identical conditions. The associated bitrate was 271 kbit/s, while in our investigations the bitrate of DVB-quality transmissions can be as high as 20 Mbit/s, where a higher number of resolvable multipath components within the dispersion-range is considered. However, the performance of various wireless tranceivers is well understood by the research community over this standard COST 207 channel. Hence, its employment is beneficial in benchmarking terms. Furthermore, since the OFDM modem has 2048 subcarriers, the subcarrier signaling rate is effectively 2000-times lower than our maximum DVB-rate of 20 Mbit/s, corresponding to 10 kbit/s. At this subchannel rate, the individual subchannel can be considered nearly frequency-flat. In summary, in conjunction with the 200 km/h vehicular speed the investigated channel conditions constitute a pessimistic scenario.

In order to facilitate unequal error protection, the data partitioning procedure of the MPEG-2 video bitstream is considered next.

14.7.5 Data Partitioning Scheme

Efficient bitstream partitioning schemes for H.263-coded video were proposed, for example, by Gharavi and Alamouti [514], and were evaluated in the context of the third-generation

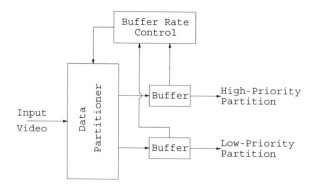

Figure 14.84: Block diagram of the data partitioner and rate controller.

mobile radio standard proposal known as IMT-2000 [389]. As portrayed in Figures 14.79 and 14.80, the corrupted variable-length coded DCT coefficients produce a high video PSNR degradation. Assuming that all MPEG-2 header information is received correctly, the fidelity of the reconstructed images at the receiver is dependent on the number of correctly decoded DCT coefficients. However, the subjective effects of losing higher spatial frequency DCT coefficients are less dramatic compared to those of the lower spatial frequency DCT coefficients. The splitting of the MPEG-2 video bitstream into two different integrity bitstreams is termed data partitioning [508]. Recall from Section 14.7.3 that the hierarchical 16-QAM and 64-QAM DVB-T transmission scheme enables us to multiplex two unequally protected MPEG-2 video bitstreams for transmission. This section describes the details of the MPEG-2 data partitioning scheme [508].

Figure 14.84 shows the block diagram of the data partitioning scheme, which splits an MPEG-2 video bitstream into two resultant bitstreams. The position at which the MPEG-2 bitstream is split is based on a variable referred to as the priority breakpoint (PBP) [508]. The PBP can be adjusted at the beginning of the encoding of every MPEG-2 image slice, based on the buffer occupancy or fullness of the two output buffers. For example, if the high-priority buffer is 80% full and the low-priority buffer is only 40% full, the rate-control module will have to adjust the PBP so that more data is directed to the low-priority partition. This measure is taken to avoid high-priority buffer overflow and low-priority buffer underflow events. The values for the MPEG-2 PBP are summarized in Table 14.19 [508].

There are two main stages in updating the PBP. The first stage involves the rate-control module of Figure 14.84 in order to decide on the preferred new PBP value for each partition based on its individual buffer occupancy and on the current value of the PBP. The second stage then combines the two desired PBPs based on the buffer occupancy of both buffers in order to produce a new PBP.

The updating of the PBP in the first stage of the rate control module is based on a heuristic approach, similar to that suggested by Aravind *et al.* [515]. The update procedure is detailed in Algorithm 1, which is discussed below and augmented by a numerical example at the end of this section.

The variable "sign" is used in Algorithm 1, in order to indicate how the PBP has to be adjusted in the high- and low-priority MPEG-2 partitions, so as to arrive at the required target buffer occupancy. More explicitly, the variable "sign" in Algorithm 1 is necessary because

Table 14.19: Priority Breakpoint Values and the Associated MPEG-2 Parameters that will be Directed to the High-priority Partition [508]. A Higher PBP Directs More Parameters to the High-priority Partition. In Contrast, for the Low-priority Partition a Higher PBP Implies Obtaining Less Data

PBP	Syntax Elements in High-priority Partition
0	Low-priority partition always has its PBP set to 0.
1	Sequence, GOP, picture and slice layer information up to extra bit slice.
2	Same as above and up to macroblock address increment.
3	Same as above plus including macroblock syntax elements but excluding coded block pattern.
4–63	Reserved for future use.
64	Same as above plus including DC DCT coefficient and the first run-length coded DCT coefficient.
65	Same as above and up to the second run-length coded DCT coefficient.
$64 + x$	Same as above and up to x run-length coded DCT coefficient.
127	Same as above and up to 64 run-length coded DCT coefficient.

the MPEG-2 PBP values [508] shown in Table 14.19 indicate the amount of information that should be directed to the high-priority partition. Therefore, if the low-priority partition requires more data, then the new PBP must be lower than the current PBP. In contrast, for the high-priority partition a higher PBP implies obtaining more data.

Once the desired PBPs for both partitions have been acquired with the aid of Algorithm 1, Algorithm 2 is used to compute the final PBP for the current MPEG-2 image slice. The inner workings of these algorithms are augmented by a numerical example at the end of this section. There are two main cases to consider in Algorithm 2. The first one occurs when both partitions have a buffer occupancy of less than 50%. By using the reciprocal of the buffer occupancy in Algorithm 2 as a weighting factor during the computation of the PBP adjustment value "delta", the algorithm will favor the new PBP decision of the less occupied buffer in order to fill the buffer with more data in the current image slice. This is simply because the buffer is closer to underflow; hence, increasing the PBP according to its instructions will assist in preventing the particular buffer from underflowing. On the other hand, when both buffers experience a buffer occupancy of more than 50%, the buffer occupancy itself is used as a weighting factor instead. Now the algorithm will instruct the buffer having a higher occupancy to adjust its desired PBP so that less data is inserted into it in the current MPEG-2 image slice. Hence, buffer overflow problems are alleviated with the aid of Algorithm 1 and Algorithm 2.

The new PBP value is then compared to its legitimate range tabulated in Table 14.19. Furthermore, we restricted the minimum PBP value so that I-, P-, and B-pictures have minimum PBP values of 64, 3, and 2, respectively. Since B-pictures are not used for future predictions, it was decided that their data need not be protected as strongly as the data for I- and P-pictures. As for P-pictures, Ghanbari and Seferidis [324] showed that correctly decoded motion vectors alone can still provide a subjectively pleasing reconstruction of the image, even if the DCT coefficients were discarded. Hence, the minimum MPEG-2

Algorithm 1 Computes the desired PBP update for the high- and low-priority partitions which is then passed to Algorithm 2, in order to determine the PBP to be set for the current image slice.

Step 1: Initialize parameters
   if High Priority Partition **then**
      sign := +1
   else
      sign := −1
   end if

Step 2:
   if buffer occupancy ≥ 80% **then**
      diff := 64 − PBP
   end if

   if buffer occupancy ≥ 70% **and** buffer occupancy < 80% **then**
      if PBP ≥ 100 **then**
         diff := −9
      end if
      if PBP ≥ 80 **and** PBP < 100 **then**
         diff := −5
      end if
      if PBP ≥ 64 **and** PBP < 80 **then**
         diff := −2
      end if
   end if

   if buffer occupancy ≥ 50% **and** buffer occupancy < 70% **then**
      diff := +1
   end if

   if buffer occupancy < 50% **then**
      if PBP ≥ 80 **then**
         diff := +1
      end if
      if PBP ≥ 70 **and** PBP < 80 **then**
         diff := +2
      end if
      if PBP ≥ 2 **and** PBP < 70 **then**
         diff := +3
      end if
   end if

Step 3:
   diff := sign × diff
   Return diff

---

bitstream splitting point or PBP for P-pictures has been set to be just before the coded block pattern parameter, which would then ensure that the motion vectors would be mapped to the high-priority partition. Upon receiving corrupted DCT coefficients, they would be set to zero, which corresponds to setting the motion-compensated error residual of the macroblock concerned to zero. For I-pictures, the fidelity of the reconstructed image is dependent on the number of DCT coefficients that can be decoded successfully. Therefore, the minimum MPEG-2 bitstream splitting point or PBP was set to include at least the first run-length-coded DCT coefficient.

14.7. DVB-T FOR MOBILE RECEIVERS

Algorithm 2 Computes the new PBP for the current image slice based on the current buffer occupancy of both partitions

Step 1:

if   Occupancy$_{HighPriority} < 50\%$ and   Occupancy$_{LowPriority} < 50\%$
or   Occupancy$_{HighPriority} = 50\%$ and   Occupancy$_{LowPriority} < 50\%$
or   Occupancy$_{HighPriority} < 50\%$ and   Occupancy$_{LowPriority} = 50\%$
or   Occupancy$_{HighPriority} < 25\%$ and $50\% <$ Occupancy$_{LowPriority} < 70\%$
or $50\% <$ Occupancy$_{HighPriority} < 70\%$ and   Occupancy$_{LowPriority} < 25\%$

then

$$\text{delta} := \frac{\text{Occupancy}_{HighPriority}^{-1} \times \text{diff}_{HighPriority} + \text{Occupancy}_{LowPriority}^{-1} \times \text{diff}_{LowPriority}}{\text{Occupancy}_{HighPriority}^{-1} + \text{Occupancy}_{LowPriority}^{-1}}$$

else

$$\text{delta} := \frac{\text{Occupancy}_{HighPriority} \times \text{diff}_{HighPriority} + \text{Occupancy}_{LowPriority} \times \text{diff}_{LowPriority}}{\text{Occupancy}_{HighPriority} + \text{Occupancy}_{LowPriority}}$$

end if

Step 2:
New_PBP := Previous_PBP + $\lceil$delta$\rceil$ where $\lceil \rceil$ means rounding up to the nearest integer
Return New_PBP

Below we demonstrate the operation of Algorithm 1 and Algorithm 2 with the aid of a simple numerical example. We will assume that the PBP prior to the update is 75 and that the buffer occupancy for the high- and low-priority partition buffers is 40% and 10%, respectively. Considering the high-priority partition, according to the buffer occupancy of 40% Algorithm 1 will set the desired PBP update difference denoted by "diff" for the PBP to $+2$. This desired update is referred to as diff$_{HighPriority}$ in Algorithm 2. For the low-priority partition, according to the buffer occupancy of 10%, Algorithm 1 will set the desired update for the PBP to -2, since the sign of diff is changed by Algorithm 1. The desired PBP update for the low-priority partition is referred to as diff$_{LowPriority}$ in Algorithm 2. Since the occupancy of both partition buffers' is less than 50%, Algorithm 2 will use the reciprocal of the buffer occupancy as the weighting factor, which will then favor the desired update of the low-priority partition due to its 10% occupancy. The final update value, which is denoted by delta in Algorithm 2, is equal to -2 (after being rounded up). Hence, according to Step 2 of Algorithm 2, the new PBP is 73. This means that for the current MPEG-2 image slice more data will be directed into the low-priority partition in order to prevent buffer underflow since PBP was reduced from 75 to 73 according to Table 14.19.

Apart from adjusting the PBP values from one MPEG-2 image slice to another to avoid buffer underflow or overflow, the output bitrate of each partition buffer must be adjusted so that the input bitrate of the inner interleaver and modulator in Figure 14.81 is properly matched between the two partitions. Specifically, in the 16QAM mode the two modem subchannels have an identical throughput of 2 bits per 4-bit symbol. In contrast, in the 64QAM mode there are three 2-bit subchannels per 6-bit 64QAM symbol, although the standard [499] recommends using a higher-priority 2-bit and a lower-priority 4-bit subchannels. Hence, it is imperative to take into account the redundancy added by forward

Table 14.20: The Bitrate Partitioning Ratios Based on the Modulation Mode and Code Rates Selected for the DVB-T Hierarchical Mode. The Line in Bold Corresponds to our Worked Example

Modulation	Conv. Code Rate (High Priority)	Conv. Code Rate (Low Priority)	Ratio (High - B1 : Low - B2)
16QAM	1/2	1/2	1 : 1
	1/2	2/3	3 : 4
	1/2	3/4	2 : 3
	1/2	5/6	3 : 5
	1/2	7/8	4 : 7
	2/3	1/2	4 : 3
64QAM	1/2	1/2	1 : 2
	1/2	2/3	3 : 8
	1/2	**3/4**	**1 : 3**
	1/2	5/6	3 : 10
	1/2	7/8	2 : 7
	2/3	1/2	2 : 3

error correction (FEC), especially when the two partitions' FECs operate at different code rates. Figure 14.85 shows a block diagram of the DVB-T system operating in the hierarchical mode and receiving its input from the video partitioner. The FEC module represents the concatenated coding system, consisting of a Reed–Solomon codec [389] and a convolutional codec [389]. The modulator can invoke both 16QAM and 64QAM [222]. We shall now use an example to illustrate the choice of the various partitioning ratios summarized in Table 14.20.

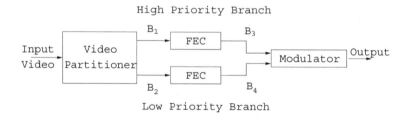

Figure 14.85: Video partitioning for the DVB-T system operating in hierarchical mode.

We shall assume that 64QAM is selected and the high-and low-priority video partitions employ $\frac{1}{2}$ and $\frac{3}{4}$ convolutional codes, respectively. This scenario is portrayed in the third line of the 64QAM section of Table 14.20. We do not have to take the Reed–Solomon code rate into account, since both partitions invoke the same Reed–Solomon codec. Based on these facts and on reference to Figure 14.85, the input bitrates B_3 and B_4 of the modulator must be in the ratio 1:2, since the two MSBs of the 64-QAM constellation are assigned to the high-priority video partition and the remaining four bits to the low-priority video partition.

Table 14.21: Summary of the Three Schemes Employed in our Investigations into the Performance of the Data Partitioning Scheme. The FEC-coded High-priority Video Bit-Stream B3, as Shown in Figure 14.85, was Mapped to the High-priority 16QAM Subchannel, while the Low-priority B4-stream to the Low Priority 16QAM Subchannel

	Modulation	Conv. Code Rate (High Prior. – B1)	Conv. Code Rate (Low Prior. – B2)	Ratio (High : Low) (B1 : B2)
Scheme 1	16QAM	1/2	1/2	1 : 1
Scheme 2	16QAM	1/3	2/3	1 : 2
Scheme 3	16QAM	2/3	1/3	2 : 1

At the same time, the ratio of B_3 to B_4 is related to the ratio of B_1 to B_2 with the FEC redundancy taken into account, requiring

$$\begin{aligned}
\frac{B_3}{B_4} &= \frac{2 \times B_1}{\frac{4}{3} \times B_2} \stackrel{64-QAM}{=} \frac{1}{2} \\
&= \frac{3}{2} \cdot \frac{B_1}{B_2} \stackrel{64-QAM}{=} \frac{1}{2} \\
\frac{B_1}{B_2} &= \frac{1}{2} \times \frac{2}{3} \\
&= \frac{1}{3}.
\end{aligned} \quad (14.25)$$

If, for example, the input video bitrate of the data partitioner module is 1 Mbit/s, the output bitrate of the high- and low-priority partition would be $B_1 = 250$ kbit/s and $B_2 = 750$ kbit/s, respectively, according to the ratio indicated by Equation 14.25.

In this section, we have outlined the operation of the data partitioning scheme which we used in the DVB-T hierarchical transmission scheme. Its performance in the context of the overall system will be characterized in Section 14.7.8. Let us, however, first evaluate the BER sensitivity of the partitioned MPEG-2 bitstream to randomly distributed bit errors using various partitioning ratios.

14.7.6 Performance of the Data Partitioning Scheme

Let us consider the 16QAM modem and refer to the equally split rate $\frac{1}{2}$ convolutional coded high- and low-priority scenario as Scheme 1. Furthermore, the 16QAM rate $\frac{1}{3}$ convolutional coded high priority data and rate $\frac{2}{3}$ convolutional coded low-priority data-based scenario is referred to here as Scheme 2. Lastly, the 16QAM rate $\frac{2}{3}$ convolutional coded high-priority data and rate $\frac{1}{3}$ coded low-priority databased partitioning scheme is termed Scheme 3. We then programmed the partitioning scheme of Figure 14.85 for maintaining the required splitting ratio B_1/B_2, as seen in Table 14.21. This was achieved by continuously adjusting the PBP using Algorithms 1 and 2. The 704×576-pixel "Football" high-definition television (HDTV) video sequence was used in these investigations.

Figures 14.86 to 14.88 show the relative frequency at which a particular PBP value occurs for each image of the "Football" video sequence for the three different schemes of Table 14.21 mentioned earlier. The reader may recall from Table 14.19 that the PBP values indicate the proportion of encoded video parameters, which are to be directed into the high-priority partition. As the PBP value increases, the proportion of video data mapped to the high-priority partition increases and vice versa. Comparing Figures 14.86 to 14.88, we observe that Scheme 3 has the most data in the high-priority partition associated with the high PBPs of Table 14.19, followed by Scheme 1 and Scheme 2. This observation can be explained as follows. We shall consider Scheme 3 first. In this scheme, the high-priority video bits are protected by a rate $\frac{2}{3}$ convolutional code and mapped to the higher integrity 16QAM subchannel. In contrast, the low-priority video bits are encoded by a rate $\frac{1}{3}$ convolutional code and mapped to the lower integrity 16QAM subchannel. Again, assuming that 16QAM is used in our experiment according to line 3 of Table 14.21, $\frac{2}{3}$ of the video bits will be placed in the high-priority 16QAM partition and the remaining video bits in the low-priority 16QAM partition, following the approach of Equation 14.25. The BER difference of the 16QAM subchannels depend on the channel error statistics, but the associated BERs are about a factor of 2–3 different [222]. In contrast to Scheme 3, Scheme 2 will have $\frac{1}{3}$ of the video bits placed in the high-priority 16QAM partition, and the remaining $\frac{2}{3}$ of the video bits mapped to the low-priority 16QAM partition, according to line 2 of Table 14.21. Lastly, Scheme 1 will have half of the video bits in the high- and low-priority 16QAM partitions, according to line 1 of Table 14.21. This explains our observation in the context of Scheme 3 in Figure 14.88, where a PBP value as high as 80 is achieved in some image frames. However, each PBP value encountered has a lower probability of being selected, since the total number of 3600 occurrences associated with investigated 3600 MPEG-2 video slices per 100 image frames is spread over a higher variety of PBPs. Hence, Scheme 3 directs about $\frac{2}{3}$ of the original video bits after $\frac{2}{3}$ rate coding to the high-priority 16QAM subchannel. This observation is in contrast to Scheme 2 of Figure 14.87, where the majority of the PBPs selected are only up to the value of 65. This indicates that about $\frac{2}{3}$ of the video bits are concentrated in the lower priority partition, as indicated in line 2 of Table 14.21.

Figures 14.89(a) to 14.91(a) show the average probability at which a particular PBP value is selected by the rate control scheme, as discussed in Section 14.7.5, during the encoding of the video sequence. Again, we observe that Scheme 3 encounters the widest range of PBP values, followed by Scheme 1 and Scheme 2, respectively. According to Table 14.21, these schemes map a decreasing number of bits to the high-priority partition in this order.

We then quantified the error sensitivity of the partitioning Schemes 1 to 3 characterized in Table 14.21, when each partition was subjected to randomly distributed bit errors, although in practice the error distribution will depend on the fading channel's characteristics. Specifically, the previously defined average PSNR degradation was evaluated for given error probabilities inflicting random errors imposed on one of the partitions, while keeping the other partition error-free. These results are portrayed in Figures 14.89(b), 14.90(b) and 14.91(b) for Schemes 1 to 3, respectively.

Comparing Figures 14.89(b) to 14.91(b), we observe that the average PSNR degradation exhibited by the three schemes of Table 14.21, when only their high-priority partitions are corrupted, is similar. The variations in the average PSNR degradation in these cases are caused by the different quantity of sensitive video bits, which resides in the high priority partition. If we compare the performance of the schemes summarized in Table 14.21 at a BER

14.7. DVB-T FOR MOBILE RECEIVERS 581

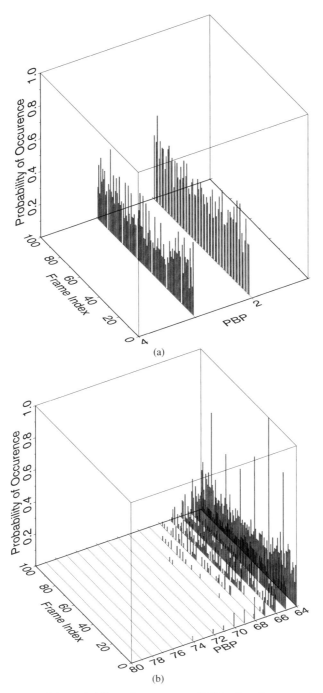

Figure 14.86: Evolution of the probability of occurrence of PBP values from one picture to another of the 704 × 576-pixel "Football" video sequence for Scheme 1 of Table 14.21.

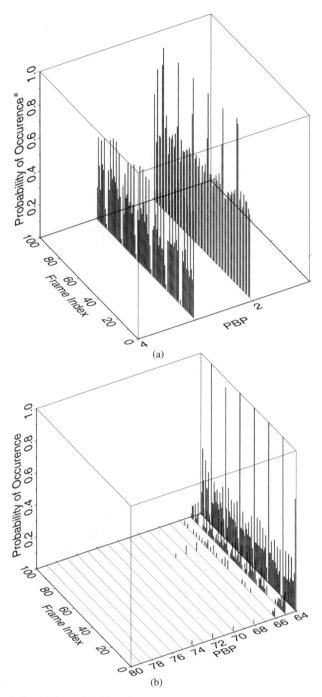

Figure 14.87: Evolution of the probability of occurrence of PBP values from one picture to another of the 704 × 576-pixel "Football" video sequence for Scheme 2 of Table 14.21.

14.7. DVB-T FOR MOBILE RECEIVERS

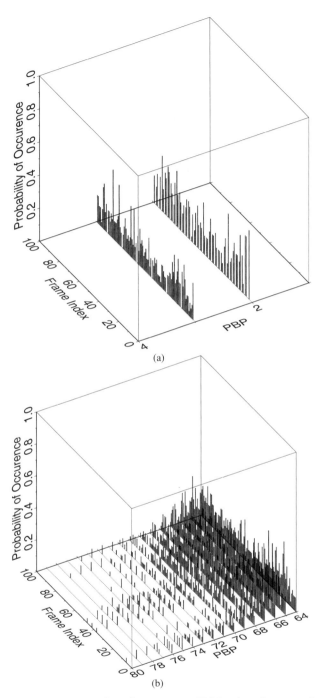

Figure 14.88: Evolution of the probability of occurrence of PBP values from one picture to another of the 704 × 576-pixel "Football" video sequence for Scheme 3 of Table 14.21.

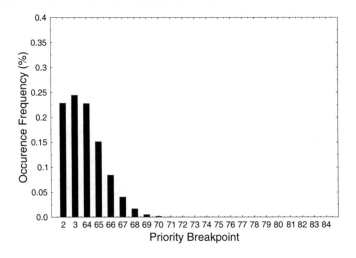

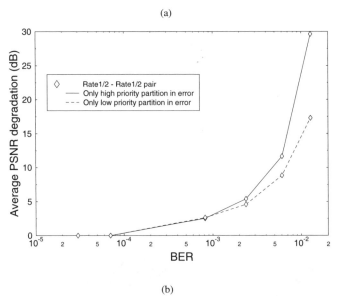

Figure 14.89: (a) Histogram of the probability of occurrence for various priority breakpoints and (b) average PSNR degradation versus BER for rate 1/2 convolutional coded high- and low-priority data in Scheme 1 of Table 14.21.

of 2×10^{-3}, Scheme 3 experienced approximately 8.8 dB average video PSNR degradation, while Schemes 1 and 2 exhibited approximately 5 dB degradation. This trend was expected, since Scheme 3 had the highest portion of the video bits, namely, $\frac{2}{3}$ residing in the high-priority partition, followed by Scheme 1 hosting $\frac{1}{2}$ and Scheme 2 having $\frac{1}{3}$ of the bits in this partition.

14.7. DVB-T FOR MOBILE RECEIVERS

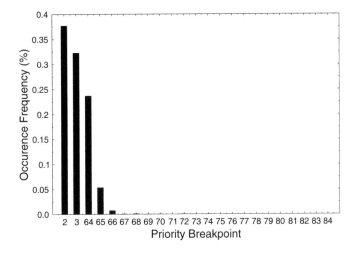

(a)

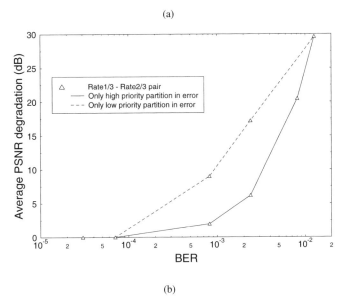

(b)

Figure 14.90: (a) Histogram of the probability of occurrence for various priority breakpoints and (b) average PSNR degradation versus BER for the rate 1/3 convolutional coded high-priority data and rate 2/3 convolutional coded low-priority data in Scheme 2 of Table 14.21.

On the other hand, we can observe a significant difference in the average PSNR degradation measured for Schemes 1 to 3 of Table 14.21, when only the low priority partitions are corrupted by comparing the curves shown as broken lines in Figures 14.89(b) to 14.91(b). Under this condition, Scheme 2 experienced approximately 16 dB average video PSNR degradation at a BER of 2×10^{-3}. In contrast, Scheme 1 exhibited an approximately 4 dB average video PSNR degradation, while Scheme 3 experienced about 7.5 dB degradation at this BER. The scheme with the highest portion of video bits in the lower priority partition

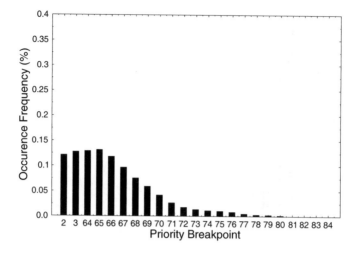

(a)

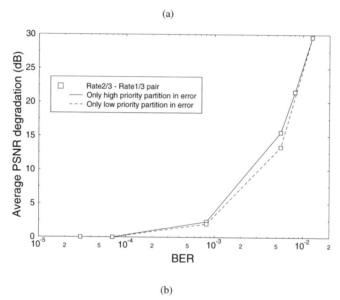

(b)

Figure 14.91: (a) Histogram of the probability of occurrence for various priority breakpoints and (b) average PSNR degradation versus BER for the rate 2/3 convolutional coded high-priority data and rate 1/3 convolutional coded low-priority data in Scheme 3 of Table 14.21.

(i.e., Scheme 2) experienced the highest average video PSNR degradation. This observation correlates well with our earlier findings in the context of the high-priority partition scenario, where the partition holding the highest portion of the video bits in the error-impaired partition exhibited the highest average PSNR degradation.

Having discussed our observations for the three schemes of Table 14.21 from the perspective of the relative amount of video bits in one partition compared to the other, we

14.7. DVB-T FOR MOBILE RECEIVERS 587

now examine the data partitioning process further in order to relate them to our observations. Figure 14.92 shows a typical example of an MPEG-2 video bitstream both prior to and after data partitioning. There are two scenarios to be considered here, namely, intra-frame and inter-frame coded macroblock partitioning. We have selected the PBP value of 64 from Table 14.19 for the intra-frame coded macroblock scenario and the PBP value of 3 for the inter-frame coded macroblock scenario, since these values have been selected frequently by the rate-control arrangement for Schemes 1 and 2. This is evident from Figures 14.86 and 14.87 as well as from Figures 14.89(a) and 14.90(a). With the aid of Table 14.19 and Figure 14.92, this implies that only the macroblock (MB) header information and a few low-frequency DCT coefficients will reside in the high-priority partition, while the rest of the DCT coefficients will be stored in the low-priority partition. These can be termed as base layer and enhancement layer, as seen in Figure 14.92. In the worst-case scenario, where the entire enhancement layer or low-priority partition data are lost due to a transmission error near the beginning of the associated low-priority bitstream, the MPEG-2 video decoder will only have the bits of the high-priority partition in order to reconstruct the encoded video sequence. Hence, the MPEG-2 decoder cannot reconstruct good-quality images. Although the results reported by Ghanbari and Seferidis [324] suggest that adequate video reconstruction is possible, provided that the motion vectors are correctly decoded, this observation is only true if the previous intra-coded frame is correctly reconstructed. If the previous intra-coded frame contains artifacts, these artifacts will be further propagated to forthcoming video frames by the motion vectors. By attempting to provide higher protection for the high-priority partition or base layer, we have indirectly forced the rate-control scheme of Section 14.7.5 to reduce the proportion of video bits directed into the high-priority partition under the constraint of a given fixed bitrate, which is imposed by the 16QAM subchannels.

In order to elaborate a little further, at a BER of 2×10^{-3}, Scheme 1 in Figure 14.89(a) exhibited a near-identical PSNR degradation for the high- and low-priority video bits. When assigning more bits to the low-priority partition, in order to accommodate a stronger FEC code in the high-priority partition, an increased proportion of error-impaired bits is inflicted in the low-priority partition. This is the reason for the associated higher error sensitivity seen in Figure 14.90(b). As such, there is a trade-off between the amount of video data protected and the code rate of the channel codec. As a comparison to the above scenarios in the context of Schemes 1 and 2, we shall now examine Scheme 3. In this scheme, more video data — namely, half the bits — can be directed into the high-priority partition, as demonstrated by Figure 14.88 due to encountering higher PBPs. This can also be confirmed with reference to Figures 14.90(b) and 14.91(b) by observing the PSNR degradations associated with the curves plotted in broken lines. If the low-priority partition is lost in Scheme 3, its effect on the quality of the reconstructed images is less detrimental than that of Scheme 2, since Scheme 3 loses only half the bits rather than 2/3. Hence, it is interesting to note that Scheme 3 experiences slightly higher average PSNR degradation than Scheme 1 at a BER of 2×10^{-3}, when only the low-priority partition is lost in both cases, despite directing only $\frac{1}{3}$ rather than $\frac{1}{2}$ of the bits to the low-priority partition. This observation can be explained as follows.

Apart from partitioning the macroblock header information and the variable-length coded DCT coefficients into the high- and low-priority partitions, synchronization information such as the picture header information [508] is replicated in the enhancement layer, as suggested by Gharavi *et al.* [514] as well as the MPEG-2 standard [508]. The purpose is to enable the MPEG-2 decoder to keep the base and enhancement layers synchronized during decoding.

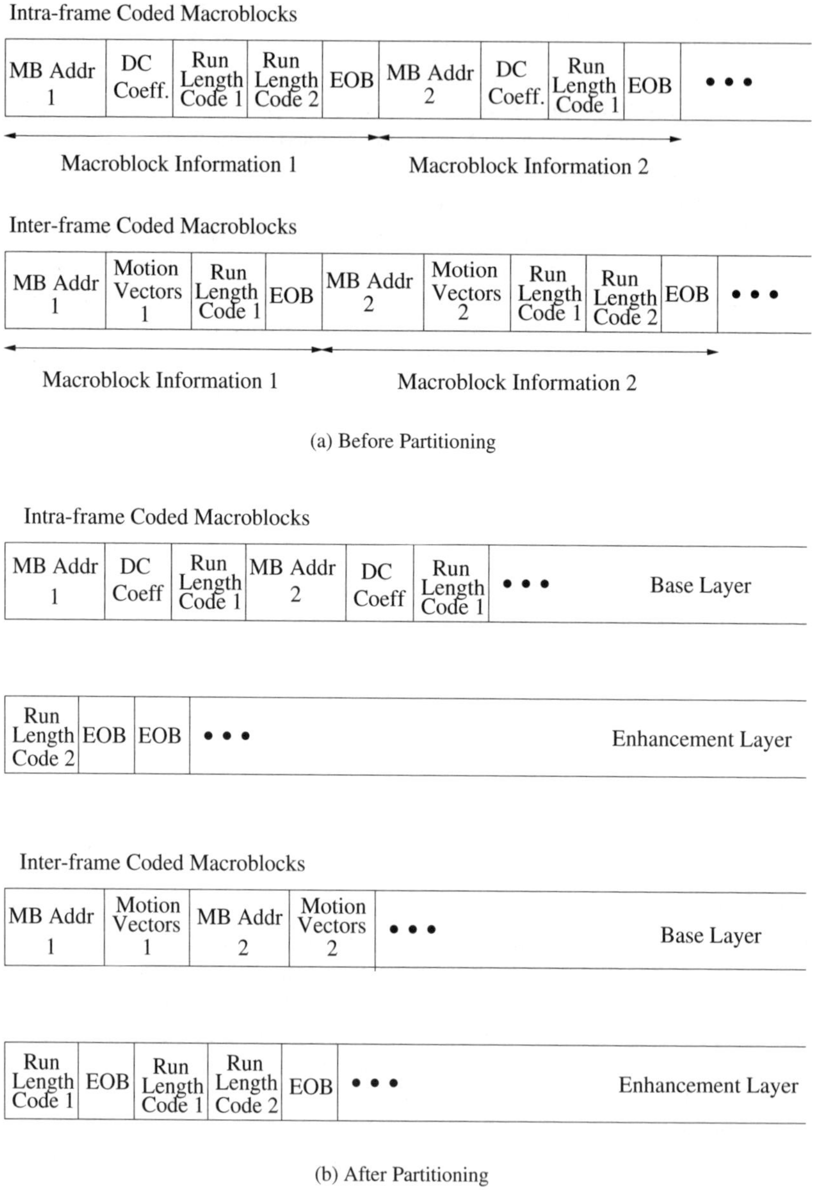

Figure 14.92: Example of video bitstream (a) before data partitioning and (b) after data partitioning for intra-frame coded macroblocks (MB) assuming a PBP of 64 and for inter-frame coded macroblocks assuming a PBP of 3.

14.7. DVB-T FOR MOBILE RECEIVERS

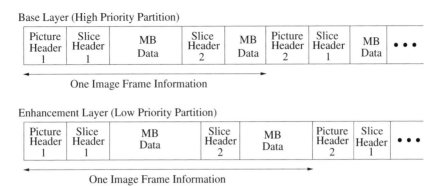

Figure 14.93: Example of high-level bitstream syntax structure of a data partitioned MPEG-2 video bitstream. The "MB data" shown in the diagram refers to the macroblock header information and to the variable-length coded DCT coefficients, which have been partitioned as shown in Figure 14.92.

An example of this arrangement is shown in Figure 14.93. This resynchronization measure is only effective when the picture start code of both the high- and low-priority partitions is received correctly. If the picture start code in the low-priority partition is corrupted, for example, the MPEG-2 decoder may not detect this PSC, and all the data corresponding to the current image frame in the low-priority partition will be lost. The MPEG-2 decoder will then interpret the bits received for the low-priority partition of the next frame as the low-priority data expected for the current frame. As expected, because of this synchronization problem, the decoded video would have a higher average PSNR degradation than for the case where picture start codes are unimpaired. This explains our observation of a higher average PSNR degradation for Scheme 3 when only its lower priority partition was corrupted by the transmission channel. On the other hand, in this specific experiment, Scheme 1 did not experience the loss of synchronization due to corruption of its picture start code. Viewing events from another perspective, by opting for allocating less useful video bits to the low-priority partition, the probability of transmission errors affecting the fixed-length PSC within the reduced-sized low priority partition becomes higher.

These findings will assist us in explaining our observations in the context of the hierarchical transmission scheme of Section 14.7.8, suggesting that the data partitioning scheme did not provide overall gain in terms of error resilience over the nonpartitioned case. Let us, however, consider first the performance of the nonhierarchical DVB-T scheme in the next section.

14.7.7 Nonhierarchical OFDM DVBP Performance

In this section, we elaborate on our findings when the convolutional code used in the standard nonhierarchical DVB scheme [499] is replaced by a turbo code. We will invoke a range of standard-compliant schemes as benchmarks. The 704×576-pixel HDTV-resolution "Football" video sequence was used in our experiments. The MPEG-2 decoder employs a simple error concealment algorithm to fill in missing portions of the reconstructed image in the event of decoding errors. The concealment algorithm will select the specific portion of the

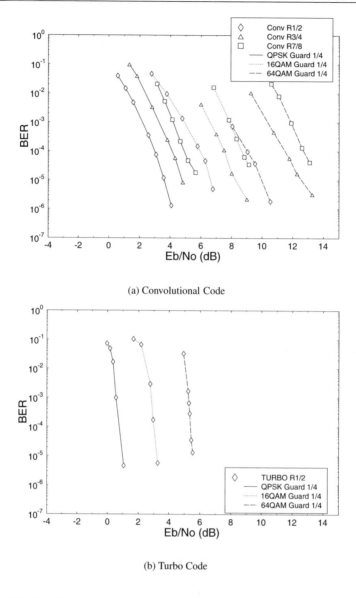

Figure 14.94: BER after (a) convolutional decoding and (b) turbo decoding for the DVB-T scheme over stationary nondispersive *AWGN* channels for *nonhierarchical transmission*.

previous reconstructed image, which corresponds to the missing portion of the current image in order to conceal the errors.

In Figure 14.94(a) and (b), the bit error rate (BER) performance of the various modem modes in conjunction with our diverse channel-coding schemes are portrayed over stationary, narrowband additive white Gaussian noise channels (AWGN), where the turbo codec exhibits a significantly steeper BER reduction in comparison to the convolutionally coded arrangements.

14.7. DVB-T FOR MOBILE RECEIVERS

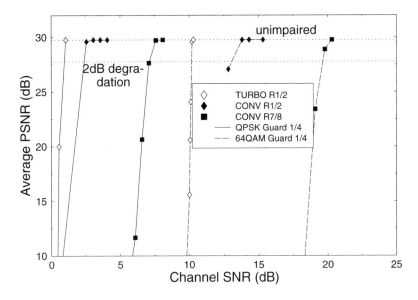

Figure 14.95: Average PSNR versus channel SNR of the DVB scheme [499] over nondispersive *AWGN* channels for *nonhierarchical transmission*.

Specifically, comparing the performance of the various turbo and convolutional codes for QPSK and 64QAM at a BER of 10^{-4}, we see that the turbo code exhibited an additional coding gain of about 2.24 dB and 3.7 dB, respectively, when using half-rate codes in Figure 14.94(a) and (b). Hence, the peak signal-to-noise ratio (PSNR) versus channel signal-to-noise ratio (SNR) graphs in Figure 14.95 demonstrate that approximately 2 dB and 3.5 dB lower channel SNRs are required in conjunction with the rate $\frac{1}{2}$ turbo codec for QPSK and 64QAM, respectively, than for convolutional coding, in order to maintain high reconstructed video quality. The term *unimpaired* as used in Figure 14.95 and Figure 14.96 refers to the condition where the PSNR of the MPEG-2 decoder's reconstructed image at the receiver is the same as the PSNR of the same image generated by the local decoder of the MPEG-2 video encoder, corresponding to the absence of channel — but not MPEG-2 coding — impairments.

Comparing the BER performance of the $\frac{1}{2}$ rate convolutional decoder in Figure 14.97(a) and the log-MAP [452] turbo decoder using eight iterations in Figure 14.97(b) for QPSK modulation over the worst-case fading mobile channel of Figure 14.83, we observe that at a BER of about 10^{-4} the turbo code provided an additional coding gain of 6 dB in comparison to the convolutional code. In contrast, for 64QAM using similar codes, a 5 dB coding gain was observed at this BER.

Similar observations were also made with respect to the average peak signal-to-noise ratio (PSNR) versus channel signal-to-noise ratio (SNR) plots of Figure 14.96. For the QPSK modulation mode and a $\frac{1}{2}$ coding rate, the turbo code required an approximately 5.5 dB lower channel SNR for maintaining near-unimpaired video quality than the convolutional code.

Comparing Figure 14.97(a) and Figure 14.98(a), we note that the Reed–Solomon decoder becomes effective in lowering the bit error probability of the transmitted data further below the BER threshold of 10^{-4}. From these figures we also observe that the rate $\frac{3}{4}$ convolutional code is unsuitable for transmission over the highly dispersive hilly terrain channel used in this

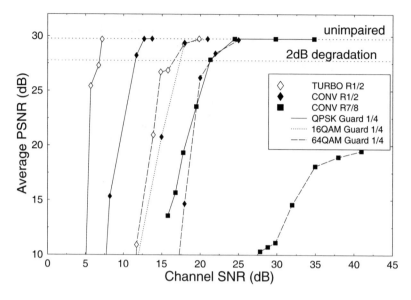

Figure 14.96: Average PSNR versus channel SNR of the DVB scheme [499] over the *wideband fading channel* of Figure 14.83 for *nonhierarchical transmission*.

Table 14.22: Summary of the *Nonhierarchical* Performance Results over Nondispersive *AWGN* Channels Tolerating a PSNR Degradation of 2 dB. The BER Measure Refers to BER after Viterbi or Turbo Decoding

Mod.	Code	CSNR (dB)	E_b/N_0	BER
QPSK	Turbo (1/2)	1.02	1.02	6×10^{-6}
64QAM	Turbo (1/2)	9.94	5.17	2×10^{-3}
QPSK	Conv (1/2)	2.16	2.16	1.1×10^{-3}
64QAM	Conv (1/2)	12.84	8.07	6×10^{-4}
QPSK	Conv (7/8)	6.99	4.56	2×10^{-4}
64QAM	Conv (7/8)	19.43	12.23	3×10^{-4}

experiment, when 64QAM is employed. When the rate $\frac{7}{8}$ convolutional code is used, both the 16QAM and 64QAM schemes perform poorly. As for the QPSK modulation scheme, a convolutional code rate as high as $\frac{7}{8}$ can still provide a satisfactory performance after Reed–Solomon decoding.

In conclusion, Tables 14.22 and 14.23 summarize the system performance in terms of the channel SNR (CSNR) required for maintaining less than 2 dB PSNR video degradation. At this PSNR degradation, decoding errors were still perceptually unnoticeable to the viewer due to the 30 frame/s refresh rate, although the typical still frame shown in Figure 14.99 in this scenario exhibits some degradation. It is important to underline once again that the $K = 3$ turbo code and the $K = 7$ convolutional code exhibited comparable complexities. The higher performance of the turbo codec facilitates, for example, the employment of

14.7. DVB-T FOR MOBILE RECEIVERS 593

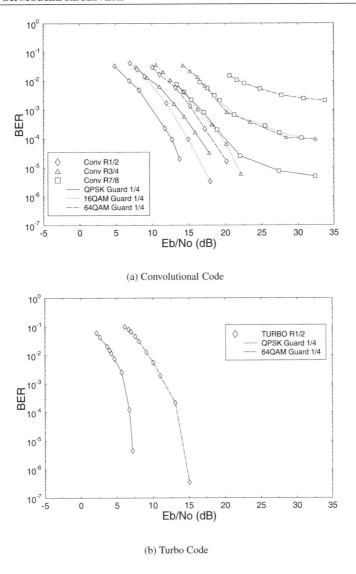

(a) Convolutional Code

(b) Turbo Code

Figure 14.97: BER after (a) convolutional decoding and (b) turbo decoding for the DVB-T scheme over the *wideband fading channel* of Figure 14.83 for *nonhierarchical transmission*.

turbo-coded 16QAM at a similar channel SNR, where convolutional-coded QPSK can be invoked. This in turn allows us to double the bitrate within the same bandwidth and thereby to improve the video quality. In the next section, we present the results of our investigations employing the DVB-T system [499] in a hierarchical transmission scenario.

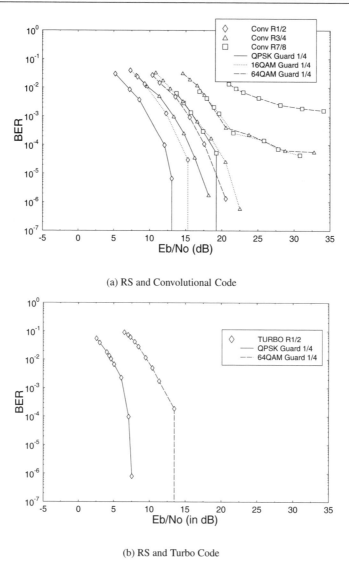

(a) RS and Convolutional Code

(b) RS and Turbo Code

Figure 14.98: BER after (a) RS and convolutional decoding and (b) RS and turbo decoding for the DVB-T scheme over the *wideband fading channel* of Figure 14.83 for *nonhierarchical transmission*.

14.7.8 Hierarchical OFDM DVB Performance

The philosophy of the hierarchical transmission mode is that the natural BER difference of a factor 2 to 3 of the 16QAM modem is exploited for providing unequal error protection for the FEC-coded video streams B3 and B4 of Figure 14.85 [222]. If the sensitivity of the video bits requires a different BER ratio between the B3 and B4 streams, the choice of the FEC

14.7. DVB-T FOR MOBILE RECEIVERS

Table 14.23: Summary of the *Nonhierarchical* Performance Results over *Wideband Fading Channels* Tolerating a PSNR Degradation of 2 dB. The BER Measure Refers to BER after Viterbi or Turbo Decoding

Mod.	Code	CSNR (dB)	E_b/N_0	BER
QPSK	Turbo (1/2)	6.63	6.63	2.5×10^{-4}
64QAM	Turbo (1/2)	15.82	11.05	2×10^{-3}
QPSK	Conv (1/2)	10.82	10.82	6×10^{-4}
64QAM	Conv (1/2)	20.92	16.15	7×10^{-4}
QPSK	Conv (7/8)	20.92	18.49	3×10^{-4}

Figure 14.99: Frame 79 of the "Football" sequence, which illustrates the visual effects of minor decoding errors at a BER of 2×10^{-4} after convolutional decoding. The PSNR degradation observed is approximately 2 dB. The sequence was coded using a rate 7/8 convolutional code and transmitted employing QPSK modulation.

codes protecting the video streams B1 and B2 of Figure 14.85 can be appropriately adjusted to equal out or to augment these differences.

Below we invoke the DVB-T hierarchical scheme in a mobile broadcasting scenario. We also demonstrate the improvements that turbo codes offer when replacing the convolutional code in the standard scheme. Hence, the convolutional codec in both the high- and low-priority partitions was replaced by the turbo codec. We have also investigated replacing only the high-priority convolutional codec with the turbo codec, pairing the $\frac{1}{2}$ rate turbo codec in the high-priority partition with the convolutional codec in the low-priority partition. Again, the "Football" sequence was used in these experiments. Partitioning was carried out using the schematic of Figure 14.85 as well as Algorithms 1 and 2. The FEC-coded high-priority video partition B3 of Figure 14.85 was mapped to the higher integrity 16QAM or 64QAM subchannel. In contrast, the low-priority partition B4 of Figure 14.85 was directed to the lower integrity 16QAM or 64QAM subchannel. Lastly, no specific mapping was required for QPSK, since it exhibits no subchannels. We note, however, that further design trade-offs become feasible when reversing the above mapping rules. This is necessary, for example, in conjunction with Scheme 2 of Table 14.21, since the high number of bits in the low-priority portion render it more sensitive than the high-priority partition. Again, the 16QAM subchannels exhibit a factor of 2 to 3 BER difference under various channel conditions, which improves the robustness of the reverse-mapped Scheme 2 of Table 14.21.

Referring to Figure 14.100 and comparing the performance of the $\frac{1}{2}$ rate convolutional code and turbo code at a BER of 10^{-4} for the low-priority partition, we find that the turbo code, employing eight iterations, exhibited a coding gain of about 6.6 dB and 5.97 dB for 16QAM and 64QAM, respectively. When the number of turbo-decoding iterations was reduced to 4, the coding gains offered by the turbo code over that of the convolutional code were 6.23 dB and 5.7 dB for 16QAM and 64QAM, respectively. We observed that by reducing the number of iterations to four halved the associated complexity, but the turbo code exhibited a coding loss of only about 0.37 dB and 0.27 dB in comparison to the eight-iteration scenario for 16QAM and 64QAM, respectively. Hence, the computational complexity of the turbo codec can be halved by sacrificing only a small amount of coding gain. The substantial coding gain provided by turbo coding is also reflected in the PSNR versus channel SNR graphs of Figure 14.101. In order to achieve transmission with very low probability of error, Figure 14.101 demonstrated that approximately 5.72 dB and 4.56 dB higher channel SNRs are required by the standard scheme compared to the scheme employing turbo coding, when using four iterations in both partitions. We have only shown the performance of turbo coding for the low-priority partition in Figures 14.100(b) and 14.102(b), since the turbo or convolutional-coded high-priority partition was received with very low probability of error after Reed–Solomon decoding for the range of SNRs used.

We also observed that the rate $\frac{3}{4}$ and rate $\frac{7}{8}$ convolutional codes in the low-priority partition were unable to provide sufficient protection to the transmitted video bits, as becomes evident from Figures 14.100(a) and 14.102(a). In these high coding rate scenarios, due to the presence of residual errors even after the Reed–Solomon decoder, the decoded video exhibited some decoding errors, which is evidenced by the flattening of the PSNR versus channel SNR curves in Figure 14.101(a), before reaching the error-free PSNR.

A specific problem when using the data partitioning scheme in conjunction with the high-priority partition being protected by the rate $\frac{1}{2}$ code and the low-priority partition protected by the rate $\frac{3}{4}$ and rate $\frac{7}{8}$ codes was that when the low-priority partition data was corrupted, the

14.7. DVB-T FOR MOBILE RECEIVERS

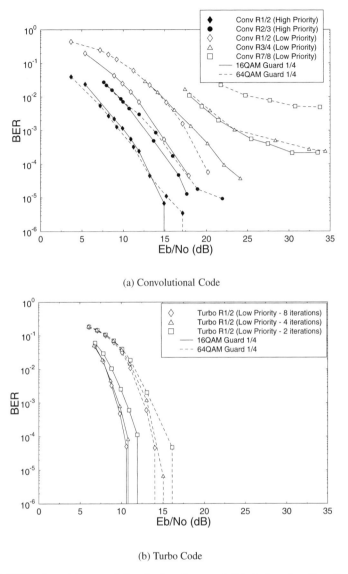

(a) Convolutional Code

(b) Turbo Code

Figure 14.100: BER after (a) convolutional decoding and (b) turbo decoding for the *DVB-T hierarchical scheme* over the *wideband fading channel* of Figure 14.83 using the schematic of Figure 14.85 as well as Algorithms 1 and 2. In (b), the BER of the turbo- or convolutional-coded high-priority partition is not shown.

error-free high priority data available was insufficient for concealing the errors, as discussed in Section 14.7.6. We have also experimented with the combination of rate $\frac{2}{3}$ convolutional coding and rate $\frac{1}{2}$ convolutional coding, in order to protect the high- and low-priority data, respectively. From Figure 14.101(a) we observed that the performance of this $\frac{2}{3}$ rate and $\frac{1}{2}$ rate combination approached that of the rate $\frac{1}{2}$ convolutional code in both partitions. This was expected, since now more data can be inserted into the high-priority partition. Hence, in the

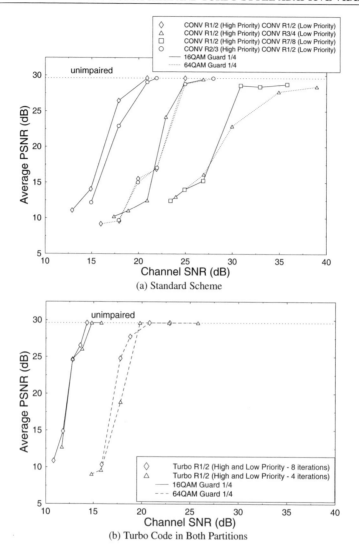

Figure 14.101: Average PSNR versus channel SNR for (a) standard DVB scheme [499] and (b) system with turbo coding employed in both partitions, for transmission over the *wideband fading channel* of Figure 14.83 for *hierarchical transmission* using the schematic of Figure 14.85 as well as Algorithms 1 and 2.

event of decoding errors in the low-priority data, we had more error-free high-priority data that could be used to reconstruct the received image.

Our last combination investigated involved using rate $\frac{1}{2}$ turbo coding and convolutional coding for the high- and low-priority partitions, respectively. Comparing Figures 14.103 and 14.101(a), the channel SNRs required for achieving unimpaired video transmission were similar in both cases. This was expected, since the turbo-convolutional combination's

14.7. DVB-T FOR MOBILE RECEIVERS

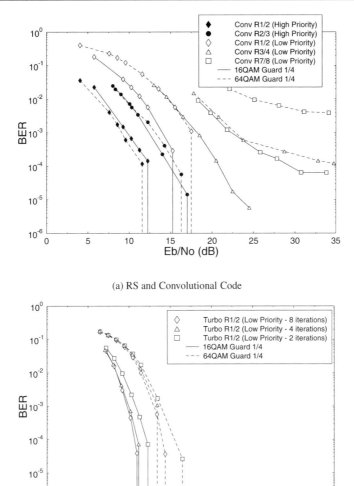

(a) RS and Convolutional Code

(b) RS and Turbo Code

Figure 14.102: BER after (a) RS and convolutional decoding and (b) RS and turbo decoding for the *DVB-T hierarchical scheme* over the *wideband fading channel* of Figure 14.83 using the schematic of Figure 14.85 as well as Algorithms 1 and 2. In (b), the BER of the turbo- or convolutional-coded high-priority partition is not shown.

video performance is dependent on the convolutional code's performance in the low-priority partition.

Lastly, comparing Figures 14.101 and 14.96, we found that the unimpaired PSNR condition was achieved at similar channel SNRs for the hierarchical and nonhierarchical schemes, suggesting that the data partitioning scheme had not provided sufficient

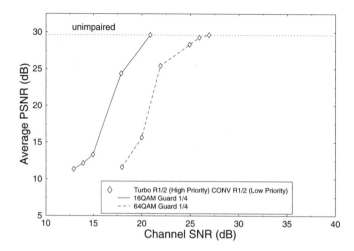

Figure 14.103: Average PSNR versus channel SNR of the DVB scheme, employing turbo coding in the high-priority partition and convolutional coding in the low-priority partition, over the *wideband fading channel* of Figure 14.83 for *hierarchical transmission* using the schematic of Figure 14.85 as well as Algorithms 1 and 2.

performance improvements in the context of the mobile DVB scheme to justify its added complexity. Again, this was a consequence of relegating a high proportion of video bits to the low integrity partition.

14.7.9 Summary and Conclusions

In this chapter, we have investigated the performance of a turbo-coded DVB system in a mobile environment. A range of system performance results was presented based on the standard DVB-T scheme as well as on an improved turbo-coded scheme. The convolutional code specified in the standard system was replaced by turbo coding, which resulted in a substantial coding gain of around 5 dB. It is important to underline once again that the $K = 3$ turbo code and the $K = 7$ convolutional code exhibited comparable complexities. The higher performance of the turbo codec facilitates, for example, the employment of turbo-coded 16QAM at a similar SNR, where convolutional-coded QPSK can be invoked. This in turn allows us to double the video bitrate within the same bandwidth and hence to improve the video quality. We have also applied data partitioning to the MPEG-2 video stream to gauge its efficiency in increasing the error resilience of the video codec. However, from these experiments we found that the data partitioning scheme did not provide substantial improvements compared to the nonpartitioned video transmitted over the nonhierarchical DVB-T system. Our future work will focus on extending this DVB-T system study to incorporate various types of channel models, as well as on investigating the effects of different Doppler frequencies on the system. Further work will also be dedicated to trellis-coded modulation (TCM) and turbo trellis-coded modulation (TTCM) based OFDM. The impact of employing various types of turbo interleavers on the system performance is also of interest. A range of further wireless video communications issues are addressed in [161, 516]. Let us now consider a variety of satellite-based turbo-coded blind-equalized multilevel modulation-assisted video broadcasting schemes.

14.8 Satellite-based Video Broadcasting[13]

14.8.1 Background and Motivation

In recent years, three harmonized digital video broadcasting (DVB) standards have emerged in Europe for terrestrial [499], cable-based [500], and satellite-oriented [501] delivery of DVB signals. The dispersive wireless propagation environment of the terrestrial system requires concatenated Reed–Solomon (RS) [389, 502] and rate-compatible punctured convolutional coding (RCPCC) [389, 502] combined with orthogonal frequency division multiplexing (OFDM)-based modulation [222]. The satellite-based system employs the same concatenated channel coding arrangement as the terrestrial scheme, while the cable-based system refrains from using concatenated channel coding, opting for RS coding only. The performance of both of the latter schemes can be improved upon invoking blind-equalized multilevel modems [222], although the associated mild dispersion or linear distortion does not necessarily require channel equalization. However, since we propose invoking turbo-coded 4-bit/symbol 16-level quadrature amplitude modulation (16QAM) in order to improve the system's performance at the cost of increased complexity, in this section we also invoked blind channel equalizers. This is further justified by the associated high video transmission rates, where the dispersion may become a more dominant performance limitation.

Lastly, the video codec used in all three systems is the Motion Pictures Expert Group's MPEG-2 codec. These standardization activities were followed by a variety of system performance studies in the open literature [517–520]. Against this background, we suggest turbo-coding improvements to the satellite-based DVB system [501] and present performance studies of the proposed system under dispersive channel conditions in conjunction with a variety of blind channel equalization algorithms. The transmitted power requirements of the standard system employing convolutional codecs can be reduced upon invoking more complex, but more powerful, turbo codecs. Alternatively, the standard quaternary or 2-bit/symbol system's bit error rate (BER) versus signal-to-noise ratio (SNR) performance can almost be matched by a turbo-coded 4-bit/symbol 16QAM scheme, while doubling the achievable bitrate within the same bandwidth and hence improving the associated video quality. This is achieved at the cost of an increased system complexity.

The remainder of this section is organized as follows. A succinct overview of the turbo-coded and standard DVB satellite scheme is presented in Section 14.8.2, while our channel model is described in Section 14.8.3. A brief summary of the blind equalizer algorithms employed is presented in Section 14.8.4. Following this, the performance of the improved DVB satellite system is examined for transmission over a dispersive two-path channel in Section 14.8.5, before our conclusions and future work areas are presented in Section 14.8.6.

---

[13]This section is based on C. S. Lee, S. Vlahoyiannatos, and L. Hanzo, "Satellite based turbo-coded, blind-equalized 4QAM and 16QAM digital video broadcasting", *IEEE Transactions on Broadcasting*, March 2000, pp. 22–34, ©2000 IEEE. Personal use of this material is permitted. However, permission to reprint/republish this material for advertising or promotional purposes or for creating new collective works for resale or redistribution to servers or lists, or to reuse any copyrighted component of this work in other works must be obtained from the IEEE.

CHAPTER 14. HSDPA-LIKE AND TURBO-STYLE ADAPTIVE VIDEO SYSTEMS

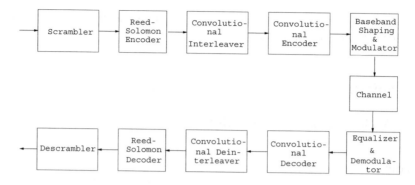

Figure 14.104: Schematic of the DVB satellite system.

Table 14.24: Parameters of the $CC(n, k, K)$ Convolutional Inner Encoder of the DVB-S Modem

Convolutional Coder Parameters	
Code rate	1/2
Constraint length	7
n	2
k	1
Generator polynomials (octal format)	171, 133

14.8.2 DVB Satellite Scheme

The block diagram of the DVB satellite (DVB-S) system [501] is shown in Figure 14.104, which is composed of a MPEG-2 video encoder (not shown in the diagram), channel-coding modules, and a quadrature phase shift keying (QPSK) modem [222]. The bitstream generated by the MPEG-2 encoder is packetized into frames 188 bytes long. The video data in each packet is then randomized by the scrambler. The details concerning the scrambler have not been included in this chapter, since these may be obtained from the DVB-S standard [501].

Because of the poor error resilience of the MPEG-2 video codec, powerful concatenated channel coding is employed. The concatenated channel codec comprises a shortened Reed–Solomon (RS) outer code and an inner convolutional encoder. The 188-byte MPEG-2 video packet is extended by the Reed–Solomon encoder [389, 502], with parity information to facilitate error recovery to form a 204-byte packet. The Reed–Solomon decoder can then correct up to 8 erroneous bytes for each 204-byte packet. Following this, the RS-coded packet is interleaved by a convolutional interleaver and is further protected by a half-rate inner convolutional encoder with a constraint length of 7 [389, 502].

Furthermore, the overall code rate of the concatenated coding scheme can be adapted by variable puncturing, not shown in the figure, which supports code rates of $\frac{1}{2}$ (no puncturing) as well as $\frac{2}{3}$, $\frac{3}{4}$, $\frac{5}{6}$, and $\frac{7}{8}$. The parameters of the convolutional encoder are summarized in Table 14.24.

In addition to implementing the standard DVB-S system as a benchmark, we have improved the system's performance with the aid of a turbo codec [401, 402]. The block

14.8. SATELLITE-BASED VIDEO BROADCASTING

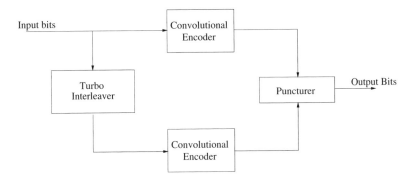

Figure 14.105: Block diagram of turbo encoder.

Table 14.25: Parameters of the Inner Turbo Encoder Used to Replace the DVB-S System's Convolutional Coder (RSC: Recursive Systematic Code)

Turbo Coder Parameters	
Turbo code rate	1/2
Input block length	17 952 bits
Interleaver type	Random
Number of turbo decoder iterations	8
Turbo Encoder Component Code Parameters	
Component code encoder type	Convolutional Encoder (RSC)
Component code decoder type	log-MAP [452]
Constraint length	3
n	2
k	1
Generator polynomials (octal format)	7, 5

diagram of the turbo encoder is shown in Figure 14.105. The turbo encoder is constructed of two component encoders. Each component encoder is a half-rate convolutional encoder whose parameters are listed in Table 14.25. The two component encoders are used to encode the same input bits, although the input bits of the second component encoder are interleaved before encoding. The output bits of the two component codes are punctured and multiplexed in order to form a single-output bitstream. The component encoder used here is known as a half-rate recursive systematic convolutional encoder (RSC) [511]. It generates one parity bit and one systematic output bit for every input bit. In order to provide an overall coding rate of one-half, half the output bits from the two encoders must be punctured. The puncturing arrangement used in our work is to transmit all the systematic bits from the first encoder and every other parity bit from both encoders.

Readers interested in further details of the DVB-S system are referred to the DVB-S standard [501]. The performance of the standard DVB-S system and the performance of

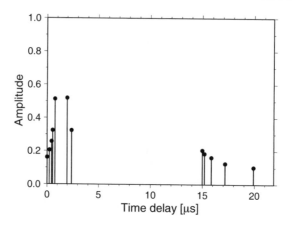

Figure 14.106: Two-path satellite channel model with either a one-symbol or two-symbol delay.

the turbo-coded system are characterized in Section 14.8.5. Let us now briefly consider the multipath channel model used in our investigations.

14.8.3 Satellite Channel Model

The DVB-S system was designed to operate in the 12 GHz frequency band (K-band). Within this frequency band, tropospheric effects such as the transformation of electromagnetic energy into thermal energy due to induction of currents in rain and ice crystals lead to signal attenuations [521, 522]. In the past 20 years, various researchers have concentrated on attempting to model the satellite channel, typically within a land mobile satellite channel scenario. However, the majority of the work conducted, for example, by Vogel and his colleagues [523–526] concentrated on modeling the statistical properties of a narrowband satellite channel in lower frequency bands, such as the 870 MHz UHF band and the 1.5 GHz L-band.

Our high bitrate DVB satellite system requires a high bandwidth, however. Hence, the video bitstream is exposed to dispersive wideband propagation conditions. Recently, Saunders *et al.* [527, 528] have proposed the employment of multipath channel models to study the satellite channel, although their study concentrated on the L-band and S-band only.

Due to the dearth of reported work on wideband satellite channel modeling in the K-band, we have adopted a simpler approach. The channel model employed in this study was the two-path (nT)-symbol spaced impulse response, where T is the symbol duration. In our studies we used $n = 1$ and $n = 2$ (Figure 14.106). This corresponds to a stationary dispersive transmission channel. Our channel model assumed that the receiver had a direct line-of-sight with the satellite as well as a second path caused by a single reflector probably from a nearby building or due to ground reflection. The ground reflection may be strong if the satellite receiver dish is only tilted at a low angle.

Based on these channel models, we studied the ability of a range of blind equalizer algorithms to converge under various path-delay conditions. In the next section, we provide a brief overview of the various blind equalizers employed in our experiments. Readers

14.8. SATELLITE-BASED VIDEO BROADCASTING

interested mainly in the system's performance may proceed directly to our performance analysis section, Section 14.8.5.

14.8.4 The Blind Equalizers

This section presents the blind equalizers used in the system. The following blind equalizers have been studied:

1. The modified constant modulus algorithm (MCMA) [529].

2. The Benveniste–Goursat algorithm (B-G) [530].

3. The stop-and-go algorithm (S-a-G) [531].

4. The per-survivor processing (PSP) algorithm [532].

We will now briefly introduce these algorithms.
First, we define the variables that we will use:

$$\mathbf{y}(n) = [y(n+N_1), \ldots, y(0), \ldots, y(n-N_2)]^T \quad (14.26)$$

$$\mathbf{c}^{(n)} = [c_{-N_1}, \ldots, c_o, \ldots, c_{N_2}]^T \quad (14.27)$$

$$z(n) = \left(\mathbf{c}^{(n)}\right)^T \mathbf{y}(n) = \mathbf{y}^T(n)\mathbf{c}^{(n)} \quad (14.28)$$

where $\mathbf{y}(n)$ is the received symbol vector at time n, containing the $N_1 + N_2 + 1$ most recent received symbols, while N_1, N_2 are the number of equalizer feedback and feedforward taps, respectively. Furthermore, $\mathbf{c}^{(n)}$ is the equalizer tap vector, consisting of the equalizer tap values, and $z(n)$ is the equalized symbol at time n, given by the convolution of the received signal with the equalizer's impulse response, while $()^T$ stands for matrix transpose. Note that the variables of Equations 14.26–14.28 assume complex values, when multilevel modulation is employed.

The modified CMA (MCMA) is an improved version of Godard's well-known *constant modulus algorithm (CMA)* [533]. The philosophy of the CMA is based on forcing the magnitude of the equalized signal to a constant value. In mathematical terms, the CMA is based on minimizing the cost function:

$$J^{(CMA)} = E\left[\left(|z(n)|^2 - R_2\right)^2\right], \quad (14.29)$$

where R_2 is a suitably chosen constant and $E[]$ stands for the expectation. Similarly to the CMA, the MCMA, which was proposed by Wesolowsky [529], forces the real and imaginary parts of the complex signal to the constant values of $R_{2,R}$ and $R_{2,I}$, respectively, according to the equalizer tap update equation of [529]:

$$\mathbf{c}^{(n+1)} = \mathbf{c}^{(n)} - \lambda \cdot \mathbf{y}^*(n) \cdot \{Re[z(n)] \cdot ((Re[z(n)])^2 - R_{2,R})$$
$$+ jIm[z(n)] \cdot ((Im[z(n)])^2 - R_{2,I})\}, \quad (14.30)$$

where λ is the step-size parameter and the $R_{2,R}$, $R_{2,I}$ constant parameters of the algorithm are defined as:

$$R_{2,R} = \frac{E\left[(Re[a(n)])^4\right]}{E\left[(Re[a(n)])^2\right]} \quad (14.31)$$

$$R_{2,I} = \frac{E\left[(Im[a(n)])^4\right]}{E\left[(Im[a(n)])^2\right]}, \quad (14.32)$$

where $a(n)$ is the transmitted signal at time instant n.

The Benveniste–Goursat (B-G) algorithm [530] is an amalgam of the Sato's algorithm [534] and the decision-directed (DD) algorithm [222]. Strictly speaking, the decision-directed algorithm is not a blind equalization technique, since its convergence is highly dependent on the channel.

This algorithm estimates the error between the equalized signal and the detected signal as:

$$\epsilon^{DD}(n) = z(n) - \hat{z}(n), \quad (14.33)$$

where $\hat{z}(n)$ is the receiver's estimate of the transmitted signal at time instant n. Similarly to the DD algorithm's error term, the Sato error [534] is defined as:

$$\epsilon^{Sato}(n) = z(n) - \gamma \cdot csgn(z(n)), \quad (14.34)$$

where γ is a constant parameter of the Sato algorithm, defined as:

$$\gamma = \frac{E\left[(Re\,[a(n)])^2\right]}{E\,[|Re\,[a(n)]|]} = \frac{E\left[(Im\,[a(n)])^2\right]}{E\,[|Im\,[a(n)]|]} \quad (14.35)$$

and $csgn(x) = sign(Re\{x\}) + jsign(Im\{x\})$ is the complex sign function. The B-G algorithm combines the above two error terms into one:

$$\epsilon^G(n) = k_1 \cdot \epsilon^{DD}(n) + k_2 \cdot |\epsilon^{DD}(n)| \cdot \epsilon^{Sato}(n), \quad (14.36)$$

where the two error terms are suitably weighted by the constant parameters k_1 and k_2 in Equation 14.36. Using this error term, the B-G equalizer updates the equalizer coefficients according to the following equalizer tap update equations [530]:

$$\mathbf{c}^{(n+1)} = \mathbf{c}^{(n)} - \lambda \cdot \mathbf{y}^*(n) \cdot \epsilon^G(n). \quad (14.37)$$

In our investigations, the weights were chosen as $k_1 = 1$, $k_2 = 5$, so that the Sato error was weighted more heavily than the DD error.

The stop-and-go (S-a-G) algorithm [531] is a variant of the decision-directed algorithm [222], where at each equalizer coefficient adjustment iteration, the update is enabled or disabled depending on whether or not the update is likely to be correct. The update equations of this algorithm are given by [531]

$$\mathbf{c}^{(n+1)} = \mathbf{c}^{(n)} - \lambda \cdot \mathbf{y}^*(n) \cdot [f_{n,R} \cdot Re\{\epsilon^{DD}(n)\} + jf_{n,I} \cdot Im\{\epsilon^{DD}(n)\}], \quad (14.38)$$

where * stands for the complex conjugate, $\epsilon^{DD}(n)$ is the decision directed error as in Equation 14.33 and the binary functions $f_{n,R}$, $f_{n,I}$ enable or disable the update of the

14.8. SATELLITE-BASED VIDEO BROADCASTING

equalizer according to the following rule. If the sign of the Sato error (the real or the imaginary part independently) is the same as the sign of the decision-directed error, then the update takes place; otherwise it does not.

In mathematical terms, this is equivalent to [531]:

$$f_{n,R} = \begin{cases} 1 & \text{if } sgn(Re[\epsilon^{DD}(n)]) = sgn(Re[\epsilon^{Sato}(n)]) \\ 0 & \text{if } sgn(Re[\epsilon^{DD}(n)]) \neq sgn(Re[\epsilon^{Sato}(n)]) \end{cases} \quad (14.39)$$

$$f_{n,I} = \begin{cases} 1 & \text{if } sgn(Im\{\epsilon^{DD}(n)\}) = sgn(Im\{\epsilon^{Sato}(n)\}) \\ 0 & \text{if } sgn(Im\{\epsilon^{DD}(n)\}) \neq sgn(Im\{\epsilon^{Sato}(n)\}). \end{cases} \quad (14.40)$$

For a blind equalizer, this condition provides us with a measure of the probability of the coefficient update being correct.

The PSP algorithm [532] is based on employing convolutional coding. Hence, it is a trellis-based sequence estimation technique in which the channel is not known *a priori*. An iterative channel estimation technique is employed in order to estimate the channel jointly with the modulation symbol. In this sense, an initial channel is used, and the estimate is updated at each new symbol's arrival.

In our case, the update was based on the *least means squares (LMS)* estimates, according to the following channel-tap update equations [532]:

$$\hat{\mathbf{h}}^{(n+1)} = \hat{\mathbf{h}}^{(n)} + \lambda \cdot \hat{\mathbf{a}}^*(n) \cdot \left(y(n) - \hat{\mathbf{a}}^T(n)\hat{\mathbf{h}}^{(n)} \right), \quad (14.41)$$

where $\hat{\mathbf{h}}^{(n)} = (\hat{h}^{(n)}_{-L_1}, \ldots, \hat{h}^{(n)}_o, \ldots, \hat{h}^{(n)}_{L_2})^T$ is the estimated (for one surviving path) channel tap vector at time instant n, $\hat{\mathbf{a}}(n) = (\hat{a}(n+L_1), \ldots, \hat{a}(0), \ldots, \hat{a}(n-L_2))^T$ is the associated estimated transmitted symbol vector, and $y(n)$ is the actually received symbol at time instant n.

Each of the surviving paths in the trellis carries not only its own signal estimation, but also its own channel estimation. Moreover, convolutional decoding can take place jointly with this channel and data estimation procedure, leading to improved bit error rate (BER) performance. The various equalizers' parameters are summarized in Table 14.26.

Having described the components of our enhanced DVB-S system, let us now consider the overall system's performance.

14.8.5 Performance of the DVB Satellite Scheme

In this section, the performance of the DVB-S system was evaluated by means of simulations. Two modulation types were used, namely, the standard QPSK and the enhanced 16QAM schemes [222]. The channel model of Figure 14.106 was employed. The first channel model had a one-symbol second-path delay, while in the second one the path-delay corresponded to the period of two symbols. The average BER versus SNR per bit performance was evaluated after the equalization and demodulation process, as well as after Viterbi [389] or turbo decoding [402]. The SNR per bit or E_b/N_o is defined as follows:

$$\text{SNR per bit} = 10 \log_{10} \frac{\bar{S}}{\bar{N}} + \delta, \quad (14.42)$$

Table 14.26: Summary of the Equalizer Parameters Used in the Simulations. The Tap Vector $(1.2, 0, \ldots, 0)$ Indicates that the First Equalizer Coefficient is Initialized to the Value 1.2, While the Others are Initialized to 0

	Step Size λ	No. of Equalizer Taps	Initial Tap Vector
Benveniste–Goursat	5×10^{-4}	10	$(1.2, 0, \ldots, 0)$
Modified CMA	5×10^{-4}	10	$(1.2, 0, \ldots, 0)$
Stop-and-go	5×10^{-4}	10	$(1.2, 0, \ldots, 0)$
PSP (1 sym delay)	10^{-2}	2	$(1.2, 0)$
PSP (2 sym delay)	10^{-2}	3	$(1.2, 0, 0)$

where $\bar{S}$ is the average received signal power, $\bar{N}$ is the average received noise power, and δ, which is dependent on the type of modulation scheme used and channel code rate (R), is defined as:

$$\delta = 10 \log_{10} \frac{1}{R \times \text{bits per modulation symbol}}. \tag{14.43}$$

Our results are further divided into two subsections for ease of discussion. First, we present the system performance over the one-symbol delay two-path channel in Section 14.8.5.1. Next, the system performance over the two-symbol delay two-path channel is presented in Section 14.8.5.2. Lastly, a summary of the system performance is provided in Section 14.8.5.3.

14.8.5.1 Transmission over the Symbol-spaced Two-path Channel

The linear equalizers' performance was quantified and compared using QPSK modulation over the one-symbol delay two-path channel model of Figure 14.107. Since all the equalizers have similar BER performance, only the modified CMA results are shown in the figure.

The equalized performance over the one symbol-spaced channel was inferior to that over the nondispersive AWGN channel. However, as expected, it was better than without any equalization. Another observation for Figure 14.107 was that the different punctured channel-coding rates appeared to give slightly different bit error rates after equalization. This was because the linear blind equalizers required uncorrelated input bits in order to converge. However, the input bits were not entirely random when convolutional coding was used. The consequences of violating the zero-correlation constraint are not generally known. Nevertheless, two potential problems were apparent. First, the equalizer may diverge from the desired equalizer equilibrium [535]. Second, the performance of the equalizer is expected to degrade, owing to the violation of the randomness requirement, which is imposed on the input bits in order to ensure that the blind equalizers will converge.

Since the channel used in our investigations was static, the first problem was not encountered. Instead, the second problem was what we actually observed. Figure 14.108 quantifies the equalizers' performance degradation due to convolutional coding. We can

14.8. SATELLITE-BASED VIDEO BROADCASTING

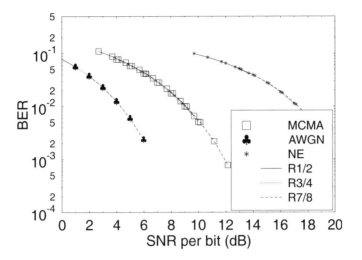

(a) After equalization and demodulation

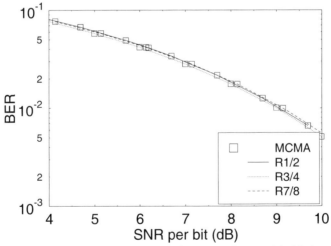

(b) Same as (a) but enlarged in order to show performance difference of the blind equalizer, when different convolutional code rates are used

Figure 14.107: Average BER versus SNR per bit performance after equalization and demodulation employing QPSK modulation and one-symbol delay channel (NE = nonequalized; MCMA = modified constant modulus algorithm).

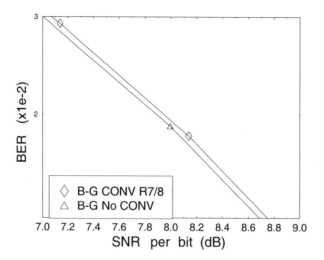

Figure 14.108: Average BER versus SNR per bit performance after equalization and demodulation employing QPSK modulation and the one-symbol delay two-path channel of Figure 14.106, for the Benveniste–Goursat algorithm, where the input bits are random (No CONV) or correlated (CONV 7/8) as a result of convolutional coding having a coding rate of 7/8.

observe a 0.1 dB SNR degradation when the convolutional codec creates correlation among the bits for this specific case.

The average BER versus SNR curves after Viterbi or turbo decoding are shown in Figure 14.109(a). In this figure, the average BER over the nondispersive AWGN channel after turbo decoding constitutes the best-case performance, while the average BER of the one-symbol delay two-path MCMA-equalized rate 7/8 convolutionally coded scenario exhibits the worst-case performance. Again, in this figure only the modified CMA was featured for simplicity. The performance of the remaining equalizers was characterized in Figure 14.109(b). Clearly, the performance of all the linear equalizers investigated was similar.

As seen in Figure 14.109(a), the combination of the modified CMA blind equalizer with turbo decoding exhibited the best SNR performance over the one-symbol delay two-path channel. The only comparable alternative was the PSP algorithm. Although the performance of the PSP algorithm was better at low SNRs, the associated curves cross over and the PSP algorithm's performance became inferior below the average BER of 10^{-3}. Although not shown in Figure 14.109, the Reed–Solomon decoder, which was concatenated to either the convolutional or the turbo decoder, became effective, when the average BER of its input was below approximately 10^{-4}. In this case, the PSP algorithm performed by at least 1 dB worse in the area of interest, which is at an average BER of 10^{-4}.

A final observation in the context of Figure 14.109(a) is that when convolutional decoding was used, the associated E_b/N_o performance of the rate 1/2 convolutional coded scheme appeared slightly inferior to that of the rate 3/4 and the rate 7/8 scenarios beyond certain E_b/N_o values. This was deemed to be a consequence of the fact that the 1/2 rate encoder introduced more correlation into the bitstream than its higher rate counterparts, and this

14.8. SATELLITE-BASED VIDEO BROADCASTING

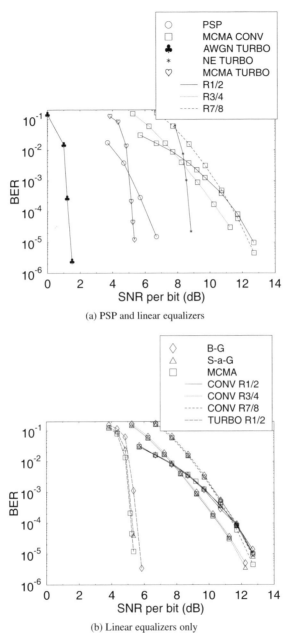

(a) PSP and linear equalizers

(b) Linear equalizers only

Figure 14.109: Average BER versus SNR per bit performance after convolutional or turbo decoding for QPSK modulation and one-symbol delay channel (NE = nonequalized; B-G = Benveniste–Goursat; S-a-G = stop-and-go; MCMA = modified constant modulus algorithm; PSP = per-survivor processing).

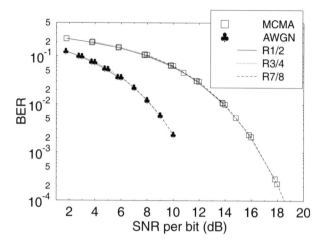

(a) After equalization and demodulation

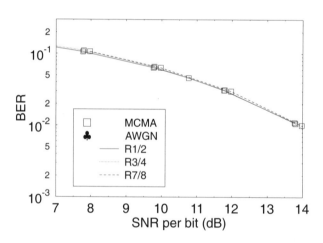

(b) Same as (a) but enlarged in order to show performance difference of the blind equalizer, when different convolutional code rates are used

Figure 14.110: Average BER versus SNR per bit after equalization and demodulation for 16QAM over the one-symbol delay two-path channel of Figure 14.106 (MCMA = modified constant modulus algorithm).

degraded the performance of the blind channel equalizers which performed best when fed with random bits.

Having considered the QPSK case, we shall now concentrate on the enhanced system which employed 16QAM under the same channel and equalizer conditions. Figures 14.110 and 14.111 present the performance of the DVB system employing 16QAM. Again, for

14.8. SATELLITE-BASED VIDEO BROADCASTING

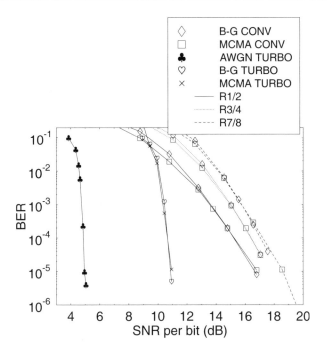

Figure 14.111: Average BER versus SNR per bit after Viterbi or turbo decoding for 16QAM over the one-symbol delay two-path channel of Figure 14.106 (B-G = Benveniste–Goursat; S-a-G = stop-and-go; MCMA = modified constant modulus algorithm; PSP = per-survivor processing).

simplicity, only the modified CMA results are given. In this case, the ranking order of the different coding rates followed our expectations more closely in the sense that the lowest coding rate of 1/2 was the best performer, followed by rate 3/4 codec, in turn followed by the least powerful rate 7/8 codec.

The stop-and-go algorithm has been excluded from these results, because it does not converge for high SNR values. This happens because the equalization procedure is activated only when there is a high probability of correct decision-directed equalizer update. In our case, the equalizer is initialized far from its convergence point, and hence the decision directed updates are unlikely to be correct. In the absence of noise, this leads to the update algorithm being permanently deactivated. If noise is present, however, then some random perturbations from the point of the equalizer's initialization activate the stop-and-go algorithm and can lead to convergence. We made this observation at medium SNR values in our simulation study. For high SNR values, the algorithm did not converge.

It is also interesting to compare the performance of the system for the QPSK and 16QAM schemes. When the one-symbol delay two-path channel model of Figure 14.106 was considered, the system was capable of supporting the use of 16QAM with the provision of an additional SNR per bit of approximately 4–5 dB. This observation was made by comparing the performance of the DVB system when employing the modified CMA and the half-rate convolutional or turbo code in Figures 14.109 and 14.111 at a BER of 10^{-4}. Although the original DVB satellite system only employs QPSK modulation, our simulations had shown that 16QAM can be employed equally well for the range of blind equalizers that we have used in our work. This allowed us to double the video bitrate and hence to substantially improve the video quality. The comparison of Figures 14.109 and 14.111 also reveals that the extra SNR requirement of approximately 4–5 dB of 16QAM over QPSK can be eliminated by employing turbo coding at the cost of a higher implementational complexity. This allowed us to accommodate a doubled bitrate within a given bandwidth, which improved the video quality.

14.8.5.2 Transmission over the Two-symbol Delay Two-path Channel

In Figure 14.112 (only for the Benveniste–Goursat algorithm for simplicity) and Figure 14.113, the corresponding BER results for the two-symbol delay two-path channel of Figure 14.106 are given for QPSK. The associated trends are similar to those in Figures 14.107 and 14.109, although some differences can be observed, as listed below:

- The "crossover point" is where the performance of the PSP algorithm becomes inferior to the performance of the modified CMA in conjunction with turbo decoding. This point is now at 10^{-4}, which is in the range where the RS decoder guarantees an extremely low probability of error.

- The rate 1/2 convolutional decoding is now the best performer, when convolutional decoding is concerned, while the rate 3/4 scheme exhibits the worst performance.

Finally, in Figure 14.114, the associated 16QAM results are presented. Notice that the stop-and-go algorithm was again excluded from the results. Furthermore, we observe a high-performance difference between the Benveniste–Goursat algorithm and the modified CMA. In the previous cases we did not observe such a significant difference. The difference in this case is that the channel exhibits an increased delay spread. This illustrated the capability of the equalizers to cope with more widespread multipaths, while keeping the equalizer order constant at 10. The Benveniste–Goursat equalizer was more efficient than the modified CMA in this case.

It is interesting to note that in this case the performance of the different coding rates was again in the expected order: the rate $\frac{1}{2}$ scheme is the best, followed by the rate $\frac{3}{4}$ scheme and then the rate $\frac{7}{8}$ scheme.

If we compare the performance of the system employing QPSK and 16QAM over the two-symbol delay two-path channel of Figure 14.106, we again observe that 16QAM can be incorporated into the DVB system if an extra 5 dB of SNR per bit is affordable in power budget terms. Here, only the B-G algorithm is worth considering out of the three linear equalizers of Table 14.26. This observation was made by comparing the performance of the DVB system when employing the Benveniste–Goursat equalizer and the half-rate convolutional coder in Figures 14.113 and 14.114.

14.8.5.3 Performance Summary of the DVB-S System

Table 14.27 provides an approximation of the convergence speed of each blind equalization algorithm of Table 14.26. PSP exhibited the fastest convergence, followed by the Benveniste–Goursat algorithm. In our simulations the convergence was quantified by observing the slope

14.8. SATELLITE-BASED VIDEO BROADCASTING

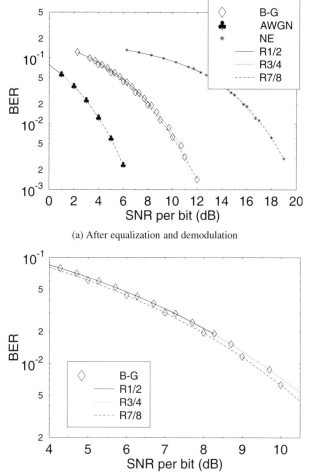

(a) After equalization and demodulation

(b) Same as (a) but enlarged in order to show performance difference of the blind equalizer, when different convolutional code rates are used

Figure 14.112: Average BER versus SNR per-bit performance after equalization and demodulation for QPSK modulation over the two-symbol delay two-path channel of Figure 14.106 (B-G = Benveniste–Goursat).

of the BER curve and finding when this curve was reaching the associated residual BER, implying that the BER has reached its steady-state value. Figure 14.115 gives an illustrative example of the equalizer's convergence for 16QAM. The stop-and-go algorithm converges significantly slower than the other algorithms, which can also be seen from Table 14.27. This happens because during start-up the algorithm is deactivated most of the time. The effect becomes more severe with increasing QAM order.

Figure 14.116 portrays the corresponding reconstructed video quality in terms of the average peak signal-to-noise ratio (PSNR) versus channel SNR (CSNR) for the one-symbol delay and two-symbol delay two-path channel models of Figure 14.106. The PSNR is defined

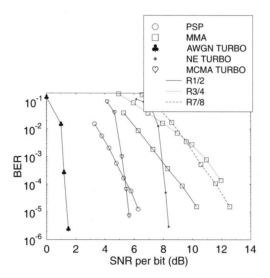

(a) PSP and linear equalizers

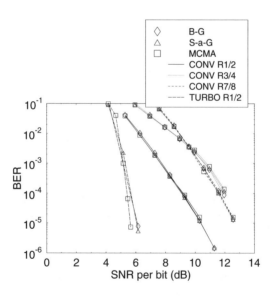

(b) Linear equalizers only

Figure 14.113: Average BER versus SNR per-bit performance after convolutional or turbo decoding for QPSK modulation over the two-symbol delay two-path channel of Figure 14.106 (B-G = Benveniste–Goursat; S-a-G = stop-and-go; MCMA = modified constant modulus algorithm; PSP = per-survivor processing).

14.8. SATELLITE-BASED VIDEO BROADCASTING

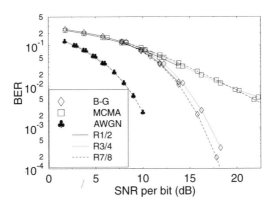

(a) After equalization and demodulation

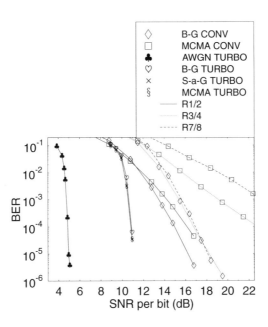

(b) After viterbi or turbo decoding

Figure 14.114: Average BER versus SNR per-bit performance (a) after equalization and demodulation and (b) after Viterbi or turbo decoding for 16QAM over the two-symbol delay two-path channel of Figure 14.106 (B-G = Benveniste–Goursat; S-a-G = stop-and-go; MCMA = modified constant modulus algorithm; PSP = per-survivor processing).

Table 14.27: Equalizer Convergence Speed (in miliseconds) Measured in the Simulations, Given as an Estimate of Time Required for Convergence When 1/2 Rate Puncturing Is Used (x sym = x-Symbol Delay Two-path Channel and x Can Take Either the Value 1 or 2)

	B-G	MCMA	S-a-G	PSP
QPSK 1 sym	73	161	143	0.139
QPSK 2 sym	73	143	77	0.139
16QAM 1 sym	411	645	1393	—
16QAM 2 sym	359	411	1320	—

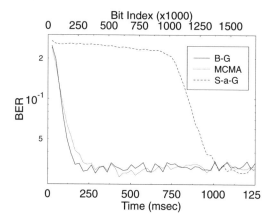

Figure 14.115: Learning curves for 16QAM, one-symbol delay two-path channel at SNR = 18 dB.

as follows:

$$\text{PSNR} = 10 \log_{10} \frac{\sum_{n=0}^{N} \sum_{m=0}^{M} 255^2}{\sum_{n=0}^{N} \sum_{m=0}^{M} \Delta^2}, \quad (14.44)$$

where Δ is the difference between the uncoded pixel value and the reconstructed pixel value. The variables M and N refer to the dimension of the image. The maximum possible 8-bit represented pixel luminance value of 255 was used in Equation 14.44 in order to mitigate the PSNR's dependence on the video material used. The average PSNR is then the mean of the PSNR values computed for all the images constituting the video sequence.

Tables 14.28 and 14.29 provide a summary of the DVB satellite system's performance tolerating a PSNR degradation of 2 dB, which was deemed to be nearly imperceptible in terms of subjective video degradations. The average BER values quoted in the tables refer to the average BER achieved after Viterbi or turbo decoding. The channel SNR is quoted in association with the 2 dB average video PSNR degradation, since the viewer will begin to perceive video degradations due to erroneous decoding of the received video around this threshold.

Tables 14.30 and 14.31 provide a summary of the SNR per bit required for the various system configurations. The BER threshold of 10^{-4} was selected here, because of this

14.8. SATELLITE-BASED VIDEO BROADCASTING

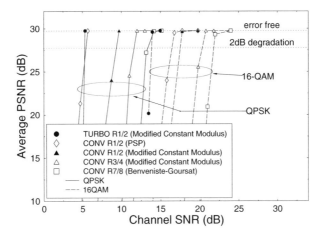

(a) One-symbol delay two-path channel model

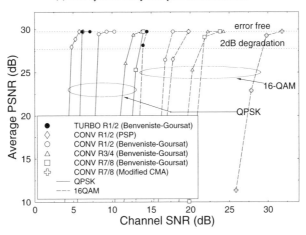

(b) Two-symbol delay two-path channel model

Figure 14.116: Average PSNR versus channel SNR for (a) the one-symbol delay two-path channel model and (b) the two-symbol delay two-path channel model of Figure 14.106 at a video bitrate of 2.5 Mbit/s using the "Football" video sequence.

average BER after Viterbi or turbo decoding, the RS decoder becomes effective, guaranteeing near error-free performance. This also translates into near unimpaired reconstructed video quality.

Finally, in Table 14.32 the QAM symbol rate or baud rate is given for different puncturing rates and for different modulation schemes, based on the requirement of supporting a video bitrate of 2.5 Mbit/sec. We observe that the baud rate is between 0.779 and 2.73 MBd, depending on the coding rate and the number of bits per modulation symbol.

Table 14.28: Summary of Performance Results over the Dispersive One-symbol Delay Two-path AWGN Channel of Figure 14.106 Tolerating a PSNR Degradation of 2 dB

Mod.	Equalizer	Code	CSNR (dB)	E_b/N_0
QPSK	PSP	R = 1/2	5.3	5.3
QPSK	MCMA	Turbo (1/2)	5.2	5.2
16QAM	MCMA	Turbo (1/2)	13.6	10.6
QPSK	MCMA	Conv (1/2)	9.1	9.1
16QAM	MCMA	Conv (1/2)	17.2	14.2
QPSK	MCMA	Conv (3/4)	11.5	9.7
16QAM	MCMA	Conv (3/4)	20.2	15.4
QPSK	B-G	Conv (7/8)	13.2	10.8
16QAM	B-G	Conv (7/8)	21.6	16.2

Table 14.29: Summary of Performance Results over the Dispersive Two-symbol Delay Two-path AWGN Channel of Figure 14.106 Tolerating a PSNR Degradation of 2 dB

Mod.	Equalizer	Code	CSNR (dB)	E_b/N_0
QPSK	PSP	R = 1/2	4.7	4.7
QPSK	B-G	Turbo (1/2)	5.9	5.9
16QAM	B-G	Turbo (1/2)	13.7	10.7
QPSK	B-G	Conv (1/2)	8.0	8.0
16QAM	B-G	Conv (1/2)	17.0	14.0
QPSK	B-G	Conv (3/4)	12.1	10.3
16QAM	B-G	Conv (3/4)	21.1	16.3
QPSK	B-G	Conv (7/8)	13.4	11.0
16QAM	MCMA	Conv (7/8)	29.2	23.8

Table 14.30: Summary of System Performance Results over the Dispersive One-symbol Delay Two-path AWGN Channel of Figure 14.106 Tolerating an Average BER of 10^{-4}, which was Evaluated after Viterbi or Turbo Decoding but before RS Decoding

Mod.	Equalizer	Code	E_b/N_0
QPSK	PSP	R = 1/2	6.1
QPSK	MCMA	Turbo (1/2)	5.2
16QAM	MCMA	Turbo (1/2)	10.7
QPSK	MCMA	Conv (1/2)	11.6
16QAM	MCMA	Conv (1/2)	15.3
QPSK	MCMA	Conv (3/4)	10.5
16QAM	MCMA	Conv (3/4)	16.4
QPSK	B-G	Conv (7/8)	11.8
16QAM	B-G	Conv (7/8)	17.2

14.8. SATELLITE-BASED VIDEO BROADCASTING

Table 14.31: Summary of System Performance Results over the Dispersive Two-symbol Delay Two-path AWGN Channel of Figure 14.106 Tolerating an Average BER of 10^{-4}, which was Evaluated after Viterbi or Turbo Decoding but before RS Decoding

Mod.	Equalizer	Code	E_b/N_0
QPSK	PSP	R = 1/2	5.6
QPSK	B-G	Turbo (1/2)	5.7
16QAM	B-G	Turbo (1/2)	10.7
QPSK	B-G	Conv (1/2)	9.2
16QAM	B-G	Conv (1/2)	15.0
QPSK	B-G	Conv (3/4)	12.0
16QAM	B-G	Conv (3/4)	16.8
QPSK	B-G	Conv (7/8)	11.7
16QAM	MCMA	Conv (7/8)	26.0

Table 14.32: The Channel Bitrate for the Three Different Punctured Coding Rates and for the Two Modulations

Punctured Rate	4QAM Baud Rate (MBd)	16QAM Baud Rate (MBd)
1/2	2.73	1.37
3/4	1.82	0.909
7/8	1.56	0.779

14.8.6 Summary and Conclusions on the Turbo-coded DVB System

In this section, we have investigated the performance of a turbo-coded DVB system in a satellite broadcast environment. A range of system performance results was presented based on the standard DVB-S scheme, as well as on a turbo-coded scheme in conjunction with blind-equalized QPSK/16QAM. The convolutional code specified in the standard system was substituted with turbo coding, which resulted in a substantial coding gain of approximately 4–5 dB. We have also shown that 16QAM can be utilized instead of QPSK if an extra 5 dB SNR per bit gain is added to the link budget. This extra transmitted power requirement can be eliminated upon invoking the more complex turbo codec, which requires lower transmitted power for attaining the same performance as the standard convolutional codecs.

Our future work will focus on extending the DVB satellite system to support mobile users for the reception of satellite broadcast signals. The use of blind turbo equalizers will also be investigated in comparison to conventional blind equalizers. Further work will also be dedicated to trellis-coded modulation (TCM) and turbo trellis-coded modulation (TTCM)-based orthogonal frequency division multiplexed (OFDM) and single-carrier equalized modems. The impact on the system performance by employing various types of turbo interleavers and turbo codes is also of interest. A range of further wireless video communications issues are addressed in [161].

14.9 Summary and Conclusions

In this concluding chapter, we attempted to provide the reader with a system-oriented overview in the context of various adaptive and/or iterative transceivers. In all the systems investigated we have employed the H.263 video codec, but the channel coding and transmission schemes were different. The channel codecs spanned from conventional low-complexity binary BCH and nonbinary RS codes to complex iteratively decoded turbo codecs.

In Section 14.1, we investigated the feasibility of real-time wireless videotelephony over GSM/GPRS links and quantified the achievable performance advantages with the advent of turbo equalization invoking joint channel decoding and channel equalization. Interactive videotelephony is feasible over GSM/GPRS even without turbo equalization and — as expected — the attainable video performance improves when the number of GPRS timeslots is increased. Apart from the higher associated bitrate the length of the interleaver can be increased, which is also beneficial in overall error resilience terms.

HSDPA-style [198] burst-by-burst adaptive AQAM systems were also employed in conjunction with direct-sequence spreading, as it was demonstrated in Section 14.2. Explicitly, in Section 14.2 multi-user detection assisted burst-by-burst adaptive AQAM/CDMA systems were studied, which attained a near-single-user performance.

In recent years OFDM has received a tremendous attention and has been standardized for employment in a whole host of applications. Specifically, the IEEE 802.11 and the High-Performance Local Area Network standard known as HIPERLAN II employ OFDM in predominantly indoor LAN environments. Similarly, both the Pan-European terrestrial Digital Video Broadcast (DVB) system and the Digital Audio Broadcast (DAB) systems opted for advocating OFDM. Motivated by these trends, we quantified the expected video performance of a LAN-type and a cellular-type high-rate, high-quality interactive OFDM video system for transmission over strongly dispersive channels.

The concept of symbol-by-symbol adaptive OFDM systems was introduced in Section 14.3, which converted the lessons of the single-carrier burst-by-burst adaptive modems to OFDM modems. It was found that in indoor scenarios — where low vehicular/pedestrian speeds prevail and hence the channel quality does not change dramatically between the instant of channel quality estimation and the instant of invoking this knowledge at the remote transmitter — symbol-by-symbol adaptive OFDM outperforms fixed-mode OFDM. More specifically, a high-quality group of subcarriers is capable of conveying a higher number of bits per subcarrier at a given target transmission frame error ratio than the low-quality subcarrier groups. Furthermore, it was found beneficial to control the number of video bits per OFDM symbol as a function of time, since in drastically faded instances the video bitrate could be reduced and hence a less heavily loaded OFDM symbol yielded fewer transmission errors. An interesting trade-off was, however, that upon loading the OFDM symbols with more bits allowed us to use a higher turbo coding interleaver and channel interleaver, which improved the system's robustness up to a limit, resulting in an optimum OFDM symbol loading.

14.10. WIRELESS VIDEO SYSTEM DESIGN PRINCIPLES

The last two sections of this chapter considered turbo-coded terrestrial and satellite based DVB broadcast systems, respectively. Specifically, Section 14.7 employed an OFDM-based system, while Section 14.8 used blind-equalized single-carrier modems. An interesting conclusion of our investigations was that upon invoking convolutional turbo coding — rather than the standard convolutional codecs — the system's robustness improves quite dramatically. Hence the required SNR is reduced sufficiently, in order to be able to support, for example, 4 bit/symbol transmissions, instead of 2 bit/symbol signaling. This then potentially allows us to double the associated video bitrate within a given bandwidth at the cost of an acceptable implementational complexity increase when replacing the standard convolutional codecs by turbo channel codecs.

14.10 Wireless Video System Design Principles

We conclude that 2G and 3G interactive cordless as well as cellular wireless videophony using the H.263 ITU codec is realistic in a packet acknowledgment assisted TDMA/TDD, CDMA/TDD, or OFDM/TDD system both with and without power control following the system design guidelines summarized below:

- In time-variant wireless environments adaptive multimode transceivers are beneficial, in particular, if the geographic distribution of the wireless channel capacity — i.e., the instantaneous channel quality — varies across the cell due to path loss, slow-fading, fast-fading, dispersion, co-channel interference, etc.

- In order to support the operation of intelligent multimode transceivers a video source codec is required that is capable of conveniently exploiting the variable-rate, time-variant capacity of the wireless channel by generating always the required number of bits that can be conveyed in the current modem mode by the transmission burst. This issue was discussed throughout this chapter.

- The H.263 video codec's local and remote video reconstruction frame buffers have to be always identically updated, which requires a reliable, strongly protected acknowledgment feedback flag. The delay of this flag has to be kept as low as possible, and a convenient mechanism for its transmission is constituted by its superposition on the reverse-direction packets in TDD systems. Alternatively, the system's control channel can be used for conveying both the video packet acknowledgment flag and the transmission mode requested by the receiver. The emerging 3G CDMA systems provide a sufficiently high bitrate to convey the necessary side information. The associated timing issues were summarized in Figure 10.21.

- The acknowledgment feedback flag can be conveniently error protected by simple majority logic decoded repetition codes, which typically provide more reliable protection than more complex similar-length error correction codes.

- For multimode operation an appropriate adaptive packetization algorithm is required that controls the transmission, storage, discarding, and updating of video packets.

- For multimode operation in cellular systems an appropriate fixed FER power-control algorithm is required, that is capable of maintaining a constant transmission burst error rate, irrespective of the modulation mode used, since it is the FER, rather than the bit error rate, which determines the associated video quality.

- We emphasize again that wireless videotelephony over 2G systems is realistic using both proprietary and standard video codecs, and the same design principles can also be applied in the context of the emerging 3G CDMA systems, as it was demonstrated in the context of our JD-ACDMA design example.

- The feasibility of using similar design principles has been shown also for high-rate WATM OFDM modems in this closing chapter.

Future video transceiver performance improvements are possible, while retaining the basic system design principles itemized above. The image compression community completed the MPEG4 video standard [536–538], while the wireless communications community has commenced research toward the fourth generation of mobile radio standards. All in all, this is an exciting era for wireless video communications researchers, bringing about new standards to serve a forthcoming wireless multimedia age.

Glossary

16CIF	Sixteen Common Intermediate Format Frames are sixteen times as big as CIF frames and contain 1408 pixels vertically and 1152 pixels horizontally
2G	Second Generation
3G	Third Generation
3GPP	Third Generation Partnership Project
4CIF	Four Common Intermediate Format Frames are four times as big as CIF frames and contain 704 pixels vertically and 576 pixels horizontally
AB	Access Burst
ACCH	Associated Control Chanel
ACELP	Algebraic Code Excited Linear Predictive (ACELP) Speech Codec
ACF	Autocorrelation Function
ACL	Autocorrelation
ACO	Augmented Channel Occupancy matrix, which contains the channel occupancy for the local and surrounding base stations. Often used by locally distributed DCA algorithms to aid allocation decisions.
ACTS	Advanced Communications Technologies and Services. The fourth framework for European research (1994–98). A series of consortia consisting of universities and industrialists considering future communications systems.
ADC	Analog-to-Digital Converter
ADPCM	Adaptive Differential Pulse Coded Modulation
AGCH	Access Grant Control Chanel
AI	Acquisition Indicator
AICH	Acquisition Indicator Chanel
ANSI	American National Standards Institute
ARIB	Association of Radio Industries and Businesses
ARQ	Automatic Repeat Request, Automatic request for retransmission of corrupted data
ATDMA	Advanced Time Division Multiple Access
ATM	Asynchronous Transfer Mode
AUC	Authentication Center

AV.26M	A draft recommendation for transmitting compressed video over error-prone channels, based on the H.263 [259] video codec
AWGN	Additive White Gaussian Noise
B-ISDN	Broadband ISDN
BbB	Burst-by-Burst
BCCH	Broadcast Control Chanel
BCH	Bose-Chaudhuri-Hocquenghem, a class of forward error correcting codes (FEC)
BCH Codes	Bose-Chaudhuri-Hocquenghem (BCH) Codes
BER	Bit Error Rate, the fraction of the bits received incorrectly
BN	Bit Number
BPSK	Binary Phase Shift Keying
BS	Base Station
BSIC	Base Station Identifier Code
BTC	Block Truncation Coding
CBER	Channel Bit Error Rate, the bit error rate before FEC correction
CBP	Coded Block Pattern, a H.261 video codec symbol that indicates which of the blocks in the macroblock are active
CBPB	A fixed-length codeword used by the H.263 video codec to convey the coded block pattern for bidirectionally predicted (B) blocks
CBPY	A variable-length codeword used by the H.263 video codec to indicate the coded block pattern for luminance blocks
CC	Convolutional Code
CCCH	Common Control Channel
CCITT	Now ITU, standardization group
CCL	Cross-correlation
CD	Code Division, a multiplexing technique whereby signals are coded and then combined in such a way that they can be separated using the assigned user signature codes at a later stage
CDF	Cumulative Density Function, the integral of the Probability Density Function (PDF)
CDMA	Code Division Multiple Access
CELL_BAR_ACCESS	Boolean flag to indicate whether the MS is permitted
CIF	Common Intermediate Format Frames containing 352 pixels vertically and 288 pixels horizontally
CIR	Carrier to Interference Ratio, same as SIR
COD	A one-bit codeword used by the H.263 video codec that indicates whether the current macroblock is empty or nonempty
CPICH	Common Pilot Channel
CT2	British Second Generation Cordless Phone Standard
CWTS	China Wireless Telecommunication Standard
DAB	Digital Audio Broadcasting
DAC	Digital-to-Analog Convertor
DAMPS	Pan-American Digital Advanced Phone System, IS-54

DB	Dummy Burst
DC	Direct Current, normally used in electronic circuits to describe a power source that has a constant voltage, as opposed to AC power in which the voltage is a sine wave. It is also used to describe things that are constant, and hence have no frequency component.
DCA	Dynamic Channel Allocation
DCH	Dedicated transport Channel
DCS1800	A digital mobile radio system standard, based on GSM but operating at 1.8 GHz at a lower power
DCT	A Discrete Cosine Transform that transforms data into the frequency domain. Commonly used for video compression by removing high-frequency components of the video frames
DECT	A Pan-European digital cordless telephone standard
DL	Down-Link
DPCCH	Dedicated Physical Control Channel
DPCH	Dedicated Physical Channel
DPCM	Differential Pulse Coded Modulation
DPDCH	Dedicated Physical Data Channel
DQUANT	A fixed-length coding parameter used to differential change the current quantizer used by the H.263 video codec
DS–CDMA	Direct Sequence Code Division Multiple Access
DSMA-CD	Digital Sense Multiple Access-Collision Detection
DTTB	Digital Terrestrial Television Broadcast
DVB-T	Terrestrial Pan-European Digital Video Broadcast Standard
EIR	Equipment Identity Register
EMC	Electromagnetic Compatibility
EOB	An End-Of-Block variable-length symbol used to indicate the end of the current block in the H.261 video codec
EREC	Error-Resilient Entropy Coding. A coding technique that improves the robustness of variable-length coding by allowing easier resynchronization after errors
ERPC	Error-Resilient Position Coding, a relative of the coding scheme known as Error-Resilient Entropy Coding (EREC)
ETSI	European Telecommunications Standards Institute
EU	European Union
FA	First Available, a simple centralized DCA scheme that allocates the first channel found that is not reused within a given preset reuse distance
FACCH	Fast Associated Control Channel
FACH	Forward Access Channel
FAW	Frame Alignment Word
FBER	Feedback Error Ratio, the ratio of feedback acknowledgment messages that are received in error
FCA	Fixed Channel Allocation
FCB	Frequency Correction Burst

FCCH	Frequency Correction Channel
FD	Frequency Division, a multiplexing technique whereby different frequencies are used for each communications link
FDD	Frequency-Division Duplex, a multiplexing technique whereby the forward and reverse links use a different carrier frequency
FDM	Frequency Division Multiplexing
FDMA	Frequency Division Multiple Access, a multiple access technique whereby Frequency Division (FD) is used to provide a set of access channels
FEC	Forward Error Correction
FEF	Frame Error Flag
FER	Frame Error Rate
FH	Frequency Hopping
FIFO	First-In First-Out, a queuing strategy in which elements that have been in the queue longest are served first
FN	TDMA Frame Number
FPLMTS	Future Public Land Mobile Telecommunication System
FRAMES	Future Radio Wideband Multiple Access System
GBSC	Group of Blocks (GOB) Start Code, used by the H.261 and H.263 video codecs to regain synchronization, playing a similar role to PSC
GEI	Functions similar to PEI but in the GOB layer of the H.261 video codec
GFID	A fixed-length codeword used by H.263 video codec to aid correct resynchronization after an error
GMSK	Gaussian Mean Shift Keying, a modulation scheme used by the Pan-European GSM standard by virtue of its spectral compactness
GN	Group of blocks Number, an index number for a GOB used by the H.261 and H.263 video codecs
GOB	Group of Blocks, a term used by the H.261 and H.263 video codecs, consisting of a number of macroblocks
GOS	Grade of Service, a performance metric to describe the quality of a mobile radio network
GP	Guard Period
GPS	Global Positioning System
GQUANT	Group of blocks Quantizer, a symbol used by the H.261 and H.263 video codecs to modify the quantizer used for the GOB
GSM	A Pan-European digital mobile radio standard, operating at 900 MHz
GSPARE	Functions similar to PSPARE but in the GOB layer of the H.261 video codec
H.261	A video coding standard [258], published by the ITU in 1990
H.263	A video coding standard [259], published by the ITU in 1996
HC	Huffman Coding
HCA	Hybrid Channel Allocation, a hybrid of FCA and DCA
HCS	Hierarchical Cell Structure
HDTV	High-Definition Television
HLR	Home Location Register

GLOSSARY

HO	Handover
HTA	Highest interference below Threshold Algorithm, a distributed DCA algorithm also known as MTA. The algorithm allocates the most interfered channel, whose interference is below the maximum tolerable interference threshold.
IF	Intermediate Frequency
IMSI	International Mobile Subscriber Identity
IMT-2000	International Mobile Telecommunications-2000
IMT2000	Intelligent Mobile Telecommunications in the Year 2000, Japanese Initiative for 3rd Generation Cellular Systems
IS-54	Pan-American Digital Advanced Phone System
IS-95	North American mobile radio standard that uses CDMA technology
ISDN	Integrated Services Digital Network, digital replacement of the analog telephone network
ISI	Intersymbol Interference, Inter-Subcarrier Interference
ITU	International Telecommunications Union, formerly the CCITT, standardization group
ITU-R	International Mobile Telecommunication Union – Radiocommunication Sector
JDC	Japanese Digital Cellular Standard
JPEG	"Lossy" DCT-based Still Picture Coding Standard
LFA	Lowest Frequency below threshold Algorithm, a distributed DCA algorithm that is a derivative of the LTA algorithm, the difference being that the algorithm attempts to reduce the number of carrier frequencies being used concurrently
LIA	Least Interference Algorithm, a distributed DCA algorithm that assigns the channel with the lowest measured interference that is available.
LODA	Locally Optimized Dynamic Assignment, a centralized DCA scheme, which bases its allocation decisions on the future blocking probability in the vicinity of the cell
LOLIA	Locally Optimized Least Interference Algorithm, a locally distributed DCA algorithm that allocates channels using a hybrid of the LIA and an ACO matrix
LOMIA	Locally Optimized Most Interference Algorithm, a locally distributed DCA algorithm that allocates channels using a hybrid of the MTA and an ACO matrix
LP filtering	Low-Pass filtering
LP-DDCA	Local Packing Dynamic Distributed Channel Assignment, a locally distributed DCA algorithm that assigns the first channel available that is not used by the surrounding base stations, whose information is contained in an ACO matrix
LPF	Low-Pass Filter
LSB	Least Significant Bit
LSR	Linear Shift Register
LTA	Least interference below Threshold Algorithm, a distributed DCA algorithm that allocates the least interfered channel, whose interference is below a preset maximum tolerable interference level
LTI	Linear Time-Invariant
MA	Abbreviation for Miss America, a commonly used head and shoulders video sequence referred to as Miss America
Macroblock	A grouping of 8 by 8 pixel blocks used by the H.261 and H.263 video codecs. Consists of four luminance blocks and two chrominance blocks

MAI	Multiple Access Interference
MAP	Maximum–A–Posteriori
MB	Macroblock
MBA	Macroblock address symbol used by the H.261 video codec, indicating the position of the macroblock in the current GOB
MBS	Mobile Broadband System
MC	Motion Compensation
MCBPC	A variable-length codeword used by the H.263 video codec to convey the macroblock type and the coded block pattern for the chrominance blocks
MCER	Motion Compensated Error Residual
MDM	Modulation Division Multiplexing
MF-PRMA	Multi-Frame Packet Reservation Multiple Access
MFlop	Mega Flop, 1 million floating point operations per second
MODB	A variable-length coding parameter used by the H.263 video codec to indicate the macroblock mode for bidirectionally predicted (B) blocks
MPEG	Motion Picture Expert Group, also a video coding standard designed by this group that is widely used
MPG	Multiple Processing Gain
MQUANT	A H.261 video codec symbol that changes the quantizer used by current and future macroblocks in the current GOB
MS	A common abbreviation for Mobile Station
MSC	Mobile Switching Center
MSE	Mean Square Error
MSQ	Mean Square centralized DCA algorithm that attempts to minimize the mean square distance between cells using the same channel
MTA	Most interference below Threshold Algorithm, a distributed DCA algorithm also known as HTA. The algorithm allocates the most interfered channel, whose interference is below the maximum tolerable interference level.
MTYPE	H.261 video codec symbol that contains information about the macroblock, such as coding mode, and flags to indicate whether optional modes are used, like motion vectors, and loop filtering
MV	Motion Vector, a vector to estimate the motion in a frame
MVD	Motion Vector Data symbol used by H.261 and H.263 video codecs
MVDB	A variable-length codeword used by the H.263 video codec to convey the motion vector data for bidirectionally predicted (B) blocks
NB	Normal Burst
NCC	Normalized Channel Capacity
NLF	Nonlinear Filtering
NMC	Network Management Center
NN	Nearest-Neighbor centralized DCA algorithm; allocates a channel used by the nearest cell, which is at least the reuse distance away
NN+1	Nearest-Neighbor-plus-one centralized DCA algorithm; allocates a channel used by the nearest cell, which is at least the reuse distance plus one cell radius away

GLOSSARY

OFDM	Orthogonal Frequency Division Multiplexing
OMC	Operation and Maintenance Center
OVSF	Orthogonal Variable Spreading Factor
P-CCPCH	Primary Common Control Physical Chanel
PCH	Paging Chanel
PCM	Pulse Code Modulation
PCN	Personal Communications Network
PCPCH	Physical Common Packet Chanel
PCS	Personal Communications System, a term used to describe third-generation mobile radio systems in North America
PDF	Probability Density Function
PDSCH	Physical Down-link Shared Chanel
PEI	Picture layer extra insertion bit, used by the H.261 video codec, indicating that extra information is to be expected
PFU	Partial Forced Update
PGZ	Peterson–Gorenstein–Zierler (PGZ) Decoder
PHP	Japanese Personal Handyphone Phone System
PI	Page Indicator
PICH	Page Indicator Chanel
PLMN	Public Land Mobile Network
PLMN_PERMITTED	Boolean flag to indicate, whether the MS is permitted
PLMR	Public Land Mobile Radio
PLR	Packet-Loss Ratio
PP	Partnership Project
PQUANT	A fixed-length codeword used by the H.263 video codec to indicate the quantizer to use for the next frame
PRACH	Physical Random Access Chanel
PRMA	Packet Reservation Multiple Access, a statistical multiplexing arrangement contrived to improve the efficiency of conventional TDMA systems, by detecting inactive speech segments using a voice activity detector, surrendering them and allocating them to subscribers contending to transmit an active speech packet
PRMA++	PRMA system allowing contention only in the so-called contention slots, which protect the information slots from contention and collisions
PSAM	Pilot Symbol-Assisted Modulation, a technique whereby known symbols (pilots) are transmitted regularly. The effect of channel fading on all symbols can then be estimated by interpolating between the pilots.
PSC	Picture Start Code, a preset sequence used by the H.261 and H.263 video codecs, that can be searched for to regain synchronization after an error
PSD	Power Spectral Density
PSNR	Peak Signal-to-Noise Ratio, noise energy compared to the maximum possible signal energy. Commonly used to measure video image quality
PSPARE	Picture layer extra information bits, indicated by a PEI symbol in H.261 video codec

PSTN	Public Switched Telephone Network
PTYPE	Picture layer information, used by H.261 and H.263 video codecs to transmit information about the picture, e.g. resolution, etc.
QAM	Quadrature Amplitude Modulation
QCIF	Quarter Common Intermediate Format Frames containing 176 pixels vertically and 144 pixels horizontally
QMF	Quadrature Mirror Filtering
QN	Quarter bit Number
QoS	Quality of Service
QT	Quad-Tree
RACE	Research in Advanced Communications Equipment Programme in Europe, from June 1987 to December 1995
RACH	Random Access Chanel
RC filtering	Raised-Cosine filtering
RF	Radio Frequency
RFCH	Radio Frequency Chanel
RFN	Reduced TDMA frame number in GSM
RING	A centralized DCA algorithm that attempts to allocate channels in one of the cells, which is at least the reuse distance away that forms a "ring" of cells
RLC	Run-Length Coding
RPE	Regular Pulse Excited
RS codes	Reed–Solomon (RS) codes
RSSI	Received Signal Strength Indicator, commonly used as an indicator of channel quality in a mobile radio network
RTT	Radio Transmission Technology
RXLEV	Received signal level: parameter used in hangovers
RXQUAL	Received signal quality: parameter used in hangovers
S-CCPCH	Secondary Common Control Physical Chanel
SAC	Syntax-based Arithmetic Coding, an alternative to variable-length coding, and a variant of arithmetic coding
SACCH	Slow Associated Control Chanel
SB	Synchronization Burst
SCH	Synchronization Chanel
SCS	Sequential Channel Search distributed DCA algorithm that searches the available channels in a predetermined order, picking the first channel found, which meets the interference constraints
SDCCH	Stand-alone Dedicated Control Chanel
SF	Spreading Factor
SINR	Signal-to-Interference plus Noise ratio, same as signal-to-noise ratio (SNR) when there is no interference
SIR	Signal-to-Interference Ratio
SNR	Signal-to-Noise Ratio, noise energy compared to the signal energy

GLOSSARY

SPAMA	Statistical Packet Assignment Multiple Access
SQCIF	Sub-Quarter Common Intermediate Format Frames containing 128 pixels vertically and 96 pixels horizontally
SSC	Secondary Synchronization Codes
TA	Timing Advance
TB	Tailing Bits
TC	Trellis Coded
TCH	Traffic Chanel
TCH/F	Full-rate traffic channel
TCH/F2.4	Full-rate 2.4 kbps data traffic channel
TCH/F4.8	Full-rate 4.8 kbps data traffic channel
TCH/F9.6	Full-rate 9.6 kbps data traffic channel
TCH/FS	Full-rate speech traffic channel
TCH/H	Half-rate traffic channel
TCH/H2.4	Half-rate 2.4 kbps data traffic channel
TCH/H4.8	Half-rate 4.8 kbps data traffic channel
TCM	Trellis Code Modulation
TCOEFF	An H.261 and H.263 video codec symbol that contains the transform coefficients for the current block
TD	Time Division, a multiplexing technique whereby several communications links are multiplexed onto a single carrier by dividing the channel into time periods, and assigning a time period to each communications link
TDD	Time-Division Duplex, a technique whereby the forward and reverse links are multiplexed in time.
TDMA	Time Division Multiple Access
TFCI	Transport-Format Combination Indicator
TIA	Telecommunications Industry Association
TN	Time slot Number
TPC	Transmit Power Control
TR	Temporal Reference, a symbol used by H.261 and H.263 video codecs to indicate the real-time difference between transmitted frames
TS	Technical Specifications
TTA	Telecommunications Technology Association
TTC	Telecommunication Technology Committee
TTIB	Transparent Tone In Band
UHF	Ultra High Frequency
UL	Up-Link
UMTS	Universal Mobile Telecommunications System, a future Pan-European third-generation mobile radio standard
UTRA	Universal Mobile Telecommunications System Terrestrial Radio Access
VA	Viterbi Algorithm
VAD	Voice Activity Detection

VAF	Voice Activity Factor, the fraction of time the voice activity detector of a speech codec is active
VE	Viterbi Equalizer
VL	Variable Length
VLC	Variable-Length Coding/codes
VLR	Visiting Location Register
VQ	Vector Quantization
W-CDMA	Wideband Code Division Multiple Access
WARC	World Administrative Radio Conference
WATM	Wireless Asynchronous Transfer Mode (ATM)
WLAN	Wireless Local Area Network
WN	White Noise
WWW	World Wide Web, the name given to computers that can be accessed via the Internet using the HTTP protocol. These computers can provide information in a easy-to-digest multimedia format using hyperlinks

Bibliography

[1] P. Howard and J. Vitter, "Arithmetic coding for data compression," *Proceedings of the IEEE*, vol. 82, pp. 857–865, June 1994.

[2] R. M. Fano, "Transmission of Information," in *M.I.T. Press* Cambridge, MA, 1949.

[3] J. B. Connell, "A Huffman-Shannon-Fano code," *Proceedings of the IEEE*, vol. 61, pp. 1046–1047, 1973.

[4] D. A. Huffman, "A method for the construction of minimum-redundancy codes," *Proceedings of IRE*, vol. 20, 9, pp. 1098–1101, September 1952.

[5] L. Hanzo, P. J. Cherriman and J. Street, *Wireless Video Communications: Second to Third Generation Systems and Beyond*. Piscataway, NJ: IEEE Press, 2001.

[6] V. Baskaran and K. Konstantinides, *Image and Video Compression Standards*. Boston: Kluwer Academic Publishers, 1995.

[7] J. L. Mitchell, W. B. Pennebaker, C. E. Fogg and D. J. LeGall, *MPEG Video Compression Standard*. New York: Chapman & Hall, 1997.

[8] B. G. Haskell, A. Puri and A. N. Netravali, *Digital Video: An Introduction to MPEG-2*. New York: Chapman & Hall, 1997.

[9] J. R. Jain, *Fundamentals of Digital Image Processing*. Englewood Cliffs, NJ: Prentice-Hall, 1989.

[10] A. Jain, *Fundamentals of Digital Image Processing*. Englewood Cliffs, NJ: Prentice-Hall, 1989.

[11] R. W. Burns, *A History of Television Broadcasting*. IEE History of Technology Series, Number 7, 1986.

[12] P. Symes, *Video Compression Demystified*. McGraw-Hill, 2001.

[13] A. Habibi, "An adaptive strategy for hybrid image coding," *IEEE Transactions on Communications*, pp. 1736–1753, December 1981.

[14] K. R. Rao and P. Yip, *Discrete Cosine Transform - Algorithms, Advantages, Applications*. San Diego, CA: Academic Press, 1990.

[15] B. Ramamurthi and A. Gersho, "Classified vector quantization of images," *IEEE Transactions on communications*, vol. COM-34, pp. 1105–1115, November 1986.

[16] C. W. Rutledge, "Vector DPCM: vector predictive coding of color images," *Proceedings of the IEEE Global Telecommunications Conference*, pp. 1158–1164, September 1986.

[17] M. Yuen and H. R. Wu, "A survey of hybrid MC/DPCM/DCT video coding distortions," *Signal Processing*, vol. 70, pp. 247–278, July 1998.

[18] J. Zhang, M. O. Ahmad and M. N. S. Swamy, "Bidirectional variable size block motion compensation," *Electronics Letters*, vol. 34, pp. 54–53, January 1998.

[19] S.-T. Hsiang and J. W. Woods, "Invertible three-dimensional analysis/synthesis system for video coding with half-pixel-accurate motion compensation," *Proceedings of SPIE 3653, Visual Communications and Image Processing'99*, January 1999.

[20] G. J. Sullivan, T. Wiegand and T. Stockhammer, "Draft H.26L video coding standard for mobile applications," in *Proceedings of IEEE International Conference on Image Processing*, vol. 3, (Thessaloniki, Greece), pp. 573–576, October 2001.

[21] A. Alatan, L. Onural, M. Wollborn, R. Mech, E. Tuncel and T. Sikora, "Image sequence analysis for emerging interactive multimedia services. The European COST211 framework," *IEEE Communications Letters*, vol. 8, pp. 802–813, November 1998.

[22] CCITT/SG XV, "Codecs for videoconferencing using primary digital group transmission," in *Recommendation H.120*, CCITT (currently ITU-T), Geneva, 1989.

[23] G. K. Wallace, "The JPEG still picture compression standard," *Communications of the Association for Computing Machinery*, vol. 34, no. 4, pp. 30–44, 1991.

[24] ITU-T/SG16/Q15, "Video coding for low bitrate communication," in *ITU-T Recommendation H.263, Version 2 (H.263+)*, ITU-T, Geneva, 1998.

[25] ISO/IEC JTC1/SC29/WG11, "Information technology - Generic coding of audio-visual objects.," in *Part 2: Visual. Draft ISO/IEC 14496-2 (MPEG-4), version 1*, ISO/IEC, Geneva, 1998.

[26] ITU-T/SG16/Q15, "Draft for "H.263++" annexes U, V, and W to recommendation H.263," in *Draft*, ITU-T, Geneva, 2000.

[27] Joint Video Team (JVT) of ISO/IEC MPEG and ITU-T VCEG, "Joint Final Committee Draft (JFCD) of joint video specification (ITU-T Rec. H.264 ISO/IEC 14496-10 AVC)," August 2002.

[28] ITU-T/SG15, "Video coding for low bitrate communication," in *ITU-T Recommendation H.263, Version 1*, ITU-T, Geneva, 1996.

[29] CCITT H.261, "Video Codec for audiovisual services at $p \times 64$ kbit/s," 1990.

[30] ISO/IEC JTC1/SC29/WG11, "Information technology - coding of moving pictures and associated audio for digital storage media at up to about 1.5 Mbits/s," in *Part 2: Video. Draft ISO/IEC 11172-2 (MPEG-1)*, ISO/IEC, Geneva, 1991.

[31] ISO/IEC JTC1/SC29/WG11, "Information technology - Generic coding of moving pictures and associated audio," in *Part 2: Video. Draft ISO/IEC 13818-2 (MPEG-2) and ITU-T Recommendation H.262*, ISO/IEC and ITU-T, Geneva, 1994.

[32] ITU-T Experts Group on very low Bitrate Visual Telephony, "ITU-T Recommendation H.263:Video coding for low bitrate communication," December 1995.

[33] MPEG Video Group, "Report of ad-hoc group on the evaluation of tools for nontested functionalities of video submissions to MPEG-4," *Munich meeting, document ISO/IEC/JTC1/SC29/WG11 N0679*, January 1996.

[34] D. E. Pearson, "Developments in model-based video coding," *Proceedings of the IEEE*, vol. 83, pp. 892–906, 5–9 December 1995.

[35] B. Liu and A. Zaccarin, "New fast algorithms for the estimation of block motion vectors," *IEEE Transactions on Circuits and Systems for Video Technology*, vol. 3, pp. 148–157, April 1993.

[36] ITU-R Recommendation BT.601-5 (10/95), "Studio encoding parameters of digital television for standard 4:3 and wide-screen 16:9 aspect ratios".

[37] AT&T, "History of AT&T: Milestone in AT&T history. Online document available at URL:http://www.att.com/history/milestones.html."

[38] D. Cohen, "Specifications for the Network Voice Protocol (NVP)," *Internet Engineering Task Force, RFC 741*, November 1977.

[39] R. Cole, "PVP - A Packet Video Protocol," *Internal Document*, USC/ISI, July 1981.

[40] E. M. Schooler, "A distributed architecture for multimedia conference control," *ISI research report ISI/RR-91-289*, November 1991.

[41] ITU-T Recommendation H.320, "Narrowband ISDN Visual telephone systems and terminal equipment," 1995.

[42] V. Jacobson, "DARTNET planning and review meeting," December 1991.

[43] T. Dorcey, "CU-SeeMe Desktop VideoConferencing Software," *Connexions*, vol. 9, March 1995.

BIBLIOGRAPHY

[44] T. Turletti, "H.261 software codec for videoconferencing over the Internet," in *Rapports de Recherche 1834*, Institut National de Recherche en Informatique et en Automatique (INRIA), Sophia-Antipolis, France, January 1993.

[45] T. Dorcey, "CU-SeeMe Desktop Videoconferencing Software," *Connexions*, pp. 42–45, March 1995.

[46] H. Schulzrinne, "RTP: The real-time transport protocol," in *MCNC 2nd Packet Video Workshop*, vol. 2, (Research Triangle Park, NC), December 1992.

[47] Cornell University, "The CU-SeeMe home page, URL: http://cu-seeme.cornell.edu/."

[48] G. Venditto, "Internet phones - the future is calling," *Internet World*, pp. 40–52, 1996.

[49] Vocaltec Communications Ltd., "The Vocaltec Telephony Gateway. Online document available at URL:http://www.vocaltec.com/products/products.htm."

[50] ITU-T Recommendation H.324, "Terminal for Low Bitrate multimedia communication," 1995.

[51] ITU-T Recommendation T.120, "Data protocols for multimedia conferencing," July 1996.

[52] ITU-T Recommendation H.323, "Visual telephone systems and equipment for LAN which provide a non-guaranteed quality of service," November 1996.

[53] CERN, "Caltech and HP open scientific datacenter," November 1997.

[54] R. Braden et al., "Resource Reservation Protocol (RSVP)," September 1997.

[55] ITU-T Recommendation H.323 Version 2, "Packet-based multimedia communication systems," January 1998.

[56] J. Lennox, J. Rosenberg and H. Schulzrinne, "Common gateway interface for SIP," June 2000.

[57] ITU-T Recommendation H.450.4, "Call hold supplementary service for H.323, Series H: Audiovisual and Multimedia Systems," May 1999.

[58] ITU-T Recommendation H.323 Version 4, "Packet-based multimedia communication systems," November 2000.

[59] S. Emani and S. Miller, "DPCM picture transmission over noisy channels with the aid of a markov model," *IEEE Transactions on Image Processing*, vol. 4, pp. 1473–1481, November 1995.

[60] M. Chan, "The performance of DPCM operating on lossy channels with memory," *IEEE Transactions on Communications*, vol. 43, pp. 1686–1696, April 1995.

[61] CCITT/SG XV, "Video codec for audiovisual services at $p \times 64$ kbit/s," in *Recommendation H.120, CCITT (currently ITU-T)*, (Geneva), 1993.

[62] ISO/IEC JTC1, "Coding of audio-visual objects - Part 2: Visual," April 1999.

[63] ITU-T Recommendation H.263, Version 2, "Video coding for low bitrate communication". International Telecommunications Union, Geneva, January 1998.

[64] M. R. Civanlar, A. Luthra, S. Wenger and W. Zhu, "Special issue on streaming video," *IEEE Transactions on Circuits and Systems for Video Technology*, vol. 11, March 2001.

[65] C. W. Chen, P. Cosman, N. Kingsbury, J. Liang and J. W. Modestino, "Special issue on error resilient image and video transmission," *IEEE Journal on Selected Area in Communications*, vol. 18, June 2001.

[66] P. List, A. Joch, J. Lainema, G. Bjontegaard and M. Karczewicz, "Adaptive deblocking filter," *IEEE Transaction on Circuits and Systems for Video Technology*, vol. 13, pp. 614–619, July 2003.

[67] Y. Wang and Q.-F. Zhu, "Error control and concealment for video communications: A review," *Proceedings of the IEEE*, vol. 86, pp. 974–997, May 1998.

[68] S. Aign and K. Fazel, "Temporal and spatial error concealment techniques for hierarchical MPEG-2 video codec," in *Proceedings IEEE International Conference on Communications ICC'95*, Seattle, WA, pp. 1778–1783, June 1995.

[69] Y. Wang, Q.-F. Zhu and L. Shaw, "Maximally smooth image recovery in transform coding," *IEEE Transactions on Communications*, vol. 41, pp. 1544–1551, October 1993.

[70] S. S. Hemami and T. H.-Y. Meng, "Transform coded image reconstruction exploiting interblock correlation," *IEEE Transactions on Image Processing*, vol. 4, pp. 1023–1027, July 1995.

[71] W.-M. Lam and A. R. Reibman, "An error concealment algorithm for images subject to channel errors," *IEEE Transactions on Image Processing*, vol. 4, pp. 533–542, May 1995.

[72] H. Sun, K. Challapali and J. Zdepski, "Error concealment in digital simulcast AD-HDTV decoder," *IEEE Transactions on Consumer Electronics*, vol. 38, pp. 108–118, August 1992.

[73] R. Aravind, M. R. Civanlar and A. R. Reibman, "Packet loss resilience of MPEG-2 scalable video coding algorithms," *IEEE Transactions on Circuits and Systems for Video Technology*, vol. 6, pp. 426–435, October 1996.

[74] W.-J. Chu and J.-J. Leou, "Detection and concealment of transmission errors in H.261 images," *IEEE Transactions on Circuits and Systems for Video Technology*, vol. 8, pp. 78–84, February 1998.

[75] J. Apostolopoulos, "Error-resilient video compression via multiple state streams," *Proceedings IEEE International Workshop on Very Low Bit Rate Video Coding*, pp. 168–171, October 1999.

[76] Q.-F. Zhu, Y. Wang and L. Shaw, "Coding and cell-loss recovery in DCT based packet video," *IEEE Transactions on Circuits and Systems for Video Technology*, vol. 3, pp. 248–258, June 1993.

[77] B. Haskell and D. Messerschmitt, "Resynchronization of motion compensated video affected by ATM cell loss," *Proceedings IEEE International Conference on Acoustics, Speech and Signal Processing, ICASSP'93*, pp. 545–548, March 1992.

[78] A. Narula and J. S. Lim, "Error concealment techniques for an all-digital high-definition television system," *Proceedings of the SPIE*, vol. 2094, pp. 304–318, November 1993.

[79] W. M. Lam, A. R. Reibman and B. Liu, "Recovery of lost or erroneously received motion vectors," *Proceedings IEEE International Conference on Acoustics, Speech and Signal Processing, ICASSP'93*, vol. 5, pp. 417–420, April 1993.

[80] J. Lu, M. L. Lieu, K. B. Letaief and J. C.-I. Chuang, "Error resilient transmission of H.263 coded video over mobile networks," *Proceedings IEEE International Symposium on Circuits and Systems*, vol. 4, pp. 502–505, June 1998.

[81] M. Ghanbari and V. Seferidis, "Cell-loss concealment in ATM video codecs," *IEEE Transactions on Circuits and Systems for Video Technology*, vol. 3, pp. 238–247, June 1993.

[82] L. H. Kieu and K. N. Ngan, "Cell-loss concealment techniques for layered video codecs in an ATM network," *IEEE Transactions on Image Processing*, vol. 3, pp. 666–677, September 1994.

[83] L. Hanzo, L. L. Yang, E. L. Kuan and K. Yen, *Single- and Multi-Carrier CDMA*. Chichester, UK: John Wiley-IEEE Press, 2003.

[84] L. Hanzo, W. Webb and T. Keller, *Single- and Multi-Carrier Quadrature Amplitude Modulation: Principles and Applications for Personal Communications, WLANs and Broadcasting*. Piscataway, NJ, USA: IEEE Press, 2000.

[85] L. Hanzo, C. H. Wong and M. S. Yee, *Adaptive Wireless Transceivers*. Chichester, UK: John Wiley-IEEE Press, 2002.

[86] L. Hanzo, T. H. Liew and B. L. Yeap, *Turbo Coding, Turbo Equalisation and Space Time Coding for Transmission over Wireless channels*. New York, USA: John Wiley-IEEE Press, 2002.

[87] P. Jung and J. Blanz, "Joint detection with coherent receiver antenna diversity in CDMA mobile radio systems," *IEEE Transactions on Vehicular Technology*, vol. 44, pp. 76–88, February 1995.

[88] L. Hanzo, M. Münster, B.-J. Choi and T. Keller, *OFDM and MC-CDMA*. Chichester, UK: John Wiley and IEEE Press, 2003.

[89] E. Steinbach, N. Färber and B. Girod, "Adaptive playout for low-latency video streaming," in *IEEE International Conference on Image Processing ICIP-01*, Thessaloniki, Greece, pp. 962–965, October 2001.

[90] M. Kalman, E. Steinbach and B. Girod, "Adaptive playout for real-time media streaming," in *IEEE International Symposium on Circuits and Systems*, vol. 1, Scottsdale, AZ, pp. 45–48, May 2002.

[91] S. Wenger, G. D. Knorr, J. Ott and F. Kossentini, "Error resilience support in H.263+," *IEEE Transactions on Circuits and Systems for Video Technology*, vol. 8, pp. 867–877, November 1998.

[92] R. Talluri, "Error-resilient video coding in the ISO MPEG-4 standard," *IEEE Communications Magazine*, vol. 2, pp. 112–119, June 1998.

[93] N. Färber, B. Girod and J. Villasenor, "Extension of ITU-T Recommendation H.324 for error-resilient video transmission," *IEEE Communications Magazine*, vol. 2, pp. 120–128, June 1998.

[94] E. Steinbach, N. Färber and B. Girod, "Standard compatible extension of H.263 for robust video transmission in mobile environments," *IEEE Transactions on Circuits and Systems for Video Technology*, vol. 7, pp. 872–881, December 1997.

[95] B. Girod and N. Färber, "Feedback-based error control for mobile video transmission," *Proceedings of the IEEE*, vol. 87, pp. 1707–1723, October 1999.

[96] G. J. Sullivan and T. Wiegand, "Rate-distortion optimization for video compression," *IEEE Signal Processing Magazine*, vol. 15, pp. 74–90, November 1998.

[97] A. Ortega and K. Ramchandran, "From rate-distortion theory to commercial image and video compression technology," *IEEE Signal Processing Magazine*, vol. 15, pp. 20–22, November 1998.

[98] T. Wiegand, X. Zhang and B. Girod, "Long-term memory motion-compensated prediction," *IEEE Transactions on Circuits and Systems for Video Technology*, vol. 9, pp. 70–84, February 1999.

[99] T. Wiegand, N. Färber and B. Girod, "Error-resilient video transmission using long-term memory motion-compensated prediction," *IEEE Journal on Selected Areas in Communications*, vol. 18, pp. 1050–1062, June 2000.

[100] P. A. Chou, A. E. Mohr, A. Wang and S. Mehrotra, "Error control for receiver-driven layered multicast of audio and video," *IEEE Transactions on Multimedia*, vol. 3, pp. 108–122, March 2001.

[101] G. Cote and F. Kossentini, "Optimal intra coding of blocks for robust video communication over the Internet," *Signal Processing: Image Communication*, vol. 15, pp. 25–34, September 1999.

[102] R. Zhang, S. L. Regunathan and K. Rose, "Video coding with optimal inter/intra-mode switching for packet loss resilience," *IEEE Journal on Selected Areas in Communications*, vol. 18, pp. 966–976, June 2000.

[103] R. Zhang, S. L. Regunathan and K. Rose, "Optimal estimation for error concealment in scalable video coding," in *Proceedings of Thirty-Fourth Asilomar Conference on Signals, Systems and Computers*, vol. 2, Pacific Grove, CA, pp. 1974–1978, 2000.

[104] R. Zhang, S. L. Regunathan and K. Rose, "Robust video coding for packet networks with feedback," in *Proceedings of Thirty-Fourth Asilomar Conference on Signals, Systems and Computers*, (Snowbird, UT), pp. 450–459, 2000.

[105] W. Tan and A. Zakhor, "Video multicast using layered FEC and scalable compression," *IEEE Transactions on Circuits and Systems for Video Technology*, vol. 11, pp. 373–387, March 2001.

[106] P. C. Cosman, J. K. Rogers, P. G. Sherwood and K. Zeger, "Image transmission over channels with bit errors and packet erasures," in *Proceedings of Thirty-Fourth Asilomar Conference on Signals, Systems and Computers*, Pacific Grove, CA, pp. 1621–1625, 1998.

[107] S. Lee and P. Lee, "Cell loss and error recovery in variable rate video," *Journal of Visual Communication and Image Representation*, vol. 4, pp. 39–45, March 1993.

[108] B. Girod, K. Stuhlmüller, M. Link and U. Horn, "Packet loss resilient Internet video streaming," in *Proceedings of Visual Communications and Image Processing VCIP-99*, San Jose, CA, pp. 833–844, January 1999.

[109] M. Khansari and M. Vetterli, "Layered transmission of signals over powerconstrained wireless channels," in *Proc. of the IEEE International Conference on Image Processing (ICIP)*, vol. 3, Washington, DC, pp. 380–383, October 1995.

[110] R. Puri and K. Ramchandran, "Multiple description source coding using forward error correction codes," in *Proc. Asilomar Conference on Signals, Systems and Computers*, Pacific Grove, CA, pp. 342–346, November 1999.

[111] R. Puri, K. Ramchandran, K. W. Lee and V. Bharghavan, "Forward error correction (FEC) codes based multiple description coding for Internet video streaming and multicast," *Signal Processing: Image Communication*, vol. 16, pp. 745–762, May 2001.

[112] P. A. Chou and K. Ramchandran, "Clustering source/channel rate allocations for receiver-driven multicast under a limited number of streams," in *Proceedings of the IEEE International Conference on Multimedia and Expo (ICME)*, vol. 3, New York, NY, pp. 1221–1224, July 2000.

[113] K. Stuhlmüller, N. Färber, M. Link and B. Girod, "Analysis of video transmission over lossy channels," *IEEE Journal on Selected Areas in Communications*, vol. 18, pp. 1012–1032, June 2000.

[114] Y. J. Liang, J. G. Apostolopoulos and B. Girod, "Model-based delay-distortion optimization for video streaming using packet interleaving," in *Proceedings of the 36th Asilomar Conference on Signals, Systems and Computers*, Pacific Grove, CA, November 2002.

[115] S. Wicker, *Error Control Systems for Digital Communication and Storage*. Prentice-Hall, 1995.

[116] B. Dempsey, J. Liebeherr and A. Weaver, "On retransmission-based error control for continuous media traffic in packet-switching networks," *Computer Networks and ISDN Systems Journal*, vol. 28, pp. 719–736, March 1996.

[117] H. Liu and M. E. Zarki, "Performance of H.263 video transmission over wireless channels using hybrid ARQ," *IEEE Journal on Selected Areas in Communications*, vol. 15, pp. 1775–1786, December 1999.

[118] C. Papadopoulos and G. M. Parulkar, "Retransmission-based error control for continuous media applications," in *Proc. Network and Operating System Support for Digital Audio and Video (NOSSDAV)*, Zushi, Japan, pp. 5–12, April 1996.

[119] G. J. Conklin, G. S. Greenbaum, K. O. Lillevold, A. F. Lippman and Y. A. Reznik, "Video coding for streaming media delivery on the Internet," *IEEE Transactions on Circuits and Systems for Video Technology*, vol. 11, pp. 269–281, March 2001.

[120] Y.-Q. Zhang, Y.-J. Liu and R. L. Pickholtz, "Layered image transmission over cellular radio channels," *IEEE Transactions on Vehicular Technology*, vol. 43, pp. 786–794, August 1994.

[121] B. Girod, N. Färber and U. Horn, "Scalable codec architectures for Internet video-on-demand," in *Proceedings of the Thirty-First Asilomar Conference on Signals, Systems and Computers*, Pacific Grove, CA, pp. 357–361, November 1997.

[122] A. Puri, L. Yan and B. G. Haskell, "Temporal resolution scalable video coding," in *Proc. of the IEEE International Conference on Image Processing (ICIP)*, vol. 2, Austin, TX, pp. 947–951, November 1994.

[123] K. M. Uz, M. Vetterli and D. J. LeGall, "Interpolative multiresolution coding of advance television with compatible subchannels," *IEEE Transactions on Circuits and Systems for Video Technology*, vol. 1, pp. 88–99, March 1991.

[124] S. Zafar, Y.-Q. Zhang and B. Jabbari, "Multiscale video representation using multiresolution motion compensation and wavelet decomposition," *IEEE Journal on Selected Areas in Communications*, vol. 11, pp. 24–35, January 1993.

[125] U. Horn, K. Stuhlmüller, M. Link and B. Girod, "Robust Internet video transmission based on scalable coding and unequal error protection," *Signal Processing: Image Communication*, vol. 15, pp. 77–94, September 1999.

[126] B. Girod and U. Horn, "Scalable codec for Internet video streaming," in *Proceedings of the International Conference on Digital Signal Processing*, Piscataway, NJ, pp. 221–224, July 1997.

[127] M. Khansari, A. Zakauddin, W.-Y. Chan, E. Dubois and P. Mermelstein, "Approaches to layered coding for dual-rate wireless video transmission," *Proceedings of the IEEE International Conference on Image Processing (ICIP)*, pp. 285–262, November 1994.

[128] G. Karlsson and M. Vetterli, "Subband coding of video for packet networks," *Optical Engineering*, vol. 27, pp. 574–586, July 1998.

[129] D. Quaglia and J. C. De Martin, "Delivery of MPEG video streams with constant perceptual quality of service," *Proceedings of the IEEE International Conference on Multimedia and Expo (ICME)*, vol. 2, pp. 85–88, August 2002.

[130] E. Masala, D. Quaglia and J. C. De Martin, "Adaptive picture slicing for distortion-based classification of video packets," *Proceedings of the IEEE Fourth Workshop on Multimedia Signal Processing*, pp. 111–116, October 2001.

[131] J. Shin, J. W. Kim and C. C. J. Kuo, "Quality-of-service mapping mechanism for packet video in differentiated services network," *IEEE Transactions on Multimedia*, vol. 3, pp. 219–231, June 2001.

[132] J. Shin, J. Kim and C. C. J. Kuo, "Relative priority based QoS interaction between video applications and differentiated service networks," in *Proceedings of the IEEE International Conference on Image Processing*, Vancouver, BC, Canada, pp. 536–539, September 2000.

[133] S. Regunathan, R. Zhang and K. Rose, "Scalable video coding with robust mode selection," *Signal Processing: Image Communication*, vol. 16, pp. 725–732, May 2001.

[134] H. Yang, R. Zhang and K. Rose, "Drift management and adaptive bit rate allocation in scalable video coding," in *Proceedings of the IEEE International Conference on Image Processing (ICIP)*, vol. 2, Rochester, NY, pp. 49–52, September 2002.

[135] S. Dogan, A. Cellatoglu, M. Uyguroglu, A. H. Sadka and A. M. Kondoz, "Error-resilient video transcoding for robust Internetwork communications using GPRS," *IEEE Transactions on Circuits and Systems for VideoTechnology - Special Issue on Wireless Video*, vol. 12, pp. 453–464, July 2002.

[136] H. G. Musmann, P. Pirsch and H. J. Grallert, "Advances in picture coding," *Proceedings of the IEEE*, vol. 73, pp. 523–548, April 1985.

[137] M. Flierl and B. Girod, "Generalized B pictures and the draft H.264/AVC video-compression standard," *IEEE Transaction on Circuits and Systems for Video Technology*, vol. 13, pp. 587–597, July 2003.

[138] T. Shanableh and M. Ghanbari, "Loss Concealment Using B-Pictures Motion Information," *IEEE Transaction on Multimedia*, vol. 5, pp. 257–266, June 2003.

[139] M. Al-Mualla, N. Canagarajah and D. Bull, "Simplex minimization for single- and multiple reference motion estimation," *IEEE Transactions on Circuits and Systems for Video Technology*, vol. 11, pp. 1029–1220, December 2001.

[140] A. Luthra, G. J. Sullivan and T. Wiegand, "Special issue on the H.264/AVC video coding standard," *IEEE Transactions on Circuits and Systems for Video Technology*, vol. 13, July 2003.

[141] S. Wenger, "H.264/AVC Over IP," *IEEE Transactions on Circuits and Systems for Video Technology*, vol. 13, pp. 587–597, July 2003.

[142] T. Stockhammer, M. M. Hannuksela and T. Wiegand, "H.264/AVC in wireless environments," *IEEE Transactions on Circuits and Systems for Video Technology*, vol. 13, pp. 587–597, July 2003.

[143] A. Arumugam, A. Doufexi, A. Nix and P. Fletcher, "An investigation of the coexistence of 802.11g WLAN and high data rate Bluetooth enabled consumer electronic devices in indoor home and office environments," *IEEE Transactions on Consumer Electronics*, vol. 49, pp. 587–596, August 2003.

[144] R. Thobaben and J. Kliewer, "Robust decoding of variable-length encoded Markov sources using a three-dimensional trellis," *IEEE communications letters*, vol. 7, pp. 320–322, July 2003.

[145] A. Murad and T. Fuja, "Joint source-channel decoding of variable-length encoded sources," in *IEEE Information Theory Workshop*, Killarney, Ireland, pp. 94–95, June 1998.

[146] M. Barnsley, "A better way to compress images," *BYTE*, pp. 215–222, January 1988.

[147] J. Beaumont, "Image data compression using fractal techniques," *BT Technology*, vol. 9, pp. 93–109, October 1991.

[148] A. Jacquin, "Image coding based on a fractal theory of iterated contractive image transformations," *IEEE Transactions on Image Processing*, vol. 1, pp. 18–30, January 1992.

[149] D. Monro and F. Dudbridge, "Fractal block coding of images," *Electronic Letters*, vol. 28, pp. 1053–1055, May 1992.

[150] D. Monro, D. Wilson and J. Nicholls, "High speed image coding with the bath fractal transform," in Damper et al. [539], pp. 23–30.

[151] J. Streit and L. Hanzo, "A fractal video communicator," in *Proceedings of IEEE VTC'94* [540], pp. 1030–1034.

[152] W. Welsh, "Model based coding of videophone images," *Electronic and Communication Engineering Journal*, pp. 29–36, February 1991.

[153] J. Ostermann, "Object-based analysis-synthesis coding based on the source model of moving rigid 3D objects," *Signal Processing: Image Communication*, vol. 6, pp. 143–161, 1994.

[154] M. Chowdhury, "A switched model-based coder for video signals," *IEEE Transactions on Circuits and Systems*, vol. 4, pp. 216–227, June 1994.

[155] G. Bozdagi, A. Tekalp and L. Onural, "3-D motion estimation and wireframe adaptation including photometric effects for model-based coding of facial image sequences," *IEEE Transactions on circuits and Systems for Video Technology*, vol. 4, pp. 246–256, June 1994.

[156] Q. Wang and R. Clarke, "Motion estimation and compensation for image sequence coding," *Signal Processing: Image Communications*, vol. 4, pp. 161–174, 1992.

[157] H. Gharavi and M. Mills, "Blockmatching motion estimation algorithms — new results," *IEEE Transactions on Circuits and Systems*, vol. 37, pp. 649–651, May 1990.

[158] J. Jain and A. Jain, "Displacement measurement and its applications in inter frame image coding," *IEEE Transactions on Communications*, vol. 29, December 1981.

[159] B. Wang, J. Yen and S. Chang, "Zero waiting-cycle hierarchical block matching algorithm and its array architectures," *IEEE Transactions on Circuits and Systems for Video Technology*, vol. 4, pp. 18–27, February 1994.

[160] P. Strobach, "Tree-structured scene adaptive coder," *IEEE Transactions on Communications*, vol. 38, pp. 477–486, April 1990.

[161] L. Hanzo, P. Cherriman and J. Streit, *Video Compression and Communications over Wireless Channels: From Second to Third Generation Systems, WLANs and Beyond.* IEEE Press, 2001. (For detailed contents please refer to http://www-mobile.ecs.soton.ac.uk.)

[162] B. Liu and A. Zaccarin, "New fast algorithms for the estimation of block motion vectors," *IEEE Transactions on Circuits and Systems*, vol. 3, pp. 148–157, April 1993.

[163] R. Li, B. Zeng and N. Liou, "A new three step search algorithm for motion estimation," *IEEE Transactions on Circuits and Systems*, vol. 4, pp. 439–442, August 1994.

[164] L. Lee, J. Wang, J. Lee and J. Shie, "Dynamic search-window adjustment and interlaced search for block-matching algorithm," *IEEE Transactions on Circuits and Systems for Video Technology*, vol. 3, pp. 85–87, February 1993.

[165] B. Girod, "Motion-compensating prediction with fractional-pel accuracy," *IEEE Transactions on Communications*, vol. 41, pp. 604–611, April 1993.

[166] J. Huang et al., "A multi-frame pel-recursive algorithm for varying frame-to-frame displacement estimation," in *Proceedings of International Conference on Acoustics, Speech, and Signal Processing, ICASSP'92* [541], pp. 241–244.

[167] N. Efstratiadis and A. Katsaggelos, "Adaptive multiple-input pel-recursive displacement estimation," in *Proceedings of International Conference on Acoustics, Speech, and Signal Processing, ICASSP'92* [541], pp. 245–248.

[168] C. Huang and C. Hsu, "A new motion compensation method for image sequence coding using hierarchical grid interpolation," *IEEE Transactions on Circuits and Systems for Video Technology*, vol. 4, pp. 42–51, February 1994.

[169] J. Nieweglowski, T. Moisala and P. Haavisto, "Motion compensated video sequence interpolation using digital image warping," in *Proceedings of the IEEE International Conference on Acoustics, Speech and Signal Processing (ICASSP'94)* [542], pp. 205–208.

[170] C. Papadopoulos and T. Clarkson, "Motion compensation using second-order geometric transformations," *IEEE Transactions on Circuits and Systems for Video Technology*, vol. 5, pp. 319–331, August 1995.

[171] C. Papadopoulos, *The use of geometric transformations for motion compensation in video data compression.* PhD thesis, University of London, 1994.

[172] M. Hoetter, "Differential estimation based on object oriented mapping parameter estimation," *Signal Processing*, vol. 16, pp. 249–265, March 1989.

[173] S. Karunaserker and N. Kingsbury, "A distortion measure for blocking artifacts in images based on human visual sensitivity," *IEEE Transactions on Image Processing*, vol. 6, pp. 713–724, June 1995.

[174] D. Pearson and M. Whybray, "Transform coding of images using interleaved blocks," *IEE Proceedings*, vol. 131, pp. 466–472, August 1984.

[175] J. Magarey and N. Kingsbury, "Motion estimation using complex wavelets," in *Proceedings of the IEEE International Conference on Acoustics, Speech and Signal Processing (ICASSP'96)* [543], pp. 2371–2374.

[176] R. Young and N. Kingsbury, "Frequency-domain motion estimation using a complex lapped transform," *IEEE Transactions on Image Processing*, vol. 2, pp. 2–17, January 1993.

[177] R. Young and N. Kingsbury, "Video compression using lapped transforms for motion estimation/compensation and coding," in *Proceedings of the SPIE Communication and Image Processing Conference*, (Boston, MA), pp. 1451–1463, SPIE, November 1992.

[178] K. Rao and P. Yip, *Discrete Cosine Transform: Algorithms, Advantages and Applications*. New York: Academic Press Ltd., 1990.

[179] A. Sharaf, *Video coding at very low bit rates using spatial transformations*. PhD thesis, Department of Electronic and Electrical Engineering, Kings College, London, 1997.

[180] R. Clarke, *Transform Coding of Images*. New York: Academic Press, 1985.

[181] A. Palau and G. Mirchandani, "Image coding with discrete cosine transforms using efficient energy-based adaptive zonal filtering," in *Proceedings of the IEEE International Conference on Acoustics, Speech and Signal Processing (ICASSP'94)* [542], pp. 337–340.

[182] H. Yamaguchi, "Adaptive DCT coding of video signals," *IEEE Transactions on Communications*, vol. 41, pp. 1534–1543, October 1993.

[183] K. Ngan, "Adaptive transform coding of video signals," *IEE Proceedings*, vol. 129, pp. 28–40, February 1982.

[184] R. Clarke, "Hybrid intra-frame transform coding of image data," *IEE Proceedings*, vol. 131, pp. 2–6, February 1984.

[185] F.-M. Wang and S. Liu, "Hybrid video coding for low bit-rate applications," in *Proceedings of the IEEE International Conference on Acoustics, Speech and Signal Processing (ICASSP'94)* [542], pp. 481–484.

[186] M. Ghanbari and J. Azari, "Effect of bit rate variation of the base layer on the performance of two-layer video codecs," *IEEE Transactions on Communications for Video Technology*, vol. 4, pp. 8–17, February 1994.

[187] N. Jayant and P. Noll, *Digital Coding of Waveforms, Principles and Applications to Speech and Video*. Englewood Cliffs, NJ: Prentice-Hall, 1984.

[188] N. Cheng and N. Kingsbury, "The ERPC: an efficient error-resilient technique for encoding positional information of sparse data," *IEEE Transactions on Communications*, vol. 40, pp. 140–148, January 1992.

[189] M. Narasimha and A. Peterson, "On the computation of the discrete cosine transform," *IEEE Transactions on Communications*, vol. 26, pp. 934–936, June 1978.

[190] L. Hanzo, F. Somerville and J. Woodard, "Voice and audio compression for wireless communications: Principles and applications for fixed and wireless channels." 2007 (For detailed contents, please refer to http://www-mobile.ecs.soton.ac.uk/).

[191] R. M. Pelz, "An un-equal error protected $p \times 8$ kbit/s video transmission for DECT," in *Proceedings of IEEE VTC'9/* [540], pp. 1020–1024.

[192] L. Hanzo, R. Stedman, R. Steele and J. Cheung, "A portable multimedia communicator scheme," in Damper et al. [539], pp. 31–54.

[193] R. Stedman, H. Gharavi, L. Hanzo and R. Steele, "Transmission of subband-coded images via mobile channels," *IEEE Transactions on Circuits and Systems for Video Technology*, vol. 3, pp. 15–27, February 1993.

[194] L. Hanzo and J. Woodard, "An intelligent multimode voice communications system for indoor communications," *IEEE Transactions on Vehicular Technology*, vol. 44, pp. 735–748, November 1995.

[195] L. Hanzo and J. Streit, "Adaptive low-rate wireless videophone systems," *IEEE Transactions on Circuits and Systems for Video Technology*, vol. 5, pp. 305–319, August 1995.

[196] L. Hanzo, S. X. Ng, W. Webb and T. Keller, *Quadrature Amplitude Modulation: From Basics to Adaptive Trellis-Coded, Turbo-Equalised and Space-Time Coded OFDM, CDMA and MC-CDMA Systems*. New York, USA: John Wiley and Sons, 2000.

[197] ETSI, *GSM Recommendation 05.05, Annex 3*, November 1988.

[198] H. Holma and A. Toskala, *HSDPA/HSUPA for UMTS*. Chichester, UK: John Wiley and Sons, 2006.

[199] G. Djuknic and D. Schilling, "Performance analysis of an ARQ transmission scheme for meteor burst communications," *IEEE Transactions on Communications*, vol. 42, pp. 268–271, February/March/April 1994.

[200] L. de Alfaro and A. Meo, "Codes for second and third order GH-ARQ schemes," *IEEE Transactions on Communications*, vol. 42, pp. 899–910, February–April 1994.

[201] T.-H. Lee, "Throughput performance of a class of continuous ARQ strategies for burst-error channels," *IEEE Transactions on Vehicular Technology*, vol. 41, pp. 380–386, November 1992.

[202] S. Lin, D. Constello Jr. and M. Miller, "Automatic-repeat-request error-control schemes," *IEEE Communications Magazine*, vol. 22, pp. 5–17, December 1984.

[203] L. Hanzo, L-L. Yang, E. L. Kuan and K. Yen, *Single- and Multi-Carrier CDMA*. New York, USA: John Wiley, IEEE Press, 2003.

[204] A. Gersho and R. Gray, *Vector Quantization and Signal Compression*. Dordrecht: Kluwer Academic Publishers, 1992.

[205] L. Torres, J. Casas and S. deDiego, "Segmentation based coding of textures using stochastic vector quantization," in *Proceedings of the IEEE International Conference on Acoustics, Speech and Signal Processing (ICASSP'94)* [542], pp. 597–600.

[206] M. Jaisimha, J. Goldschneider, A. Mohr, E. Riskin and R. Haralick, "On vector quantization for fast facet edge detection," in *Proceedings of the IEEE International Conference on Acoustics, Speech and Signal Processing (ICASSP'94)* [542], pp. 37–40.

[207] P. Yu and A. Venetsanopoulos, "Hierarchical finite-state vector quantisation for image coding," *IEEE Transactions on Communications*, vol. 42, pp. 3020–3026, November 1994.

[208] C.-H. Hsieh, K.-C. Chuang and J.-S. Shue, "Image compression using finite-state vector quantization with derailment compensation," *IEEE Transactions on Circuits and Systems for Video Technology*, vol. 3, pp. 341–349, October 1993.

[209] N. Nasrabadi, C. Choo and Y. Feng, "Dynamic finite-state vector quantisation of digital images," *IEEE Transactions on Communications*, vol. 42, pp. 2145–2154, May 1994.

[210] V. Sitaram, C. Huang and P. Israelsen, "Efficient codebooks for vector quantisation image compression with an adaptive tree search algorithm," *IEEE Transactions on Communications*, vol. 42, pp. 3027–3033, November 1994.

[211] W. Yip, S. Gupta and A. Gersho, "Enhanced multistage vector quantisation by joint codebook design," *IEEE Transactions on Communications*, vol. 40, pp. 1693–1697, November 1992.

[212] L. Po and C. Chan, "Adaptive dimensionality reduction techniques for tree-structured vector quantisation," *IEEE Transactions on Communications*, vol. 42, pp. 2246–2257, June 1994.

[213] L. Lu and W. Pearlman, "Multi-rate video coding using pruned tree-structured vector quantization," in *Proceedings of the IEEE International Conference on Acoustics, Speech and Signal Processing (ICASSP'93)*, vol. 5, Minneapolis, MN, pp. 253–256, IEEE, 27–30 April 1993.

[214] F. Bellifemine and R. Picco, "Video signal coding with DCT and vector quantisation," *IEEE Transactions on Communications*, vol. 42, pp. 200–207, February 1994.

[215] K. Ngan and K. Sin, "HDTV coding using hybrid MRVQ/DCT," *IEEE Transactions on Circuits and Systems for Video Technology*, vol. 3, pp. 320–323, August 1993.

[216] D. Kim and S. Lee, "Image vector quantiser based on a classification in the DCT domain," *IEEE Transactions on Communications*, pp. 549–556, April 1991.

[217] L. Torres and J. Huguet, "An improvement on codebook search for vector quantisation," *IEEE Transactions on Communications*, vol. 42, pp. 208–210, February 1994.

[218] W. Press, S. Teukolsky, W. Vetterling and B. Flannery, *Numerical Recipes in C*. Cambridge: Cambridge University Press, 1992.

[219] J. Streit and L. Hanzo, "Dual-mode vector-quantised low-rate cordless videophone systems for indoors and outdoors applications," *IEEE Transactions on Vehicular Technology*, vol. 46, pp. 340–357, May 1997.

[220] Telcomm. Industry Association (TIA), Washington, DC, *Dual-mode subscriber equipment — Network equipment compatibility specification, Interim Standard IS-54*, 1989.

[221] Research and Development Centre for Radio Systems, Japan, *Public Digital Cellular (PDC) Standard, RCR STD-27*.

[222] L. Hanzo, W. Webb and T. Keller, *Single- and Multi-Carrier Quadrature Amplitude Modulation: Principles and Applications for Personal Communications, WLANs and Broadcasting*. IEEE Press, 2000.

[223] C. E. Shannon, "A Mathematical Theory of Communication," *The Bell system Technical Journal*, vol. 27, pp. 379–656, July 1948.

[224] L. Hanzo, P. J. Cherriman and J. Street, *Wireless Video Communications: Second to Third Generation Systems and Beyond*. New York, USA: IEEE Press, 2001.

[225] S. X. Ng, R. G. Maunder, J. Wang, L-L. Yang and L. Hanzo, "Joint Iterative-Detection of Reversible Variable-Length Coded Constant Bit Rate Vector-Quantized Video and Coded Modulation," in *European Signal Processing Conference (EUSIPCO)*, Vienna, Austria, pp. 2231–2234, September 2004.

[226] ISO/IEC 14496-2:2004, *Information Technology – Coding of Audio-Visual Objects – Part 2: Visual*.

[227] S. X. Ng, J. Y. Chung, F. Guo and L. Hanzo, "A Turbo-Detection Aided Serially Concatenated MPEG-4/TCM Videophone Transceiver," in *IEEE Vehicular Technology Conference (VTC)*, Los Angeles, CA, September 2004.

[228] Q. Chen and K. P. Subbalakshmi, "Joint source-channel decoding for MPEG-4 video transmission over wireless channels," *IEEE Journal on Selected Areas in Communications*, vol. 21, no. 10, pp. 1780–1789, 2003.

[229] S. Benedetto, D. Divsalar, G. Montorsi and F. Pollara, "Serial Concatenation of Interleaved Codes: Performance Analysis, Design and Iterative Decoding," *IEEE Transactions on Information Theory*, vol. 44, pp. 909–926, May 1998.

[230] J. Hagenauer and N. Görtz, "The Turbo Principal in Joint Source-Channel Coding," in *Proceedings of the IEEE Information Theory Workshop*, Paris, France, pp. 275–278, March 2003.

[231] S. X. Ng and L. Hanzo, "Space-Time IQ-Interleaved TCM and TTCM for AWGN and Rayleigh Fading Channels," *IEE Electronics Letters*, vol. 38, pp. 1553–1555, November 2002.

[232] R. Bauer and J. Hagenauer, "Symbol-by-Symbol MAP Decoding of Variable Length Codes," in *ITG Conference on Source and Channel Coding*, Munich, Germany, pp. 111–116, January 2000.

[233] M. W. Marcellin and T. R. Fischer, "Trellis Coded Quantization of Memoryless and Gauss-Markov Sources," *IEEE Transactions on Communications*, vol. 38, pp. 82–93, January 1990.

[234] L. R. Bahl, J. Cocke, F. Jelinek and J. Raviv, "Optimal decoding of linear codes for minimizing symbol error rate," *IEEE Transactions on Information Theory*, vol. 20, pp. 284–287, March 1974.

[235] J. Kliewer and R. Thobaben, "Iterative Joint Source-Channel Decoding of Variable-Length Codes Using Residual Source Redundancy," *IEEE Transactions on Wireless Communications*, vol. 4, May 2005.

[236] J. Hagenauer, E. Offer and L. Papke, "Iterative Decoding of Binary Block and Convolutional Codes," *IEEE Transactions on Information Theory*, vol. 42, pp. 429–445, March 1996.

[237] L. Hanzo, T. H. Liew and B. L. Yeap, *Turbo Coding, Turbo Equalisation and Space Time Coding for Transmission over Wireless Channels*. Chichester, UK: Wiley, 2002.

[238] Y. Takishima, M. Wada and H. Murakami, "Reversible Variable Length Codes," *IEEE Transactions on Communications*, vol. 43, pp. 158–162, February–April 1995.

[239] D. A. Huffman, "A Method for the Construction of Minimum-Redundancy Codes," *Proceedings of the IRE*, vol. 40, no. 9, pp. 1098–1101, 1951.

[240] Y. Linde, A. Buzo and R. Gray, "An Algorithm for Vector Quantizer Design," *IEEE Transactions on Communications*, vol. 28, pp. 84–95, January 1980.

[241] P. Robertson, E. Villebrun and P. Höher, "A comparison of optimal and sub-optimal MAP decoding algorithms operating in Log domain," in *Proceedings of the International Conference on Communications*, pp. 1009–1013, June 1995.

[242] V. Franz and J. B. Anderson, "Concatenated Decoding with a Reduced-Search BCJR Algorithm," *IEEE Journal on Selected Areas in Communications*, vol. 16, pp. 186–195, February 1998.

[243] L. Hanzo, S. X. Ng, T. Keller and W. Webb, *Quadrature Amplitude Modulation*. Chichester, UK: Wiley, 2004.

[244] S. ten Brink, "Convergence Behaviour of Iteratively Decoded Parallel Concatenated Codes," *IEEE Transactions on Communications*, vol. 49, pp. 1727–1737, October 2001.

[245] "Feature topic: Software radios," *IEEE Communications Magazine*, vol. 33, pp. 24–68, May 1995.

[246] X. Zhang, M. Cavenor and J. Arnold, "Adaptive quadtree coding of motion -compensated image sequences for use on the broadband ISDN," *IEEE Transactions on Circuits and Systems for Video Technology*, vol. 3, pp. 222–229, June 1993.

[247] J. Vaisey and A. Gersho, "Image compression with variable block size segmentation," *IEEE Transactions on Signal Processing*, vol. 40, pp. 2040–2060, August 1992.

[248] M. Lee and G. Crebbin, "Classified vector quantisation with variable block-size DCT models," *IEE Proceedings, Vision, Image and Signal Processing*, pp. 39–48, February 1994.

[249] E. Shustermann and M. Feder, "Image compression via improved quadtree decomposition algorithms," *IEEE Transactions on Image Processing*, vol. 3, pp. 207–215, March 1994.

[250] F. DeNatale, G. Desoli and D. Giusto, "A novel tree-structured video codec," in *Proceedings of the IEEE International Conference on Acoustics, Speech and Signal Processing (ICASSP'94)* [542], pp. 485–488.

[251] M. Hennecke, K. Prasad and D. Stork, "Using deformable templates to infer visual speech dynamics," in *Proceedings of the 28th Asilomar Conference on Signals, Systems and Computers*, vol. 1, Pacific Grove, CA, pp. 578–582, 30 October – 2 November 1994.

[252] G. Wolf et al., "Lipreading by neural networks: Visual preprocessing, learning and sensory integration," *Proceedings of the neural information processing systems*, vol. 6, pp. 1027–1034, 1994.

[253] J. Streit and L. Hanzo, "Quad-tree based parametric wireless videophone systems," *IEEE Transactions Video Technology*, vol. 6, pp. 225–237, April 1996.

[254] E. Biglieri and M. Luise, "Coded modulation and bandwidth-efficient transmission," in *Proceedings of the Fifth Tirrenia International Workshop*, Netherlands, 8–12 September 1991.

[255] L.-F. Wei, "Trellis-coded modulation with multidimensional constellations," *IEEE Transactions on Information Theory*, vol. IT-33, pp. 483–501, July 1987.

[256] L. Hanzo and J. Stefanov, "The Pan-European Digital Cellular Mobile Radio System — known as GSM," in Steele and Hanzo [322], ch. 8, pp. 677–765.

[257] ITU-T, *ISO/IEC-CD-11172 — Coding of moving pictures and associated audio for digital storage*.

[258] ITU-T, *Recommendation H.261: Video codec for audiovisual services at $p \times 64$ Kbit/s*, March 1993.

[259] ITU-T, "Recommendation H.263: Video Coding for Low Bitrate communication," March 1998.

[260] D. Redmill and N. Kingsbury, "Improving the error resilience of entropy encoded video signals," in *Proceedings of the Conference on Image Processing: Theory and Applications (IPTA)*, (Netherlands), pp. 67–70, Elsevier, 1993.

[261] N. Jayant, "Adaptive quantization with a one-word memory," *Bell System Technical Journal*, vol. 52, pp. 1119–1144, September 1973.

[262] L. Zetterberg, A. Ericsson and C. Couturier, "DPCM picture coding with two-dimensional control of adaptive quantisation," *IEEE Transactions on Communications*, vol. 32, no. 4, pp. 457–642, 1984.

[263] C. Hsieh, P. Lu and W. Liou, "Adaptive predictive image coding using local characteristics," *IEE Proceedings*, vol. 136, pp. 385–389, December 1989.

[264] P. Wellstead, G. Wagner and J. Caldas-Pinto, "Two-dimensional adaptive prediction, smoothing and filtering," *Proceedings of the IEE*, vol. 134, pp. 253–266, June 1987.

[265] O. Mitchell, E. Delp and S. Carlton, "Block truncation coding: A new approach to image compression," in *IEEE International Conference on Communications (ICC)*, pp. 12B.1.1–12B.1.4, 1978.

[266] E. Delp and O. Mitchell, "Image compression using block truncation coding," *IEEE Transactions on Communications*, vol. 27, pp. 1335–1342, September 1979.

[267] D. Halverson, N. Griswold and G. Wiese, "A generalized block truncation coding algorithm for image compression," *IEEE Transactions Acoustics, Speech and Signal Processing*, vol. 32, pp. 664–668, June 1984.

[268] G. Arce and N. Gallanger, "BTC image coding using median filter roots," *IEEE Transactions on Communications*, vol. 31, pp. 784–793, June 1983.

[269] M. Noah, "Optimal Lloyd-Max quantization of LPC speech parameters," in *Proceedings of International Conference on Acoustics, Speech, and Signal Processing, ICASSP'84*, San Diego, CA, pp. 1.8.1–1.8.4, IEEE, 19–21 March 1984.

[270] R. Crochiere, S. Webber and J. Flangan, "Digital coding of speech in sub-bands," *Bell System Technology Journal*, vol. 52, pp. 1105–1118, 1973.

[271] R. Crochiere, "On the design of sub-band coders for low bit rate speech communication," *Bell System Technology Journal*, vol. 56, pp. 747–770, 1977.

[272] J. Woods and S. O'Neil, "Subband coding of images," *IEEE Transactions on Acoustic, Sound and Signal Processing*, vol. 34, pp. 1278–1288, October 1986.

[273] J. Woods, ed., *Subband Image Coding*. Dordrecht: Kluwer Academic Publishers, March 1991.

[274] H. Gharavi and A. Tabatabai, "Subband coding of digital images using two-dimensional quadrature mirror filtering," in *Proceedings of SPIE*, 1986.

[275] H. Gharavi and A. Tabatabai, "Subband coding of monochrome and color images," *IEEE Transactions on Circuits and Systems*, vol. 35, pp. 207–214, February 1988.

[276] H. Gharavi, "Subband coding algorithms for video applications: Videophone to HDTV-conferencing," *IEEE Transactions on Circuits and Systems for Video Technology*, vol. 1, pp. 174–183, February 1991.

[277] A. Alasmari, "An adaptive hybrid coding scheme for HDTV and digital video sequences," *IEEE Transactions on consumer electronics*, vol. 41, no. 3, pp. 926–936, 1995.

[278] K. Irie et al., "High-quality subband coded for HDTV transmission," *IEEE Transactions on Circuits and Systems for Video Technology*, vol. 4, pp. 195–199, April 1994.

[279] E. Simoncelli and E. Adelson, "Subband transforms," in Woods [273], pp. 143–192.

[280] K. Irie and R. Kishimoto, "A study on perfect reconstructive subband coding," *IEEE Transactions on Circuits and Systems for Video Technology*, vol. 1, pp. 42–48, January 1991.

[281] J. Woods and T. Naveen, "A filter based bit allocation scheme for subband compression of HDTV," *IEEE Transactions on Image Processing*, vol. 1, pp. 436–440, July 1992.

[282] D. Esteban and C. Galand, "Application of quadrature mirror filters to split band voice coding scheme," in *Proceedings of International Conference on Acoustics, Speech, and Signal Processing, ICASSP'77*, Hartford, CT, pp. 191–195, IEEE, 9–11 May 1977.

[283] J. Johnston, "A filter family designed for use in quadrature mirror filter banks," in *Proceedings of International Conference on Acoustics, Speech, and Signal Processing, ICASSP'80*, Denver, CO, pp. 291–294, IEEE, 9–11 April 1980.

[284] H. Nussbaumer, "Complex quadrature mirror filters," in *Proceedings of International Conference on Acoustics, Speech, and Signal Processing, ICASSP'83*, Boston, MA, pp. 221–223, IEEE, 14–16 April 1983.

[285] C. Galand and H. Nussbaumer, "New quadrature mirror filter structures," *IEEE Transactions on Acoustic Speech Signal Processing*, vol. ASSP-32, pp. 522–531, June 1984.

[286] R. Crochiere and L. Rabiner, *Multirate Digital Processing*. Englewood Cliffs, NJ: Prentice-Hall, 1993.

[287] S. Aase and T. Ramstad, "On the optimality of nonunitary filter banks in subband coders," *IEEE Transactions on Image Processing*, vol. 4, pp. 1585–1591, December 1995.

[288] V. Nuri and R. Bamberger, "Size limited filter banks for subband image compression," *IEEE Transactions on Image Processing*, vol. 4, pp. 1317–1323, September 1995.

[289] H. Gharavi, "Subband coding of video signals," in Woods [273], pp. 229–271.

[290] O. Egger, W. Li and M. Kunt, "High compression image coding using an adaptive morphological subband decomposition," *Proceedings of the IEEE*, vol. 83, pp. 272–287, February 1995.

[291] P. Westerink and D. Boekee, "Subband coding of color images," in Woods [273], pp. 193–228.

[292] Q. Nguyen, "Near-perfect-reconstruction pseudo-QMF banks," *IEEE Transactions on signal processing*, vol. 42, pp. 65–76, January 1994.

[293] S.-M. Phoong, C. Kim, P. Vaidyanathan and R. Ansari, "A new class of two-channel biorthogonal filter banks and wavelet bases," *IEEE Transactions on Signal Processing*, vol. 43, pp. 649–665, March 1995.

[294] E. Jang and N. Nasrabadi, "Subband coding with multistage VQ for wireless image communication," *IEEE Transactions in Circuit and Systems for Video Technology*, vol. 5, pp. 347–253, June 1995.

[295] P. Cosman, R. Gray and M. Vetterli, "Vector quantisation of image subbands: A survey," *IEEE Transactions on Image Processing*, vol. 5, pp. 202–225, February 1996.

[296] ITU, *Joint Photographic Experts Group ISO/IEC, JTC/SC/WG8, CCITT SGVIII. JPEG technical specifications, revision 5. Report JPEG-8-R5*, January 1990.

[297] P. Franti and O. Nevalainen, "Block truncation coding with entropy coding," *IEEE Transactions on Communications*, vol. 43, no. 4, pp. 1677–1685, 1995.

[298] V. Udpikar and J. Raina, "BTC image coding using vector quantisation," *IEEE Transactions on Communications*, vol. 35, pp. 353–359, March 1987.

[299] International Standards Organization, *ISO/IEC 11172 MPEG 1 International Standard, Coding of moving pictures and associated audio for digital storage media up to about 1.5 Mbit/s, Parts 1–3*.

[300] International Standards Organization, *ISO/IEC CD 13818 MPEG 2 International Standard, Information Technology, Generic Coding of Moving Video and Associated Audio Information, Parts 1–3*.

[301] Telenor Research and Development, P.O. Box 83, N-2007 Kjeller, Norway, *Video Codec Test Model 'TMN 5', ITU Study Group 15, Working Party 15/1*.

[302] D. Choi, "Frame alignment in a digital carrier system — a tutorial," *IEEE Communications Magazine*, vol. 28, pp. 46–54, February 1990.

[303] ITU (formerly CCITT), *ITU Recommendation X25*, 1993.

[304] M. Al-Subbagh and E. Jones, "Optimum patterns for frame alignment," *IEE Proceedings*, vol. 135, pp. 594–603, December 1988.

[305] T. Turletti, "A H.261 software codec for videoconferencing over the internet," *Technical Report* 1834, INRIA, 06902 Sophia-Antipolis, France, January 1993.

[306] N. Kenyon and C. Nightingale, eds., *Audiovisual Telecommunications*. London: Chapman and Hall, 1992.

[307] N. MacDonald, "Transmission of compressed video over radio links," *BT Technology Journal*, vol. 11, pp. 182–185, April 1993.

[308] M. Khansari, A. Jalali, E. Dubois and P. Mermelstein, "Robust low bit-rate video transmission over wireless access systems," in *Proceedings of International Communications Conference (ICC)*, pp. 571–575, 1994.

[309] M. Khansari, A. Jalali, E. Dubois and P. Mermelstein, "Low bit-rate video transmission over fading channels for wireless microcellular systems," *IEEE Transactions on Circuits and Systems for Video Technology*, vol. 6, pp. 1–11, February 1996.

[310] N. Cheng, *Error resilient video coding for Noisy Channels*. PhD thesis, Department of Engineering, University of Cambridge, 1991.

[311] D. Redmill, *Image and Video Coding for Noisy Channels*. PhD thesis, Signal Processing and Communication Laboratory, Department of Engineering, University of Cambridge, November 1994.

[312] Y. Matsumura, S. Nakagawa and T. Nakai, "Very low bit rate video coding with error resilience," in *VLBV'95* [544], pp. L–1.

[313] K. Ngan and D. Chai, "Enhancement of image quality in VLBR coding," in *VLBV'95* [544], pp. L–3.

[314] K. Ngan and D. Chai, "Very low bit rate video coding using H.263 coder," *IEEE Transactions on Circuits and Systems for Video Technology*, vol. 6, pp. 308–312, June 1996.

[315] W. Webb and L. Hanzo, "Square QAM," in *Modern Quadrature Amplitude Modulation: Principles and Applications for Wireless Communications* [222], ch. 5, pp. 156–169.

[316] IBM Corp., White Plains, NY, *General Information: Binary Synchronous Communication, IBM Publication GA27-3004*, 1969.

[317] S. Lin and D. Constello Jr., *Error Control Coding: Fundamentals and Applications*. Englewood Cliffs, NJ: Prentice-Hall, October 1982.

[318] S. Sampei, S. Komaki and N. Morinaga, "Adaptive modulation/TDMA scheme for large capacity personal multi-media communication systems," *IEICE Transactions on Communications (Japan)*, vol. E77-B, pp. 1096–1103, September 1994.

[319] J. Torrance and L. Hanzo, "Upper bound performance of adaptive modulation in a slow Rayleigh fading channel," *Electronics Letters*, vol. 32, pp. 718–719, 11 April 1996.

[320] W. Webb and L. Hanzo, "Variable rate QAM," in *Modern Quadrature Amplitude Modulation: Principles and Applications for Wireless Communications* [222], ch. 13, pp. 384–406.

[321] R. Steele and W. Webb, "Variable rate QAM for data transmission over Rayleigh fading channels," in *Proceedings of Wireless'91*, (Calgary, Alberta), pp. 1–14, IEEE, 1991.

[322] R. Steele and L. Hanzo, eds., *Mobile Radio Communications*. Piscataway, NJ: IEEE Press, 1999.

[323] P. Skelly, M. Schwartz and S. Dixit, "A histogram-based model for video traffic behavior in a ATM multiplexer," *IEEE/ACM Transactions Networking*, vol. 1, pp. 446–459, August 1993.

[324] M. Ghanbari and V. Seferidis, "Cell-loss concealment in ATM video codecs," *IEEE Transactions on Circuits and Systems for Video Technology*, vol. 3, pp. 238–247, June 1993.

[325] W. Chung, F. Kossentini and M. Smith, "An efficient motion estimation technique based on a rate-distortion criterion," in *Proceedings of the IEEE International Conference on Acoustics, Speech and Signal Processing (ICASSP'96)* [543], pp. 1977–1980.

[326] M. Whybray and W. Ellis, "H.263 - video coding recommendation for PSTN videophone and multimedia," in *IEE Colloquium (Digest)*, pp. 6/1–6/9, IEE, June 1995.

[327] Telenor Research and Development, P.O. Box 83, N-2007 Kjeller, Norway, *H.263 Software Codec*. http://www.nta.no/brukere/DVC.

[328] N. Färber, E. Steinbach and B. Girod, "Robust H.263 video transmission over wireless channels," in *Proceedings of International Picture Coding Symposium (PCS)*, Melbourne, Australia, pp. 575–578, March 1996.

[329] W. Ding and B. Liu, "Rate control of MPEG video coding and recording by rate-quantization modeling," *IEEE Transactions on Circuits and Systems for Video Technology*, vol. 6, pp. 12–20, February 1996.

[330] G. Schuster and A. Katsaggelos, "A video compression scheme with optimal bit allocation between displacement vector field and displaced frame difference," in *Proceedings of the IEEE International Conference on Acoustics, Speech and Signal Processing (ICASSP'96)* [543], pp. 1967–1970.

[331] F. Martins, W. Ding and E. Feig, "Joint control of spatial quantization and temporal sampling for very low bitrate video," in *Proceedings of the IEEE International Conference on Acoustics, Speech and Signal Processing (ICASSP'96)* [543], pp. 2074–2077.

[332] T. Wiegand, M. Lightstone and D. Mukherjee, "Rate-distortion optimized mode selection for very low bit rate video coding and the emerging H.263 standard," *IEEE Transactions on Circuits and Systems for Video Technology*, vol. 6, pp. 182–190, April 1996.

[333] K. Wong and L. Hanzo, "Channel coding," in Steele and Hanzo [322], ch. 4, pp. 347–488.

[334] A. Paulraj, "Diversity techniques," in Gibson [545], ch. 12, pp. 166–176.

[335] A. Mämmelä, *Diversity receivers in a fast fading multipath channel*. PhD thesis, Department of Electrical Engineering, University of Oulu, Finland, 1995.

[336] R. Steele, "Digital European Cordless Telecommunications (DECT) systems," in Steele and Hanzo [322], ch. 1.7.2, pp. 79–83.

[337] P. Crespo, R. M. Pelz and J. Cosmas, "Channel error profile for DECT," *IEE Proceedings on Communications*, vol. 141, pp. 413–420, December 1994.

[338] S. Asghar, "Digital European Cordless Telephone," in Gibson [545], ch. 30, pp. 478–499.

[339] L. Chiariglione, "The development of an integrated audiovisual coding standard: MPEG," *Proceedings of the IEEE*, vol. 83, pp. 151–157, February 1995.

[340] R. Schäfer and T. Sikora, "Digital video coding standards and their role in video communications," *Proceedings of the IEEE*, vol. 83(10), pp. 907–924, June 1995.

[341] D. J. Le Gall, "The MPEG video compression algorithm," *Signal Processing: Image Communication*, vol. 4, pp. 129–140, 1992.

[342] T. Sikora, "MPEG-4 very low bit rate video," *Proceedings of IEEE ISCAS Conference*, Hong Kong, pp. 1440–1443, February 1997.

[343] T. Sikora, "The MPEG-4 video standard verification model," *IEEE Transactions on Circuit and Systems for Video Technology*, vol. 7, pp. 19–31, February 1997.

[344] ISO/IEC JTC1/SC29/WG11 N0702 Rev., "Information technology - Generic coding of moving pictures and associated audio, Recommendation H.262. Draft International Standard," vol. 83, March 1994.

[345] ISO/IEC 11172-2 Information technology, "Coding of moving pictures and associated audio for digital storage media at up to about 1.5Mbit/s - Video". Standards Organization/International Electrotechnical (in German). International Commission, 1993.

[346] MPEG AOE Group, "Proposal package description (PPD)-Revision 3," July 1995.

[347] ISO/IEC JTC1/SC29/WG11, "Information technology - Coding of moving pictures and associated audio for digital storage media at up to 1.5 Mbits/s. Part 2: Video. Draft ISO/IEC 11172-2 (MPEG-1)," *ISO/IEC*, 1991.

[348] ISO/IEC JTC1/SC29/WG11, "Information technology - Generic coding of moving pictures and associated audio. Part 2: Video. Draft ISO/IEC 13818-2 (MPEG-2) and ITU-T Recommendation H.262, ISO/IEC and ITU-T," *ISO/IEC*, 1994.

[349] ISO/IEC JTC1/SC29/WG11, "Information technology - Generic coding of audio-visual objects. Part 2: Visual. Draft ISO/IEC 14496-2 (MPEG-4), version 1," 1998.

[350] A. Jain, *Fundamentals of Digital Image Processing*. Englewood Cliffs, NJ: Prentice-Hall, 1989.

[351] O. Avaro, A. Eleftheriadis, C. Herpel, G. Rajan and L. Ward, "MPEG-4 systems: overview," *Signal Processing: Image Communication*, vol. 15, pp. 281–298, 2000.

[352] G. Franceschini, "The delivery layer in MPEG-4," in *Signal Processing: Image Communication*, vol. 15, pp. 347–363, 2000.

[353] C. Herpel, "Architectural considerations for carriage of MPEG-4 over IP network," *ISO/IEC JTC1/SC29/WG11 N2615*, December 1998.

[354] C. Herpel and A. Eleftheriadis, "MPEG-4 systems: elementary stream management," *Signal Processing: Image Communication*, vol. 15, pp. 299–320, 2000.

[355] R. Talluri, "Error-resilient video coding in the ISO MEPG-4 standard," *IEEE Communications Magazine*, pp. 112–119, June 1998.

[356] H. Schulzrinne, S. Casner, R. Frederick and V. Jacobson, "RTP: A transport protocol for real-time applications," *RFC 1889*, January 1996.

[357] L. Chiariglione, "MPEG and multimedia communications," *IEEE Transactions on Circuits and Systems for Video Technology*, vol. 7, pp. 5–18, February 1997.

[358] K. N. Ngan, T. Sikora, M.-T. Sun and S. Pamchanathan, "Segmentation, description and retrieval of video content," *IEEE Transactions on Circuits and Systems for Video Technology, special issue*, vol. 8(5), pp. 521–524, September 1998.

[359] K. N. Ngan, T. Sikora, M.-T. Sun and S. Pamchanathan, "Representation and coding of images and video," *IEEE Transactions on Circuits and Systems for Video Technology, special issue*, vol. 8, pp. 797–801, November 1998.

[360] ISO/IEC 13818-2 MPEG-2 Video Coding Standard, "Information technology - Generic coding of moving pictures and associated audio information: Video," March 1995.

[361] T. Sikora and L. Chiariglione, "MPEG-4 Video and its potential for future multimedia services," *Proceedings of IEEE ISCAS Conference*, Hong Kong, vol. 2, pp. 1468–1471, June 1997.

[362] F. Bossen and T. Ebrahimi, "A simple and efficient binary shape coding technique based on bitmap representation," *Proceedings of the International Conference on Acoustics, Speech and Signal Processing (ICASSP'97)*, Munich, Germany, vol. 4, pp. 3129–3132, April 1997.

[363] T. Ebrahimi, C. Horne, "MPEG-4 natural video coding - An overview," *Signal Processing: Image Communication*, vol. 15, no. 4, pp. 365–385, 2000.

[364] ISO/IEC JTC1/SC29/WG11 N1902, "Information technology - coding of audio visual objects: visual," October 1997.

[365] Recommendation ITU-T BT.500-11, "Methodology for the subjective assessment of the quality of television pictures," ITU-T, 2002.

[366] ITU-T/SG 16/VCEG(formerly Q.15 now Q.6), "H.26L test model long term number 7 (TML-7), Doc. VCEG-M81," April 2001.

[367] Y. Zeng, L. Cheng, G. Bi and A. Kot, "Integer DCTs and fast algorithms," *IEEE Transactions on Signal Processing*, vol. 49, pp. 2774–2782, November 2001.

[368] W. Choi and B. Jeon, "Dynamic UVLC codeword remapping with fixed re-association table for H.26L," in *Picture Coding Symposium(PCS)*, (Seoul, Korea), pp. 167–170, April 2001.

[369] D. Marpe, G. Blättermann, G. Heising and T. Wiegand, "Further results for CABAC entropy coding scheme," Document VCEG-M59, ITU-T Video Coding Experts Group, Apr. 2001, http://standards.pictel.com/ftp/video-site/0104Aus/VCEG-M59.doc."

[370] R. J. Clarke, "Transform coding of images," in *Microelectronics and Signal Processing*. London: Academic Press, 1985.

[371] T. N. N. Ahmed and K. Rao, "Discrete Cosine Transform," *IEEE Transactions on Computers*, pp. 90–93, January 1974.

[372] ITU-T Rec. H.26L/ISO/IEC 11496-10, "Advanced video coding," *Final Committee Draft, Document JVT-E022*, September 2002.

[373] D. Marpe, G. Blättermann, G. Heising and T. Wiegand, "Adaptive codes for H.26L," *ITU-T SG16/Q.6 VCEG-L-13*, January 2001.

[374] ITU-T, "Video coding for low bitrate communication," *ITU-T Recommendation H.263; version 1*, November 1995.

[375] J. Signes, Y. Fisher and A. Eleftheriadis, "MPEG-4's binary format for scene description," in *Signal Processing: Image Communication, Special issue on MPEG-4*, vol. 15, pp. 312–345, January 2000.

[376] A.M. Tekalp and J. Ostermann, "Face and 2-D mesh animation in MPEG-4," in *Signal Processing: Image Communication, Special issue on MPEG-4*, vol. 15, pp. 387–421, January 2000.

[377] ISO/IEC JTC1/SC29/WG11 N1902, "Information technology - coding of audio visual objects: visual," November 1998.

[378] ISO/IEC JTC1/SC29/WG11, "Adhoc group on core experiments on error resilience aspects of MPEG-4 video, description of error resilience aspects of MPEG-4 video," *Description of Error Resilience Core Experiments*, November 1996.

[379] J. G. Proakis, *Digital Communication*, 3rd ed. McGraw-Hill, New York, 1995.

[380] S. B. Wicker, *Error Control Systems for Digital Communication and Storage*. Englewood Cliffs, NJ: Prentice-Hall, 1994.

[381] A. Andreadis, G. Benelli, A. Garzelli and S. Susini, "FEC coding for H.263 compatible video transmission," *Proceedings of International Conference on Image Processing*, Santa Barbara, CA, pp. 579–581, October 1997.

[382] J. Wen and J. D. Villasenor, "A class of reversible variable length codes for robust image and video coding," *Proceedings 1997 IEEE International Conference on Image Processing*, vol. 2, pp. 65–68, October 1997.

[383] H. Sun, J. W. Zdepski, W. Kwok and D. Raychaudhuri, "Error concealment for robust decoding of MPEG compressed video," in *Signal Processing: Image Communication*, vol. 10(4), pp. 249–268, September 1997.

[384] A. Li, S. Kittitornkun, Y. H. Hu, D. S. Park and J. Villasenor, "Data partitioning and reversible variable length codes for robust video communications," in *IEEE Data Compression Conference Proceedings*, Snowbird, UT, pp. 460–469, March 2000.

[385] B. L. Montgomery and J. Abrahams, "Synchronization of binary source codes," in *IEEE Transactions on Information Theory*, vol. 32, pp. 849–854, November 1996.

[386] Y. Takishima, M. Wada and H. Murakami, "Reversible variable length codes," *IEEE Transactions on Communications*, vol. 43, pp. 158–162, February 1995.

[387] R. Talluri, I. Moccagatta, Y. Nag and G. Cheung, "Error concealment by data partitioning," *Signal Processing Magazine*, vol. 14, pp. 505–518, May 1999.

[388] C. Bormann, L. Cline, G. Deisher, T. Gardos, C. Maciocco, D. Newell, J. Ott, G. Sullimendation, S. Wenger and C. Zhu, "RTP Payload format for the 1998 version of ITU-T recommendation H.263 video (H.263+)"; *Request for Comments 2429l*, May 1998.

[389] R. Steele and L. Hanzo, eds., *Mobile Radio Communications* 2nd ed., New York: IEEE Press-John Wiley, 1999.

[390] L. Hanzo, W. Webb and T. Keller, *Single- and Multi-carrier Quadrature Amplitude Modulation*. New York: John Wiley-IEEE Press, April 2000.

[391] L. Hanzo, F. Somerville and J. Woodard, "Voice compression and communications: Principles and applications for fixed and wireless channels." 2001 (For detailed contents, please refer to http://www-mobile.ecs.soton.ac.uk.).

[392] P. Cherriman and L. Hanzo, "Programmable H.263-based wireless video transceivers for interference-limited environments," *IEEE Transactions on Circuits and Systems for Video Technology*, vol. 8, pp. 275–286, June 1998.

[393] C. Douillard, A. Picart, M. Jézéquel, P. Didier, C. Berrou and A. Glavieux, "Iterative correction of intersymbol interference: Turbo-equalization," *European Transactions on Communications*, vol. 6, pp. 507–511, 1995.

[394] M. Gertsman and J. Lodge, "Symbol-by-symbol MAP demodulation of CPM and PSK signals on Rayleigh flat-fading channels," *IEEE Transactions on Communications*, vol. 45, pp. 788–799, July 1997.

[395] I. Marsland, P. Mathiopoulos and S. Kallel, "Non-coherent turbo equalization for frequency selective Rayleigh fast fading channels," in *Proceedings of the International Symposium on Turbo Codes & Related Topics*, Brest, France, pp. 196–199, 3–5 September 1997.

[396] Q. Dai and E. Shwedyk, "Detection of bandlimited signals over frequency selective Rayleigh fading channels," *IEEE Transactions on Communications*, pp. 941–950, February–April 1994.

[397] G. Bauch, H. Khorram and J. Hagenauer, "Iterative equalization and decoding in mobile communications systems," in *European Personal Mobile Communications Conference*, pp. 301–312, 1997.

[398] M. Moher, "Decoding via cross-entropy minimization," in *Proceedings of the IEEE Global Telecommunications Conference 1993*, Houston, TX, pp. 809–813, 29 November – 2 December 1993.

[399] G. Bauch and V. Franz, "Iterative equalisation and decoding for the GSM-system," in *Proceedings of IEEE Vehicular Technology Conference (VTC'98)* [546], pp. 2262–2266.

[400] D. Raphaeli and Y. Zarai, "Combined turbo equalization and turbo decoding," *IEEE Communications Letters*, vol. 2, pp. 107–109, April 1998.

[401] C. Berrou, A. Glavieux and P. Thitimajshima, "Near Shannon Limit Error-Correcting Coding and Decoding: Turbo Codes," in *Proceedings of the International Conference on Communications*, Geneva, Switzerland, pp. 1064–1070, May 1993.

[402] C. Berrou and A. Glavieux, "Near optimum error correcting coding and decoding: turbo codes," *IEEE Transactions on Communications*, vol. 44, pp. 1261–1271, October 1996.

[403] K. Narayanan and G. Stuber, "A serial concatenation approach to iterative demodulation and decoding," *IEEE Transactions on Communications*, vol. 47, pp. 956–961, July 1999.

[404] B. Yeap, T. Liew, J. Hamorsky and L. Hanzo, "Comparative study of turbo equalisers using convolutional codes and block-based turbo-codes for GMSK modulation," in *Proceeding of VTC'99 (Fall)*, Amsterdam, Netherlands, pp. 2974–2978, IEEE, 19–22 September 1999.

[405] A. Klein, R. Pirhonen, J. Skoeld and R. Suoranta, "FRAMES multiple access mode 1 — wideband TDMA with and without spreading," in *Proceedings of IEEE International Symposium on Personal, Indoor and Mobile Radio Communications, PIMRC'97*, vol. 1, Marina Congress Centre, Helsinki, Finland, pp. 37–41, IEEE, 1–4 September 1997.

[406] E. Kuan and L. Hanzo, "Joint detection CDMA techniques for third-generation transceivers," in *Proceeding of ACTS Mobile Communication Summit'98* [547], pp. 727–732.

[407] "COST 207: Digital land mobile radio communications, final report." Office for Official Publications of the European Communities, Luxembourg, 1989.

[408] H. Matsuoka, S. Sampei, N. Morinaga and Y. Kamio, "Adaptive modulation system with variable coding rate concatenated code for high quality multi-media communications systems," in *Proceedings of IEEE VTC'96* [548], pp. 487–491.

[409] S.-G. Chua and A. Goldsmith, "Variable-rate variable-power mQAM for fading channels," in *Proceedings of IEEE VTC'96* [548], pp. 815–819.

[410] J. Torrance and L. Hanzo, "Latency and networking aspects of adaptive modems over slow indoors rayleigh fading channels," *IEEE Transactions on Vehicular Technology*, vol. 48, no. 4, pp. 1237–1251, 1998.

[411] J. Torrance, L. Hanzo and T. Keller, "Interference aspects of adaptive modems over slow rayleigh fading channels," *IEEE Transactions on Vehicular Technology*, vol. 48, pp. 1527–1545, September 1999.

[412] T. Liew, C. Wong and L. Hanzo, "Block turbo coded burst-by-burst adaptive modems," in *Proceedings of Microcoll'99*, Budapest, Hungary, pp. 59–62, 21–24 March 1999.

[413] V. Lau and M. Macleod, "Variable rate adaptive trellis coded QAM for high bandwidth efficiency applications in rayleigh fading channels," in *Proceedings of IEEE Vehicular Technology Conference (VTC'98)* [546], pp. 348–352.

[414] A. Goldsmith and S. Chua, "Variable-rate variable-power MQAM for fading channels," *IEEE Transactions on Communications*, vol. 45, pp. 1218–1230, October 1997.

[415] C. Wong, T. Liew and L. Hanzo, "Turbo coded burst by burst adaptive wideband modulation with blind modem mode detection," in *Proceeding of ACTS Mobile Communication Summit'99*, Sorrento, Italy, pp. 303–308, ACTS, 8–11 June 1999.

[416] M. Yee and L. Hanzo, "Upper-bound performance of radial basis function decision feedback equalised burst-by-burst adaptive modulation," in *Proceedings of ECMCS'99*, Krakow, Poland, 24–26 June 1999.

[417] T. Keller and L. Hanzo, "Adaptive orthogonal frequency division multiplexing schemes," in *Proceeding of ACTS Mobile Communication Summit'98* [547], pp. 794–799.

[418] E. Kuan, C. Wong and L. Hanzo, "Burst-by-burst adaptive joint detection CDMA," in *Proceeding of VTC'99 (Spring)*, (Houston, TX), IEEE, 16–20 May 1999.

[419] A. Czylwik, "Adaptive OFDM for wideband radio channels," in *Proceeding of IEEE Global Telecommunications Conference, Globecom 96* [549], pp. 713–718.

[420] R. Fischer and J. Huber, "A new loading algorithm for discrete multitone transmission," in *Proceeding of IEEE Global Telecommunications Conference, Globecom 96* [549], pp. 713–718.

[421] P. Chow, J. Cioffi and J. Bingham, "A practical discrete multitone transceiver loading algorithm for data transmission over spectrally shaped channels," *IEEE Transactions on Communications*, vol. 48, pp. 772–775, 1995.

[422] H. Rohling and R. Grünheid, "Performance of an OFDM-TDMA mobile communication system," in *Proceeding of IEEE Global Telecommunications Conference, Globecom 96* [549], pp. 1589–1593.

[423] K. Fazel, S. Kaiser, P. Robertson and M. Ruf, "A concept of digital terrestrial television broadcasting," *Wireless Personal Communications*, vol. 2, pp. 9–27, 1995.

[424] H. Sari, G. Karam and I. Jeanclaude, "Transmission techniques for digital terrestrial TV broadcasting," *IEEE Communications Magazine*, pp. 100–109, February 1995.

[425] J. Borowski, S. Zeisberg, J. Hübner, K. Koora, E. Bogenfeld and B. Kull, "Performance of OFDM and comparable single carrier system in MEDIAN demonstrator 60GHz channel," in *Proceeding of ACTS Mobile Communication Summit'97* [550], pp. 653–658.

[426] I. Kalet, "The multitone channel," *IEEE Transactions on Communications*, vol. 37, pp. 119–124, February 1989.

[427] P. Cherriman, T. Keller and L. Hanzo, "Constant-rate turbo-coded and block-coded orthogonal frequency division multiplex videophony over UMTS," in *Proceeding of Globecom'98* [551], pp. 2848–2852.

[428] J. Woodard, T. Keller and L. Hanzo, "Turbo-coded orthogonal frequency division multiplex transmission of 8 kbps encoded speech," in *Proceeding of ACTS Mobile Communication Summit'97* [550], pp. 894–899.

[429] Y. Li and N. Sollenberger, "Interference suppression in OFDM systems using adaptive antenna arrays," in *Proceeding of Globecom'98* [551], pp. 213–218.

[430] F. Vook and K. Baum, "Adaptive antennas for OFDM," in *Proceedings of IEEE Vehicular Technology Conference (VTC'98)* [546], pp. 608–610.

[431] G. Ungerboeck, "Channel coding with multilevel/phase signals," *IEEE Transactions on Information Theory*, vol. IT-28, pp. 55–67, January 1982.

[432] E. Zehavi, "8-PSK trellis codes for a Rayleigh fading channel," *IEEE Transactions on Communications*, vol. 40, pp. 873–883, May 1992.

[433] G. Caire, G. Taricco and E. Biglieri, "Bit-Interleaved Coded Modulation," *IEEE Transactions on Information Theory*, vol. IT-44, pp. 927–946, May 1998.

[434] S. L. Goff, A. Glavieux and C. Berrou, "Turbo-codes and high spectral efficiency modulation," in *Proceedings of IEEE International Conference on Communications*, pp. 645–649, 1994.

[435] P. Robertson and T. Worz, "Bandwidth-Efficient Turbo Trellis-Coded Modulation Using Punctured Component Codes," *IEEE Journal on Selected Areas in Communications*, vol. 16, pp. 206–218, February 1998.

[436] U. Wachsmann and J. Huber, "Power and bandwidth efficient digital communications using turbo codes in multilevel codes," *European Transactions on Telecommunications*, vol. 6, pp. 557–567, September–October 1995.

[437] D. J. Costello, A. Banerjee, T. E. Fuja and P. C. Massey, "Some Reflections on the Design of Bandwidth Efficient Turbo Codes," in *Proceedings of 4th ITG Conference on Source and Channel Coding*, no. 170 in ITG Fachbericht, (Berlin), pp. 357–363, VDE–Verlag, 28–30 January 2002.

[438] L. Hanzo, C. H. Wong and M. S. Yee, *Adaptive Wireless Transceivers: Turbo-Coded, Turbo-Equalized and Space-Time Coded TDMA, CDMA and OFDM Systems*. New York, USA: John Wiley, IEEE Press, 2002.

[439] C. Wong and L. Hanzo, "Upper-bound performance of a wideband burst-by-burst adaptive modem," *IEEE Transactions on Communications*, vol. 48, pp. 367–369, March 2000.

[440] S. M. Alamouti and S. Kallel, "Adaptive Trellis-Coded Multiple-Phased-Shift Keying Rayleigh fading channels," *IEEE Transactions on Communications*, vol. 42, pp. 2305–2341, June 1994.

[441] K. J. Hole, H. Holm and G. E. Oien, "Adaptive multidimensional coded modulation over flat fading channels," *IEEE Journal on Selected Areas in Communications*, vol. 18, pp. 1153–1158, July 2000.

[442] A. Goldsmith and S. Chua, "Adaptive coded modulation for fading channels," *IEEE Transactions on Communications*, vol. 46, pp. 595–602, May 1998.

[443] D. Goeckel, "Adaptive coding for fading channels using outdated fading estimates," *IEEE Transactions on Communications*, vol. 47, pp. 844–855, June 1999.

[444] V. K. N. Lau and M. D. Macleod, "Variable-Rate Adaptive Trellis Coded QAM for Flat-Fading Channels," *IEEE Transactions on Communications*, vol. 49, pp. 1550–1560, September 2001.

[445] P. Ormeci, X. Liu, D. Goeckel and R. Wesel, "Adaptive bit-interleaved coded modulation," *IEEE Transactions on Communications*, vol. 49, pp. 1572–1581, September 2001.

[446] V. K. N. Lau, "Performance analysis of variable rate: symbol-by-symbol adaptive bit interleaved coded modulation for Rayleigh fading channels," *IEEE Transactions on Vehicular Technology*, vol. 51, pp. 537–550, May 2002.

[447] S. X. Ng, C. H. Wong and L. Hanzo, "Burst-by-Burst Adaptive Decision Feedback Equalized TCM, TTCM, BICM and BICM-ID," *International Conference on Communications (ICC)*, 11–15 June 2001. Helsinki, Finland, pp. 3031–3035.

[448] P. Cherriman, C. Wong and L. Hanzo, "Turbo- and BCH-coded wide-band burst-by-burst adaptive H.263-assisted wireless video telephony," *IEEE Transactions on Circuits and Systems for Video Technology*, vol. 10, pp. 1355–1363, December 2000.

[449] Special Mobile Group of ETSI, "UMTS: Selection procedures for the choice of radio transmission technologies of the UMTS," *Technical Report*, European Telecommunications Standard Institute (ETSI), France, 1998.

[450] A. Duel-Hallen, S. Hu and H. Hallen, "Long range prediction of fading signals," *IEEE Signal Processing Magazine*, vol. 17, pp. 62–75, May 2000.

[451] L. Hanzo, M. Münster, B. J. Choi and T. Keller, *OFDM and MC-CDMA for Broadcasting Multi-User Communications, WLANs and Broadcasting*. New York, USA: John Wiley, IEEE Press, 2003.

[452] P. Robertson, E. Villebrun and P. Hoeher, "A comparison of optimal and sub-optimal MAP decoding algorithms operating in the log domain," in *Proceedings of the International Conference on Communications*, pp. 1009–1013, June 1995.

[453] A. Klein, G. Kaleh and P. Baier, "Zero forcing and minimum mean square error equalization for multiuser detection in code division multiple access channels," *IEEE Transactions on Vehicular Technology*, vol. 45, pp. 276–287, May 1996.

[454] G. Golub and C. van Loan, *Matrix Computations*. North Oxford Academic, 1983.

[455] B. J. Choi and L. Hanzo, "Optimum Mode-Switching-Assisted Constant-Power Single- and Multicarrier Adaptive Modulation," *IEEE Transactions on Vehicular Technology*, vol. 52, pp. 536–560, May 2003.

[456] V. Tarokh, N. Seshadri and A. R. Calderbank, "Space-time codes for high rate wireless communication: Performance analysis and code construction," *IEEE Transactions on Information Theory*, vol. 44, pp. 744–765, March 1998.

[457] M. Tao and R. S. Cheng, "Diagonal Block Space-time Code Design for Diversity and Coding Advantage over Flat Rayleigh Fading Channels," *IEEE Transactions on Signal Processing*, to appear in April 2004.

[458] S. X. Ng, F. Guo and L. Hanzo, "Iterative Detection of Diagonal Block Space Time Trellis Codes, TCM and Reversible Variable Length Codes for Transmission over Rayleigh Fading Channels," in *IEEE Vehicular Technology Conference*, Los Angeles, CA, 26–29 September 2004.

[459] U. Wachsmann, R. F. H. Fischer and J. B. Huber, "Multilevel Codes: Theoretical Concepts and Practical Design Rules," *IEEE Transactions on Information Theory*, vol. 45, pp. 1361–1391, July 1999.

[460] R. H. Morelos-Zaragoza, M. P. C. Fossorier, L. Shu and H. Imai, "Multilevel Coded Modulation for Unequal Error Protection and Multistage Decoding – Part I: Symmetric Constellations," *IEEE Transactions on Communications*, vol. 48, pp. 204–213, February 2000.

[461] ISO/IEC JTC1/SC29/WG11 W2502, "ISO/IEC 14496-2.," in *Final Draft International Standard. Part 2: Visual*, Atlantic City, NJ, 1998.

[462] J. Kliewer, S. X. Ng and L. Hanzo, "Efficient Computation of EXIT Functions for Non-Binary Iterative Decoding," *To appear in IEEE Transactions on Communications*, 2005.

[463] G. J. Foschini, Jr., "Layered Space-time architecture for wireless communication in a fading environment when using multi-element antennas," *Bell Labs Technical Journal*, pp. 41–59, 1996.

[464] J. Proakis, *Digital Communications*. New York: McGraw-Hill, 1987.

[465] R. Gallager, *Information Theory and Reliable Communication*. John Wiley and Sons, 1968.

[466] E. Telatar, "Capacity of multi-antenna Gaussian channels," *European Transactions on Telecommunication*, vol. 10, pp. 585–595, Nov–Dec 1999.

[467] B. Vucetic and J. Yuan, *Space-Time Coding*. New York: John Wiley-IEEE Press, May 2003.

[468] S. M. Alamouti, "A simple transmit diversity technique for wireless communications," *IEEE Journal on Selected Areas in Communications*, vol. 16, pp. 1451–1458, October 1998.

[469] V. Tarokh, H. Jafarkhani and A. Calderbank, "Space-time block codes from orthogonal designs," *IEEE Transactions on Information Theory*, vol. 45, pp. 1456–1467, May 1999.

[470] S. X. Ng and L. Hanzo, "Space-Time IQ-interleaved TCM and TTCM for AWGN and Rayleigh Fading Channels," *IEE Electronics Letters*, vol. 38, pp. 1553–1555, November 2002.

[471] S. ten Brink, "Convergence behaviour of iteratively decoded parallel concatenated codes," *IEEE Transactions on Communications*, vol. 49, pp. 1727–1737, October 2001.

[472] H. Chen and A. Haimovich, "EXIT charts for turbo trellis-coded modulation," *IEEE Communications Letters*, vol. 8, pp. 668–670, November 2004.

[473] A. Grant, "Convergence of non-binary iterative decoding," in *Proceedings of the IEEE Global Telecommunications Conference (GLOBECOM)*, San Antonio, TX, pp. 1058–1062, November 2001.

[474] M. Tüchler, "Convergence prediction for iterative decoding of threefold concatenated systems," in *Proceedings of Global Telecommunications Conference — Globecom'02*, vol. 2, Taipei, Taiwan, pp. 1358–1362, IEEE, 17–21 November 2002.

[475] M. Tüchler and J. Hagenauer, "EXIT charts of irregular codes," in *Conference on Information Sciences and Systems*, Princeton, NJ, pp. 748–753, March 2002.

[476] ten Brink, S., "Convergence of iterative decoding," *Electronics Letters*, vol. 35, no. 10, pp. 806–808, 1999.

[477] A. Ashikhmin, G. Kramer and S. ten Brink, "Extrinsic information transfer functions: model and erasure channel properties," *IEEE Transactions on Information Theory*, vol. 50, pp. 2657–2673, November 2004.

[478] J. Wang, S. X. Ng, A. Wolfgang, L-L. Yang, S. Chen and L. Hanzo, "Near-capacity three-stage MMSE turbo equalization using irregular convolutional codes," in *International Symposium on Turbo Codes*, Munich, Germany, April 2006. Electronic publication.

[479] L. Hanzo, T. H. Liew and B. L. Yeap, *Turbo Coding, Turbo Equalisation and Space-Time Coding for Transmission over Fading Channels*. John Wiley-IEEE Press, 2002.

[480] C. Shannon, *Mathematical Theory of Communication*. University of Illinois Press, 1963.

[481] D. A. Huffman, "A method for the construction of minimum-redundancy codes," *Proceedings of the IRE*, vol. 40, pp. 1098–1101, September 1952.

[482] V. Buttigieg and P. G. Farrell, "Variable-length error-correcting codes," *IEE Proceedings on Communications*, vol. 147, pp. 211–215, August 2000.

[483] S. Benedetto and G. Montorsi, "Serial concatenation of block and convolutional codes," *Electronics Letters*, vol. 32, no. 10, pp. 887–888, 1996.

[484] S. Benedetto and G. Montorsi, "Iterative decoding of serially concatenated convolutional codes," *Electronics Letters*, vol. 32, no. 13, pp. 1186–1188, 1996.

[485] R. G. Maunder, J. Wang, S. X. Ng, L-L. Yang and L. Hanzo, "Irregular Variable Length Coding for Near-Capacity Joint Source and Channel coding." Submitted to *IEEE Workshop on Signal Processing Systems*, Shanghai, China, October 2007.

[486] G. Ungerboeck, "Channel coding with multilevel/phase signals," *IEEE Transactions on Information Theory*, vol. 28, no. 1, pp. 55–67, 1982.

[487] R. Bauer and J. Hagenauer, "Symbol by symbol MAP decoding of variable length codes," in *3rd ITG Conference on Source and Channel Coding*, Munich, Germany, pp. 111–116, January 2000.

[488] J. Kliewer and R. Thobaben, "Iterative joint source-channel decoding of variable-length codes using residual source redundancy," *IEEE Transactions on Wireless Communications*, vol. 4, no. 3, pp. 919–929, 2005.

[489] V. B. Balakirsky, "Joint source-channel coding with variable length codes," in *IEEE International Symposium on Information Theory*, Ulm, Germany, p. 419, June 1997.

[490] S. Lloyd, "Least squares quantization in PCM," *IEEE Transactions on Information Theory*, vol. 28, no. 2, pp. 129–137, 1982.

[491] J. Max, "Quantizing for minimum distortion," vol. 6, pp. 7–12, March 1960.

[492] L. Bahl, J. Cocke, F. Jelinek and J. Raviv, "Optimal decoding of linear codes for minimizing symbol error rate (Corresp.)," *IEEE Transactions on Information Theory*, vol. 20, no. 2, pp. 284–287, 1974.

[493] J. Hagenauer, E. Offer and L. Papke, "Iterative decoding of binary block and convolutional codes," *IEEE Transactions on Information Theory*, vol. 42, no. 2, pp. 429–445, 1996.

[494] J. Wang, L-L. Yang and L. Hanzo, "Iterative construction of reversible variable-length codes and variable-length error-correcting codes," *IEEE Communications Letters*, vol. 8, pp. 671–673, November 2004.

[495] R. Thobaben and J. Kliewer, "Low-complexity iterative joint source-channel decoding for variable-length encoded Markov sources," *IEEE Transactions on Communications*, vol. 53, pp. 2054–2064, December 2005.

[496] L. Hanzo, S. X. Ng, T. Keller and W. Webb, *Quadrature Amplitude Modulation*. Chichester, UK: Wiley, 2004.

[497] M. Tüchler, "Design of serially concatenated systems depending on the block length," *IEEE Transactions on Communications*, vol. 52, pp. 209–218, February 2004.

[498] R. Bauer and J. Hagenauer, "On variable length codes for iterative source/channel decoding," in *Data Compression Conference*, Snowbird, UT, pp. 273–282, March 2001.

[499] ETSI, *Digital Video Broadcasting (DVB); Framing structure, channel coding and modulation for digital terrestrial television*, August 1997. EN 300 744 V1.1.2.

[500] ETSI, *Digital Video Broadcasting (DVB); Framing structure, channel coding and modulation for cable systems*, December 1997. EN 300 429 V1.2.1.

[501] ETSI, *Digital Video Broadcasting (DVB); Framing structure, channel coding and modulation for 11/12 GHz Satellite Services*, August 1997. EN 300 421 V1.1.2.

[502] A. Michelson and A. Levesque, *Error Control Techniques for Digital Communication*. New York: Wiley-Interscience, 1985.

[503] S. O'Leary and D. Priestly, "Mobile broadcasting of DVB-T signals," *IEEE Transactions on Broadcasting*, vol. 44, pp. 346–352, September 1998.

[504] W.-C. Lee, H.-M. Park, K.-J. Kang and K.-B. Kim, "Performance analysis of viterbi decoder using channel state information in COFDM system," *IEEE Transactions on Broadcasting*, vol. 44, pp. 488–496, December 1998.

[505] S. O'Leary, "Hierarchical transmission and COFDM systems," *IEEE Transactions on Broadcasting*, vol. 43, pp. 166–174, June 1997.

[506] L. Thibault and M. Le, "Performance evaluation of COFDM for digital audio broadcasting Part I: parametric study," *IEEE Transactions on Broadcasting*, vol. 43, pp. 64–75, March 1997.

[507] B. Haskell, A. Puri and A. Netravali, *Digital Video: An Introduction To MPEG-2*. Digital Multimedia Standards Series, London: Chapman and Hall, 1997.

[508] *ISO/IEC 13818-2: Information Technology — Generic Coding of Moving Pictures and Associated Audio Information — Part 2: Video*, March 1995.

[509] L. Hanzo and J. Woodard, "An intelligent voice communications system for indoors communications," in *Proceedings of IEEE Vehicular Technology Conference (VTC'95)*, vol. 4, Chicago, IL, pp. 735–749, IEEE, 15–28 July 1995.

[510] P. Shelswell, "The COFDM modulation system: the heart of digital audio broadcasting," *Electronics & Communication Engineering Journal*, vol. 7, pp. 127–136, June 1995.

[511] S. Wicker, *Error Control Systems for Digital Communication and Storage*. Englewood Cliffs, NJ: Prentice-Hall, 1994.

[512] A. Barbulescu and S. Pietrobon, "Interleaver design for turbo codes," *IEE Electronic Letters*, pp. 2107–2108, December 1994.

[513] M. Failli, "Digital land mobile radio communications COST 207," *Technical Report*, European Commission, 1989.

[514] H. Gharavi and S. Alamouti, "Multipriority video transmission for third-generation wireless communication system," in Gharavi and Hanzo [516], pp. 1751–1763.

[515] A. Aravind, M. Civanlar and A. Reibman, "Packet loss resilience of MPEG-2 scalable video coding algorithms," *IEEE Transaction on Circuits And Systems For Video Technology*, vol. 6, pp. 426–435, October 1996.

[516] H. Gharavi and L. Hanzo, eds., *Proceedings of the IEEE*, vol. 87, October 1999.

[517] G. Reali, G. Baruffa, S. Cacopardi and F. Frescura, "Enhancing satellite broadcasting services using multiresolution modulations," *IEEE Transactions on Broadcasting*, vol. 44, pp. 497–506, December 1998.

[518] Y. Hsu, Y. Chen, C. Huang and M. Sun, "MPEG-2 spatial scalable coding and transport stream error concealment for satellite TV broadcasting using Ka-band," *IEEE Transactions on Broadcasting*, vol. 44, pp. 77–86, March 1998.

[519] L. Atzori, F. D. Natale, M. D. Gregario and D. Giusto, "Multimedia information broadcasting using digital TV channels," *IEEE Transactions on Broadcasting*, vol. 43, pp. 383–392, December 1997.

[520] W. Sohn, O. Kwon and J. Chae, "Digital DBS system design and implementation for TV and data broadcasting using Koreasat," *IEEE Transactions on Broadcasting*, vol. 44, pp. 316–323, September 1998.

[521] J. Griffiths, *Radio Wave Propagation and Antennas — An Introduction*. Englewood Cliffs, NJ: Prentice-Hall, 1987.

[522] M. Karaliopoulos and F.-N. Pavlidou, "Modelling the land mobile satellite channel: a review," *Electronics and Communication Engineering Journal*, vol. 11, pp. 235–248, October 1999.

[523] J. Goldhirsh and W. Vogel, "Mobile satellite system fade statistics for shadowing and multipath from roadside trees at UHF and L-band," *IEEE Transactions on Antennas and Propagation*, vol. 37, pp. 489–498, April 1989.

[524] W. Vogel and J. Goldhirsh, "Multipath fading at L band for low elevation angle, land mobile satellite scenarios," *IEEE Journal on Selected Areas in Communications*, vol. 13, pp. 197–204, February 1995.

[525] W. Vogel and G. Torrence, "Propagation measurements for satellite radio reception inside buildings," *IEEE Transactions on Antennas and Propagation*, vol. 41, pp. 954–961, July 1993.

[526] W. Vogel and U. Hong, "Measurement and modelling of land mobile satellite propagation at UHF and L-band," *IEEE Transactions on Antennas and Propagation*, vol. 36, pp. 707–719, May 1988.

[527] S. Saunders, C. Tzaras and B. Evans, "Physical statistical propagation model for mobile satellite channel," *Technical Report*, European Commission, 1998.

[528] S. Saunders, *Antennas and Propagation for Wireless Communication Systems Concept and Design*. New York: John Wiley and Sons, 1999.

[529] K. Wesolowsky, "Analysis and properties of the modified constant modulus algorithm for blind equalization," *European Transactions on Telecommunication*, vol. 3, pp. 225–230, May–June 1992.

[530] M. Goursat and A. Benveniste, "Blind equalizers," *IEEE Transactions on Communications*, vol. COM–28, pp. 871–883, August 1984.

[531] G. Picchi and G. Prati, "Blind equalization and carrier recovery using a "stop–and–go" decision–directed algorithm," *IEEE Transactions on Communications*, vol. COM–35, pp. 877–887, September 1987.

[532] A. Polydoros, R. Raheli and C. Tzou, "Per–survivor processing: a general approach to MLSE in uncertain environments," *IEEE Transactions on Communications*, vol. COM–43, pp. 354–364, February–April 1995.

[533] D. Godard, "Self–recovering equalization and carrier tracking in two–dimensional data communication systems," *IEEE Transactions on Communications*, vol. COM–28, pp. 1867–1875, November 1980.

[534] Y. Sato, "A method of self–recovering equalization for multilevel amplitude–modulation systems," *IEEE Transactions on Communications*, vol. COM–23, pp. 679–682, June 1975.

[535] Z. Ding, R. Kennedy, B. Anderson and R. Johnson, "Ill-convergence of Godard blind equalizers in data communications systems," *IEEE Transactions on Communications*, vol. COM-39, pp. 1313–1327, September 1991.

[536] Y.-Q. Zhang, F. Pereira, T. Sikora and C. Reader (Guest Editors), "Special issue on MPEG-4," *IEEE Transactions on Circuits and Systems for Video Technology*, vol. 7, February 1997.

[537] L. Chiariglione, "MPEG and multimedia communication," *IEEE Transaction On Circuits And Systems For Video Technology*, vol. 7, pp. 5–18, February 1997.

[538] T. Sikora, "The MPEG-4 video standard verification model," *IEEE Transaction On Circuits And Systems For Video Technology*, vol. 7, pp. 19–31, February 1997.

[539] R. Damper, W. Hall and J. Richards, eds., *Proceedings of IEEE International Symposium of Multimedia Technologies and Future Applications*. London: Pentech Press, April 1993.

[540] IEEE, *Proceedings of IEEE VTC'94*, Stockholm, Sweden, 8–10 June 1994.

[541] IEEE, *Proceedings of International Conference on Acoustics, Speech, and Signal Processing, ICASSP'92*, March 1992.

[542] IEEE, *Proceedings of the IEEE International Conference on Acoustics, Speech and Signal Processing (ICASSP'94)*, Adelaide, Australia, 19–22 April 1994.

[543] IEEE, *Proceedings of the IEEE International Conference on Acoustics, Speech and Signal Processing (ICASSP'96)*, Atlanta, GA, 7–10 May 1996.

[544] *Proceedings of International Workshop on Coding Techniques for Very Low Bit-rate Video (VLBV'95)*, Shinagawa, Tokyo, Japan, 8–10 November 1995.

[545] J. Gibson, ed., *The Mobile Communications Handbook*. Boca Raton FL: CRC Press and IEEE Press, 1996.

[546] IEEE, *Proceedings of IEEE Vehicular Technology Conference (VTC'98)*, Ottawa, Canada, 18–21 May 1998.

[547] ACTS, *Proceeding of ACTS Mobile Communication Summit'98*, Rhodes, Greece, 8–11 June 1998.

[548] IEEE, *Proceedings of IEEE VTC'96*, Atlanta, GA, 28 April–1 May 1996.

[549] IEEE, *Proceeding of IEEE Global Telecommunications Conference, Globecom 96*, London, 18–22 November 1996.

[550] ACTS, *Proceeding of ACTS Mobile Communication Summit'97*, Aalborg, Denmark, 7–10 October 1997.

[551] IEEE, *Proceeding of Globecom'98*, Sydney, Australia, 8–12 November 1998.

Index

ACF, 39
Active/passive concept, 61
Adaptive differential pulse code modulation, 177
Adaptive turbo-coded OFDM-based
        videotelephony, 486–507
Adaptive vector quantization, 100–102
Adaptive video systems, 459–628
Adaptive VQ, 100
ADPCM, 177
Affine transformation, 21
Algorithmic complexity, 103–104
Aliasing distortion, 185
AMR, 66, 70, 92
Analysis filtering, 185–188
Antenna diversity, 361–363
AOFDM modem mode adaptation and signaling,
        487–488
AOFDM subband BER estimation, 488
Architecture of system 1, 109–111
Architecture of system 2, 111–112
Architecture of systems 3–6, 112–113
Arithmetic coding, 311, 312
ARQ, 16, 73, 257, 287, 290, 291, 293, 341,
        357–359, 367, 370, 371
Automatic repeat request, 79–80, 257, 341
AV.26M, 340
AVC, 14
AVQ, 100
AWGN, 71, 403, 404

B-Codes, 220
Basic differential pulse code modulation, 173–175
BbB, 18
BbB-AQAM, 16
BER, 404
Bidirectional coding, 13
Bit sensitivity, 71, 106–107, 157–158
Bit sensitivity of codec I and II, 71–73
Bit-allocation, 155–157, 229–230
Bit-allocation strategy, 67–68, 105–106

Bitrate, 403, 452
Bitstream, 410
    DCT parameters, 452
    Hierarchial, 437
    Parameters, 445, 451, 452, 454
    Parameters index, 451
    Start code, 440
    Start code list, 440
    Start code prefix, 440
    VO parameters, 451
    VOL parameters, 451
    VOP parameters, 451
Blind equalizers, 608–611
Block truncation algorithm, 177–179
Block truncation codec implementations, 180
Block Truncation Coding (BTC), 177–183
BTC, 177
    Algorithm, 177
    Implementation, 180
    Inter-coding, 182
    Intra-coding, 180
    Performance-intra, 182, 183
    Quantizers, 180, 181
    Upper bounds (Intra), 179, 183
Burst-by-burst adaptive videophone system,
        480–485

Camera shake, 310
CCITT, 8
CDMA
    JD-CDMA, 16, 18
Channel coding and modulation, 158–159
Choice of modem, 73–74
Choice of modulation, 107–108
Chrominance, 5
Classified vector quantization, 102–103
Classified VQ, 102
Codebook design, 93–95
Codec outline, 58–60
Codec performance, 230–232

Coded Modulation, 18
Combined source/channel coding, 257, 342
Comparison
    Bitrate, 167, 168
    Error sensitivity, 169
    Performance, 168
Comparison of subband-adaptive OFDM and fixed-mode OFDM transceivers, 489–493
Complex quadrature mirror filters, 191
Complexity considerations and reduction techniques, 65–66
Contractive affine transforms, 21
Cost/gain, 59
Cr_u, 5
Cr_v, 5
CVQ, 102

Data compression, 1
Data partition, 443, 445
    DCT data, 445
    Motion data, 445
Data partitioning scheme, 576–581
DCT, 8, 35, 205, 383, 392, 397, 409, 415, 441
    B-encoded, 221
    Bit-allocation, 67, 230
    Codec
        Active/passive, 61, 62
        Bit sensitivity, 71, 72
        Bit-allocation, 67
        Complexity, 65
        Cost gain controlled, 64
        Degradation profile, 72
        Erroneous conditions, 70
        Error sensitivity, 73
        Gain controlled, 60
        Intra-block sizes, 60
        Intra-frame, 60
        Performance, 68
        PFU degradation, 64
        Preselection, 66
        Results, 68
        Results – "Lab" sequence, 69, 70
        Results – "Miss America" sequence, 69
    Codec outline, 58
    Codec schematic, 59
    Complexity, 233
    DCT coefficients, 15, 392, 410
        AC coefficients, 392
        DC coefficients, 392
    Efficiency
        Inter, 219
        Intra, 218
    Error sensitivity, 233, 234
    Integer Transform, 415
        4 × 4 integer transform, 416
        Example, 421
    Inter frame, 215
    Inter/Intra codec, 226
    Inter/Intra decision, 226
    Intra bit-allocation, 209
    Intra performance, 210
    Introduction, 205
    Multiclass, 58
    Optimum mask, 208
    Performance, 230, 231
        Erroneous conditions, 232, 233
    PFU, 63
    Proposed codec, 224
    Quantizer allocation, 56, 57
    Quantizer training, 55, 205
    Quantizer types, 58
    Schematic, 224
    Single class, 55
    Zonal mask, 206
DCT basis images, 53
DCT codec performance under erroneous conditions, 70–73
DCT codecs, 35
DCT coefficients, 409, 445
DCT-based low-rate video transceivers, 73–80
De-blocking filter, 427
Decimated signals, 185
Decomposition algorithmic issues, 145–148
DECT, 362, 364–366
DECT channels, 364–366
DECT frame structure, 364
Degradation, 15
DFD, 39
Differential pulse code modulation, 173–177
Distance measures, 40–42
Domain block, 22
DPCM, 8, 173, 177
    Adaptive, 177
    ADPCM-performance, 178
    Basic, 173
    Inter/Intra, 175
    PDF, 175
    Performance, 176
    Schematic, 174
DVB satellite scheme, 605–607
DVB terrestrial scheme, 572–574
DVB-T for mobile receivers, 560–604
DVD, 12

Energy compaction coefficient, 217
Enhanced QT codec, 153–154
Entropy coding
    CABAC, 423, 424
    UVLC, 423
EREC, 340
ERPC, 169
Error control background, 255
Error control mechanisms, 255
Error correction coding, 275
Error mechanisms, 253

INDEX

Error resilient, 441, 442
    Data recovery, 441
    Error concealment, 441, 442
    Resynchronization, 441–443
    Resynchronization marker, 442, 443, 447
    Synchronization, 443, 444
Error sensitivity and complexity, 32–34, 233–235
EXIT charts, 545
Extrinsic information transfer charts, 545
Eye and mouth detection, 149–151

FAW, 65, 157, 227
    Autocorrelation, 230
    Data-emulating probability, 229
FDD, 367
FEC, 16, 73, 169, 273, 275, 284, 286, 287, 345, 441
FER performance, 84–86
FIFO, 101
First-order intensity match, 144–145
Forced updating, 63
Forward error correction (FEC), 75, 109
Fractal codec design, 27–28
Fractal codec performance, 28–32
Fractal encoding, 21
    Affine transformation, 21, 22
    Bit sensitivity, 33
    Bit-allocation, 33
    Block types, 30
    Codec comparison, 29
    Codec design, 27
    Collage theorem, 27
    Complexity, 32
    Conclusion, 34
    DB-RB mapping example, 23
    Domain block (DB), 22
    Error sensitivity, 32, 34
    IFS
        Encoding, 26
        Example, 26
        Example decoding, 27
    Iterated function system (IFS), 23
    Iterative reconstruction, 32
    One-dimensional, 24
    One-dimensional example, 25
    PDF, 29
    Performance, 28, 31
    Principles, 21
    Random collage theorem, 23
Fractal image codecs, 21–34
Fractal principles, 21–23
Frame alignment, 227–229
Frame-differencing, 240
Full or exhaustive motion search, 42–43

Gain-controlled motion compensation, 60–61
Gain-cost-controlled motion compensation, 46–48
Gain/cost-controlled inter-frame codec, 64–66
Gaussian channels, 283, 287–291, 356–359

GOB, 443, 444
GPRS, 17
Gradient constraint equation, 44
Gradient-based motion estimation, 43–44
Grid interpolation techniques, 49

H.261
    Bitrate adjustment, 273
    Block diagram, 240, 296
    Block layer, 247–249
    Coded block pattern, 247, 248, 261, 262, 302
    Coding control, 242
    Comparison with H.263, 295–337
    Effects of errors, 259
    End of block marker, 247, 249, 262
    Error degradation
        Inter, 265, 266, 268–272
        Intra, 263, 264
        Motion vectors, 268, 270
        Quantizer, 267–269
    Error recovery, 258
    Errors in inter blocks, 264
    Errors in intra blocks, 263
    Extra data bits (GSPARE), 246, 247, 260, 262
    Extra data bits (PSPARE), 245, 260, 262, 264, 267
    Extra insertion bit (GEI), 246, 247, 260, 262
    Extra insertion bit (PEI), 245, 260, 262–264, 266, 271
    Frame divided into GOBs, 246
    GOB index, 245, 246, 260, 262–264, 266, 267, 271
    GOB quantizer, 245, 246, 260, 262, 263, 267, 268, 271, 272, 275, 301
    GOB start code, 245, 246, 259, 260, 262
    Group of blocks, 244, 245
    Group of blocks layer, 245, 246
    Hierarchical structure, 244
    Intra DC parameters, 261, 262
    Macroblock, 240, 247
    Macroblock address, 247, 248, 261–264, 266, 267, 301, 302
    Macroblock layer, 247, 248
    Macroblock motion vector, 247, 248, 261, 262, 269, 301, 302
    Macroblock quantizer, 245, 247, 248, 261, 262, 267, 275, 301, 302
    Macroblock size, 274, 276, 277, 286
    Macroblock type, 247, 248, 261, 262, 266, 267, 269, 301, 302
    Motion compensation, 239, 240, 247, 248, 261, 269, 295, 306
    Overview, 239
    Performance, 319–322
    Picture layer, 244, 245
    Picture layer header, 244
    Picture start code, 244, 245, 259, 262
    Picture type word, 245, 259, 260, 262

Pixel sampling, 241
Qualitative error effects, 259
Quantitative error effects, 262
Quantizers, 243, 274
Run-length coding, 249, 250
Source encoder, 240, 241
Statistics, 250
Temporal reference, 245, 260, 262
Transform coefficients, 247, 249, 262
Transmission coder, 250
Video multiplex coder, 243
Video-coding standard, 239–253
Videophone system, 272–294
Zig-zag run-length coding, 249
H.261 performance, 319–322
    Comparison to H.263, 322–325
    Fixed bitrate, 321
    Fixed PSNR, 321
    PSNR versus bitrate, 319
    PSNR versus compression, 320
H.261/H.263 comparison, 295–337
    Coding algorithms, 297–317
    Conclusions, 335
    Introduction, 295
    Motion compensation, 306–308
    Overview, 295
    Performance, 322–325
        Fixed bitrate, 324
        Fixed PSNR, 323
        PSNR versus bitrate, 323
        PSNR versus compression, 324
        Results, 318–335
    Source encoder, 297–298
    Video multiplex coder, 298–306
H.263, 222, 232
    16CIF format, 296, 297, 299
    4CIF format, 296, 297, 299
    Advanced prediction, 296, 298, 312–315
    Bidirectional prediction, 318
    Bilinear interpolation, 308
    Bit stuffing, 300
    Bitrate control, 343
    Block diagram, 296
    Block layer, 305
    CBPB coding parameter, 304
    CBPY coding parameter, 304
    CIF format, 296, 297, 299
    COD coding parameter, 302, 305
    Coded block pattern
        B-Blocks, 304
        Chrominance, 302, 305
        Luminance, 304
    Coded macroblock indicator, 302, 305
    Coding algorithms, 297
    Coding control, 298
    Comparison with H.261, 295–337
    Differential quantizer, 304, 305
    DQUANT coding parameter, 304, 305
    Four motion vectors per MB, 313, 314
    GOB frame ID, 301
    GOB index, 300, 301
    GOB quantizer, 301
    GOB start code, 300, 301
    GOB structure, 299
    Group of Blocks layer, 300
    Half-pixel interpolation, 308
    Half-pixel motion vectors, 308
    In a mobile environment, 339
    Macroblock layer, 301, 302
    Macroblock layer quantizer, 304, 305
    Macroblock mode (B-Blocks), 304
    Macroblock motion vector, 304, 305
    Macroblock type, 302, 305
    MCBPC coding parameter, 302, 305
    MODB coding parameter, 304
    Motion compensation, 295, 297, 306, 307
    Motion vector (B-Blocks), 305
    Motion vector predictor, 307, 308
    Motion vectors, 307
    MVD coding parameter, 304, 305
    MVDB coding parameter, 305
    Negotiable options, 296, 309
    Overlapped motion vectors, 313, 315
    P-B mode, 296–298, 305, 315, 316, 318
    Performance, 325–335
    Picture layer, 300
    Picture layer quantizer, 300
    Picture start code, 300
    Picture type word, 300
    Prediction, 297
    QCIF format, 296, 297, 299
    Quantization, 298
    Run-length coding, 306
    Source encoder, 297
    SQCIF format, 296, 297, 299
    Syntax-based arithmetic coding, 296, 310
    Temporal reference, 300
    Transform coding, 298
    Unrestricted motion vectors, 296, 298, 309
    Video multiplex coder, 298
    Video-coding standard, 295–337
    Videophone system, 343–377
    Wireless problems, 339
    Wireless solutions, 340
    Zig-zag run-length coding, 306
H.263 performance, 325–335
    4CIF, 16CIF, 335
    Comparison to H.261, 322–325
    Different resolutions, 328–335
    Fixed bitrate, 328, 330, 333, 335
    Fixed PSNR, 327, 330, 332, 334
    Gray versus color, 325–328
    PSNR versus bitrate, 326, 329, 331, 333, 336
    PSNR versus compression, 327, 329, 331, 334, 336
    QCIF comparison, 328

INDEX

SQCIF, QCIF, 4CIF, 332–335
SQCIF, QCIF, CIF, 328–332
Video sequences, 325
H.264, 407
    Decoder block diagram, 409
    Encoder block diagram, 408
    Inter frame prediction, 412
    Intra frame prediction, 410
        16×16 description, 412
        16×16 diagram, 412
        4×4 description, 411
        4×4 diagram, 411
HDTV, 379
Header Extension Code, 447
Hierarchical OFDM DVB performance, 595–604
Hierarchical or tree search, 44–45
Hierarchical search, 44
High-resolution DCT coding, 205–235
High-resolution image coding, 171–236
Huffman coding, 1

Image block, 391
Image resolution
    CIF, 11, 12, 381
    QCIF, 11, 381, 386, 443, 452
Initial intra-frame coding, 60
Inter coding, 390
    coding sequence, 390
    error, 452
Inter-frame, 17
Inter-frame block truncation coding, 182–183
Inter-frame correlation, 212
Inter-frame DCT coding, 215–223
Inter/Intra-DCT codec, 226–227
Interlacing, 2
Interpolation, 185
Intra coding, 390
    coding sequence, 390
    error, 452
Intra-frame, 16
Intra-frame block truncation coding, 180–181
Intra-frame coding, 60, 256, 341
Intra-frame correlation, 216
Intra-frame quantizer training, 205–209
Intra/Inter-frame differential pulse code modulation, 175–176
Irregular variable-length codes (IRVLC), 545
ISO, 379
Iterated function system, 22
ITU, 11
    H.261, 11
    H.263, 12, 443
    H.263++, 13
ITU (International Telecommunication Union), 35

Joint motion compensation and residual encoding, 222–223

JPEG, 8, 200
JVT, 14

Leakage factor, 63
Local decoder, 37
Low-complexity techniques, 173–203
Luminance, 5

Macroblock (MB), 391, 409, 443–445, 451
    Diagram, 392
    MB coefficients, 409
MAD, 41
Max–Lloyd-based subband coding, 197–202
MC, 37
MC in the transform domain, 50
MC using higher order transformations, 49–50
MCER, 35, 202, 298, 317, 409, 415
MCER active/passive concept, 61–63
Mean absolute difference, 41
Mean and shape gain vector quantization, 99–100
Mean squared error (MSE), 40
Model-based parametric enhancement, 148–153
Modulation
    BPSK, 404
Motion compensation, 35, 37, 209, 225–226
    ACF, 41
    Algorithm comparison, 46
    Comparison, 213
    Conclusion, 50
    Distance measures, 40, 43
    Example, 40
    Exhaustive motion search, 42
    Frame difference, 36
    Full search, 42
    Gain cost controlled, 46
    Gradient based, 43
    Grid interpolation, 49
    Grid interpolation example, 50
    Higher order, 49
    In transform domain, 50
    Inter-frame correlation, 212
    Joint, 222, 223
    MCER entropy, 47
    MV bitrate, 48
    MV correlation (Inter), 214
    MV correlation (Intra), 214, 215
    Other techniques, 48
    Pel recursive, 49
    Post-processing, 46
    Principle, 37
    Schematic, 39
    Search algorithms, 42
    Search techniques, 212
    Simple codec, 36
    Subsampling search, 45
    Tree search, 44, 45
Motion compensation (MC), 11, 12, 15, 383, 384, 392

Motion compensation for high-quality images, 209–215
Motion estimation techniques, 48–50
Motion search algorithms, 42–48
Motion vector (MV), 15, 38, 46, 384, 392
Motivation and video transceiver overview, 473–476
MPEG, 17, 345, 379
    MPEG-1, 11
    MPEG-1+, 12
    MPEG-2, 12
    MPEG-3, 12
    MPEG-4, 13, 379
        BIFS, 438
        Bit sensitivity, 448
        Bitsream structure, 437
        Bitstream, 388, 439, 443
        Content based, 380, 387
        Decoder block diagram, 383, 393
        Elementary stream, 383
        Encoder block diagram, 383, 393
        Flexmuxed stream, 383
        Functionalities, 379
        Object based, 384, 387
        Resynchronization, 444
        Scalability, 396
        Shape encoding, 393
        shape encoding, 394
        Standardization, 379
        Synchronization layer, 383
        Tools, 380
        Transmux stream, 384
MPEG-2 bit error sensitivity, 561–571
MPEG-X, 207
MPEG2, 297
MSE, 40, 399
MSVQ, 99
Multimode video system performance, 476–480
Multiple modulations schemes, 341

Nonhierarchical OFDM DVB performance, 592–595

One-dimensional fractal coding, 24–32
One-dimensional transform coding, 51–52

Parametric codebook training, 151–152
Parametric encoding, 152–153
Partial forced update, 63–64
PDC, 41
PDF, 39
Pel difference classification, 41
Pel-recursive displacement estimation, 49
Perfect reconstruction quadrature mirror filtering, 185–191
Performance and considerations under erroneous conditions, 154–158
Performance of system 1, 80–84
Performance of system 2, 84–88

Performance of systems 1 and 3, 114–115
Performance of systems 2 and 6, 117–118
Performance of systems 3–5, 88–89
Performance of systems 4 and 5, 115–117
Performance of the data partitioning scheme, 581–592
Performance of the DVB satellite scheme, 611–623
Performance summary of the DVB-S system, 618–623
Performance under erroneous conditions, 105–107
PFU, 63
PIC, 18
Pilot symbol-assisted modulation, 284
Pixel, 8
pixel, 391
PNN, 94
Post-processing of motion vectors, 46
Practical QMF design constraints, 189–191
Practical quadrature mirror filters, 191–195
Predicted coding, 13
Preface, xxi–xxiii
Principle of motion compensation, 37–51
Properties, 215
Properties of the DCT transformed MCER, 215–222
Proposed codec, 224–235
Pseudo quantizer, 56
PSNR, 42, 399
    degradation, 452
    performance, 87–88
PSTN, 379

QAM, 257
QAM subchannels, 277
QCIF, 35
QMF, 185
    Analysis-synthesis schematic, 186
    Band-splitting, 186
QMF design, 189
QoS, 17
QT codecs, 139
QT decomposition, 139
QT-based transceiver architectures, 159–162
QT-based video-transceiver performance, 162–165
QT-codec-based video transceivers, 158–162
Quad-tree, 139
    Algorithmic issues, 145
    Bit sensitivity, 157
    Bit-allocation, 155, 157
    Bitrate, 146
    Decomposition, 139, 140
    Decomposition example, 140
    Erroneous conditions, 154
    Intensity match, 142
        Comparison, 145
        First order, 144
        Zero order, 142
    Introduction, 139
    Nodes/leaves, 147

Parametric enhancement, 148
   Algorithm, 149
   Bit-allocation, 153, 157
   Codebook, 152
   Codebook training, 151
   Comparison, 156
   Detection, 149
   Encoding, 152
   Example, 153
   Performance, 156
   Process example, 150
   Schematic, 154
   Template, 151
   PDF, 143
   Performance, 144, 148, 155
   Quantizer ranges, 144
   Regular decomposition, 140
   Segmentation, 142
Quad-tree based codecs, 139–169
Quad-tree decomposition, 139–142
Quad-tree intensity match, 142–148
Quadrature mirror filtering, 185
Quadrature mirror filters, 185, 191
Quality criterion, 225
Quantizer, 392, 400, 409, 445
   Quantizer step, 403
Quantizer training for multi-class DCT, 55–58
Quantizer training for single-class DCT, 55

Random collage theorem, 23
Range block, 22
Rate distortion, 16
Rayleigh channels, 283, 287, 289–291, 293, 356–359, 361–363
Reconfigurable modulation schemes, 257, 341
Reconfigurable videophone system, 272, 339, 343
RGB, 5
RL, 71
RS, 16
RTP, 384, 447
Run-length based intra-frame subband coding, 195–202

Satellite channel model, 607–608
Satellite-based video broadcasting, 604–626
Scalable coding, 16
Search algorithms, 42
Sensitivity-matched modulation, 75
Simulation environment, 113–114
Slot occupancy performance, 86–87
SNR, 17, 42
Source sensitivity, 75
Source-matched transceiver, 74–80
Spatial filtering, 240
Subband coding, 183–195
   10 split band example, 194
   Corrupted, 202
   Dead zone, 198

Dead zone quantizer, 197
Error sensitivity, 201
Max–Lloyd based, 197
PDF, 196
Performance "Susie", 200
Run-length based, 195
Split-band codec, 184
Split-band filters, 192, 193
Without quantization, 195
Subband-adaptive OFDM transceivers having different target bitrates, 493–497
Subsampling search, 45–46
Summary of low-rate codecs/transceivers, 166–169
Summary of QT-based video transceivers, 165–166
Synthesis filtering, 188–189
System 1, 74–77
System 2, 77–80
System concept, 74–75
System performance, 80–89
Systems 3–5, 80
System performance, 113–118

TDD, 367
Terrestrial broadcast channel model, 574–575
Test sequences, 4
Texture, 390
Time division multiple access (TDMA), 371
Time-variant target bitrate OFDM transceivers, 497–504
TMN5, 221
Transform coding, 51–58
Transform coding efficiency test, 216
transmission error, 452
Transmission format, 75–77
Transmission over the symbol-spaced two-path channel, 612–617
Transmission over the two-symbol delay two-path channel, 618
Tree search, 44
Two-dimensional transform coding, 52–55

UEP, 16, 18
UMTS-like videophone system, 473–485
Universal access, 380

Variable Length Code (VLC), 392
   RVLC, 443, 447
   RVLC diagram, 447
Variable-length coding, 195, 441
VCEG, 14
   H.264, 14
   H.26L, 14
Vector quantization, 93
   127-entry codebook, 96
   16-entry codebook, 96
   Adaptive, 100
      Performance, 102
      Schematic, 101

Basic schematic, 97
Bit sensitivity, 106–108
Bit-allocation, 105, 106
Classified, 102
    Active/passive, 105
    Complexity, 103
    Performance, 104
    Schematic, 103
Codebook design, 93
Codebook sizes, 99
Design, 95
Erroneous conditions, 105
Introduction, 93
MSVQ, 99
    Performance, 101
    Schematic, 100
PDF, 94
Performance, 98
Vector quantizer design, 95–104
Very low bitrate DCT codecs and multimode videophone transceivers, 35–92
Very low bitrate VQ codecs and multimode videophone transceivers, 93–138
Video
    Frame rate, 8
    Block, 8
Video codec outline, 35–37
Video coding standard, H.261, 239–253
Video communications, 15
Video compression, 1
Video compression and transmission aspects, 488
Video formats, 2–5
Video object
    Video Object (VO), 385, 386, 389, 441, 451
    Video Object Layer (VOL), 386, 387, 451
    Video Object Plane (VOP), 386, 389, 390, 394, 398, 440, 451
    Video object scene, 384
Video quality measure, 398
    Objective measure, 399
    Subjective measure, 398
Video sequence
    Akiyo, 394, 400, 403
    Foreman, 400, 428
    Miss America, 394, 400, 401, 428
    Suzi, 428
Video systems based on proprietary video codecs, 21–171
Video systems based on standard video codecs, 239–629
Video transmission, 15
Videophone system, 272, 339, 343
VLC, 195, 215
VQ, 8, 93
VQ-based low-rate video transceivers, 107–113

Wireless videophone, 272–294, 343–377
    ARQ, 341, 371

Bitrate adjustment, 273
Bitrate control, 343
Combined source/channel coding, 342
Conclusions, 376
Encoding history, 278, 279, 347, 348
End of frame effect, 281
Error correction coding, 275
Error resilience, 273
Error-free performance, 288, 354, 355
Features, 292, 343, 353
Feedback channel, 273, 282
Frame errors versus channel SNR, 358, 361
Gaussian channels, 360
H.261/H.263 comparison, 359, 360
Intra-frame coding, 341
Macroblock compounding, 279–284
Macroblock truncation, 282–284
Modulation schemes, 285, 287
Objectives, 273
Packet size, 287
Packet structure, 346, 347
Packetization, 273, 278, 346, 349
Packetization algorithm, 349, 350
Packetization examples, 349
Performance, 283, 286, 287, 292, 352–354, 360–362, 365, 366, 369
Performance over DECT channels, 362, 365, 366
Performance versus channel SNR, 290, 291, 293, 356, 357, 359, 363
Performance with antenna diversity, 361–363
Performance with errors, 288, 354–356
Problems, 339
PSNR versus channel SNR, 360
Rayleigh channels, 360
Reconfigurable, 273
Reconfigurable modulation, 341
Resynchronization, 273
Solutions, 340
Subchannel equalization, 277
Switching levels, 366
System architecture, 283
System environment, 352
Transmission feedback, 282, 367–371, 375, 376
    Majority logic codes, 369, 373, 374
Transmission feedback timing, 370
Use of ARQ, 257, 287, 290, 291, 293, 357–359, 371
Use of FEC, 273, 275, 286, 287, 345

YUV, 5, 381, 391, 428
    Chrominance, 391
    Luminance, 391
    YUV representation, 391

Zero-order intensity match, 142–144
Zig-zag, 392, 409, 445

Author Index

Symbols
261, CCITT-H. [29] 9, 11, 408

A
Aase, S.O. [287] 192, 203
Abrahams, J. [385] 443
Adelson, E.H. [279] 184
Ahmed, N. [371] 416
Aign, S. [68] 15
Al-Mualla, M. [139] 17
Al-Subbagh, M. [304] 229
Alamouti, S.M. [514] 573, 587
Alamouti, S.M. [468] 533
Alasmari, A.K. [277] 184
Anderson, B.D.O. [535] 608
Anderson, J.B. [242] 132
Andreadis, A. [381] 441
Ansari, R. [293] 194
Anti Toskala, [198] 76, 92, 108, 158, 159, 161, 165, 368, 622
Apostolopoulos, J.G. [114] 16
Aravind, A. [515] 574
Aravind, R. [73] 15, 16, 396
Arce, G.R. [268] 179
Arnold, J.F. [246] 139, 145, 196
Arumugam, A. [143] 17
Asghar, S. [338] 363
Ashikhmin, A. [477] 543
Atzori, L. [519] 601
Avaro, O. [351] 383
Azari, J. [186] 54

B
Bahl, L. [492] 547
Bahl, L.R. [234] 119, 122, 130
Balakirsky, V.B. [489] 545, 547, 555
Bamberger, R.H. [288] 192
Barbulescu, A.S [512] 571
Barnsley, M.F. [146] 21, 22, 27, 28
Baruffa, G. [517] 601

Baskaran, V. [6] 1
Bauch, G. [399] 462
Bauch, G. [397] 462, 464
Bauer, R. [232] 119, 127
Bauer, R. [487] 546, 555, 557
Baum, K.L. [430] 486
Beaumont, J.M. [147] 21, 24, 27, 28
Bellifemine, F. [214] 93
Benedetto, S. [229] 119
Benedetto, S. [483] 547
Benedetto, S. [484] 547
Benelli, G. [381] 441
Benveniste, A. [530] 605, 606
Berrou, C. [401] . 463, 473, 474, 485, 488, 506, 571, 602
Berrou, C. [393] 462, 463
Berrou, C. [402] 463, 571, 602, 607
Bharghavan, V. [111] 16
Bi, G. [367] 410
Biglieri, E. [254] 158
Bingham, J.A.C. [421] 486
Bjontegaard, G. [66] 15, 410
Blanz, J. [87] 16
Blättermann, G. [373] 424, 427, 431
Blattermann, G. [369] 410
Boekee, D.E. [291] 193
Bogenfeld, E. [425] 486
Bormann, C. [388] 448
Borowski, J. [425] 486
Bossen, F. [362] 393, 394
Bozdagi, G. [155] 38
Braden, R. [54] 10
Bull, D. [139] 17
Buttigieg, V. [482] 543
Buzo, A. [240] 125

C
Cacopardi, S. [517] 601
Caldas-Pinto, J.R. [264] 177
Calderbank, A.R. [456] 527, 532

Calderbank, A.R. [469] 533
Canagarajah, N. [139] 17
Carlton, S.G. [265] 177
Casas, J.R. [205] 93
Casner, S. [356] 384, 448
Cavenor, M.C. [246] 139, 145, 196
CCITT/SG, XV [22] 8, 9
CCITT/SG, XV [61] 8
Cellatoglu, A. [135] 17
CERN, [53] 10
Chae, J.S. [520] 601
Chai, D. [314] 274, 313
Chai, D. [313] 274, 313
Challapali, K. [72] 15
Chan, C. [212] 93
Chan, M. [60] 8, 173
Chan, W.-Y. [127] 17
Chang, S. [159] 44
Chen, C.W. [65] 15
Chen, Q. [228] 119
Chen, S. [478] 543
Chen, Y.C. [518] 601
Cheng, L. [367] 410
Cheng, N.T. [188] 62, 72, 169, 272, 340
Cheng, N.T. [310] 272, 340
Cheng, R.S. [457] 527
Cherriman, P. [392] 462, 465, 473, 488
Cherriman, P. [161] . 47, 61, 70, 108, 111, 112, 196, 475, 482, 487, 600, 622
Cherriman, P. [427] 486
Cherriman, P.J. [224] 118–120, 132, 133
Cherriman, P.J. [5] 1, 12, 15, 16, 384, 392, 409, 427, 539
Cheung, G. [387] 447
Cheung, J. [192] 70
Chiariglione, L. [339] 379
Chiariglione, L. [357] 387
Chiariglione, L. [361] 393
Choi, [88] 16
Choi, D.W. [302] 227
Choi, W.I. [368] 410, 423, 427
Choo, C. [209] 93
Chou, P.A. [100] 16
Chou, P.A. [112] 16
Chow, P.S. [421] 486
Chowdhury, M.F. [154] 38
Chu, W.-J. [74] 15
Chua, S-G. [409] 485
Chua, S. [414] 485
Chuang, J.C.-I. [80] 15
Chuang, K-C. [208] 93
Chung, J.Y. [227] 118–120, 134, 135
Chung, W.C. [325] 306
Cioffi, J.M. [421] 486
Civanlar, M.R. [515] 574
Civanlar, M.R. [73] 15, 16, 396
Civanlar, M.R. [64] 15
Clarke, R.J. [180] 54, 184

Clarke, R.J. [184] 54
Clarke, R.J. [370] 415
Clarke, R.J. [156] 38, 46
Clarkson, T.G. [170] 50
Cline, L. [388] 448
Cocke, J. [234] 119, 122, 130
Cocke, J. [492] 547
Cole, R. [39] 9
Conklin, G.J. [119] 16, 396
Connell, J.B. [3] 1
Constello, D.J. Jr [202] . 78, 112, 161, 282, 341, 371
Constello, D.J. Jr [317] 282, 371
Cornell University, [47] 9, 10
Cosman, P. [65] 15
Cosman, P.C. [106] 16
Cosman, P.C. [295] 196
Cosmas, J. [337] 362–364
Couturier, C. [262] 177
Crebbin, G. [248] 139
Crespo, P. [337] 362–364
Crochiere, R.E. [270] 183
Crochiere, R.E. [271] 183
Crochiere, R.E. [286] 192
Czylwik, A. [419] 486

D

Dai, Q. [396] 462
Danny Cohen, [38] 9
de Alfaro, L. [200] 78, 112, 161
De Martin, J.C. [130] 17
De Martin, J.C. [129] 17
De Natale, F.G.B. [519] 601
deDiego, S. [205] 93
Deisher, G. [388] 448
Delp, E.J. [265] 177
Delp, E.J. [266] 178–181
Dempsey, B. [116] 16
DeNatale, F.G.B. [250] 139
Desoli, G.S. [250] 139
Di Gregario, M. [519] 601
Didier, P. [393] 462, 463
Ding, W. [329] 344
Ding, W. [331] 345
Ding, Z. [535] 608
Divsalar, D. [229] 119
Dixit, S. [323] 285
Djuknic, G.M. [199] 78, 112, 161
Dogan, S. [135] 17
Dorcey, T. [45] 9
Doufexi, A. [143] 17
Douillard, C. [393] 462, 463
Dubois, E. [309] 272, 340
Dubois, E. [127] 17
Dubois, E. [308] 272, 340
Dudbridge, F. [149] 21, 27, 28

AUTHOR INDEX

E

Ebrahimi, T. [362] 393, 394
Ebrahimi, T. [363] 394, 396
Efstratiadis, N. [167] 49
Egger, O. [290] 193
Eleftheriadis, A. [351] 383
Eleftheriadis, A. [354] 383
Eleftheriadis, A. [375] 438
Elke Offer, [236] 119, 121, 122, 131, 133
Ellis, W. [326] 310, 313, 317
Emani, S. [59] 8, 173
Ericsson, A. [262] 177
Esteban, D. [282] 185, 187, 189
Evans, B.G. [527] 604
Experts Group on very low Bitrate Visual
      Telephony, ITU-T [32] 390

F

Failli, M. [513] 572
Fano, R.M. [2] 1
Färber, N. [328] 340
Färber, N. [95] 16
Farrell, P.G. [482] 543
Fazel, K. [68] 15
Fazel, K. [423] 486
Feder, M. [249] 139, 145
Feig, E. [331] 345
Feng, Y. [209] 93
Fischer, R.F.H. [420] 486
Fischer, R.F.H. [459] 527
Fischer, T.R. [233] 119, 130
Fisher, Y. [375] 438
Flangan, J.L. [270] 183
Flannery, B.P. [218] 106
Fletcher, P. [143] 17
Flierl, M. [137] 17
Fogg, C.E. [7] 1
Foschini, G.J. [463] 532
Fossorier, M.P.C. [460] 527
Franceschini, G. [352] 383
Franti, P. [297] 203
Franz, V. [399] 462
Franz, V. [242] 132
Frederick, R. [356] 384, 448
Frescura, F. [517] 601
Fuja, T. [145] 17
Farber, N. [113] 16

G

Galand, C. [282] 185, 187, 189
Galand, C.R. [285] 191
Gallanger, N. [268] 179
Gardos, T. [388] 448
Garzelli, A. [381] 441
Gersho, A. [247] 139
Gersho, A. [204] 93–95, 103
Gersho, A. [15] 8, 27, 28, 30, 98, 102, 103
Gersho, A. [211] 93

Gertsman, M.J. [394] 462
Ghanbari, M. [81] 15
Ghanbari, M. [138] 17
Ghanbari, M. [324] 289, 351, 575, 587
Ghanbari, M. [186] 54
Gharavi, H. [193] 70, 158, 193
Gharavi, H. [514] 573, 587
Gharavi, H. [157] 41
Gharavi, H. [516] 600
Gharavi, H. [274] 183
Gharavi, H. [275] 183, 195, 196
Gharavi, H. [289] 192, 220
Gharavi, H. [276] 184
Girod, B. [137] 17
Girod, B. [93] 16
Girod, B. [95] 16
Girod, B. [98] 16, 410
Girod, B. [90] 16
Girod, B. [114] 16
Girod, B. [108] 16
Girod, B. [121] 16, 17, 396
Girod, B. [126] 17
Girod, B. [165] 49, 295
Girod, B. [328] 340
Giusto, D.D. [250] 139
Giusto, D.D. [519] 601
Glavieux, A. [401] 463, 473, 474, 485, 488, 571, 602
Glavieux, A. [393] 462, 463
Glavieux, A. [402] 463, 571, 602, 607
Godard, D.N. [533] 605
Goldhirsh, J. [523] 604
Goldhirsh, J. [524] 604
Goldschneider, J.R. [206] 93
Goldsmith, A.J. [414] 485
Goldsmith, A.J. [409] 485
Görtz, N. [230] 119
Goursat, M. [530] 605, 606
Grallert, H.J. [136] 17
Gray, R. [240] 125
Gray, R.M. [295] 196
Gray, R.M. [204] 93–95, 103
Greenbaum, G.S. [119] 16, 396
Griffiths, J. [521] 604
Griswold, N.C. [267] 179
Group, MPEGAOE [346] 380, 382
Grünheid, R. [422] 486
Guo, F. [458] 527
Guo, F. [227] 118–120, 134, 135
Gupta, S. [211] 93

H

Haavisto, P. [169] 49
Hagenauer, J. [493] 547
Hagenauer, J. [397] 462, 464
Hagenauer, J. [232] 119, 127
Hagenauer, J. [230] 119
Hagenauer, J. [475] 543, 550
Hagenauer, J. [487] 546, 557, 558

Halverson, D.R. [267] 179
Hamorsky, J. [404] 464, 465
Hannuksela, M.M. [142] 17
Hanzo, L. [406] 473
Hanzo, L. [428] 486
Hanzo, L. [417] 486, 487
Hanzo, L. [415] 485
Hanzo, L. [85] 16, 482
Hanzo, L. [392] 462, 465, 473, 488
Hanzo, L. [195] . 72, 77, 80–91, 166, 257, 272, 340, 341, 561
Hanzo, L. [193] 70, 158, 193
Hanzo, L. [389] . 459–461, 464, 558, 570, 573, 578, 601, 602, 607
Hanzo, L. [322] 285, 473, 474
Hanzo, L. [88] 16
Hanzo, L. [222] . 114, 158, 341, 343, 352, 485, 558, 570, 578, 580, 594, 601, 602, 606, 607
Hanzo, L. [161] 47, 61, 70, 108, 111, 112, 196, 475, 482, 487, 600, 622
Hanzo, L. [190] 66, 70, 91, 92, 136
Hanzo, L. [391] 460
Hanzo, L. [192] 70
Hanzo, L. [416] 485
Hanzo, L. [418] 486
Hanzo, L. [243] 133
Hanzo, L. [237] 122, 130, 132
Hanzo, L. [224] 118–120, 132, 133
Hanzo, L. [509] 561
Hanzo, L. [194] 71
Hanzo, L. [203] 91, 161
Hanzo, L. [256] 159, 363
Hanzo, L. [315] 277, 345
Hanzo, L. [320] 284, 341
Hanzo, L. [333] 352
Hanzo, L. [478] 543
Hanzo, L. [485] 544
Hanzo, L. [410] 485
Hanzo, L. [462] 527, 534
Hanzo, L. [390] 460
Hanzo, L. [319] 284, 341
Hanzo, L. [412] 485
Hanzo, L. [253] 158, 160, 162–164, 166
Hanzo, L. [470] 533
Hanzo, L. [458] 527
Hanzo, L. [225] 118–120, 134, 135
Hanzo, L. [231] 119–122, 133
Hanzo, L. [227] 118–120, 134, 135
Hanzo, L. [151] 33, 166
Hanzo, L. [427] 486
Hanzo, L. [516] 600
Hanzo, L. [196] 72–76, 79, 108, 112, 158, 277, 532, 535
Hanzo, L. [84] 16
Hanzo, L. [496] 549
Hanzo, L. [83] 16, 66, 70, 76, 78, 108, 109, 111, 112, 137, 165, 460, 482
Hanzo, L. [86] 16, 75, 109, 158, 526, 528

Hanzo, L. [411] 485
Hanzo, L. [219] 107, 110, 114–118, 166
Hanzo, L. [494] 549
Hanzo, L. [5]. 1, 12, 15, 16, 384, 392, 409, 427, 539
Hanzo, L. [404] 464, 465
Haralick, R.M. [206] 93
Harri Holma, [198] 76, 92, 108, 158, 159, 161, 165, 368, 622
Haskell, B. [77] 15
Haskell, B.G. [507] 559, 560
Haskell, B.G. [8] 1, 12
Haskell, B.G. [122] 17
Heising, G. [373] 424, 427, 429
Heising, G. [369] 410
Hemami, S.S. [70] 15
Hennecke, M.E. [251] 149
Herpel, C. [351] 383
Herpel, C. [354] 383
Herpel, C. [353] 383, 384
Hoeher, P. [452] 571, 591, 603
Hoetter, M. [172] 50
Höher, P. [241] 132
Hong, U.S. [526] 604
Horn, U. [126] 17
Horn, U. [125] 17, 396
Horne, C. [363] 394, 396
Howard, P.G. [1] 310–312
Hsieh, C-H. [208] 93
Hsieh, C. [263] 177
Hsu, C. [168] 49
Hsu, Y.F. [518] 601
Hu, Y.H. [384] 443
Huang, C. [168] 49
Huang, C. [210] 93
Huang, C.J. [518] 601
Huang, J. [166] 49
Huber, J.B. [420] 486
Huber, J.B. [459] 527
Hübner, J. [425] 486
Huffman, D.A. [239] 125
Huffman, D.A. [4] 1
Huffman, D.A. [481] 543
Huguet, J. [217] 104

I

Imai, H. [460] 527
Irie, K. [280] 184
Irie., K. [278] 184
ISO/IEC 11172-2 Information technology, [345]379
ISO/IEC 13818-2 MPEG-2 Video Coding Standard, [360] 391
ISO/IEC JTC1, [62] 14, 425
ISO/IEC JTC1/SC29/WG11 N0702 Rev., [344] 379, 391, 397
ISO/IEC JTC1/SC29/WG11 N1902, [364] 394, 396, 397, 437, 438, 440, 441
ISO/IEC JTC1/SC29/WG11 N1902, [377] 441
ISO/IEC JTC1/SC29/WG11 W2502, [461] 527

ISO/IEC JTC1/SC29/WG11, [30] .. 11, 12, 17, 389, 392, 408
ISO/IEC JTC1/SC29/WG11, [31] . 12, 17, 389, 392, 408
ISO/IEC JTC1/SC29/WG11, [25] ... 10, 13, 15, 17, 380, 382, 397, 406, 408, 527
ISO/IEC JTC1/SC29/WG11, [378] 441
ISO/IEC JTC1/SC29/WG11, [347] .. 380, 381, 389, 393
ISO/IEC JTC1/SC29/WG11, [348] .. 380, 381, 389, 393
ISO/IEC JTC1/SC29/WG11, [349] .. 381, 386, 391, 424, 439, 445, 454
Israelsen, P. [210] 93
ITU-T, [374] 424, 425
ITU-T, [259]166, 205, 207, 215, 220, 221, 295, 302, 304, 337, 473
ITU-T/SG 16/VCEG(formerly Q.15 now Q.6), [366]408, 412, 416–418, 425–427
ITU-T/SG15, [28] 10, 12, 392, 408
ITU-T/SG16/Q15, [24] 10, 13, 17
ITU-T/SG16/Q15, [26] 13

J

Jabbari, B. [124] 17
Jacobson, V. [42] 9
Jacobson, V. [356] 384, 448
Jacquin, A.E. [148] 21, 27–30
Jafarkhani, H. [469] 533
Jain, A. [350] 381, 391, 416
Jain, A.K. [10] . 5, 51, 52, 54, 55, 65, 139, 166, 175, 184, 206, 295
Jain, A.K. [158] 43, 44
Jain, J.R. [9] 1
Jain, J.R. [158] 43, 44
Jaisimha, M.Y. [206] 93
Jalali, A. [309] 272, 340
Jalali, A. [308] 272, 340
Jang, E. [294] 196
Jayant, N.S. [261] 177
Jayant, N.S. [187] . . 55, 63, 100, 143, 145, 192, 209, 295
Jeanclaude, I. [424] 486
Jelinek, F. [234] 119, 122, 130
Jelinek, F. [492] 547
Jeon, B.W. [368] 410, 424, 427
Jézéquel, M. [393] 462, 463
Joachim Hagenauer, [236] .. 119, 121, 122, 131, 133
Joch, A. [66] 15, 410
Johnson, R.C. [535] 608
Johnston, J.D. [283] 185, 191, 192, 194
Joint Video Team (JVT) of ISO/IEC MPEG, [27]11, 15, 17, 406
Jones, E.V. [304] 229
Jung, P. [87] 16

K

Kaiser, S. [423] 486
Kalet, I. [426] 486
Kallel, S. [395] 462
Kalman, M. [90] 16
Kamio, Y. [408] 485
Kang, K-J. [504] 558
Karaliopoulos, M.S. [522] 604
Karam, G. [424] 486
Karczewicz, M. [66] 15, 410
Karlsson, G. [128] 17
Karunaserker, S.A. [173] 50
Katsaggelos, A. [167] 49
Katsaggelos, A.K. [330] 345
Keller, [88] 16
Keller, T. [428] 486
Keller, T. [417] 486, 487
Keller, T. [222] .. 114, 158, 341, 343, 352, 485, 558, 570, 571, 578, 580, 594, 601, 602, 606, 607
Keller, T. [243] 133
Keller, T. [390] 460
Keller, T. [427] 486
Keller, T. [196] . 72–76, 79, 108, 112, 158, 277, 532, 535
Keller, T. [84] 16
Keller, T. [496] 549
Keller, T. [411] 485
Kennedy, R.A. [535] 608
Kenyon, N.D. [306] 240
Khansari, M. [309] 272, 340
Khansari, M. [127] 17
Khansari, M. [109] 16, 396
Khansari, M. [308] 272, 340
Khorram, H. [397] 462, 464
Kieu, L.H. [82] 15
Kim, C.W. [293] 194
Kim, D.S. [216] 93
Kim, J. [132] 17
Kim, J.W. [131] 17
Kim, K-B. [504] 558
Kingsbury, N. [65] 15
Kingsbury, N.G. [188] 62, 72, 169, 272, 340
Kingsbury, N.G. [260] 169, 205
Kingsbury, N.G. [173] 50
Kingsbury, N.G. [176] 50
Kingsbury, N.G. [177] 50
Kingsbury, N.G. [175] 50
Kishimoto, R. [280] 184
Kittitornkun, S. [384] 443
Klein, A. [405] 473, 474
Kliewer, J. [144] 17
Kliewer, J. [495] 549
Kliewer, J. [462] 527, 534
Kliewer, J. [235] 119, 130, 131
Knorr, G.D. [91] 16
Komaki, S. [318] 284
Kondoz, A.M. [135] 17
Konstantinides, K. [6] 1
Koora, K. [425] 486

Kossentini, F. [91] 16
Kossentini, F. [325] 306
Kot, A. [367] 410
Kramer, G. [477] 543
Kuan, E.L. [406] 473
Kuan, E.L. [418] 486
Kuan, E.L. [203] 91, 161
Kuan, E.L. [83] ... 16, 66, 70, 76, 78, 108, 109, 111, 112, 137, 165, 460, 482
Kull, B. [425] 486
Kunt, M. [290] 193
Kuo, C.C.J. [132] 17
Kuo, C.C.J. [131] 17
Kwok, W. [383] 442
Kwon, O.H. [520] 601

L

Lainema, J. [66] 15, 410
Lam, W.-M. [71] 15
Lam, W.M. [79] 15
Lau, V.K.N. [413] 485
Le Gall, D.J. [341] 379
Le, M.T. [506] 558
Lee, J. [164] 48
Lee, K.W. [111] 16
Lee, L. [164] 48
Lee, M.H. [248] 139
Lee, S.U. [216] 93
Lee, T-H. [201] 78, 112, 161
Lee, W-C. [504] 558
LeGall, D.J. [7] 1
LeGall, D.J. [123] 17
Lennox, J. [56] 10
Leou, J.-J. [74] 15
Letaief, K.B. [80] 15
Levesque, A.H. [502] 558, 570, 601, 602
Li, A. [384] 443
Li, R. [163] 48
Li, W. [290] 193
Li, Y. [429] 486
Liang, J. [65] 15
Liang, Y.J. [114] 16
Liebeherr, J. [116] 16
Lieu, M.L. [80] 15
Liew, T.H. [415] 485
Liew, T.H. [237] 122, 130, 132
Liew, T.H. [412] 485
Liew, T.H. [86] 16, 75, 109, 158, 526, 529
Liew, T.H. [404] 464, 465
Lightstone, M. [332] 345
Lillevold, K.O. [119] 16, 396
Lim, J.S. [78] 15
Lin, S. [202] 78, 112, 161, 282, 341, 371
Lin, S. [317] 282, 371
Linde, Y. [240] 125
Link, M. [113] 16
Link, M. [108] 16
Link, M. [125] 17, 396

Liou, N. [163] 48
Liou, W. [263] 177
Lippman, A.F. [119] 16, 396
List, P. [66] 15, 410
Liu, B. [79] 15
Liu, B. [329] 344
Liu, B. [162] 48
Liu, S. [185] 54, 142
Liu, Y.-J. [120] 16, 396
Lloyd, S. [490] 546
Lodge, J.L. [394] 462
Lu, J. [80] 15
Lu, L. [213] 93
Lu, P. [263] 177
Luise, M. [254] 158
Luthra, A. [140] 17
Luthra, A. [64] 15
Lutz Papke, [236] 119, 121, 122, 131, 133

M

MacDonald, N. [307] 257, 272, 340, 341
Maciocco, C. [388] 448
Macleod, M.D. [413] 485
Magarey, J. [175] 50
Mämmelä, A. [335] 361
Mann Pelz, R. [337] 362–364
Mann Pelz, R. [191] 70, 272, 340, 346
Marcellin, M.W. [233] 119, 130
Marpe, D. [373] 424, 427, 431
Marpe, D. [369] 410
Marsland, I.D. [395] 462
Martins, F.C.M. [331] 345
Masala, E. [130] 17
Mathiopoulos, P.T. [395] 462
Matsumura, Y. [312] 272, 340
Matsuoka, H. [408] 485
Maunder, R.G. [485] 544
Maunder, R.G. [225] 118–120, 134, 135
Mehrotra, S. [100] 16
Meng, T.H.-Y. [70] 15
Meo, A.R. [200] 78, 112, 161
Mermelstein, P. [309] 272, 340
Mermelstein, P. [127] 17
Mermelstein, P. [308] 272, 340
Messerschmitt, D. [77] 15
Michelson, A.M. [502] 558, 570, 601, 602
Miller, M.J. [202] 78, 112, 161, 282, 341, 371
Miller, S. [59] 8, 173
Mills, M. [157] 41
Mirchandani, G. [181] 54
Mitchell, J.L. [7] 1
Mitchell, O.R. [265] 177
Mitchell, O.R. [266] 178–181
Moccagatta, I. [387] 447
Modestino, J.W. [65] 15
Moher, M. [398] 462
Mohr, A.E. [100] 16
Mohr, A.E. [206] 93

Moisala, T. [169] 49
Monro, D.M. [149] 21, 27, 28
Monro, D.M. [150] 21, 27, 28
Montgomery, B.L. [385] 443
Montorsi, G. [229] 119
Montorsi, G. [483] 547
Montorsi, G. [484] 547
Morelos-Zaragoza, R.H. [460] 527
Morinaga, N. [318] 284
Morinaga, N. [408] 485
Müenster, [88] 16
Mukherjee, D. [332] 345
Murad, A. [145] 17
Murakami, H. [386] 444, 447
Murakami, H. [238] 124, 125, 132
Musmann, H.G. [136] 17

N

Nag, Y. [387] 447
Nakagawa, S. [312] 272, 340
Nakai, T. [312] 272, 340
Narasimha, M. [189] 65
Narayanan, K.R. [403] 464
Narula, A. [78] 15
Nasrabadi, N. [294] 196
Nasrabadi, N. [209] 93
Natarajan, T. [371] 416
Naveen, T. [281] 184
Netravali, A.N. [507] 559, 560
Netravali, A.N. [8] 1, 12
Nevalainen, O. [297] 203
Newell, D. [388] 448
Ng, S.X. [243] 133
Ng, S.X. [478] 543
Ng, S.X. [485] 544
Ng, S.X. [462] 527, 534
Ng, S.X. [470] 533
Ng, S.X. [458] 527
Ng, S.X. [225] 118–120, 134, 135
Ng, S.X. [231] 119–122, 133
Ng, S.X. [227] 118–120, 134, 135
Ng, S.X. [196] . 72–76, 79, 108, 112, 158, 277, 532, 535
Ng, S.X. [496] 549
Ng, K.N. [82] 15
Ng, K.N. [314] 274, 313
Ng, K.N. [359] 388
Ng, K.N. [358] 388
Ng, K.N. [313] 274, 313
Ng, K.N. [183] 54
Ng, K.N. [215] 93
Nguyen, Q.T. [292] 194
Nicholls, J.A. [150] 21, 27, 28
Nieweglowski, J. [169] 49
Nightingale, C. [306] 240
Nix, A. [143] 17
Noah, M.J. [269] 180, 208
Noll, P. [187] . . 55, 63, 100, 143, 145, 192, 209, 295

Nuri, V. [288] 192
Nussbaumer, H.J. [285] 191
Nussbaumer, H.J. [284] 191

O

O'Leary, S. [503] 558
O'Leary, S. [505] 558
O'Neil, S.D. [272] 183
Offer, E. [493] 547
Onural, L. [155] 38
Ortega, A. [97] 16
Ostermann, J. [153] 38
Ostermann, J. [376] 439
Ott, J. [91] 16
Ott, J. [388] 448

P

Palau, A. [181] 54
Pamchanathan, S. [359] 388
Pamchanathan, S. [358] 388
Papadopoulos, C. [118] 16
Papadopoulos, C.A. [170] 50
Papadopoulos, C.A. [171] 50
Papke, L. [493] 547
Park, D.S. [384] 443
Park, H-M. [504] 558
Parulkar, G.M. [118] 16
Paulraj, A. [334] 361
Pavlidou, F-N. [522] 604
Pearlman, W.A. [213] 93
Pearson, D.E. [34] 8
Pearson, D.E. [174] 50
Pennebaker, W.B. [7] 1
Pereira, F. [536] 624
Peterson, A. [189] 65
Phoong, S-M. [293] 194
Picart, A. [393] 462, 463
Picchi, G. [531] 605, 606
Picco, R. [214] 93
Pickholtz, R.L. [120] 16, 396
Pietrobon, S.S. [512] 571
Pirhonen, R. [405] 473, 474
Pirsch, P. [136] 17
Po, L. [212] 93
Pollara, F. [229] 119
Polydoros, A. [532] 605, 606
Prasad, K.V. [251] 149
Prati, G. [531] 605, 606
Press, W.H. [218] 106
Priestly, D. [503] 558
Proakis, J.G. [464] 532, 533
Proakis, J.G. [379] 441
Puri, A. [507] 559, 560
Puri, A. [8] 1, 12
Puri, A. [122] 17
Puri, R. [111] 16
Puri, R. [110] 16

Q

Quaglia, D. [130] 17
Quaglia, D. [129] 17

R

Rabiner, L.R. [286] 192
Raheli, R. [532] 605, 606
Raina, J. [298] 203
Rajan, G. [351] 383
Ramamurthi, B. [15] 8, 27, 28, 30, 98, 102, 103
Ramchandran, K. [97] 16
Ramchandran, K. [112] 16
Ramchandran, K. [111] 16
Ramchandran, K. [110] 16
Ramstad, T.A. [287] 192, 203
Rao, K.R. [371] 416
Rao, K.R. [14] 8, 415, 416
Rao, K.R. [178] 52, 54, 65, 563
Raphaeli, D. [400] 463
Raviv, J. [234] 119, 122, 130
Raviv, J. [492] 547
Raychaudhuri, D. [383] 442
Reader, C. [536] 624
Reali, G. [517] 601
Rec. H.26L/ISO/IEC 11496-10, ITU-T [372]
    418–420
Recommendation BT.601-5 (10/95), ITU-R [36] . . 8
Recommendation H.263, ITU-T [63] 15, 443
Recommendation H.320, ITU-T [41] 9
Recommendation H.323 Version 2, ITU-T [55] . 10, 12
Recommendation H.323 Version 4, ITU-T [58] . . 10
Recommendation H.323, ITU-T [52] 10
Recommendation H.324, ITU-T [50] 10
Recommendation H.450.4, ITU-T [57] 10
Recommendation ITU-T BT.500-11, [365] 398
Recommendation T.120, ITU-T [51] 10
Redmill, D.W. [260] 169, 205
Redmill, D.W. [311] 272, 340
Regunathan, S. [133] 17
Regunathan, S.L. [102] 16
Regunathan, S.L. [103] 16
Regunathan, S.L. [104] 16
Reibman, A.R. [515] 574
Reibman, A.R. [73] 15, 16, 396
Reibman, A.R. [71] 15
Reibman, A.R. [79] 15
Reznik, Y.A. [119] 16, 396
Riskin, E.A. [206] 93
Robertson, P. [423] 486
Robertson, P. [452] 571, 591, 603
Robertson, P. [241] 132
Rogers, J.K. [106] 16
Rohling, H. [422] 486
Rose, K. [102] 16
Rose, K. [133] 17
Rose, K. [134] 17
Rose, K. [103] 16
Rose, K. [104] 16
Rosenberg, J. [56] 10
Ruf, M.J. [423] 486

S

Sadka, A.H. [135] 17
Sampei, S. [318] 284
Sampei, S. [408] 485
Sari, H. [424] 486
Sato, Y. [534] 606
Saunders, S. [528] 604
Saunders, S.R. [527] 604
Schäfer, R. [340] 379, 380, 389–391
Schilling, D.L. [199] 78, 112, 161
Schooler, E.M. [40] 9
Schulzrinne, H. [56] 10
Schulzrinne, H. [46] 9
Schulzrinne, H. [356] 384, 448
Schuster, G.M. [330] 345
Schwartz, M. [323] 285
Seferidis, V. [81] 15
Seferidis, V. [324] 289, 351, 575, 587
Seshadri, N. [456] 527, 532
Shanableh, T. [138] 17
Shannon, C.E. [223] 118, 134, 137
Shannon, C.E. [480] 543
Sharaf, A. [179] 53
Shaw, L. [69] 15
Shaw, L. [76] 15
Shelswell, P. [510] 570
Sherwood, P.G. [106] 16
Shie, J. [164] 48
Shin, J. [132] 17
Shin, J. [131] 17
Shu, L. [460] 527
Shue, J-S. [208] 93
Shustermann, E. [249] 139, 145
Shwedyk, E. [396] 462
Signes, J. [375] 438
Sikora, T. [538] 624
Sikora, T. [359] 388
Sikora, T. [358] 388
Sikora, T. [536] 624
Sikora, T. [340] 379, 380, 389–391
Sikora, T. [342] 379, 384, 387
Sikora, T. [343] 379, 454
Sikora, T. [361] 393
Simoncelli, E.P. [279] 184
Sin, K.K. [215] 93
Sitaram, V. [210] 93
Skelly, P. [323] 285
Skoeld, J. [405] 473, 474
Smith, M.J.T. [325] 306
Sohn, W. [520] 601
Sollenberger, N.R. [429] 486
Somerville, F.C.A. [190] 66, 70, 91, 92, 136
Somerville, F.C.A. [391] 460
Stedman, R. [193] 70, 158, 193

Stedman, R. [192] 70
Steele, R. [193] 70, 158, 193
Steele, R. [389]..459–461, 464, 558, 570, 573, 574, 578, 601, 602, 607
Steele, R. [322] 285, 473, 474
Steele, R. [192].............................70
Steele, R. [336]....................... 362–364
Steele, R. [321] 284, 341, 485
Stefanov, J. [256] 159, 363
Steinbach, E. [94] 16
Steinbach, E. [90] 16
Steinbach, E. [89] 16
Steinbach, E. [328] 340
Stockhammer, T. [142] 17
Stockhammer, T. [20] 10, 11, 407, 408, 424
Stork, D.G. [251] 149
Street, J. [224] 118–120, 132, 133
Street, J. [5] . . 1, 12, 15, 16, 384, 392, 409, 427, 539
Streit, J. [195] ...72, 77, 80–91, 166, 257, 272, 340, 341, 561
Streit, J. [161]..47, 61, 70, 108, 111, 112, 196, 475, 482, 487, 600, 622
Streit, J. [253] 158, 160, 162–164, 166
Streit, J. [151] 33, 166
Streit, J. [219] 107, 110, 114–118, 166
Strobach, P. [160]....45, 46, 56, 139, 141, 145, 216
Stuber, G.L. [403] 464
Subbalakshmi, K.P. [228] 119
Sullimendation, G. [388] 448
Sullivan, G.J. [140] 17
Sullivan, G.J. [96] 16
Sullivan, G.J. [20] 10, 11, 407, 408, 424
Sun, H. [72]................................15
Sun, H. [383].............................. 442
Sun, M.-T. [359] 388
Sun, M.-T. [358]........................... 388
Sun, M.J. [518] 601
Suoranta, R. [405] 473, 474
Susini, S. [381]............................ 441

T

Tabatabai, A. [274] 183
Tabatabai, A. [275] 183, 195, 196
Takishima, Y. [386] 444, 447
Takishima, Y. [238]............... 124, 125, 132
Talluri, R. [387] 447
Talluri, R. [355] 383, 443
Tan, W. [105] 16
Tao, M. [457] 527
Tarokh, V. [456] 527, 532
Tarokh, V. [469] 533
Tekalp, A.M. [155] 38
Tekalp, A.M. [376] 439
Telatar, E. [466] 533
ten Brink, S. [471] 535
ten Brink, S. [244] 133
ten Brink, S. [476] 543, 550
ten Brink, S. [477] 543

Teukolsky, S.A. [218] 106
Thibault, L. [506] 558
Thitimajshima, P. [401] ... 463, 473, 474, 485, 488, 506, 571, 602
Thobaben, R. [144] 17
Thobaben, R. [495] 549
Thobaben, R. [235] 119, 130, 131
Tim Dorcey, [43] 9
Torrance, J.M. [410] 485
Torrance, J.M. [319] 284, 341
Torrance, J.M. [411] 485
Torrence, G.W. [525] 604
Torres, L. [205] 93
Torres, L. [217] 104
Tüchler, M. [497] 552
Tüchler, M. [475] 543, 550
Tüchler, M. [474] 534, 536
Turletti, T. [305].......................... 240
Turletti, T. [44] 9
Tzaras, C. [527] 604
Tzou, C. [532] 605, 607

U

Udpikar, V. [298] 203
Ungerboeck, G. [486] 545, 547
Ungerboeck, G. [431] 526
Uyuroglu, M. [135]......................... 17
Uz, K.M. [123] 17

V

Vaidyanathan, P.P. [293] 194
Vaisey, J. [247] 139
VCEG, ITU-T [27] 11, 15, 17, 406
Venditto, G. [48] 10
Venetsanopoulos, A. [207].....................93
Vetterli, M. [128] 17
Vetterli, M. [109] 16, 396
Vetterli, M. [295] 196
Vetterli, M. [123] 17
Vetterling, W.T. [218] 106
Video Group, MPEG [33] 381
Villasenor, J. [384] 443
Villasenor, J.D. [382] 442, 443, 447
Villebrun, E. [452] 571, 591, 603
Villebrun, E. [241].......................... 132
Vitter, J.S. [1] 310–312
Vocaltec Communications Ltd., [49] 10
Vogel, W.J. [523] 604
Vogel, W.J. [524] 604
Vogel, W.J. [526] 604
Vogel, W.J. [525] 604
Vook, F.W. [430] 486
Vucetic, B. [467] 533

W

Wachsmann, U. [459] 527
Wada, M. [386] 444, 447
Wada, M. [238] 124, 125, 132

Wagner, G.R. [264] 177
Wallace, G.K. [23] 8, 10
Wang, A. [100] 16
Wang, B.M. [159] 44
Wang, F-M. [185] 54, 142
Wang, J. [478] 543
Wang, J. [485] 544
Wang, J. [164] 48
Wang, J. [225] 118–120, 134, 135
Wang, J. [494] 549
Wang, Q. [156] 38, 46
Wang, Y. [67] 15
Wang, Y. [69] 15
Wang, Y. [76] 15
Ward, L. [351] 383
Weaver, A. [116] 16
Webb, W. [243] 133
Webb, W. [196] 72–76, 79, 108, 112, 158, 277, 532, 535
Webb, W. [84] 16
Webb, W. [496] 549
Webb, W.T. [222]114, 158, 341, 343, 352, 485, 558, 570, 571, 578, 580, 594, 601, 602, 606, 607
Webb, W.T. [315] 277, 345
Webb, W.T. [320] 284, 341
Webb, W.T. [390] 460
Webb, W.T. [321] 284, 339, 485
Webber, S.A. [270] 183
Wei, L-F. [255] 158
Wellstead, P.E. [264] 177
Welsh, W.J. [152] 38
Wen, J. [382] 442, 443, 447
Wenger, S. [141] 17
Wenger, S. [91] 16
Wenger, S. [388] 448
Wenger, S. [64] 15
Wesolowsky, K. [529] 605, 606
Westerink, P.H. [291] 193
Whybray, M.W. [326] 310, 313, 317
Whybray, M.W. [174] 50
Wicker, S.B. [115] 16
Wicker, S.B. [511] 571, 603
Wicker, S.B. [380] 441
Wiegand, T. [140] 17
Wiegand, T. [142] 17
Wiegand, T. [96] 16
Wiegand, T. [99] 16
Wiegand, T. [98] 16, 410
Wiegand, T. [332] 345
Wiegand, T. [373] 424, 427, 431
Wiegand, T. [20] 10, 11, 407, 408, 424
Wiegand, T. [369] 410
Wiese, G.L. [267] 179
Wilson, D.L. [150] 21, 27, 28
Wolf., G.J. [252] 149
Wolfgang, A. [478] 543
Wong, C.H. [415] 485

Wong, C.H. [85] 16, 482
Wong, C.H. [418] 486
Wong, C.H. [412] 485
Wong, K.H.H. [333] 352
Woodard, J.P [428] 486
Woodard, J.P. [190] 66, 70, 91, 92, 136
Woodard, J.P. [391] 460
Woodard, J.P. [509] 561
Woodard, J.P. [194] 71
Woods, J.W. [273] 183
Woods, J.W. [281] 184
Woods, J.W. [272] 183

Y

Yamaguchi, H. [182] 54
Yan, L. [122] 17
Yang, H. [134] 17
Yang, L-L. [203] 91, 161
Yang, L-L. [478] 543
Yang, L-L. [485] 544
Yang, L-L. [225] 118–120, 134, 135
Yang, L-L. [494] 549
Yang, L.L. [83] . . . 16, 66, 70, 76, 78, 108, 109, 111, 112, 137, 165, 460, 482
Yeap, B.L. [237] 122, 130, 132
Yeap, B.L. [86] 16, 75, 109, 158, 526, 529
Yeap, B.L. [404] 464, 465
Yee, M.S. [85] 16, 482
Yee, M.S. [416] 485
Yen, J.C. [159] 44
Yen, K. [203] 91, 161
Yen, K. [83] . 16, 66, 70, 76, 78, 108, 109, 111, 112, 137, 165, 460, 482
Yip, P. [14] 8, 415, 416
Yip, P. [178] 52, 54, 65, 563
Yip, W. [211] 93
Young, R.W. [176] 50
Young, R.W. [177] 50
Yu, P. [207] 93
Yuan, J. [467] 533

Z

Zaccarin, A. [162] 48
Zafar, S. [124] 17
Zakauddin, A. [127] 17
Zakhor, A. [105] 16
Zarai, Y. [400] 463
Zdepski, J. [72] 15
Zdepski, J.W. [383] 442
Zeger, K. [106] 16
Zehavi, E. [432] 526
Zeisberg, S. [425] 486
Zeng, B. [163] 48
Zeng, Y. [367] 410
Zetterberg, L. [262] 177
Zhang, R. [102] 16
Zhang, R. [133] 17
Zhang, R. [134] 17

AUTHOR INDEX

Zhang, R. [103] 16
Zhang, R. [104] 16
Zhang, X. [98] 16, 410
Zhang, X. [246] 139, 145, 196
Zhang, Y.-Q. [536] 624
Zhang, Y.-Q. [124] 17

Zhang, Y.-Q. [120] 16, 396
Zhu, C. [388] 448
Zhu, Q.-F. [67] 15
Zhu, Q.-F. [69] 15
Zhu, Q.-F. [76] 15
Zhu, W. [64] 15